Protein Engineering Handbook

Volume 2

Edited by
Stefan Lutz and
Uwe T. Bornscheuer

Further Reading

Cox, M. M., Phillips, G. N. (eds.)

Handbook of Proteins

Structure, Function and Methods. 2 Volume Set

2008

Hardcover

ISBN: 978-0-470-06098-8

Miller, L. W. (eds.)

Probes and Tags to Study Biomolecular Function

2008

Hardcover

ISBN: 978-3-527-31566-6

Lengauer, T. (ed.)

Bioinformatics – From Genomes to Therapies

2007

Hardcover

ISBN: 978-3-527-31278-8

Schreiber, S. L., Kapoor, T., Wess, G. (eds.)

Chemical Biology

From Small Molecules to Systems Biology and Drug Design

2007

Hardcover

ISBN: 978-0-470-84984-2

Aehle, W. (ed.)

Enzymes in Industry

Production and Applications

2007

Hardcover

ISBN: 978-3-527-31689-2

Meyers, R. A. (ed.)

Proteins

From Analytics to Structural Genomics

2006

Hardcover

ISBN: 978-3-527-31608-3

Protein Engineering Handbook

Volume 2

Edited by
Stefan Lutz and Uwe T. Bornscheuer

WILEY-
VCH

WILEY-VCH Verlag GmbH & Co. KGaA

The Editors

Prof. Dr. Stefan Lutz
Dept. of Chemistry
Emory University
1515 Dickey Drive
Atlanta GA 30322
USA

Prof. Dr. Uwe T. Bornscheuer
Dept. of Biotechnology and Enzyme Catalysis
Institute of Biochemistry
Greifswald University
Felix-Hausdorff-Str. 4
17487 Greifswald

Library of Congress Card No.: applied for

British Library Cataloguing-in-Publication Data
A catalogue record for this book is available from the British Library.

Bibliographic information published by the Deutsche Nationalbibliothek
The Deutsche Nationalbibliothek lists this publication in the Deutsche Nationalbibliografie; detailed bibliographic data are available on the Internet at <http://dnb.d-nb.de>.

Typesetting SNP Best-set Typesetter Ltd., Hong Kong
Printing betz-druck GmbH, Darmstadt
Binding Litges & Dopf GmbH, Heppenheim

Printed in the Federal Republic of Germany
Printed on acid-free paper

ISBN: 978-3-527-31850-6

Contents

Volume 1

Protein Engineering Handbook. Edited by Stefan Lutz and Uwe T. Bornscheuer

ISBN: 978-3-527-31850-6

Preface

Protein engineering is pursued by scientists from many different disciplines. Chemists, biochemists, biologists, and engineers alike are engaged in tailoring enzymes. As diverse as their intellectual background is their motivation to do so, varying from a desire to understand the fundamentals of biocatalysis such as the intimate relationship of structure, dynamics and function to questions of evolution, from a need to adjust enzyme properties for industrial processes to the challenge of generating novel proteins for therapeutic and biomedical applications. To meet their objectives, researchers are using highly creative and innovative approaches to introduce beneficial changes to enzymes, focusing on – among other properties – greater activity, altered substrate specificity, improved enantioselectivity, and increased stability.

As a field of research, protein engineering has made significant contributions towards a better understanding of the physical and chemical properties of proteins. In return, it has benefited from advances in traditional areas of biochemistry and biophysics. Insights into the role of protein structure from x-ray crystallography and NMR spectroscopy experiments have been rapidly growing and, together with clever mechanistic studies by enzymologists, have greatly contributed towards a better rationale for function. Separately, the emerging appreciation for protein dynamics, as well as the implementation of single-molecule studies has given us an intimate look at the performance of not just bulk catalyst but individual molecules as they move along the reaction coordinate. Paralleling advances in our understanding of the fundamentals, the last two decades have brought three paradigm shifts on the technological side of protein engineering. Starting with the introduction of the polymerase chain reaction and recombinant gene technology, progress in the field has empowered researchers to manipulate amino acid sequences in a relatively straightforward fashion and obtain vast quantities of selected polypeptides in heterologous expression systems. Next, the recreation of protein evolution processes in the laboratory, using random mutagenesis and *in vitro* recombination techniques, has opened up exciting and powerful new opportunities for protein engineers in all disciplines. Lastly, the recent development of predictive computer algorithms has added an important new tool, complementing experimental approaches by guiding the design and, in some cases, allowing for complete *de novo* construction of enzymes.

Protein Engineering Handbook. Edited by Stefan Lutz and Uwe T. Bornscheuer

ISBN: 978-3-527-31850-6

Capturing these exciting developments, Volume 1 of this book series focuses on fundamental aspects of protein engineering. While the opening chapter by Edmondson defines some of the terminology related to the characterization of engineered enzymes and the comparison to its natural parents, the contributions of Kazlauskas, Hollfelder and Miller concentrate on the active site, exploring enantioselectivity and substrate promiscuity. An often neglected yet critical aspect of protein engineering is folding of the polypeptide chain. While Fersht highlights the application of protein engineering for the studies of the folding process, DeLisa recapitulates some of the strategies to identify properly folded proteins. Along the same line, protein dynamics is another largely overlooked aspect of protein engineering. The contribution by Romesberg introduces a series of spectrophotometric techniques to capture protein motion while the article by Pelletier summarizes recent findings by NMR spectroscopy and x-ray crystallography. In one example for putting protein folding and dynamics data to work in the context of enzyme engineering, Sagermann presents a simple yet elegant method to explore and exploit conformational changes for creating functional protein switches.

Giving thought to the observation that not all proteins are equally suitable for laboratory evolution, Höcker provides a more practical perspective on the selection of protein frameworks as starting points for enzyme engineering. New activity and promising templates for engineering can also be found in the vastness of the metagenome. Many new opportunities in this emerging research area are discussed in the chapters by Ferrer and Eck. Separately, the contributions by Wood and Horswill review the utilization of intein sequences as protein engineering tools. Finally, the application of computational methods to guide protein engineering and *de novo* design is examined in the section by Saven.

In Volume 2, mutagenesis and shuffling strategies for generating libraries are described in the contributions by Reetz, Jäger and Tawfik while computational and experimental tools for chimeragenesis are reviewed in the chapters by Arnold and Lutz. Less conventional but highly useful, Ostermeier discusses the impact of circular permutation on the structure and function of proteins.

As library generation represents only half the challenge in directed evolution, effective methods for searching the often substantial library diversity are necessary. Such screening or selection protocols are commonly performed in vivo or with the help of display systems as reported by Hilvert, Withers, Soumillion and Wittrup. The combination of such systems with spectroscopic assays offers a highly versatile screening strategy as outlined by Reymond and Coco. Alternatively, Ueda and Tawfik describe elegant in vitro strategies for library analysis. More product-oriented, Zhao and Bornscheuer discuss the application of protein engineering towards altering substrate and cofactor specificity, as well as enantioselectivity in individual enzymes. These strategies are not limited to single-enzyme systems. The chapters by Khosla, Steinbüchel, and Schmidt-Dannert demonstrate their application towards the manipulation of entire pathways. Similarly, protein engineering also offers new opportunities for tailoring biomaterials as described in the contributions of Taguchi, Conticello and Kaplan.

In summary, this book series attempts to capture some of the diverse interests and approaches in protein engineering, reflecting the many different disciplines and individual motivations and objectives in this area. We hope that it offers solutions to existing protein engineering problems and inspires new ideas to tackle the challenges in the field. In today's fast-moving world, it is unrealistic to expect an all-inclusive, up-to-date collection of knowledge and methods in any printed media. The current research literature is a more appropriate source for the latest hypotheses and technology. Aiming for scientists new to the field, we instead emphasize a review of the basics in the field, as well as introduce selected new and promising strategies for protein engineering. We hope that this will provide readers with a comprehensive overview of this highly interdisciplinary research topic. For the experienced protein engineer, the book series might offer some new inspiration as well.

A book project such as this would never succeed without the wonderful support of many individuals that inspired, encouraged, and assisted in its assembly. In addition to thanking all of the authors for their efforts, we would like to acknowledge our colleagues and students at Emory University and the University of Greifswald for their advice in managing such a project, as well as their willingness to review and proof-read the pages that make up the two volumes. Finally, our special thanks also extend to the people at Wiley Publisher, namely Dr. Frank Weinreich and Dr. Heike Nöthe for their editorial assistants, as well as Claudia Zschernitz and Nele Denzau for their help during the printing stage of the books.

Atlanta/Greifswald, July 2008 *Stefan Lutz & Uwe T. Bornscheuer*

List of contributors

Amir Aharoni
Department of Life Science and the NIBN
University of Ben Gurion in the Negev
POB 653
Beer-Sheva
84105
Israel
aaharoni@chem.ubc.ca

Miguel Alcalde
CSIC, Institute of Catalysis
Dept. of Applied Biocatalysis
28049 Madrid
Spain

Jennifer N. Andexer
Institute of Molecular Enzyme Technology
Heinrich-Heine-University Düsseldorf
Research Centre Jülich
52426 Jülich
Germany

Frances H. Arnold
Division of Chemistry and Chemical Engineering
California Institute of Technology
210-41
1200 E California Blvd
Pasadena
CA 91125
USA
frances@cheme.caltech.edu

Ana Beloqui
CSIC, Institute of Catalysis
Dept. of Applied Biocatalysis
28049 Madrid
Spain

Blaise R. Boles
Department of Internal Medic
Roy J. and Lucille A. Carver College of Medicine
University of Iowa 440 EMBR
Iowa City
IA 52242
USA

Protein Engineering Handbook. Edited by Stefan Lutz and Uwe T. Bornscheuer

ISBN: 978-3-527-31850-6

Uwe T. Bornscheuer
Department of Biotechnology and Enzyme Catalysis
Institute of Biochemistry
Greifswald University
Felix-Hausdorff-Str. 4
17487 Greifswald
Germany
uwe.bornscheuer@uni-greifswald.de

Dominique Böttcher
Department of Biotechnology and Enzyme Catalysis
Institute of Biochemistry
Greifswald University
Felix-Hausdorff-Str. 4
17487 Greifswald
Germany

Martina Carbone
Division of Chemistry and Chemical Engineering
California Institute of Technology
210-41
1200 E California Blvd
Pasadena
CA 91125
USA
martina@cheme.caltech.edu

Peggy Cebe
Department of Physics
Tufts University
Science and Technology Center
Room 208
4 Colby Street
Medford
MA 02155
USA

Alice Y. Chen
Department of Chemical Engineering
Stanford University
Stauffer III
381 North-South Mall
Stanford
CA 94305
USA

Wayne M. Coco
DIREVO Biotech AG
Nattermannalle 1
50829 Cologne
Germany
coco@direvo.com

Vincent P. Conticello
Department of Chemistry
Emory University
1515 Dickey Drive
Atlanta
GA 30322
USA
vcontic@emory.edu

Matthew P. DeLisa
120 Olin Hall
Department of Chemical and Biomolecular Engineering
Cornell University
Ithaca
NY 14853
USA
md255@cornell.edu

Kevin K. Desai
Department of Chemistry and Biochemistry
The Florida State University
213 Dittmer Laboratory
Tallahassee
FL 32306-4390
USA

Matthew DeSieno
Department of Chemical and Biomolecular Engineering
University of Illinois at Urbana-Champaign
600 South Mathews Avenue
Urbana
IL 61801
USA

Nicolas Doucet
Université de Montréal
Département de biochimie
CP 6128
Succursale Centre-Ville
Montréal
Québec
H3C 3J7 Canada

Jing Du
Department of Chemical and Biomolecular Engineering
University of Illinois at Urbana-Champaign
600 South Mathews Avenue
Urbana
IL 61801
USA

Jürgen Eck
B·R·A·I·N AG
Darmstaedter Straße 34-36
64673 Zwingenberg
Germany

Dale E. Edmondson
Departments of Biochemistry and Chemistry
Emory University
Atlanta
GA 30322-4098
USA

Thorsten Eggert
Institute of Molecular Enzyme Technology
Heinrich-Heine-University Düsseldorf
Research Centre Jülich
52426 Jülich
Germany
Present address
evocatal GmbH
Merowingerplatz 1a
40225 Düsseldorf
Germany

Neil Ferguson
Medical Research Council Centre for Protein Engineering
Hills Road
Cambridge CB2 0QH
United Kingdom
and
Cambridge University
Chemical Laboratory
Lensfield Road
Cambridge CB2 1EW
United Kingdom

Manuel Ferrer
CSIC, Institute of Catalysis
Dept. of Applied Biocatalysis
28049 Madrid
Spain
mferrer@icp.csic.es

Alan R. Fersht
Medical Research Council Centre for Protein Engineering
Hills Road
Cambridge CB2 0QH
United Kingdom
and
Cambridge University
Chemical Laboratory
Lensfield Road
Cambridge CB2 1EW
United Kingdom
arf25@cam.ac.uk

Adam C. Fisher
120 Olin Hall
Department of Chemical and Biomolecular Engineering
Cornell University
Ithaca
NY 14853
USA

Susanne A. Funke
Institute of Molecular Enzyme Technology
Heinrich-Heine-University Düsseldorf
Research Centre Jülich
52426 Jülich
Germany
Present address
Institute of Neuroscience and Biophysics, Molecular Biophysics
Research Center Jülich
52426 Jülich
Germany

Esther Gabor
B·R·A·I·N AG
Darmstaedter Straße 34-36
64673 Zwingenberg
Germany

Giovanni Gadda
Departments of Chemistry and Biology
Georgia State University
The Center for Biotechnology and Drug Design
Atlanta
GA 30302-4098
USA

Alison R. Gillies
Department of Chemical Engineering
Princeton University
Princeton
NJ 08544
USA

Peter N. Golyshin
Division of Microbiology
HZI – Helmholtz Centre for Infection Research
38124 Braunschweig
Germany
and
Department of Biological Sciences
University of Wales
Bangor LL57 2DG
United Kingdom

Benjamin J. Hackel
Massachusetts Institute of Technology
Building E19-563
50 Ames Street
Cambridge
MA 02142
USA

Ulrich Haupts
DIREVO Biotech AG
Nattermannalle 1
50829 Cologne
Germany

Asael Herman
Department of Biological Chemistry
Weizmann Institute of Science
Rehovot 76100
Israel
and
Department of Pathology
University of Washington School of Medicine
HSB K-058, BOX 357 705
Seattle
WA 98195-7705
USA

Oliver Hesse
DIREVO Biotech AG
Nattermannalle 1
50829 Cologne
Germany

Donald Hilvert
Laboratory of Organic Chemistry
E.T.H. Zurich
Altwiesenstrasse 64
CH-8093 Zurich
Switzerland
hilvert@org.chem.ethz.ch

Birte Höcker
Max-Planck-Institute for
Developmental Biology
Spemannstrasse 35
72076 Tübingen
Germany
birte.hoecker@tuebingen.mpg.de

Florian Hollfelder
University of Cambridge
Department of Biochemistry
Cambridge CB2 1GA
United Kingdom
fh111@cam.ac.uk

Alexander R. Horswill
Department of Microbiology
Roy J. and Lucille A. Carver
College of Medicine
431 Newton Rd
540 F. Eckstein
Medicinal Research Building
University of Iowa
Iowa City
IA 52242
USA
alex-horswill@uiowa.edu

Karl-Erich Jaeger
Institute of Molecular Enzyme
Technology
Heinrich-Heine-University Düsseldorf
Research Centre Jülich
52426 Jülich
Germany

Stefanie Jonas
University of Cambridge
Department of Biochemistry
Cambridge CB2 1GA
United Kingdom

Manu Kanwar
Department of Chemical and
Biomolecular Engineering
Johns Hopkins University
3400 N. Charles St.
Baltimore
MD 21218-2681
USA

David. L. Kaplan
Department of Biomedical
Engineering
Tufts University
Medford
MA 02155
USA

Peter Kast
Laboratory of Organic Chemistry
E.T.H. Zurich
Altwiesenstrasse 64
CH-8093 Zurich
Switzerland

Romas Kazlauskas
Department of Biochemistry,
Molecular Biology
and Biophysics, and The Biotechnology Institute,
University of Minnesota
1479 Gortner Avenue
Saint Paul
MN 55108
USA

Chaitan Khosla
Departments of Chemical
Engineering, Chemistry and
Biochemistry
Stanford University, Keck 337
Stanford
CA 94305
USA
khosla@stanford.edu

Andreas C. Kleeb
Laboratory of Organic Chemistry
E.T.H. Zurich
Altwiesenstrasse 64
CH-8093 Zurich
Switzerland

Marco Landwehr
Division of Chemistry and
Chemical Engineering
California Institute of Technology
210-41
1200 E California Blvd
Pasadena
CA 91125
USA

Yougen Li
Division of Chemistry and Chemical
Engineering
California Institute of Technology
210-41
1200 E California Blvd
Pasadena
CA 91125
USA

Klaus Liebeton
B·R·A·I·N AG
Darmstaedter Straße 34-36
64673 Zwingenberg
Germany

Stefan Lutz
Department of Chemistry
Emory University
1515 Dickey Drive
Atlanta
GA 30322
USA
sal2@emory.edu

Thomas J. Mansell
120 Olin Hall
Department of Chemical and
Biomolecular Engineering
Cornell University
Ithaca
NY 14853
USA

Glenna E. Meister
Department of Chemical and
Biomolecular Engineering
Johns Hopkins University
3400 N. Charles St.
Baltimore
MD 21218-2681
USA

Guido Meurer
B·R·A·I·N AG
Darmstaedter Straße 34-36
64673 Zwingenberg
Germany

Michelle Meyer
Division of Chemistry and
Chemical Engineering
California Institute of Technology
210-41
1200 E California Blvd
Pasadena
CA 91125
USA

Brian G. Miller
Department of Chemistry and
Biochemistry
The Florida State University
213 Dittmer Laboratory
Tallahassee
FL 32306-4390
USA
miller@chem.fsu.edu

Philippe Minard
Laboratoire de Modelisation et
Ingénierie des Protéines
Institut de Biochimie et
Biophysique Moléculaire et
Cellulaire
Université Paris-Sud - Bat. 430
91405 Orsay
France

Martin Neuenschwander
Laboratory of Organic Chemistry
E.T.H. Zurich
Altwiesenstrasse 64
CH-8093 Zurich
Switzerland

Frank Niehaus
B·R·A·I·N AG
Darmstaedter Straße 34-36
64673 Zwingenberg
Germany

Marc Ostermeier
Department of Chemical and Biomo-
lecular Engineering
Johns Hopkins University
3400 N. Charles St.
Baltimore
MD 21218-2681
USA
oster@jhu.edu

Melissa Patterson
Department of Chemistry
Emory University
1515 Dickey Drive
Atlanta
GA 30322
USA

Sonha C. Payne
Department of Chemistry
Emory University
1515 Dickey Drive
Atlanta
GA 30322
USA

Joelle N. Pelletier
Université de Montréal
Département de chimie &
Département de biochimie
CP 6128
Succursale Centre-Ville
Montréal
Québec
H3C 3J7 Canada
joelle.pelletier@umontreal.ca

Alexander Pisarchik
Department of Biochemistry,
Molecular Biology and Biophysics
University of Minnesota
1479 Gortner Avenue
St. Paul
MN 55108
USA

Manfred T. Reetz
Max-Planck-Institut für
Kohlenforschung
Kaiser-Wilhelm-Platz 1
45470 Mülheim an der Ruhr
Germany
reetz@mpi-muelheim.mpg.de

Jean-Louis Reymond
University of Berne
Department of Chemistry and
Biochemistry
Freiestrasse 3
3012 Berne
Switzerland
jean-louis.reymond@ioc.unibe.ch

Floyd E. Romesberg
The Scripps Research Institute
Department of Chemistry
10550 N. Torrey Pines Road
La Jolla
California
USA
floyd@scripps.edu

Gloria Saab-Rincon
Departmento de Ingenieria
Celular y Biocatalisis Instituto de
Biotecnologia Universidad
Nacional Autonoma de Mexico
Apdo Postal 510-3 Cueernavaca
Morelos 62250
Mexico

Martin Sagermann
Department of Chemistry and
Biochemistry and
Interdepartmental Program in
Biomolecular Science and Engineering
University of California Santa Barbara
Santa Barbara 4649 B PSB North
California 93106-9510
USA
sagermann@chem.ucsb.edu

Jeffery G. Saven
Department of Chemistry
University of Pennsylvania
231 South 34th Street
Philadelphia
PA 19104-6323
USA
saven@sas.upenn.edu

Marlen Schmidt
Department of Biotechnology and
Enzyme Catalysis
Institute of Biochemistry
Greifswald University
Felix-Hausdorff-Str. 4
17487 Greifswald
Germany

Claudia Schmidt-Dannert
Department of Biochemistry,
Molecular Biology and Biophysics
University of Minnesota
1479 Gortner Avenue
St. Paul
MN 55108
USA
schmi232@umn.edu

Patrice Soumillion
Laboratoire d'Ingénierie des
Protéines et des Peptides
Institut des Sciences de la Vie
Université catholique de Louvain
Place Croix du Sud 4-5, bte 3
1348 Louvain-la-Neuve
Belgium
patrice.soumillion@uclouvain.be

Alexander Steinbüchel
Institut für Molekulare Mikrobiologie und Biotechnologie der
Westfälischen
Wilhelms-Universität
Corrensstraße 3
48149 Münster
Germany
steinbu@uni-muenster.de

Anna Steinle
Institut für Molekulare
Mikrobiologie und Biotechnologie
der Westfälischen
Wilhelms-Universität
Corrensstraße 3
48149 Münster
Germany

Michael Strerath
DIREVO Biotech AG
Nattermannalle 1
50829 Cologne
Germany

Seiichi Taguchi
Division of Biotechnology and
Macromolecular Chemistry
Graduate School of Engineering
Hokkaido University
N13W8, Kita-ku
Sapporo 060-8628
Japan
staguchi@isc.meiji.ac.jp

Dan S. Tawfik
Department of Biological Chemistry
Weizmann Institute of Science
Ullmann Bldg
Room 201a
PO BOX 26
Rehovot 76100
Israel
tawfik@weizmann.ac.il

Megan C. Thielges
The Scripps Research Institute
Department of Chemistry
10550 N. Torrey Pines Road
La Jolla
California
USA

Manuela Trani
Department of Chemistry
Emory University
1515 Dickey Drive
Atlanta
GA 30322
USA

Takeharu Tsuge
Department of Innovative and Engineered Materials
Tokyo Institute of Technology
4529 Nagatsuta
Midori-ku
Yokohama 226-8502
Japan

Takuya Ueda
Department of Medical Genome
Sciences
Graduate School of Frontier Sciences
The University of Tokyo
FSB-401
5-1-5 Kashiwanoha
Kashiwa
Chiba 277-8562
Japan
ueda@k.u-tokyo.ac.jp

Agathe Urvoas
Laboratoire de Modelisation et
Ingénierie des Protéines
Institut de Biochimie et Biophysique Moléculaire et Cellulaire
Université Paris-Sud – Bat. 430
91405 Orsay
France

Peter J. Walla
Technische Universität
Braunschweig
Institute for Physical and
Theoretical Chemistry
Dept. for Biophysical Chemistry
Hans-Sommerstr. 10
38106 Brunswick
Germany
and
Max-Planck-Institute for Biophysical Chemistry
Am Faßberg 11
37077 Göttingen
Germany

Xiaoqin Wang
Department of Biomedical
Engineering
Tufts University
Science Technic Center
4 Colby Street
Medford
MA 02155
USA

Stephen G. Withers
Department of Chemistry
University of British Columbia
2036 Main Mall
Vancouver
British Columbia V6T 1Z1
Canada

K. Dane Wittrup
Massachusetts Institute of Technology
Building E19-551
77 Massachusettes Ave
Cambridge
MA 02139
USA
wittrup@mit.edu

David W. Wood
Department of Chemical Engineering
and
Department of Molecular Biology
Princeton University
Princeton
NJ 08544
USA
dwood@princeton.edu

Bei-Wen Ying
Department of Bioinformatic
Engineering
Graduate School of Information
Science and Technology
Osaka University
2-1 Yamadaoka
Suita
Osaka 565-0871
Japan

Wayne Yu
The Scripps Research Institute
Department of Chemistry
10550 N. Torrey Pines Road
La Jolla
California
USA

Huimin Zhao
Departments of Chemical and
Biomolecular Engineering,
Chemistry and Bioengineering
University of Illinois at
Urbana-Champaign
600 South Mathews Avenue
Urbana
IL 61801
USA
zhao5@uiuc.edu

Jörg Zimmermann
The Scripps Research Institute
Department of Chemistry
10550 N. Torrey Pines Road
La Jolla
California
USA

Miren Zumárraga
CSIC, Institute of Catalysis
Dept. of Applied Biocatalysis
28049 Madrid
Spain

16
A Method for Rapid Directed Evolution

Manfred T. Reetz

16.1
Introduction

Directed evolution [1] has emerged as a powerful means to improve essentially any property of an enzyme, including thermostability [2], robustness in hostile organic solvents [3] and stereoselectivity [4]. It is based on the appropriate combination of gene mutagenesis and expression coupled with high-throughput screening (HTS) or selection [5]. Some of the most important methods for gene mutagenesis are error-prone polymerase chain reaction (epPCR) [6], saturation mutagenesis [7] and DNA shuffling [8]. Although numerous studies have been reported which are based on these or related mutagenesis methods, until now it is not clear how these methods should be applied optimally. For example, what mutation rate should be used when applying epPCR? Which sites in an enzyme are best suited for saturation mutagenesis, and which genes are optimally shuffled? Also, should these methods be combined and, if so, how? Actually, any approach in directed evolution is likely to result in some progress, as long as a reasonable degree of evolutionary pressure is exerted by going through several cycles of mutagenesis/expression/screening (selection). Whilst the 1990s witnessed the 'rise' of directed evolution of functional enzymes, efficiency was not always the focus of interest. Moreover, very few studies were reported in which methods or combinations of methods were critically compared [1]. In recent years, however, it has become clear that new methods and strategies for probing protein sequence space efficiently are needed [1, 9]. The importance of methodology development was emphasized as early as 2004 in a review [9a] entitled 'Novel methods for directed evolution: quality, not quantity'. In this article, and in other summaries, new and older mutagenesis methods were critically reviewed [1, 9] and only a few points are reiterated here before turning to the actual subject of this chapter.

To this day the most often-used mutagenesis method is epPCR, as first reported by Leung *et al.* in 1989 [6a] and later improved by Joyce [6b]. The reason for this method's popularity relates to the fact that it is easy to perform in the laboratory,

Protein Engineering Handbook. Edited by Stefan Lutz and Uwe T. Bornscheuer

ISBN: 978-3-527-31850-6

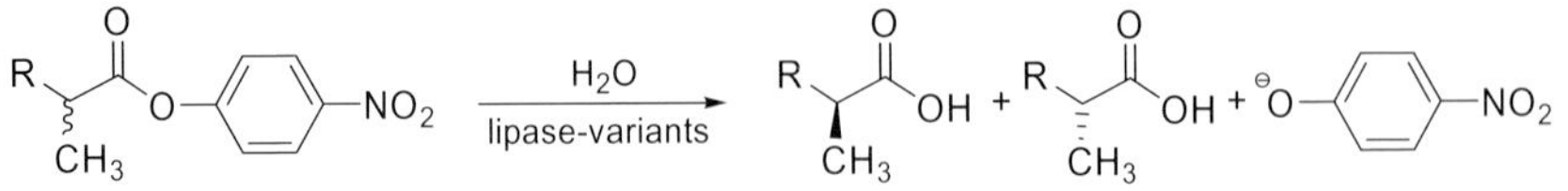

Scheme 16.1

although the problem of HTS must also be considered. Typically, five to ten cycles of epPCR are transversed, with at each stage between 1000 and 10 000 (or more) transformants being screened. During the 1990s, essentially all research groups adhered to a low mutation rate averaging one amino acid exchange per enzyme molecule, as for example in the first study regarding the directed evolution of an enantioselective enzyme [10]. In that study the selectivity factor *E* in the hydrolytic kinetic resolution of the chiral ester *rac*-**1**, catalyzed by the lipase from *Pseudomonas aeruginosa*, was increased stepwise from 1.1 to 11 by applying four cycles of epPCR at low mutation rate (Scheme 16.1). In each round one amino acid substitution occurred, resulting in the accumulation of four mutations [10].

Later, it was demonstrated that epPCR at higher error-rate with the simultaneous introduction of three amino acid substitutions leads to considerably better results [11], at least in the early stage of directed evolution. This may have surprised researchers in the field, but subsequently it was shown by a statistical analysis that even higher mutation rates may be beneficial [12]. Crucial in all real applications is, of course, the availability of an efficient screening system [5]. It is important to illuminate once more the limitations of epPCR [1, 13]. Due to the degeneracy of the genetic code, this mutagenesis method is not truly random. The bias of epPCR has been recognized in many critical reports [1, 13]. For example, at best one nucleotide of a given codon will be exchanged, thereby leading to just nine (instead of 64 possible) different codons encoding four to seven (instead of 20) different amino acids. For example, as illustrated for the lipase from *Bacillus subtilis* (Lip A), a 181 amino-acid protein, the number of amino acid substitutions achieved by epPCR depends on the type of the original codon [13a]. The number of enzyme variants obtained in reality using low error-rate (one mutation) is only one-third of the total theoretically possible variants [13a]. Obtaining variants having, for example, two amino acid substitutions at two neighboring sites is essentially impossible due to statistical reasons. This is unfortunate, because synergistic effects often occur at such neighboring positions (see Sections 16.3 and 16.4).

DNA shuffling [8], especially family shuffling [8b], is a powerful method for creating high diversity, but it also has limitations [14]. Relatively high homology (>60%) is usually required, but this disadvantage has been overcome to some extent by shuffling variations such as RACHITT [15] or synthetic shuffling [16]. However, the experimental effort may be extensive. The relative merits of the various approaches have been reviewed [9].

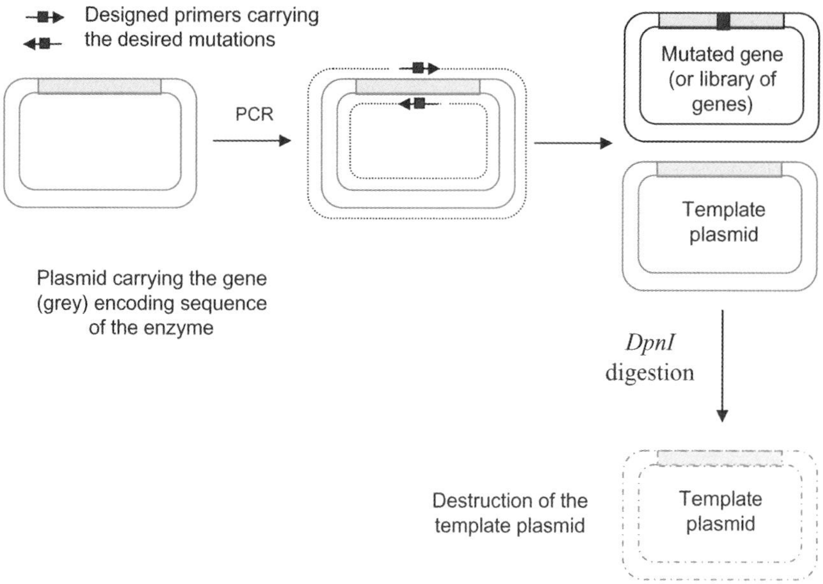

Figure 16.1 Schematic illustration of QuikChange (Stratagene) [17].

Saturation mutagenesis (sometimes termed cassette mutagenesis) allows for randomization at one, two, three, or more positions in the enzyme [1]. In general, this means the simultaneous introduction of all 20 proteinogenic amino acids at predefined positions, leading to 20, 400, 8000 and so on, enzyme variants, respectively. Several different molecular biological approaches to achieving this type of mutagenesis [7, 17] with the formation of focused libraries [9, 11a, 18] have been described. When applying this mutagenesis method, a rational decision as to the choice of appropriate sites in the enzyme to be randomized is required. The currently most often-used molecular biological method for saturation mutagenesis is the QuikChange protocol [17], which is based on earlier investigations [7b, c] and which has been extended [7d, e]. The general scheme is shown in Figure 16.1. It forms the molecular biological basis of the subject highlighted in this chapter, namely Iterative Saturation Mutagenesis (ISM) [19], as described in Section 16.3.

When applying saturation mutagenesis, it is important to consider codon degeneracy (which determines the use of the amino acid alphabet) and its relationship to oversampling [20]. Using statistical models published in 2003/2004 [21], data of the type shown in Table 16.1 were calculated [20], which are useful when designing saturation mutagenesis experiments. It can be seen that NNK codon degeneracy (N: Ade/Cyt/Gua/Thy; K: Gua/Thy), meaning all 20 proteinogenic amino acids as building blocks, requires the highest degree of oversampling when striving for 95% coverage. For example, when randomizing a site comprised of three amino acids, 98 163 bacterial colonies (clones) need to be evaluated. In sharp contrast for example, NDT degeneracy (N: Ade/Cyt/Gua/Thy; D: Ade/Gua/Thy; T: Thy),

Table 16.1 Statistical analysis of codon usage (selected examples) [20].

Codon degeneracy	No. of codons	No. of AA	No. of stops	AA encoded	95% coverage for 2 pos.	95% coverage for 3 pos.
NNK	32	20	1	All 20	3066	98 163
NDT	12	12	0	RNDCGHILF SYV	430	5175
DBK	18	12	0	ARCGILMFS TWV	969	17 470
NRT	8	8	0	RNDCGHSY	190	1532

meaning utilization of the 12 amino acids Arg, Asn, Asp, Cys, Gly, His, Ile, Leu, Phe, Ser, Tyr, Trp and Val as building blocks at the three defined amino acid positions, requires only 5175 clones to be evaluated if 95% coverage is desired [20].

This raises a strategic question which may be crucial to success in a given directed evolution study based on saturation mutagenesis. For example, if a choice is made to restrict the screening effort to about 5000 clones, then in the case of simultaneous randomization at a site composed of three amino acid positions, the choice of NNK would mean that only about 15% of the respective protein sequence space is covered. Compare this to NDT degeneracy, which restricts structural diversity to 12 amino acids, but ensures 95% coverage when screening 5000 clones. In both cases the same amount of laboratory work and screening effort is involved, but which choice is better? Although different amino acid alphabets have been employed for other purposes [22a], such as binary patterning [22b, c], this question has not been posed previously in the present context [20]. It might be suspected that NDT is the better choice, because the 'cocktail' of the 12 amino acids contains polar, nonpolar (lipophilic), charged and aromatic building blocks which should suffice (as a start in a directed evolution study). The results of a preliminary study (see Section 16.3) corroborate this conclusion, although it is currently not clear how general this is. Questions of this type do not touch on the mutagenesis method as such, but on its strategic use, which in turn affects the overall efficiency.

Questions of efficiency also need to be posed when considering other mutagenesis methods, especially when comparing them [1, 4c, 9, 14]. In this regard it is useful to consider the above-mentioned lipase from *P. aeruginosa* as a catalyst in the hydrolytic kinetic resolution of *rac*-**1** [10, 11, 23]. No other enzyme has been studied by directed evolution so systematically as this lipase; the most important results of these studies are summarized in Figure 16.2 [4c, 4d, 11a].

The best variant X was obtained by a combination of high error-rate epPCR, simultaneous saturation mutagenesis at two amino acid positions and DNA shuffling [11a]. Accordingly, the first step involved epPCR with the introduction of three amino acid substitutions, leading to mutants IV and V. The respective genes were considered for DNA shuffling in a special manner. Since positions 155 and 162 had been previously identified by epPCR and saturation mutagenesis to be

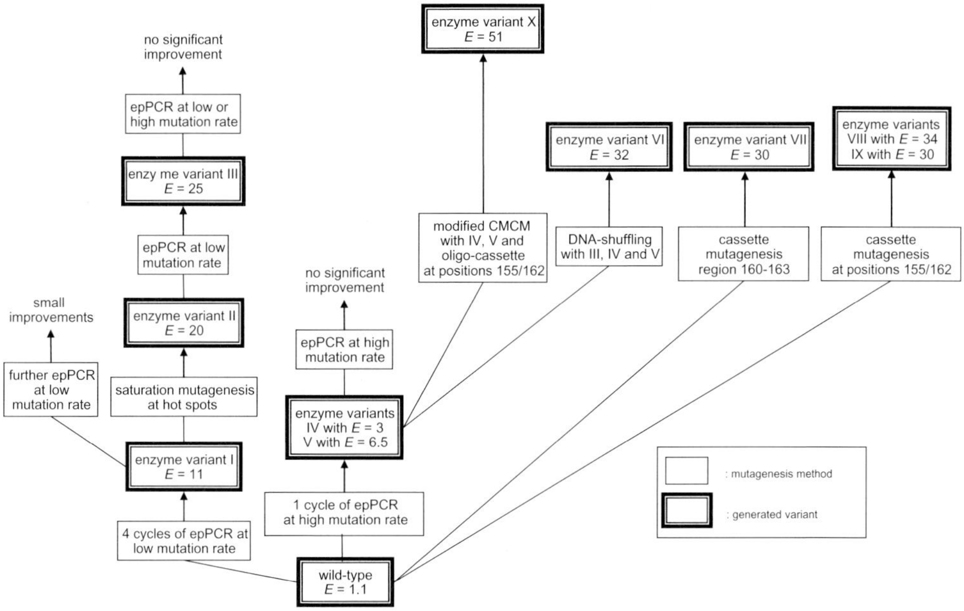

Figure 16.2 Schematic summary of the directed evolution of enantioselective lipase-variants originating from the wild-type (WT) *Pseudomonas aeruginosa* lipase (PAL) used as catalysts in the hydrolytic kinetic resolution of ester *rac*-1. CMCM = Combinatorial multiple-cassette mutagenesis [4c, d, 11a].

'hot spots', genes D and E and a mutagenic oligocassette inducing randomization at 155 and 162 were subjected to DNA shuffling, giving rise to the best mutant X. It leads to an *E*-value of 51, and is characterized by six amino acid substitutions. Five of the mutations are remote and only one mutation occurs near the binding pocket [4, 11]. Remote effects had been reported previously in directed evolution studies [1], mainly with respect to thermostability [2] which involves a fundamentally different phenomenon. In contrast, researchers traditionally associate enantioselectivity and/or substrate acceptance with the binding pocket along the lines of Fischer's lock-and-key hypothesis and/or Koshland's induced-fit model [24]. A detailed quantum mechanics/molecular mechanics (QM/MM) study predicted that of the six mutations only three are required for high enantioselectivity, two of them being particularly important [25]. One of these, Ser53Pro, is located on the surface of the enzyme, the other one (Leu162Gly) being next to the binding pocket. A relay mechanism explaining the high enantioselectivity was shown to be likely, coupled with the prediction of a cooperative (synergistic) effect [25a]. Having learned from directed evolution, especially when accompanied by a sound theoretical analysis, further experiments using site specific mutagenesis were performed which fully corroborated the theoretical predictions. For example, the double mutant Ser53Pro/Leu162Gly led to a similar enantioselectivity ($E = 64$) [25b].

This analysis is not only a triumph of theory but also raises another important question: Why did the evolutionary process accumulate three to four mutations which are not required for the evolved property (enantioselectivity)? Superfluous mutations have been noted in other studies [1], although in general research groups investigating directed evolution do not consider deconvoluting the mutations of the final improved mutant. The simplest explanation is an inefficiency of the overall evolutionary process due to low-quality libraries, and it is likely that this pertains to the case described. A total of 40 000–50 000 bacterial colonies was transformed and picked, and the same number of clones screened. The extent of molecular biological and microbiological investigations, as well as the screening effort, is formidable yet tolerable. An inversion of enantioselectivity with the formation of *(R)*-**2** was also achieved by using a similar process [11b]. Nevertheless, the question of efficiency remains to be answered. The same applies to numerous other successful studies regarding the directed evolution of enantioselectivity [4] or thermostability [2, 3] which utilize similar strategies as described here. Especially in industrial applications speed is crucial, which means that rapid directed evolution is desired [9, 20].

16.2 Focused Libraries Generated by Saturation Mutagenesis

Saturation mutagenesis, irrespective of the particular molecular biological version used [7, 17], produces what today are called focused libraries [1, 9, 11a, 18–20] (see Section 16.1). This immediately raises the crucial question regarding the choice of the amino acid positions to be randomized. Three strategies are possible:

I. Perform epPCR, identify the 'hot spots', and then choose these for saturation mutagenesis [1, 23, 26].
II. Perform saturation mutagenesis systematically at all amino acid positions of a given enzyme [27].
III. Choose sites for saturation mutagenesis on the basis of structural/mechanistic data [7, 11a, 17–20].

Strategy I was introduced in 1999/2000 for studies investigating the thermostability of an esterase [26] and the aforementioned enantioselectivity of the lipase from *P. aeruginosa* [23b] (see Section 16.1). While this approach was certainly successful and has since been used in numerous studies [1], not all experiments of this type lead to improved enzyme variants, nor can they be expected to do so [4]. Strategy II was developed by Diversa [27a] and by the Mülheim Jülich groups [27b], specifically in the quest to increase the enantioselectivity of a nitrilase and of a lipase, respectively. Improved variants were identified, but the experimental effort was formidable. Moreover, such a scan does not allow for cooperative effects since, by nature, only single mutants are possible. Strategy III has the potential of being extremely powerful, provided that the appropriate criteria for choosing the correct sites to be saturated can be developed [7, 11a, 17–20]. Certainly, when improving such different properties as thermostability, pH stability, enantioselectivity or substrate acceptance, different criteria are necessary.

When evolving enantioselectivity or substrate acceptance (rate) using saturation mutagenesis, one logical possibility is to focus on the binding pocket. The first example regarding enantioselectivity involves the above-mentioned hydrolytic kinetic resolution of *rac*-**1** catalyzed by mutants of the lipase from *P. aeruginosa* [11a] (Section 16.1). Based on the X-ray structure [28], it was concluded that the site comprising four amino acids (positions 160–163) next to the binding pocket (Figure 16.3) could be a logical place to perform saturation mutagenesis (NNK codon usage). Indeed, upon screening only 5000 clones, a mutant (Glu160Ala/Ser161Asp/Leu162Gly/Asn163Phe) was identified which showed fairly high enantioselectivity in the model reaction *rac*-**1** $\rightarrow$ **2** ($E = 30$) [11a]. For 95% coverage, about 3×10^6 clones would have been necessary [20], which shows that complete scanning is not mandatory. However, it is likely that even better mutants are possible, and that these were actually missed. The real weakness of this study is a different one, namely that other sites around the binding pocket–which can be expected to be equally or even more important–were not considered [11a].

The above critical comment applies to many other studies in which focused libraries were created by saturation mutagenesis [7, 17, 18]. Again, improvements in the catalytic profile of enzymes were achieved, but the choice of only one or two amino acid sites seems somewhat arbitrary, leading to the conclusion that numerous hits were missed by such a procedure. This central issue is treated in Section 16.3 and constitutes the starting point for the development of a more efficient strategy.

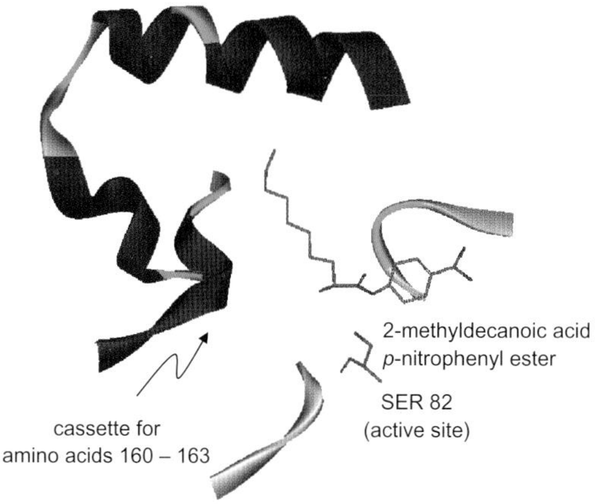

Figure 16.3 Binding pocket of *Pseudomonas aeruginosa* lipase (PAL) for the acid part of *rac*-**1**, showing the geometric position of amino acids 160–163 which were randomized simultaneously by saturation mutagenesis [11a].

Another important point to be considered concerns the question of remote versus close mutations, specifically when attempting to evolve enantioselectivity. The aforementioned lipase studies (see Section 16.1) showed that the best mutants have at least one remote mutation (Ser53Pro) on the surface and one close mutation (Leu160Gly) next to the binding pocket [4, 11a, 25]. However, this does not mean that remote mutations are required when evolving enantioselectivity. Statistically there are more remote sites than those next to the binding pocket. This is the reason why the overall process, which included epPCR, led to a hot spot at a remote site. This point was also clearly stressed in a study regarding an esterase, the conclusion being that close mutations are more important [18a]. However, the same authors showed in a later study, using a different enzyme, that remote mutations can be important [18b], while still other studies have likewise uncovered distal effects when evolving enantioselectivity or substrate acceptance (activity) [1, 29].

Although this situation may appear to be confusing, it need not be so. It is legitimate to allow remote mutations to occur, for example, by epPCR, and then to illuminate the source of enhanced stereo- or regioselectivity on a molecular level by sound theoretical analyses. Novel effects can be discovered by using such a strategy. In apparent 'contradiction' to this, it also makes sense to ignore epPCR (which in repeating rounds addresses the whole enzyme over and over again) and to restrict mutagenesis to properly chosen sites next to the binding pocket. From a practical point of view, this may be more efficient, and novel structural effects can also be unveiled by theoretical analyses. Finally, it is also clear that in any study regarding thermostability, a different criterion for choosing sites for saturation mutagenesis must be developed.

16.3 Iterative Saturation Mutagenesis

16.3.1 General Concept

Iterative saturation mutagenesis (ISM) constitutes the *systematic* application of this type of gene mutagenesis method in an evolutionary manner. It was first illustrated in the directed evolution of enantioselectivity [19] (see Section 16.3.2) and thermostability [20, 30] (Section 16.3.3). The first step in applying ISM is to analyze the enzyme of interest, generally on the basis of its three-dimensional (3-D) structure (X-ray analysis or homology model), and to identify *all* sites which can be expected to be important for saturation mutagenesis. Naturally, this will depend upon the nature of the catalytic property which is being addressed, an issue that will be treated in Sections 16.3.2 and 16.3.3. Consider, for example, an enzyme for which the respective analysis suggests four sites A, B, C and D (in other cases more or less sites are conceivable). These are comprised of two or three amino acid positions (only one or more than three are likewise conceivable). Following this analysis, the sites are randomized with the creation of four focused libraries, which are then screened for the property of interest. The decision as to the choice of codon usage (see Section 16.2) depends upon several factors, including the capacity of the screening system that is available. The best hit from each library A, B, C and D is then sequenced. Thereafter, the respective mutant genes are used as templates for further rounds of saturation mutagenesis, and the process is continued iteratively as shown in Figure 16.4 [19, 20].

ISM exerts evolutionary pressure in a defined region of protein sequence space. In every mutagenesis/screening round, the enzyme mutates exclusively at sensitive positions, and in each subsequent cycle the enzyme is allowed to 'respond' to the structural changes at the respective other sites. This not only allows for possible additivity, but also for cooperative effects in predefined regions.

The iterative process illustrated in Figure 16.4 is not convergent. However, convergency comes about automatically when stipulating that in a given upward pathway each site is considered only once (Figure 16.5) [19, 20]. So far, all ISM studies have adhered to this simplified version. In this case a total of 64 saturation mutagenesis libraries are possible (Figure 16.5), although of course if more than four sites are chosen this number rises. As before, it is not necessary to perform all of the experiments, which raises the question as to the optimal upward pathway. A pathway such as A → B → C → D can lead to a mutant which is different from one that emerges by transversing a different pathway, such as the reverse D → C → B → A. Extensive exploration of protein sequence space along these lines constitutes an intriguing endeavor. Thus far, the decision as to the order in visiting the sites has been somewhat arbitrary (Sections 16.3.2 and 16.3.3), for example, by taking a hierarchical approach. A conceptual precursor of ISM is the idea of DNA shuffling and simultaneous randomization at two amino acid positions previously identified by epPCR and earlier saturation mutagenesis [11a], as described in Section 16.1. Related to this is also the engineering of an epoxide hydrolase for enhancing the aerobic mineralization of *cis*-1,2-dichlorethylene by saturation

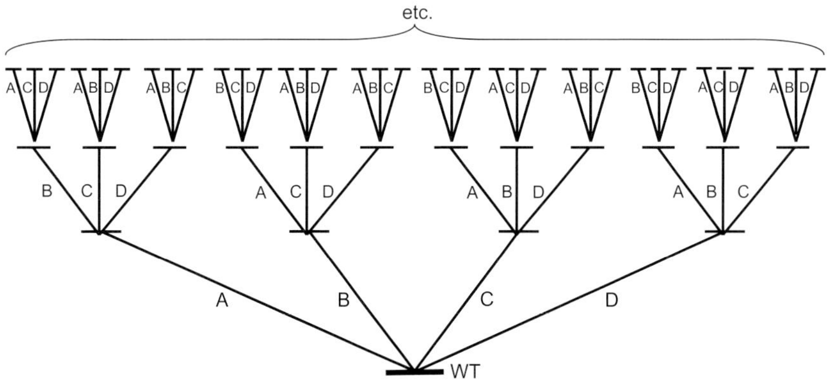

Figure 16.4 ISM employing four sites A, B, C and D [19, 20].

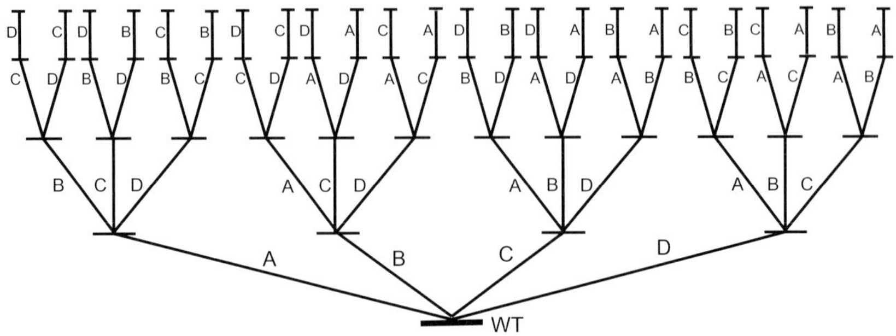

Figure 16.5 ISM employing four sites A, B, C and D, each site in a given upward pathway being visited only once [19, 20].

mutagenesis and co-expression in a DNA-shuffled toluene σ-monooxygenase [31]. Related also is a previous study of selective binding of zinc finger proteins to DNA, in which each domain was randomized and assembled individually, although catalysis was not involved [32].

16.3.2
Combinatorial Active-Site Saturation Test (CAST) as a Means to Control Substrate Acceptance and/or Enantioselectivity

As described in Section 16.3.1, the successful application of ISM requires reliable criteria for choosing the appropriate sites at which saturation mutagenesis is to be performed iteratively. In the case of expanding or shifting the scope of substrate acceptance or the enantioselectivity of an enzyme by ISM, the so-called Combinatorial Active-Site Saturation Test (CAST) [33] was applied [19, 20]. This requires knowledge of the 3-D structure of the enzyme under study, via either X-ray data or a homology model. Appropriate sites composed of one to three amino acids, the side chains of which reside near the binding pocket, are then identified [19, 20, 33]. CASTing is thus the systematic design and screening of focused libraries around the complete binding pocket, preferably with bound substrate or inhibitor [33]. In more general terms, the catalytically active center is used as a reference point of the enzyme, around which the Cartesian space within a radius of about 10 or more is partitioned into defined regions (sites) to be randomized [20]. Figure 16.6 features the case of four sites, each harboring two or three amino acids (more are possible).

CASTing was first proposed and illustrated in the quest to enlarge the substrate scope of the lipase from *P. aeruginosa*, specifically in an attempt to increase the rate of hydrolysis of the bulky *p*-nitrophenyl esters **4–14** (Scheme 16.2) [33]. Most of the sterically encumbered esters are not accepted by the wild-type lipase, and only a slow background reaction occurs.

The CAST analysis of the enzyme led to five sites, A, B, C, D and E, each harboring two amino acid positions (Figure 16.7) [33]. Such a systematization can be

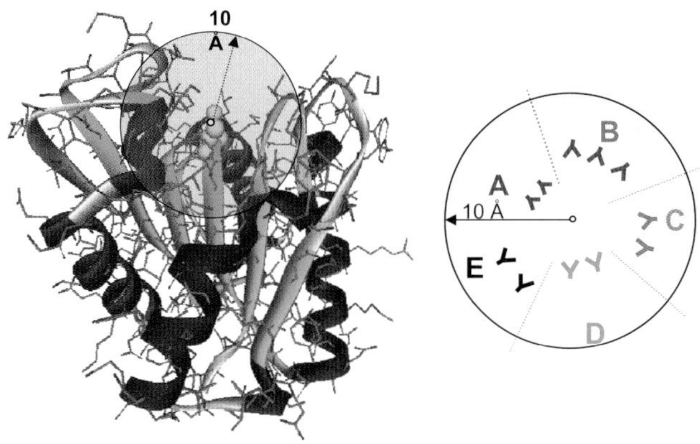

Figure 16.6 Generalization of CASTing [20]. For illustrative purposes the lipase from *Bacillus subtilis* [34] is shown, the yellow residue representing the catalytically active amino acid (left). An arbitrary dissection leads to four sites harboring, for example, two or three amino acid positions (right) [20].

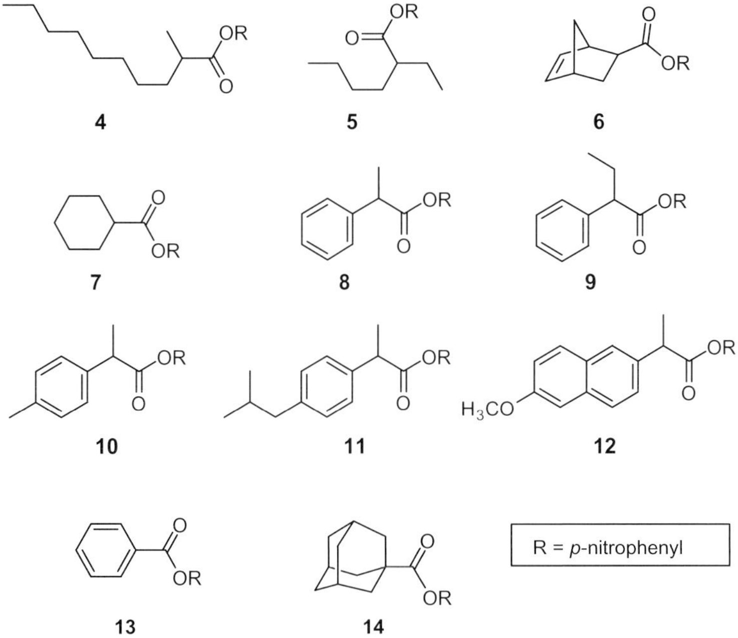

Scheme 16.2

compared to the earlier arbitrary choice of amino acid positions 160–163, as illustrated in Figure 16.3 [11a].

The results of this multisubstrate screening are remarkable, and only part of the data is shown here (Figure 16.8). Improved mutants (hits) were found for all of

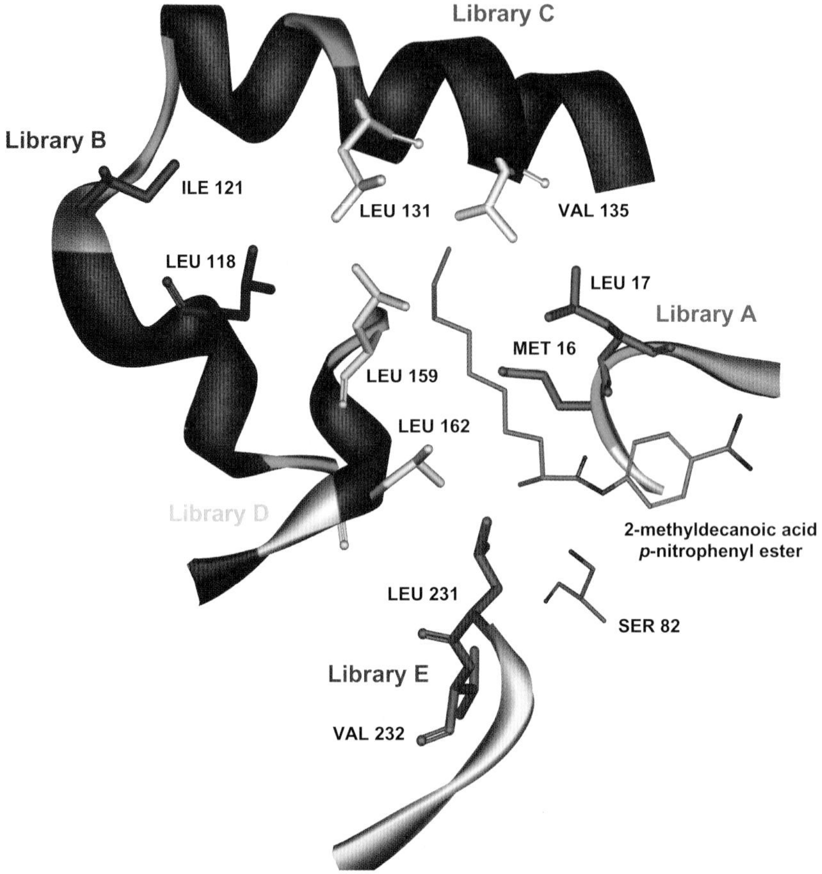

Figure 16.7 CASTing of the lipase from *Pseudomonas aeruginosa* (PAL) leading to the construction of five libraries of mutants (A–E) produced by simultaneous randomization at sites composed of two amino acids. (For illustrative purposes, the binding of substrate **4** is shown) [33, 35].

the substrates in these initial CAST libraries [33]. Most of the hits originated from libraries A and D, but interestingly many of the best ones were due to amino acid substitutions in which the side chains were sterically *larger* than those in the wild-type, which may be counterintuitive. However, molecular dynamics (MD) calculations show that H-bonding or π-π stacking effects are operating, which place the respective side chains away from the active center, resulting in actual enlargement of the binding pocket. In a follow-up study, mutations from hits originating from libraries A and D were combined, leading to further rate enhancement (up to a factor of 250) [35]. Such a traditional approach provides new mutants, but not any new mutations, which limits structural diversity and thus the degree of enzyme improvement. The application of CASTing iteratively in the form of ISM remains

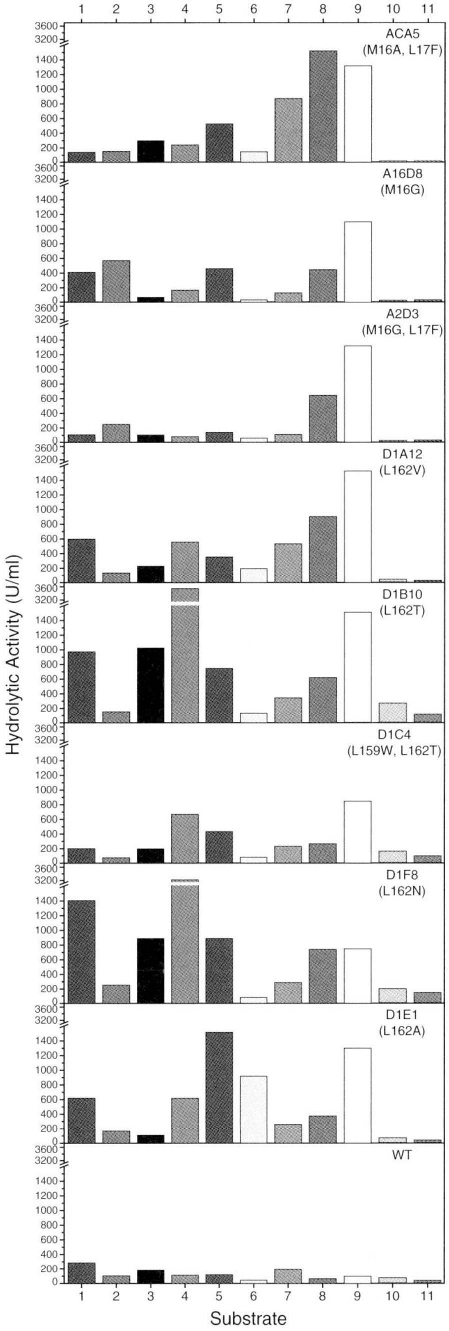

Figure 16.8 Substrate profile of lipase variants produced by CASTing [33]. Hydrolytic activity of selected mutants and the wild-type (WT) was measured photometrically in a continuous assay by following the absorbance at 405 nm for 8 min in a microtiter plate.

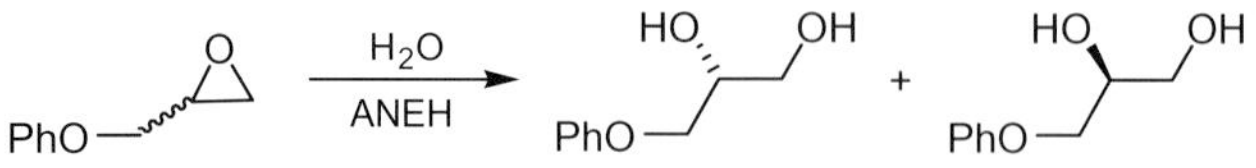

rac-**15** (*R*)-**16** (*S*)-**16**

Scheme 16.3

to be performed systematically in this particular case, but can be expected to be more efficient. Indeed, preliminary experiments applying iterative CASTing in the quest to enhance the enantioselectivity and/or to influence substrate acceptance of lipase have proved to be successful (M.T. Reetz, Y. Gumulya and S. Prasad, unpublished results).

Iterative CASTing was first applied in a study directed towards enhancing the enantioselectivity of the hydrolytic kinetic resolution of *rac*-**15** with preferential formation of (*S*)-**16**, catalyzed by the epoxide hydrolase from *Aspergillus niger* (ANEH) (Scheme 16.3) [19]. The wild-type shows a slight preference for (*S*)-**16** ($E = 4.6$).

In an earlier study, the traditional approach based on libraries produced by epPCR at different mutation rates had provided a triple mutant showing only moderately enhanced enantioselectivity ($E = 11$) [36]. This somewhat disappointing result had required the production and screening of 20 000 clones, and suggested that the ANEH is difficult to evolve. In the CAST study, six potentially sensitive sites – A, B, C, D, E and F – were identified on the basis of the X-ray structure [37] and by modeling the substrate into the somewhat narrow binding pocket [19] (Figure 16.9). These sites are comprised of either two or three amino acid positions.

Preliminary saturation mutagenesis at several sites showed that the best hit originates from library B, giving rise to $E = 14$, which is already better than the results obtained in the earlier study using epPCR [36]. In the ensuing ISM experiments (NNK codon degeneracy), the upward pathway B $\rightarrow$ C $\rightarrow$ D $\rightarrow$ F $\rightarrow$ E led to a dramatic increase in enantioselectivity, the final variant LW202 with nine amino acid substitutions showing an *E*-value of 115 (Figure 16.10).

These remarkable results deserve several comments. First, the overall effort required only 20 000 clones to be picked and screened. This happened to be approximately the same number evaluated in the earlier study using epPCR [36], yet the enantioselectivity was much higher ($E = 115$ versus $E = 11$). This comparison strongly suggests that ISM is more efficient than the traditional use of epPCR, at least in the present case. Moreover, site A was not even considered. It is also important to note that the libraries were considerably smaller than what would be required for 95% coverage of the respective protein sequence space. In summary, iterative CASTing appears to be a powerful approach to enhancing enantioselectivity because it provides 'smart' libraries. It is likely that other highly enantioselective ANEH mutants can be found if the iterative CAST search were to be performed more systematically, and/or if a different codon usage were to be considered.

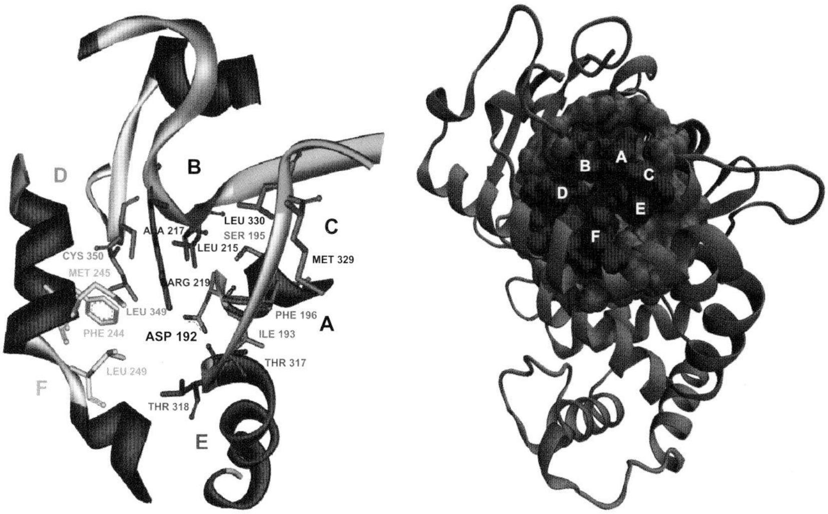

Figure 16.9 CASTing [19] of the epoxide hydrolase from *Aspergillus niger* (ANEH) based on the X-ray structure of the wild-type (WT) [37]. Left: Defined randomization sites A–E; Right: Top view of tunnel-like binding pocket showing sites A–E (blue) and the catalytically active Asp192 (red).

Two fundamentally different questions arise from this study. First, are there other upward pathways which likewise result in highly enantioselective ANEH mutants, and which are they? (These investigations are currently in progress; M.T. Reetz and J. Sanchis, unpublished results.) Second, is the specific order B $\rightarrow$ C $\rightarrow$ D $\rightarrow$ F $\rightarrow$ E using the observed *sets* of mutations the only one leading to the *specific* mutant LW202. Preliminary deconvolution of these sets shows that B $\rightarrow$ C $\rightarrow$ D $\rightarrow$ F $\rightarrow$ E is not unique, and that almost half of 120 constructed pathways are energetically feasible (M.T. Reetz and J. Sanchis, in press). This is of considerable theoretical interest, because another study had previously concluded that "Darwinian evolution can follow only few mutational pathways to fitter proteins" [38]. A question of a different theoretical nature concerns the source of enhanced enantioselectivity of variant LW202. The mechanism of ANEH activity was known to involve two tyrosines (251 and 314) which bind and activate the epoxide, followed by S_N2 reaction of Asp192 and hydrolysis of the covalently bound enzyme ester intermediate [37, 39] (Figure 16.11).

A preliminary MD study in which the evolutionary progress at each upward step in the sequence B $\rightarrow$ C $\rightarrow$ D $\rightarrow$ F $\rightarrow$ E was modeled, uncovered an interesting phenomenon (Figure 16.11) (M.T. Reetz and M. Bocola, unpublished results). In the wild-type the distance d between the O-atom of Asp192 and the C-atom of the activated substrate is very similar for *(R)*-**15** and *(S)*-**15**; that is, Δd is only 0.7 with the favored *(S)*-enantiomer being closer to the attacking nucleophile. As iterative CASTing progresses along with enhanced enantioselectivity, Δd

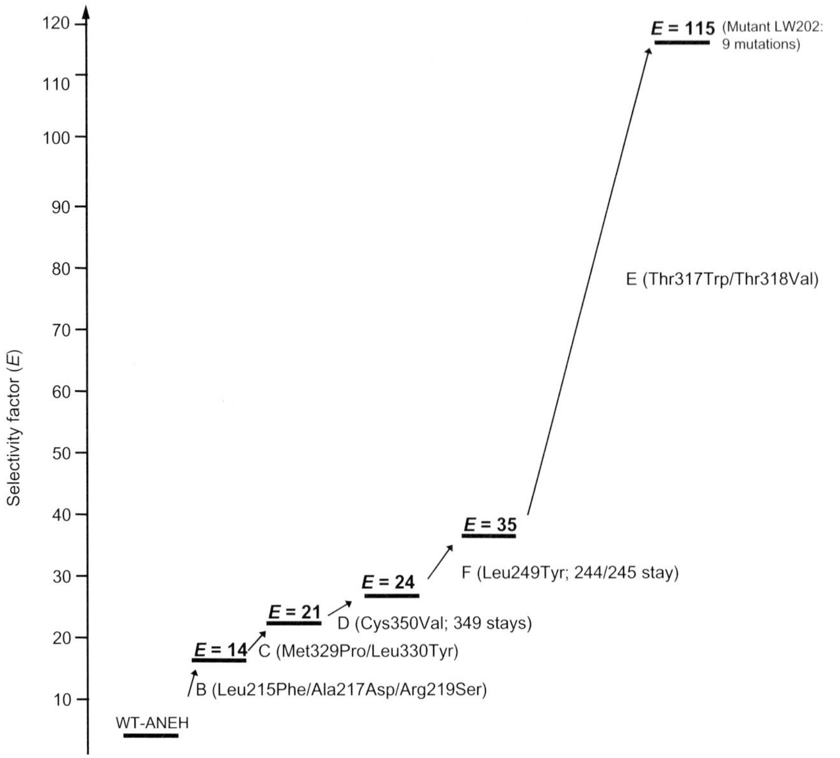

Figure 16.10 Iterative CASTing in the evolution of enantioselective ANEH mutants as catalysts in the hydrolytic kinetic resolution of *rac*-**15** [19].

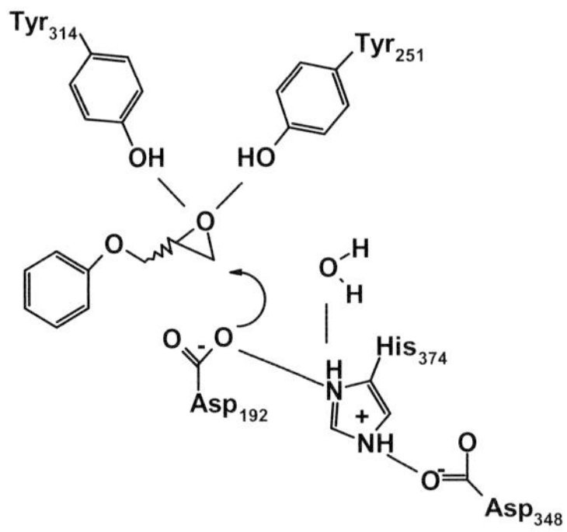

Figure 16.11 Mechanism of ANEH catalysis (the O-C distance d is indicated by the arrow).

increases (up to 2.4 Å in the case of LW202). Thus, the mutations lead to structures in which the disfavored enantiomer *(R)*-**15** is oriented farther away from Asp192 (~6 Å). In other words, the favored *(S)*-**15** takes on a near-attack conformer [40], which is not possible for disfavored *(R)*-**15**. Thus, it can be expected that the mutational changes responsible for this effect will slow down the reaction rate of *(R)*-**15**, leading to a high *E*-value. Although these calculations are crude, they are in line with kinetic studies [41].

More recently, it has been shown that the use of a reduced set of amino acids as specified by the use of the appropriate codon degeneracy provides the experimenter with an additional powerful tool for designing and generating focused libraries (M.T. Reetz, D. Kahakeaw and R. Lohmer, in press).

CASTing has also been utilized in the directed evolution of an enantioselective Baeyer-Villigerase (Cyclopentanone Monooxygenase), although only the initial saturation mutagenesis libraries were considered [42]. In the directed evolution of hybrid catalysts [43], in which a Rh/ligand entity was anchored to a host protein, iterative CASTing led to a significant improvement of enantioselectivity in olefin-hydrogenation [44]. Iterative CASTing was applied successfully to the original model reaction regarding the hydrolytic kinetic resolution of *rac*-**1** catalyzed by the lipase from *P. aeruginosa* (M.T. Reetz, Y. Gumulya and S. Prasad, unpublished results). Finally, CASTing has already been applied successfully by other groups, as in the directed evolution of an esterase [45] and a xylose reductase [46].

16.3.3
B-Factor Iterative Test (B-FIT) as a Means to Increase Thermostability

When using enzymes as catalysts in organic chemistry, polymer technology and pollution cleanup, as components in detergents, as diagnostic tools (sensors) or as bionanotechnological devices, thermostability is a crucial issue. A great deal of effort has been invested in the quest to increase thermostability by considering protein engineering, post-translational modification, immobilization or the use of additives. Methods based on protein engineering utilize either site-specific mutagenesis guided by rational design [47] or directed evolution relying on evolutionary principles [1, 2]. Typically, the increase in thermostability, as measured by T_m or T_{50}^{60} values, amounted to 5–15 °C, although in rare cases larger effects were observed (up to 35 °C) [48]. Most directed evolution studies regarding thermostabilization rely on multiple rounds of epPCR and/or DNA shuffling [1, 2]. For example, in an early study regarding the thermostabilization of an esterase from *Bacillus subtilis*, six to seven rounds of epPCR and DNA shuffling provided an improved mutant showing a ΔT_m-value of 14 °C [49]. In a recent study of another esterase, two rounds of epPCR were transversed, a process in which 1.5 million clones were screened, leading to an enhanced thermostability by 13 °C [50]. DNA shuffling was reported to fail in this case, which could not be explained unambiguously.

Recently, a novel approach to increase the thermostability of proteins was proposed and shown to be unusually efficient [20, 30]. This is based on ISM and, as detailed in the previous section, in order for ISM to be applicable it is necessary to develop criteria for choosing appropriate sites for mutagenesis [20]. In the case of thermostability, the B-Factor Iterative Test (B-FIT) was proposed [30].

Previously, numerous structural studies regarding mesophilic and thermophilic enzymes had shown that the latter are characterized by higher rigidity due to such interactions as salt bridges, H-bonding and π-π effects [2, 51]. The strategy behind B-FIT is to increase rigidity at those sites showing the greatest degree of flexibility [20, 30]. Although such sites in an enzyme can, in principle, be identified by NMR studies [52] or by computational means [53], a straightforward criterion was proposed: Sites showing high B-factors, which are available from X-ray data [54], should be chosen for saturation mutagenesis. They reflect smearing of atomic electron densities relative to their equilibrium positions as a result of thermal motion and positional disorder. The conventional X-ray data provides B-factors of all atoms, but this is of little use in B-FIT. Therefore, a computer-aided (B-FITTER) was developed with which it is possible automatically to calculate the average B-factor of each amino acid in the protein and thereby to obtain a hierarchical overview of the whole protein, beginning with the amino acid having the highest B-factor [20].

In the original study the lipase from *B. subtilis* (Lip A) [34] was chosen as the enzyme for enhancing thermostability [20, 30]. This is a mesophilic lipase composed of 181 amino acids, and has been characterized several times by using X-ray crystallography [55]. Only the resolved amino acids were considered, which means that in the particular case at hand the non-resolved C and N termini were not addressed. The 10 amino acids that showed the highest average B factors were then chosen as sites for randomization (Arg33: average B-factor = 50.9; Lys69: 44.1; Gln164: 40.7; Asp34: 39.9; Lys112: 39.6; Lys35: 38.9; Met134: 38.5; Tyr139: 37.9; Ile157: 37.4; Gly13: 37.0). Eight libraries were subsequently constructed by saturation mutagenesis, namely at sites A (Gly13), B (Arg33, Asp34, Lys35), C (Lys69), D (Lys112), E (Met134), F (Tyr139), G (Ile157) and H (Gln164) (see Figure 16.12).

Saturation mutagenesis was performed using the QuikChange mutagenesis method from Stratagene [17a]. NNK codon usage was chosen and appropriate oversampling performed. Owing to the difficulties in performing high-throughput assays with triglycerides (which are the natural substrates), *p*-nitrophenyl caprylate was employed as the substrate [20, 30]. Fast screening for improved thermostability in the initial eight libraries was performed in 96-well microtiter plates by heating the enzyme solutions (diluted supernatants) at 54 °C for 15 min using a PCR thermocycler. The reactions were monitored by a UV/Vis plate reader (405 nm). Under these test conditions the wild-type (WT) enzyme displayed a residual activity of only 9% in the hydrolysis of *p*-nitrophenyl caprylate. Thermostability was assessed by measuring the residual activity subsequent to the exposure to high temperature. In the literature, the so-called T_{50}-value is often used to quantitatively characterize thermostability [2, 24, 47, 51]; this is the temperature required to reduce the initial enzymatic activity by 50% within a given period of time. T_{50} is near or at the critical temperature of denaturation. In order to speed

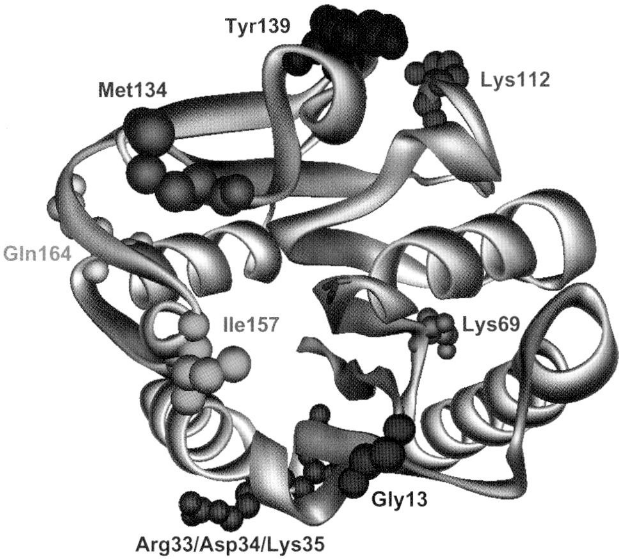

Figure 16.12 Sites in Lip A chosen for saturation mutagenesis [20, 30]. The picture shows a structural model based on the X-ray structure [55b]. Library A (dark blue); Library B (red); Library C (dark green); Library D (violet); Library E (brown); Library F (light blue); Library G (light green); Library H (yellow). Ser77 is the catalytically active site.

up the screening, a 15 min heat treatment, resulting in T_{50}^{15} values, was used. The initial eight libraries A, B, C, D, E, F, G and H were screened for thermostability, followed by a hierarchical upward pathway [20, 30] (Figure 16.13). Two sites C and D were not 'visited' again, due to the poor harvest in the initial mutagenesis round. In the last round, ISM along the pathway E → G → F → D → A (red letters in Figure 16.13), two variants of high thermostability were identified, X and XI [20, 30].

Following the isolation and purification of both variants, the respective residual activity curves, the real T_{50}^{60} values (60 min heating time), the kinetic constants, as well as the substrate and selectivity profiles were determined. Figure 16.14 shows the residual activity curves of the WT and of variants X and XI, demonstrating the enormous differences in thermostability. The half-lives ($t_{1/2}$) at 55 °C were measured as 905 and 980 min, respectively, whereas the $t_{1/2}$ of the WT was <2 min [20, 30]. The enhanced thermostability was also revealed by the T_{50}^{60} values of variants X and XI, which were 89 and 93 °C, respectively, compared to 48 °C for the WT. This means that ΔT_{50}^{60} values of 41 °C and 45 °C, respectively, had been achieved, which had no precedence. Most importantly, only 8000 clones had to be picked and screened.

It was observed that thermostabilization is achieved without compromising the catalytic profile at room temperature, which is a crucial issue [20, 30]. Kinetic

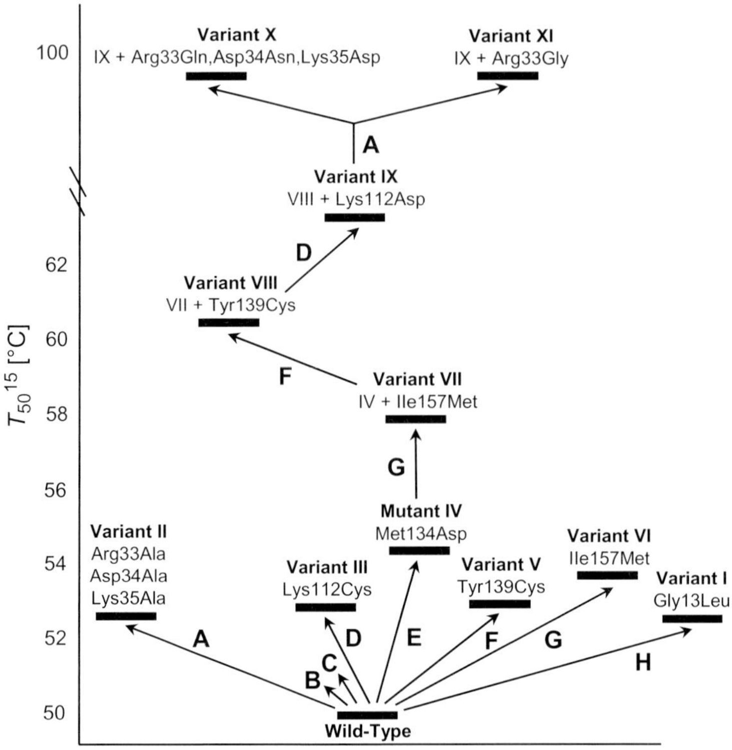

Figure 16.13 Enhancing the thermostability of the lipase from *Bacillus subtilis* (Lip A) by ISM using the B-FIT method [20, 30]. The sites A–H are defined in Figure 16.12.

measurements were made by using *p*-nitrophenyl acetate (PNPA) as the substrate at 25 °C in potassium phosphate buffer solution (10 m*M*, pH 7.0). The values of K_m and k_{cat} of the WT and of the purified variants X and XI were derived from the corresponding Michaelis–Menten plots. The data showed that no significant differences in catalytic parameters existed (WT: $K_m = 1.1\,\mathrm{m}M$; $k_{cat} = 6.5 \times 10^{-2}\,\mathrm{min}^{-1}$; $k_{cat}/K_m = 5.8 \times 10^{-2}\,\mathrm{m}M^{-1}\,\mathrm{min}^{-1}$. Variant X: $K_m = 0.62\,\mathrm{m}M$, $k_{cat} = 5.8 \times 10^{-2}\,\mathrm{min}^{-1}$; $k_{cat}/K_m = 9.4 \times 10^{-2}\,\mathrm{m}M^{-1}\,\mathrm{min}^{-1}$. Variant XI: $K_m = 0.73\,\mathrm{m}M$; $k_{cat} = 6.3 \times 10^{-2}\,\mathrm{min}^{-1}$; $k_{cat}/K_m = 8.6 \times 10^{-2}\,\mathrm{m}M^{-1}\,\mathrm{min}^{-1}$). Thus, the thermostability of the variants has been increased significantly without influencing enzyme activity. This conclusion was corroborated by measuring the activity and, when relevant, the enantioselectivity of the WT and variants X and XI as catalysts in the hydrolysis of 12 other *p*-nitrophenyl esters, some of which were chiral [20, 30]. Significant differences in the catalytic profile were not observed in any of the cases. For example, in the hydrolytic kinetic resolution of *p*-nitrophenyl 2-[4-(2-methylpropyl)phenyl] propanoate, the selectivity factor *(E)* in favor of the *(R)* enantiomer remained constant when going from the WT to the variants (WT: $E = 8.1$; variant X: $E = 8.2$; variant XI: $E = 8.5$) [20, 30].

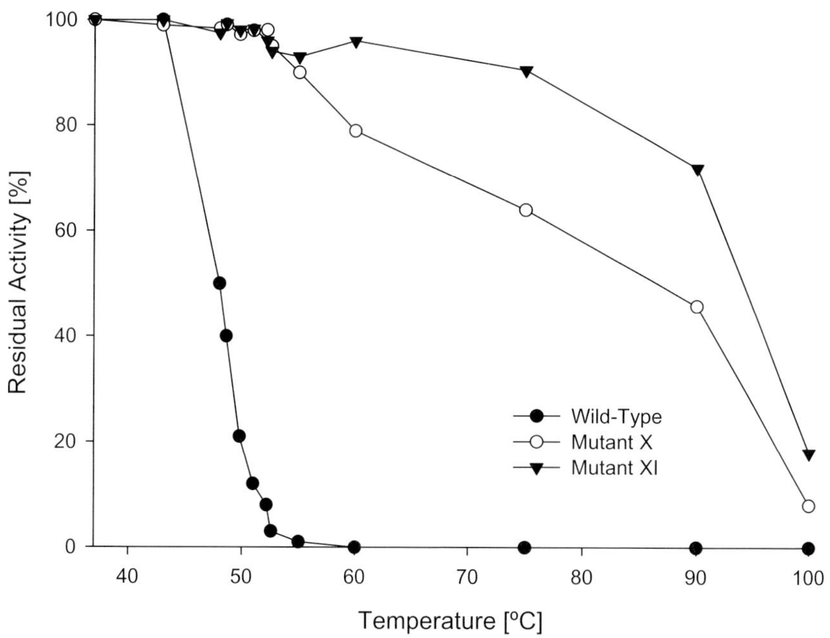

Figure 16.14 Thermostability of purified wild-type (WT) and mutants X and XI of Lip A as displayed by the residual activity curves. Enhancing the thermostability of the lipase from *Bacillus subtilis* (Lip A) by ISM using the B-FIT method [20, 30]. The sites A–H are defined in Figure 16.13 [20, 30]. Activity in the hydrolysis of *p*-nitrophenyl caprylate was measured after the enzyme solutions were treated at various temperatures (37–100 °C) for 1 h.

Previous X-ray studies of Lip A had shown that the enzyme has a compact minimal α/β hydrolase fold with a six-stranded parallel β sheet flanked by five α-helices, two being on one side of the sheet and three being on the other side [55]. The catalytic triad (Ser77, Asp133 and His156), as well as the amino acids forming the oxyanion hole, reside in positions very similar to those of other lipases. Most of the mutations (positions 13, 33, 34, 35, 112, 134 and 157) lie in surface loops (Figure 16.15) where flexibility is expected to be highest. The respective amino acid exchange helps in stabilizing these loops – and thus the protein – through a variety of interactions [1, 2, 24, 47, 51]. Solvent exposure of the randomized positions may also play a role [20, 30].

It is thus clear that although a hyperthermophilic enzyme has been evolved, its catalytic profile is that of a mesophile. This is important for long-term function under operating conditions at room temperature, and a detailed theoretical study is required in order to explain, on a molecular level, the source of this enhanced thermostability. A further noteworthy conclusion was made in this study, namely the suggestion that the directed evolution of thermostabilization should precede attempts to increase enantioselectivity or to influence substrate acceptance [30].

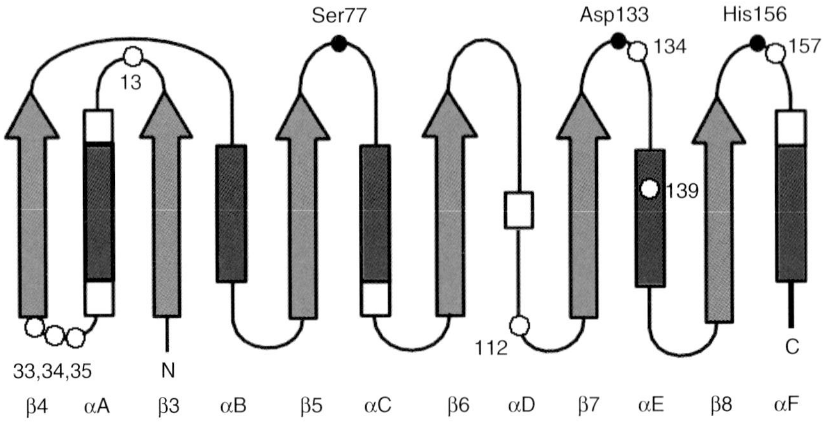

Figure 16.15 Secondary-structure topology of Lip A, modified from the representation published elsewhere [54], and showing the mutations [20, 30]. The nomenclature of the strands is taken from the canonical α/β hydrolase fold. All the positions that lead to thermostable variants are represented as white circles, the amino acids of the triad as black circles, and the termini of the protein as C and N. The white boxes represent 3_{10} helical turns [30].

Indeed, this goes hand in hand with the conclusion that enhanced thermostability promotes the ease of protein evolvability [56].

16.3.4
Practical Hints for Applying ISM

As highlighted above, several molecular biological methods for performing saturation mutagenesis are available [1, 7, 17]. The QuikChange protocol from Stratagene [17], which is based on earlier studies [7d, e], is used most often, but details of other extensions have also been published [17d, e]. In order to design and generate CAST libraries in rapid time [19, 33, 35, 42, 44], the computer aid (CASTER) is available from the literature [20] or directly from the author (http://www.mpi-muelheim.mpg.de/reetz.html). The same applies to the design and generation of an appropriate saturation mutagenesis for the purpose of increasing the thermostability of proteins using the B-FIT method [20, 30], and potential users should employ B-FITTER, which is also available from the literature [20] or from the author (http://www.mpi-muelheim.mpg.de/reetz.html). Details relating to the design of libraries, to molecular biological investigations and to screening along all steps of B-FIT are likewise accessible [20].

16.4
Conclusions

The directed evolution of enzymes leading to increased thermostability, different substrate scope and enhanced enantioselectivity, constitutes a powerful form of protein engineering. Traditional methods based on repeating rounds of epPCR, saturation mutagenesis at 'hot spots' and/or DNA shuffling are successful in these endeavors, as demonstrated by numerous earlier studies [1, 2, 4]. However, the question of efficiency is a pressing issue which touches not only on the actual mutagenesis methods but also on improved strategies for probing protein sequence space. In particular, industry requires rapid directed evolution, and ISM–in its two presently known embodiments, namely iterative CASTing and B-FIT–represents a respectable contribution towards achieving this important goal [19, 20, 33, 35, 42, 44]. Nonetheless, further studies are clearly required in the quest to generalize this approach.

In all of the studies considered so far, ISM has been unusually successful, although the reasons for such high efficiency remain to be disclosed. However, it can be anticipated that one important factor relates to the appropriate choice regarding the sites at which saturation mutagenesis is performed. If this choice is properly made–as in CASTing or B-FIT, which rely on structural data–then the chances of success are high. For statistical reasons, the probability of obtaining the CAST or B-FIT mutants by repeating cycles of epPCR are extremely low (or even close to zero). The limitations of ISM arise when no structural data (X-ray, NMR, homology model) whatsoever are available, even after much effort has been expended in these directions. Although, fortunately, this is rarely the case, ISM can still be applied even then, one possibility being to perform epPCR and then to apply ISM at the hot spots. By completing cycles of saturation mutagenesis it should soon become apparent which of the sites are not real hot spots. Bioinformatics may also be helpful in such cases, there being a variety of different approaches possible.

In future developments ISM must be studied more closely as a function of codon degeneracy. The correct choice of an amino acid alphabet can be expected to have a huge impact on the quality of libraries. For example, the use of NDT degeneracy, which encodes only 12 amino acids (Phe, Leu, Ile, Val, Tyr, His, Asn, Asp, Cys, Arg, Ser, Gly), appears to deliver libraries of superior quality (M.T. Reetz, D. Kahakeaw and R. Lohmer, in press), while genetic algorithms should also be helpful in these endeavors.

Finally, the application of ISM is apparent not only in the improvement of enzymes as catalysts in organic chemistry, but also for practical applications in such fields as polymer technology, pollution clean-up, detergents and bionanotechnological devices such as sensors. Iterative CASTing is also predestined to be useful not only in the study of binding phenomena but also in pathway engineering.

References

1 (a) Arnold, F.H. and Georgiou, G. (eds) (2003) *Directed Enzyme Evolution: Screening and Selection Methods*, Vol. 230, Humana Press, Totowa, New Jersey. (b) Brakmann, S. and Johnsson, K. (eds) (2002) *Directed Molecular Evolution of Proteins or How to Improve Enzymes for Biocatalysis*, Wiley-VCH Verlag GmbH, Weinheim, Germany. (c) Brakmann, S. and Schwienhorst, A. (eds) (2004) *Evolutionary Methods in Biotechnology Clever Tricks for Directed*

Evolution, Wiley-VCH Verlag GmbH, Weinheim, Germany.
(d) Taylor, S.V., Kast, P. and Hilvert, D. (2001) Investigating and engineering enzymes by genetic selection. *Angewandte Chemie*, **113**, 3408–36; *Angewandte Chemie – International Edition*, **40**, 3310–35.
(e) Powell, K.A., Ramer, S.W., Del Cardayré, S.B., Stemmer, W.P.C., Tobin, M.B., Longchamp, P.F. and Huisman, G.W. (2001) Directed evolution and biocatalysis. *Angewandte Chemie*, **113**, 4068–80; *Angewandte Chemie – International Edition*, **40**, 3948–59.
(f) Rubin-Pitel, S.B. and Zhao, H. (2006) Recent advances in biocatalysis by directed enzyme evolution. *Combinatorial Chemistry and High Throughput Screening*, **9**, 247–57.
(g) Kaur, J. and Sharma, R. (2006) Directed evolution: An approach to engineer enzymes. *Critical Reviews in Biotechnology*, **26**, 165–99.
(h) Wang, T.-W., Zhu, H., Ma, X.-Y., Zhang, T., Ma, Y.-S. and Wei, D.-Z. (2006) Mutant library construction in directed molecular evolution. *Molecular Biotechnology*, **34**, 55–68.
(i) Yuan, L., Kurek, I., English, J. and Keenan, R. (2005) Laboratory-directed protein evolution. *Microbiology and Molecular Biology Reviews*, **69**, 373–92.
(j) Ling, M.M. and Robinson, B.H. (1997) Approaches to DNA mutagenesis: An overview. *Analytical Biochemistry*, **254**, 157–78.
(k) Stevenson, J.D. and Benkovic, S.J. (2002) Combinatorial approaches to engineering hybrid enzymes. *Journal of the Chemical Society – Perkin Transactions*, **2**, 1483–93.
(l) Hibbert, E.G., Baganz, F., Hailes, H. C., Ward, J.M., Lye, G.J., Woodley, J.M. and Dalby, P.A. (2005) Directed evolution of biocatalytic processes. *Biomolecular Engineering*, **22**, 11–19.
(m) Bershtein, S. and Tawfik, D. (2008) Advances in laboratory evolution of enzymes. *Current Opinion in Chemistry and Biology*, **12**, 1–8.
(n) Fox, R.J. and Huisman, G.W. (2008) Enzyme optimization: moving from blind evolution to statistical exploration of sequence-function space. *Trends in Biotechnology*, **26**, 132–8.

2 (a) Wintrode, P.L. and Arnold, F.H. (2001) Temperature adaptation of enzymes: Lessons from laboratory evolution. *Advances in Protein Chemistry*, **55**, 161–225.
(b) Eijsink, V.G.H., Gåseidnes, S., Borchert, T.V. and van den Burg, B. (2005) Directed evolution of enzyme stability. *Biomolecular Engineering*, **22**, 21–30.

3 Arnold, F.H. (1998) Design by directed evolution. *Accounts of Chemical Research*, **31**, 125–31.

4 Reviews of directed evolution of enantioselective enzymes: (a) Reetz, M.T. (2002) Directed evolution of selective enzymes and hybrid catalysts. *Tetrahedron*, **58**, 6595–602.
(b) Reetz, M.T. (2004) Changing the enantioselectivity of enzymes of directed evolution, in *Methods in Enzymology* (eds D.E. Robertson and J.P. Noel), Elsevier Academic Press, San Diego, California, 388, pp. 238–56.
(c) Reetz, M.T. (2004) Controlling the enantioselectivity of enzymes by directed evolution: Practical and theoretical ramifications. *Proceedings of the National Academy of Sciences of the United States of America*, **101**, 5716–22.
(d) Reetz, M.T. (2006) Directed evolution of enantioselective enzymes as catalysts for organic synthesis, in *Advances in Catalysis*, Vol. 49 (eds B.C. Gates and H. Knözinger), Elsevier, San Diego, pp. 1–69.
(e) Turner, N.J. (2003) Directed evolution of enzymes for applied biocatalysis. *Trends in Biotechnology*, **21**, 474–8.

5 Reviews of high-throughput *ee*-screens: (a) Reetz, M.T. (2002) New methods for the high-throughput screening of enantioselective catalysts and biocatalysts. *Angewandte Chemie*, **114**, 1391–94; *Angewandte Chemie – International Edition*, **41**, 1335–8.
(b) Reetz, M.T. (2003) Select protocols of high-throughput *ee*-screening for assaying enantioselective enzymes, in *Methods in Molecular Biology*, Vol. 230 (eds F.H. Arnold and G. Georgiou), Humana Press, Totowa, New Jersey, pp. 283–90.
(c) Reetz, M.T. (2004) Screening for enantioselective enzymes, in *Enzyme Functionality – Design, Engineering and*

Screening (ed A. Svendsen), Marcel Dekker, New York, USA, pp. 559–98.
(d) Reetz, M.T. (2004) High-throughput screening of enantioselective industrial biocatalysts, in *Evolutionary Methods in Biotechnology* (eds S. Brakmann and A. Schwienhorst), Wiley-VCH Verlag GmbH, Weinheim, pp. 113–41.
(e) Reetz, M.T. (2006) High-throughput screening systems for assaying the enantioselectivity of enzymes, in *Enzyme Assays – High-throughput Screening, Genetic Selection and Fingerprinting* (ed J.-L. Reymond), Wiley-VCH Verlag GmbH, Weinheim, pp. 41–76.
See also: (f) Reymond, J.-L. (2005) *Enzyme Assays – High-throughput Screening, Genetic Selection and Fingerprinting*, Wiley-VCH Verlag GmbH, Weinheim.

6 (a) Leung, D.W., Chen, E. and Goeddel, D.V. (1989) A method for random mutagenesis of a defined DNA segment using a modified polymerase chain reaction. *Technique (Philadelphia)*, **1**, 11–15.
(b) Cadwell, R.C. and Joyce, G.F. (1992) Randomization of genes by PCR mutagenesis. *PCR Methods and Applications*, **2**, 28–33.

7 Numerous different molecular biological approaches to saturation (or cassette) mutagenesis are known, in addition to the currently most often used QuikChange protocol by Stratagene [17], e.g.: (a) Dominy, C.N. and Andrews, D.W. (2003) Site-directed mutagenesis by inverse PCR, in *Methods in Molecular Biology*, Vol. 235 (eds N. Casali and A. Preston), Humana Press, Totowa, New Jersey, pp. 209–23.
(b) Vandeyar, M.A., Weiner, M.P., Hutton, C.J. and Batt, C.A. (1988) A simple and rapid method for the selection of oligodeoxynucleotide-directed mutants. *Gene*, **65**, 129–33.
(c) Kirsch, R.D. and Joly, E. (1998) An improved PCR-mutagenesis strategy for two-site mutagenesis or sequence swapping between related genes. *Nucleic Acids Research*, **26**, 1848–50.
(d) Zheng, L., Baumann, U. and Reymond, J.-L. (2004) An efficient one-step site-directed and site-saturation mutagenesis protocol. *Nucleic Acids Research*, **32**, e115.
(e) Smith, M. (1985) In vitro mutagenesis. *Annual Review of Genetics*, **19**, 423–62.
(f) Georgescu, R., Bandara, G. and Sun, L. (2003) Saturation Mutagenesis, in *Directed Evolution Library Creation* (eds F.H. Arnold and G. Georgiou), Humana Press, Totowa, New Jersey, pp. 75–84.
(g) Wells, J.A., Vasser, M. and Powers, D. B. (1985) Cassette mutagenesis: an efficient method for generation of multiple mutations at defined sites. *Gene*, **34**, 315–23.
(h) Barettino, D., Feigenbutz, M., Valcarcel, R. and Stunnenberg, H.G. (1994) Improved method for PCR-mediated site-directed mutagenesis. *Nucleic Acids Research*, **22**, 541–2.
(i) Horwitz, M.S.Z. and Loeb, L.A. (1986) Promoters selected from random DNA sequences. *Proceedings of the National Academy of Sciences of the United States of America*, **83**, 7405–9.
(j) Oliphant, A.R., Nussbaum, A.L. and Struhl, K. (1986) Cloning of random-sequence oligodeoxynucleotides. *Gene*, **44**, 177–83.
(k) Reidhaar-Olson, J.F. and Sauer, R.T. (1988) Combinatorial cassette mutagenesis as a probe of the informational content of protein sequences. *Science (Washington, DC)*, **241**, 53–7.
(l) Hermes, J.D., Parekh, S.M., Blacklow, S.C., Köster, H. and Knowles, J.R. (1989) A reliable method for random mutagenesis: The generation of mutant libraries using spiked oligodeoxyribonucleotide primers. *Gene*, **84**, 143–51.
(m) Sarkar, G. and Sommer, S.S. (1990) The "megaprimer" method of site-directed mutagenesis. *BioTechniques*, **8**, 404–7.
(n) Martin, A., Toselli, E., Rosier, M.-F., Auffray, C. and Devignes, M.-D. (1995) Rapid and high efficiency site-directed mutagenesis by improvement of the homologous recombination technique. *Nucleic Acids Research*, **23**, 1642–3.

8 (a) Stemmer, W.P.C. (1994) Rapid evolution of a protein in vitro by DNA shuffling. *Nature (London, UK)*, **370**, 389–91.
(b) Stemmer, W.P.C. (1994) DNA shuffling

by random fragmentation and reassembly: In vitro recombination for molecular evolution. *Proceedings of the National Academy of Sciences of the United States of America*, **91**, 10747–51.
(c) Crameri, A., Raillard, S.-A., Bermudez, E. and Stemmer, W.P.C. (1998) DNA shuffling of a family of genes from diverse species accelerates directed evolution. *Nature*, **391**, 288–91.

9 (a) Lutz, S. and Patrick, W.M. (2004) Novel methods for directed evolution of enzymes: quality, not quantity. *Current Opinion in Biotechnology*, **15**, 291–7.
(b) Neylon, C. (2004) Chemical and biochemical strategies for the randomization of protein encoding DNA sequences: Library construction methods for directed evolution. *Nucleic Acids Research*, **32**, 1448–59.
(c) Polizzi, K.M., Parikh, M., Spencer, C. U., Matsumura, I., Lee, J.H., Realff, M.J. and Bommarius, A.S. (2006) Pooling for improved screening of combinatorial libraries for directed evolution. *Biotechnology Progress*, **22**, 961–7.

10 Reetz, M.T., Zonta, A., Schimossek, K., Liebeton, K. and Jaeger, K.-E. (1997) Creation of enantioselective biocatalysts for organic chemistry by in vitro evolution. *Angewandte Chemie*, **109**, 2961–3; *Angewandte Chemie – International Edition*, **36**, 2830–2.

11 (a) Reetz, M.T., Wilensek, S., Zha, D. and Jaeger, K.-E. (2001) Directed evolution of an enantioselective enzyme through combinatorial multiple cassette mutagenesis. *Angewandte Chemie*, **113**, 3701–3; *Angewandte Chemie – International Edition*, **40**, 3589–91.
(b) Zha, D., Wilensek, S., Hermes, M., Jaeger, K.-E. and Reetz, M.T. (2001) Complete reversal of enantioselectivity of an enzyme-catalyzed reaction by directed evolution. *Chemical Communications*, 2664–5.

12 Drummond, D.A., Iverson, B.L., Georgiou, G. and Arnold, F.H. (2005) Why high-error-rate random mutagenesis libraries are enriched in functional and improved proteins. *Journal of Molecular Biology*, **350**, 806–16.

13 (a) Eggert, T., Reetz, M.T. and Jaeger, K.-E. (2004) Directed evolution by random mutagenesis: a critical evaluation, in *Enzyme Functionality – Design, Engineering, and Screening* (ed A. Svendsen), Marcel Dekker, New York, USA, pp. 375–90.
(b) Cirino, P.C., Mayer, K.M. and Umeno, D. (2003) Generating mutant libraries using error-prone PCR, in *Directed Evolution Library Creation: Methods and Protocols*, Vol. 231 (eds G. Georgiou and F.H. Arnold), Humana Press, Totowa, New Jersey, pp. 3–9.
(c) Wong, T.S., Tee, K.L., Hauer, B. and Schwaneberg, U. (2004) Sequence saturation mutagenesis (SeSaM): a novel method for directed evolution. *Nucleic Acids Research*, **32**, e26.

14 (a) Kikuchi, H., Ohnishi, K. and Harayama, S. (1999) An effective family shuffling method using single-stranded DNA. *Gene*, **243**, 133–7.
(b) Rowe, L.A., Geddie, M.L., Alexander, O.B. and Matsumura, I. (2003) A comparison of directed evolution approaches using the β-glucuronidase model system. *Journal of Molecular Biology*, **332**, 851–60.
(c) Parikh, M.R. and Matsumura, I. (2005) Site-saturation mutagenesis is more efficient than DNA shuffling for the directed evolution of β-fucosidase from β-galactosidase. *Journal of Molecular Biology*, **352**, 621–8.

15 (a) Coco, W.M., Levinson, W.E., Crist, M.J., Hektor, H.J., Darzins, A., Pienkos, P.T., Squires, C.H. and Monticello, D.J. (2001) DNA shuffling method for generating highly recombined genes and evolved enzymes. *Nature Biotechnology*, **19**, 354–9.
(b) Coco, W.M. (2003) RACHITT: Gene family shuffling by random chimeragenesis on transient templates, in *Directed Evolution Library Creation: Methods and Protocols*, Vol. 231 (eds F.H. Arnold and G. Georgiou), Humana Press, Totowa, New Jersey, pp. 111–27.

16 (a) Ness, J.E., Kim, S., Gottman, A., Pak, R., Krebber, A., Borchert, T.V., Govindarajan, S., Mundorff, E.C. and Minshull, J. (2002) Synthetic shuffling expands functional protein diversity by allowing amino acids to recombine

independently. *Nature Biotechnology*, **20**, 1251–5.
(b) Coco, W.M., Encell, L.P., Levinson, W.E., Crist, M.J., Loomis, A.K., Licato, L.L., Arensdorf, J.J., Sica, N., Pienkos, P.T. and Monticello, D.J. (2002) Growth factor engineering by degenerate homoduplex gene family recombination. *Nature Biotechnology*, **20**, 1246–50.
(c) Zha, D., Eipper, A. and Reetz, M.T. (2003) Assembly of designed oligonucleotides as an efficient method for gene recombination: A new tool in directed evolution. *ChemBioChem*, **4**, 34–9.

17 Hogrefe, H.H., Cline, J., Youngblood, G.L. and Allen, R.M. (2002) Creating randomized amino acid libraries with the QuikChange® multi site-directed mutagenesis kit. *BioTechniques*, **33**, 1158–65.

18 (a) Horsman, G.P., Liu, A.M.F., Henke, E., Bornscheuer, U.T. and Kazlauskas, R.J. (2003) Mutations in distant residues moderately increase the enantioselectivity of *Pseudomonas fluorescens* esterase towards methyl 3-bromo-2-methylpropanoate and ethyl 3-phenylbutyrate. *Chemistry–A European Journal*, **9**, 1933–9.
(b) Schmidt, M., Hasenpusch, D., Kähler, M., Kirchner, U., Wiggenhorn, K., Langel, W. and Bornscheuer, U.T. (2006) Directed evolution of an esterase from *Pseudomonas fluorescens* yields a mutant with excellent enantioselectivity and activity for the kinetic resolution of a chiral building block. *ChemBioChem*, **7**, 805–9.

19 Reetz, M.T., Wang, L.-W. and Bocola, M. (2006) Directed evolution of enantioselective enzymes: Iterative cycles of CASTing for probing protein-sequence space. *Angewandte Chemie*, **118**, 1258–63; *Erratum*, 2556; *Angewandte Chemie – International Edition* (2006) **45**, 1236–41; *Erratum*, 2494.

20 Reetz, M.T. and Carballeira, J.D. (2007) Iterative Saturation Mutagenesis (ISM) for rapid directed evolution of functional enzymes. *Nature Protocols*, **2**, 891–903.

21 (a) Bosley, A.D. and Ostermeier, M. (2005) Mathematical expressions useful in the construction, description and evaluation of protein libraries. *Biomolecular Engineering*, **22**, 57–61.
(b) Patrick, W.M. and Firth, A.E. (2005) Strategies and computational tools for improving randomized protein libraries. *Biomolecular Engineering*, **22**, 105–12.
See also: (c) Denault, M. and Pelletier, J.N. (2007) Protein library design and screening. Working out the probabilities, in *Protein Engineering Protocols* (eds K.M. Arndt and K.M. Müller), Humana Press, Totowa, New Jersey, pp. 127–54.

22 Examples of different applications of amino acid alphabet usage: (a) Kamtekar, S., Schiffer, J.M., Xiong, H.Y., Babik, J.M. and Hecht, M.H. (1993) Protein design by binary patterning of polar and nonpolar amino-acids. *Science*, **262**, 1680–5.
(b) Davidson, A.R., Lumb, K.J. and Sauer, R.T. (1995) Cooperative folded proteins in random sequence libraries. *Nature Structural Biology*, **2**, 856–64.
(c) Walter, K.U., Vamvaca, K. and Hilvert, D. (2005) An active enzyme constructed from a nine amino acid alphabet. *The Journal of Biological Chemistry*, **280**, 37742–6.

23 (a) Reetz, M.T. (1999) Strategies for the development of enantioselective catalysts. *Pure and Applied Chemistry*, **71**, 1503–9.
(b) Liebeton, K., Zonta, A., Schimossek, K., Nardini, M., Lang, D., Dijkstra, B.W., Reetz, M.T. and Jaeger, K.-E. (2000) Directed evolution of an enantioselective lipase. *Chemistry and Biology*, **7**, 709–18.
(c) Reetz, M.T. (2000) Application of directed evolution in the development of enantioselective enzymes. *Pure and Applied Chemistry*, **72**, 1615–22.

24 Fersht, A. (1999) *Structure and Mechanism in Protein Science*, W.H Freeman, New York.

25 (a) Bocola, M., Otte, N., Jaeger, K.-E., Reetz, M.T. and Thiel, W. (2004) Learning from directed evolution: Theoretical investigations into cooperative mutations in lipase enantioselectivity. *ChemBioChem*, **5**, 214–23.
(b) Reetz, M.T., Puls, M., Carballeira, J.D., Vogel, A., Jaeger, K.-E., Eggert, T., Thiel, W., Bocola, M. and Otte, N. (2007) Learning from directed evolution: Further

lessons from theoretical investigations into cooperative mutations in lipase enantioselectivity. *ChemBioChem*, **8**, 106–12.

26 (a) Miyazaki, K. and Arnold, F.H. (1999) Exploring nonnatural evolutionary pathways by saturation mutagenesis: Rapid improvement of protein function. *Journal of Molecular Evolution*, **49**, 716–20.
(b) Arnold, F.H., Wintrode, P.L., Miyazaki, K. and Gershenson, A. (2001) How enzymes adapt: Lessons from directed evolution. *Trends in Biochemical Sciences*, **26**, 100–6.

27 (a) Desantis, G., Wong, K., Farwell, B., Chatman, K., Zhu, Z., Tomlinson, G., Huang, H., Tan, X., Bibbs, L., Chen, P., Kretz, K. and Burk, M.J. (2003) Creation of a productive, highly enantioselective nitrilase through Gene Site Saturation Mutagenesis (GSSM). *Journal of the American Chemical Society*, **125**, 11476–7.
(b) Funke, S.A., Eipper, A., Reetz, M.T., Otte, N., Thiel, W., Van Pouderoyen, G., Dijkstra, B.W., Jaeger, K.-E. and Eggert, T. (2003) Directed evolution of an enantioselective *Bacillus subtilis* lipase. *Biocatalysis and Biotransformation*, **21**, 67–73.
(c) Brissos, V., Eggert, T., Cabral, J.M.S. and Jaeger, K.-E. (2008) Improving activity and stability of cutinase towards the anionic detergent AOT by complete saturation mutagenis. *Protein Engineering, Design & Selection*, **21**, 387–93.

28 Nardini, M., Lang, D.A., Liebeton, K., Jaeger, K.-E. and Dijkstra, B.W. (2000) Crystal structure of *Pseudomonas aeruginosa* lipase in the open conformation. *The Journal of Biological Chemistry*, **275**, 31219–25.

29 See for example: (a) Fong, S., Machajewski, T.D., Mak, C.C. and Wong, C.-H. (2000) Directed evolution of D-2-keto-3-deoxy-6-phosphogluconate aldolase to new variants for the efficient synthesis of D- and L-sugars. *Chemistry and Biology*, **7**, 873–83.
(b) Wang, L.-F., Goodey, N.M., Benkovic, S.J. and Kohen, A. (2006) Coordinated effects of distal mutations on environmentally coupled tunneling in dihydrofolate reductase. *Proceedings of the National Academy of Sciences of the United States of America*, **103**, 15753–8.
(c) Oelschlaeger, P., Schmid, R.D. and Pleiss, J. (2003) Modeling domino effects in enzymes: molecular basis of the substrate specificity of the bacterial metallo-β-lactamases IMP-1 and IMP-6. *Biochemistry*, **42**, 8945–56.
(d) Fowler, S.M., Taylor, J.M., Friedberg, T., Wolf, C.R. and Riley, R.J. (2002) CYP3A4 active site volume modification by mutagenesis of leucine 211. *Drug Metabolism and Disposition*, **30**, 452–6.
(e) Chin, J.K. and Klinman, J.P. (2000) Probes of a role for remote binding interactions on hydrogen tunnelling in the horse liver alcohol dehydrogenase reaction. *Biochemistry*, **39**, 1278–84.

30 Reetz, M.T., Carballeira, J.D. and Vogel, A. (2006) Iterative saturation mutagenesis on the basis of B factors as a strategy for increasing protein thermostability. *Angewandte Chemie*, **118**, 7909–15; *Angewandte Chemie – International Edition*, **45**, 7745–51.

31 Rui, L., Cao, L., Chen, W., Reardon, K.F. and Wood, T.K. (2004) Active site engineering of the epoxide hydrolase from *Agrobacterium radiobacter* AD1 to enhance aerobic mineralization of cis-1,2-dichloroethylene in cells expressing an evolved toluene ortho-monooxygenase. *The Journal of Biological Chemistry*, **279**, 46810–17.

32 Greisman, H.A. and Pabo, C.O. (1997) A general strategy for selecting high-affinity zinc finger proteins for diverse DNA target sites. *Science*, **275**, 657–61.

33 Reetz, M.T., Bocola, M., Carballeira, J.D., Zha, D. and Vogel, A. (2005) Expanding the range of substrate acceptance of enzymes: Combinatorial active-site saturation test. *Angewandte Chemie*, **117**, 4264–8; *Angewandte Chemie – International Edition*, **44**, 4192–6.

34 (a) Dartois, V., Baulard, A., Schanck, K. and Colson, C. (1992) Cloning, nucleotide-sequence and expression in *Escherichia coli* of a lipase gene from *Bacillus subtilis* 168. *Biochimica et Biophysica Acta*, **1131**, 253–60.
(b) Jaeger, K.-E., Dijkstra, B.W. and Reetz, M.T. (1999) Bacterial biocatalysts:

molecular biology, three-dimensional structures, and biotechnological applications of lipases. *Annual Review of Microbiology*, **53**, 315–51.

35 Reetz, M.T., Carballeira, J.D., Peyralans, J.J.-P., Höbenreich, H., Maichele, A. and Vogel, A. (2006) Expanding the substrate scope of enzymes: combining mutations obtained by CASTing. *Chemistry–A European Journal*, **12**, 6031–8.

36 Reetz, M.T., Torre, C., Eipper, A., Lohmer, R., Hermes, M., Brunner, B., Maichele, A., Bocola, M., Arand, M., Cronin, A., Genzel, Y., Archelas, A. and Furstoss, R. (2004) Enhancing the enantioselectivity of an epoxide hydrolase by directed evolution. *Organic Letters*, **6**, 177–80.

37 Zou, J.Y., Hallberg, B.M., Bergfors, T., Oesch, F., Arand, M., Mowbray, S.L. and Jones, T.A. (2000) Structure of *Aspergillus niger* epoxide hydrolase at 1.8 Å resolution: implications for the structure and function of the mammalian microsomal class of epoxide hydrolases. *Structure*, **8**, 111–22.

38 Weinreich, D.M., Delaney, N.F., Depristo, M.A. and Hartl, D.L. (2006) Darwinian evolution can follow only very few mutational paths to fitter proteins. *Science*, **312**, 111–14.

39 (a) Faber, K. and Orru, R.V.A. (2002) Hydrolysis of Epoxides, in *Enzyme Catalysis in Organic Synthesis: A Comprehensive Handbook* (eds K. Drauz and H. Waldmann), Vol. II, Wiley-VCH Verlag GmbH, Weinheim, pp. 579–608.
(b) Morisseau, C. and Hammock, B.D. (2005) Epoxide hydrolases: mechanisms, inhibitor designs, and biological roles. *Annual Review of Pharmacology and Toxicology*, **45**, 311–33.
(c) Archelas, A. and Furstoss, R. (2001) Synthetic applications of epoxide hydrolases. *Current Opinion in Chemical Biology*, **5**, 112–19.

40 Bruice, T.C. (2002) A view at the millennium: the efficiency of enzymatic catalysis. *Accounts of Chemical Research*, **35**, 139–48.

41 Wang, L.-W. (2006) Directed evolution of the *Aspergillus niger* epoxide hydrolase. Dissertation, Ruhr-Universität Bochum, Germany.

42 Clouthier, C.M., Kayser, M.M. and Reetz, M.T. (2006) Designing new Baeyer-Villiger monooxygenases using restricted CASTing. *The Journal of Organic Chemistry*, **71**, 8431–7.

43 (a) Reetz, M.T. (2001) Patent DE-A 101 29 187.6.
(b) Reetz, M.T. (2002) Directed evolution of selective enzymes and hybrid catalysts. *Tetrahedron*, **58**, 6595–602.

44 (a) Reetz, M.T., Peyralans, J.J.-P., Maichele, A., Fu, Y. and Maywald, M. (2006) Directed evolution of hybrid enzymes: evolving enantioselectivity of an achiral Rh-complex anchored to a protein. *Chemical Communications*, 4318–20.
(b) Reetz, M.T., Rentzsch, M., Pletsch, A., Maywald, M., Maiwald, P., Peyralans, J.J.-P., Maichele, A., Fu, Y., Jiao, N., Hollmann, F., Mondière, R. and Taglieber, A. (2007) Directed evolution of enantioselective hybrid catalysts: a novel concept in asymmetric catalysis. *Tetrahedron*, **63**, 6404–14.
(c) Reetz, M.T., Rentzsch, M., Pletsch, A., Taglieber, A., Hollmann, F., Mondière, R. J.G., Dickmann, N., Höcker, B., Cerrone, S., Haeger, M.C. and Sterner, R. (2008) A robust protein host for anchoring chelating ligands and organocatalysts. *ChemBioChem*, **9**, 552–64.

45 Bartsch, S., Kourist, R. and Bornscheuer, U.T. (2008) Complete inversion of enantioselectivity towards acetylated tertiary alcohols by a double mutant of a *Bacillus subtilis* esterase. *Angewandte Chemie*, **120**, 1531–4; *Angewandte Chemie–International Edition*, **47**, 1508–11.

46 Liang, L., Zhang, J. and Lin, Z. (2007) Altering coenzyme specificity of *Pichia stipitis* xylose reductase by the semi-rational approach CASTing. *Microbial Cell Factories*, **6**, 36.

47 (a) Oshima, T. (1994) Stabilization of proteins by evolutionary molecular engineering techniques. *Current Opinion in Structural Biology*, **4**, 623–8.
(b) Ó'Fágáin, C. (2003) Enzyme stabilization–recent experimental progress. *Enzyme and Microbial Technology*, **33**, 137–49.
(c) Eijsink, V.G.H., Bjørk, A., Gåseidnes, S., Sirevåg, R., Synstad, B., van den Burg, B. and Vriend, G. (2004) Rational

engineering of enzyme stability. *Journal of Biotechnology*, **113**, 105–20.
48 Palackal, N., Brennan, Y., Callen, W.N., Dupree, P., Frey, G., Goubet, F., Hazlewood, G.P., Healey, S., Kang, Y.E., Kretz, K.A., Lee, E., Tan, X., Tomlinson, G.L., Verruto, J., Wong, V.W.K., Mathur, E.J., Short, J.M., Robertson, D.E. and Steer, B.A. (2004) An evolutionary route to xylanase process fitness. *Protein Science*, **13**, 494–503.
49 Giver, L., Gershenson, A., Freskgard, P.O. and Arnold, F.H. (1998) Directed evolution of a thermostable esterase. *Proceedings of the National Academy of Sciences of the United States of America*, **95**, 12809–13.
50 Valinger, G., Hermann, M., Wagner, U.G. and Schwab, H. (2007) Stability and activity improvement of cephalosporin esterase EstB from *Burkholderia gladioli* by directed evolution and structural interpretation of muteins. *Journal of Biotechnology*, **129**, 98–108.
51 (a) Matthews, B.W. (1993) Structural and genetic analysis of protein stability. *Annual Review of Biochemistry*, **62**, 139–60.
(b) Jaenicke, R. and Böhm, G. (1998) The stability of proteins in extreme environments. *Current Opinion in Structural Biology*, **8**, 738–48.
(c) Vieille, C. and Zeikus, G.J. (2001) Hyperthermophilic enzymes: sources, uses, and molecular mechanisms for thermostability. *Microbiology and Molecular Biology Reviews*, **65**, 1–43.
(d) Sterner, R. and Liebl, W. (2001) Thermophilic adaption of proteins. *Critical Reviews in Biochemistry and Molecular Biology*, **36**, 39–106.
(e) Lazaridis, T., Lee, I. and Karplus, M. (1997) Dynamics and unfolding pathways of a hyperthermophilic and a mesophilic rubredoxin. *Protein Science*, **6**, 2589–605.
(f) Kumar, S. and Nussinov, R. (2001) How do thermophilic proteins deal with heat? *Cellular and Molecular Life Sciences*, **58**, 1216–33.
(g) Podar, M. and Reysenbach, A.-L. (2006) New opportunities revealed by biotechnological explorations of extremophiles. *Current Opinion in Biotechnology*, **17**, 1–6.
(h) Demirjian, D.C., Morís-Varas, F. and Cassidy, C.S. (2001) Enzymes from extremophiles. *Current Opinion in Chemical Biology*, **5**, 144–51.
(i) Rodriguez-Larrea, D., Minning, S., Borchert, T.V. and Sanchez-Ruiz, J.M. (2006) Role of solvation barriers in protein kinetic stability. *Journal of Molecular Biology*, **360**, 715–24.
(j) Atomi, H. (2005) Recent progress towards the application of hyperthermophiles and their enzymes. *Current Opinion in Biotechnology*, **9**, 166–73.
(k) Vogt, G., Woell, S. and Argos, P. (1997) Protein thermal stability, hydrogen bonds, and ion pairs. *Journal of Molecular Biology*, **269**, 631–43.
(l) Buchner, J. and Kiefhaber, T. (2005) *Protein Folding Handbook*, Wiley-VCH Verlag GmbH, Weinheim.
52 See for example: (a) Schlessinger, A. and Rost, B. (2005) Protein flexibility and rigidity predicted from sequence. *Proteins: Structure, Function, and Bioinformatics*, **61**, 115–26.
(b) Wang, C., Karpowich, N., Hunt, J.F., Rance, M. and Palmer, A.G. (2004) Dynamics of ATP-binding cassette contribute to allosteric control, nucleotide binding and energy transduction in ABC transporters. *Journal of Molecular Biology*, **342**, 525–37.
53 (a) Schlessinger, A., Yachdav, G. and Rost, B. (2006) PROFbval: predict flexible and rigid residues in proteins. *Bioinformatics*, **22**, 891–3.
(b) Jin, Y. and Dunbrack, R.L. Jr (2005) Assessment of disorder predictions in CASP6. *Proteins: Structure, Function, and Bioinformatics*, **61**, 167–75.
54 (a) Karplus, P.A. and Schulz, G.E. (1985) Prediction of chain flexibility in proteins. *Naturwissenschaften*, **72**, 212–13.
(b) Vihinen, M. (1987) Relationship of protein flexibility of thermostability. *Protein Engineering*, **1**, 477–80.
(c) Parthasarathy, S. and Murthy, M.R.N. (2000) Protein thermal stability: insights from atomic displacement parameters (B values). *Protein Engineering*, **13**, 9–13.
(d) Yuan, Z., Zhao, J. and Wang, Z.-X. (2003) Flexibility analysis of enzyme active

sites by crystallographic temperature factors. *Protein Engineering*, **16**, 109–14.
(e) Radivojac, P., Obradovic, Z., Smith, D.K., Zhu, G., Vucetic, S., Brown, C.J., Lawson, J.D. and Dunker, A.K. (2004) Protein flexibility and intrinsic disorder. *Protein Science*, **13**, 71–80.
(f) Trueblood, K.N., Bürgi, H.-B., Burzlaff, H., Dunitz, J.D., Gramaccioli, C.M., Schulz, H.H., Shmueli, U. and Abrahams, S.C. (1996) Atomic displacement parameter nomenclature report of a subcommittee on atomic displacement parameter nomenclature. *Acta Crystallographica*, **A52**, 770–81.

55 (a) Van Pouderoyen, G., Eggert, T., Jaeger, K.-E. and Dijkstra, B.W. (2001) The crystal structure of *Bacillus subtilis* lipase: a minimal a/β hydrolase fold enzyme. *Journal of Molecular Biology*, **309**, 215–26.
(b) Kawasaki, K., Kondo, H., Suzuki, M., Ohgiya, S. and Tsuda, S. (2002) Alternative conformations observed in catalytic serine of *Bacillus subtilis* lipase determined at 1.3 Å resolution. *Acta Crystallographica Section D. Biological Crystallography*, **58**, 1168–74.

56 Bloom, J.D., Labthavikul, S.T., Otey, C.R. and Arnold, F.H. (2006) Protein stability promotes evolvability. *Proceedings of the National Academy of Sciences of the United States of America*, **103**, 5869–74.

17
Evolution of Enantioselective *Bacillus subtilis* Lipase

Thorsten Eggert, Susanne A. Funke, Jennifer N. Andexer, Manfred T. Reetz and Karl-Erich Jaeger

17.1
Introduction

Lipases represent a very important class of enzymes for biotechnological applications as they can catalyze both hydrolysis and synthesis reactions in aqueous as well as in organic solvents. Furthermore, they usually do not require cofactors, and they exhibit broad substrate specificity and high enantioselectivity. Although, currently lipases are used widely as detergent additives, their application in organic synthesis to produce enantiopure compounds is of major importance. Both academia and industry are today pushing forward the development of novel biocatalytic and chemocatalytic routes leading to enantiopure compounds, as the U.S. Federal Drug Administration (FDA) approves novel pharmaceuticals only if the manufacturer can prove that they are produced as single enantiomers. Unfortunately, many industrial relevant substrates are non-natural and therefore, new enzymes must be identified or existing ones tailored to allow the efficient and enantioselective conversion of these artificial substrates [1–3].

During the past few years, two major methodologies, namely metagenomics and directed evolution, have proven to be effective in the quest to identify new and optimize existing enzymes.

Metagenomics comprises the isolation of DNA directly from the environment, its cloning, expression and the identification of novel enzyme activities, usually by high-throughput screening (HTS) methods (see Chapters 9, 11) [4, 5]. In theory, metagenomics provides access to a huge genetic diversity, without the need to culture organisms in the laboratory. In practice, the chance to find and identify novel biocatalysts is severely restricted by the fact that many of the newly isolated genes cannot be functionally expressed in standard host strains such as *Escherichia coli*. Also, only a few of the HTS systems currently available allow for the functional screening of novel enzymatic activities (see Chapters 28, 30) [6].

Directed evolution methods have shown their potential to optimize existing enzymes with respect to industrially relevant properties including activity, sub-

Protein Engineering Handbook. Edited by Stefan Lutz and Uwe T. Bornscheuer

ISBN: 978-3-527-31850-6

strate specificity, thermostability, pH-stability or solvent stability, and enantioselectivity [7–10]. An array of different methods has been developed to:

- create genetic diversity by random mutagenesis
- combine advantageous mutations by mimicking recombination-like events
- allow for selection or HTS to identify improved variants.

Despite the huge number of available methods, error-prone polymerase chain reaction (epPCR) is still the most widely used method to create libraries of protein variants. This method is simple, easy to perform, and can therefore be established also in laboratories which possess only limited experience with sophisticated molecular biological techniques. However, it has been demonstrated that epPCR has major drawbacks, in particular concerning the number and type of mutations that can be created [11]. The probability to substitute a given amino acid depends on the DNA sequence of its respective codon(s) (Figure 17.1a), and the exchange of more than one base within a single codon is highly unlikely. Therefore, only a limited number of theoretically possible point mutations will practically occur in a library generated with epPCR (see Figure 17.1). This mutational bias is further amplified by the natural bias of DNA polymerases used for PCR amplification. For example, *Taq*-polymerase preferentially introduces A→T, T→A transversions and A→G, T→C transitions when used in $MnCl_2$ buffer; thereby further reducing the number of amino acid exchanges to about 20% (Figure 17.1b). The probability to generate G→T, C→A transversions is very low, and C→G, G→C transversions occur only rarely [11, 12]. Recently, a detailed study has evaluated mutational efficiencies by comparing different random mutagenesis methods [13]. These authors also developed a statistical analysis tool named MAP (Mutagenesis Assistant

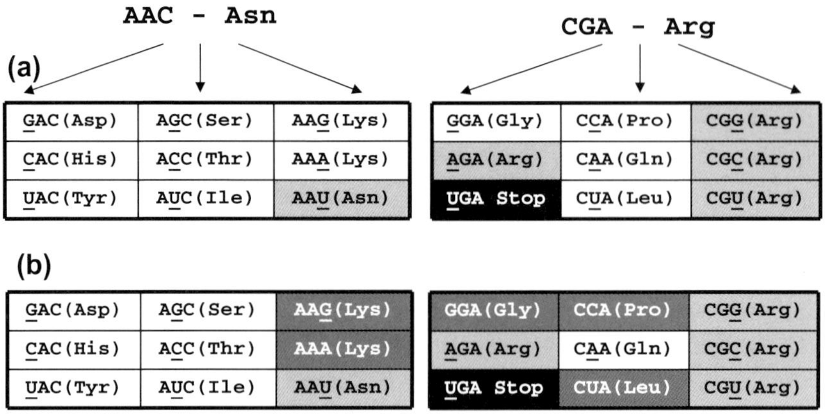

Figure 17.1 Mutational bias of epPCR. (a) Theoretical number of possible mutations of codons AAC encoding Asn and CGA (Arg); (b) additional mutational bias caused by low frequencies for transversions of G→T, C→A, G→C and C→G. Newly created amino acids are shown in white, silent mutations in light gray, mutations unlikely to occur in dark gray, and mutations creating a stop codon in black.

Program) which allows the comparison of 19 different mutagenesis methods with respect to their efficacy by analyzing amino acid substitution patterns at each amino acid position of a single gene [14]. A statistical analysis of single and double nucleotide exchanges in each of the 64 codons revealed that single nucleotide exchanges generate only five to seven amino acid substitutions per codon, corresponding to only about 40% of natural diversity, whereas a simultaneous exchange of both the first and the second nucleotide of a codon generates a diverse amino acid substitution pattern, achieving 83% of the natural diversity. Furthermore, transversions were shown to generate a more complex substitution pattern with chemically more diverse amino acids than can be achieved by transitions [15]. Alternative protocols have therefore been developed for random mutagenesis, among them sequence saturation mutagenesis (SeSaM) as a novel method which truly randomizes a protein at each amino acid position [16]. Meanwhile, SeSaM – which is more complex and time-consuming than epPCR – was further optimized, and this resulted in increased numbers of double and triple mutants to be generated by tuning the experimental conditions [17].

Apart from the development of more efficient mutagenesis methods, a persisting problem relates to the numbers of variants that can, in theory, be created. A calculation using Equation 17.1 [18] reveals a theoretical number of 3800 variants to be screened if just one out of 200 amino acids of a given protein is substituted.

$$N = \frac{E^M X!}{(X - M)!M!} \tag{17.1}$$

where N is the maximal number of variants, E is the number of amino acids exchanged per position, M is the total number of amino acids exchanged per molecule, and X is the number of amino acids per molecule.

However, if two amino acids are substituted, this number already increases to 7 183 900, while for three amino acid substitutions within the same protein the number of theoretically possible variants exceeds 9 000 000 000 and can thus no longer be screened by methods available to date.

How can these limitations be overcome? Complete site-saturation mutagenesis was proven to represent a valuable starting point for directed evolution experiments [19–22]. This method allows each amino acid of a given protein to be replaced by the remaining 19 amino acids using site-specific mutagenesis with randomized primers (Figure 17.2). A library consisting of about 200 variants is created for each amino acid position. We have used this method to evolve an enantioselective enzyme using *Bacillus subtilis* lipase LipA (BSLA).

The Gram-positive soil bacterium *B. subtilis* is used widely in the biotechnology industries and possesses at least two extracellular lipases, BSLA and BSLB, both of which are secreted into the culture medium. These lipases share an unusual consensus pentapeptide (Ala-X-Ser-X-Gly), with the first glycine being replaced by an alanine. When both enzymes were biochemically characterized, BSLB was

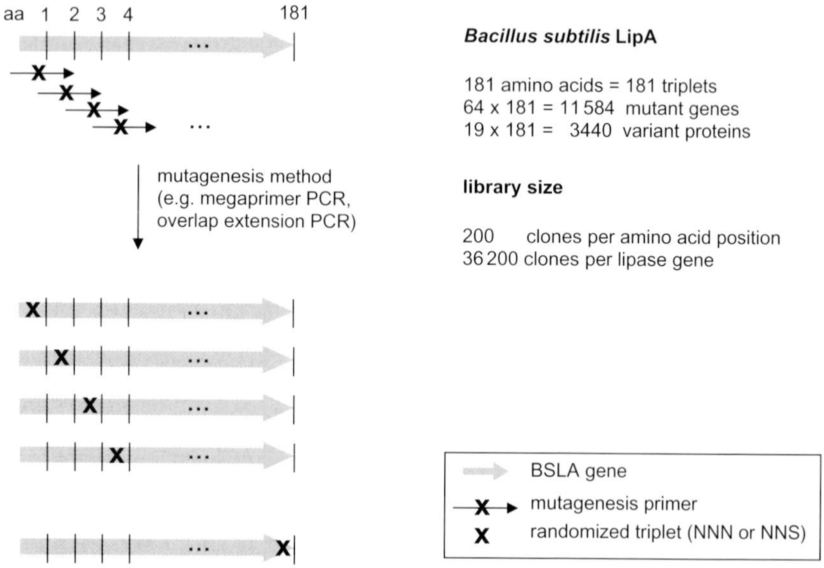

Figure 17.2 Strategy for complete saturation mutagenesis exemplified for *Bacillus subtilis* lipase A (BSLA). Each amino acid position is randomized using mutagenesis primers; this results in 181 different libraries, each representing one saturated amino acid position.

found to be an esterase which showed specificity for short-chain fatty acid ester substrates. BSLA is the smallest lipase known and can easily be expressed not only in *B. subtilis* but also in *E. coli*. It does not show interfacial activation at oil–water interfaces, and also exhibits a high pH-stability in the alkaline range, with an optimum at pH 10.0 [23, 24]. The X-ray structure of BSLA revealed an α/β-hydrolase fold [25] with the residues forming the catalytic triad located close to the protein surface. A lid-like structure covering the active site was missing, which may explain the lack of interfacial activation. Its small size of only 181 amino acids ($M_r = 19.4\,kDa$) and the order of its secondary structural elements led us to conclude that BSLA represents a minimal α/β-hydrolase fold enzyme, thus making it an interesting candidate for directed evolution experiments [24, 26].

17.2
Directed Evolution of Enantioselective Lipase from *Bacillus subtilis*

The asymmetric hydrolysis of *pseudo-meso*-1,4-diacetoxy-cyclopentene catalyzed by BSLA was used as the model reaction. The deuterium-labeled substrate is hydrolyzed to yield two enantiomers of the product, which can be distinguished by their mass difference using a high-throughput electrospray ionization-mass spectrometry (ESI-MS) method (see Figure 17.3) [27]. The wild-type enzyme hydrolyzes

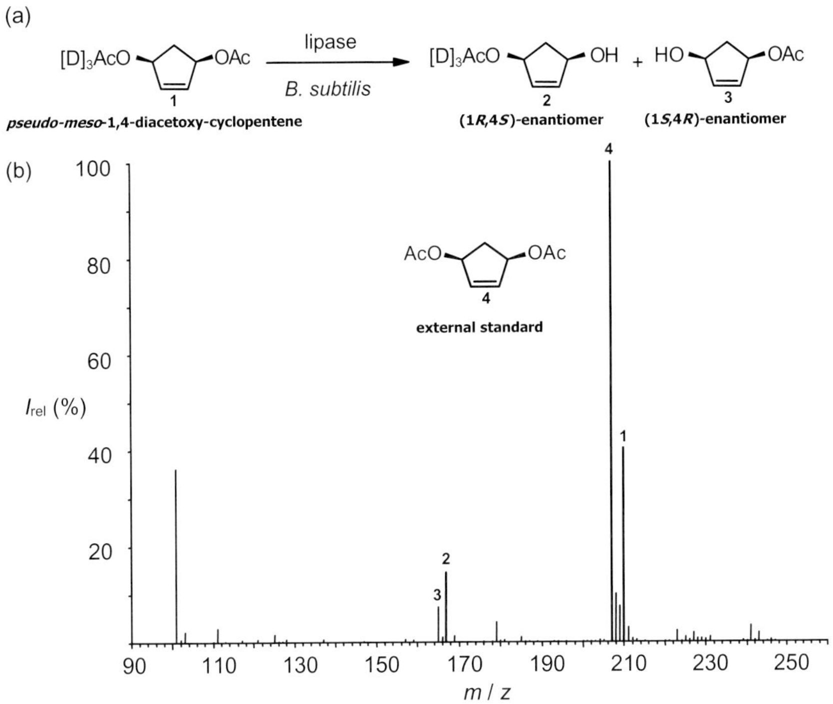

Figure 17.3 (a) Hydrolysis of *pseudo-meso*-1,4-diacetoxy-cyclopentene as a model reaction; (b) The two enantiomers of the product can be distinguished by their mass difference using a high-throughput ESI-MS method.

the model substrate with only moderate enantioselectivity [enantiomeric excess (ee) = 38%] to form the (1*R*,4*S*) product. In these studies, we attempted to evolve BSLA variants with improved enantioselectivity.

17.3 Directed Evolution by Error-Prone PCR

As a starting point, we applied epPCR at different mutation rates, leading to variant libraries with either one or two to three amino acid substitutions per lipase molecule. The libraries were screened in the model reaction and five variants were identified which showed enantioselectivities of up to 58% ee. These variants were subjected to a second round of epPCR with a high error rate, and four additional variants with ee-values up to 69% were identified. Interestingly, a third round of epPCR did not result in the identification of variants with further improved enantioselectivity, although 9000 variants were screened. The results of the epPCR experiments are summarized in Table 17.1.

Obviously, epPCR resulted in the formation of BSLA variants which showed improved enantioselectivity towards the model substrate; however, iterative rounds

Table 17.1 Evolution of BSLA with epPCR.

Generation	Error rate	Number of variants screened	Active variants (%)	Enantioselectivity[c] (% ee)
0 (wild-type)			100	38
1	Low[a]	1000	85	48
1	High[b]	4000	60	58
2	High[b]	15000	45	69
3	High[b]	9000	10	65–69

a Corresponding to an average of one amino acid substitution per lipase molecule.
b Corresponding to an average of two to three amino acid substitutions per lipase molecule.
c In formation of the (1*R*,4*S*) product.

of mutagenesis did not result in further improvements. It was concluded from these results that the small size of BSLA may hamper or even prevent the generation of large numbers of functional BSLA variants. Exhaustive random mutagenesis leading to the accumulation of several amino acid substitutions at a time may cause drastic reduction or loss of enzyme stability and/or lipolytic activity. Indeed, it was observed that only a few variants obtained from the third round of epPCR were still enzymatically active when tested by lipase standard assays such as the tributyrin agar plate assay [28]. We presently assume that natural evolution has designed the compact structure of BSLA as an optimal adaptation to its function, inferring that only very few additional amino acid substitutions are tolerated without disturbing the enzyme's structural integrity. Hence, the decision was taken to apply complete site-saturation mutagenesis to scan the entire BSLA sequence and separately to evaluate the contribution of each amino acid to the enantioselectivity of BSLA.

17.4 Complete Site-Saturation Mutagenesis

A first-generation library constructed by epPCR can serve to identify so-called 'hot-spots' – that is, amino acid positions which obviously affect the searched enzyme property. These positions can subsequently be subjected to site-saturation mutagenesis. For BSLA, we have sequenced the genes of several variants identified within epPCR libraries. Six amino acid positions were found at which substitutions resulted in improved enantioselectivities, namely Ile22, Tyr49, Asn50, Gen60, Leu124 and Gen164. Additionally, multiple site-saturation libraries were established at positions Asn50, Phe58, Gen60 and Ile157, Leu160 and Gen164. Interestingly, we also identified a single variant which showed inverted enantioselectivity (15% ee, reverse) yielding the (1*S*,4*R*) enantiomer of the product [21].

Based on these findings, we decided to subject BSLA to complete site-saturation mutagenesis. Libraries for each of the 181 amino acid positions of BSLA consisting of about 300 variants were constructed and screened to ensure that all possible amino acid substitutions were covered. This approach led us to identify amino acid position Asn18 as a 'hot-spot' for reversal of enantioselectivity (Figures 17.4c and 17.5). This position had never been touched by our previous mutagenesis experiments using epPCR, clearly indicating the advantage of the complete site-saturation mutagenesis method. It is hardly possible to decide whether or not amino acid Asn18 would have been identified also within an epPCR library. On the other hand, all hot-spot positions previously determined by the epPCR approach were also identified by site-saturation mutagenesis. The BSLA variant with the highest enantioselectivity for the (1*R*,4*S*) product showed an ee-value of 69%, and better variants were not found in the site-saturation libraries. Typical results which were obtained for most of the amino acid positions are shown in Figure 17.4a and b, for positions Gly30 and Ser127, respectively. Amino acid substitutions at position 30 produced many inactive variants, whereas at position 127, mostly variants with wild-type activity and enantioselectivity were identified. Additionally, variant

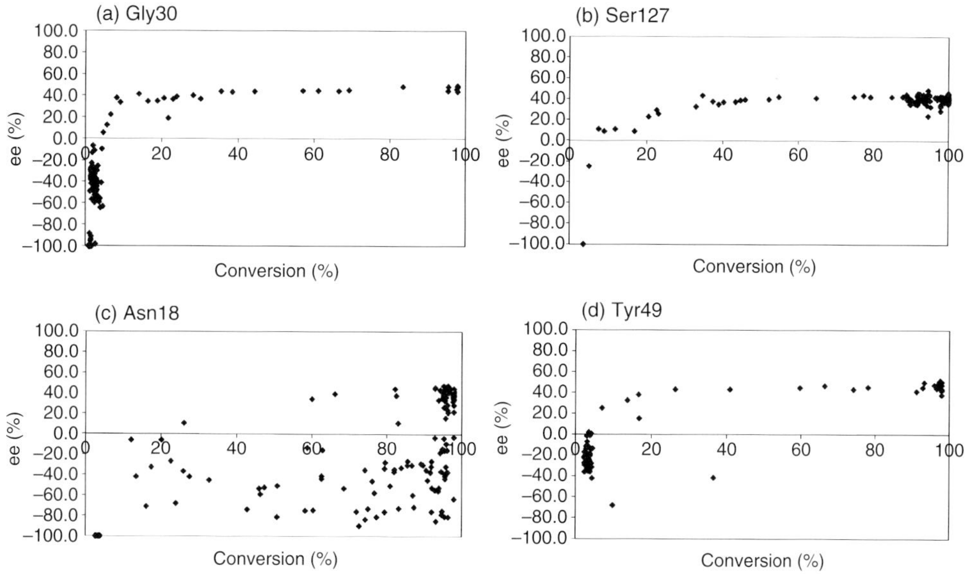

Figure 17.4 Representative results obtained by complete site-saturation mutagenesis of BSLA. Enantioselectivities and activities of selected variants were identified by HTS using ESI-MS. (a) Saturation mutagenesis at position Gly30 revealed many inactive enzyme variants; (b) variants at position Ser127 mostly showed wild-type activity (90–100% substrate conversion) and enantioselectivity (ee = 40%); (c) amino acid positions Asn18 and (d) Tyr49 were identified as 'hot-spots' for determining the enantioselectivity of BSLA.

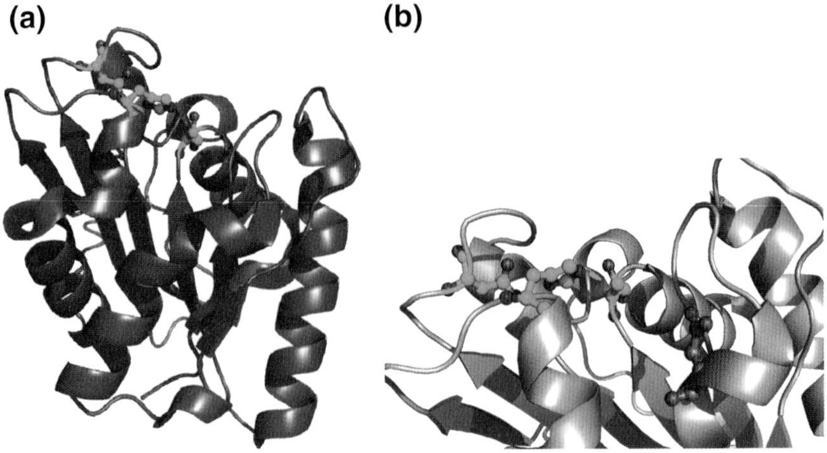

Figure 17.5 Structural model of BSLA. (a) X-ray structure (pdb 1I6W), the catalytic triad residues Ser77, Asp133 and His156 are shown in green; (b) amino acid Asn18 highlighted in orange.

Tyr49Ile with inverted enantioselectivity (ee = 15% reverse) was identified (Figure 17.4d), but this variant was expressed only at very low levels in *E. coli*.

Presently, we cannot provide a feasible explanation for these results because no structural data are available for BSLA with bound substrate. However, structure determination as well as molecular modeling studies of BSLA variant Asn18 appear to be particularly interesting with respect to the molecular mechanism underlying enantiodiscrimination of the substrate. By using *Pseudomonas aeruginosa* lipase as a model enzyme, we have previously demonstrated that the knowledge of an enzymes' three- dimensional structure in combination with quantum mechanics/molecular mechanics (QM/MM) calculations can provide reasonable explanations for the effect of several amino acid substitutions on enantioselectivity [29, 30].

17.5 Conclusions

Several different experimental approaches were applied in the quest to evolve enantioselective variants of BSLA. The results obtained clearly indicated that complete site-saturation mutagenesis is a promising and meaningful approach to identify hot-spot positions within the primary structure of a given enzyme. The size of the respective libraries is comparatively small, which results in a reasonable number of variants to be screened. The hot-spot substitutions can either be combined using recombinative methods or, alternatively, improved variants identified within site-saturation libraries can be subjected to further rounds of random mutagenesis, followed by optimization via consecutive rounds of saturation mutagenesis at newly identified amino acid positions.

Another important result which has emerged from our studies relates to the model enzyme BSLA. This lipase is a minimal α/β-hydrolase which seems to be particularly suited to the study of structure–function relationships concerning not only enantioselectivity but also thermostability. Recently, a phage-display system was described which allowed the selection of enantioselective BSLA variants [31, 32]. Furthermore, the thermostability of BSLA was significantly increased using two initial rounds of epPCR, followed by a computational analysis of the best variants to identify positive point mutations and to separate them from those with negative effects [33]. In a mechanistically different approach, B-factors were extracted from X-ray data to identify amino acids responsible for thermal instability of BSLA. Subsequent iterative saturation mutagenesis resulted in a hyperthermophilic BSLA variant showing a T_{50} value of 93 °C as compared to T_{50} = 48 °C for the wild-type enzyme [34, 35].

At present, a general protocol is not available to optimize an enzyme by directed evolution. Complete site-saturation mutagenesis appears to be an excellent starting point to scan the entire amino acid sequence of a given enzyme and to assign to each amino acid a specific role for the property to be optimized.

References

1 Hatti-Kaul, R., Tornvall, U., Gustafsson, L. and Borjesson, P. (2007) Industrial biotechnology for the production of bio-based chemicals – a cradle-to-grave perspective. *Trends in Biotechnology*, **25**, 119–24.

2 Jaeger, K.E. and Eggert, T. (2002) Lipases for biotechnology. *Current Opinion in Biotechnology*, **13**, 390–7.

3 Schmid, A., Dordick, J.S., Hauer, B., Kiener, A., Wubbolts, M. and Witholt, B. (2001) Industrial biocatalysis today and tomorrow. *Nature*, **409**, 258–68.

4 Lorenz, P. and Eck, J. (2005) Metagenomics and industrial applications. *Nature Reviews. Microbiology*, **3**, 510–16.

5 Streit, W.R., Daniel, R. and Jaeger, K.E. (2004) Prospecting for biocatalysts and drugs in the genomes of non-cultured microorganisms. *Current Opinion in Biotechnology*, **15**, 285–90.

6 Reymond, J.-L. (2006) *Enzyme Assays: High-Throughput Screening, Genetic Selection and Fingerprinting*, Wiley-VCH Verlag GmbH, Weinheim.

7 Arnold, F.H. and Volkov, A.A. (1999) Directed evolution of biocatalysts. *Current Opinion in Chemical Biology*, **3**, 54–9.

8 Jaeger, K.E., Eggert, T., Eipper, A. and Reetz, M.T. (2001) Directed evolution and the creation of enantioselective biocatalysts. *Applied Microbiology and Biotechnology*, **55**, 519–30.

9 Otten, L.G. and Quax, W.J. (2005) Directed evolution: selecting today's biocatalysts. *Biomolecular Engineering*, **22**, 1–9.

10 Sen, S., Dasu, V.V. and Mandal, B. (2007) Developments in directed evolution for improving enzyme functions. *Applied Biochemistry and Biotechnology*, **143**, 212–23.

11 Eggert, T., Jaeger, K.E. and Reetz, M.T. (2004) Directed evolution by random mutagenesis: a critical evaluation, in *Enzyme Functionality: Design, Engineering, and Screening* (ed. A. Svendsen), Marcel Dekker, New York, pp. 375–390.

12 Vartanian, J.P., Henry, M. and WainHobson, S. (1996) Hypermutagenic PCR involving all four transitions and a sizeable proportion of transversions. *Nucleic Acids Research*, **24**, 2627–31.

13 Wong, T.S., Zhurina, D. and Schwaneberg, U. (2006) The diversity challenge in directed protein evolution. *Combinatorial Chemistry and High Throughput Screening*, **9**, 271–88.

14 Wong, T.S., Roccatano, D., Zacharias, M. and Schwaneberg, U. (2006) A statistical analysis of random mutagenesis methods used for directed protein evolution. *Journal of Molecular Biology*, **355**, 858–71.

15 Wong, T.S., Roccatano, D. and Schwaneberg, U. (2007) Challenges of the genetic code for exploring sequence space in directed protein evolution. *Biocatalysis and Biotransformation*, **25**, 229–41.

16 Wong, T.S., Tee, K.L., Hauer, B. and Schwaneberg, U. (2004) Sequence saturation mutagenesis (SeSaM): a novel method for directed evolution. *Nucleic Acids Research*, **32**, e32.

17 Wong, T.S., Tee, K.L., Hauer, B. and Schwaneberg, U. (2005) Sequence saturation mutagenesis with tunable mutation frequencies. *Analytical Biochemistry*, **341**, 187–9.

18 Arnold F.H. (1996) Directed evolution: creating biocatalysts for the future. *Chemical Engineering Science*, **51**, 5091–102.

19 Brissos, V., Eggert, T., Cabral, J.M.S. and Jaeger, K.E. (2008) Improving activity and stability of cutinase towards the anionic detergent AOT by complete saturation mutagenesis. *Protein Engineering, Design and Selection*, **21**, 387–93.

20 DeSantis, G., Wong, K., Farwell, B., Chatman, K., Zhu, Z.L., Tomlinson, G., Huang, H.J., Tan, X.Q., Bibbs, L., Chen, P., Kretz, K. and Burk, M.J. (2003) Creation of a productive, highly enantioselective nitrilase through gene site saturation mutagenesis (GSSM). *Journal of the American Chemical Society*, **125**, 11476–7.

21 Funke, S.A., Eipper, A., Reetz, M.T., Otte, N., Thiel, W., Van Pouderoyen, G., Dijkstra, B.W., Jaeger, K.E. and Eggert, T. (2003) Directed evolution of an enantioselective *Bacillus subtilis* lipase. *Biocatalysis and Biotransformation*, **21**, 67–73.

22 Gray, K.A., Richardson, T.H., Kretz, K., Short, J.M., Bartnek, F., Knowles, R., Kan, L., Swanson, P.E. and Robertson, D.E. (2001) Rapid evolution of reversible denaturation and elevated melting temperature in a microbial haloalkane dehalogenase. *Advanced Synthesis Catalysis*, **343**, 607–17.

23 Eggert, T., van Pouderoyen, G., Dijkstra, B.W. and Jaeger, K.E. (2001) Lipolytic enzymes LipA and LipB from *Bacillus subtilis* differ in regulation of gene expression, biochemical properties, and three-dimensional structure. *FEBS Letters*, **502**, 89–92.

24 Eggert, T., van Pouderoyen, G., Pencreac'h, G., Douchet, I., Verger, R., Dijkstra, B.W. and Jaeger, K.E. (2002) Biochemical properties and three-dimensional structures of two extracellular lipolytic enzymes from *Bacillus subtilis*. *Colloids and Surfaces B: Biointerfaces*, **26**, 37–46.

25 van Pouderoyen, G., Eggert, T., Jaeger, K.E. and Dijkstra, B.W. (2001) The crystal structure of *Bacillus subtilis* lipase: a minimal alpha/beta hydrolase fold enzyme. *Journal of Molecular Biology*, **309**, 215–26.

26 Eggert, T., Leggewie, C., Puls, M., Streit, W., van Pouderoyen, G., Dijkstra, B.W. and Jaeger, K.E. (2004) Novel biocatalysts by identification and design. *Biocatalysis and Biotransformation*, **22**, 139–44.

27 Reetz, M.T. (2004) High-throughput screening of enantioselective industrial biocatalysts, in *Evolutionary Methods in Biotechnology* (eds S. Brakmann and A. Schwienhorst), Wiley-VCH Verlag GmbH, Weinheim, p. 214.

28 Kok, R.G., Christoffels, V.M., Vosman, B. and Hellingwerf, K.J. (1993) Growth-phase-dependent expression of the lipolytic system of *Acinetobacter calcoaceticus* Bd413 – cloning of a gene encoding one of the esterases. *Journal of General Microbiology*, **139**, 2329–42.

29 Bocola, M., Otte, N., Jaeger, K.E., Reetz, M.T. and Thiel, W. (2004) Learning from directed evolution: theoretical investigations into cooperative mutations in lipase enantioselectivity. *ChemBioChem*, **5**, 214–23.

30 Reetz, M.T., Puls, M., Carballeira, J.D., Vogel, A., Jaeger, K.E., Eggert, T., Thiel, W., Bocola, M. and Otte, N. (2007) Learning from directed evolution: further lessons from theoretical investigations into cooperative mutations in lipase enantioselectivity. *ChemBioChem*, **8**, 106–12.

31 Boersma, Y.L., Dröge, M.J. and Quax, W.J. (2007) Selection strategies for improved biocatalysts. *The FEBS Journal*, **274**, 2181–95.

32 Droge, M.J., Boersma, Y.L., van Pouderoyen, G., Vrenken, T.E., Ruggeberg, C.J., Reetz, M.T., Dijkstra, B.W. and Quax, W.J. (2006) Directed evolution of *Bacillus subtilis* lipase A by use of enantiomeric phosphonate inhibitors: crystal structures and phage display selection. *ChemBioChem*, **7**, 149–57.

33 Acharya, P., Rajakumara, E., Sankaranarayanan, R. and Rao, N.M. (2004) Structural basis of selection and thermostability of laboratory evolved *Bacillus subtilis* lipase. *Journal of Molecular Biology*, **341**, 1271–81.

34 Reetz, M.T. and Carballeira, J.D. (2007) Iterative saturation mutagenesis (ISM) for rapid directed evolution of functional enzymes. *Nature Protocols*, **2**, 891–903.

35 Reetz, M.T., Carballeira, J.D. and Vogel, A. (2006) Iterative saturation mutagenesis on the basis of B factors as a strategy for increasing protein thermostability. *Angewandte Chemie – International Edition*, **45**, 7745–51.

18
Circular Permutation of Proteins

Glenna E. Meister, Manu Kanwar and Marc Ostermeier

18.1
Introduction

The circular permutation of a protein can be defined as the intramolecular relocation of its N and C termini. Such relocations have been identified in natural proteins through comparison of the primary and tertiary structures of protein families. Hence, circular permutation is a natural evolutionary mechanism for proteins, although the mechanism by which it occurs and its utility are not fully understood. The artificial circular permutation of a protein is most easily achieved through manipulation of its gene. Conceptually, at the protein level, circular permutation involves ligation of the N and C termini – usually using a small peptide linker – and breaking the peptide bond at another site in the molecule, thereby generating new N and C termini (Figure 18.1). In general, the N and C termini of the original protein must be sufficiently close (~5–10 Å) to enable joining the two termini. As 50% of single-domain proteins have their termini within 5 Å of each other [3], a wide variety of proteins are potentially amenable to circular permutation.

Circular permutation affects a protein in three fundamental ways. First, the continuity of the polypeptide chain is altered. Second, new termini and a linker between the old termini are introduced and must be accommodated in the structure. Third, some structures adjacent in the primary sequence are moved apart, whereas others are brought together. These changes have the potential of altering a protein's folding, stability, structure and function. Yet, proteins are surprisingly tolerant to circular permutation at a wide variety of locations, including within secondary structure elements [4–8]. Circular permutation can occur with very little change in overall structure (Figure 18.1). In this chapter, the details of natural circularly permuted proteins are provided and their evolution outlined, together with a description of the use of artificial circular permutation as a protein-engineering tool. Circular permutation as a tool for understanding protein folding, structure and function will be addressed only briefly.

Protein Engineering Handbook. Edited by Stefan Lutz and Uwe T. Bornscheuer

ISBN: 978-3-527-31850-6

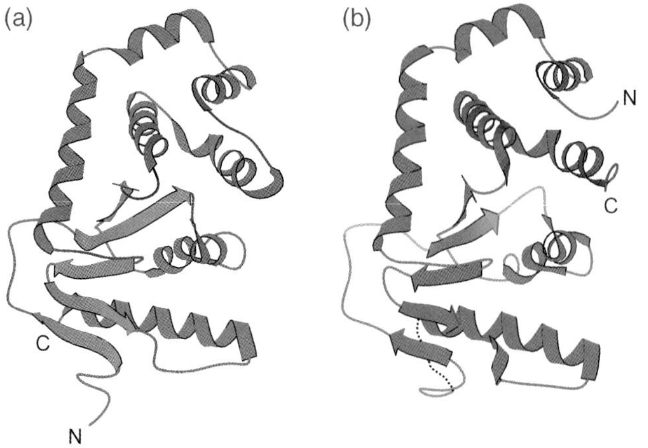

Figure 18.1 Circular permutation illustrated with the crystal structures of (a) DsbA [1] and (b) circularly permuted DsbA [2]. DsbA was circularly permuted by joining the natural termini of DsbA with a GGGTG peptide linker (disordered in the structure but shown as a dotted line for illustrative purposes) and the introduction of new termini (N and C) in the helical domain. The two proteins show very small differences in structure.

18.2
Evolution of Circular Permutations in Nature

18.2.1
Naturally Occurring Circular Permutations

Interest in natural circular permutations – which once were considered quite rare – has grown during recent years as the number of examples has increased. Improvements in sequence alignment and structure comparison software, in addition to an increase in the number of searchable proteins in databases, have led to a rise in the number of known circular permutations. Several protein families exhibit circular permutation, including lectins, methyltransferases, β-glucanases, saposins, surface layer homology domains, transolidases, transhydrogenases, adenosine triphosphate-binding cassette, glycosyl hydrolases and bacterial phosphocarrier proteins [9, 10]. Circular permutation can occur at either the peptide or genetic level, although genetic rearrangements are far more common.

The first circular permutation was found through a comparison of the amino acid sequences of two lectins, favin and concanavalin A (Con A) [11]. Con A is composed of one polypeptide chain, while favin is a heterodimeric protein with an alpha chain and a beta chain. When the favin and Con A polypeptides were sequenced, a comparison of the two showed that the N-terminal regions of the favin beta chain were homologous to the C-terminal region of Con A, starting at residue 123. At first, it was postulated that the circular permutation occurred at the genetic level, but it was later determined to be the result of post-translational modifications [12]. Several other proteins have been shown to undergo post-translational modifications involving the breaking and joining of peptide bonds. For example, favin is spliced into two polypeptide chains after its synthesis. However, Con A is the only known circular permutation arising from this mechanism. The CyBase database contains information for backbone-cyclized proteins [13], many of which result from post-translational modifications analogous to that of Con A. This suggests that conditions in nature allow circular permutations to arise by this mechanism.

The first known natural circular permutation arising from genetic rearrangement was identified in saposins. There are four saposin functional domains, which are normally seen as tandem repeats separated by a linker sequence in prosaposin. Each domain is composed of four alpha helices (identified as α1–α4). A saposin-like domain was identified in a plant aspartic proteinase [14], where the saposins-like domain, called swaposins, had the same four alpha helices, although the alpha helices were in the order α3–α4–α1–α2, which is a circular permutation of the order typically found in saposins [15].

DNA methyltransferases belong to a protein family which exhibits extensive circular permutations caused by genetic rearrangement. Methyltransferases are responsible for transferring methyl groups to DNA. Methylation patterns are important for the epigenetic control of gene expression in higher eukaryotes, as well as forming part of the restriction/modification system in prokaryotes to protect against foreign DNA invasion. Methyltransferases have ten conserved motifs (I–X) and a target recognition domain (TRD). There are two groups of methyltransferases that differ in the location of the TRD region in relation to other motifs. In one group, the TRD falls between motifs II and III, whereas in the second group the TRD falls between motifs VIII and X (motif IX is not conserved in some methyltransferases). Both groups have different classes of circular permutations [16]. Methyltransferases are assigned to one of seven classes of circular permutations (α, δ, ε, m^5C, β, γ, ζ), the classes being determined by the order of the catalytic region (motifs IV–VIII), the substrate binding region (motifs I–III and X) and the target recognition domain. These classes are described in detail by Vogel and Morea [17]. When first proposed, the δ, ε and ζ classes were considered hypothetical, but two ζ class (M.*Bss*HI and M.*Tvo*ORF1413P) and one δ class (M.*Mwo*I) methyltransferases have been found [18], and a new class (η) was discovered using directed evolution experiments [19].

18.2.2
Identification of Natural Circular Permutations

The first natural circular permutations were found by manual comparison of related sequences. As many of the currently known circular permutations occur in closely related protein families (many maintaining the same or fairly similar function), a comparison of these sequences is routine. However, the manual identification of nonhomologous proteins that are related by circular permutation is difficult, and an automated screening method is required in order to probe databases for these proteins.

Two main approaches have been used for screening databases. The first approach is to search and align protein sequences from protein or gene databases, while the second is to search the structure database and compare structures. Each search method has its own merits and complications. As the unique entries in sequence databases vastly outnumber the unique entries in the PDB database, a sequence comparison approach provides access to many proteins that a structure-based approach does not. However, as most circular permutations have a conserved structure, despite significant sequence divergence, structural comparisons can in theory detect circular permutations that a sequence alignment would miss. In addition, a structural comparison approach can identify circular permutations resulting from post-translational modifications (e.g. Con A).

Problems with using standard algorithms for sequence alignments of circular permutations have been reviewed previously [20]. Many algorithms align proteins by identifying homologous sequences in a sequential order. In circular permutations, the domains are – by definition – in a different sequence order, thereby causing difficulties in identifying matches. In some cases there could be a match based on similarity between one domain, but in many cases circular permutations have diverged substantially so that the algorithms cannot detect sufficient similarities. Database search methods such as BLAST also have their problems. BLAST breaks down sequences into short fragments and tries to match fragments to other proteins. These small fragments may detect similarities in circular permutations, but reassembling the small fragments could be problematic.

Uliel *et al.* addressed these problems with an algorithm that compared sequences to a tandemly duplicated version of the queried protein [20, 21]. By using this simple procedure, a protein sequence and all possible circular permutations thereof can then be screened for similarities to sequences in the database. This approach was able to determine new circular permutations in the Swissprot database, but required a manual comparison to check the results. Also, the strict criterion of the algorithm defines circular permutations as a case where the C termini are similar to the N termini of the other protein, and *vice versa*. This is problematic for identifying circular permutations in proteins of differing sizes (i.e. those that have an additional domain in the protein). The algorithm was able to find several known circular permutations such as lectins and β-glucanases, but did not detect some known circular permutations such as saposin. Another more recent algorithm developed by Weiner *et al.* searches for circular permutations based on domains or motifs in the sequence [22]. Using the ProDom database that identifies several conserved protein domains, the algorithm breaks the protein sequences into a linear sequence of domains. This method is more tolerant to the presence of extra domains and is faster than the previously described method. Novel circular permutations were found using this method between chitinases from different *Bacillus* species and between protozoan transhydrogenases and transhydrogenases from higher eukaryotes.

Structural alignment algorithms have been developed by Jung and Le [23] and Chen *et al.* [24]. In one method, the circular permutations are found by aligning structures which have had a new cut site in the middle of the sequence [23], and the structural alignment then matches as many residues as possible. Based on whether the number of residues matched before or after the cut are greater, the cut site is then moved a residue away from the last aligned residue. The protein is then realigned and evaluated for structure similarity. A total of 412 circularly permutated structures was found, including several novel pairs with remote homology. The SAMO software uses a simple method of aligning the greatest number of residues while minimizing the rms in a sequence order-independent manner [24]. This allows for the detection of circular permutations, although the software does not screen specifically for these cases.

18.2.3
Mechanisms of Circular Permutation

There are three main mechanisms believed to result in the natural circular permutation of proteins (recently reviewed by Weiner and Bornberg-Bauer [10]). These methods are: (i) independent fusion; (ii) duplication/deletion; and (iii) cut and paste (Figure 18.2).

The independent fusion model posits that proteins which are related by circular permutation can arise by independent assembly of protein modules (from independent genes) in different orders. For example, if we imagine the existence of two independent genes that code for two independent domains A and B, then potentially these domains could be fused in either of two orders: AB or BA. This model is more likely applicable to simple two-domain proteins for which there is evidence of the domains existing independently elsewhere in the genome. A variation of this method is the fusion/fission model, where a fission event separates a promiscuous domain from the protein, and later undergoes a fusion event to create a circularly permuted form of the gene [10].

Natural tandem domain repeats are well-known; indeed, this is the functional form of some proteins. Such genes should be prone to circular permutation by extraction of a single gene fragment corresponding to the end of one domain and the beginning of the next [15]. This mechanism is believed to be responsible for circular permutation in saposins [15]. The duplication/deletion model extends this mechanism to proteins for which the functional form does not have tandem duplications of domains [22, 25]. The duplication/deletion model posits an initial duplication of the protein, making a multi-domain protein, such as AB into ABAB. Following this, independent deletions of superfluous domains creates a circularly permutated protein (i.e. BA). As it is unlikely for all un-needed domains to be eliminated in one step, the model supposes that intermediate circular permutations (iCPs) occur, such as BAB or ABA. This mechanism is supported by the identification of several iCPs in protein databases [22].

A seeming requirement of the duplication/deletion model is that the fused tandemly duplicated gene as well as the iCPs should remain functional. Directed evolution experiments on M.*Hae*III, a 5-methylcytosine-class methyltransferase, have provided convincing experimental evidence of the feasibility of the model. These experiments evolved a circularly permuted M.*Hae*III through each of the

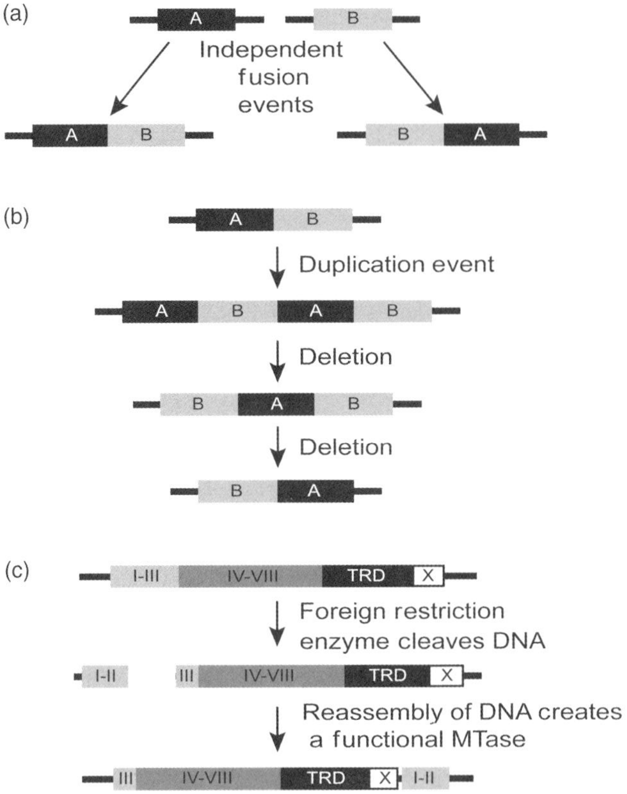

Figure 18.2 Evolutionary mechanisms for circular permutation. In the independent fusion mechanism (a) two domains become fused independently in opposite orders, resulting in two proteins the sequences of which are related by circular permutation. The duplication/deletion mechanism (b) posits that a gene first becomes tandemly duplicated whereupon successive introduction of new start and stop signals (i.e. deletion) results in a circularly permuted gene. The cut-and-paste mechanism (c) is illustrated with a gene for a DNA methyltransferase. The gene first becomes fragmented by restriction enzymes. The strong selective pressure to reconstitute an active DNA methyltransferase can result in a reassembled, circularly permuted gene.

three steps of the model [19]. First, the fused, tandemly duplicated M.*Hae*III was shown to be functional, after which separate libraries of N- and C-terminal truncations of the gene were subjected to selection for functional methyltransferases. Several functional methyltransferases were found which consisted of an entire M.*Hae*III domain fused to a partial M.*Hae*III domain. Next, libraries derived from these iCP genes were created by truncation from the other end. Several active circular permutants were identified through selection. Interestingly, libraries in which the tandemly duplicated gene was truncated simultaneously from both ends yielded similar permutants to those of the stepwise process. This suggests that the requirement for functional iCPs will not constrain the number of circular permutants that are evolutionary achievable–that is, this discounts the possibility that a potential viable circular permutant cannot be evolved because its corresponding iCP is not functional.

The third mechanism for the evolution of circular permutation of proteins relates specifically to DNA methyltransferases [18]. In bacteria, it is possible for the host to be invaded by a foreign plasmid encoding a restriction/modification system which is different from that of the host. The foreign restriction endonuclease could cut the host methyltransferase gene into segments which are then rearranged and pasted back together in a different order (possibly creating a circular permutation). Due to the intense selective pressure of maintaining a functional methyltransferase in the presence of an active endonuclease, active circular permutations will be selected. This mechanism seems confined to the specific example of methyltransferases, and is supported by the specific circular permutation of M.*Tha*I and M.*Tvo*ORF1413P and the rearrangement of M.*Sfi*I and M.*Mwo*I.

18.3 Artificial Circular Permutations

18.3.1 Early Studies

Artificial circular permutations were first employed to study the folding properties of proteins. In the classic study of Goldenberg and Creighton [26], bovine pancreatic trypsin inhibitor (BPTI) was circularly permuted to study the effect of circular permutation on BPTI's folding and stability. The N and C termini of BPTI are in close proximity and were chemically crosslinked by treatment with 1-ethyl-3-(3-dimethylamino-propyl) carbodiimide. This cyclic protein was purified from unlinked proteins by ion-exchange chromatography and subsequently treated with trypsin. The trypsin specifically cleaved BPTI between Lys15 and Ala16 to create a functional circular permutant. This method of circular permutation is limited to *in vitro* studies and requires a suitable cut site for a protease to cleave.

Circular permutation of the gene represents an easier approach, whereby the protein can then be expressed *in vivo* or *in vitro* in the permuted form. Modification at the genetic level also allows new termini to be produced at any residue in the sequence. The first example of artificial circular permutation of a gene involved the rearrangement of a two-domain protein, yeast phosphoribosyl anthranilate isomerase (yPRAI) using recombinant DNA technology [27]. Unique restriction sites were used to remove the 3′ end of the gene and reinsert it at the 5′ end, creating a circularly permuted gene. Two different surface loops were selected for the new amino and carboxyl termini. Both circular permutants were found to be active.

Circular permutation of dihydrofolate reductase (DHFR) was performed to study the kinetic and folding changes it caused [28]. The authors utilized a slightly different approach in which the gene was duplicated in tandem, after which new start and stop codons were introduced (in an outer loop) to form the circular permutant. Such an approach is much like the tandem repeat or duplication/deletion method proposed for the evolution of natural circularly permuted proteins. The k_{cat} and k_m of the circular permutant and the wild-type DHFR were almost indistinguishable, with the only difference between the two proteins being a slightly lower stability and solubility. Wild-type activity was also observed with one of the circular permutants of yPRAI [27]. These two studies firmly established that artificial circular permutation was not devastating to protein function and could result in proteins with essentially wild-type properties. This no doubt encouraged future studies to be conducted a decade later which showed that circular permutation could improve protein activity. The early investigations on artificial circular permutation were comprehensively reviewed by Heinemann and Hahn [29].

18.3.2
Systematic and Random Circular Permutation

In the early studies of circular permutations, new termini were introduced in surface loops of the proteins in order to minimize disruptions of internal interactions within the protein. In most circular permutation studies, only a few constructs were created and tested as a systematic probe of an entire protein for circular permutations would be tedious. Despite this, just such a mammoth study was conducted on DHFR [6]. DHFR is a small monomeric protein of 159 amino acids, and by using PCR methods all circular permutations of the protein were cloned and expressed. Of the 158 permutants, 61 could not be produced (possibly due to folding or stability problems), while another 12 formed inclusion bodies and could not be refolded. However, the remaining 85 permutants were able to fold properly into an active enzyme. Thus, at least in DHFR, over 50% of the potential circular permutation sites are compatible with enzyme activity. Many sites permissible to circular permutation were within loop regions, although circular permutation sites were also found within secondary structure elements. Clearly, some proteins are fairly tolerant of circular permutations and a variety of new termini are feasible.

Although the creation of a particular circular permutation of a gene is relatively easy using recombinant DNA methodologies, such an approach is not efficient to evaluate all potential cut sites. Hence, the method of random circular permutation was developed to address this problem [4]. The method creates a library comprised of nominally all-circular permutations of a gene (Figure 18.3). In order to create the library, the gene is prepared in such a manner that DNA cyclization results in the fusion of DNA coding for a suitable linker between the original N and C termini (with the start and stop codon removed). Typically, this linker DNA contains a restriction enzyme site used in the ligation. Cyclization is achieved by DNA ligation performed under very dilute DNA concentrations so that cyclization is favored over linear dimerization. Digestion with an exonuclease can be used to remove any noncyclized DNA [30]. Next, the DNA is digested with very low levels of a nonspecific endonuclease such as DNaseI such that, on average, one double-

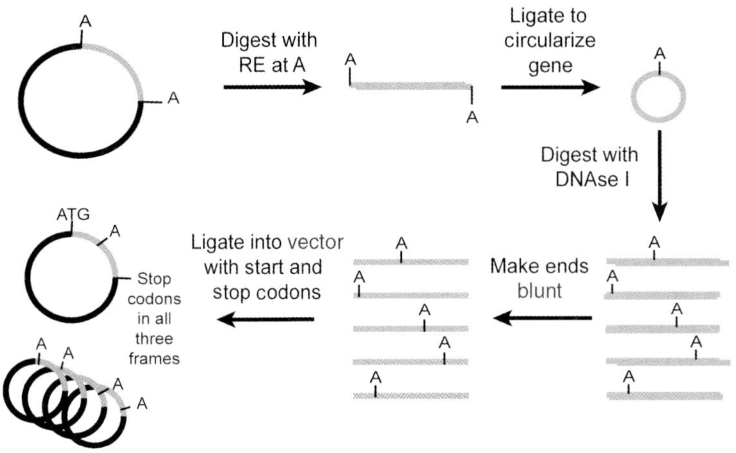

Figure 18.3 Creation of a random circular permutation library. The gene to be circularly permuted is first excised from a plasmid using a restriction enzyme (RE) that digests at both A sites. The excised gene has previously been designed such that the start codon and stop codons have been removed, and DNA coding for a linker peptide has been introduced. The cyclization step occurs in the presence of high concentrations of DNA ligase and very low concentrations of DNA, such that cyclization is favored over dimerization. The cyclized DNA is isolated and digested with dilute concentrations of DNaseI such that on average one double-strand break is created per molecule. These double-strand breaks can produce blunt ends, but predominantly produce 3′ or 5′ overhangs that must be repaired to make blunt-ended DNA. This repaired DNA is then ligated into a vector such that it is flanked by an ATG start codon on one side and a series of three stop codons in all three frames on the other side.

stranded break occurs per molecule. The ends of the DNA are then repaired to form blunt ends. The ends of the DNA must be repaired, since DNaseI digestion predominantly produces 3′ and 5′ overhangs, although perfect blunt ends can occur. The repair of 5′ overhangs results in tandem duplications of sequences, whereas the repair of 3′ overhangs results in deletions of sequences in the region of the new termini. The library will also consist of variants with short C-terminal extensions, depending on the stop codon with which the gene is in frame. The library of circularly permuted genes is then inserted into an expression vector containing a 5′ start site and a series of 3′ stop codons in all three reading frames.

Aspartate transcarbamoylase (ATCase) was the model protein used in the first demonstration of random circular permutation [4]. ATCase had been shown previously to tolerate circular permutations at a few locations [31]. Random circular permutation resulted in the identification of several new permutants, many with new termini in loop regions but others with termini in the alpha helices. Random circular permutations have since been used to identify functional circular permutants of DsbA [5], 5-aminolevulinate synthase [32], GFP [8], *Candida antarctica* Lipase B [7] and TEM1 β-lactamase [33]. The most striking conclusion of these

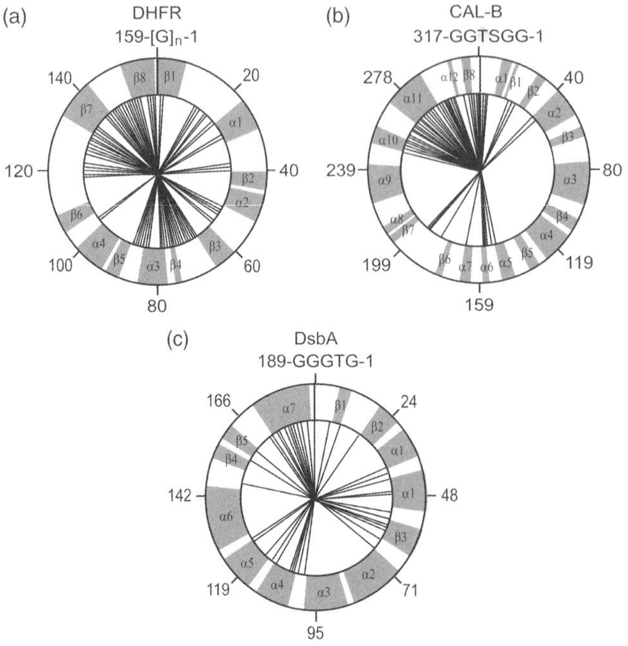

Figure 18.4 Permissive sites for circular permutation in (a) *E. coli* DHFR [6], (b) *Candida antarctica* Lipase B (CAL-B) [7], and (c) *E. coli* DsbA [5]. For CAL-B and DsbA some variants had tandem duplications at the beginning and end of the gene, making the marking of the permutation site arbitrary within the duplicated sequence. The radial lines indicate the locations of the new N-terminus in active permutants, whether or not the permutant contained tandem duplications of sequences at the ends of the protein. The locations of secondary structure elements are indicated in gray.

studies is the high tolerance of proteins to circular permutation in a variety of locations (Figure 18.4). Furthermore, for several of these proteins, random circular permutation enabled identification of permutants with improved catalytic activity, as will be discussed in Section 18.4.3.

18.3.3
Protein Folding and Stability

Many circular permutation studies have been undertaken to address questions on protein folding and stability [29]; however, some highlights will only be briefly touched upon here. Permutation can alter the rate of folding as well as the folding pathway itself [34, 35], perhaps by changing the spatial distance between structural blocks in the unfolded protein. In general, permutants are less stable than their unpermuted parent. In the case of a circular permutation of DsbA, a crystal structure of the permutant (Figure 18.1b) revealed that decreased stability arose for small structural changes, primarily near the new termini and the circularizing linker [2]. Interestingly, the active site of DsbA also was slightly perturbed, displaying increased flexibility. Circular permutation has been used to define folding elements – contiguous amino acids within which circular permutation is not tolerated [6]. Circular permutation, combined with segmental deletion has been used to show that no one segment in *Bacillus licheniformis* β-lactamase is essential to the folded structure [36].

18.4 Circular Permutation and Protein Engineering

Circular permutation can alter the folding, stability, structure, function and topology of a protein. These changes can be beneficial for applications, as a variety of studies have demonstrated.

18.4.1 Alteration of the Spatial Arrangement of Protein Fusions

Protein fusion has a wide variety of applications, the most common type being end-to-end fusion. Although only two types of end-to-end fusion are possible between two proteins A and B, namely AB and BA, this limited topology can be overcome if, instead, one protein is inserted into the other. Although proteins have been shown to tolerate insertions of short peptides and proteins at a variety of locations, there are significant limitations to this approach. The acceptor protein must be able to tolerate insertion at the desired site, and the inserted protein presumably must have its N and C termini proximal, but circular permutation of the acceptor sequence can potentially circumvent these limitations. If the acceptor sequence can tolerate circular permutation at the desired insertion site, then an end-to-end fusion can be made to this circular permuted variant.

Changes in the spatial orientation of fusions can have a significant impact on their biological properties. As interleukin-4 (IL-4) receptors are overexpressed on a variety of hematological and solid tumors, fusions of toxins to IL-4 could be of potential therapeutic benefit for the treatment of cancers. For this strategy to work, the fusions must have an appropriate affinity for the IL-4 receptor. End-to-end fusion of a circularly permuted variant of IL-4 and a truncated *Pseudomonas* exotoxin (PE) showed improved IL-4 receptor binding over that of an end-to-end fusion of unpermuted IL-4 and PE [37] which in turn resulted in a raised antitumor activity in mice [38]. These fusions have been shown to have potential for the treatment of human glioma [39], Hodgkin lymphoma [40] and ovarian cancer [41]. Similarly, the relative orientation of fusions of interleukin-3 (IL-3) and granulocyte colony-stimulating factor (GCSF) affects the ratio of IL-3 receptor and GCSF receptor agonist activities [42].

Bacterial display technologies generally rely on fusion of the desired displayed protein/peptide to a protein that naturally localizes to the inner or outer membrane of the bacterium. Predominantly, peptides are displayed as insertions in a surface loop, but the necessary step of transferring the selected peptides to a soluble scaffold can often result in significant losses in affinity [43]. An alternative approach would be to display the peptide as an N-terminal or C-terminal fusion, thus minimizing the impact of the host protein on the properties of the peptide library. However, this requires a suitable outer membrane protein that has at least one of its termini on the outside of the outer membrane. The bacterial protein OmpX does not have surface termini, but its proximal N and C termini and beta-barrel structure [44] make it an ideal candidate for circular permutation. Consequently, Rice *et al.* developed a new scaffold for display in which both OmpX termini were relocated to the surface by circular permutation [45]. Peptides could be displayed on the surface of *E. coli* as either N-terminal or C-terminal fusions. The N-terminal-displayed peptides were selected for binding to streptavidin, and the N-terminal display was found to yield sequences with greater diversity, affinity and modularity than the insertional display [45].

18.4.2
Oligomeric State Modification

The ability to change the oligomeric state of a protein without affecting its desired properties can be beneficial for some applications. For example, the Caf1 protein from *Yersinia pestis* has potential for use in a vaccine for the organism. However, recombinant Caf1 forms polymers of indefinite size instead of the desired homogeneous quaternary structure. Chalton *et al.* were able to create a monomeric form of Caf1 through circular permutation [46]. This example is an interesting exception to the rule that N and C termini must be proximal for circular permutation. Although the N and C termini of a monomeric unit of oligomeric Caf1 are not near each other, the N terminus of one unit is near the C terminus of the next unit. Thus, removal of the N-terminal sequence that interacts with the next subunit, and placing it at the C terminus, is feasible and prevents oligomerization [46]. As a result, the permuted Caf1 was monomeric yet retained the tertiary structure and immunogenic characteristics of polymeric Caf1.

Circular permutation has also been used to convert tetrameric human hemoglobin into a dimer of essentially the same molecular weight. Cell-free hemoglobin solutions have potential use as a blood substitute, but hemoglobin dissociates into dimers in the bloodstream and is rapidly filtered by the kidneys. As an alternative to stabilizing the tetramer by chemical crosslinking, Sanders *et al.* [47] devised a circular permutation of one hemoglobin α subunit that allowed its fusion with the second, unpermuted α subunit via a short linker. The resulting engineered hemoglobin displayed a fivefold higher affinity for oxygen but a decreased oxygen-binding cooperativity. In a similar study, the fusion of two circular permuted variants of chicken avidin, which normally forms a homotetramer, resulted in a protein that is presumably a homodimer of equivalent molecular weight to the homotetramer [48]. The biological activity and structural characteristics of the homodimer were similar to those of the homotetramer. Such a fusion allows mutations in only one of the two subunits of the fusion–mutations that are not possible in the homotetramer.

18.4.3
Improvement of Function

A reasonable expectation of even successful circular permuted proteins is that their stability and function will be no better than the unpermuted parent, and will often be compromised. After all, the amino acid composition of the unpermuted parent and the permuted protein are the same – aside from the linker, which is often located far from the active site and viewed as only having a connecting function. Only the linear order of the circularly permutated protein has been altered, and the alteration is in a very simple block-like manner. Perhaps for these reasons circular permutation as a method for creating improved protein properties has been only minimally addressed until recently. However, like random mutagenesis – for which a few mutations result in a gain of function and most mutations are either neutral or negative – there is no reason *a priori* why some circular permutants may not have improved properties relative to their parents. Indeed, although the examples so far are few in number, circular permutation has been shown to result in proteins with improved catalytic activity, ligand-binding properties and thermodynamic stability.

The circular permutation of eye lens βB2-crystallin resulted in a small increase in stability [49], while the circular permutation and subunit fusion of hemoglobin resulted in a fivefold increased affinity for oxygen [47], as described above. These two examples were found serendipitously in the sense that the goal of the circular permutation was other than improvement in these properties. The improved properties were only discovered through characterization of the permutants. What is desirable to evaluate the potential of circular permutation as a tool for improving protein function is a systematic or combinatorial search for improved variants.

The landmark construction of all possible circular permutations of DHFR and the subsequent kinetic characterization of all active permutants demonstrated the potential of circular permutation for the improvement of catalytic activity [6]. One permutant had a sixfold increase in k_{cat} and several had 1.5- to 2.5-fold increases in k_{cat}/k_m. However, even in this study, the goal was not to address the potential of circular permutation as a tool for diversity generation for directed evolution but rather to identify the essential folding elements of the protein. In particular, only one circular permutant of each potential cut point was tested. As both the linker identity [50, 51] and the addition/deletion of peptides on the new N and C termini [52] can potentially modulate the permutant's properties, there is a very large number of potential variants for any one cut-point.

Fortunately, the method of random circular permutation [4] creates libraries of permutants comprised of members that have deletions of sequences and tandem duplications of sequences at the two new termini and C-terminal appendages of short sequences. This added diversity has the potential to be beneficial to both folding and function. Studies on the random circular permutation of aspartate transcarbamoylase [4], DsbA [5] and GFP [8] employed convenient selections for function but did not address whether permutants with improved function existed among the functional variants.

The first demonstration that random circular permutation can produce variants with improved function occurred with the enzyme 5-aminolevulinate synthase (ALAS) [32]. The goal of this study was not to create improved variants but rather to assess the importance of continuity in the polypeptide's chain for the enzyme's function. However, one permutant among the nine active permutants discovered exhibited a 3.5-fold increase in k_{cat} and a 10-fold increase in k_{cat}/k_m for glycine (the improvement in k_{cat}/k_m for the second substrate of the reaction, succinyl-CoA, was about twofold). The high frequency of improved variants of ALAS was encouraging.

A high frequency of improved variants was also found among circular permutants of *Candida antarctica* Lipase B (CAL-B) [7]. Among the 63 functional variants (Figure 18.4b) identified through a functional screen, eight were chosen for kinetic characterization based on their structural location. Several had improvements in kinetic parameters, including one permutant that had a 29-fold increase in k_{cat}/k_m and a 60-fold increase in k_{cat} for one substrate. Interestingly, removal of the amino acid duplications and a poly-histidine tag positively affected the catalytic activity of some permutants, but not of others [53]. After removal of these sequences, two permutants had a catalytic efficiency that was over 100-fold better than that of the wild-type enzyme. Most importantly, such improvements were not achieved at the expense of enantioselectivity, and the largest improvements in catalysis were seen in variants in which the new termini were located proximal to the active site. This was postulated to result from an improved active-site accessibility and increased active-site flexibility [7].

Improved circular permutants of TEM1 β-lactamase (BLA) also had new N and C termini near the active site [33]. The search for catalytically improved permutants of BLA grew out of the observation that certain fusion proteins of BLA and maltose binding protein (MBP), in which BLA was circularly permuted in the Ω-loop and inserted inside of MBP, conferred a much higher resistance to cefotaxime than would be predicted from the resistance conferred to other penicillins and cephalosporins [54]. A library of circular permutants of BLA was subjected to selection for conferring improved resistance to four different β-lactam antibiotics, but only in the case of cefotaxime were clones with improved resistance identified. Sequencing revealed that, like the BLA-MBP fusions, the improved variants were permuted in BLA's Ω-loop. As the improved variants conferred a fourfold higher minimum inhibitory concentration (MIC) for cefotaxime, but a several-fold reduced resistance to ampicillin compared with unpermuted BLA, at a minimum the permutation resulted in a change in substrate specificity in the protein. As the expression of the permutants was poorer than that of wild-type BLA, the catalytic activity of the permutants must be improved over wild-type BLA, though this is yet to be confirmed with *in vitro* kinetic studies.

18.4.4 Creation of Protein Switches

The defining property of a protein switch is the modulation of function by external signals, usually ligand-binding. As circular permutation can be used to vary the spatial arrangement between two protein domains in a protein fusion, it has great potential for creating hybrid protein switches comprised of protein domains with the respective input and output functions of the desired switch. Baird *et al.* used random circular permutation of green fluorescent protein (GFP) to identify sites that may be tolerant to insertion [55]. Insertion of the Ca^{2+} binding protein calmodulin into one such site resulted in a fusion protein the fluorescence intensity of which was increased sevenfold in the presence of Ca^{2+}.

Guntas *et al.* also utilized random circular permutation to create enzymatic protein switches [54, 56]. A combination of random circular permutation of the TEM1 BLA gene and random insertion of the circularly permuted library into the MBP gene led to the creation of a family of maltose-activated β-lactamase switches. It is possible to conceptualize the structural space that is accessed by a combination of circular permutation and insertion as 'rolling' the two proteins across each other's surface and fusing them through two peptide bonds at the points where their surfaces meet. In the best switches, activity was severely compromised in the absence of maltose, which acted as a positive effector (an agonist), increasing β-lactamase activity up to 600-fold. In contrast, β-cyclodextrin acted as an antagonist, preventing maltose-activation. Cassette mutagenesis at the maltose-binding site, combined with a genetic selection scheme, illustrated the modularity of this design through to the creation of variants that responded to sucrose as a positive effector [54]. Unexpectedly, in one switch the combination of circular permutation and insertion created a new binding site for Zn^{2+}, which functioned as a negative effector for turning off the enzyme activity [50]. The role of circular permutation in this simultaneous emergence of a ligand-binding site and allostery is not clear, as the circular permutation of BLA alone did not create such a site and the Zn^{2+}-binding site has not yet been identified. However, it is tempting to speculate that the structural perturbations and fusion geometries uniquely enabled by circular permutation may prove useful in the creation of new functions, particularly in the case of protein switches.

18.4.5 Protein Crystallization

A routine trick in protein crystallization is to remove unstructured regions from the N and C termini of the domain to be crystallized. This facilitates crystallization by reducing its entropic cost, but the approach is not readily applicable when the unstructured region is internal to the domain. Circular permutation can be used to relocate the N and C termini within the flexible region, thereby allowing the deletion of the region. This approach was successfully applied to the β subunit of the eukaryotic signal recognition particle [57], whereby circular permutation allowed the deletion of an internal 26-residue flexible loop and enabled crystallization.

18.5
Perspective

Whilst circular permutation is unique for the type of mutagenesis involved, it can also occur without changing either the amino acid content or the molecular weight of a protein (if no additional amino acids are added in the linker between the N and C termini). As with the noncyclic rearrangements of a secondary structure [58–60], the hallmark of circular permutation is an altered connectivity of the peptide backbone and an altered topology, but this – along with perturbations of the newly introduced termini – can result in an altered structure and, in turn, a change in protein fitness.

The role of circular permutation in the evolution of natural proteins is unknown, and it is unclear whether the observed circular permutations have derived from an event that led to a gain of function, or were merely an accident of evolutionary drift. Artificial circular permutation experiments have revealed the ability of the mechanism to improve protein fitness, but have not yet addressed whether such improvements might be achieved more easily by other mutagenic mechanisms that are, presumably, more common. It follows that circular permutation may be best employed for applications that utilize the unique ability of circular permutation to alter protein topology, such as in the creation of altered protein fusions, protein switches and topologically changed protein platforms for nanotechnological applications [61].

Acknowledgments

These studies were supported in part by grants from NIH and NSF.

References

1 Guddat, L.W., Bardwell, J.C. and Martin, J.L. (1998) Crystal structures of reduced and oxidized DsbA: investigation of domain motion and thiolate stabilization. *Structure*, **6**, 757–67.

2 Manjasetty, B.A., Hennecke, J., Glockshuber, R. and Heinemann, U. (2004) Structure of circularly permuted DsbA(Q100T99): preserved global fold and local structural adjustments. *Acta Crystallographica. Section D, Biological Crystallography*, **60**, 304–9.

3 Krishna, M.M. and Englander, S.W. (2005) The N-terminal to C-terminal motif in protein folding and function. *Proceedings of the National Academy of Sciences of the United States of America*, **102**, 1053–8.

4 Graf, R. and Schachman, H.K. (1996) Random circular permutation of genes and expressed polypeptide chains: application of the method to the catalytic chains of aspartate transcarbamoylase. *Proceedings of the National Academy of Sciences of the United States of America*, **93**, 11591–6.

5 Hennecke, J., Sebbel, P. and Glockshuber, R. (1999) Random circular permutation of DsbA reveals segments that are essential for protein folding and stability. *Journal of Molecular Biology*, **286**, 1197–215.

6 Iwakura, M., Nakamura, T., Yamane, C. and Maki, K. (2000) Systematic circular permutation of an entire protein reveals essential folding elements. *Nature Structural Biology*, **7**, 580–5.

7 Qian, Z. and Lutz, S. (2005) Improving the catalytic activity of *Candida antarctica* lipase B by circular permutation. *Journal of the American Chemical Society*, **127**, 13466–7.

8 Topell, S. and Glockshuber, R. (2002) Circular permutation of the green fluorescent protein. *Methods in Molecular Biology (Clifton, N.J.)*, **183**, 31–48.

9 Lindqvist, Y. and Schneider, G. (1997) Circular permutations of natural protein sequences: structural evidence. *Current Opinion in Structural Biology*, **7**, 422–7.

10 Weiner, J. III and Bornberg-Bauer, E. (2006) Evolution of circular permutations in multidomain proteins. *Molecular Biology and Evolution*, **23**, 734–43.

11 Cunningham, B.A., Hemperly, J.J., Hopp, T.P. and Edelman, G.M. (1979) Favin versus concanavalin A: circularly permuted amino acid sequences. *Proceedings of the National Academy of Sciences of the United States of America*, **76**, 3218–22.

12 Carrington, D.M., Auffret, A. and Hanke, D.E. (1985) Polypeptide ligation occurs during post-translational modification of concanavalin A. *Nature*, **313**, 64–7.

13 Mulvenna, J.P., Wang, C. and Craik, D.J. (2006) CyBase: a database of cyclic protein sequence and structure. *Nucleic Acids Research*, **34**, D192–4.

14 Guruprasad, K., Tormakangas, K., Kervinen, J. and Blundell, T.L. (1994) Comparative modeling of barley-grain aspartic proteinase: a structural rationale for observed hydrolytic specificity. *FEBS Letters*, **352**, 131–6.

15 Ponting, C.P. and Russell, R.B. (1995) Swaposins: circular permutations within genes encoding saposin homologues. *Trends in Biochemical Sciences*, **20**, 179–80.

16 Malone, T., Blumenthal, R.M. and Cheng, X. (1995) Structure-guided analysis reveals nine sequence motifs conserved among DNA amino-methyltransferases, and suggests a catalytic mechanism for these enzymes. *Journal of Molecular Biology*, **253**, 618–32.

17 Vogel, C. and Morea, V. (2006) Duplication, divergence and formation of novel protein topologies. *BioEssays*, **28**, 973–8.

18 Bujnicki, J.M. (2002) Sequence permutations in the molecular evolution of DNA methyltransferases. *BMC Evolutionary Biology*, **2**, doi: 10.1186/1471-2148-2-3.

19 Peisajovich, S.G., Rockah, L. and Tawfik, D.S. (2006) Evolution of new protein topologies through multistep gene rearrangements. *Nature Genetics*, **38**, 168–74.

20 Uliel, S., Fliess, A. and Unger, R. (2001) Naturally occurring circular permutations in proteins. *Protein Engineering*, **14**, 533–42.

21 Uliel, S., Fliess, A., Amir, A. and Unger, R. (1999) A simple algorithm for detecting circular permutations in proteins. *Bioinformatics (Oxford, England)*, **15**, 930–6.

22 Weiner, J. III, Thomas, G. and Bornberg-Bauer, E. (2005) Rapid motif-based prediction of circular permutations in multi-domain proteins. *Bioinformatics (Oxford, England)*, **21**, 932–7.

23 Jung, J. and Lee, B. (2001) Circularly permuted proteins in the protein structure database. *Protein Science*, **10**, 1881–6.

24 Chen, L., Wu, L.Y., Wang, Y., Zhang, S. and Zhang, X.S. (2006) Revealing divergent evolution, identifying circular permutations and detecting active-sites by protein structure comparison. *BMC Structural Biology*, **6**, 18.

25 Jeltsch, A. (1999) Circular permutations in the molecular evolution of DNA methyltransferases. *Journal of Molecular Evolution*, **49**, 161–4.

26 Goldenberg, D.P. and Creighton, T.E. (1983) Circular and circularly permuted forms of bovine pancreatic trypsin inhibitor. *Journal of Molecular Biology*, **165**, 407–13.

27 Luger, K., Hommel, U., Herold, M., Hofsteenge, J. and Kirschner, K. (1989) Correct folding of circularly permuted variants of a beta alpha barrel enzyme in vivo. *Science (New York, N.Y.)*, **243**, 206–10.

28 Buchwalder, A., Szadkowski, H. and Kirschner, K. (1992) A fully active variant of dihydrofolate reductase with a circularly permuted sequence. *Biochemistry*, **31**, 1621–30.

29 Heinemann, U. and Hahn, M. (1995) Circular permutation of polypeptide chains: implications for protein folding and stability. *Progress in Biophysics and Molecular Biology*, **64**, 121–43.

30 Ostermeier, M. and Benkovic, S.J. (2001) Construction of hybrid gene libraries involving the circular permutation of DNA. *Biotechnology Letters*, **23**, 303–10.

31 Yang, Y.R. and Schachman, H.K. (1993) Aspartate transcarbamoylase containing circularly permuted catalytic polypeptide chains. *Proceedings of the National Academy of Sciences of the United States of America*, **90**, 11980–4.

32 Cheltsov, A.V., Barber, M.J. and Ferreira, G.C. (2001) Circular permutation of 5-aminolevulinate synthase. Mapping the polypeptide chain to its function. *The Journal of Biological Chemistry*, **276**, 19141–9.

33 Guntas, G. (2005) Creation of molecular switches by combinatorial protein engineering, PhD Thesis, Johns Hopkins University, Baltimore.

34 Bulaj, G., Koehn, R.E. and Goldenberg, D.P. (2004) Alteration of the disulfide-coupled folding pathway of BPTI by circular permutation. *Protein Science*, **13**, 1182–96.

35 Lindberg, M., Tangrot, J. and Oliveberg, M. (2002) Complete change of the protein folding transition state upon circular permutation. *Nature Structural Biology*, **9**, 818–22.

36 Gebhard, L.G., Risso, V.A., Santos, J., Ferreyra, R.G., Noguera, M.E. and Ermacora, M.R. (2006) Mapping the distribution of conformational information throughout a protein sequence. *Journal of Molecular Biology*, **358**, 280–8.

37 Kreitman, R.J., Puri, R.K. and Pastan, I. (1994) A circularly permuted recombinant interleukin 4 toxin with increased activity. *Proceedings of the National Academy of Sciences of the United States of America*, **91**, 6889–93.

38 Kreitman, R.J., Puri, R.K. and Pastan, I. (1995) Increased antitumor activity of a circularly permuted interleukin 4-toxin in mice with interleukin 4 receptor-bearing human carcinoma. *Cancer Research*, **55**, 3357–63.

39 Kawakami, M., Kawakami, K. and Puri, R.K. (2003) Interleukin-4-*Pseudomonas* exotoxin chimeric fusion protein for malignant glioma therapy. *Journal of Neuro-Oncology*, **65**, 15–25.

40 Kawakami, M., Kawakami, K., Kioi, M., Leland, P. and Puri, R.K. (2005) Hodgkin lymphoma therapy with interleukin-4 receptor-directed cytotoxin in an infiltrating animal model. *Blood*, **105**, 3707–13.

41 Kioi, M., Takahashi, S., Kawakami, M., Kawakami, K., Kreitman, R.J. and Puri, R.K. (2005) Expression and targeting of interleukin-4 receptor for primary and advanced ovarian cancer therapy. *Cancer Research*, **65**, 8388–96.

42 McWherter, C.A., Feng, Y., Zurfluh, L.L., Klein, B.K., Baganoff, M.P., Polazzi, J.O., Hood, W.F., Paik, K., Abegg, A.L., Grabbe, E.S., Shieh, J.J., Donnelly, A.M. and McKearn, J.P. (1999) Circular permutation of the granulocyte colony-stimulating factor receptor agonist domain of myelopoietin. *Biochemistry*, **38**, 4564–71.

43 Bessette, P.H., Rice, J.J. and Daugherty, P.S. (2004) Rapid isolation of high-affinity protein binding peptides using bacterial display. *Protein Engineering, Design and Selection*, **17**, 731–9.

44 Vogt, J. and Schulz, G.E. (1999) The structure of the outer membrane protein OmpX from *Escherichia coli* reveals possible mechanisms of virulence. *Structure*, **7**, 1301–9.

45 Rice, J.J., Schohn, A., Bessette, P.H., Boulware, K.T. and Daugherty, P.S. (2006) Bacterial display using circularly permuted outer membrane protein OmpX yields high affinity peptide ligands. *Protein Science*, **15**, 825–36.

46 Chalton, D.A., Musson, J.A., Flick-Smith, H., Walker, N., McGregor, A., Lamb, H.K., Williamson, E.D., Miller, J., Robinson, J.H. and Lakey, J.H. (2006) Immunogenicity of a *Yersinia pestis* vaccine antigen monomerized by circular permutation. *Infection and Immunity*, **74**, 6624–31.

47 Sanders, K.E., Lo, J. and Sligar, S.G. (2002) Intersubunit circular permutation of human hemoglobin. *Blood*, **100**, 299–305.

48 Nordlund, H.R., Laitinen, O.H., Hytonen, V.P., Uotila, S.T., Porkka, E. and Kulomaa, M.S. (2004) Construction of a dual chain pseudotetrameric chicken avidin by combining two circularly permuted avidins. *Journal of Biological Chemistry*, **279**, 36715–19.

49 Wieligmann, K., Norledge, B., Jaenicke, R. and Mayr, E.M. (1998) Eye lens betaB2-crystallin: circular permutation does not influence the oligomerization state but enhances the conformational stability. *Journal of Molecular Biology*, **280**, 721–9.

50 Liang, J., Kim, J.R., Boock, J.T., Mansell, T.J. and Ostermeier, M. (2007) Ligand binding and allostery can emerge simultaneously. *Protein Science*, **16**, 929–37.

51 Osuna, J., Perez-Blancas, A. and Soberon, X. (2002) Improving a circularly permuted TEM-1 beta-lactamase by directed evolution. *Protein Engineering*, **15**, 463–70.

52 Kojima, M., Ayabe, K. and Ueda, H. (2005) Importance of terminal residues on circularly permutated *Escherichia coli* alkaline phosphatase with high specific activity. *Journal of Bioscience and Bioengineering*, **100**, 197–202.

53 Qian, Z., Fields, C.J. and Lutz, S. (2007) Investigating the structural and functional consequences of circular permutation on lipase B from *Candida antarctica*. *ChemBioChem*, **8**, 1989–96.

54 Guntas, G., Mansell, T.J., Kim, J.R. and Ostermeier, M. (2005) Directed evolution of protein switches and their application to the creation of ligand-binding proteins. *Proceedings of the National Academy of Sciences of the United States of America*, **102**, 11224–9.

55 Baird, G.S., Zacharias, D.A. and Tsien, R.Y. (1999) Circular permutation and receptor insertion within green fluorescent proteins. *Proceedings of the National Academy of Sciences of the United States of America*, **96**, 11241–6.

56 Guntas, G., Mitchell, S.F. and Ostermeier, M. (2004) A molecular switch created by in vitro recombination of nonhomologous genes. *Chemistry and Biology*, **11**, 1483–7.

57 Schwartz, T.U., Walczak, R. and Blobel, G. (2004) Circular permutation as a tool to reduce surface entropy triggers crystallization of the signal recognition particle receptor beta subunit. *Protein Science*, **13**, 2814–18.

58 Bittker, J.A., Le, B.V., Liu, J.M. and Liu, D.R. (2004) Directed evolution of protein enzymes using nonhomologous random recombination. *Proceedings of the National Academy of Sciences of the United States of America*, **101**, 7011–16.

59 MacBeath, G., Kast, P. and Hilvert, D. (1998) Redesigning enzyme topology by directed evolution. *Science*, **279**, 1958–61.

60 Tabtiang, R.K., Cezairliyan, B.O., Grant, R.A., Cochrane, J.C. and Sauer, R.T. (2005) Consolidating critical binding determinants by noncyclic rearrangement of protein secondary structure. *Proceedings of the National Academy of Sciences of the United States of America*, **102**, 2305–9.

61 Paavola, C.D., Chan, S.L., Mazzarella, K.M., McMillan, R.A. and Trent, J.D. (2006) A versatile platform for nanotechnology based on circular permutation of a chaperonin protein. *Nanotechnology*, **17**, 1171–6.

19
Incorporating Synthetic Oligonucleotides via Gene Reassembly (ISOR): A Versatile Tool for Generating Targeted Libraries

Asael Herman and Dan S. Tawfik

19.1
Introduction

The addition of synthetic oligonucleotides to a mixture of gene fragments prior to DNA shuffling was suggested in Stemmer's original report [1]. Perhaps due to the lack of a systematic, well-established protocol, this approach has been only very rarely applied [2, 3]. In a recent report we described the optimization of this method, which we refer to as ISOR (Incorporating Synthetic Oligonucleotides via Gene Reassembly) [4], and its application towards the generation of a range of different targeted libraries while incorporating base substitutions, insertions and deletions. Here, we will describe the advantages of this method and its applications, and provide a general protocol for its implementation.

19.1.1
Background

Rational design and directed evolution are the two conceptually contrasting strategies that underpin protein engineering. Whilst directed evolution requires no prior knowledge of the target protein, it does rely on selection capabilities that sample only a miniscule fraction of all possible permutations. Rational and computational designs greatly minimize the number of sequence permutations that are explored (often down to one sequence), but are hampered by the complexity of proteins and a limited knowledge of sequence–function relationships. An awareness of the relative strengths and weaknesses of these approaches has led workers to combine them, for example in 'semi-rational' protein engineering [5–7] and in targeting library diversity to a given set of residues (for example a defined set of active site positions). Recent examples of this targeted library approach include the modification of the substrate specificity of a Cre DNA recombinase [8], an epoxide hydrolase [9, 10], a lipase [11], an esterase [10] and, most recently, a blue fluorescent protein [12]. Other 'targeted library' approaches utilize computational methods and protein design algorithms. Examples include methods that perform a 'virtual screening'

Protein Engineering Handbook. Edited by Stefan Lutz and Uwe T. Bornscheuer

ISBN: 978-3-527-31850-6

of otherwise impossibly large libraries [13], computational methods for the design of enzyme active sites [14], and algorithms that direct recombination by predicting optimal crossover loci [15] or predict optimal combinations of beneficial mutations [16].

'Targeted libraries' are constructed primarily by directing randomization (by saturation mutagenesis) to specific positions within the gene. Saturation mutagenesis uses synthetic oligonucleotides that encode the desired diversity at the specified positions (e.g. see Refs [8, 9, 17–19]). The diversified oligonucleotides are incorporated by PCR, or directly cloned into the gene of interest as a cassette. Saturation mutagenesis is obviously limited as the simultaneous diversification of many residues creates library sizes that are beyond any available screening capabilities; even high-throughput technologies that allow 10^{10} variants to be screened can accommodate only six fully randomized positions. One potential solution to the above obstacle is parsimonious mutagenesis [20]. As the name suggests, this technique provides a means of partial diversification by using oligonucleotides in which the diversified codons comprise a small proportion of mutating bases among an excess of wild-type bases. However, this technique has not been used extensively, most likely because of the high cost of 'doped' oligonucleotides and their limited purity.

In our search for techniques that allow a systematic design of gene libraries informed by inputs from rational, or computational design, we had made attempts to perform gene assembly from long synthetic oligonucleotides (60–80 bp) that were designed to introduce the targeted diversified residues in a manner similar to the synthetic shuffling method [21, 22]. It was found, however, that the libraries constructed in this way were very sensitive to oligonucleotide quality, and that long oligonucleotides contain a significant fraction of $(n-1)$ and $(n+1)$ products. Moreover, the purification of these oligonucleotides using polyacrylamide gel electrophoresis (PAGE) resulted in an even higher frequency of frameshifts. Chastened by these experiences, we directed our efforts at developing a general and versatile technique that targets diversity to predefined and specific positions, thereby creating the desired gene libraries with high precision. ISOR is the result of this effort.

19.1.2
Overview of the Method

ISOR is a simple adaptation of gene shuffling and allows the diversification – by substitution, insertion or deletion – of large sets of residues. Each library variant carries a random and different subset of mutated residues, with the entire set represented in the complete library. A biotinylated PCR product (or a purified restriction fragment) of the target gene is subjected to fragmentation by digestion with DNaseI. The DNaseI fragments are then mixed with a set of synthetic oligonucleotides, and assembled in a process of self-primed extension by *Taq* polymerase. The assembled genes can be enriched by capture on streptavidin-coated magnetic beads, thereby maintaining the diversity created in the assembly reaction, or amplified directly. The product can be recloned and transformed into

E. coli, or used directly for selections using *in-vitro* compartmentalization (IVC) [23, 24]. The oligonucleotides used in such an approach are typically short (~30 bp), and can encode substitutions, insertions or deletions at any given position with a high degree of precision. The frequency of errors in such short oligonucleotides and their cost are much lower than those of their long counterparts; moreover, they do not require any chromatographic purification that increases costs and biases the library content. ISOR, therefore, begins from a reliable starting point, yet is extremely versatile and adaptable. Once a set of oligonucleotides has been synthesized, it can be used for the assembly of various libraries with different rates of diversification, or libraries created with different subsets of the same oligonucleotides. As most of the gene sequence is reassembled from DNaseI fragments, and because the oligonucleotides used are short, the method is not very sensitive to oligonucleotide quality.

19.1.3
Applications

The major advantage of ISOR is its 'tuneability', in that it allows a 'parsimonious' representation of diversity in many positions, while affording the opportunity to control the mutation rate at each targeted residue. It simply allows the input of any structural or functional data into gene libraries. We utilized the benefits of ISOR in the preparation of several targeted gene libraries. The procedure was used to randomize a set of 45 noncontiguous amino acids of a DNA methylase that were identified using bioinformatics analysis, and generate a library where each variant carries a different subset of the complete set of 45 possible changes. Further, the number of average randomized positions per gene could be tuned in the range of one to six [4]. ISOR was also used to input structural, functional and homology data to create gene libraries of a serum paraoxonase by targeting indels [4] and substitutions [24] to the protein's various structural elements. A clear demonstration of how functional data and computation could be combined in each round of directed evolution was recently shown in the evolution of improved halohydrin halogenase [16]. Fox *et al.* used the ProSAR algorithm to evaluate the contribution toward enzyme performance of individual mutations within multiply mutated sequences. They then used synthetic oligonucleotides, in similar fashion to the procedure described here, to incorporate beneficial mutations into their libraries.

19.2
Materials

19.2.1
DNaseI Digestion

1. 6 μg of purified, biotinylated (see Note 1) PCR product of the target gene.
2. 500 m*M* Tris–HCl buffer pH 7.5.

3. 100 m*M* $MnCl_2$.
4. 0.05 U μl^{-1} DNaseI.
5. 0.5 *M* EDTA.
6. 2% agarose gel.
7. QIAEX II Gel Extraction Kit (Qiagen).

19.2.2
Assembly

1. Diversity-encoding oligonucleotides (see Note 2).
2. *Pfu* Turbo DNA polymerase (Stratagene) or another DNA polymerase (Note 3).
3. 10× *Pfu* buffer.
4. 10 m*M* dNTPs (10 m*M* each).

19.2.3
Magnetic Separation and Product Amplification

1. M280 streptavidin-coated magnetic beads (Dynal).
2. Washing and binding buffer: 10 m*M* Tris–HCl buffer, pH 7.4 containing 1 *M* NaCl, 25 m*M* EDTA and 15 m*M* EGTA.
3. 50 m*M* Tris–HCl (pH 8).
4. 10 pmol μl^{-1} of each two 'nested' oligonucleotide primers.
5. *Pfu* Turbo DNA polymerase (Stratagene) or other high-fidelity polymerase.
6. 10× *Pfu* buffer.
7. 10 m*M* dNTPs (10 m*M* each).

19.3
Methods

19.3.1
DNaseI Digestion (see Note 4)

1. Mix 6 μg of the target gene with 5 μl of 500 m*M* Tris–HCl buffer pH 7.5 and 5 μl of 100 m*M* $MnCl_2$, in a final volume of 50 μl and equilibrate at 20 °C in a thermocycler.
2. Prepare a thin-walled PCR tube with 1 μl of DNaseI, and equilibrate at 20 °C in a thermocycler.
3. Transfer the 50 μl DNA mixture to the tube with DNaseI (in the thermocycler), mix by pippeting up and down twice. Incubate for 5 min at 20 °C.
4. Stop the reaction by adding 15 μl 0.5 *M* EDTA and heating at 90 °C for 10 min.
5. Separate the digestion products in a 2% agarose gel and excise the DNA fragments of 70 to 100 bp in size.

6. Purify the fragments using the QIAEX II Gel Extraction Kit according to the manufacturer's instructions.

19.3.2
Assembly

1. Make an assembly reaction mix by combining 100 ng of purified DNA fragments with oligonucleotides (see Notes 5 and 6 and Figure 19.1) in a 50 μl reaction mixture that contain 2.5 U Pfu Turbo DNA polymerase (Stratagene) in the supplied buffer, and 0.4 m*M* of each dNTP.
2. Use the following cycle conditions
   a. Denaturation step at 96 °C for 1.5 min
   b. 35 cycles composed of:
      i. Denaturation step at 94 °C for 30 s.
      ii. Nine successive hybridization steps separated by 3 °C each, from 65 to 41 °C for 1.5 min each (total 13.5 min).
      iii. Elongation step of 1.5 min at 72 °C.
   c. A final 7 min elongation step at 72 °C.

19.3.3
Magnetic Separation and Product Amplification (see Note 1)

1. Bind 50 μl of the assembly reaction to 2.5 μl M280 streptavidin-coated magnetic beads (Dynal) in 50 μl washing and binding buffer, and incubate at ambient temperature for 1 h.

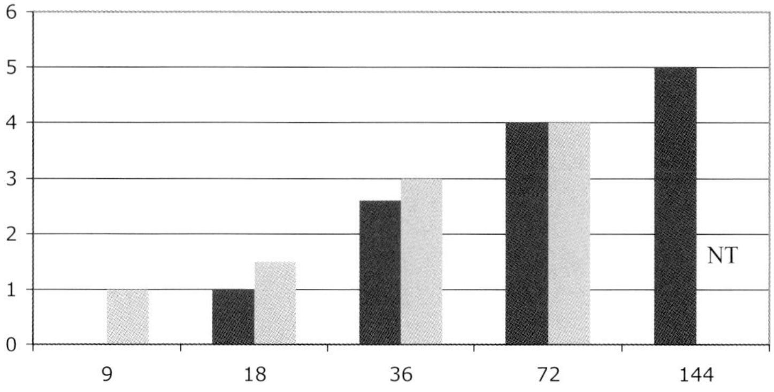

Figure 19.1 The rate of oligonucleotide incorporation into gene-libraries. A set of libraries of the DNA methyltransferase M.*Hae*III (black bars), or the serum paraoxonase PON1 (gray bars), were created using ISOR, with different oligonucleotide concentrations as indicated. The oligonucleotide mix applied contained 45 different oligonucleotides (M.*Hae*III libraries), or 90 different oligonucleotides (PON1 libraries) encoding substitutions (M.*Hae*III), or indels (PON1). Clones from each library were sequenced and the average frequency of oligonucleotides incorporated per gene was determined. NT = not tested. For details, see Ref. [4].

2. Rinse the beads three times with the same buffer, and three times with 50 m*M* Tris–HCl (pH 8). Use the magnetic particles separator (do not centrifuge).
3. Prepare the nested PCR mixture containing DNA polymerase, dNTPs and nested forward and back primers in the polymerase buffer.
4. Resuspend the beads in a 50 μl nested PCR mixture and cycle in a thermocycler.
5. The purified PCR product could be cloned to an expression vector or used directly in *in-vitro* transcription and translation mixtures.

19.4 Notes

1. In most standard DNA shuffling protocols the assembly reaction is highly diluted (typically 1 : 100) before amplification. If not diluted, excess of DNA fragments and short assembly products may inhibit the amplification. This dilution compromises the diversity created in the assembly reaction. Here, we describe an alternative that maintains the diversity. In this modified protocol, the assembled genes can be enriched by capture on streptavidin-coated magnetic beads, thereby minimizing mispriming and the amplification of short products. In cases of relatively small libraries, magnetic bead separation need not be applied, and the DNA need not be biotinylated as described above.
2. The standard oligonucleotide described here was a 33 mer carrying a codon substitution, insertion or deletion. Larger oligonucleotides could also be used to incorporate multiple substitutions or insertions. High-performance liquid chromatography (HPLC) or PAGE purification of oligonucleotides is generally unnecessary.
3. The method of DNA shuffling, which is the 'heart' of ISOR, is mutagenic and can introduce random mutations at frequencies as high as 0.7% [1]. Random mutations can be introduced in the steps of target gene preparation, DNaseI digestion and product assembly or amplification. Whilst in some cases this additional source of diversity may be desirable, in other cases it is not. The rate of these mutations can be controlled with few modifications to the original protocol, as was previously suggested [25]. Here, we used manganese ions in the DNaseI digestion step and *Pfu* Turbo DNA polymerase in the assembly and all amplification steps; this allowed us to keep the random substitution frequency at 0.05%.
4. An optimal DNaseI digestion will result in a mixture of DNA fragments of around the same size. In order to optimize the reaction, the amount of DNaseI, reaction time and the temperature may be adjusted.
5. The concentration of each oligonucleotide in the assembly reaction in respect to the gene fragments will determine the frequency of its incorporation (Figure 19.1). A very high amount of oligonucleotides (five- to 10-fold higher then gene-fragments) will inhibit the assembly reaction. Based on our experience when

using large oligonucleotides sets (45–90), the dynamic range for oligonucleotides incorporation was between 1/10, and twofold the amount of gene-fragments.
6. In several cases we used oligonucleotides that either introduce a new restriction-enzyme recognition sequence, or abolish one [4]. We then compared the pattern of restriction-fragment lengths of the assembled genes to the naïve gene and estimated the rate of oligonucleotides incorporation.

Acknowledgments

The authors gratefully acknowledge research grants from the Estate of Fannie Sherr, and the Israel Science Foundation.

References

1 Stemmer, W.P. (1994) DNA shuffling by random fragmentation and reassembly: *in vitro* recombination for molecular evolution. *Proceedings of the National Academy of Sciences of the United States of America*, **91** (22), 10747–51.

2 Stutzman-Engwall, K., Conlon, S., Fedechko, R., McArthur, H., Pekrun, K., Chen, Y., Jenne, S., La, C., Trinh, N., Kim, S., Zhang, Y.X., Fox, R., Gustafsson, C. and Krebber, A. (2005) Semi-synthetic DNA shuffling of aveC leads to improved industrial scale production of doramectin by *Streptomyces avermitilis*. *Metabolic Engineering*, **7** (1), 27–37.

3 van den Beucken, T., van Neer, N., Sablon, E., Desmet, J., Celis, L., Hoogenboom, H.R. and Hufton, S.E. (2001) Building novel binding ligands to B7.1 and B7.2 based on human antibody single variable light chain domains. *Journal of Molecular Biology*, **310** (3), 591–601.

4 Herman, A. and Tawfik, D.S. (2007) incorporating synthetic oligonucleotides via gene reassembly (ISOR): a versatile tool for generating targeted libraries. *Protein Engineering, Design and Selection*, **20** (5), 219–26.

5 Chica, R.A., Doucet, N. and Pelletier, J.N. (2005) Semi-rational approaches to engineering enzyme activity: combining the benefits of directed evolution and rational design. *Current Opinion in Biotechnology*, **16** (4), 378–84.

6 Minshull, J., Govindarajan, S., Cox, T., Ness, J.E. and Gustafsson, C. (2004) Engineered protein function by selective amino acid diversification. *Methods*, **32** (4), 416–27.

7 Patrick, W.M. and Firth, A.E. (2005) Strategies and computational tools for improving randomized protein libraries. *Biomolecular Engineering*, **22** (4), 105–12.

8 Santoro, S.W. and Schultz, P.G. (2002) Directed evolution of the site specificity of Cre recombinase. *Proceedings of the National Academy of Sciences of the United States of America*, **99** (7), 4185–90.

9 Rui, L., Cao, L., Chen, W., Reardon, K.F. and Wood, T.K. (2004) Active site engineering of the epoxide hydrolase from *Agrobacterium radiobacter* AD1 to enhance aerobic mineralization of cis-1,2-dichloroethylene in cells expressing an evolved toluene ortho-monooxygenase. *Journal of Biological Chemistry*, **279** (45), 46810–17.

10 Park, S., Morley, K.L., Horsman, G.P., Holmquist, M., Hult, K. and Kazlauskas, R.J. (2005) Focusing mutations into the *P. fluorescens* esterase binding site increases enantioselectivity more effectively than distant mutations. *Chemistry and Biology*, **12** (1), 45–54.

11 Reetz, M.T., Bocola, M., Carballeira, J.D., Zha, D. and Vogel, A. (2005) Expanding the range of substrate acceptance of enzymes: combinatorial active-site saturation test. *Angewandte Chemie – International Edition in English*, **44** (27), 4192–6.

12 Mena, M.A., Treynor, T.P., Mayo, S.L. and Daugherty, P.S. (2006) Blue fluorescent proteins with enhanced brightness and photostability from a structurally targeted library. *Nature Biotechnology*, **24** (12), 1569–71.

13 Hayes, R.J., Bentzien, J., Ary, M.L., Hwang, M.Y., Jacinto, J.M., Vielmetter, J., Kundu, A. and Dahiyat, B.I. (2002) Combining computational and experimental screening for rapid optimization of protein properties. *Proceedings of the National Academy of Sciences of the United States of America*, **99** (25), 15926–31.

14 Dwyer, M.A., Looger, L.L. and Hellinga, H.W. (2004) Computational design of a biologically active enzyme. *Science*, **304** (5679), 1967–71.

15 Voigt, C.A., Martinez, C., Wang, Z.G., Mayo, S.L. and Arnold, F.H. (2002) Protein building blocks preserved by recombination. *Nature Structural and Molecular Biology*, **9** (7), 553–8.

16 Fox, R.J., Davis, S.C., Mundorff, E.C., Newman, L.M., Gavrilovic, V., Ma, S.K., Chung, L.M., Ching, C., Tam, S., Muley, S., Grate, J., Gruber, J., Whitman, J.C., Sheldon, R.A. and Huisman, G.W. (2007) Improving catalytic function by ProSAR-driven enzyme evolution. *Nature Biotechnology*, **25** (3), 338–44.

17 Antikainen, N.M., Hergenrother, P.J., Harris, M.M., Corbett, W. and Martin, S.F. (2003) Altering substrate specificity of phosphatidylcholine-preferring phospholipase C of *Bacillus cereus* by random mutagenesis of the headgroup binding site. *Biochemistry*, **42** (6), 1603–10.

18 Reetz, M.T. (2004) Controlling the enantioselectivity of enzymes by directed evolution: practical and theoretical ramifications. *Proceedings of the National Academy of Sciences of the United States of America*, **101** (16), 5716–22.

19 Reetz, M.T., Wilensek, S., Zha, D. and Jaeger, K.E. (2001) Directed evolution of an enantioselective enzyme through combinatorial multiple-cassette mutagenesis. *Angewandte Chemie – International Edition in English*, **40** (19), 3589–91.

20 Balint, R.F. and Larrick, J.W. (1993) Antibody engineering by parsimonious mutagenesis. *Gene*, **137** (1), 109–18.

21 Zha, D., Eipper, A. and Reetz, M.T. (2003) Assembly of designed oligonucleotides as an efficient method for gene recombination: a new tool in directed evolution. *Chembiochem*, **4** (1), 34–9.

22 Ness, J.E., Kim, S., Gottman, A., Pak, R., Krebber, A., Borchert, T.V., Govindarajan, S., Mundorff, E.C. and Minshull, J. (2002) Synthetic shuffling expands functional protein diversity by allowing amino acids to recombine independently. *Nature Biotechnology*, **20** (12), 1251–5.

23 Miller, O.J., Bernath, K., Agresti, J.J., Amitai, G., Kelly, B.T., Mastrobattista, E., Taly, V., Magdassi, S., Tawfik, D.S. and Griffiths, A.D. (2006) Directed evolution by *in vitro* compartmentalization. *Nature Methods*, **3** (7), 561–70.

24 Aharoni, A., Amitai, G., Bernath, K., Magdassi, S. and Tawfik, D.S. (2005) High-throughput screening of enzyme libraries: thiolactonases evolved by fluorescence-activated sorting of single cells in emulsion compartments. *Chemistry and Biology*, **12** (12), 1281–9.

25 Zhao, H. and Arnold, F.H. (1997) Optimization of DNA shuffling for high fidelity recombination. *Nucleic Acids Research*, **25** (6), 1307–8.

20
Protein Engineering by Structure-Guided SCHEMA Recombination

Gloria Saab-Rincon, Yougen Li, Michelle Meyer, Martina Carbone, Marco Landwehr and Frances H. Arnold

20.1
Introduction

20.1.1
SCHEMA Recombination of Proteins: Theoretical Framework

With data from chimeric and randomly mutated β-lactamases, Drummond *et al.* [1] showed that recombination is much more conservative than random mutation, leading to a folding probability that is many orders of magnitude greater at the highest mutation levels. By exploiting the conservative nature of mutations introduced into a structure that has already proven to tolerate them, recombination creates chimeric enzymes that are distant from one another in sequence with minimal loss in their probability of folding.

When designing site-directed recombination libraries, the primary goal is simultaneously to maximize the mutation level of the chimeras and the probability of folding in order to promote functional evolution without disrupting structure. Increasing the number, of parents, their sequence divergence and the number of crossovers can increase mutation levels. However, as the parents increase in number the homologous mutations become less conservative because they are recruited into a less-native environment and participate in novel interactions that usually jeopardize the structural integrity of the chimeras. In fact, randomly recombining sequences with less than ~70% identity generates mostly unfolded proteins [2, 3]. Structure-guided recombination attempts to overcome this hurdle by directing crossovers to the least-disruptive locations, enabling the recombination of more or more distant parents and/or the creation of more diversified libraries (see also Chapter 21).

Thus, we developed an optimization algorithm that would select crossovers to minimize the average disruption, *E*, of the library, subject to constraints on the length of each fragment (and therefore to mutation levels) [4, 3]. SCHEMA disruption *E* counts the number of interactions that are broken by recombination. In

Protein Engineering Handbook. Edited by Stefan Lutz and Uwe T. Bornscheuer

ISBN: 978-3-527-31850-6

order to calculate E, we use the 3-D structure of at least one of the proteins and identify pairs of interacting residues (those within 4.5 Å of each other) to construct a contact matrix that represents these interactions (see Figure 20.1). The contact matrix is then adjusted for the identity between the two proteins: contacts where at least one of the amino acids of an interacting pair is conserved in the parental proteins cannot be broken by recombination and are effectively removed from the matrix. The generation of a chimera by recombination breaks a certain number of the remaining contacts, and this number is designated E, the SCHEMA energy or disruption. The SCHEMA disruption is therefore given by

$$E = \sum_{i}\sum_{j>i} C_{ij}\Delta_{ij} \qquad (20.1)$$

where C_{ij} is the contact matrix element and has a value of 1 if residues i and j are in contact [4]. The SCHEMA delta function $\Delta_{ij} = 0$ if the contact made by amino acid pair i and j is already present in any parent; otherwise, $\Delta_{ij} = 1$.

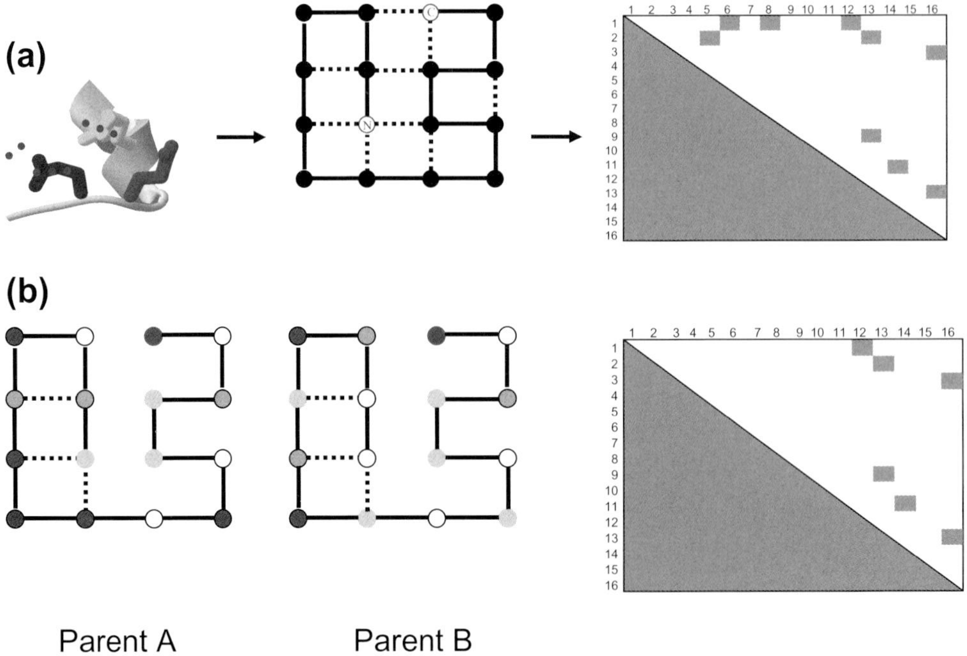

Figure 20.1 (a) SCHEMA disruption is based upon a contact matrix representing interactions between amino acids in the three-dimensional structure of a protein (illustrated here with a simplified model); (b) This contact matrix is then adjusted for the sequence identity of the parent proteins. Contacts that cannot be broken by recombination (where one or both of the interacting amino acids are conserved in the two parents) are removed from the contact matrix. The remaining contacts can be broken and are counted in Equation 20.1.

The crossovers that minimize the average energy, E, of the library are found by the optimization algorithm RASPP [5]. By design, these crossovers partition the structure into a set of fragments that maximize the number of conserved amino acids at their interfaces upon recombination, thus minimizing the number of broken interactions.

The algorithm can be modified to take any pairwise energy function (e.g. Equation 20.1) as input and calculate optimal libraries in regard to folding and diversity (RASPP curve; see Ref. [5]). Such pairwise energy functions may be as simple as Equation 20.1, or they may incorporate detailed biochemical and physical properties of the amino acids.

20.1.2
Comparison of SCHEMA with Other Guided-Recombination Methods [6]

To minimize the number of broken native interactions, crossovers should partition the structure into structurally and functionally independent 'building blocks'. Several groups have proposed that 'contiguous peptide chains forming compact structures' or secondary structural elements could act as independent, recombinable structural units [7–9]. With SCHEMA, however, it has become clear that such elements cannot be defined without consideration of the parent sequences. SCHEMA analysis on several proteins showed that, in certain cases, crossovers are less disruptive in the middle rather than at the ends of an alpha helix [10]; this shows that the best partitions are not intrinsic to the fold, but rather depend on the spatial organization of conserved residues. According to this framework, the interfaces between independent structural units are defined by interfaces composed primarily of conserved residues.

Hernandez and LeMaster developed the algorithm HybNat, which partitions residues into mutually exclusive clusters of interacting amino acids [11]. Unlike SCHEMA, HybNat interactions are atom-based rather than residue-based, and conservative mutations do not contribute to disruption. The interfaces between the clusters are perfectly conserved, eliminating disruption upon recombination, and the authors showed that in this limit, the fragments make independent contributions to protein stability [12]. The creation of diverse libraries using HybNat is difficult because, as the desired level of mutational diversity increases, the identification of perfectly conserved interfaces becomes nearly impossible and, when found, they will partition the structure into very large or small fragments.

Algorithms such as SCHEMA and HybNat only use structural information to minimize disruption. The evolutionary information contained in a natural multiple sequence alignment (MSA) can also be used to minimize structural disruption for the recombination of distant sequences [13–16]. Structure-guided and evolutionary-based recombination algorithms both seek to identify and preserve native interactions. The latter use the statistical covariance, or the pair-wise conservation of amino acids, within a natural MSA to identify important interactions. Evolutionary information in natural MSAs has been used to identify potential stabilizing mutations [17] in 'consensus stabilization' approaches, which assume that highly conserved residue positions represent energetically important sites and that the most frequent amino acids at those sites are the most stabilizing [18, 19]. Similarly, coevolving or highly conserved pairs of residues in a MSA may represent an evolutionary trace of energetically important interactions. Double-mutant thermodynamic cycles have in fact verified energetic coupling between coevolving residues [20]. The FamClash algorithm developed by Maranas and coworkers uses a natural MSA to identify pairs of positions that exhibit conserved amino acids properties (hydrophobicity, charge and volume; HCV) and then scores chimeras according to the total number of 'clashes', where a clash designates a pair of residues with physical properties that deviate from those found at the corresponding positions in the MSA [13]. Recently, Pantazes *et al.* [14] developed a hybrid of SCHEMA and FamClash which penalizes broken interactions, defined by a structural distance cut-off, in proportion to how dissimilar the HCV properties of their components are from those in the family sequence alignment. Ye *et al.* developed another algorithm that uses both evolutionary and structural information to score chimeras and design chimeric libraries [16]. Unlike other methods, this system takes into account higher-than-second-order-interactions. All of these algorithms can be used to direct crossovers to locations that optimize their scoring metric.

Ranganathan and coworkers [15] developed an evolution-based algorithm for generating libraries the residue statistics of which resemble those of the natural MSA. Coevolving residues are identified by a perturbation method known as statistical coupling analysis (SCA), which stores the information in the SCA matrix [20]. Design proceeds by shuffling the columns of the natural MSA while minimizing the differences between the natural and artificial SCA matrix, thus creating protein libraries with conserved independent and pair-wise residue statistics. The resulting proteins contain homologous substitutions and are chimeras, albeit with many crossovers.

Algorithms that maximize the conservation of interactions enable the creation of more diversified libraries. Although several such algorithms have been described here, there are no rigorous studies comparing the performance of each one for different sets of proteins. For algorithms that require structural information it is important to have either one high-resolution crystal structure and a reliable parental sequence alignment, or all the parental crystal structures so that structural alignments can be used. When parental homology is low, structural alignments may provide a better starting point for structure-guided recombination. For algorithms that use evolutionary information, it is important to have a large, high-quality family sequence alignment such that the sequence statistics are well-representative of that fold. Methods that do not fix crossover locations (Ranganathan's SCA-based design) are advantageous when recombining many parents, and may allow for greater mutational diversity arising from the freedom of choosing new crossovers for each new chimeric sequence. However, this method has only been tested on a single very small protein domain (35 amino acids), and it is not clear how it will perform on larger proteins. Furthermore, the experimental characterization of libraries with nonfixed crossovers is inherently more difficult and expensive.

20.1.3
Practical Guidelines for SCHEMA Recombination

For SCHEMA recombination it is necessary to have at least one high-resolution crystal structure and a reliable parental sequence alignment. As the parents diverge in sequence it may be useful to have the crystal structures of all the parents so that a structural alignment can be used instead. SCHEMA software and documentation is available at http://www.che.caltech.edu/groups/fha/index.html. A pdb file containing the atom coordinates of a single parent and a text file containing the parental sequence alignment is all that is required to execute the program and design SCHEMA libraries. The software currently available online is suited to designing chimeras from three parents based on Equation 20.1. However, the code can be modified to accommodate different numbers of parents and different pairwise energy functions.

20.2
Examples of Chimeric Libraries Designed Using the SCHEMA Algorithm

20.2.1
SCHEMA Recombination of β-Lactamases

A family of β-lactamase chimeras was constructed by recombining SCHEMA-designed gene fragments from three homologues: TEM-1, PSE-4 and SED-1 [21]. These proteins are approximately 265 amino acids in length and share between 34 and 42% amino acid sequence identity. We chose to use eight blocks (seven crossovers), giving $3^8 = 6561$ possible chimeras. To ensure that a significant fraction of the chimeras would fold, we used RASPP to minimize the library average SCHEMA energy ($<E>$). The gene fragments were combinatorially assembled using SISDC (Sequence Independent Site-Directed Chimeragenesis) [22] to create the library of 6561 possible chimeras. These genes were expressed in *Escherichia coli*, and the sequences and functional status of more than 500 unique chimeras were determined by high-throughput probe hybridization and screening for the ability to confer ampicillin resistance, a function shared by all three parents. Of 553 unique chimeras chosen at random, 111 (20%) retained at least a low level of β-lactamase activity. An additional 50 functional chimeras were sequenced, generating a family of 161 functional lactamases and 442 nonfunctional lactamases [21].

The functional β-lactamases are highly mosaic and have up to 92 mutations to the closest parental sequence (m). Similar to previous observations for β-lactamase chimeras [22, 23], most functional chimeras (80%) retain the N- and C- terminal fragments from the same parent. Because there are many interactions between these sequence elements, recombining them from different parents is highly disruptive. The functional β-lactamases have a lower SCHEMA disruption than the nonfunctional β-lactamases ($E = 23 \pm 17$ versus $E = 49 \pm 14$) and fewer mutations

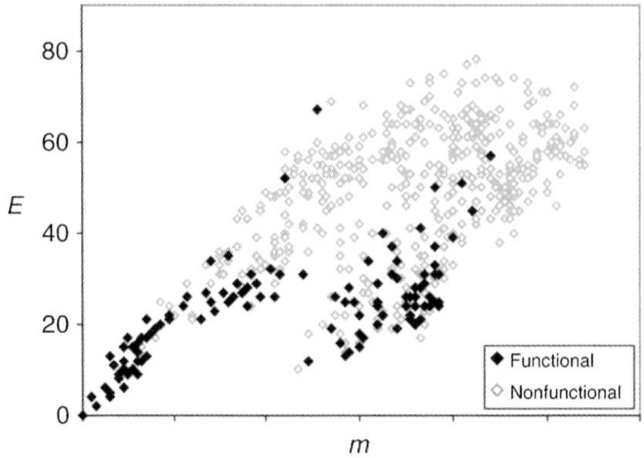

Figure 20.2 SCHEMA disruption value E versus m (number of mutations from the closest parent sequence) for 161 functional (◆) and 442 nonfunctional (◇) β-lactamase chimeras. Functional β-lactamase chimeras contain up to 92 mutations to the closest parental sequence, but have lower SCHEMA disruption (E) than nonfunctional chimeras. E is a good predictor of chimera function [23].

($m = 44 \pm 29$ versus $m = 71 \pm 31$). The average number of crossovers in the functional sequences (4.1) is only slightly less than the average crossovers in the nonfunctional sequences (4.5). An examination of E and m for functional and nonfunctional chimeras shows that, at the same level of mutation, chimeras with a lower E are much more likely to function and fold (Figure 20.2) [21].

20.2.2 SCHEMA-Guided Recombination of Cytochrome P450 Heme Domains

Natural evolution has created a multitude of cytochrome P450s with activities on an amazing variety of substrates. These homologues retain a common overall fold despite their low sequence identity. The aim was to take advantage of the evolvability of this enzyme scaffold to study SCHEMA-directed recombination and functional evolution. We used SCHEMA to recombine three bacterial cytochrome P450 heme domains, creating thousands of new and diverse P450s [10]. A synthetic family of 6561 cytochrome P450 heme domain genes was constructed by swapping eight sequence blocks from three bacterial parents sharing 61–63% sequence identity: CYP102A1, CYP102A2 and CYP102A3 (A1, A2 and A3) (Figure 20.3a). A1 (also known as P450 BM-3 from *Bacillus megaterium*) and its *Bacillus subtilis* homologues A2 and A3 are soluble proteins and are expressed in *E. coli* [24, 25].

Of 955 unique chimeras that were sequenced and partially characterized, 620 encode proteins that fold properly and incorporate the heme cofactor, while another 335 chimeric sequences do not encode folded P450s. (Of the chimeras picked at random, 47% were properly folded; this data set of 955 includes additional folded proteins.) More than 70% of the folded P450 chimeras exhibit per-

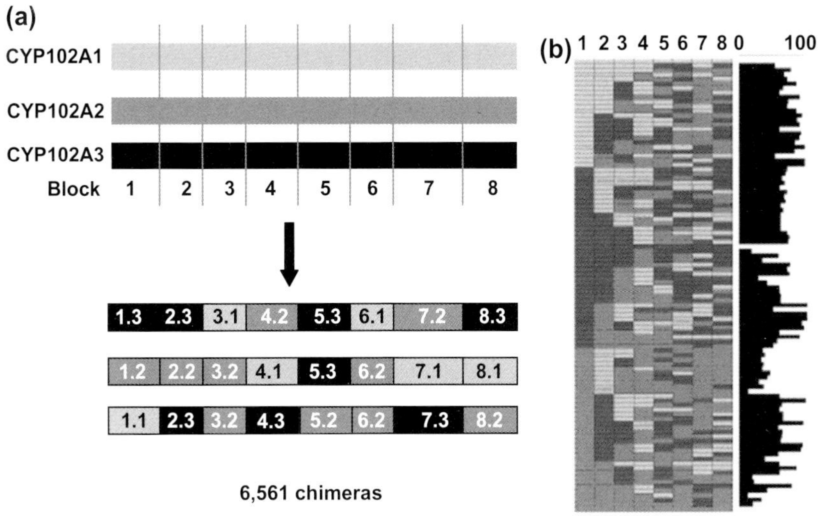

Figure 20.3 A synthetic family of P450s by SCHEMA recombination. (a) Site-directed recombination of three bacterial cytochromes P450 using seven crossover sites chosen to maximize the number of folded yet highly active diverse sequences; (b) Examples of folded chimeric P450s showing sequence changes relative to the closest parent (black bar on right) [10].

oxygenase activity on at least one of a limited set of substrates. These novel P450s are diverse, having as many as 109 and an average of 72 amino acid substitutions with respect to the closest parent sequence [10] (Figure 20.3b).

Although recombination can lead to the creation of many mutations with relatively little structural disruption, the degree of functional diversity that is accessible to a process which only explores combinations of mutations already accepted during natural evolution is not known. Previous studies have shown that chimeric enzymes can acquire catalytic activities not exhibited by the parents [26] motivating us to explore the functional evolution of the chimeric P450 library generated by SCHEMA. Enzymes of the CYP102 family are comprised of a reductase domain and a heme domain connected by a flexible linker. With a single amino acid substitution, the heme domains can function alone as peroxygenases, catalyzing oxygen insertion in the presence of hydrogen peroxide. The synthetic CYP102A family was constructed from parental sequences containing this mutation; all of the chimeric proteins can therefore potentially function as peroxygenases. They can also be reconstituted into functional monooxygenases that utilize NADPH and molecular oxygen for catalysis by fusion to a parental reductase domain. Any of the parental reductases can be used because the key interactions between the heme and reductase domains are conserved in the parents [27].

Hence, a set of 14 chimeric heme domains was selected, and these were reconstituted with all three parental reductase domains. Peroxygenase and monooxygenase activities on the 11 substrates shown in Figure 20.4 were then determined

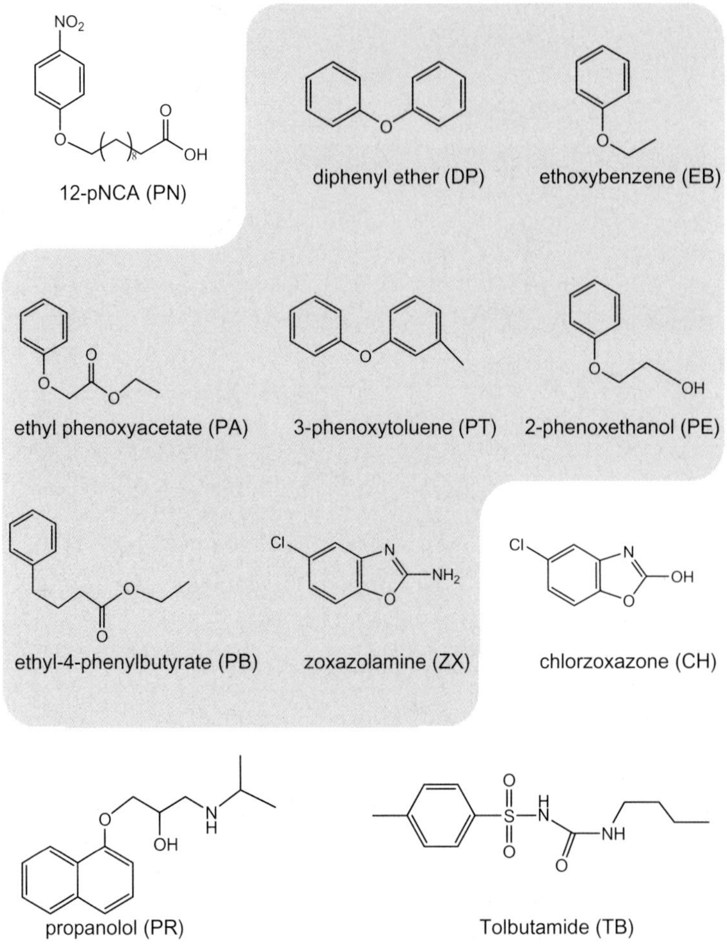

Figure 20.4 Chemical structures and abbreviations. Substrates are grouped according to the pairwise correlations. Members of a group are highly correlated; intergroup correlations are low [27].

[27]. All of the chimeras were successfully reconstituted into functional monooxygenases, and nearly all of the chimera fusions outperformed even the best parent holoenzyme. The chimeric peroxygenases also consistently outperformed the parent peroxygenases; in fact, the best enzyme for each substrate was always a chimera. The chimeric enzymes also exhibited distinct specificities and could be partitioned into clusters based on their specificities [27].

By examining the thermostabilities of the heme domain chimeras, two important observations were made: (i) a significant number of chimeras was found that were more stable than any of the parents; and (ii) a remarkably simple relationship could be identified between chimera sequence and stability. The thermostabilities

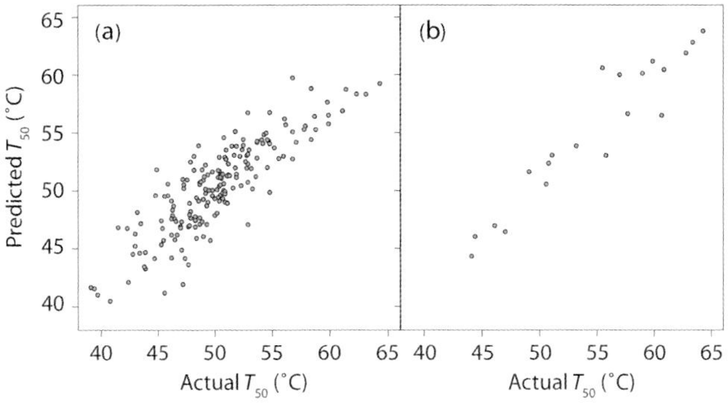

Figure 20.5 Sequence elements contribute additively to thermostabilities of chimeric cytochromes P450. (a) Predicted T_{50} from a simple linear model correlates with the measured T_{50} for 185 P450 chimeras with $r = 0.857$; (b) Linear model accurately predicts stabilities of 20 new chimeras, including the most-stable P450 (MTP) [30].

of 185 folded cytochromes P450 chimeras were measured in the form of T_{50} (the temperature at which 50% of the protein is irreversibly denatured after incubation for 10 min) [28]. Thermostability measured in this way correlates linearly with irreversible half-maximal protein denaturation with urea [29]. The chimera T_{50} values were distributed over a wide range (Figure 20.5), with a number of the chimeric proteins sampled being more stable than any of the three parents, at 54.9 °C (A1), 43.6 °C (A2) and 49.1 °C (A3).

The linear regression of protein thermostability (T_{50}) against chimera fragment composition revealed a strong linear correlation between predicted and observed T_{50} over all 185 chimeras (Figure 20.5a). The next step was to measure the T_{50} of 12 chimeras selected randomly from the subset of 620 folded chimeras reported previously [10], in addition to seven chimeras predicted to be highly thermostable. Predicted and measured T_{50}-values for these 20 previously unstudied proteins correlated very closely ($r = 0.949$) (Figure 20.5b). Over all of the chimeras, the root-mean-square deviation of predicted T_{50} from the measured value was 2.3 °C [30]. It is believed that this linearity is a direct consequence of the SCHEMA design, and reflects the high frequency of conserved residues at the interfaces of the recombined fragments. Conserved interfaces act as insulating walls and partition the structure into independent structural units that make additive contributions to stability, as observed by Le Master *et al.* [12].

The most thermostable chimeras also contain sequence fragments that are more frequent in folded chimeras compared with unfolded chimeras. This observation has precedence in the theory of consensus stabilization [18, 19]. In order to determine whether the highest-frequency (consensus) residues (or in this case fragments) are over-represented in highly stable chimeras, we analyzed the sequences of the nine most thermostable ($T_{50} > 60$ °C) P450 chimeras. Consensus fragments dominated every position in these nine chimeras, and the most thermostable P450 consisted entirely of consensus fragments (2 131 2333, where the number 1, 2 or 3 denotes the parent from which that fragment was derived) (Figure 20.6). These

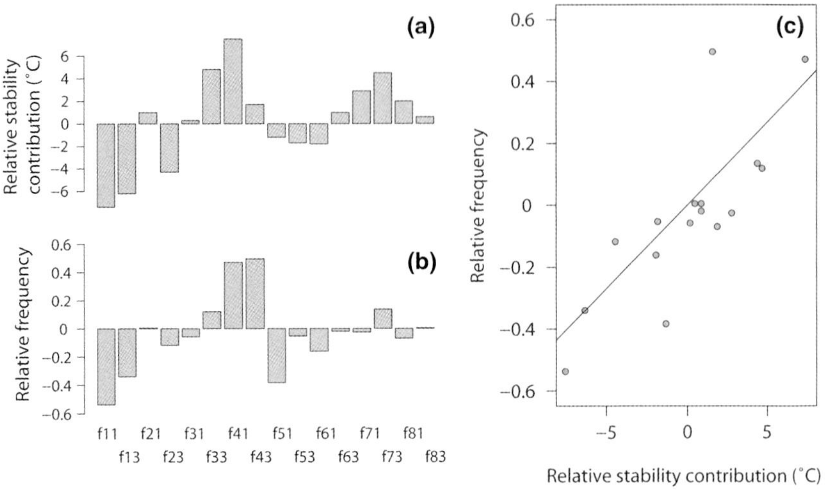

Figure 20.6 Correlation between relative frequencies of sequence elements among folded chimeras and relative stability contributions. (a) Thermostability contributions of fragments from parents A1 and A3 relative to those from parent A2, obtained by linear regression analysis of 205 folded chimeras with measured T_{50}; (b) Frequencies of fragments from parents A1 and A3 relative to those from parent A2 among folded chimeras; (c) Relative fragment thermostability contributions correlate with their relative frequencies among folded chimeras [30].

results show that sequence statistics alone can be used to make nontrivial predictions about chimeric protein folding and relative stability.

These results were subsequently used to predict the most stable chimeras in the 6561-member SCHEMA library. In total, 44 predicted-stable chimeras were constructed, tested, and all were found to be more stable than the most stable parent [30]. The half-lives of inactivation at 57 °C were shown to be up to 108 times that of the most stable parent.

The thermostable chimeras were then tested for activity on 2-phenoxyethanol, a substrate common to all three parent enzymes. All 44 stable P450 chimeras were active, ranking from 0.2 to 9-fold relative to the most active parent, A1 [30]. The chimeras were also tested for activity on the human drugs verapamil and astemizole, substrates for which none of the parents showed any activity. Three of the stable chimeras produced authentic metabolites as produced by the human Cyp3A4, 1A2, 2C and 2D6 enzymes [30].

20.3
Conclusions

Recombination methods that are annealing-based [31], or that are sequence-independent and random, generate libraries of chimeras with unknown sequences and often limited diversity. This restricts the ability to evaluate how efficiently a recombination strategy traverses protein sequence space and contributes to functional evolution. Functional analyses of the resultant libraries have not provided data that can be used to systematically determine which chimeras are most likely to retain structure and function, or exhibit new functional properties. The implementation of SCHEMA has allowed site-directed recombination libraries to be generated with high sequence diversity, and a substantially higher proportion of folded proteins with which to explore the sequence determinants of stability and the functional diversity that is accessible to recombination. The construction and analysis of such proteins has shown that significant sequence and functional diversity can be generated by using this approach. Due to the block-additivity of the SCHEMA fragments, the analysis of a small subset of a chimeric library allows an accurate prediction to be made of the thermostable members, and these can then be rapidly assembled to create an entire family of stable chimeras with high sequence diversity.

References

1. Drummond, D.A., Silberg, J.J., Meyer, M.M., Wilke, C.O. and Arnold, F.H. (2005) On the conservative nature of intragenic recombination. *Proceedings of the National Academy of Sciences of the United States of America*, **102**, 5380–5.
2. Ostermeier, M., Shim, J.H. and Benkovic, S.J. (1999) A combinatorial approach to hybrid enzymes independent of DNA homology. *Nature Biotechnology*, **17**, 1205–9.
3. Lutz, S., Ostermeier, M., Moore, G.L., Maranas, C.D. and Benkovic, S.J. (2001) Creating multiple-crossover DNA libraries independent of sequence identity. *Proceedings of the National Academy of Sciences of the United States of America*, **98**, 11248–53.
4. Voigt, C.A., Martinez, C., Wang, Z.G., Mayo, S.L. and Arnold, F.H. (2002) Protein building blocks preserved by recombination. *Nature Structural Biology*, **9**, 553–8.
5. Endelman, J.B., Silberg, J.J., Wang, Z.G. and Arnold, F.H. (2004) Site-directed protein recombination as a shortest-path problem. *Protein Engineering, Design & Selection*, **17**, 589–94.
6. Carbone, M.N. and Arnold, F.H. (2007) Engineering by homologous recombination: exploring sequence and function within a conserved fold. *Current Opinion in Structural Biology*, **17**, 454–9.
7. Tsuji, T., Onimaru, M. and Yanagawa, H. (2006) Towards the creation of novel proteins by block shuffling. *Combinatorial Chemistry & High Throughput Screening*, **9**, 259–69.
8. O'Maille, P.E., Tsai, M.D., Greenhagen, B.T., Chappell, J. and Noel, J.P. (2004) Gene library synthesis by structure-based combinatorial protein engineering. *Methods in Enzymology*, **388**, 75–91.
9. Go, M. (1985) Protein structures and split genes. *Advances in Biophysics*, **19**, 91–131.
10. Otey, C.R., Landwehr, M., Endelman, J.B., Hiraga, K., Bloom, J.D. and Arnold, F.H. (2006) Structure-guided recombination creates an artificial family of cytochromes P450. *PLoS Biology*, **4**, e112.
11. Hernandez, G. and LeMaster, D.M. (2005) Hybrid native partitioning of interactions among nonconserved residues in chimeric proteins. *Proteins*, **60**, 723–31.
12. LeMaster, D.M. and Hernandez, G. (2005) Additivity in both thermodynamic stability and thermal transition temperature for rubredoxin chimeras via hybrid native partitioning. *Structure*, **13**, 1153–63.
13. Saraf, M.C., Horswill, A.R., Benkovic, S.J. and Maranas, C.D. (2004) FamClash: a method for ranking the activity of

engineered enzymes. *Proceedings of the National Academy of Sciences of the United States of America*, **101**, 4142–7.

14 Pantazes, R., Saraf, M.C. and Maranas, C.D. (2007) Optimal protein library design using recombination or point mutations based on sequence based scoring functions. *Protein Engineering, Design & Selection*, **20**, 361–73.

15 Socolich, M., Lockless, S.W., Russ, W.P., Lee, H., Gardner, K.H. and Ranganathan, R. (2005) Evolutionary information for specifying a protein fold. *Nature*, **437**, 512–18.

16 Ye, X., Friedman, A. and Bailey-Kellogg, C. (2006) Hypergraph model of multi-residue interactions in proteins: sequentially-constrained partitioning algorithms for optimization of site-directed recombination. *Lecture Notes in Computer Science*, 15–29.

17 Lehmann, M., Kostrewa, D., Wyss, M., Brugger, R., D'Arcy, A., Pasamontes, L. and van Loon, A.P. (2000) From DNA sequence to improved functionality: using protein sequence comparisons to rapidly design a thermostable consensus phytase. *Protein Engineering*, **13**, 49–57.

18 Steipe, B., Schiller, B., Pluckthun, A. and Steinbacher, S. (1994) Sequence statistics reliably predict stabilizing mutations in a protein domain. *Journal of Molecular Biology*, **240**, 188–92.

19 Finkelstein, A.V., Badretdinov, A. and Gutin, A.M. (1995) Why do protein architectures have Boltzmann-like statistics? *Proteins*, **23**, 142–50.

20 Lockless, S.W. and Ranganathan, R. (1999) Evolutionarily conserved pathways of energetic connectivity in protein families. *Science*, **286**, 295–9.

21 Meyer, M.M., Hochrein, L. and Arnold, F.H. (2006) Structure-guided SCHEMA recombination of distantly related beta-lactamases. *Protein Engineering, Design & Selection*, **19**, 563–70.

22 Hiraga, K. and Arnold, F.H. (2003) General method for sequence-independent site-directed chimeragenesis. *Journal of Molecular Biology*, **330**, 287–96.

23 Meyer, M.M., Silberg, J.J., Voigt, C.A., Endelman, J.B., Mayo, S.L., Wang, Z.G. and Arnold, F.H. (2003) Library analysis of SCHEMA-guided protein recombination. *Protein Science*, **12**, 1686–93.

24 Gustafsson, M.C., Roitel, O., Marshall, K.R., Noble, M.A., Chapman, S.K., Pessegueiro, A., Fulco, A.J., Cheesman, M.R., von Wachenfeldt, C. and Munro, A.W. (2004) Expression, purification, and characterization of *Bacillus subtilis* cytochromes P450 CYP102A2 and CYP102A3: flavocytochrome homologues of P450 BM3 from *Bacillus megaterium*. *Biochemistry*, **43**, 5474–87.

25 Munro, A.W., Coggins, J.R., Lindsay, J.G., Kelly, S. and Price, N.C. (1996) Deflavination of cytochrome P450 BM3 by treatment with guanidinium chloride. *Biochemical Society Transactions*, **24**, 19S.

26 Raillard, S., Krebber, A., Chen, Y., Ness, J.E., Bermudez, E., Trinidad, R., Fullem, R., Davis, C., Welch, M., Seffernick, J. *et al.* (2001) Novel enzyme activities and functional plasticity revealed by recombining highly homologous enzymes. *Chemistry & Biology*, **8**, 891–8.

27 Landwehr, M., Carbone, M., Otey, C.R., Li, Y. and Arnold, F.H. (2007) Diversification of catalytic function in a synthetic family of chimeric cytochrome P450s. *Chemistry & Biology*, **14**, 269–78.

28 Salazar, O., Cirino, P.C. and Arnold, F.H. (2003) Thermostabilization of a cytochrome P450 peroxygenase. *ChemBioChem*, **4**, 891–3.

29 Bloom, J.D., Labthavikul, S.T., Otey, C.R. and Arnold, F.H. (2006) Protein stability promotes evolvability. *Proceedings of the National Academy of Sciences of the United States of America*, **103**, 5869–74.

30 Li, Y., Drummond, D.A., Otey, C.R., Snow, C.D., Bloom, J.D. and Arnold, F.H. (2007) A diverse family of thermostable cytochrome P450s created by recombination of stabilizing fragments. *Nature Biotechnology*, **25**, 1051–6.

31 Crameri, A., Raillard, S.A., Bermudez, E. and Stemmer, W.P. (1998) DNA shuffling of a family of genes from diverse species accelerates directed evolution. *Nature*, **391**, 288–91.

32 Silberg, J.J., Endelman, J.B. and Arnold, F.H. (2004) SCHEMA-guided protein recombination. *Methods in Enzymology*, **388**, 35–42.

21
Chimeragenesis in Protein Engineering

Manuela Trani and Stefan Lutz

21.1
Introduction

Major advances in natural evolution occur by means of dramatic gene reorganization such as DNA swapping and juxtaposition. Evidence for such events include the divergence of entire organisms such as *Escherichia coli*, evolving from *Salmonella* through multiple horizontal gene transfer [1] and, on a smaller scale, the emergence of bacterial resistance to penicillin [2]. The major role played by genomic recombination in the natural evolution of organisms and protein function has inspired molecular biologists to develop *in vitro* recombination strategies, mimicking Nature's astonishing ability to adjust to changing environments and to create new function through Darwinian evolution.

The introduction of DNA shuffling marks a pivotal moment in protein engineering using *in vitro* recombination by its ability to simulate Nature's apparent random tinkering on existing protein scaffolds via a more directed and faster strategy to engineer protein in the laboratory [3, 4]. In its original version, the two-step procedure uses limited DNaseI digestion of a gene or gene collection to generate random oligonucleotides. Multiple rounds of self-primed PCR then reassemble the parent-size DNA sequence by homologous recombination. Such a procedure allows the combination, elimination and redistribution of sequence variations found in individual genes with other members in the sequence pool. When applied iteratively, DNA shuffling – in combination with a selection or high-throughput screening (HTS) technique – has the potential to yield recombinants with enhanced functional properties. Since its introduction in the mid 1990s, DNA shuffling has enabled experimentalists to improve enzyme activity and substrate specificity, as well as increase protein robustness to changes in the reaction environment such as elevated temperature, pH changes and chelating agents. Subsequently, a number of alternate techniques have been described in the literature, including the staggered extension process (StEP) [5], random chimeragenesis on transient templates (RACHITT) [6] and nucleotide exchange and excision technology (NExT) [7] to name only a few.

Protein Engineering Handbook. Edited by Stefan Lutz and Uwe T. Bornscheuer

ISBN: 978-3-527-31850-6

Although these DNA shuffling protocols have proven highly successful in enhancing enzyme function, the approach faces inherent limitations as the creation of diversity relies on homologous recombination. Attempts to shuffle parental genes with less than 70% DNA sequence identity have been challenging, as gene fragments from parents with DNA sequence identity below this threshold tend to prefer reassembly with cognate oligonucleotides. This leads to libraries with crossovers heavily biased towards local regions of higher sequence identity or, more frequently, results in reconstruction of the native DNA sequences [8, 9].

Such limitations are relevant to protein engineers as multiple sequence alignments of many protein families show that a significant fraction of target enzymes fall below the 70% minimum. Two solutions developed to extend the working range of the shuffling protocols are family DNA shuffling, also called molecular breeding and gene optimization. In the former, the insufficient DNA sequence identity between two parent genes can be bridged by one or more additional genes with intermediate sequence identity [10, 11]. Alternatively, computational tools can redesign genes for optimal DNA sequence identity without altering the polypeptide sequence, taking advantage of the degeneracy of the genetic code [12]. The latter approach has certainly gained popularity as whole-gene synthesis has become a low-cost routine procedure. Nevertheless, even these strategies are typically limited to effectively sampling proteins above 50% DNA sequence identity, preventing the shuffling of too many interesting proteins in the twilight zone of 25–40% DNA sequence identity.

21.1.1
Homology-Independent *in vitro* Recombination (Chimeragenesis)

In response, several research groups have engaged in the development of novel methods for homology-independent random *in vitro* recombination, also termed chimeragenesis (see also Chapter 20). While experimentally more demanding, the potential benefits of chimeragenesis were first outlined by the theoretical studies of Bogorad and Deem [13]. Their results suggested that domain swapping in proteins is more efficient in generating new function than homologous recombination and point mutations alone. In their Monte Carlo simulations, the evolution of a small, 100-amino acid polypeptide suffered tremendous limitations when restricted to point mutations and swaps within peptides belonging to the same secondary structure pools. On the other hand, the random juxtaposition of different types of secondary structures and domains resulted in dramatic changes, leading to proteins with new tertiary structures and improved properties.

21.1.1.1 Homology-Independent Random Gene Fusion

When testing these predictions in the laboratory, early experimental strategies focused on creating random hybrid proteins with a single crossover; that is, fusing the N-terminal protein sequence of parent A to the C-terminal portion of parent B (Figure 21.1). In practice, these hybrids were generated by blunt-end ligation of

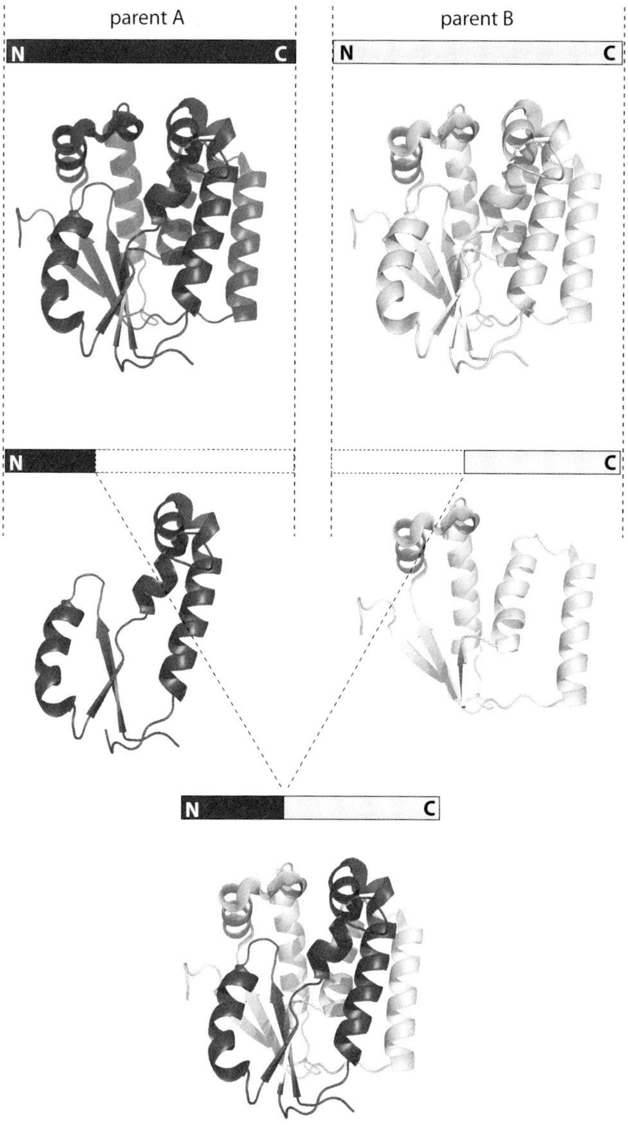

Figure 21.1 Hybrid enzyme formation by fusion of gene fragments. The concept of chimeragenesis involves the substitution of short peptide sequences, secondary structures, or entire domains between two parents. The fusion of a gene fragment, encoding the N-terminal portion of parent protein A with the gene fragment, corresponding to the C-terminal region of parent protein B, results in a hybrid gene with a single crossover. Expression in a host organism subsequently yields the corresponding chimeric protein.

the corresponding gene fragments, which were created using either incremental truncation or circular permutation (Figure 21.2a and b) [14–16]. In both cases, the methods successfully increased the diversity of chimeras by facilitating the introduction of crossovers in regions beyond the reach of homologous recombination, yielding novel proteins with interesting functional properties. By using incremental truncation for the creation of hybrid enzymes (ITCHY), Ostermeier and coworkers created hybrid enzymes of human glycinamide ribonucleotide formyltransferase (hGART) and its *E. coli* equivalent, PurN. With 50% DNA and 41% protein sequence identity (SI) (50/41% SI), these two enzymes are not amenable to traditional DNA shuffling, yet domains from these proteins can be swapped by chimeragenesis. Similarly, Sieber and coworkers used sequence homology-independent protein recombination (SHIPREC) to generate hybrids of human P450 (1A2) and the soluble heme domain of *Bacillus megaterium* P450 that share only 43/16% SI. Nevertheless, both approaches can only introduce a single crossover and alternative protocols are needed to introduce multiple crossovers per gene.

21.1.1.2 **Homology-Independent Recombination with Multiple Crossovers**

One of the first methods to raise the number of homology-independent crossover in a hybrid gene library was SCRATCHY, a chimeragenesis protocol that uses two complementary ITCHY libraries in combination with DNA shuffling (Figure 21.2c) [8]. Although functional multicrossover hybrids of hGART and PurN could be isolated using SCRATCHY, thus verifying the feasibility of protein chimeragenesis, the catalytic efficiency of these hybrids was generally lower than that of the parental enzymes [17]. The limited success of these initial experiments was probably linked to deficiencies in the selection system and library quality. Functional chimeras were identified by the genetic complementation of an *E. coli* auxotroph with a low threshold for GART activity and little room for adjustments of the selection pressure. In addition, DNA sequence analysis of the naive SCRATCHY library showed only a small fraction of multicrossover hybrids in a gene pool that was dominated by reassembled wild-type sequence and single crossover hybrids (see discussion below). While the choice or selection or screening system is project-specific, improvements in library quality can be accomplished by general technical advances in library preparation. Two such advanced protocols are the enhanced-crossover [18] and forced-crossover SCRATCHY methods [19, 20]. These approaches enable the enrichment of gene libraries with members containing more than one crossover and, in the latter case, give the experimentalist strict control over the number and location of fusion points in a target gene. The technical details, as well as the advantages of these higher-quality hybrid libraries in regards to enzyme function, will be discussed in greater detail in Section 21.2.

Alternatively, non-homologous random recombination (NRR), originally developed by the Liu group for raising nucleic acid aptamers [21], can generate protein chimeras with multiple crossovers (Figure 21.2d). Bittker and coworkers applied this method to the directed evolution of chorismate mutase from *Methanococcus jannaschii* (mMjCM), identifying a number of active clones with significant inser-

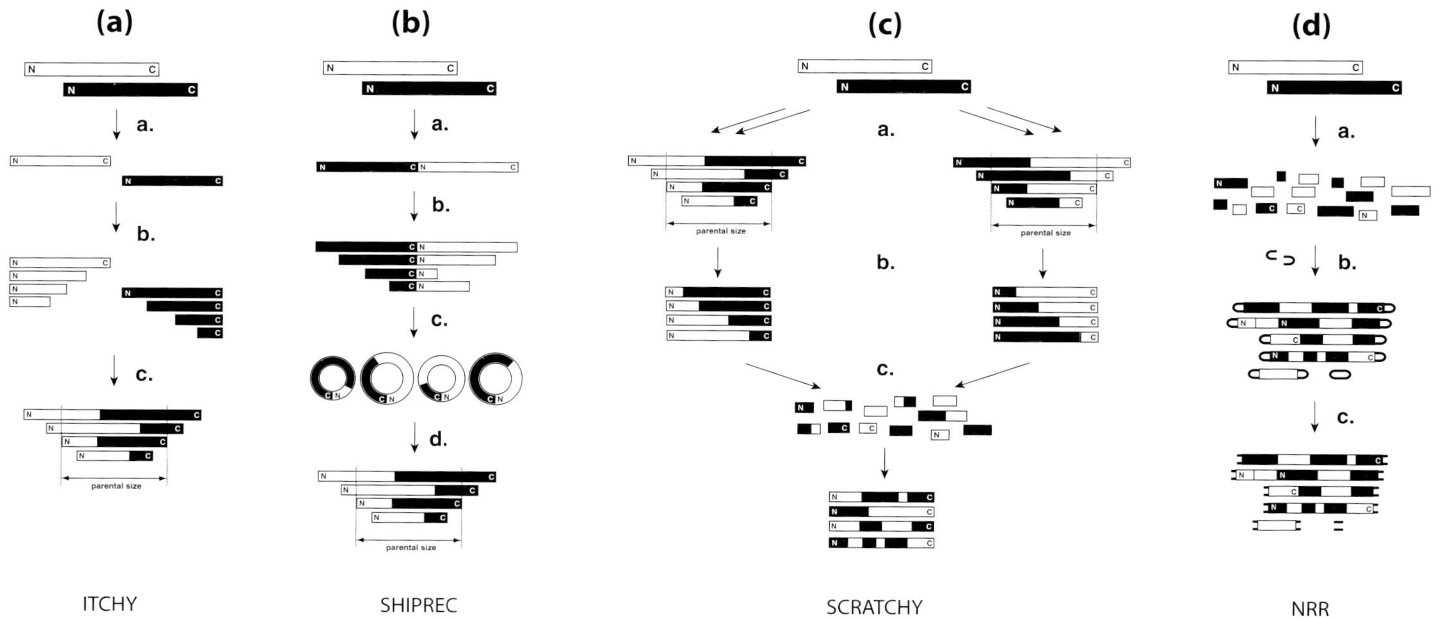

Figure 21.2 An overview of chimeragenesis methods. In these schematic drawings, the two parental DNA sequences are represented by black and white bars. (a) Incremental truncation for the creation of hybrid enzymes (ITCHY) is a method to create gene fusions, using random-length fragments of the N- and C-terminal regions of the two parents. As shown, the protocol creates incremental truncation libraries of both parents, targeting either the C terminus (white parent) or N terminus (black parent). In the subsequent ligation step, a randomly picked DNA fragment encoding the N-terminal portion of the white parent protein is joined with a randomly picked DNA sequence encoding the C-terminal region of the black parent. The resulting hybrid gene library consists of fusion constructs of parental size, as well as members with duplications and deletions of coding sequence; (b) Sequence homology-independent protein recombination (SHIPREC) is an alternative strategy to create libraries of random gene fusions, using a circular permutation step to invert the two parental sequences after random truncation to yield hybrid genes with a single crossover; (c) SCRATCHY is a combination of ITCHY and DNA shuffling, enabling the creation of hybrid genes with more than one fusion point. The two-step protocol starts with the preparation of two complementary ITCHY libraries (white-black and black-white). In the subsequent DNA shuffling step, the combined ITCHY libraries are partially digested by DNaseI and the resulting oligodeoxyribonucleotides (including fragments containing the fusion points) are reassembled via homologous recombination; (d) Nonhomologous random recombination (NRR) distinguishes itself from most other chimeragenesis protocols in that it does not attempt to preserve the original order of DNA fragments during reassembly. The parental genes are fragmented by limited digestion, followed by random reassembly of the oligodeoxyribonucleotides in the presence of short hairpin DNA elements that function as ligation terminators. The resulting 'capped' DNA sequences are subsequently treated with a restriction endonuclease to free the ends of the chimera for integration into a DNA vector.

tions and deletions (in some cases in multiple locations) throughout the parental sequence [22]. In the same study, the chimeragenesis of mMjCM with the structurally and functionally unrelated fumarase from *E. coli* yielded hybrid sequences with several deletions and insertions as well as up to 11 crossovers per gene. Unfortunately, all characterized chimeras exhibited chorismate mutase activity which was five to 2000-fold lower than the wild-type enzyme. In summary, NRR probably produces the highest level of diversity of all reported chimeragenesis protocols, yet the library's functional content ranks among the lowest as reading frame shifts and noncoding fragments lead to a larger percentage of insoluble, misfolded and inactive chimeras.

21.1.2
Predictive Algorithms in Chimeragenesis

Together with the development of new and improved experimental chimeragenesis protocols, *in silico* approaches to predict the outcome of chimeragenesis and guide the design of higher-quality gene libraries have emerged. Probably the most widely used predictive algorithm is SCHEMA [23–25] (see also Chapter 20). Briefly, SCHEMA uses information on noncovalent interactions between individual amino acids in a protein chain, obtained from multiple sequence alignments and crystal structures, to identify suitable crossover points between two or more parental enzymes. The algorithm ranks the quality of crossovers based on their potential to disturb the protein's structural integrity. The experimentalist is supplied with a sequence profile that highlights regions that cause minimal disruption of the three-dimensional structure upon fragment swapping. The computational data can be used in conjunction with sequence-independent site-directed chimeragenesis (SISDS). SISDS generates chimera libraries via recombination of gene fragments, obtained by the restriction endonuclease (RE) digestion of parental genes that carry unique endonucleases cleavage sites at selected SCHEMA-guided crossover sites [26]. Several examples of successful chimeragenesis with the help of SCHEMA have been reported, including the recombination of β-lactamases [24–27] and cytochrome P450s [9, 28–30].

Alternatively, Maranas and coworkers developed a series of predictive frameworks based on residue-specific interactions. Early computational models focused on the correlation of enzyme function with compatibility of swapped protein fragments in chimeras, taking into consideration electrostatic, steric and hydrogen-bonding interactions [31, 32]. While successful in predicting function *per se*, the model was insufficient for estimating the level of activity. An elegant solution to the problem is the successor algorithm FamClash, which combines information on structural compatibility with correlated mutation patterns among residues in the protein sequence [33]. When applied to hybrids of dihydrofolate reductase from *E. coli* and *Bacillus subtilis*, the computational predictions agree well with relative activity levels measured for the corresponding fusion enzymes in the laboratory.

More recently, the same group reported the development of OPTCOMB (Optimal Pattern of Tiling for COMBinatorial library design), a protocol for determining the optimal composition of hybrids, using protein fragments from multiple parents [34]. Similar to FamClash, the algorithm is designed to minimize the number of clashes in hybrid proteins, yet can further optimize the chimeras by choosing the most suitable protein fragment from among the parental sequence pool and adjusting the crossover position(s). Furthermore, OPTCOMB delivers valuable information on optimal library diversity by correlating library size with structural degeneracy due to clashes. Following additional refinements of the scoring method, the authors successfully tested their predictive framework against experimental data for dihydrofolate reductases and cytochrome P450s [35].

In summary, the development of computational tools to guide the design of protein-engineering libraries has made tremendous advances during recent years. While it is important to remember that their success can vary with the number of available sequences and structure information, these algorithms can be powerful allies for directed evolution in general, and for chimeragenesis in particular. In combination with changes and improvements of experimental protocols, as outlined below for SCRATCHY, methods such as SCHEMA and OPTCOMB enable practitioners to focus their efforts and accelerate the search of the vastness of protein sequence space by helping to eliminate potential structural incompatibilities and preserve critical residues and contacts throughout the protein scaffold.

21.2 Experimental Aspects of the SCRATCHY Protocol

Since its first implementation, the SCRATCHY protocol has undergone significant modifications to improve the library quality and simplify the experimental aspects of the method.

21.2.1 Creation of ITCHY Libraries

Several strategies for the creation of ITCHY libraries have been reported [15, 36, 37]. Due to its simplicity, our laboratory employs mostly the PCR amplification of the target vector in the presence of α-phosphothioate dNTPs (Thio-ITCHY). Briefly, a DNA plasmid carrying the two target genes or gene fragments in series (Figure 21.3a) is linearized by restriction digestion and the template is amplified via PCR. In contrast to a regular PCR, the reaction contains a mixture of dNTP and α-S dNTPs. Unable to effectively distinguish a native 2′-deoxyribonucleotide from its thiophosphate analogue, the DNA polymerase will randomly incorporate the analogue during template-directed DNA synthesis. While a regular phosphodiester linkage in the DNA backbone is susceptible to exonuclease digestion, the incorporation of a thiophosphate creates a phosphodiester bond which is resistant to enzymatic hydrolysis. Our protocol takes advantage of this nuclease resistance as the PCR product is subsequently treated with *Exo*III, a 3′–5′-exonuclease which

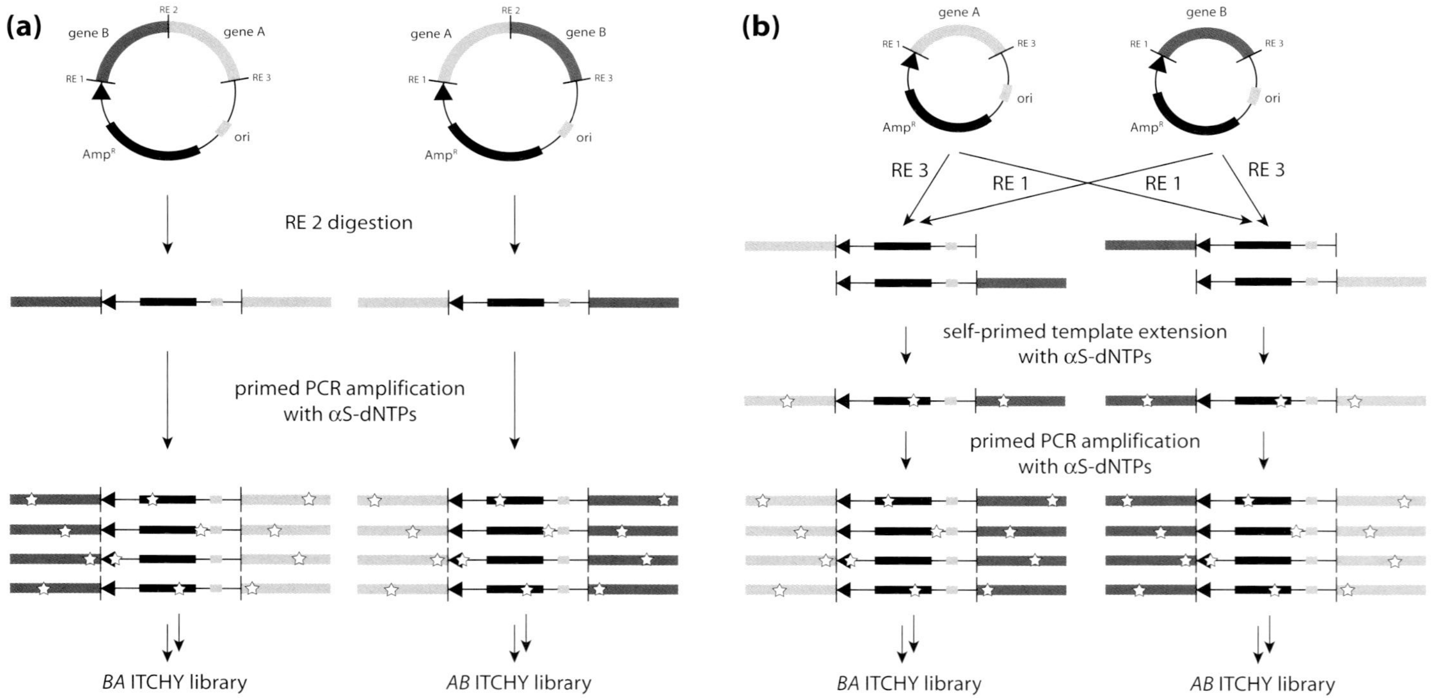

Figure 21.3 Alternative strategies for creating incremental truncation libraries by Thio-ITCHY. Complementary ITCHY libraries (*BA* and *AB* libraries) are necessary for the preparation of SCRATCHY libraries. (a) The original protocol for Thio-ITCHY requires two separate DNA plasmids, carrying either the full-length *BA* or *AB* genes. Upon linearization by restriction endonuclease (RE) digestion, the DNA sequences serve as templates in the PCR with α-phosphothioate dNTPs. In minimizing time-consuming gene cloning and maximizing flexibility in recombining genes of interest, an alternative strategy was implemented to generate Thio-ITCHY libraries; (b) Rapid Thio-ITCHY starts by cloning all parental genes into a common DNA vector, followed by linearization at either the 5′ or 3′ end of the target gene (RE1 and RE3). After mixing the appropriate parental DNA sequences, initial rounds of self-primed primer extension generate the full-length, two-gene template which is directly used in a second, primed PCR reaction in the presence of α-phosphothioate dNTPs.

will stop degradation of the double-stranded DNA template upon encountering the first thiophosphate linker, thereby creating a collection of incrementally truncated DNA fragments. In the final step, *intra*-molecular ligation of the plasmid fuses the two gene fragments and generates the ITCHY library. A detailed technical description of the approach can be found elsewhere [38].

While the Thio-ITCHY approach has successfully been applied on a variety of DNA sequences, assembly of the initial vector with the two target genes cloned in series is cumbersome. We recently developed and applied a new strategy to expedite the template assembly process (Figure 21.3b) [19]. In the latter scheme, both parental genes (*A* and *B*) are inserted separately in the same DNA plasmid, using identical restriction sites. To create an *AB* ITCHY library, *A* vector is linearized with RE3, while the *B* plasmid is digested with RE1. Following agarose gel purification, the two linearized plasmids are mixed and amplified over four rounds with polymerase and dNTPs/α-S dNTPs in the absence of primer. An aliquot of that reaction mixture is then used directly as the template for a regular PCR amplification with primers, yielding the phosphothiolate-labeled PCR product. The complementary *BA* ITCHY library can easily be assembled, starting with the same *A* and *B* vectors but switching the restriction endonucleases. Besides reducing the number of required cloning steps, the protocol can easily be adapted to experiments using more than two parental genes, allowing for a multigene chimeragenesis (family-SCRATCHY).

21.2.2 Size and Reading Frame Selection

The creation of comprehensive ITCHY and SCRATCHY libraries is limited for practical reasons and likely unproductive from the standpoint of enzyme structure and function. The immense diversity of chimera containing all possible gene fragment sizes including insertions and deletions in multiple crossover combinations fare exceeds experimental limitations such as transformation efficiency and screening or selection capacity. Furthermore, it is unlikely that hybrid proteins with extensive sequence deletions and insertions will retain the structural integrity and stability and therefore the function of its parents. As such, the preselection of gene fusions of approximately parental size seems sensible and can experimentally be accomplished in two ways: (i) by using methods which favor the formation of parental-size libraries [37]; or (ii) by separating DNA fragments via agarose gel electrophoresis, followed by excision and recovery of nucleic acids of the desired size range [16, 19, 39].

The quality of chimeragenesis libraries can further be improved by eliminating gene fusions with scrambled reading frames. Statistically, two-thirds of ITCHY library members lose the correct reading frame, resulting in random sequences and early termination. The selection of library members with an intact reading frame across the gene fusion can significantly enhance the quality of an ITCHY library. The SCRATCHY protocol is particularly affected by out-of-frame sequences as ITCHY libraries serve as templates for multiple crossover hybrids, causing a rapid decline in the fraction of hybrid sequences with an overall correct reading frame.

In order to isolate DNA sequences with the correct reading frame, we developed two DNA vector-based selection systems called pSALect and pInSALect (Figure 21.4) [40, 41]. In pSALect, the target DNA is cloned between an N-terminal TAT signal sequence and a C-terminal lactamase gene. Upon expression, target peptides that preserve the correct reading frame act as linkers between the periplasmic export signal and the antibiotic resistance marker. Following secretion into the periplasm, the trifunctional fusion protein renders the host organism resistant to ampicillin or carbenicillin in the growth media. In contrast, out-of-frame constructs are eliminated as they fail to translate the β-lactamase portion of the fusion protein. Furthermore, the selection via TAT and lactamase eliminates false positives that can originate from translational initiation at internal ribosomal binding sites, observed in reading frame selections utilizing traditional strategies such as N-terminal fusion to chloramphenicol and green fluorescent protein (GFP) [16, 42].

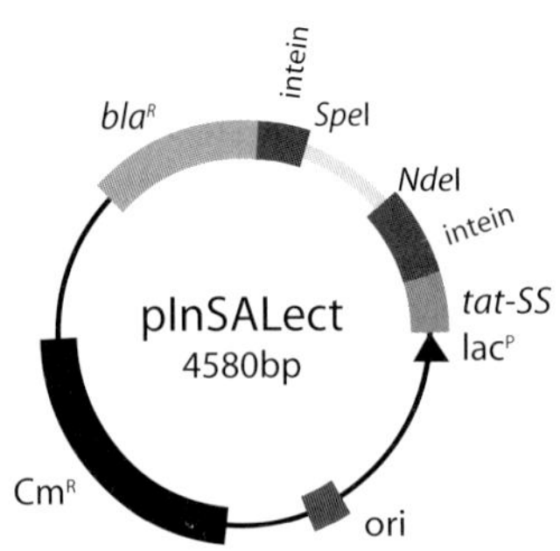

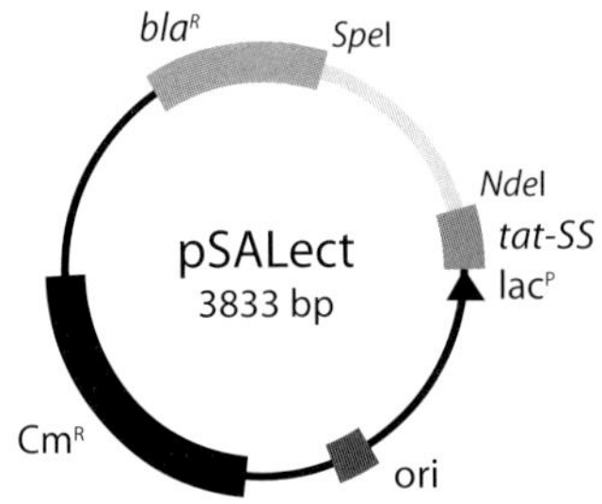

Figure 21.4 The pSALect and pInSALect vectors for reading frame selection. Both DNA vectors carry the chloramphenicol resistance gene (Cm^R) as a selection marker. The gene of interest is cloned via the *NdeI*/*SpeI* restriction sites. In pSALect, the cloning site is flanked by the N-terminal *tat* signal sequence and the ampicillin-resistance gene (*bla*). In pInSALect, the N- and C-terminal fragments of the cis-splicing VMA intein from *Saccharomyces cerevisiae* has been added, allowing for post-translational excision of the target protein to minimize library biases due to solubility.

Besides reading-frame selection, pSALect also functions as a solubility probe for the target peptide. Target sequences with a tendency to aggregate or misfold will coprecipitate the fused lactamase, effectively eliminating the host's ability to grow in the presence of antibiotics. While potentially useful as a selection system for soluble proteins (see also Chapter 5), this secondary effect is undesirable for the purpose of creating diverse libraries of in-frame gene fusions with random crossover points. A particular crossover by itself may result in an insoluble hybrid protein, yet its combination with a second, third or more crossovers elsewhere in the protein sequence can potentially stabilize the structure and once again yield a functional biocatalyst. Given that the target peptide's sole function after translation is to link the export signal sequence to the antibiotic resistance marker, we designed a second-generation vector called pInSALect to remove the target peptide post-translationally with the help of inteins [40]. In pInSALect, the gene for the cis-splicing VMA intein from *Saccharomyces cerevisiae* is inserted between the N-terminal TAT signal sequence and the C-terminal antibiotic resistance marker. Unique *Nde*I and *Spe*I restriction sites in the endonuclease portion of the intein serve as the new cloning sites for target DNA. Following translation of the entire fusion complex, the intein will excise not only itself but also the target peptide, effectively preventing the latter from interfering with lactamase function. Concomitantly with target excision, the flanking protein moieties are covalently linked, fusing the TAT-signal sequence directly onto the lactamase, which ensures efficient periplasmic export and antibiotic resistance for the host organism. DNA sequence analysis of currently over 1000 selected library members (~750 000 bp) from various projects suggests that pInSALect performs reading frame selection with >99% reliability and independent of the amino acid composition or solubility of the target peptides.

The protocols for size and reading frame selection can easily be combined to streamline the procedure. Following the isolation of plasmid DNA containing the ITCHY library, the vector is treated with restriction enzymes that cleave in positions flanking the gene fusion. The digestion mixture is separated using agarose gel electrophoresis, and hybrid genes of the desired size can be excised with a razor and the DNA recovered from the gel matrix. These DNA fragments are then directly ligated into linearized pInSALect and transformed into the *E. coli* expression system. Reading frame selection is performed by culturing the host on growth media containing carbenicillin, as described in detail elsewhere [40]. Following colony harvest, the plasmid-based library is recovered and can directly be used as template for the enhanced-SCRATCHY protocol.

21.2.3
Enhanced SCRATCHY via Forced Crossovers

The construction of hybrids with multiple crossovers at positions independent of DNA sequence homology is accomplished in the second step of the SCRATCHY protocol. In the original protocol, we used a simple DNA shuffling step that resulted in libraries heavily skewed towards revertant sequences [8]. More specifically, the bias is a consequence of the low fraction of DNA fragments carrying a gene fusion: for example, DNaseI-treatment of a 1000 bp hybrid gene into fragments of 100 bp average size generates only one chimeric gene fragment, but nine DNA fragments from either one of the two parental genes. Upon reassembly of these fragment mixtures, the low ratio of fragments containing a gene fusion to fragments consisting entirely of wild-type sequence, as well as more favorable annealing properties of the latter due to greater complementarity, will disfavor the formation of multicrossover hybrid genes.

As a means to compensate for these biases, the selective enrichment of DNA fragments containing gene fusions was introduced to skew the reassembly process in favor of multicrossover hybrids [18–20]. As outlined in Figure 21.5, ITCHY

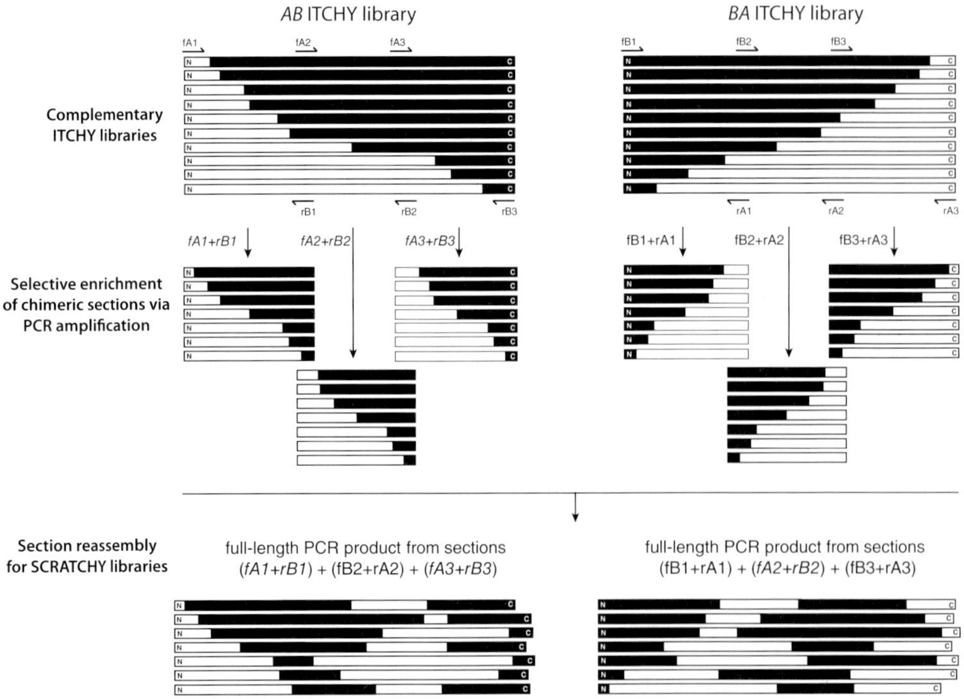

Figure 21.5 Chimeragenesis by enhanced and forced crossover SCRATCHY. The selective enrichment of ITCHY library sections containing the fusion point can significantly raise the level of multicrossover hybrids in the subsequent DNA shuffling step (enhanced crossover SCRATCHY). Alternatively, multiple overlapping sections that cover the entire length of the target gene(s) can be used to reassemble chimera libraries in a semi-rational fashion (forced crossover SCRATCHY). Hybrid genes, containing a fusion point in a particular section, are easily amplified from the random library by using parent-specific primers. For example, the enrichment of DNA fragments with crossovers in the first (N-terminal) section of the *AB* library can be accomplished by using an *A*-specific forward primer (fA1) in combination with a *B*-specific reverse primer (rBIX1). The selection of suitable primer binding sites for these two approaches can be arbitrary, but might benefit from consideration of structural and functional features.

libraries can serve as templates for the PCR amplification with forward primers, complementary to the N-terminal parental sequence, and reverse primers matching the C-terminal parental sequence. The location of the primer binding sites might be chosen arbitrarily, dividing the DNA sequence into equal portions. Alternatively, primer design can be guided by structural features and function of the parental proteins such as domain boundaries and active site residues. Finally, predictive frameworks such as SCHEMA [25], FamClash [33] and other computational tools offer a semi-rational strategy to define suitable primer binding sites.

Following the amplification of selected regions of the *AB* and *BA* ITCHY libraries, the PCR products can be mixed with the complementary ITCHY libraries for regular DNA shuffling, creating an enhanced-crossover SCRATCHY library [18]. In the case of glutathione transferase chimeragenesis, the spiking of the oligonucleotide pool with chimeras from the PCR reactions raised the percentage of reassembled genes with two or more crossovers to >40%, approximately doubling the fraction of hybrids with more than one crossover compared to the classical SCRATCHY protocol. Further library analysis also revealed a threefold lower number of revertant sequences (~12%) and a slight increase in the average crossover frequency.

An even more stringent method for the generation of multicrossover hybrid libraries is the assembly of enhanced SCRATCHY libraries via forced crossovers [19, 20]. The method requires a complete set of chimeric sections that can be obtained from two complementary ITCHY libraries (*AB* and *BA*) (Figure 21.5). By mixing the N-terminal section (fA1+rBIX1) and C-terminal section (fA3+rBIX3) from the *AB* ITCHY library with the center section (fBIX2+rA2) from the *BA* ITCHY library, stepwise annealing of adjacent fragments via primer overlap extension leads to the reconstruction of hybrid genes of parental size. Accordingly, the complementary triple-crossover SCRATCHY library is built from fragments (fBIX1+rA1), (fA2+rBIX2) and (fBIX3+rA3). The approach provides precise control over the number of fusion points in the final SCRATCHY library. While the example in Figure 21.5 generates three crossovers per gene, simply redefining the chimera sections by relocating the primer binding sites enables the creation of SCRATCHY libraries with two, three, four or more fusion points. DNA sequence analysis of native libraries has confirmed that typical SCRATCHY libraries with forced crossovers are composed of >98% hybrids of the desired complexity (S. Lutz *et al.*, unpublished results). Although the predetermined number of crossovers might initially be perceived as a limitation, the strategy does enable the assembly of more comprehensive libraries with defined composition. Library complexity cannot only be controlled at the crossover level but can be adjusted by combining homology-independent and homology-dependent methods [20] and targeting specific regions within a gene for chimeragenesis (Y. Liu, unpublished data). Furthermore, the simplicity of the experimental protocol, which basically is a series of PCRs, makes it feasible to generate multiple SCRATCHY libraries in parallel using only template DNA quantities of ITCHY libraries; this is in sharp contrast to the large amount of nucleic acids required for DNA shuffling experiments.

21.3
Future Trends in Chimeragenesis

In recent years, chimeragenesis has become a well-established technology and one of the standard procedures in the 'molecular toolbox' of protein engineers. Method improvements have made the experimental protocols more reliable and user-friendly, allowing for a wider distribution of the strategy in the research community. Often misconceived as an alternative to methods such as DNA shuffling, homology-independent recombination is not so much designed to compete rather than to complement the array of existing, directed evolution techniques. Experienced protein engineers will agree that the strategy is an unlikely choice for recombining proteins with sequence identities of 80% or higher, simply because it is more laborious than methods based on homologous recombination. Nevertheless, by enabling experimentalists to explore hybrid proteins without constraints of sequence identity, chimeragenesis provides a relatively easy access to proteins in the 'sequence twilight zone' that are largely beyond the reach of traditional homology-based recombination methods.

Venturing into the twilight zone of sequence identity has potential functional benefits, as has been demonstrated in several recent reports of successful protein engineering by chimeragenesis. Paralleling the technological advances, the results from chimeragenesis applications have also demonstrated the method's potential for modulating, improving and creating novel enzyme function. Recently, several groups have reported the successful remodeling of protein frameworks, using parents with high structure homology but low sequence identity. Working with distantly related glutathione *S*-transferases from human and rat (63/54% SI) – members of a large enzyme family that play a key role in cellular detoxification – Griswold *et al.* identified functional hybrid enzymes with up to 300-fold increased turnover rates for 7-amino-4-chloromethyl coumarin compared to the two parental enzymes [20]. The rate enhancement was not limited to this substrate, as an additional kinetic analysis of the chimeras showed elevated k_{cat}/K_m values for a number of alternate electrophiles.

In our laboratory, the homology-independent recombination of 2′-deoxyribonucleoside kinases – which are key enzymes for the activation of nucleoside analogue prodrugs used in antiviral and cancer therapy – identified variants with not only an elevated substrate specificity but also a novel activity for substrates not recognized by either parental enzyme [19]. Chimeragenesis of the human thymidine kinase 2 and the 2′-deoxyribonucleoside kinase from *Drosophila melanogaster* (39/41% SI) yielded hybrid enzymes with not only a slightly reduced activity for the native substrates (the four natural 2′-deoxyribonucleosides), but also an elevated specific activity for nucleoside analogues such as 2′,3′-dideoxythymidine and 2′,3′-dideoxycytidine. More importantly, the two characterized hybrids showed substantial catalytic activity for 2′,3′-didehydro-2′,3′-dideoxy thymidine (d4T), a nucleoside analogue for which neither one of the two parents shows any detectable activity.

Finally, the Arnold laboratory has utilized site-directed homologous recombination guided by their SCHEMA algorithm for creating chimera libraries of cytochromes P450 with improved thermostability and altered substrate specificity [28–30]. Cytochromes P450 comprise a large superfamily of heme-cofactor-carrying enzymes involved in drug metabolism and xenobiotics breakdown, as well as in the biosynthesis of steroid hormones and secondary metabolites. With three parental sequences (67/65% SI), the authors used SCHEMA to fragment the enzymes into eight portions whilst minimizing the disruption of structural contacts. The success of the design was verified upon construction of the 3^8 library members (6561 chimeras), followed by screening for folded hybrid proteins via CO-difference spectroscopy. The results from this high-throughput screen, which relied on the absorbance at 450 nm which is characteristic of a reduced CO-saturated heme group, suggested that almost 50% of the proteins (~3000 members) assumed a ternary structure capable of cofactor binding [30].

In a follow-up study, Landwehr *et al.* chose a subset of 14 P450 chimeras, as well as the three parent enzymes, for an in-depth analysis of their catalytic activities and substrate specificities [28]. When selected based on their above-average activities in preliminary tests, the candidates were evaluated for peroxygenase activity, the hydrogen peroxide-dependent oxygenation of substrates. Furthermore, the authors generated fusion proteins of all 17 heme domains with either one of the three parental reductase domains to reconstitute the native monooxygenase function. The relative activity of each enzyme variant for 11 different substrates was subsequently determined in 96-well microtiterplate assays, confirming the emergence of novel and improved substrate specificities among members of the library.

More recently, Li and coworkers used members of the same library as a training set for the prediction of a hybrid's thermostability [29]. While the thermostability of proteins is often compromised as a result of chimeragenesis, earlier results by Otey *et al.* [30] suggested the presence of few P450 hybrids with thermostability greater than the parental wild-type enzymes. In the subsequent, more systematic, analysis of 184 chimeras, Li found a direct correlation between the hybrids' fragment composition and their T_{50}, the temperature at which 50% of the protein retains an ability to bind heme cofactor after 10 min of incubation. In complementing the sequence–stability correlation data, the team successfully employed an alternative approach, using multiple sequence alignments to compute chimera stability based on consensus energies. The two methods were subsequently combined to guide the construction of several hybrids predicted to represent the most stable variants, yielding chimeras with an over-100-fold extended half-life at 57 °C. The increase in stability did not compromise enzyme activity, as the selected library members showed elevated peroxygenase activity for 2-phenoxyethanol by up to 50-fold and specificity for substrates not recognized by any one of the parent enzymes.

21.3.1 Combining SCRATCHY and SCHEMA

The creation of 'high-quality' protein libraries, rich in folded and stable structures, is desirable as it can significantly accelerate the discovery process of enzyme variants with novel properties. More so, in cases of limited screening capacity, the preparation of libraries with a high functional content can be critical to the success of the engineering project. The generation of high-quality libraries not only considers preserving critical intramolecular protein contacts but also enables the extensive sampling of regions of interest by containing broad diversity within a focus region. In the context of chimeragenesis, SCRATCHY has been shown to generate very diverse libraries of chimeric genes, enabling the construction of gene fusions with one or more crossover points, without any apparent sequence bias. Nevertheless, the protocol completely disregards structural information, resulting in a large percentage of chimeras with compromised foldability and structural stability which potentially forces the experimentalist to analyze large numbers of library members in search for improved protein variants.

By introducing the new protocol for creating SCRATCHY libraries with forced crossovers, we are now in the position to not only control the number of fusion points per gene but also to utilize structural information to guide our primer binding site selection. When used in combination with SCHEMA to identify regions important for structural integrity, we can eliminate gene fusions with crossovers that might disrupt chimera folding and stability, while enriching the library for hybrids with crossovers at or near locations suggested by the algorithm (Figure 21.6a and b). The idea was put to the test by calculating the SCHEMA crossover pattern for the deoxyribonucleoside kinase sequences, *Dm*dNK and human TK2, for which extensive experimental chimeragenesis data are available ([19]; also Y. Liu, unpublished data). When selecting for three crossovers, the SCHEMA prediction shows an overall good correlation with the gene fragment fusions of functional hybrid kinases, selected from SCRATCHY libraries, even though the latter experiments were conducted without prior knowledge of the predicted crossover data (Figure 21.6c). It is envisioned that future chimeragenesis library designs will increasingly take advantage of the predictive frameworks such as SCHEMA and OPTCOMB. The integration of these computational data in the analysis of experimental findings might also aid in interpreting and rationalizing partiality for selected chimeras.

21.3.2 The Future of Chimeragenesis

Finally, it might be interesting to take a closer look at Nature's mechanisms for some perspective on the current chimeragenesis methods and ideas for future protein-engineering strategies by homology-independent recombination. Particularly inspiring is the impressive performance of the *Deinococcus radiodurans* DNA repair system, which is capable of efficiently and accurately repairing very large

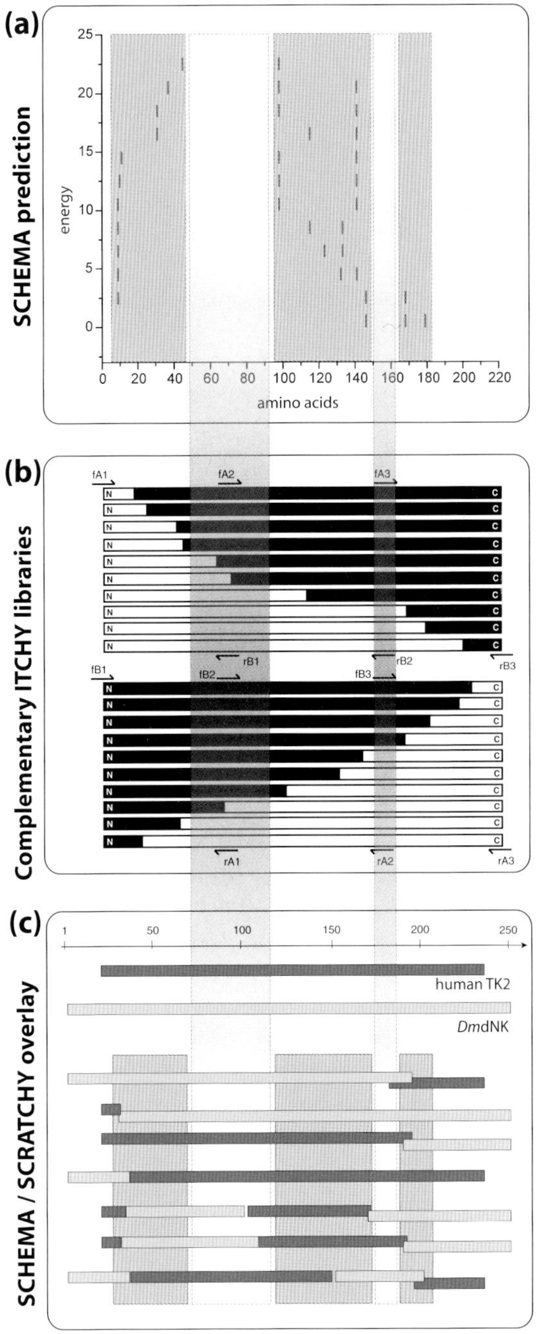

Figure 21.6 Combining SCHEMA predictions with SCRATCHY experiments. The design of enhanced and forced crossover SCRATCHY libraries can greatly benefit from computational algorithms that capture intramolecular protein contacts, which are important to protein stability and function. (a) The SCHEMA analysis of a protein scaffold provides a predictive framework, highlighting regions marked in orange that might be suitable for crossovers. On the opposite, blank sections in the same plot, marked by gray boxes, indicate functionally conserved sections which are most likely not suitable for the introduction of crossovers; (b) Functionally conserved sections can make excellent primer binding sites as the oligonucleotide protects the region from chimeragenesis by discriminating against sequences with gene fusions in that particular section; (c) The comparison of DNA sequence data of functional SCRATCHY library members with an independently calculated SCHEMA plot shows a good match of predicted and experimentally found crossover regions.

numbers of double-strand breaks in its genomic DNA following exposure to high doses of radiation [43]. Recruiting such highly efficient enzyme machinery for facilitating the reassembly of laboratory-made DNA fragments could be very attractive, not only from a practical perspective but also from a viewpoint of understanding the basic principles of DNA repair. The two major pathways for repairing double-strand breaks in Nature are homologous recombination and non-homologous end joining (NHEJ). While the former concept has already been very successfully adapted to laboratory evolution in the form of DNA shuffling, it is NHEJ that might offer some interesting new tricks for creating chimeragenesis libraries.

The NHEJ pathway is found in prokaryotes and eukaryotes and consists, in its simplest form, of two components, the DNA-end-binding protein Ku and a complementary DNA ligase [44]. Mycobacteria and *Pseudomonas aeruginosa* carry such minimal NHEJ systems and have been models for studying this DNA repair mechanism [45, 46]. Although many details of the pathway remain unclear, the working model predicts that each subunit of the homodimeric Ku protein interacts with one of the DNA termini regions, positioning the ends in proximity to each other. Separately, DNA ligase D (LigD), which in addition to the ligase function carries a phosphoesterase and polymerase domain, docks to a specific site on Ku and the ternary complex performs ATP-dependent linkage of the two bound DNA fragments.

The Ku/LigD system can efficiently ligate DNA fragments with blunt ends, as well as 5′- and 3′-overhangs. Nevertheless, one of the challenges for using the NHEJ machinery in chimeragenesis might be infidelity. Studies in mycobacteria have shown that joining DNA fragments with blunt-ends and 5′-overhangs generates approximately 50% frame-shifted ligation products [46]. Whilst linking DNA with blunt ends tends to result in template-independent single nucleotide addition, DNA fragments with 5′-overhangs show signs of polymerase-catalyzed fill-in of the overhang prior to ligation of the DNA termini. The potential benefit of such a high frequency of frame-shifts to Nature is unclear and would seriously compromise the quality of a chimeragenesis library. Nevertheless, recent findings indicate that the ligation fidelity can be enhanced by disabling the polymerase function in LigD with two active-site mutations [47]. In the absence of primer extension, the accuracy of the Ku/LigD system increases to >90% for blunt-end DNA fragments and >80% for DNA strands with 5′-overhangs, respectively. Alternatively, the creation of chimeragenesis libraries via NHEJ could be accomplished with DNA fragments carrying 3′-overhangs. In contrasting the infidelity of blunt-end and 5′-overhang ligation, experiments in mycobacterial cells have shown highly faithful linkage (>97%) of DNA with 3′-overhangs [47]. Although strain mutation and deletion experiments by the same authors suggest a more complex *in vivo* scenario, potentially involving other cellular polymerases and ligases as functional back-ups, the findings are encouraging and justify investigating the system for *in vitro* applications.

In the context of chimeragenesis, the use of overexpressed and purified Ku and LigD, in combination with synthetic double-stranded oligodeoxyribonucleotides, could significantly enhance the efficiency of DNA fragment reassembly. Furthermore, Ku protein could minimize potential biases towards recombination of DNA pieces with greater sequence homology. While initial studies would have to focus on establishing a better understanding of the factors that determine recombination fidelity – or a lack thereof – it seems sensible to contemplate the idea of incorporating these mechanisms as part of the reassembly and library diversification process.

21.4 Conclusions

In summary, the results of recent studies have demonstrated conclusively that a chimeric enzyme can in fact be more than just the sum of its parts [48]. Improved library design and preparation methods for chimeragenesis, combined with clever screening and selection protocols, have yielded biocatalysts with new and improved properties that range from catalytic activity to substrate specificity and protein stability. In addition, the integration of computational algorithms with laboratory experiments has afforded an opportunity to test and refine current hypotheses on the structure–function relationship in proteins and enzymes.

Acknowledgments

The authors would like to acknowledge in part the financial support provided by the National Institutes of Health, and also the members of the Lutz laboratory for their many critical and helpful comments during the preparation of this chapter.

References

1 Lawrence, J.G. (1997) Selfish operons and speciation by gene transfer. *Trends in Microbiology*, **5**, 355–9.
2 Shapiro, J.A. (1997) Genome organization, natural genetic engineering and adaptive mutation. *Trends in Genetics*, **13**, 98–108.
3 Stemmer, W.P. (1994) DNA shuffling by random fragmentation and reassembly: in vitro recombination for molecular evolution. *Proceedings of the National Academy of Sciences of the United States of America*, **91**, 10747–51.
4 Stemmer, W.P. (1994) Rapid evolution of a protein in vitro by DNA shuffling. *Nature*, **370**, 389–91.
5 Zhao, H., Giver, L., Shao, Z., Affholter, J.A. and Arnold, F.H. (1998) Molecular evolution by staggered extension process (StEP) in vitro recombination. *Nature Biotechnology*, **16**, 258–61.
6 Coco, W.M., Levinson, W.E., Crist, M.J., Hektor, H.J., Darzins, A., Pienkos, P.T., Squires, C.H. and Monticello, D.J. (2001) DNA shuffling method for generating highly recombined genes and evolved enzymes. *Nature Biotechnology*, **19**, 354–9.
7 Muller, K.M., Stebel, S.C., Knall, S., Zipf, G., Bernauer, H.S. and Arndt, K.M. (2005) Nucleotide exchange and excision technology (NExT) DNA shuffling: a robust method for DNA fragmentation

and directed evolution. *Nucleic Acids Research*, **33**, e117.

8 Lutz, S., Ostermeier, M., Moore, G.L., Maranas, C.D. and Benkovic, S.J. (2001) Creating multiple-crossover DNA libraries independent of sequence identity. *Proceedings of the National Academy of Sciences of the United States of America*, **98**, 11248–53.

9 Otey, C.R., Silberg, J.J., Voigt, C.A., Endelman, J.B., Bandara, G. and Arnold, F.H. (2004) Functional evolution and structural conservation in chimeric cytochromes P450: calibrating a structure-guided approach. *Chemistry and Biology*, **11**, 309–18.

10 Chang, C.C., Chen, T.T., Cox, B.W., Dawes, G.N., Stemmer, W.P., Punnonen, J. and Patten, P.A. (1999) Evolution of a cytokine using DNA family shuffling. *Nature Biotechnology*, **17**, 793–7.

11 Ness, J.E., Welch, M., Giver, L., Bueno, M., Cherry, J.R., Borchert, T.V., Stemmer, W.P. and Minshull, J. (1999) DNA shuffling of subgenomic sequences of subtilisin. *Nature Biotechnology*, **17**, 893–6.

12 Moore, G.L. and Maranas, C.D. (2002) eCodonOpt: a systematic computational framework for optimizing codon usage in directed evolution experiments. *Nucleic Acids Research*, **30**, 2407–16.

13 Bogorad, L.D. and Deem, M.W.A. (1999) A hierarchical approach to protein molecular evolution. *Proceedings of the National Academy of Sciences of the United States of America*, **96**, 2591–5.

14 Ostermeier, M., Nixon, A.E. and Benkovic, S.J. (1999) Incremental truncation as a strategy in the engineering of novel biocatalysts. *Bioorganic and Medicinal Chemistry*, **7**, 2139–44.

15 Ostermeier, M., Shim, J.H. and Benkovic, S.J. (1999) A combinatorial approach to hybrid enzymes independent of DNA homology. *Nature Biotechnology*, **17**, 1205–9.

16 Sieber, V., Martinez, C.A. and Arnold, F.H. (2001) Libraries of hybrid proteins from distantly related sequences. *Nature Biotechnology*, **19**, 456–60.

17 Lee, S.G., Lutz, S. and Benkovic, S.J. (2003) On the structural and functional modularity of glycinamide ribonucleotide formyltransferases. *Protein Science*, **12**, 2206–14.

18 Kawarasaki, Y., Griswold, K.E., Stevenson, J.D., Selzer, T., Benkovic, S.J., Iverson, B.L. and Georgiou, G. (2003) Enhanced crossover SCRATCHY: construction and high-throughput screening of a combinatorial library containing multiple non-homologous crossovers. *Nucleic Acids Research*, **31**, e126.

19 Gerth, M.L. and Lutz, S. (2007) Non-homologous recombination of deoxyribonucleoside kinases from human and *Drosophila melanogaster* yields human-like enzymes with novel activities. *Journal of Molecular Biology*, **370**, 742–51.

20 Griswold, K.E., Kawarasaki, Y., Ghoneim, N., Benkovic, S.J., Iverson, B.L. and Georgiou, G. (2005) Evolution of highly active enzymes by homology-independent recombination. *Proceedings of the National Academy of Sciences of the United States of America*, **102**, 10082–7.

21 Bittker, J.A., Le, B.V. and Liu, D.R. (2002) Nucleic acid evolution and minimization by nonhomologous random recombination. *Nature Biotechnology*, **20**, 1024–9.

22 Bittker, J.A., Le, B.V., Liu, J.M. and Liu, D.R. (2004) Directed evolution of protein enzymes using nonhomologous random recombination. *Proceedings of the National Academy of Sciences of the United States of America*, **101**, 7011–16.

23 Endelman, J.B., Silberg, J.J., Wang, Z.-G. and Arnold, F.H. (2004) Site-directed protein recombination as a shortest-path problem. *Protein Engineering Design and Selection*, **17**, 589–94.

24 Meyer, M.M., Silberg, J.J., Voigt, C.A., Endelman, J.B., Mayo, S.L., Wang, Z.G. and Arnold, F.H. (2003) Library analysis of SCHEMA guided protein recombination. *Protein Science*, **12**, 1686–93.

25 Voigt, C.A., Martinez, C., Wang, Z.-G., Mayo, S.L. and Arnold, F.H. (2002) Protein building blocks preserved by recombination. *Nature Structural Biology*, **9**, 553–8.

26 Higara, K. and Arnold, F.H. (2003) General method for sequence-independent site-directed chimeragenesis. *Journal of Molecular Biology*, **330**, 287–96.

27 Meyer, M.M., Hochrein, L. and Arnold, F.H. (2006) Structure-guided SCHEMA recombination of distantly related β-lactamases. *Protein Engineering Design and Selection*, **19**, 563–70.

28 Landwehr, M., Carbone, M., Otey, C.R., Li, Y. and Arnold, F.H. (2007) Diversification of catalytic function in a synthetic family of chimeric cytochrome P450s. *Chemistry and Biology*, **14**, 269–78.

29 Li, Y., Drummond, D.A., Sawayama, A.M., Snow, C.D., Bloom, J.D. and Arnold, F.H. (2007) A diverse family of thermostable cytochrome P450s created by recombination of stabilizing fragments. *Nature Biotechnology*, **25**, 1051–6.

30 Otey, C.R., Landwehr, M., Endelman, J.B., Higara, K., Bloom, J.D. and Arnold, F.H. (2006) Structure-guided recombination creates an artificial family of cytochromes P450. *PLoS Biology*, **4**, e112.

31 Moore, G.L. and Maranas, C.D. (2003) Identifying residue-residue clashes in protein hybrids by using a second-order mean-field approach. *Proceedings of the National Academy of Sciences of the United States of America*, **100**, 5091–6.

32 Saraf, M.C., Moore, G.L. and Maranas, C.D. (2003) Using multiple sequence correlation analysis to characterize functionally important protein regions. *Protein Engineering*, **16**, 397–406.

33 Saraf, M.C., Horswill, A.R., Benkovic, S.J. and Maranas, C.D. (2004) FamClash: a method for ranking the activity of the engineered enzymes. *Proceedings of the National Academy of Sciences of the United States of America*, **101**, 4142–7.

34 Saraf, M.C., Gupta, A. and Maranas, C.D. (2005) Design of combinatorial protein libraries of optimal size. *Proteins*, **60**, 769–77.

35 Pantazes, R.J., Saraf, M.C. and Maranas, C.D. (2007) Optimal protein library design using recombination or point mutations based on sequence-based scoring functions. *Protein Engineering Design and Selection*, **20**, 361–73.

36 Lutz, S., Ostermeier, M. and Benkovic, S.J. (2001) Rapid generation of incremental truncation libraries for protein engineering using alpha-phosphothioate nucleotides. *Nucleic Acids Research*, **29**, e16.

37 Ostermeier, M. and Benkovic, S.J. (2001) Construction of hybrid gene libraries involving the circular permutation of DNA. *Biotechnology Letters*, **23**, 303–10.

38 Ostermeier, M. and Lutz, S. (2003) *The Creation of ITCHY Hybrid Protein Libraries*, Vol. 231 (eds G. Georgiou and F.H. Arnold), Humana Press Inc., Totowa, NJ, pp. 129–41.

39 Lutz, S. and Ostermeier, M. (2003) *Preparation of SCRATCHY Hybrid Protein Libraries*, Vol. 231 (eds G. Georgiou and F. H. Arnold), Humana Press Inc., Totowa, NJ, pp. 143–51.

40 Gerth, M.L., Patrick, W.M. and Lutz, S. (2004) A second-generation system for unbiased reading frame selection. *Protein Engineering Design and Selection*, **17**, 595–602.

41 Lutz, S., Fast, W. and Benkovic, S.J. (2002) A universal, vector-based system for nucleic acid reading-frame selection. *Protein Engineering*, **15**, 1025–30.

42 Cabantous, S. and Waldo, G.S. (2006) In vivo and in vitro protein solubility assays using split GFP. *Nature Methods*, **3**, 845–54.

43 Zahrandka, K., Slade, D., Bailone, A., Sommer, S., Averbeck, D., Petranovic, M., Lindner, A.B. and Radman, M. (2006) Reassembly of shattered chromosomes in *Deinococcus radiodurans*. *Nature*, **443**, 569–73.

44 Shuman, S. and Glickman, M.S. (2007) Bacterial DNA repair by non-homologous end joining. *Nature Reviews Microbiology*, **5**, 852–61.

45 Della, M., Palmbos, P.L., Tseng, H.M., Tonkin, L.M., Daley, J.M., Topper, L.M., Pitcher, R.S., Tomkinson, A.E., Wilson, T. E. and Doherty, A.J. (2004) Mycobacterial Ku and ligase proteins constitute a two-component NHEJ repair machine. *Science*, **306**, 683–5.

46 Gong, C., Bongiorno, P., Martins, A., Stephanou, N.C., Zhu, H., Shuman, S. and Glickman, M.S. (2005) Mechanism of nonhomologous end-joining in mycobacteria: a low-fidelity repair system driven by Ku, ligase D and ligase C. *Nature*

Structural and Molecular Biology, **12**, 304–12.

47 Aniukwu, J., Glickman, M.S. and Shuman, S. (2008) The pathways and outcomes of mycobacterial NHEJ depend on the structure of the broken DNA ends. *Genes and Development*, **22**, 512–27.

48 Beguin, P. (1999) Hybrid enzymes. *Current Opinion in Biotechnology*, **10**, 336–40.

22
Protein Generation Using a Reconstituted System

Bei-Wen Ying and Takuya Ueda

22.1
Introduction

Cell-free methodologies for the generation of proteins have been widely utilized in protein science studies. These so-called cell-free translation systems are typically based on crude-cell extracts which comprise all of the necessary components for transcription and translation processes. Current representative strategies include those using wheat-germ extract [1, 2], the S30 fraction of *E. coli* [3–5], and rabbit reticulocyte lysate [6]. In order to avoid rapid depletion of the energy source and the degradation of proteins and nucleic acids in batch-formed systems, a variety of alternative techniques such as continuous-flow or continuous-exchange [5, 7–9] have been developed, and this has resulted in considerably higher productivities.

Compared to conventional *in vivo* expression methods, cell-free translation systems are especially feasible for high-throughput and genome-wide production [1, 9–12]. The application of cell-free translation systems to structural proteomics is also more effective because of the huge number of protein sequences with unknown functions that have been identified in genome projects. Among all these cell extract systems, the wheat-germ cell-free expression system particularly excels in the efficient synthesis of eukaryotic proteins [9]. Indeed, a platform that utilizes a wheat-germ cell-free pipeline to produce protein samples for structural determination by NMR has recently reviewed [11]. Techniques for the functional production of integral membrane proteins – which are perhaps the most intriguing targets in protein generation – have been recently introduced [13, 14]. Screening techniques such as ribosome display and mRNA display exhibit great potential and have been developed using these systems [15–19]. Indeed, the progress in these *in vitro* selection technologies can also be exploited for drug discovery and therapeutics [14, 20].

One crucial problem in many biotechnological applications of cell-free translation is the generation of proteins with accurate protein folding [21–25]. Further studies on protein translation, folding and maturation are therefore essential to

Protein Engineering Handbook. Edited by Stefan Lutz and Uwe T. Bornscheuer

ISBN: 978-3-527-31850-6

generate better cell-free protein production. In addition, the removal of proteinases and nucleases from the expression system is required. Here, we present a highly controllable reconstituted expression system that is ideally built to meet these demands. The system, which is known as PURE (Protein synthesis Using Recombinant Elements) [26, 27], comprises only a minimum set of purified factors and enzymes that are required for protein synthesis. The PURE system and its related approaches which are introduced here will open new doors that, undoubtedly, will enhance future research in protein engineering and protein science.

22.2 The PURE System

22.2.1 Concept and Strategy

Protein synthesis is one of the most complicated biological processes occurring in cells, as it involves hundreds of biochemical reactions and multiple subsystems. It is well known that the genesis of proteins includes transcription from DNAs to mRNAs, as well as translation initiation, elongation, termination and recycling steps. In total, more than 100 factors take part in the whole process. Recently, a new method of generating proteins has been developed which is based on a constructive concept. For this, a minimal set of components that are essential for protein synthesis was purified and reconstituted into a biologically functional system, the PURE system [26–28], as illustrated in Figure 22.1. This protein generation system is composed of recombinant enzymes and chemicals, and was created by employing an approach whereby all of the recombinant enzymes contain a His-tag. The His-tag is especially useful for the purification of the recombinant enzymes as well as during purification of the synthesized protein products. More-

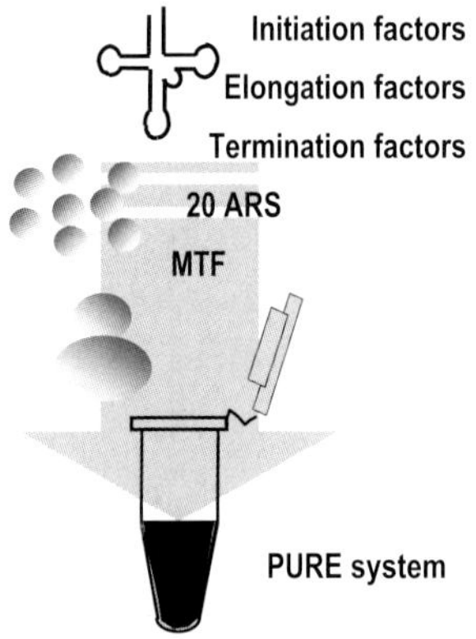

Figure 22.1 Schematic representation of the PURE system. ARS, aminoacyl-tRNA synthetase; MTF, methionyl-tRNA transformylase.

over, an energy recycling scheme is embedded in the system to maintain the protein synthetic process, which is highly energy-consuming in nature. The information provided in the following sections demonstrates that such a complex metabolic system can be reconstructed relatively easily from its purified components. It is also clear that such a constructive approach represents an efficient and practical strategy for protein generation.

22.2.2 The Composition of PURE

In cells, the translation process comprises four steps: initiation, elongation, termination, and recycling. In *E. coli*, the proteinaceous factors responsible for the translation process consist of three initiation factors (IF1, IF2 and IF3), three elongation factors (EF-G, EF-Tu and EF-Ts), three releasing factors (RF1, RF2 and RF3), and a ribosome recycling factor (RRF). In order to initiate translation, methionyl-tRNA transformylase (MTF) catalyzes the formylation of Met-RNA_f. Furthermore, to ensure that the amino acids polymerize in the proper order as a correct peptide, 20 different aminoacyl-tRNA synthetases (ARSs) are needed to activate the corresponding tRNAs. The transcription from DNA to mRNA requires a corresponding RNA polymerase. Thus, 32 essential catalysts and cofactors, IF1, IF2, IF3, EF-G, EF-Tu, EF-Ts, RF1, RF2, RF3, RRF, 20 ARSs, MTF and T7 RNA polymerase, were constructed with an extra His-tag at either their C or N termini. These factors were overexpressed in *E. coli* and purified using a Ni^{2+} column. Ribosomes – the vital element for protein synthesis – were purified from wild-type bacteria. Additional components, which were also crucial to establish the PURE system, included 46 tRNAs, ATP, GTP, CTP, TTP, creatine phosphate, 10-formyl-5,6,7,8-tetrahydrofolic acid (FD), 20 proteinogenic amino acids, creatine kinase (CK), myokinase, nucleoside-diphosphate kinase (NDK) and pyrophosphatase (PPiase). Detailed information regarding to how to create the PURE system is provided in Table 22.1.

22.2.3 Advantages of PURE

The first major advantage of the PURE system is that synthesized proteins can easily be purified in their native form within just a few hours. As described above, the proteinaceous factors are fused with a His-tag so that they can be eliminated by passage through a Ni^{2+} column. The target protein can therefore be affinity-purified without any tags and produced in its native form. In addition, ribosomes are very large molecules that can be removed by ultrafiltration using a membrane with a molecular weight cutoff of 100 kDa. This advantage has enabled the rapid purification of highly purified dihydrofolate reductase (DHFR). Such rapid and simple purification procedures will benefit high-throughput protein preparations and will clearly offer major advantages for protein structural analyses using NMR and X-ray diffraction.

Table 22.1 Composition of the PURE system.

E. coli tRNA mix[a]	10–56 A260 ml^{-1}	EF-TS	0.66 μ*M*	MetRS	2.08 μg ml^{-1}
Magnesium acetate[a]	9–13 m*M*	EF-Tu	0.92 μ*M*	AlaRS	68.79 μg ml^{-1}
HEPES–KOH (pH 7.6)	50 m*M*	EF-G	0.26 μ*M*	IleRS	39.53 μg ml^{-1}
Potassium acetate[b]	18.7 m*M*	RF-1	0.25 μ*M*	AspRS	7.97 μg ml^{-1}
Potassium glutamate[a]	100–157 m*M*	RF-2	0.24 μ*M*	HisRS	0.8 μg ml^{-1}
20 Amino acids[a]	0.1–0.3 m*M* each	RF-3	0.17 μ*M*	GluRS	12.63 μg ml^{-1}
Spermidine	2 m*M*	RRF	0.5 μ*M*	GlyRS	9.6 μg ml^{-1}
Creatine phosphate	20 m*M*	IF-1	2.7 μ*M*	ProRS	10.24 μg ml^{-1}
NDK	1.08 μg ml^{-1}	IF-2	0.4 μ*M*	ThrRS	6.29 μg ml^{-1}
Ribosome	1 μ*M*	IF-3	1.5 μ*M*	ArgRS	2 μg ml^{-1}
T7 RNA polymerase	10 μg ml^{-1}	PPiase	0.1 μg ml^{-1}	CysRS	1.23 μg ml^{-1}
FD	10 μg ml^{-1}	CK	4 μg ml^{-1}	GlnRS	3.79 μg ml^{-1}
DTT	1 m*M*	Myokinase	3 μg ml^{-1}	AsnRS	22 μg ml^{-1}
GTP	2 m*M*	MTF	20 μg ml^{-1}	TrpRS	1.05 μg ml^{-1}
UTP	1 m*M*	LeuRS	4.02 μg ml^{-1}	TyrRS	0.61 μg ml^{-1}
CTP	1 m*M*	LysRS	6.4 μg ml^{-1}	SerRS	1.87 μg ml^{-1}
ATP	2 m*M*	PheRS	16.52 μg ml^{-1}	ValRS	1.81 μg ml^{-1}

a Varied concentrations reported by the relative research groups.
b Omitted in commercial kits (produced by PGI) of the PURE system.
CK, creatine kinase; DTT, dithiothreitol; FD, 10-formyl-5,6,7,8-tetrahydrofolic acid; MTF, methionyl-tRNA transformylase; NDK, nucleoside diphosphate kinase; PPiase, pyrophosphatase.

The second advantage of the PURE system is that, since it is reconstructed with purified elements, there are very few RNase or protease contaminants, which can often cause degradation of the mRNAs and proteins in cell-extracted systems. The stability of both mRNAs and proteins is therefore greatly increased in the PURE system compared to other approaches. The PURE system can also express target proteins directly from linear DNA fragments (PCR amplification), which allows the time-consuming cloning step to be omitted. Taking these merits into account, the PURE system represents an excellent technique that can be applied to *in vitro* selection systems, such as ribosome display, mRNA display and other screening approaches [16, 19, 29, 30].

A third advantage of the PURE system is the freedom to omit or add any element into the system. This highly controllable property provides tremendous flexibility and precision by allowing adjustments in the concentrations of the various components. It also offers considerable benefits for studies on translation mechanisms and protein maturation processes. We have reported a number of studies on translation initiation and translation termination mechanisms [31, 32], trans-translation processes [33, 34], the chaperone-mediated protein-folding process [21, 35, 36] and secretion pathways [37], as well as applications development, for example, the incorporation of unnatural amino acids [27, 38]. Easy manipulation is feasible for a wide range of research applications.

22.2.4 Preparation of the Components

Although the PURE system is commercially available, the methods used to construct the purified translation system are introduced here, together with additional information relating to its specific applications. It is hoped that these methods will be employed by research groups specifically to tailor the PURE system to their studies.

22.2.4.1 Overexpression and Purification of Translation Factors

The protein factors (IF1, IF2, IF3, EF-G, EF-Tu, EF-Ts, RF1, RF2, RF3, RRF, 20 ARSs, MTF and T7 RNA polymerase) were cloned as His-tagged fusion forms into the pQE or pET vectors, and overexpressed in the *E. coli* strain BL21 (pREP4 or DE3, respectively) as described previously [26, 27]. These factors were purified, usually from a 1 liter culture, using a 5 ml HiTrap chelating column (Amersham Pharmacia) precharged with Ni^{2+}. Chromatography was performed using AKTA (Amersham Pharmacia) with a linear gradient buffer system containing 10 m*M* or 400 m*M* imidazole. The purified proteins were stocked in universal buffer containing 50 m*M* HEPES–KOH (pH 7.6), 100 m*M* KCl, 10 m*M* $MgCl_2$, 7 m*M* β-mercaptoethanol and 30% glycerol. The purified factors and enzymes were verified to be biologically active, despite the presence of the fused His-tag. Other biochemical components, including 20 proteinogenic amino acids, NTPs, tRNA mix, spermidine, creatine phosphate, myokinase, NDK, PPiase, CK and FD, were acquired commercially. All of the components were stocked separately in small aliquots at −30 °C for normal use, or at −80 °C for long-term storage.

22.2.4.1.1 Tips for Overexpression and Purification

- Problems of overexpressing EF-G in bacteria can often be resolved by the addition of GTP to the medium.
- GDP usually increases the stability of EF-Tu; thus, a low concentration (~10 μ*M*) of GDP in the stock buffer and/or the dialysis buffer is recommended.
- IF-3 is easily aggregated during the dialysis step after purification; therefore it is better to increase the concentration of salt (up to 200 m*M* is fine) in the final stock.
- PheRS is typically prone to aggregation and requires a chaperonin (GroEL-GroES) to facilitate its folding *in vivo*, which causes the copurification of GroEL during PheRS preparation. Additional ATP in the purification buffer system helps to eliminate GroEL contamination.

22.2.4.2 Preparation of Ribosomes

Our laboratory has used two major methods to isolate active ribosomes, both of which are suitable for the PURE system. Either method can be used, depending on the purpose of the research or the facilities available. Initially, *E. coli* cells are collected during the mid-log phase and disrupted using a French Press (7000–10 000 psi). After removing the debris by centrifugation (20 000 *g*, 30 min), the separation procedures are performed in a number of ways (see Table 22.2 for detailed protocols).

The conventional method of ribosome preparation is ideal for isolating small quantities of ribosomes of high purity. However, as this is combined with additional washing steps the entire process takes at least four days. Based on personal experience, ribosomes required for mechanistic studies are better when prepared in this manner. In contrast, the rapid method of ribosome preparation is quite good for large-scale purification, and is used in commercial PURE system kits. In general, ribosomes separated using this method provide better productivity for protein synthesis.

22.2.4.2.1 Tips for Ribosome Preparation

- For mechanistic studies, washing steps are essential to clear away the membrane debris, ribosome-associated proteins or chaperones.
- If a washing step is carried out, most of the 30S ribosomes are eliminated. It is necessary to use either a weaker centrifugal force or a shorter ultracentrifugation time if the 30S fraction is required.
- For a rough (70S) ribosome preparation, the ribosome formats can be collected visually by using a pipette with good illumination.
- In order to obtain a better separation, it is better to load only 100–120 OD of crude or washed ribosomes onto the sucrose gradient layer per tube (SW 28 rotor; see Table 22.2); otherwise the 50S and 70S portions will overlap.

Table 22.2 Preparation of the 70S ribosome.

Conventional preparation

- 100 000 g ultracentrifugation

Pellet as crude ribosomes

- Wash steps (optional):
  Suspense the crude ribosome in the Wash buffer (final volume 0.5–1 ml)
  Load onto the Wash cushion buffer (20% sucrose, 1.5 ml)
  Ultracentrifugation (Beckman rotor 100.3 Ti, 55 000 rpm (~12 500 g), 4 h)
  Dissolve the pellet (clear part only) in the Wash buffer
  Repeat wash step again
- Suspend the crude or washed ribosomes in the Rbs buffer
- Separate the ribosomes by sucrose density gradient ultracentrifugation
  (6–38% sucrose gradient, SDG buffer, Beckman rotor SW28, 18 000 rpm (~43 000 g), 15.5 h)
- Collect the 70S fractions using Fractionator
- Pellet down the 70S ribosomes by ultracentrifugation
  (Beckman rotor 45 Ti, 40 000 rpm (~12 500 g), 24 h)
- Dissolve the final pellet (70S ribosome) with the Rbs buffer
- Store at –80 °C.

Quick preparation

- Ammonium precipitation (final conc. 1.5 *M*)
- Remove the precipitates by centrifugation (20 000 g, 30 min)
- Keep the supernatant for purification
- Separate the ribosomes by hydrophobic chromatography
  (~1000 OD for a 10-ml HiTrap Butyl FF column, Buffer A and B)
- Elute the ribosome fractions (0.75 *M* $(NH_4)_2SO_4$)
- Overlay onto the Cushion buffer (30%, equal volume)
- Recovery of the 70S ribosomes by ultracentrifugation
  (Beckman rotor 70Ti, 36 000 rpm (~95 000 g), 16 h)
- Dissolve the final pellet (70S ribosome) with the Rbs buffer
- Store at –80 °C.

Buffers

Wash buffer:
  20 m*M* HEPES–KOH (pH 7.6)
  10 m*M* $Mg(OAc)_2$
  500 m*M* NH_4Cl
  7 m*M* β-mercaptoethanol

SDG buffers (6% or 38%):
  20 m*M* HEPES–KOH (pH 7.6)
  10 m*M* $Mg(OAc)_2$
  30 m*M* KCl
  6% or 38% sucrose
  7 m*M* β-mercaptoethanol

Wash cushion buffer (20%):
  20 m*M* HEPES–KOH (pH 7.6)
  10 m*M* $Mg(OAc)_2$
  500 m*M* NH4Cl
  20% sucrose
  7 m*M* β-mercaptoethanol

Table 22.2 *Continued*

Buffer A:
20 m*M* HEPES–KOH (pH 7.6)
10 m*M* $Mg(OAc)_2$
1.5 *M* $(NH4)_2SO_4$
7 m*M* β-mercaptoethanol
Buffer B:
20 m*M* HEPES–KOH (pH 7.6)
10 m*M* $Mg(OAc)_2$
7 m*M* β-mercaptoethanol
Rbs buffer:
20 m*M* HEPES–KOH (pH 7.6)
10 m*M* $Mg(OAc)_2$
30 m*M* KCl
7 m*M* β-mercaptoethanol
Cushion buffer (30%):
20 m*M* HEPES–KOH (pH 7.6)
10 m*M* $Mg(OAc)_2$
30 m*M* NH4Cl
30% sucrose
7 m*M* β-mercaptoethanol

- In order to obtain tight-coupled 70S ribosomes, a lower concentration (6 m*M*) of Mg^{2+} is recommended during the purification process (SDG buffers; see Table 22.2). In this case it is better if the final stock buffer contains 10–13 m*M* Mg^{2+}.

22.2.5
Set-Up of the Translation Reaction

An active transcription/translation-coupled system is easily reconstituted by blending the purified components described in Table 22.1. All of the reagents (purified factors) should be stored on ice during manipulation. For convenience, several solutions in the PURE system can be pre-prepared and stored at −30 °C for at least several months. The pre-prepared solutions can be divided into different groups: the EF mix (EF-Tu, EF-G and EF-Ts), RF mix (RF1, RF2, RF3 and RRF), enzyme mix (all other enzymes), buffer mix and ribosomes. As the commercial products from the PGI (Post Genome Institute Co. Ltd.) allow a translation reaction to be started simply by mixing two solutions, it is also possible to simplify these groups into an enzyme mix and a buffer mix.

Both, DNA and mRNA can be used as the template for the PURE system. The competitive way to prepare a template for PURE translation is simply to induce a T7 promoter and a Shine–Dalgarno (SD) sequence upstream of the target open reading frame (ORF) by PCR amplification. The PCR products, even without the commonly performed clear-up step, can be easily used as templates in the PURE system.

22.3
Current Applications

22.3.1
Protein Generation

The primary task of any cell-free translation system is to generate proteins. To date, a large number of proteins have been successfully synthesized using the PURE system. The first reported application of the PURE system described the production of the full-length proteins DHFR, lysozyme, green fluorescent protein (GFP) and the T7 gene 10 product in their active conformations, with a productivity of more than $100\,\mu g\,ml^{-1}\,h^{-1}$ reaction [27]. Thereafter, a variety of proteins, from prokaryotic to eukaryotic targets, from natural polypeptides to semi-artificial ones, and even those incorporating non-natural amino acids, have been successfully generated using this purified system (see Table 22.3). These reports have shown that the PURE system could productively generate active proteins derived from a diverse range of other species. The PURE system contains the essential translation factors that are both crucial and sufficient for producing active proteins; however, slight modifications of the system are recommended for better expression. Some such examples are included in the following.

Cell-free translation systems are often utilized more frequently than conventional *in vivo* expression techniques due to their efficient and cost-effective reputation; however, problems during protein maturation process, such as protein aggregation, commonly occur. For instance, single-chain antibodies are frequently employed as target proteins to test *in vitro* translation systems because they are semi-artificial polypeptides with crucial folding problems and are always central players in protein engineering [17, 23, 39, 40]. When using the PURE system, we observed that the relative amounts of soluble product were significantly higher than those obtained using a bacterial S30 extract system, whereas the total amount of product was relatively higher with the S30 extracts [21, 28]. These results indicated that the crowded and condensed environment of the extract system may cause aggregation and hinder the folding process, whereas the intrinsic chaperones in the cell extracts could assist protein folding. It was hypothesized, therefore, that selectively supplementing the PURE system may enhance functional protein production.

In vivo, a subset of proteins requires the presence of other proteins (chaperones or isomerases) to facilitate their correct folding and proper maturation [41, 42]. In order to develop a production system for biologically active proteins, the following proteins have been optionally added to the PURE system to facilitate protein folding: DnaK system (DnaK, DnaJ and GrpE), trigger factor (TF), a chaperonin system (GroEL and GroES), HSP104 (ClpB), SecA, SecB, protein disulfide

Table 22.3 Proteins synthesized using the PURE system. Target proteins were synthesized using the PURE system under standard conditions at 37 °C for 1–2 h.

Proteins	MW (KDa)	Productivity	Solubility	Helpful additives	Reference(s)
2-Phosphoglycerate dehydratase (Enolase)	47	+++	+++		[35, 26, 28]
Glutamate decarboxylase (GAD)	53	++	+	DnaK system	[35]
Galactitol-1-phosphate 5-dehydrogenase (GatD)	38	+	++		[35, 28]
Threonine-3-dehydrogenase (TDH)	35	+	++	DnaK system	[35, 28]
5,10-Methylenetetrahydrofolate reductase (MetF)	33	++	+	DnaK and GroEL systems	[35, 28]
S-Adenosylmethionine synthetase (MetK)	42	++	++	GroEL, GroES	[35, 28]
D-Tagatose-1,6-bisphosphate aldolase (GatY)	31	+++	+	DnaK system	[35, 28]
Dihyodipicolinate synthetase (DapA)	38	++	+	GroEL, GroES	[35, 28, 26]
Triosephosphate isomerase (TIM)	27	++	+++		[35, 28]
Phosphomethylpyrimidine kinase (ThiD)	29	+	++	DnaK system	[35, 28]
Uracil phosphoribosyltransferase (UPRTase)	20	+++	+++		[35]
DNA-directed RNA polymerase α (RpoA)	36	++	+++		[35]
Malate dehydrogenase (MDH)	33	+	+++		[28]
Glucose-6-phosphate dehydrogenase (G6PDH)	56	+++	+++		[28]
Lactate dehydrogenase (LDH)	36	+++	+++		[28]
D-Lactate dehydrogenase (DLD)	65	+++	+++		[28]
Maltose binding protein (double mutant, MBP5)	41	++	+++		[28]
Dihydrofolate reductase (DHFR)	18	++	+++		[27, 28]

Table 22.3 *Continued*

Proteins	MW (KDa)	Productivity	Solubility	Helpful additives	Reference(s)
Rhodanese	33	+	+++	DnaK system	Unpublished
Green fluorescent protein (wild-type)	27	+	n.d.		Unpublished
Green fluorescent protein (mutant, UV)	27	++	n.d.	GroEL, GroES	Unpublished
Luciferase	61	+	n.d.	DnaK system	Unpublished
Ribulose bisphosphate carboxylase (Rubisco)	55	+	n.d.		Unpublished
pOmpA	37	+++	+	SecB, DnaK system	[37]
MtlA	65	++	++		[37]
FtsQ	31	+	++		[37]
HyHEL10 scFv	37	++	++	DnaK and GroEL systems	[21, 28]
anti-BSA scFv	35	++	++	DnaK and GroEL systems	[21, 28]
β-Galactosidase (Gal)	100	+	n.d.		[26]
ClpB (HSP104)	95	+	n.d.		Unpublished
DnaJ (HSP40)	40	++	n.d.		[26]
Chloramphenicol acetyltransferase (CAT)	26	+++	n.d.		[26]
MAP kinase	44	++	n.d.		[26]
Interleukin-8 (IL8)	7	+++	n.d.		[26]
Ras protein	22	+	n.d.		[26]
Lysozyme	14	++	n.d.		[27]
Glutathione S-transferase (GST)	26	+	n.d.		[27]
Bacteriophage T7 gene 10	32	++	n.d.		[27]

+++, ++, + and n.d. refer to high, medium, low and not determined, respectively.

isomerase (PDI), and so on (see Table 22.3). These factors can be supplied either separately or simultaneously to the PURE system to reconstruct a combined translation–maturation system. For instance, as shown in Figure 22.2, the additional DnaK system significantly prevented the newly synthesized HyHEL10 scFv from aggregation. In general, the reaction conditions of the translation–maturation systems are the same as the PURE system standard protocol, with the exception of the additional supplements. The following conditions are commonly used for individual additions: 1–2 μ*M* DnaK, 0.4 μ*M* DnaJ, 0.4 μ*M* GrpE, 0.1 μ*M* GroEL, 0.2 μ*M* GroES, 2 μ*M* TF, 0.1 $\mu g\,\mu l^{-1}$ SecA, 0.3 $\mu g\,\mu l^{-1}$ SecB, 1 μ*M* PDI, 1 μ*M* ClpB, and so on in a 20–50 μl PURE reaction with 1 μ*M* ribosomes. Translation is usually performed at 37 °C for 1–2 h, followed by centrifugation at 14 000 r.p.m. (14 000–15 000 *g*) for 10–20 min at 4 °C to eliminate any possible aggregates and to estimate protein solubility. Methods for detection and calculations have been clearly described elsewhere [21, 28, 35, 37].

When investigating scFv antibodies translated in the presence or absence of chaperones, it was found that a single addition of either the DnaK system or TF greatly increased the soluble amount of two scFv antibodies; whereas the addition of ClpB alone hardly suppressed the aggregation of either of the newly synthesized scFvs [28]. The addition of multiple chaperones – for example TF and the DnaK system – increased the amount of soluble scFv antibody more than either chaperone alone [21, 28]. Of note, the two single-chain antibodies (anti-BSA scFv and HyHEL10 scFv) exhibited different properties with respect to protein folding compared to similar scFv proteins.

We also observed a relatively high solubility for prokaryotic proteins synthesized in the PURE system, and most products showed a high productivity compared to that of the model protein DHFR. More than 20 target proteins [28, 35], particularly

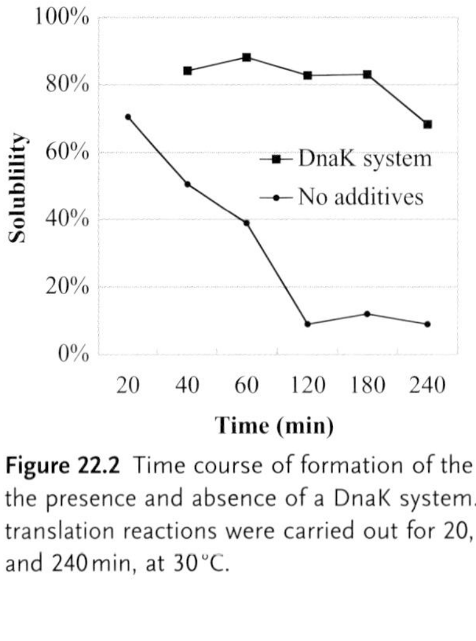

Figure 22.2 Time course of formation of the HyHEL1o scFv in the presence and absence of a DnaK system. The cell-free translation reactions were carried out for 20, 40, 60, 120, 180 and 240 min, at 30 °C.

of prokaryotic origin, were synthesized using both the standard PURE system and the folding-enhanced systems (the PURE system coupled with different folding helpers). The results showed that significantly more soluble proteins could be produced in the presence of folding helpers, and that this resulted in a more efficient process. Although the efficient generation of active protein was improved by the addition of folding helpers, the simultaneous addition of chaperones or enzymes appeared to be unnecessary [28, 30, 35].

We also investigated the production of membrane proteins using the PURE system, as they are considered to constitute difficult targets for protein generation. Instead of the folding-assistant elements, a membrane integration and/or translocation system was combined with the PURE system to produce a soluble integral membrane or presecretory proteins [37, 38]. The method used was similar to that described above, except for the addition of inverted membrane vesicles (INVs) and other related purified factors. In addition to SecA, SecB and the chaperones (DnaK system and TF) mentioned above, other ancillary factors, such as Ffh and FtsY (SRP/SR), were also evaluated [37]. The preparation of these components has been described previously. The working concentrations for these factors were SecA 96 $\mu g\,ml^{-1}$, SecB 300 $\mu g\,ml^{-1}$, Ffh 5 $\mu g\,ml^{-1}$, FtsY 144 $\mu g\,ml^{-1}$, TF 96 $\mu g\,ml^{-1}$, DnaK 90 $\mu g\,ml^{-1}$, DnaJ 35 $\mu g\,ml^{-1}$, GrpE 18 $\mu g\,ml^{-1}$ and INVs 2 $mg\,ml^{-1}$ in a 50 μl reaction carried out under standard PURE conditions. As a result, this secretory-enhanced PURE system is capable of generating soluble membrane proteins. Moreover, the system permitted an efficient membrane translocation of the presecretory protein, pOmpA (outer membrane protein A), and integration of the inner-membrane protein MtlA (mannitol permease) into INVs [37, 38].

As a reference, most of the substrate proteins tested are summarized in Table 22.3. It is hoped that these results will help other investigators to optimize and manage the PURE system easily, according to their particular targets. In this way, the reaction conditions can easily be customized in order to achieve increased productivity and/or activity.

Another convenient and creative application of the PURE system is the production of non-natural peptides. The successful introduction of non-natural amino acids into proteins by suppressing an amber codon using Val-$tRNA^{sup}$ in the PURE system has been reported previously [27, 38]. The latest reports ascertained that, when combined with the PURE system, the *de novo* tRNA acylation system (the flexizyme (Fx) system) is highly efficient for the initiation of translation with designated N-terminal acyl amino acids [43, 44]. This combination of systems results in a highly flexible and efficient method for non-natural polypeptide synthesis.

In conclusion, all of these applications benefit from the purity, flexibility and controllable characteristics of the PURE system. Normally, cell-extract systems demonstrate several-fold higher productivities than the PURE system, and this seems to be a major weak point of the latter approach. However, as productivity is not often equal to efficiency (i.e. functional production), further optimizations currently being investigated by various groups should provide the chance of better performances with the PURE system.

22.3.2
In vitro Selection

In vitro screening methodologies for the applied evolution of protein structure and function have made rapid progress in recent years. Indeed, ribosome display [15, 18, 45, 46] is today becoming one of the most widely used tools for general selection, as well as being a novel approach to drug discovery. By using this technology, the screening of super-high-affinity antibodies and synthetic biocatalysts, as well as selecting for structure and stability, has been the subject of intense research scrutiny [17–19, 47]. Although currently, cell extracts are the most widely utilized *in vitro* translation format for the ribosome display approach, the inherent problems such as intrinsic proteinases and RNases and selection complex (mRNA-ribosome-peptide) instability still need to be resolved. With this in mind, the PURE format has been applied with the aim of establishing a more stable screening system.

The key point of ribosome display (RD) is stabilization of the genotype–phenotype complex composed of the mRNA, ribosome and nascent polypeptide. Although the use of prokaryotic or eukaryotic crude cell extract systems to generate functional polysome complexes has been reported, the use of an entirely purified system – such as PURE – is expected to form ternary complexes much more stably, as it lacks proteinases and RNases.

Several studies have recently reported screening systems using the RD method together with the PURE system [30, 48, 49]. In these studies, the ternary complexes were shown to be efficiently generated from a variety of sequences at the 3' end of RNAs, and even from those with a termination codon. The stable complexes could be maintained even if the temperature was raised to 50 °C. In addition, studies using the so-called pure ribosome display (PRD) showed that the mRNA–ribosome–polypeptide ternary complex was more stable than that obtained in the conventional RD system [30]. Surprisingly, the PRD system was able to condense the target molecule (an scFv cDNA) up to 12 000-fold in just a single round of selection [30], and in so doing overcame the experimental limitations of the conventional RD system. Incidentally, it has been reported that the commercially available PURE system is highly advantageous compared to the conventional S30 extract system when used for RD [48].

The PRD system is specifically optimized for the efficient selection of single-chain antibodies; thus, additional elements (1 m*M* oxidized glutathione, 0.1 m*M* reduced glutathione and 1 μ*M* PDI) are supplied together with the PURE system. The release factors are also omitted from the reaction so as allow translation pausing of target protein at a termination codon. Single-round translation is performed at 37 °C for 10–20 min, depending on the length of the gene. In addition, preincubation (37 °C, 5 min) of the translation mixture (absent of mRNA) is favorable. The translation is subsequently quenched by adding chilled buffer containing 50 m*M* Tris–acetate (pH 7.5), 150 m*M* NaCl, 50 m*M* $Mg(OAc)_2$, 0.5% Tween 20 and 2.5 mg ml^{-1} heparin (sodium). The following selection process is performed under similar conditions with an additional blocking reagent such as bovine serum albumin (BSA) in the buffers.

Another successful application of the PURE system is that of self-binding selection. Some proteins are known to interact with their own mRNAs to regulate their translation by means of product inhibition. For instance, the expression of ribosomal proteins is autogenously regulated at the level of translation. A novel selection method was designed to identify proteins that exhibit self-mRNA binding properties. The *E. coli* ribosomal protein S15, which is known to interact with its own mRNA to regulate its translation, was employed as a model molecule in this screening system. No more than 2% of the wild-type S15 mRNA in the starting mixture was selectively isolated from the mutant mRNA lacking the secondary structure responsible for S15 binding, and was remarkably concentrated through multiple rounds of the selection procedure [29]. Strategic principles and detailed methods have been introduced elsewhere. The direct and functional selection of self-mRNA targeting proteins would provide a platform for the study of protein–nucleic acid interactions in a systematic manner.

22.3.3
Extensive Relevance in Mechanism Studies

In addition to the productive generation of polypeptides, the PURE system is appropriate for the precise investigation of translation-related mechanisms. For example, it has provided precise and conclusive evidence of the effectiveness and function of folding helpers. The PURE system is a completely reconstituted system and contains no intrinsic chaperones [21]; thus, a precise study of the chaperone-mediated folding pathway can be carried out through the simple addition of exogenous factors.

In order to illustrate the correlation between individual chaperone systems and their substrate proteins in cells, candidate substrate proteins were synthesized using the PURE system in the presence or absence of various chaperone systems. The influence of the major chaperone systems, DnaK and GroEL, on the solubility of these newly synthesized proteins was evaluated [21, 28, 35, 37]. The results showed that chaperone addition increased the proportion of soluble protein products, and that each chaperone system was individually responsible for the efficient production of particular substrate proteins.

One new application of the PURE system is the study of the cotranslational folding pathway, which is apparently quite difficult when using other extracts or chemical systems. Cotranslational and/or post-translational folding processes have been studied using the PURE system by simply arresting translation with antibiotics or chemicals. In this way it is possible to determine whether chaperones recognize their substrates during or after translation. By using this strategy, the cotranslational and/or post-translational actions of chaperones on newly synthesized polypeptides were precisely evaluated, and this led to a new understanding of chaperone function. In particular, the flexibility of this reconstituted translation system helped to demonstrate that GroEL could be involved in the folding of its stringent substrate MetK [35, 36]. One technical point here was that the elimination of release factors from the PURE system resulted in an effective preparation of the mRNA–ribosome–polypeptide translation complex, which has been employed to detect the cotranslational association of chaperones.

The secretion pathway is related to the folding process and would also be easy to investigate using the PURE system. In addition to the production of membrane-related proteins, the PURE system provides an improved platform for the biological study of protein translocation and integration mechanisms. Three types of membrane protein have been tested as model substrates for SecA/B-dependent translocation, SRP/SR-dependent membrane integration and SecA/B- and SRP/SR-dependent processes [37]. The coupling of translation and membrane translocation was successfully accomplished and post-translational or cotranslational translocation processes could be easily assessed.

Another feasible study is that of the translation process. The PURE system allows the concentration of each component to be adjusted and the translation reaction to be paused at a particular step by omitting a component that is essential for that step. By using the IF-free PURE system, the process initiated by leaderless mRNA (mRNA without an SD sequence) translation was clarified [32]. The results showed that the translation of leaderless mRNA is initiated by a direct binding of the mRNA to the intact 70S ribosome, and that dissociation of the subunit followed by an association of mRNA with the 30S subunit was not involved [32]. We have also used the RF-free PURE system to study the ribosome termination process in detail. Interestingly, during our investigations on the ribosome recycling pathway using this technique, we learned that spermidine inhibits transient and stable ribosome subunit dissociation [31]. In addition, the facility with which this system allows an evaluation of the involvement of a particular factor in translation has allowed the function of SmpB (small protein B) in the *trans*-translation procedure to be determined [34]. It was found that SmpB, together with tmRNA (transfer messenger RNA), is sufficient to complete the trans-translation process and trigger GTP hydrolysis by compensating for the lack of codon–anticodon interaction during trans-translation initiation [33]. If the PURE system were to be combined with the advanced technology of single-molecule detection, then a more detailed and precise understanding of this field would be possible.

22.4
Prospective Research

Transcription from genetic DNA to mRNAs and the subsequent translation to proteins is a fundamental phenomenon. The PURE system, due to its reconstructing characteristics, provides a powerful tool for both engineering applications and fundamental research.

22.4.1 Modifications and Developments

Efficient protein synthesis was reconstituted from purified natural components several decades ago, and this process can now be improved by using recombinant His-tagged translation factors, as described here. Both systems involve the study of molecular mechanisms of the translation process. Although these approaches predominantly use *E. coli* and are not yet available for eukaryotes, there may be a need for a eukaryotic PURE system in the future. Indeed, a eukaryotic PURE system would be highly advantageous for research involving eukaryotic mechanisms, such as post-translational modification.

As the PURE system (current version) consists of only a minimal set of components for transcription–translation, versatile PURE systems can be set up by simply adding various factors according to the specific purpose(s) of the research. For example, if the mRNA has a very strong secondary structure, then translation could be inhibited and the introduction of an RNA helicase into the system in this situation could be helpful. The production of condensed proteins using the purified system remains a challenge in most instances.

The only element in the PURE system that is not easily controllable is the ribosome, the large size of which in the RNA–protein structure makes it the complex compartment in the purified system. Whilst the careful preparation of purified ribosomes has been explained above, it is difficult to obtain homogenous 70S ribosomes without traces of contaminating ribosome binding factors. A uniformly assembled ribosome may contribute to a more tightly controlled and precisely designed PURE system. Indeed, it has been reported that an *in vitro* self-assembled ribosome was functional. Yet, the use of reconstructed ribosomes in the purified translation system would offer much more specificity and control.

22.4.2 Artificial Cells

The PURE system provides a reconstructive way in which to mimic the most original process of life. Another PURE approach, also referred to as the 'bottom-up strategy', is being developed that allows the construction of artificial cells in a test tube. This artificial system comprises biological reactions and chemical components.

Artificial life was defined as having up to 12 requirements [50], the properties of which can be summarized as self-maintenance, self-reproduction and evolvability [51]. The most fundamental process – protein synthesis – has been extensively studied with the aim of creating a living cell. Protein expression within compartments has been reported and the synthesis of polyPhe and expression of GFP inside lipid vesicles have both been achieved [52, 53]. Recently, a simple genetic network was reported in liposomes, and a self-nutrient expression system within giant vesicles was also successfully described [54, 55]. Taken together, these achievements suggest that minimal living cells may be semi-synthetically reconstituted in the near future.

To date, the success of constructing cell-like compartments has been largely dependent on *in vitro* translation systems that are based on the use of cell extracts. In comparison, the PURE system provides a more controllable design and could greatly enhance the development of a synthetic system of artificial life. Previously, we have verified that the PURE system is capable of generating soluble membrane proteins; therefore, if a biological secretion pathway (e.g. a secretory translocon, or a sec-independent pathway, etc.) were to be combined with the PURE translation system, we would be closer to attaining a semi-synthetic living system. Advanced chemical and physical techniques are also required to accomplish an artificial cell; therefore, further development of reconstructive protein engineering will be necessary to achieve this goal. In order to address the demands of synthetic cell engineering, the next challenge for the PURE system will be, perhaps, to self-produce the PURE system using the PURE system.

22.4.3
Complexity and Network

The PURE system is a cell-free expression system that is comprised of a minimal set of components (~100) that are essential for transcription (from DNA to RNA) and translation (from RNA to protein). Hundreds of chemical and physical reactions occur among these components, the result being a biologically productive network – the protein synthesis circuit. Although this system is more complex than the chemical amino acid condensation reaction, it is much simpler than a living network of cells. It would be interesting to use a constructive approach to tackle the robustness of a living network against perturbations, to explore why disturbances do not cause cells to fail catastrophically, and to investigate how evolution shapes the network. Living network circuits appear to be highly stable, despite containing thousands of nonlinear reactions. For example, the stability theory, which is based on differences between living and nonliving networks, can be exploited as the design principle of robust control in network applications. Furthermore, new concepts and definitions of life will be generated, which may provide further insight into life-like mechanisms.

22.5
Concluding Remarks

The efficient and rapid expression–purification of biologically active proteins is necessary for proteomic research and high-throughput screening strategies. The enhanced productivity and activity of the PURE system illustrates the capabilities of such a reconstituted protein generation system. The absence of unessential components (e.g. chaperones) provides ideal conditions for the identification of additional strictly required components and for elucidating the processes in which these components are involved. Highly controllable translation–folding and translation–secretion systems have been successfully developed for both efficient production and fundamental studies. The systematic synthesis and global analysis of a protein bank using this system is certainly achievable. Furthermore, the PURE system is a powerful tool for the emerging research fields of systems biology and synthetic biology.

References

1 Endo, Y. and Sawasaki, T. (2005) Advances in genome-wide protein expression using the wheat germ cell-free system. *Methods in Molecular Biology*, **310**, 145–67.

2 Endo, Y., Otsuzuki, S., Ito, K. and Miura, K. (1992) Production of an enzymatic active protein using a continuous flow cell-free translation system. *Journal of Biotechnology*, **25**, 221–30.

3 Spirin, A.S., Baranov, V.I., Ryabova, L.A., Ovodov, S.Y. and Alakhov, Y.B. (1988) A continuous cell-free translation system capable of producing polypeptides in high yield. *Science*, **242**, 1162–4.

4 Kigawa, T. and Yokoyama, S. (1991) A continuous cell-free protein synthesis system for coupled transcription-translation. *Journal of Biochemistry (Tokyo)*, **110**, 166–8.

5 Spirin, A.S. (2004) High-throughput cell-free systems for synthesis of functionally active proteins. *Trends in Biotechnology*, **22**, 538–45.

6 Beckler, G.S., Thompson, D. and Van Oosbree, T. (1995) In vitro translation using rabbit reticulocyte lysate. *Methods in Molecular Biology*, **37**, 215–32.

7 Ryabova, L.A., Morozov, I. and Spirin, A.S. (1998) Continuous-flow cell-free translation, transcription-translation, and replication-translation systems. *Methods in Molecular Biology*, **77**, 179–93.

8 Endo, Y., Oka, T., Ogata, K. and Natori, Y. (1993) Production of dihydrofolate reductase by an improved continuous flow cell-free translation system using wheat germ extract. *Tokushima Journal of Experimental Medicine*, **40**, 13–17.

9 Endo, Y. and Sawasaki, T. (2006) Cell-free expression systems for eukaryotic protein production. *Current Opinion in Biotechnology*, **17**, 373–80.

10 Yokoyama, S. (2003) Protein expression systems for structural genomics and proteomics. *Current Opinion in Chemical Biology*, **7**, 39–43.

11 Vinarov, D.A., Loushin Newman, C.L. and Markley, J.L. (2006) Wheat germ cell-free platform for eukaryotic protein production. *The FEBS Journal*, **273**, 4160–9.

12 Hoffmann, M., Nemetz, C., Madin, K. and Buchberger, B. (2004) Rapid translation system: a novel cell-free way from gene to protein. *Biotechnology Annual Review*, **10**, 1–30.

13 Schwarz, D., Klammt, C., Koglin, A., Lohr, F., Schneider, B., Dotsch, V. and Bernhard, F. (2007) Preparative scale cell-free expression systems: new tools for the large scale preparation of integral membrane proteins for functional and structural studies. *Methods*, **41**, 355–69.

14 Klammt, C., Schwarz, D., Lohr, F., Schneider, B., Dotsch, V. and Bernhard, F. (2006) Cell-free expression as an emerging technique for the large scale production of integral membrane protein. *The FEBS Journal*, **273**, 4141–53.

15 Hanes, J. and Pluckthun, A. (1997) In vitro selection and evolution of functional proteins by using ribosome display. *Proceedings of the National Academy of Sciences of the United States of America*, **94**, 4937–42.

16 Lipovsek, D. and Pluckthun, A. (2004) In-vitro protein evolution by ribosome display and mRNA display. *Journal of Immunological Methods*, **290**, 51–67.

17 Irving, R.A., Coia, G., Roberts, A., Nuttall, S.D. and Hudson, P.J. (2001) Ribosome display and affinity maturation: from antibodies to single V-domains and steps

towards cancer therapeutics. *Journal of Immunological Methods*, **248**, 31–45.

18 Jermutus, L., Honegger, A., Schwesinger, F., Hanes, J. and Pluckthun, A. (2001) Tailoring in vitro evolution for protein affinity or stability. *Proceedings of the National Academy of Sciences of the United States of America*, **98**, 75–80.

19 Schaffitzel, C., Hanes, J., Jermutus, L. and Pluckthun, A. (1999) Ribosome display: an in vitro method for selection and evolution of antibodies from libraries. *Journal of Immunological Methods*, **231**, 119–35.

20 Swartz, J.R. (2001) Advances in *Escherichia coli* production of therapeutic proteins. *Current Opinion in Biotechnology*, **12**, 195–201.

21 Ying, B.W., Taguchi, H., Ueda, H. and Ueda, T. (2004) Chaperone-assisted folding of a single-chain antibody in a reconstituted translation system. *Biochemical and Biophysical Research Communications*, **320**, 1359–64.

22 Ellis, R.J. and Hartl, F.U. (1999) Principles of protein folding in the cellular environment. *Current Opinion in Structural Biology*, **9**, 102–10.

23 Kawasaki, T., Gouda, M.D., Sawasaki, T., Takai, K. and Endo, Y. (2003) Efficient synthesis of a disulfide-containing protein through a batch cell-free system from wheat germ. *European Journal of Biochemistry*, **270**, 4780–6.

24 Hoyer, W., Ramm, K. and Pluckthun, A. (2002) A kinetic trap is an intrinsic feature in the folding pathway of single-chain Fv fragments. *Biophysical Chemistry*, **96**, 273–84.

25 Komar, A.A., Kommer, A., Krasheninnikov, I.A. and Spirin, A.S. (1997) Cotranslational folding of globin. *Journal of Biological Chemistry*, **272**, 10646–51.

26 Shimizu, Y., Kanamori, T. and Ueda, T. (2005) Protein synthesis by pure translation systems. *Methods*, **36**, 299–304.

27 Shimizu, Y., Inoue, A., Tomari, Y., Suzuki, T., Yokogawa, T., Nishikawa, K. and Ueda, T. (2001) Cell-free translation reconstituted with purified components. *Nature Biotechnology*, **19**, 751–5.

28 Ying, B.W., Shimizu, Y. and Ueda, T. (2006) The PURE system: a minimal cell-free translation system, in *Cell-Free Expression* (edsT., Kudlicki, F., Katzen, R., Bennett), Landes Bioscience, Austin, pp. 76–83.

29 Ying, B.W., Suzuki, T., Shimizu, Y. and Ueda, T. (2003) A novel screening system for self-mRNA targeting proteins. *Journal of Biochemistry (Tokyo)*, **133**, 485–91.

30 Ohashi, H., Shimizu, Y., Ying, B.W. and Ueda, T. (2007) Efficient protein selection based on ribosome display system with purified components. *Biochemical and Biophysical Research Communications*, **352**, 270–6.

31 Umekage, S. and Ueda, T. (2006) Spermidine inhibits transient and stable ribosome subunit dissociation. *FEBS Letters*, **580**, 1222–6.

32 Udagawa, T., Shimizu, Y. and Ueda, T. (2004) Evidence for the translation initiation of leaderless mRNAs by the intact 70S ribosome without its dissociation into subunits in eubacteria. *Journal of Biological Chemistry*, **279**, 8539–46.

33 Shimizu, Y. and Ueda, T. (2006) SmpB triggers GTP hydrolysis of EF-Tu on ribosome by compensating for the lack of codon-anticodon interaction during trans-translation initiation. *Journal of Biological Chemistry*, **281**, 15987–96.

34 Shimizu, Y. and Ueda, T. (2002) The role of SmpB protein in trans-translation. *FEBS Letters*, **514**, 74–7.

35 Ying, B.W., Taguchi, H., Kondo, M. and Ueda, T. (2005) Co-translational involvement of the chaperonin GroEL in the folding of newly translated polypeptides. *Journal of Biological Chemistry*, **280**, 12035–40.

36 Ying, B.W., Taguchi, H. and Ueda, T. (2006) Co-translational binding of GroEL to nascent polypeptides is followed by post-translational encapsulation by GroES to mediate protein folding. *Journal of Biological Chemistry*, **281**, 21813–19.

37 Kuruma, Y., Nishiyama, K., Shimizu, Y., Muller, M. and Ueda, T. (2005) Development of a minimal cell-free translation system for the synthesis of presecretory and integral membrane proteins. *Biotechnology Progress*, **21**, 1243–51.

38 Shimizu, Y., Kuruma, Y., Ying, B.W., Umekage, S. and Ueda, T. (2006) Cell-free translation systems for protein engineering. *The FEBS Journal*, **273**, 4133–40.

39 Merk, H., Stiege, W., Tsumoto, K., Kumagai, I. and Erdmann, V.A. (1999) Cell-free expression of two single-chain monoclonal antibodies against lysozyme: effect of domain arrangement on the expression. *Journal of Biochemistry (Tokyo)*, **125**, 328–33.

40 Ryabova, L.A., Desplancq, D., Spirin, A.S. and Pluckthun, A. (1997) Functional antibody production using cell-free translation: effects of protein disulfide isomerase and chaperones. *Nature Biotechnology*, **15**, 79–84.

41 Young, J.C., Agashe, V.R., Siegers, K. and Hartl, F.U. (2004) Pathways of chaperone-mediated protein folding in the cytosol. *Nature Reviews Molecular Cell Biology*, **5**, 781–91.

42 Hartl, F.U. and Hayer-Hartl, M. (2002) Molecular chaperones in the cytosol: from nascent chain to folded protein. *Science*, **295**, 1852–8.

43 Murakami, H., Ohta, A., Ashigai, H. and Suga, H. (2006) A highly flexible tRNA acylation method for non-natural polypeptide synthesis. *Nature Methods*, **3**, 357–9.

44 Goto, Y., Ashigai, H., Sako, Y., Murakami, H. and Suga, H. (2006) Translation initiation by using various N-acylaminoacyl tRNAs. *Nucleic Acids Symposium Series (Oxford)*, 293–4.

45 He, M. and Taussig, M.J. (2007) Eukaryotic ribosome display with in situ DNA recovery. *Nature Methods*, **4**, 281–8.

46 Hanes, J., Schaffitzel, C., Knappik, A. and Pluckthun, A. (2000) Picomolar affinity antibodies from a fully synthetic naive library selected and evolved by ribosome display. *Nature Biotechnology*, **18**, 1287–92.

47 Schaffitzel, C., Berger, I., Postberg, J., Hanes, J., Lipps, H.J. and Pluckthun, A. (2001) In vitro generated antibodies specific for telomeric guanine-quadruplex DNA react with *Stylonychia lemnae* macronuclei. *Proceedings of the National Academy of Sciences of the United States of America*, **98**, 8572–277.

48 Villemagne, D., Jackson, R. and Douthwaite, J.A. (2006) Highly efficient ribosome display selection by use of purified components for in vitro translation. *Journal of Immunological Methods*, **313**, 140–8.

49 Matsuura, T., Yanagida, H., Ushioda, J., Urabe, I. and Yomo, T. (2007) Nascent chain, mRNA, and ribosome complexes generated by a pure translation system. *Biochemical and Biophysical Research Communications*, **352**, 372–7.

50 Deamer, D. (2005) A giant step towards artificial life? *Trends in Biotechnology*, **23**, 336–8.

51 Luisi, P.L. (2002) Toward the engineering of minimal living cells. *The Anatomical Record*, **268**, 208–14.

52 Oberholzer, T., Nierhaus, K.H. and Luisi, P.L. (1999) Protein expression in liposomes. *Biochemical and Biophysical Research Communications*, **261**, 238–41.

53 Nomura, S.M., Tsumoto, K., Hamada, T., Akiyoshi, K., Nakatani, Y. and Yoshikawa, K. (2003) Gene expression within cell-sized lipid vesicles. *Chembiochem*, **4**, 1172–5.

54 Noireaux, V. and Libchaber, A. (2004) A vesicle bioreactor as a step toward an artificial cell assembly. *Proceedings of the National Academy of Sciences of the United States of America*, **101**, 17669–74.

55 Noireaux, V., Bar-Ziv, R., Godefroy, J., Salman, H. and Libchaber, A. (2005) Toward an artificial cell based on gene expression in vesicles. *Physical Biology*, **2**, P1–8.

23
Equipping *in vivo* Selection Systems with Tunable Stringency

Martin Neuenschwander, Andreas C. Kleeb, Peter Kast and Donald Hilvert

23.1
Genetic Selection in Directed Evolution Experiments

Directed enzyme evolution exploits random mutagenesis in combination with high-throughput screening (HTS) or selection to tailor the properties of protein catalysts [1–6]. This approach provides insight into protein folding, structure and catalytic mechanism, but it can also be employed for the design of new catalysts with novel activities and selectivities.

Selection methods are particularly effective for directed evolution because they enable the exhaustive analysis of much larger protein libraries than standard screening techniques (10^7–10^{15} versus 10^2–10^6 members, respectively). The classical *in vivo* approach involves genetic complementation of a microbial auxotroph. Functional gene variants are identified by their ability to correct a metabolic defect that otherwise prevents cell growth [3]. However, selection can also be based on the resistance of cells to toxins (such as mutagens, antibiotics or metabolic analogues) or environmental stress (such as temperature variation or UV radiation) [7]. A variety of *in vitro* formats, ranging from phage [8], cell surface [9, 10], ribosome [11, 12] and RNA [13] display to compartmentalization technologies [14], significantly extend the range of functions and environmental conditions that can be explored by selection.

Traditional *in vivo* selection strategies are readily applied to catalysts with natural metabolic counterparts. Construction of a selection host is relatively straightforward in such cases, as the corresponding wild-type gene can simply be disabled. Selection is not restricted to metabolic transformations, however. By linking the outcome of a chemical reaction to the transcription of a reporter gene that complements a host defect, or that provides resistance to an antibiotic, selection can be readily extended to other processes. For example, a chemosensing transcriptional activator that turns on the expression of an antibiotic resistance gene upon binding the product of a biocatalytic reaction has been successfully used to couple signal detection with cell survival [15]. Yeast three-hybrid systems, in which a small molecule is used to reconstitute a functional transcriptional activator by bridging

Protein Engineering Handbook. Edited by Stefan Lutz and Uwe T. Bornscheuer

ISBN: 978-3-527-31850-6

between DNA-binding and activation domains [16], have similarly been adapted to select enzymes that catalyze the synthesis or cleavage of the adaptor ligand [17, 18].

Nevertheless, designing *in vivo* selection systems is nontrivial. Living cells are highly complex, and modification of the host genome often leads to unanticipated phenotypes. In addition, the threshold of activity needed for the survival of an auxotrophic host and the dynamic range of selectable activities are usually difficult to predict. Methods for adjusting selection pressure to the task at hand would consequently be very valuable. They would facilitate the identification of weakly active catalysts, which generally require a low selection hurdle, and they could also be used at higher stringency to distinguish the most active enzymes from less-effective variants. Selection systems with tunable stringency would also benefit directed evolution experiments, as improvements in catalyst performance could be achieved over multiple rounds of mutation and increasingly demanding selection, without switching host strains.

In principle, there are two fundamentally different strategies to adjust selection pressure *in vivo*. Control of selection stringency can be achieved by regulating either the concentration of the catalyst or, alternatively, the concentration of the substrate. Intracellular catalyst concentration can be manipulated by altering promoter strength [19, 20] or plasmid copy number [21] for the catalyst gene, or by targeting the encoded enzyme for degradation [20]. Substrate concentration can be readily varied if the relevant molecule is provided in the growth medium [19], as is the case for detoxification-based systems involving antibiotics [22, 23]; if the substrate is produced inside the cell, metabolic engineering methods can be employed [24]. In this chapter, we discuss these two general approaches for tuning selection stringency *in vivo*, and provide specific examples to show how they can be applied for directed evolution.

23.2
Inducible Promoters for Controlling Selection Stringency

In genetic complementation assays, growth of the host depends directly on the metabolic flux at the selectable target step and, more generally, on the intracellular concentration of specific key metabolites. By regulating the concentration of the enzyme catalysts that act on these metabolites it is possible to control flux and thus growth rate. Intracellular enzyme concentrations depend on a variety of factors, including the rate of gene transcription. Differential rates of transcription have been successfully achieved using sets of constitutive promoters that vary in strength [19, 25]. However, for many applications inducible promoter systems, in which gene expression strength is regulated by the change of an external stimulus, represent a more convenient alternative.

For bacterial hosts, several inducible promoter systems have been described that exhibit little or no gene expression in the absence of inducer and strong overexpression as soon as the inducer molecule is added [26, 27]. Although both very low and very high protein levels can be attained with such systems, these concentration extremes may not be physiologically relevant or useful for genetic selection experiments. In order to sample a continuum of intermediate protein concentrations, intermediate gene expression levels are required, necessitating promoter systems that show a graded dose–response profile. If enzymatic activity is to correlate with growth phenotype, it is also important that each individual cell in a culture experiences the same level of induction. Ensuring discrete and homogeneous protein production levels in all cells in a population is, unfortunately, nontrivial. Furthermore, induction levels should remain stable over time. Even a small decrease in induction due to degradation of the inducer molecule would be disadvantageous, as genetic complementation experiments are typically carried out over periods of several days or even weeks. For these reasons, many well-established promoter systems are ill-suited for regulating genetic selection.

23.2.1
Problems Associated with Commonly Used Inducible Promoter Systems

In principle, one might expect to obtain adjustable levels of gene expression by adding subsaturating concentrations of an inducer to systems under the control of an inducible promoter. Although macroscopic protein yields can be systematically varied in this way, changing the inducer concentration frequently alters the fraction of induced cells, rather than the induction level in individual cells. For instance, Novick and Weiner showed in early studies of the *lac* operon that inducer concentrations below those needed for maximum induction resulted in the full induction of some cells in the culture but no induction of the remainder [28].

This effect is referred to as 'binary response' [29], 'all-or-none', or 'autocatalytic' induction [28]. It arises if the inducer controls the expression of the genes responsible for its own import into the cell. Many commonly used carbohydrate-responsive operons exploit this phenomenon to achieve low-level expression in the absence of substrate, but a rapid, high-level response as soon as the substrate is detected [30]. In the case of the *lac* operon, the lactose metabolite 1,6-allo-lactose induces the expression of the gene for lactose permease, which facilitates lactose uptake. The *ara* operon exhibits autocatalytic behavior due to the inducible L-arabinose transport system [31].

Autocatalytic induction is not the only cause of binary expression. Competition between transactivators and transrepressors for the same operator site on DNA can also convert a graded output into a binary all-or-nothing response [32]. For example, a model promoter system that is controlled by exogenously added doxycycline was found to exhibit homogeneous gene induction in eukaryotic cells in the presence of either a transactivator or transrepressor alone, but a binary response when transrepressor and transactivator were simultaneously present [32]. Differences in the dose-dependent response profiles were also evident. Under conditions that afforded homogeneous induction, the dose-dependent response profiles showed weak cooperativity, with Hill coefficients between 1.6 and 1.8. Under conditions leading to a binary response, the cooperativity of the dose-dependent response profile increased, as judged by Hill coefficients larger than 3.

Based on mathematical simulations, Louis and Becskei proposed that cooperativity itself is sufficient to elicit a binary all-or-nothing pattern, especially at induction levels close to half-maximal expression [29]. This observation is of potential practical relevance. Although relatively few promoter systems have been tested with respect to the homogeneity of gene expression at the single cell level, dose-dependent response curves are generally available and can be used to assess the cooperativity of induction. Strongly cooperative induction behavior should be avoided in selection experiments not only to steer clear of all-or-none effects, but also because small changes in the applied inducer concentration can lead to large changes in promoter activity, decreasing experimental reproducibility.

23.2.2
Engineering Graded Homogeneous Gene Expression

The problems associated with autocatalytic induction and loss of induction over time can frequently be solved by genetically reengineering the host strain. For instance, deletion of the genes for arabinose transport (*araE* and *araFGH*) and arabinose degradation (*araBAD*) from the *E. coli* genome affords an arabinose-dependent promoter system that shows favorable, slightly negative-cooperative induction behavior [30, 33, 34], with tight repression in the absence of inducer and a 1000-fold induction range [35] (Table 23.1). Induction is stably maintainable at low arabinose concentrations due to the greater metabolic stability of the inducer in the modified strain [34]. Nevertheless, the fact that the host must be re-engineered represents a drawback for directed evolution experiments that require specific selection strains or additional chromosomal modifications.

Stable inducer analogues, which diffuse passively across the cell membrane independent of a transport system and are not metabolized by the cell, can also provide homogeneous graded gene expression. A salient example is isopropyl β-D-1-thiogalactopyranoside (IPTG), a synthetic analogue of 1,6-allo-lactose which induces the natural *lac* operon. IPTG has been claimed to diffuse across the cell membrane without the aid of the lactose transporter [37]. In contrast to the natural inducer, IPTG has been reported to induce homogeneous gene expression from the P_{tac} and P_{trc} promoters [37]. Nevertheless, flow cytometry data indicate that the induction range is only fourfold. This small dynamic range, together with the high basal expression level in the absence of IPTG, limits the use of this system as a means of adjusting selection pressure.

Another highly regulatable promoter system, based on propionate, was recently described that may be useful for directed evolution experiments [36], since it was claimed to provide homogeneous gene expression over a 1500-fold range. In this case, adding glucose to the medium helped to lower the basal activity observed in the absence of inducer. The stability of the propionate inducer in this system has not been examined, however, and metabolic conversion may conceivably affect gene expression over prolonged time periods.

Table 23.1 Properties of selected gene expression systems.

Promoter	Inducer compound	Cooperativity (Hill coefficient)[a,b]	Induction range[b]	Analyzed by flow cytometry	Reference(s)
P_{prpB}	Propionate	ND	1500	+	[36]
P_{tac}, P_{trc}	IPTG	ND	4	+	[37]
P_{BAD}	Arabinose	0.3	1000	+	[30, 33]
P_{tetA}	Anhydrotetracycline	ND	ND	–	[38]
$P_{LtetO\text{-}1}$	Anhydrotetracycline	6	5000	–	[39]
P_{tetA}	Tetracycline	1.5 (1.9 with TetM)	40 (1000 with TetM)	+	[40, 41]
P_{tetA}	Tetracycline	1.6 (0.96 with Tn*10*)	220 (370 with Tn*10*)	+	This chapter

a Estimates of the Hill coefficients were derived graphically [42] from the published dose–response profiles, where available.
b *ND* = not determined.

Metabolically inert inducers, such as the antibiotic tetracycline, which diffuse passively across the cellular membrane, have proven particularly useful for achieving stably regulated, homogeneous expression of genes in *E. coli* [40]. Production of the tetracycline resistance determinant of the Tn*10* transposon is controlled by a well-characterized regulatory circuit [43], and expression systems based on Tn*10* regulatory elements generally show very low basal activity [38, 39]. The protein components of this system are the tetracycline resistance protein TetA and the repressor TetR (Figure 23.1a). The corresponding *tetA* and *tetR* genes are transcribed from divergent promoters, both of which are repressed when TetR binds to its DNA operator sites. The tetracycline resistance protein TetA is an antiporter, which couples tetracycline efflux to proton import. As too much TetA causes the membrane potential to collapse [43], and excess TetR leads to inhibition of cell growth [48], tightly controlled expression of both genes is essential for Tn*10*-containing cells.

Both Skerra [38] and Lutz and Bujard [39] have described promoter systems based on Tn*10* regulatory elements that utilize anhydrotetracycline as the inducer. This compound binds TetR 35 times more strongly than tetracycline, and its antibiotic activity is 100-fold lower [39]. Full induction of the promoter requires about 50–200 $ng\,ml^{-1}$ anhydrotetracycline, which is well below the threshold for cytotoxicity. In the system reported by Skerra, TetR is overproduced from the target gene expression plasmid using a weak, constitutive promoter. Unfortunately, neither quantitative dose-dependent response profiles nor flow cytometry data

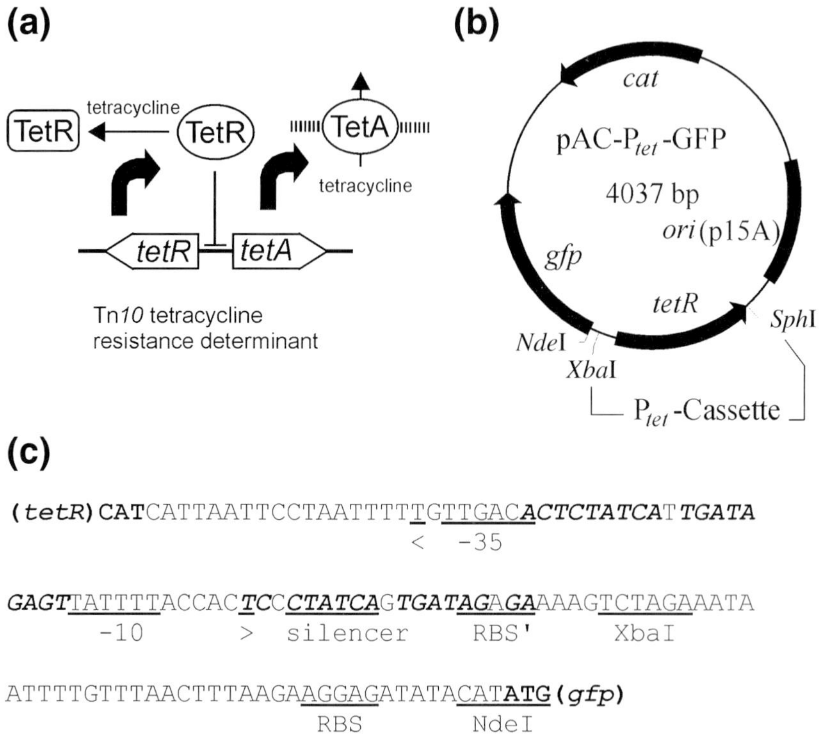

Figure 23.1 Tn*10* regulatory elements. (a) The tetracycline resistance determinant of Tn*10* can be located either in the chromosome or on an F′-episome. Release of the TetR repressor from DNA by binding to tetracycline leads to transcription of the *tetR* and *tetA* genes. The tetracycline resistance protein TetA is an antiporter in the cytoplasmic membrane that exports tetracycline to the surrounding medium; (b) pAC-P$_{tet}$-GFP is a moderate-copy number plasmid based on pACYC184 (p15A derivative) [44]; it encodes the *cat* gene for antibiotic selection with chloramphenicol. The P$_{tet}$ cassette, which consists of the *tetR* coding sequence (in which the *Nde*I and *Xba*I restriction sites were removed by introduction of silent mutations) and the adjacent regulatory region, controls expression of the target gene *gfp* as well as of *tetR*; (c) Detailed sequence of the P$_{tet}$ control region on pAC-P$_{tet}$-GFP. Highlighted in bold type are the TetR operator binding sites (italics) and the start codons of the *tetR* and *gfp* genes (roman). Underlined are the transcription start of the *tetR* gene (<), the −35 and −10 regions and the transcription start of the *tetA* promoter (>) [45], the silencer and the original, relatively inefficient ribosomal binding site of the *tetA* gene (RBS′) [46], the new efficient ribosomal binding site placed upstream of the target gene *gfp* (RBS) [47], and the restriction sites *Xba*I and *Nde*I.

were reported. In the system described by Lutz and Bujard, *tetR* is located on the host chromosome and controlled by a strong promoter, while target gene expression is driven by a synthetic promoter with engineered TetR operator sites. Dose–response profiles reveal strong cooperativity (with a Hill coefficient close to 6; Table 23.1), which was attributed to cooperative binding of anhydrotetracycline to TetR. The system has a very large (5000-fold) induction range and shows very tight repression in the absence of inducer. However, given its cooperativity, this system is unlikely to afford homogeneous gene expression at intermediate expression levels.

Bahl *et al.* [40] have exploited the autoregulatory production of TetR to construct plasmids that function as effective biosensors for tetracycline in environmental samples. The gene for TetR and its adjacent control region were placed upstream of a reporter gene. In this configuration, TetR regulates its own synthesis as well as that of the reporter protein. Autoregulatory expression of *tetR* ensures that all operator sites on the expression plasmids can be occupied, while maintaining TetR at moderate levels. Cytotoxicity is thus prevented and the system responds to tetracycline in a concentration-dependent fashion. Flow cytometry data confirm homogeneous gene expression, and the dose–response profile shows much lower cooperativity than is observed in the system described by Lutz and Bujard [39]. Introduction of the TetM gene, which further reduces cytotoxicity by blocking access of tetracycline to the ribosome, allows the use of higher inducer concentrations, effectively increasing the induction range of the system to 1000-fold [40].

23.2.3
An Optimized Tetracycline-Based Promoter System for Directed Evolution

The tetracycline biosensor system has been adapted by Neuenschwander *et al.* [20] for *in vivo* selection experiments. Specifically, a selection vector containing a P_{tet} cassette, which contains the *tetR* gene and the adjacent promoter sites, was constructed (Figure 23.1b). This cassette can be easily cloned into other vectors as a single *Sph*I-*Xba*I or *Sph*I-*Nde*I DNA fragment. An efficient ribosomal binding site was also introduced immediately upstream of the target gene (RBS in Figure 23.1c) to enable high-level gene expression, if desired. Protein overproduction directly from the selection plasmid facilitates the *in vitro* characterization of catalysts selected *in vivo*, saving an additional cloning step. Direct comparisons show that the P_{tet} cassette, when fully induced, affords similar yields of green fluorescent protein (GFP) as the strong standard T7 promoter (M. Neuenschwander, unpublished results). This system eliminates the need for tandem promoters which have been exploited for the same purpose [49].

GFP production was used to evaluate the P_{tet} promoter system in tetracycline-sensitive DH5α cells and tetracycline-resistant XL1-Blue cells. Flow cytometric analysis (Figure 23.2a and b) showed that induction was homogeneous at all inducer concentrations tested (M. Neuenschwander, unpublished results). Moreover, the presence of Tn*10* on a single-copy F′-episome in XL1-Blue cells increased the highest tolerated (i.e. nontoxic) tetracycline concentration to 5000 ng ml^{-1}, compared to 100 ng ml^{-1} for DH5α cells. The induction range, which was estimated from the global mean fluorescence recorded in the flow cytometry experiments (Figure 23.2c), spans two to three orders of magnitude, and is slightly increased in the presence of Tn*10*. The dose–response profile in the absence of the tetracycline resistance determinant shows relatively little cooperativity (Hill coefficient of 1.6; Table 23.1), similar to that described for the tetracycline biosensors [40],

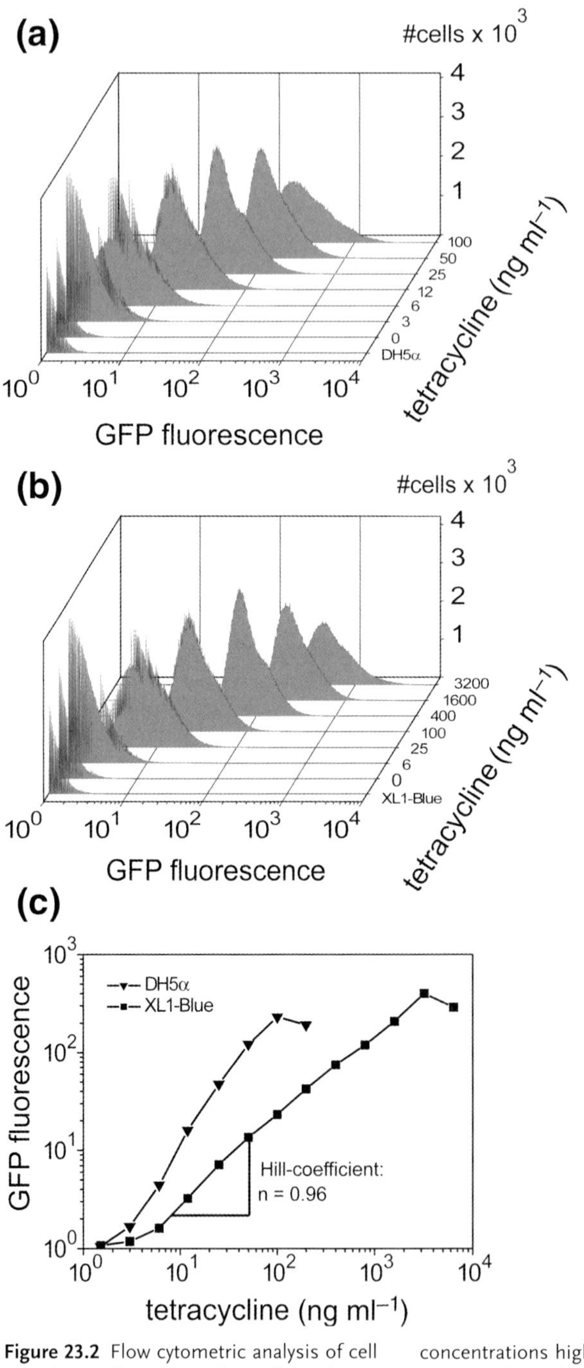

Figure 23.2 Flow cytometric analysis of cell samples producing GFP from plasmid pAC-P_{tet}-GFP induced with different concentrations of tetracycline. (a) Distribution of cells of the nontetracycline-resistant strain DH5α. Reduced cell growth was observed at tetracycline concentrations higher than 100 ng ml^{-1}; (b) Distribution of cells of the tetracycline-resistant strain XL1-Blue, which contains Tn*10* on the F′-episome. Reduced cell growth was observed at tetracycline concentrations higher than 5000 ng ml^{-1}; (c) Dose-dependent response profiles. The GFP fluorescence values correspond to the global mean fluorescence values of the histograms shown in (a) and (b). The Hill coefficient, which indicates the degree of cooperativity, corresponds to the slope of the response curve at 50% saturation [42]. The steepest part of the response curve, indicated by the triangle, was used to approximate the maximum cooperativity value.

whereas in the presence of Tn*10* induction was fully noncooperative (Hill coefficient of 0.96; Table 23.1). As outlined in the next two sections, the P_{tet} promoter system has been successfully employed in genetic selection experiments, where it provides fine-tuned control over selection stringency.

23.3 Controlling Catalyst Concentration

The ability of a protein to replace a missing enzyme in a bacterial auxotroph depends on its concentration as well as its specific activity. In an ideal selection system, it should be possible to match the intracellular concentration of the growth-limiting catalyst with the level of activity needed to overcome the selection hurdle. Thus, weak catalysts should confer a growth advantage to the host strain only when produced at very high levels, whereas highly active catalysts would complement the genetic defect even at very low concentrations. A robust method for regulating protein production would also facilitate the evolutionary improvement of mediocre catalysts, as selection pressure could be steadily increased after each round of mutagenesis simply by reducing the catalyst concentration to a level that only allows survival of cells harboring the most efficient variants.

23.3.1 Reducing Catalyst Concentration by Switching to Weaker Promoters

The rate of protein production depends on promoter strength, gene dose and the efficiency of the ribosomal binding site. By manipulating these factors, it is possible to control catalyst concentration within a cell. The conversion of an aspartate aminotransferase (AspAT) into an efficient valine aminotransferase illustrates the efficacy of such an approach [19]. Variants of the starting enzyme were selected for their ability to complement an auxotrophic *E. coli* strain lacking the gene for the branched-chain amino acid aminotransferase, which is required for the synthesis of valine, leucine and isoleucine. Selection pressure was increased during the optimization process by sequential cloning of the library of AspAT variants into a series of vectors containing constitutive promoters of steadily decreasing strength. Although this strategy was ultimately successful in generating an enzyme with the desired specificity, it was experimentally tedious. Not only are the many recloning steps time consuming and laborious, but the effect of switching promoters on selection stringency is usually only qualitatively predictable. Moreover, a loss of library diversity at each step is a potential problem.

In another study, a dimeric helical bundle chorismate mutase (EcCM) was successfully converted into a weakly active hexamer (hEcCM) [50]. This variant was identified by its ability to complement a chorismate mutase-deficient *E. coli* strain [51]. Because overproduction of chorismate mutases from the strong, fully induced *trc* promoter unit resulted in protein levels that were to high [51, 52], allowing even mediocre catalysts to fully complement the host strain, efforts to evolve more active versions of hEcCM exploited the weaker *bla* promoter [53] (Figure 23.3a). The k_{cat}/K_m parameter of the best variant, tEcCM, which contained three mutations and possessed a trimeric quaternary structure, was improved some 400-fold. Nevertheless, its k_{cat} value was still 14-fold lower than that of the parent EcCM dimer. Unfortunately, attempts to optimize this catalyst by further selection were unsuccessful, because tEcCM conferred wild-type levels of growth to the host. The *bla*

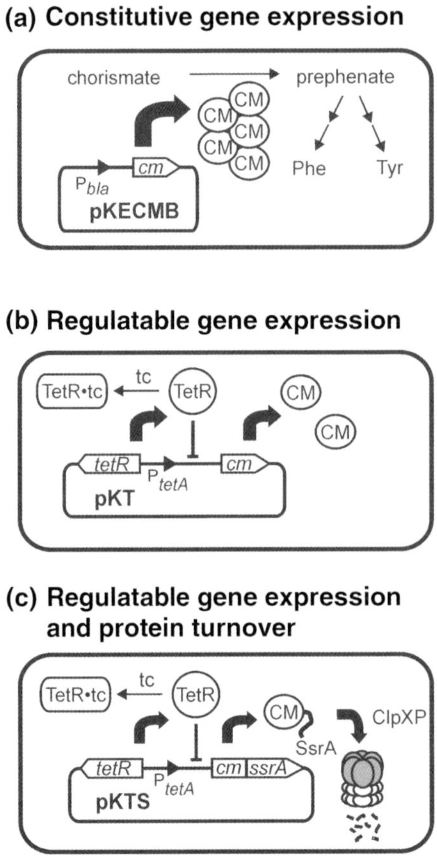

Figure 23.3 Selection strategies for controlling intracellular catalyst concentrations [20]. (a) The catalyst gene (*cm*) on plasmid pKECMB [52] is constitutively expressed from the relatively weak *bla* promoter. If the resulting enzyme (CM) complements the CM deficiency of the *E. coli* host strain, the transformed cells will grow on minimal medium [51]; (b) The selection plasmid pKT provides graded and homogeneous transcriptional control of catalyst production. The *tetR* gene and its promoter region are located upstream of *cm*, so that the TetR repressor simultaneously regulates transcription of catalyst and TetR repressor genes in response to added tetracycline (tc) [43]; (c) The selection plasmid pKTS permits graded transcriptional control and lowers enzyme half-life. The *ssrA* sequence is incorporated as a downstream genetic fusion to the catalyst gene. The resulting enzyme carries the SsrA degradation tag at its C terminus, and is directed to the ClpXP protease, where it is degraded.

promoter evidently produced sufficient quantities of the suboptimal enzyme to fully satisfy the metabolic needs of the cell. This problem, which is typical for conventional genetic selection systems, underscores the need for alternative gene expression strategies.

23.3.2
Reducing Catalyst Concentration through Graded Transcriptional Control

Regulatable promoter systems that achieve homogeneous gene expression represent a potentially attractive means of controlling catalyst concentration *in vivo*. To test this idea, the genes for the chorismate mutase variants described in the previous section were expressed from an inducible P_{tet} promoter cassette, rather than from the constitutive *bla* promoter (Figure 23.3b) [20]. In the absence of tetracycline, the weakly active hEcCM enzyme did not complement the chorismate mutase-deficient host strain under selective conditions, but prototrophy was restored as the inducer concentration was raised. However, cells containing the more active tEcCM and EcCM variants grew even in the absence of inducer. Transcriptional control thus provides adequate discrimination between good and poor catalysts in this system, but it fails to distinguish the most active enzymes from their somewhat less active counterparts.

Although gene expression systems based on the TetR repressor reportedly have a very low background in the absence of inducer [38, 39], the highly active chorismate mutases are apparently produced abundantly enough under the most stringent selection conditions to complement the chorismate mutase deficiency. In the case of wild-type EcCM, growth was even observed when the entire P_{tet} promoter cassette was replaced with the *trpA* transcriptional termination signal [54] (M. Neuenschwander, unpublished results). Because the growth of cells containing the genes for these active enzymes could not be suppressed by blocking transcription originating from all upstream promoters, a new strategy was developed to reduce intracellular catalyst concentration. It couples graded transcriptional control of protein production with targeted protein degradation [20].

23.3.3
Combining Graded Transcriptional Control and Protein Degradation

The eleven-amino acid SsrA sequence is a C-terminal degradation signal that targets proteins to the intracellular ClpXP protease, where they are rapidly degraded [55]. This peptide has been successfully used as a tag to decrease the intracellular lifetime of fluorescent reporter proteins [56, 57]. When fused to the C terminus of chorismate mutases produced with the P_{tet} system (Figure 23.3c), it also effectively reduces the intracellular enzyme concentration [20]. In contrast to the results obtained under selective conditions with the untagged proteins, no cell growth was observed for EcCM and tEcCM bearing the SsrA sequence in the absence of tetracycline. Moreover, addition of the inducer to these constructs led to dose-dependent growth that correlated with the specific activity of the enzyme that was produced. Fastest growth was observed with wild-type EcCM, and the slowest with hEcCM. At intermediate tetracycline concentrations, it was even possible to distinguish the mediocre tEcCM from EcCM. The possibility of identifying the optimal induction level for discriminating among catalysts with similar activities through experiment represents a significant advantage of this system.

This tunable transcription/degradation system was subsequently tested in a directed evolution experiment with hEcCM [20]. A library of randomized hEcCM genes cloned into the pKTS vector (Figure 23.3c) served as the starting point for selection. As expected, the percentage of clones that complement the chorismate-mutase deficiency of the selection strain decreased with decreasing inducer concentration, that is, with increasing stringency (Figure 23.4a). DNA sequence analysis of single colonies revealed a strong correlation between complementation frequency and the occurrence of specific mutations. At high complementation frequencies only a weak consensus was evident, but at intermediate complementation frequencies a significant bias for specific mutations was observed at two positions (Figure 23.4b). At low complementation frequencies, the consensus mutations disappeared, and only 'false positives' were selected. The latter had lost the degradation sequence tag, enabling their escape from the ClpXP degradation machinery. In this system, the optimal complementation frequency can be readily determined simply by varying the concentration of the inducer molecule tetracycline. By adjusting selection pressure in this way, the frequency of productive variants can be maximized while simultaneously minimizing the appearance of artifacts.

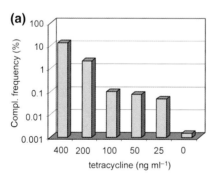

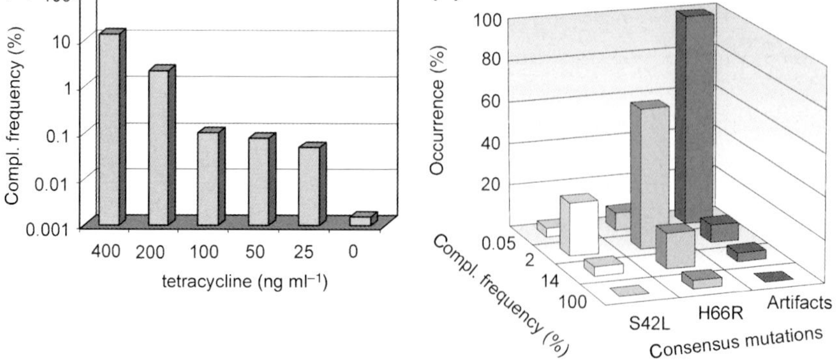

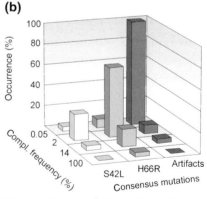

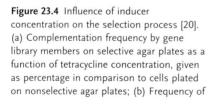

Figure 23.4 Influence of inducer concentration on the selection process [20]. (a) Complementation frequency by gene library members on selective agar plates as a function of tetracycline concentration, given as percentage in comparison to cells plated on nonselective agar plates; (b) Frequency of DNA mutations in individual selected clones as a function of the complementation frequency. The occurrence of false positives lacking the degradation tag and of the frequent consensus mutations His66Arg and Ser42Leu is plotted for different selection stringencies.

The hEcCM variants selected at the optimal complementation frequency were produced with a hexahistidine tag in place of the SsrA sequence, affinity-purified on a nickel resin, and tested for their ability to catalyze the rearrangement of chorismate to prephenate [20]. Significantly, all were catalytically superior to the best previously selected mutants, exhibiting V_{max} values similar to that of wild-type EcCM. The steady-state parameters for the best variant, which contained both consensus mutations and an additional mutation, were $k_{cat} = 12\,s^{-1}$ and $K_m = 270\,\mu M$. These values are comparable to those obtained with EcCM ($k_{cat} = 14\,s^{-1}$ and $K_m = 350\,\mu M$) and represent a 75-fold improvement in k_{cat} over hEcCM and a 10-fold improvement relative to tEcCM.

Apparently, by pairing a tunable promoter with a degradation tag, substantially more active enzymes can be evolved than was previously possible using a weak constitutive promoter. When the gene library in this system is subjected to selective conditions, the dynamic induction range translates into a range of observed complementation frequencies. While the exact correlation between the two will differ for each selection experiment (as it depends on the frequency of active catalysts in the library and on the metabolic needs of the auxotrophic strain), the general trend of decreasing complementation frequency with decreasing induction level should always hold. The graded transcription/degradation strategy described here provides a simple and effective means of setting the complementation frequency, the true gauge of selection stringency. By optimizing this parameter, it is possible to increase the probability of finding improved catalysts among the many artifacts and otherwise uninteresting variants that arise if the selection pressure is either too high or too low.

23.3.4 General Considerations

Although genetic selection can greatly facilitate the search for rare catalysts in large protein libraries, relatively inefficient enzymes produced by standard expression systems are often sufficient to achieve wild-type growth, precluding the evolution of more active agents. Even if a poor catalyst is produced at very low rates, significant amounts can accumulate over time–provided that the protein is stable–leading eventually to concentrations sufficient to complement the metabolic defect. As a consequence, attempts to control catalyst concentration solely via tight transcriptional control may prove ineffective, as seen with the mediocre chorismate mutase variant tEcCM described above. In contrast, appropriate steady-state protein concentrations can be achieved by balancing protein production with degradation [58]. The specific combination of graded gene expression with an encoded degradation tag provides finely tuned control over protein levels over a broad dynamic range, from very high to extremely low concentrations, and hence excellent control over selection pressure. As both genetic elements are located on a single plasmid and do not require dedicated host strains, this system should be applicable in a wide variety of selection formats.

23.4 Controlling Substrate Concentrations

Selection pressure does not depend solely on catalyst concentration. It is also influenced by the concentration of critical metabolites within the cell. In laboratory evolution experiments, the latter parameter can be manipulated either by controlling the supply of relevant compounds in the growth medium, or by modifying genetic and regulatory mechanisms or metabolic pathways within the cell to steer the production (or destruction) of a key substance.

Operationally, it is relatively easy to adjust selection stringency in evolutionary experiments involving an exogenously added substance that is critical for growth. For example, if cell survival depends on the production of an enzyme capable of destroying a toxic antibiotic, increasing the antibiotic concentration in the environment will favor those organisms harboring the most efficient degradative enzymes, assuming a constant catalyst concentration in all cells in the population. The activity of TEM-1 β-lactamase was dramatically improved in this way [22, 59]. Multiple cycles of DNA shuffling followed by selection on increasing concentrations of cefotaxime resulted in a 32 000-fold increase in the minimum inhibitory concentration (MIC) of this antibiotic. Providing limiting amounts of a biosynthetic precursor to an essential metabolite can similarly be exploited to power the evolution of enzymes. This strategy was successfully exploited for the conversion of AspAT into ValAT [19], mentioned above. In the initial selection round, 2-oxovaline, a precursor of valine and the substrate for ValAT, was added to the growth medium to reduce selection stringency and increase the probability of finding AspAT variants with weak ValAT activity. Once a viable catalyst emerged, however, the exogenously added metabolite could be excluded, forcing the cells to evolve more efficient enzyme variants capable of exploiting the limited pool of endogenous 2-oxovaline.

Unfortunately, in many instances the key substrate for a selection experiment cannot be added to the growth medium, either because it lacks sufficient solubility or stability, or because it is not taken up by the host strain, or simply because it is unavailable. In such cases, other strategies are needed. If the compound in question is a natural metabolite and can be produced biosynthetically, modern metabolic engineering techniques [60, 61] can be exploited to control its production. Such methods have been effectively utilized to reroute flux down alternative branches of a metabolic pathway, alterning the concentration of compounds downstream of a metabolic branch point [62]. The optimization of microbial strains for the production of aromatic amino acids from glucose illustrates the power of this approach [63]. As outlined in the next section, metabolic engineering can also be effectively used to tune the stringency of genetic selection systems.

23.4.1
Engineering a Tunable Selection System Controlled by Substrate Concentration

The shikimate pathway links carbohydrate metabolism to the biosynthesis of aromatic compounds [64]. With its multiple branch points, this pathway provides an excellent testing ground for exploring the utility of metabolic engineering for the construction of sophisticated selection systems. The creation of a bacterial selection strain for prephenate dehydratase (PDT) with adjustable stringency shows the basic strategy (Figure 23.5).

PDT enzymes catalyze the transformation of prephenate to phenylpyruvate, the penultimate step in the production of phenylalanine. In *E. coli*, PDT is fused to a

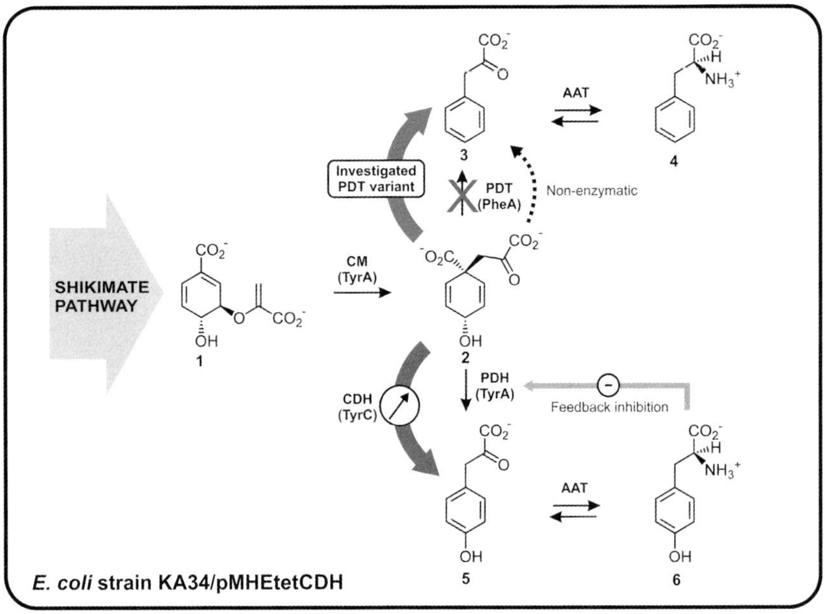

Figure 23.5 The PDT selection system for tunable stringency [24]. The chorismate mutase-prephenate dehydratase (PheA) is absent in the *E. coli* KA34 host strain. As a consequence, prephenate (**2**), generated from chorismate (**1**) by chorismate mutase (TyrA), is enzymatically converted only to 4-hydroxyphenylpyruvate (**5**) and L-tyrosine (**6**) by the successive action of PDH (prephenate dehydrogenase; TyrA) and AAT (aromatic amino acid aminotransferase). However, due to feedback inhibition of TyrA by L-tyrosine, prephenate ultimately accumulates within the cell. Spontaneous, nonenzymatic decomposition of prephenate to phenylpyruvate (**3**), which is converted in turn to L-phenylalanine (**4**) by AAT, enables KA34 to grow slowly under selective conditions. Introduction of a tyrosine feedback-resistant cyclohexadienyl dehydrogenase (CDH; TyrC) from *Zymomonas mobilis*, which acts on prephenate and reduces its intracellular concentration, eliminates this background growth, allowing proteins with PDT activity to be identified by complementation. Furthermore, by regulating the production of CDH, the stringency of the PDT selection can be systematically adjusted.

chorismate mutase. Deletion of the *pheA* gene encoding this bifunctional enzyme from the bacterial genome was accomplished by a homologous recombination technique [65] to give the phenylalanine auxotroph KA34 [24] (Figure 23.6). This strain can still produce prephenate because it possesses a second chorismate mutase, encoded by the *tyrA* gene, but it lacks the enzyme needed to convert prephenate to phenylpyruvate. As a consequence, growth on medium lacking phenylalanine is severely compromised. Complementation of the genetic defect with a functional PDT gene restores growth, as expected.

Although KA34 is suitable for identifying highly active PDTs, the isolation of very weak catalysts – and their optimization by directed evolution – is complicated by relatively high background growth under selective conditions [24]. Thus, even in the absence of exogenously added phenylalanine, the auxotroph is able to grow slowly when plated at high density (>10^7 cells per agar plate), at low pH (<6.0), or

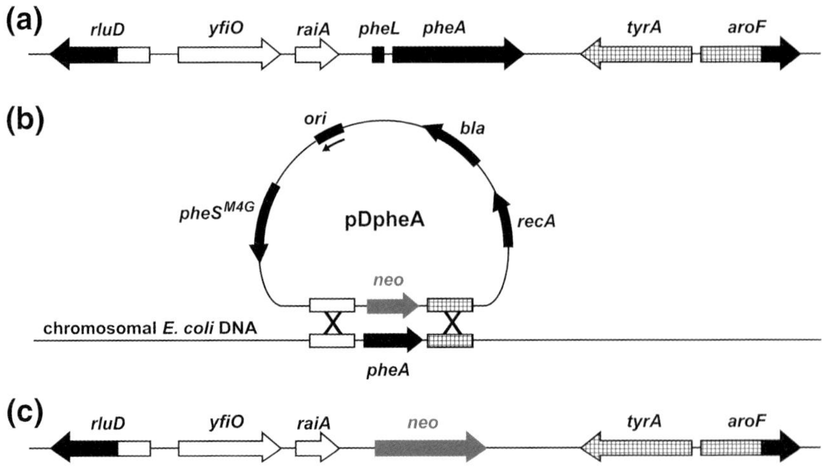

Figure 23.6 Chromosomal deletion of the chorismate mutase-prephenate dehydratase gene (*pheA*). (a) Chromosomal location of the *pheA* gene in KA10. The white and grid-patterned loci indicate the DNA sequences that were used for homologous recombination. Genes in the chromosomal region proximal to *pheA* include: *rluD, yfiO, raiA, pheL, tyrA*, and *aroF*, which encode the 23S rRNA pseudouridine synthase, a predicted lipoprotein, a putative sigma 54 modulator, the *pheA* leader peptide, chorismate mutase-prephenate dehydrogenase, and DAHP synthase, respectively; (b) Double homologous recombination between plasmid pDpheA [24] and the chromosomal DNA of the *recA*-deficient *E. coli* strain KA10 was performed using the strategy of Gamper and Kast [65] to give strain KA34. The *recA* gene encodes recombinase A, which catalyzes ATP-dependent strand exchange during homologous recombination. Its presence on pDpheA is essential for this chromosomal gene-targeting strategy in RecA-deficient strains [65]. *bla* is the ampicillin resistance gene, *ori* the origin of replication, and *pheS*^{M4G} encodes the phenylalanyl-tRNA synthetase α subunit variant used for *p*-chlorophenylalanine counter-selection; (c) Chromosomal map of KA34 in which *pheA* has been replaced by *neo*, a gene specifying kanamycin resistance.

in the presence of tyrosine (>30 μg ml^{-1}). Overproduction of an intracellular chorismate mutase also results in 'leaky' growth. The relatively facile formation of phenylpyruvate from prephenate in solution ($\tau_{1/2}$ = 26 h at pH 8 and 37 °C [66]) provides a likely explanation for these observations.

Deletion of the *pheA* gene would be expected to cause prephenate to accumulate within the cell [67, 68]. Nonenzymatic conversion of some of this material to phenylpyruvate, followed by reaction to phenylalanine, would allow the host to grow slowly. High intracellular chorismate mutase concentration should significantly increase prephenate and hence phenylalanine levels. If prephenate were to leak into the periplasmic or extracellular space, a drop in the pH of the medium would accelerate its decomposition [69] which, in turn, would boost cell growth due to rapid uptake of phenylpyruvate (P. Kast, unpublished results). In principle, the TyrA prephenate dehydrogenase (PDH), which converts prephenate to 4-hydroxyphenylpyruvate and normally competes with PDT for this branchpoint metabolite, might have been expected to siphon off additional prephenate, essentially preventing its accumulation. However, TyrA is feedback-inhibited by tyrosine [70, 71], the end product of this branch of the pathway, ultimately blocking this possibility; this feedback inhibition also explains why tyrosine-supplemented medium leads to enhanced background growth.

If background growth is due to a build-up of prephenate, reduction of intracellular prephenate concentration should help to circumvent this problem. A cyclohexadienyl dehydrogenase (CDH) from *Zymomonas mobilis* [72] was therefore added to KA34 to convert prephenate irreversibly to 4-hydroxyphenylpyruvate rather than to phenylpyruvate (Figure 23.5). Importantly, and in contrast to the endogenous prephenate dehydrogenase (TyrA), CDH is not feedback-inhibited by tyrosine. Consequently, the modified selection strain should accumulate tyrosine rather than prephenate. In order to be able to regulate the relative carbon flux towards tyrosine as opposed to phenylalanine, the CDH-encoding *tyrC* gene was placed under the control of the tetracycline-dependent P_{tet} system described in Section 23.2 [24]. In the absence of tetracycline, the intracellular CDH concentration is extremely low, and the growth phenotype of the *tyrC* transformant is similar to that of the parent KA34 strain under selective conditions. However, when inducer is added to the medium, CDH is produced and leakiness is substantially diminished, consistent with consumption of excess prephenate. Above a tetracycline concentration of 0.4 μg ml^{-1}, background growth was suppressed completely [24].

Because CDH and PDT utilize a common resource – namely prephenate – increasing CDH levels demands a higher catalytic activity from any competing dehydratase provided to the selection strain. Thus, the use of a regulatable metabolic shunt to control intracellular prephenate levels provides a simple mechanism for adjusting PDT selection stringency. The dynamic range of this system was calibrated with a series of catalysts covering a broad range of activities. Cells were transformed with plasmids encoding different PDTs and grown on selection plates lacking phenylalanine but containing different concentrations of tetracycline. At low tetracycline concentrations (<0.5 μg ml^{-1}) it was possible to distinguish dehydratases with k_{cat}/K_m values as low as $24\,M^{-1}s^{-1}$ from background. These variants complemented the genetic defect less well as the stringency of selection was increased by adding higher concentrations of the inducer, and were easily differentiated from more active PDTs. Under the most stringent conditions tested ($5\,\mu g\,ml^{-1}$ tetracycline) it was even possible to discriminate between wild-type PDTs ($k_{cat}/K_m = 10^4$–$10^6\,M^{-1}s^{-1}$) that differ only by a factor of approximately 50 in specific activity [24]. These results show that the stringency of the PDT selection system can be varied over a >50 000-fold range simply by adjusting the relative metabolic flux into different branches of the shikimate biosynthetic pathway through the action of an exogenous inducer molecule.

23.4.2
Applications

The utility of the regulatable PDT selection system for directed evolution has been demonstrated in model selection experiments. For example, plasmids encoding weakly active PDTs were successfully recovered from mixtures containing an excess of a control plasmid lacking PDT. The selection system has also been used to investigate potential roles for conserved residues in the PDT family [24]. For instance, using the monofunctional and thermostable PDT from *Methanocaldococcus jannaschii* (MjPDT) [66] as a template, the tripeptide sequence Thr172-Arg173-Phe174, which is found in >95% of all known PDTs, was targeted for randomizing mutagenesis and selection. Only wild-type threonine and phenylalanine residues were selected at positions 172 and 174, consistent with their location in the putative active site pocket. In contrast, Arg173, which is the only conserved cation in the entire protein and hence a possible binding partner for the dianionic substrate, was found to be surprisingly tolerant to substitution. However, subsequent crystallographic studies showed that this residue is located on the protein surface, excluding a role in substrate recognition or catalysis.

The ability to adjust selection pressure can be useful for identifying positions in a protein that may be mutationally tolerant with respect to weak activity (i.e. they permit growth under the least stringent conditions), but are important to achieve high catalytic efficiency (i.e. they confer a survival advantage under the most stringent conditions). In a proof-of-principle experiment, three MjPDT positions at which aromatic residues are >85% conserved in all known PDT sequences were randomized. The library was subjected to selection at three different stringencies. As the selection pressure for PDT activity increased, aromatic residues were strongly favored at position 14 of viable variants, but not at positions 7 and 114 (Figure 23.7). This bias presumably reflects the fact that residue 14 is located in the active site, where it may play a structural or mechanistic role.

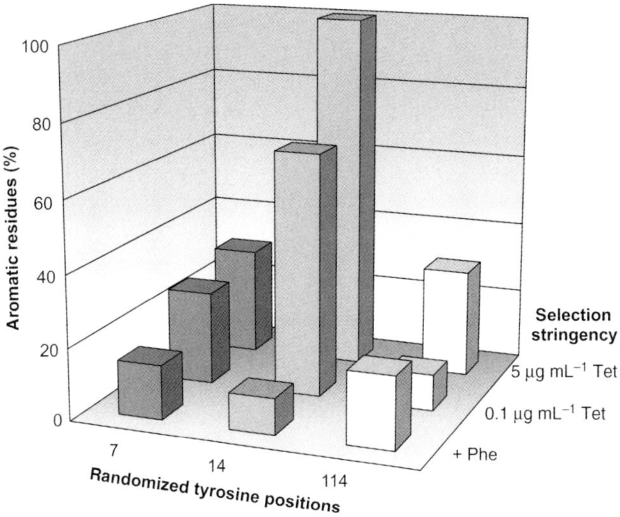

Figure 23.7 Selection results from a library with three randomized tyrosines in *Methanocaldococcus jannaschii* PDT [24]. The graph shows the frequency of aromatic amino acids (including histidine) at the three randomized positions in clones selected at different stringencies. Under nonselective conditions (i.e. in the presence of phenylalanine), no significant preference for aromatic amino acids is seen at any of the positions. In contrast, under selective conditions, aromatic residues are enriched at position 14 as the selection pressure increases due to lowered intracellular prephenate concentrations as a result of elevated concentrations of tetracycline added to the growth medium.

23.4.3
Advantages of Metabolic Engineering Approaches

The deletion of a key enzyme in a metabolic pathway can severely perturb carbon flux within a cell, leading to an accumulation of biosynthetic intermediates upstream of the defect [62]. As seen in the PDT selection system, elevated concentrations of the substrate for the missing enzyme can pose a problem if they increase the nonenzymatic background reaction. The provision of a metabolic shunt that reduces substrate levels by channeling the substrate away from the blocked step can solve this problem. The resultant decrease in intracellular substrate concentrations not only diminishes background growth but also raises the demands on enzymes that are capable of complementing the genetic defect. By regulating the production of the shunt enzyme, it is thus simultaneously possible to adjust the stringency of the selection system over a very broad dynamic range.

As forks are abundant in cellular metabolism, this basic strategy offers a simple and potentially general mechanism for varying selection pressure in genetic selection systems. It is not limited to intervention directly at the selection step. Metabolic flux can be controlled, in principle, at any branchpoint in the biosynthetic reaction sequence, before or even after the target reaction and, as a consequence, it should be possible to create regulatable selection regimes for virtually any metabolic reaction. The fact that additional chromosomal modifications of the host organism are not required is an attractive feature.

23.5
Perspectives

Genetic selection systems are often considered to be nonquantitative and incapable of differentiating alleles of variable strength [73]. As the examples discussed in this chapter show, however, selection systems are capable of sorting catalysts according to their activity – provided that they can be equipped with efficient mechanisms to control selection pressure.

The tight regulation of catalyst concentration provides one means of influencing selection stringency in a directed evolution experiment, and this can be readily achieved by balancing protein production with degradation. The specific combination of a tunable promoter system, which tightly regulates gene transcription, and a peptide sequence tag, which directs the translated enzyme to a housekeeping protease, allows systematic variation of intracellular protein concentration from very high to very low levels. As a consequence, substantially more active enzymes can be evolved [20]. The isolation of variants with high k_{cat} values – which are important for industrial biocatalysts – is particularly notable. Nonetheless, there is still room for improvement. Because the SsrA degradation tag is C-terminal, loss of the tag through frameshifts or the introduction of stop codons gives rise to artifacts that escape degradation, and these were found to dominate the population at the highest stringency conditions. Fusion of the constructs to an antibiotic resistance gene to enforce correct in-frame translation [74] may reduce this problem. Recently described N-terminal degradation tags, which direct proteins to the ClpXP protease complex [75], represent another possibility for minimizing such artifacts.

The control of intracellular substrate concentrations is a complementary approach to modulating selection pressure *in vivo*. As noted above, the use of regulated metabolic shunts to divert key intermediates away from the selectable step is likely to be generally applicable in any genetic selection system based on the conversion of an endogenous metabolite. This strategy provides a simple mechanism for reducing (or even eliminating) leakiness in auxotrophs arising from undesired background reactions, and should also facilitate the isolation of improved catalysts over multiple rounds of mutagenesis and selection. In contrast to controlling catalyst concentration, which tends to promote the evolution of enzymes with high k_{cat} values, reduction in substrate levels should favor the emergence of catalysts with high k_{cat}/K_m values. The regulation of both catalyst and substrate concentrations independently in the same system would afford an extremely versatile means of optimizing catalyst properties.

In the future, efforts to improve existing enzymes or to augment the activity of primitive catalysts, either designed *de novo* or generated by the redesign of an existing protein [76], stand to profit greatly from selection systems with tunable selection stringency. The large dynamic selection range that can be accessed by controlling either substrate or catalyst concentration is likely to be useful in this context. Systematic optimization of selection strains through rational engineering promises to open up new avenues for the directed evolution of enzymes.

References

1 Farinas, E.T., Bulter, T. and Arnold, F.H. (2001) Directed enzyme evolution. *Current Opinion in Biotechnology*, **12**, 545–51.

2 Powell, K.A., Ramer, S.W., del Cardayré, S.B., Stemmer, W.P.C., Tobin, M.B., Longchamp, P.F. and Huisman, G.W. (2001) Directed evolution and biocatalysis. *Angewandte Chemie – International Edition*, **40**, 3948–59.

3 Taylor, S.V., Kast, P. and Hilvert, D. (2001) Investigating and engineering enzymes by genetic selection. *Angewandte Chemie – International Edition*, **40**, 3310–35.

4 Kazlauskas, R.J. (2005) Enhancing catalytic promiscuity for biocatalysis. *Current Opinion in Chemical Biology*, **9**, 195–201.

5 Woycechowsky, K.J., Vamvaca, K. and Hilvert, D. (2007) Novel enzymes through design and evolution. *Advances in Enzymology and Related Areas of Molecular Biology*, **75**, 241–94.

6 Jaeger, K.E., Eggert, T., Eipper, A. and Reetz, M.T. (2001) Directed evolution and the creation of enantioselective biocatalysts. *Applied Microbiology and Biotechnology*, **55**, 519–30.

7 Jestin, J.-L. and Kaminski, P.A. (2004) Directed enzyme evolution and selections for catalysis based on product formation. *Journal of Biotechnology*, **113**, 85–103.

8 Winter, G., Griffiths, A.D., Hawkins, R.E. and Hoogenboom, H.R. (1994) Making antibodies by phage display technology. *Annual Review of Immunology*, **12**, 433–55.

9 Boder, E.T. and Wittrup, K.D. (1997) Yeast surface display for screening combinatorial polypeptide libraries. *Nature Biotechnology*, **15**, 553–7.

10 Ståhl, S. and Uhlén, M. (1997) Bacterial surface display: trends and progress. *Trends in Biotechnology*, **15**, 185–92.

11 Mattheakis, L.C., Bhatt, R.R. and Dower, W.J. (1994) An *in vitro* polysome display system for identifying ligands from very large peptide libraries. *Proceedings of the National Academy of Sciences of the United States of America*, **91**, 9022–6.

12 Schaffitzel, C., Hanes, J., Jermutus, L. and Plückthun, A. (1999) Ribosome display: an *in vitro* method for selection and evolution of antibodies from libraries. *Journal of Immunological Methods*, **231**, 119–35.

13 Roberts, R.W. and Szostak, J.W. (1997) RNA-peptide fusions for the *in vitro* selection of peptides and proteins. *Proceedings of the National Academy of Sciences of the United States of America*, **94**, 12297–302.

14 Tawfik, D.S. and Griffiths, A.D. (1998) Man-made cell-like compartments for molecular evolution. *Nature Biotechnology*, **16**, 652–6.

15 van Sint Fiet, S., van Beilen, J.B. and Witholt, B. (2006) Selection of biocatalysts for chemical synthesis. *Proceedings of the National Academy of Sciences of the United States of America*, **103**, 1693–8.

16 Licitra, E.J. and Liu, J.O. (1996) A three-hybrid system for detecting small ligand-protein receptor interactions. *Proceedings of the National Academy of Sciences of the United States of America*, **93**, 12817–21.

17 Baker, K., Bleczinski, C., Lin, H., Salazar-Jimenez, G., Sengupta, D., Krane, S. and Cornish, V.W. (2002) Chemical complementation: a reaction-independent genetic assay for enzyme catalysis. *Proceedings of the National Academy of Sciences of the United States of America*, **99**, 16537–42.

18 Lin, H., Tao, H. and Cornish, V.W. (2004) Directed evolution of a glycosynthase via chemical complementation. *Journal of the American Chemical Society*, **126**, 15051–9.

19 Yano, T., Oue, S. and Kagamiyama, H. (1998) Directed evolution of an aspartate aminotransferase with new substrate specificities. *Proceedings of the National Academy of Sciences of the United States of America*, **95**, 5511–15.

20 Neuenschwander, M., Butz, M., Heintz, C., Kast, P. and Hilvert, D. (2007) A simple selection strategy for evolving highly efficient enzymes. *Nature Biotechnology*, **25**, 1145–7.

21 Yano, T. and Kagamiyama, H. (2001) Directed evolution of ampicillin-resistant activity from a functionally unrelated DNA fragment: a laboratory model of molecular evolution. *Proceedings of the National Academy of Sciences of the United States of America*, **98**, 903–7.

22 Crameri, A., Raillard, S.-A., Bermudez, E. and Stemmer, W.P.C. (1998) DNA shuffling of a family of genes from diverse species accelerates directed evolution. *Nature*, **391**, 288–91.

23 Park, H.-S., Nam, S.-H., Lee, J.K., Yoon, C.N., Mannervik, B., Benkovic, S.J. and Kim, H.-S. (2006) Design and evolution of new catalytic activity with an existing protein scaffold. *Science*, **311**, 535–8.

24 Kleeb, A.C., Edalat, M.H., Gamper, M., Haugstetter, J., Giger, L., Neuenschwander, M., Kast, P. and Hilvert, D. (2007) Metabolic engineering of a genetic selection system with tunable stringency. *Proceedings of the National Academy of Sciences of the United States of America*, **104**, 13907–12.

25 Alper, H., Fischer, C., Nevoigt, E. and Stephanopoulos, G. (2005) Tuning genetic control through promoter engineering. *Proceedings of the National Academy of Sciences of the United States of America*, **102**, 12678–83.

26 Jana, S. and Deb, J.K. (2005) Strategies for efficient production of heterologous proteins in *Escherichia coli*. *Applied Microbiology and Biotechnology*, **67**, 289–98.

27 Makrides, S.C. (1996) Strategies for achieving high-level expression of genes in *Escherichia coli*. *Microbiological Reviews*, **60**, 512–38.

28 Novick, A. and Weiner, M. (1957) Enzyme induction as an all-or-none phenomenon. *Proceedings of the National Academy of Sciences of the United States of America*, **43**, 553–66.

29 Louis, M. and Becskei, A. (2002) Binary and graded responses in gene networks. *Science's STKE*, **143**, pe33.

30 Khlebnikov, A., Risa, Ø., Skaug, T., Carrier, T.A. and Keasling, J.D. (2000) Regulatable arabinose-inducible gene expression system with consistent control in all cells of a culture. *Journal of Bacteriology*, **182**, 7029–34.

31 Siegele, D.A. and Hu, J.C. (1997) Gene expression from plasmids containing the *araBAD* promoter at subsaturating inducer concentrations represents mixed populations. *Proceedings of the National Academy of Sciences of the United States of America*, **94**, 8168–72.

32 Rossi, F.M.V., Kringstein, A.M., Spicher, A., Guicherit, O.M. and Blau, H.M. (2000) Transcriptional control: rheostat converted to on/off switch. *Molecular Cell*, **6**, 723–8.

33 Khlebnikov, A., Datsenko, K.A., Skaug, T., Wanner, B.L. and Keasling, J.D. (2001) Homogeneous expression of the P_{BAD} promoter in *Escherichia coli* by constitutive expression of the low-affinity high-capacity AraE transporter. *Microbiology (UK)*, **147**, 3241–7.

34 Morgan-Kiss, R.M., Wadler, C. and Cronan, J.E. (2002) Long-term and homogeneous regulation of the *Escherichia coli araBAD* promoter by use of a lactose transporter of relaxed specificity.

Proceedings of the National Academy of Sciences of the United States of America, **99**, 7373–7.

35 Guzman, L.-M., Belin, D., Carson, M.J. and Beckwith, J. (1995) Tight regulation, modulation, and high-level expression by vectors containing the arabinose P_{BAD} promoter. *Journal of Bacteriology*, **177**, 4121–30.

36 Lee, S.K. and Keasling, J.D. (2005) A propionate-inducible expression system for enteric bacteria. *Applied and Environmental Microbiology*, **71**, 6856–62.

37 Khlebnikov, A. and Keasling, J.D. (2002) Effect of *lacY* expression on homogeneity of induction from the P_{tac} and P_{trc} promoters by natural and synthetic inducers. *Biotechnology Progress*, **18**, 672–4.

38 Skerra, A. (1994) Use of the tetracycline promoter for the tightly regulated production of a murine antibody fragment in *Escherichia coli*. *Gene*, **151**, 131–5.

39 Lutz, R. and Bujard, H. (1997) Independent and tight regulation of transcriptional units in *Escherichia coli* via the LacR/O, the TetR/O and AraC/ I_1-I_2 regulatory elements. *Nucleic Acids Research*, **25**, 1203–10.

40 Bahl, M.I., Hansen, L.H. and Sørensen, S.J. (2005) Construction of an extended range whole-cell tetracycline biosensor by use of the *tet*(M) resistance gene. *FEMS Microbiology Letters*, **253**, 201–5.

41 Hansen, L.H., Ferrari, B., Sørensen, A.H., Veal, D. and Sorensen, S.J. (2001) Detection of oxytetracycline production by *Streptomyces rimosus* in soil microcosms by combining whole-cell biosensors and flow cytometry. *Applied and Environmental Microbiology*, **67**, 239–44.

42 Fersht, A. (1998) *Structure and Mechanism in Protein Science: A Guide to Enzyme Catalysis and Protein Folding*, W. H. Freeman and Company, New York.

43 Hillen, W. and Berens, C. (1994) Mechanisms underlying expression of Tn*10* encoded tetracycline resistance. *Annual Review of Microbiology*, **48**, 345–69.

44 Chang, A.C.Y. and Cohen, S.N. (1978) Construction and characterization of amplifiable multicopy DNA cloning vehicles derived from the P15A cryptic miniplasmid. *Journal of Bacteriology*, **134**, 1141–56.

45 Bertrand, K.P., Postle, K., Wray, L.V. and Reznikoff, W.S. (1983) Overlapping divergent promoters control expression of Tn*10* tetracycline resistance. *Gene*, **23**, 149–56.

46 Flache, P., Baumeister, R. and Hillen, W. (1992) The Tn*10*-encoded tetracycline resistance mRNA contains a translational silencer in the 5′ nontranslated region. *Journal of Bacteriology*, **174**, 2478–84.

47 Studier, F.W., Rosenberg, A.H., Dunn, J.J. and Dubendorff, J.W. (1990) Use of T7 RNA polymerase to direct expression of cloned genes. *Methods in Enzymology*, **185**, 60–89.

48 Oehmichen, R., Klock, G., Altschmied, L. and Hillen, W. (1984) Construction of an *Escherichia coli* strain overproducing the Tn*10*-encoded TET repressor and its use for large scale purification. *The EMBO Journal*, **3**, 539–43.

49 Gamper, M., Hilvert, D. and Kast, P. (2000) Probing the role of the C-terminus of *Bacillus subtilis* chorismate mutase by a novel random protein-termination strategy. *Biochemistry*, **39**, 14087–94.

50 MacBeath, G., Kast, P. and Hilvert, D. (1998) Probing enzyme quaternary structure by combinatorial mutagenesis and selection. *Protein Science*, **7**, 1757–67.

51 Kast, P., Asif-Ullah, M., Jiang, N. and Hilvert, D. (1996) Exploring the active site of chorismate mutase by combinatorial mutagenesis and selection: the importance of electrostatic catalysis. *Proceedings of the National Academy of Sciences of the United States of America*, **93**, 5043–8.

52 MacBeath, G., Kast, P. and Hilvert, D. (1998) Exploring sequence constraints on an interhelical turn using in vivo selection for catalytic activity. *Protein Science*, **7**, 325–35.

53 Vamvaca, K., Butz, M., Walter, K.U., Taylor, S.V. and Hilvert, D. (2005) Simultaneous optimization of enzyme activity and quaternary structure by directed evolution. *Protein Science*, **14**, 2103–14.

54 Christie, G.E., Farnham, P.J. and Platt, T. (1981) Synthetic sites for transcription termination and a functional comparison with tryptophan operon termination sites *in vitro*. *Proceedings of the National Academy of Sciences of the United States of America*, **78**, 4180–4.

55 Karzai, A.W., Roche, E.D. and Sauer, R.T. (2000) The SsrA-SmpB system for protein tagging, directed degradation and ribosome rescue. *Nature Structural and Molecular Biology*, **7**, 449–55.

56 Andersen, J.B., Sternberg, C., Poulsen, L.K., Bjørn, S.P., Givskov, M. and Molin, S. (1998) New unstable variants of green fluorescent protein for studies of transient gene expression in bacteria. *Applied and Environmental Microbiology*, **64**, 2240–6.

57 DeLisa, M.P., Samuelson, P., Palmer, T. and Georgiou, G. (2002) Genetic analysis of the twin arginine translocator secretion pathway in bacteria. *Journal of Biological Chemistry*, **277**, 29825–31.

58 Leveau, J.H.J. and Lindow, S.E. (2001) Predictive and interpretive simulation of green fluorescent protein expression in reporter bacteria. *Journal of Bacteriology*, **183**, 6752–62.

59 Stemmer, W.P.C. (1994) Rapid evolution of a protein *in vitro* by DNA shuffling. *Nature*, **370**, 389–91.

60 Bailey, J.E. (1991) Toward a science of metabolic engineering. *Science*, **252**, 1668–75.

61 Raab, R.M., Tyo, K. and Stephanopoulos, G. (2005) Metabolic engineering, in *Advances in Biochemical Engineering/ Biotechnology*, Vol. 100 (ed. J. Nielsen), Springer, Berlin/Heidelberg, pp. 1–17.

62 Stephanopoulos, G. and Vallino, J.J. (1991) Network rigidity and metabolic engineering in metabolite overproduction. *Science*, **252**, 1675–81.

63 Bongaerts, J., Krämer, M., Müller, U., Raeven, L. and Wubbolts, M. (2001) Metabolic engineering for microbial production of aromatic amino acids and derived compounds. *Metabolic Engineering*, **3**, 289–300.

64 Haslam, E. (1993) *Shikimic Acid: Metabolism and Metabolites*, John Wiley & Sons, Inc., New York.

65 Gamper, M. and Kast, P. (2005) Strategy for chromosomal gene targeting in RecA-deficient *Escherichia coli* strains. *Biotechniques*, **38**, 405–8.

66 Kleeb, A.C., Kast, P. and Hilvert, D. (2006) A monofunctional and thermostable prephenate dehydratase from the archaeon *Methanocaldococcus jannaschii*. *Biochemistry*, **45**, 14101–10.

67 Davis, B.D. (1953) Autocatalytic growth of a mutant due to accumulation of an unstable phenylalanine precursor. *Science*, **118**, 251–2.

68 Katagiri, M. and Sato, R. (1953) Accumulation of phenylalanine by a phenylalanineless mutant of *Escherichia coli*. *Science*, **118**, 250–1.

69 Hermes, J.D., Tipton, P.A., Fisher, M.A., O'Leary, M.H., Morrison, J.F. and Cleland, W.W. (1984) Mechanisms of enzymatic and acid-catalyzed decarboxylations of prephenate. *Biochemistry*, **23**, 6263–75.

70 Christopherson, R.I. (1985) Chorismate mutase-prephenate dehydrogenase from *Escherichia coli*: cooperative effects and inhibition by L-tyrosine. *Archives of Biochemistry and Biophysics*, **240**, 646–54.

71 Turnbull, J., Morrison, J.F. and Cleland, W.W. (1991) Kinetic studies on chorismate mutase-prephenate dehydrogenase from *Escherichia coli*: models for the feedback inhibition of prephenate dehydrogenase by L-tyrosine. *Biochemistry*, **30**, 7783–8.

72 Zhao, G., Xia, T., Ingram, L.O. and Jensen, R.A. (1993) An allosterically insensitive class of cyclohexadienyl dehydrogenase from *Zymomonas mobilis*. *European Journal of Biochemistry*, **212**, 157–65.

73 Link, A.J., Jeong, K.J. and Georgiou, G. (2007) Beyond toothpicks: new methods for isolating mutant bacteria. *Nature Reviews of Microbiology*, **5**, 680–8.

74 Cox, J.C., Lape, J., Sayed, M.A. and Hellinga, H.W. (2007) Protein fabrication automation. *Protein Science*, **16**, 379–90.

75 Flynn, J.M., Neher, S.B., Kim, Y.-I., Sauer, R.T. and Baker, T.A. (2003) Proteomic discovery of cellular substrates of the ClpXP protease reveals five classes of ClpX-recognition signals. *Molecular Cell*, **11**, 671–83.

76 Jäckel, C., Kast, P. and Hilvert, D. (2008) Design and redesign of proteins by directed evolution. *Annual Review of Biophysics*, **37**, 153–73.

24
Protein Engineering by Phage Display

Agathe Urvoas, Philippe Minard and Patrice Soumillion

24.1
Introduction

Since the first description of phage display by Georges Smith in 1985 [1], several thousands of research articles reporting the applications of this technology have been published. As the result of a simple genetic fusion between a chosen open reading frame (ORF) and a gene encoding a filamentous phage coat protein, the corresponding peptide, protein or antibody fragment can be expressed at the surface of a phage particle, physically linked to its genetic information via the coat. Using large libraries of variants, the selective capture of a phage based on its specific interaction with an immobilized ligand allows the concomitant selection of the genetic information which can be amplified by simple phage infection (Figure 24.1). This chapter will focus on phage display for protein engineering, and will not describe other applications such as peptide or whole-phage engineering, nor all of the alternative display methods that have been developed to circumvent some of the limitations of phage technology [2]. Several interesting reviews and books should also be of interest for those readers who are using, or who plan to use, phage display [3–12].

24.2
The State of the Art

24.2.1
Engineering Protein Binders by Phage Display

24.2.1.1 Antibodies and Antibody Fragments

One of the most successful applications of the phage display technology is the selection of engineered antibodies binding to various antigenic targets including proteins, peptides or haptens.

Protein Engineering Handbook. Edited by Stefan Lutz and Uwe T. Bornscheuer

ISBN: 978-3-527-31850-6

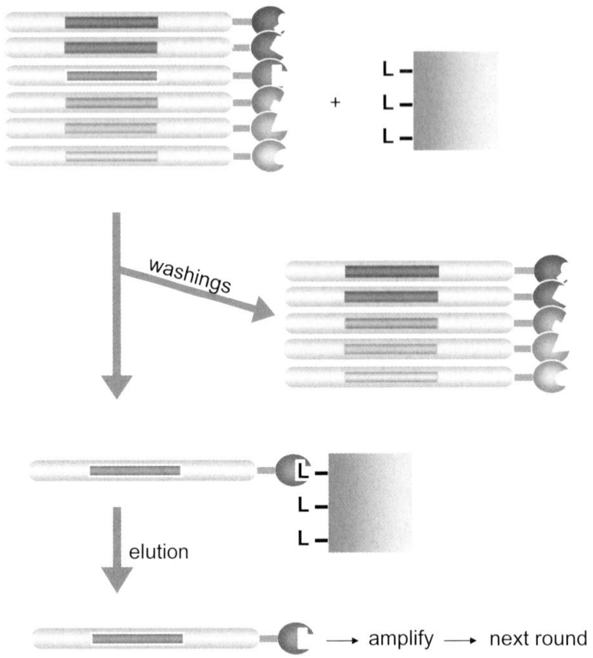

Figure 24.1 The principle of phage display is based on *in vitro* selection of peptides or proteins expressed at the surface of filamentous phages. In a 'biopanning' experiment, specific binders can be captured from a library using an immobilized target ligand and their genes amplified by simple phage infection.

24.2.1.1.1 **Engineering the Antibody Format for Phage Display** Whole antibodies are too difficult to handle as recombinant proteins and thus to expose on the phage surface because of their large size and their structure, which is composed of four polypeptide chains. Phage display of antibody is performed either on Fab or scFv fragments [7]. Fab fragments which are composed of two chains (VL-CL and VH-CH1) are more difficult to handle but tend to be more stable than ScFv in their soluble forms, and are easier to convert into full-length immunoglobulin G (IgG) for further applications. A recently described single-chain Fab offers new perspectives for phage display selection of bivalent antibody fragments [13]. ScFv fragments are composed of the variable domain of the light chain (VL) and of the heavy chain (VH), linked by a flexible peptide. Those fragments are easier to handle for recombinant expression in *Escherichia coli*, but can be subject to dimerization effects (diabodies). Depending on the size of the linker, dimers, trimers and higher multiples have been created and used as multivalent binders. Such multimers with an increased avidity for their targets (cancer cells, viruses) and a longer half-life *in vivo* are promising for the development of therapeutic molecules [14–16].

Camelidae antibodies are homodimeric heavy-chain antibodies which contrast with classical heteromeric antibodies. The variable region (VHH) of camelidae antibodies is the smallest antigen-binding fragment (11 to 15 kDa). The low complexity of the binding site (only three hypervariable loops), their good production levels in both prokaryotic and eukaryotic hosts, and their good thermodynamic stability are favorable properties for engineering in the perspective of various therapeutic or diagnostic applications [17, 18]. The grafting of specific VHH residues onto a human (or mammalian) single-domain VH can render the domain more soluble and improve its biophysical and functional properties; this process is called camelization [19–21]. Phage display was also used to select anti-IgG VHH fragments bearing a functional mimicry of a DNA fragment [22]. An immune library of llama antibodies was used to select, by phage display, those VHH domains which bind a hapten (methotrexate) with an affinity in the high nanomolar range [23].

Recently, stable monomeric human VHs were identified by phage display using a selection strategy based on proteolytic susceptibility (see Section 24.2.2). These VHs should also serve as good scaffolds for the development of immunotherapeutics [24].

24.2.1.1.2 **The Phage Antibody Libraries Used for the Selection of Antigen Binders**

In order to select an antibody fragment recognizing a given target, two types of recombinant antibody library can be used, namely naïve or immune (for reviews, see Refs [25–27]). Naïve libraries are derived from natural nonimmunized human V genes, while immune libraries are derived from V genes from immunized humans or mice. Immune libraries present a bias of specificity for a specific antigen, and have been created using immunoglobulin genes from immunized animals with an antigen, or from infected patients by pathogen viruses [28–30]. The affinity obtained from an immune library is generally superior to that obtained from a naïve library for the same size of library. The dissociation constant K_D of the selected fragments is often proportional to the size of library, and ranges from 10^{-6} to $10^{-9}\,M$. The construction of large naïve libraries (10^9 to 10^{10} clones) has allowed the selection of antibody fragments with nanomolar affinities for various targets [31–35]. An alternative to the use of naïve or immune libraries is the design of synthetic or semi-synthetic libraries [36]. Barbas *et al.* first developed this strategy by grafting degenerated fragments of the variable complementarity-determining region (CDR) loops on a pre-existing Fab scaffold [37]. This strategy was further improved and allowed the selection of nanomolar binders against different antigens (peptides, haptens, proteins) [38]. The resulting actual Fab library, named n-CoDeR (1.5×10^{10} clones), is now commercially available. Following similar principles, other large synthetic libraries were designed and are commercialized for various clinical applications such as HuCal (Human Combinatorial Antibody Library; 1.6×10^{10} clones) [39] or 'fdDOG-2lox' (6.5×10^{10} clones) (for reviews, see Refs [36, 40]). The results of recent experiments indicated that a range of tight binders can be selected from a phage library of a stable Fab for which CDR loops were randomized, leaving only serine or tyrosine as the allowed residues (for a review, see Ref. [41]).

24.2.1.1.3 **Tuning Affinity and Specificity by Antibody Engineering** If selected antibody fragments have insufficient affinities for therapeutic or diagnostic applications, they can be improved by affinity maturation using directed evolution *in vitro*. This process was developed to mimic natural evolution, and involves three steps: (i) diversity introduction in the pre-existing sequence; (ii) selection for a higher affinity; and (iii) screening for the best variants. Mutations can be introduced in the pre-existing antibody fragment using several approaches, whether random or localized: localized rational saturation mutagenesis [42–46]; systematic mutagenesis [47]; random mutations [error-prone polymerase chain reaction (epPCR), mutator strains] [48–50]; or loop or DNA shuffling [51–56]. Random approaches are generally better suited to improve low-affinity antibodies, whereas directed approaches are employed for the fine-affinity tuning, with the randomization of five to six targeted residues in contact with the antigen. Several phage display selection rounds are performed with increased stringency conditions (k_{off} selection, addition of a soluble competitor. etc.). Improved antibody fragments have been obtained by phage display with affinities in the picomolar range (10 to 1200-fold affinity increase) [43, 45, 46, 56–59]. Directed evolution was also used to improve the specificity [60–63] and stability of antibody fragments [64, 65], or specific cell-targeting properties for therapeutic applications [66–68].

24.2.1.1.4 **Applications of Engineered Antibodies** The development of *in vitro* display technologies, including phage display and directed evolution, the construction of large diverse antibody libraries and the various antibody formats explored, have contributed to the emergence of numerous diagnostic and clinical applications of engineered antibodies (for reviews, see Refs [3, 25, 27, 69]). As the risk of immunogenicity may be reduced by using humanized antibodies, these molecules represent a potential for novel therapeutic approaches [70–73]. The application formats explored include delivering therapeutic molecules to specific targets (radioisotopes, fusion proteins, chemicals, SiRNA) [74] or targeting and killing cancer cells or viral particles with multivalent antibody fragments [14, 72]. Recombinant antibody-derived molecules represent half of the molecules in clinical phase development, and the first antibody evolved by phage display (named Humira) was approved for arthritis therapy in 2002 by the American Food and Drug Administration (FDA) [27, 75]. The development of high-throughput selection and screening methodologies should increase the potential of those molecules for therapeutic and research applications (e.g. for proteome analysis) [40].

24.2.1.2 Alternative Scaffolds

Monoclonal antibodies (mAbs) produced either by hybridomas or, more recently, selected from phage libraries, have long been a unique source of specific reagents, and still are by far the most commonly used source of defined specific proteins. Unfortunately, antibody molecules suffer from a number of drawbacks, the major problem being that they are large molecules composed of several polypeptide chains that are stabilized by both intrachain and interchain disulfide bonds. This large size, complex quaternary structure and requirements for disulfide bonds and post-translational modification (glycosylation) render mAbs unfavorable for their efficient expression as recombinant proteins. Consequently, most antibodies cannot be efficiently expressed in *E. coli*, and high-level expression can only be achieved in systems based on animal cell cultures [76], which involves the investment of substantial time and costs. The application of antibodies that require significant amounts of pure protein (especially when production costs are critical) is clearly limited when expression in these simple systems is poor. The expression of engineered antibody derivatives, such as fusion with fluorescent protein or reporters, is also inefficient in common expression systems due to the unfavorable expression properties of the antibody moiety. Furthermore, due to disulfide bond requirements, most antibodies cannot fold in the reducing environment of the cell cytoplasm, which in turn limits the use of most antibodies as effectors/modulators of biological function within living cells. Finally, the intricate situation among patents in the field of therapeutic antibodies may lead to problems in the commercial development of molecules generated using protected technologies.

In order to overcome these limitations, specific binding proteins are being developed which are derived not from proteins related to antibodies but from so-called 'alternative scaffolds'. In the near future, these latter types of protein may replace antibody molecules in some diagnostics, research and biotechnological applications, and perhaps even as therapeutic agents; indeed, alternative scaffold-derived proteins are already undergoing clinical development [77, 78]. Moreover, new protein scaffolds have begun to open new intellectual properties on the basis of which a number of new biotechnology companies have recently emerged [79].

24.2.1.2.1 **Creating a Binding Site in a Protein Scaffold** The general idea is that, in any protein, the binding site for its biological partners involves a restricted number of side chains that are located on the surface of the molecule and brought into close proximity by the overall folding of the scaffold protein. The results of natural protein evolution and protein engineering experiments have suggested that all proteins have some degree of tolerance to sequence substitution, particularly for solvent-exposed amino acids. The strategy is therefore to randomly substitute a spatially defined subset of side chains in order to create a huge diversity of potential binding surfaces, but without any drastic destabilization of the protein fold [80]. The residues selected to be randomized are often those naturally involved in the binding sites for the cognate ligands, although in other cases side chains are selected simply based on their location in the same local patch of the protein surface. The library is typically built by complete random substitution of selected positions using NNK/NNS degenerate codons, but biased codons or restricted diversity can occasionally be selected on the basis of structural or evolutionary data. Large libraries made in phage display format are then used to identify specific binders against any target of interest. Although other display technologies such as ribosome display or mRNA display have been used, phage libraries are the most common [81, 82].

24.2.1.2.2 **An Increasing Diversity of Alternative Scaffolds** Although initially relatively marginal with respect to antibody libraries and restricted to a small number of protein architectures, the 'scaffold' approach has now been applied to an increasing diversity of natural proteins. A recently produced series of reviews has listed more than 40 different proteins proposed as possible scaffolds, each at various stages of development [79, 83–86].

These were selected generally as small monomeric proteins, with favorable expression properties. Most of the proposed scaffolds have been single-domain proteins, but those prepared by the concatenation of single domains have also resulted in tight-binding proteins [87, 88]. As the starting protein is subjected to extensive sequence modification, which might have a destabilizing effect, a high initial stability may be an important criterion [89]. A disulfide-free protein scaffold may be desirable if the scaffold is to be used for intracellular applications. Some of these scaffolds have been selected to be of human origin, in order to minimize the immunogenicity of the resultant binders in terms of their therapeutics applications. Proteins with very different structures have been used successfully, and different structural features have been randomized to create new binding surfaces. In scaffolds with a beta-sandwich topology, the loops connecting the beta strands have been most often used as variable surfaces [90], by analogy with the immunoglobulin domain organization. However, the solvent-exposed surfaces of beta strands have also been used successfully in proteins made by beta sheets [91, 92]. In other types of structure, it is the external surfaces of the helices involved in helical bundles [93, 94], or the loops connecting the stacks of repeated modules [95], that were used as supports for side-chain diversity. All of these options have met with some success, however, and it appears that the binding sites of engineered scaffolds proteins will be as diverse as in natural proteins.

24.2.1.2.3 **Scaffold Success Stories** As most proposed protein scaffolds have been described only recently, many are currently at the proof-of-concept stage. Only a limited subset of protein architectures has effectively produced tight and specific binding proteins against a range of different targets. Most of these proteins were developed initially in academic research groups and later transferred to biotechnology companies, often with therapeutic applications as the main objective. Details of some of the best-characterized and most successful scaffolds are described in the following sections, while a more comprehensive list of references can be found in recent reviews [79, 83–86] (Figure 24.2).

24.2.1.2.3.1 **Affibodies** Affibodies are small helical proteins derived from bacterial protein domains similar to Protein A. The side chains located on the outside surface of the three helix bundles have been randomized to create large phage display libraries [93, 94]. In the first attempts, binders were obtained in the micro-

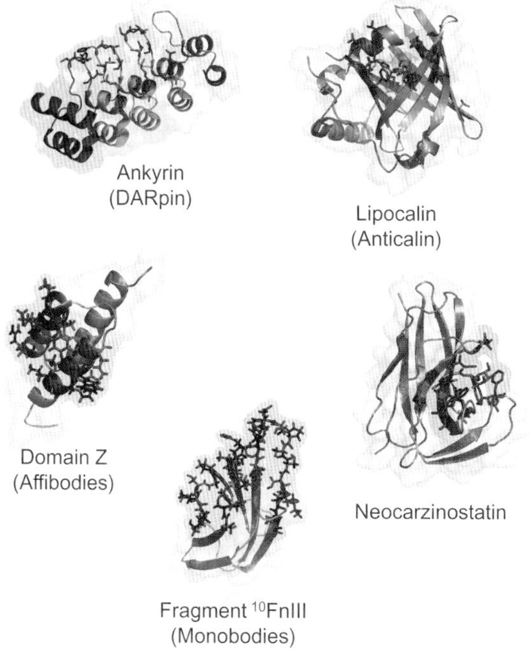

Figure 24.2 Proteins engineered as scaffolds for molecular recognition. The positions of sides chains where molecular diversity was introduced are highlighted. Ankyrin (DARpins), domain Z (affibodies) and fibronectin-based scaffolds have been developed mainly for protein recognition. Neocarzinostatin and lipocalins (anticalins) have also been engineered for small-molecule ('hapten') recognition.

molar range, but subsequent results showed that high-affinity binders could be selected [96] and that moderate affinities could be improved either by subsequent shuffling of helices [97], multimerization or affinity maturation. Structural data have been described explaining how this small and regular architecture can adapt to a range of binding surfaces [98–100]. One structural study has identified an affibody which is partially folded in its secondary structure, has a fluctuating tertiary structure, and becomes fully folded only when it binds to its target [100]. As the small size of affibodies results in their rapid clearance and good penetration into solid tumors, this scaffold represents a promising candidate as a source of specific reagents for solid tumor imaging [101]. The good expression properties of these disulfide-free small helical proteins opens a wide range of applications ranging from microarray [87] to specific binding reagents for affinity chromatography [102].

24.2.1.2.3.2 **Zinc Fingers** Zinc fingers are classical DNA recognition elements commonly found in transcription factors, and the rules governing DNA specificity have been efficiently addressed using zinc fingers domains libraries exposed on phage [103, 104]. Engineered zinc fingers specific for a range of trinucleotide sequences have been obtained [105]. These domains can be combined to produce polydactyl-specific recognition domains adapted to alter specifically the regulation of any specific gene in a complex genome [106]. Similarly, a combination of zinc finger domains can be associated with a recombinase or a nuclease for genome engineering. The rules governing sequence-specific recognition in this zinc finger scaffold are now relatively well understood, up to the point where phage display of the scaffold is no longer necessary, and a specific binder can be generated by combining recognition rules previously identified from phage libraries [107, 108].

24.2.1.2.3.3 **Fibronectin** The fibronectin III domain has the characteristic fold of the immunoglobulin domain, but without disulfide bonds. By randomizing two loops that are structurally equivalent to the CDR in antibodies, binders against ubiquitin were initially selected [90]. This first success was not expanded by using phage display libraries, and tight binders against other proteins (tumor necrosis factor-alpha; TNF-α) were obtained using RNA display technology. The much larger libraries produced with RNA display experiments were presumably the key to selecting tighter binders than those found when using phage display [109]. Recently, by employing a strategy of restricted diversity that had been successfully applied to antibodies [41], binders in the submicromolar range were selected from a library of fibronectin mutants for which the three pseudo-CDR loops were randomized with only serine or tyrosine [110]. These studies demonstrated that the amazing simplification of the explored sequence space initially discovered with antibody libraries could be successfully used on a simpler binder scaffold, with a reduced binding surface relative to the antibody derivatives.

24.2.1.2.3.4 **Lipocalins and Anticalins** Lipocalins are a family of proteins found in many organisms (including human) that naturally bind specifically small hydrophobic molecules. These proteins have a classical up- and -down beta-barrel topology, and bind their specific ligands in a funnel-shaped binding site located in the center of the beta strand at one end of the barrel. These folds have been investigated by the group of Skerra as a potential small-molecule binding platform [111]. In fact, this is one of a very small number of binding proteins that are specifically engineered to bind small-molecule targets. Interestingly, the same scaffold family has also been used successfully to bind proteins targets. In this case, the loops connecting the beta strands on one side of the barrel have been simultaneously randomized, and the resultant library used to select (by phage display) tight and specific binders [112–114]. A variety of different proteins of the lipocalin family can potentially be engineered in the same way, including proteins of human origin (tear lipocalin), in order to minimize immunological responses in medical applications.

24.2.1.2.3.5 **Protein Repeats and Designed Ankyrin Repeat Proteins (DARPINS)** Several families of repeated proteins are currently being investigated as very promising alternative scaffold candidates [115]. Although diverse, these protein families have all been used widely in natural evolution processes to create a broad diversity of very tight-binding proteins. This remarkable evolutionary success is a direct result of their structure, which results from the juxtaposition of self-compatible modules. Within each module, those residues which are important for the module's stability and for its interactions with neighboring modules, are relatively well conserved. These structurally important residues can therefore be identified from a consensus analysis of a structurally homogeneous sequence collection that can be extracted from sequence databases [116]. The remaining part of the sequence of each module is highly variable between modules, and often involves residues with solvent-exposed side chains. In the folded protein, the juxtaposition of modules with variable outside surfaces creates potentially large and chemically diverse interaction surfaces. This modular organization opens new routes for the combinatorial engineering of artificial repeat proteins; such examples are bifunctional or multifunctional binding by combining compatible modules, or straightforward affinity maturation processes by the stepwise selection/combination of modules to an initial binder.

Large libraries of artificial repeat proteins based on ankyrin repeats have been particularly successful in generating very tight-binding proteins against a number of targets [95, 117]. Some of these proteins have been shown to act as specific intracellular inhibitors [118] or as scaffolding components in cocrystallization experiments with soluble proteins, or even with membrane proteins [119]. Although the ankyrin libraries were initially selected by ribosome display (a very powerful but not widely used selection methodology), the same type of library can now also be used successfully to generate tight binders from phage display libraries [120]. Besides ankyrin repeats, other repeat protein families (Heat, Armadillo, leucine-rich repeats) have been, or are currently being, explored [120, 121].

It has been realized only recently that the very same logic which underlies repeat scaffold engineering has also been used by Nature. Here, the primordial version of an adaptive immune system, which is still employed in jawless vertebrates, is not based on immunoglobulin architecture but rather on leucine-rich repeat proteins [122].

24.2.2
Engineering Protein Stability by Phage Display

If a phage-displayed protein is inserted between the structural domains of the pIII coat protein, or fused to the C-terminal of pIII but with an N-terminal affinity tag, its proteolytic degradation will result respectively in the loss of phage infectivity or in loss of the tag (Figure 24.3). As filamentous phages are resistant to many proteases (e.g. trypsin, Factor Xa, IgA protease, Asp-N, chymotrypsin, Arg-C, Glu-C, thrombin, thermolysin [123]), the treatment of a library with a protease will target only the displayed protein. Hence, after protease incubation, simple infection or affinity capture of the tag allows the easy selection of proteins that are properly folded as they are more resistant to proteolysis than their unfolded counterparts. Moreover, as filamentous phages can tolerate temperatures of up to 60 °C,

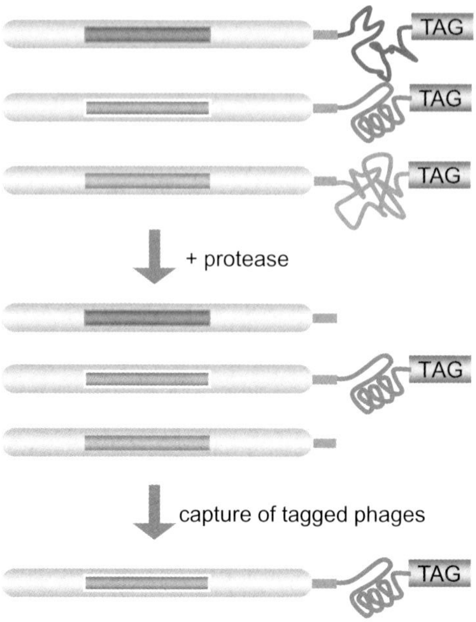

Figure 24.3 Selection strategy for improving protein stability. The incubation of a library of phage-displayed proteins with a protease results in degradation of the less-stable and misfolded mutants. If the protein is fused to a tag at the N terminus, the phage-displaying nondegraded proteins can be selected by affinity capture with a tag binder. The tag can be replaced by a pIII domain necessary for phage infection. In this case, proteolytic degradation will result in a loss of infectivity. Selection pressure can be increased by changing the conditions (pH, temperature, denaturing agents, etc.) before or during protease incubation.

high concentrations of denaturants (urea up to 10 *M*, guanidinium up to 4 *M*), reducing agents, organic solvents [dimethylformamide (DMF) up to 50%] and a large pH range (between 2 and 12), the selection pressure on a particular fold can be easily increased [123]. This selection strategy was first demonstrated in 1998 by stabilizing barnase [123] and ribonuclease T1 [124] as model proteins. It was then successfully used to improve the stability of various proteins and to create new stable protein folds (for reviews, see Refs [4, 125]). It should be noted, however, that an increase in the proteolytic susceptibility without any loss in global thermodynamic stability is always possible, for example if the propensity to local unfolding is increased. This can lead to the selection of nonstabilized mutants [126].

Nevertheless, the results of recent comparative studies have indicated that this experimental phage display strategy affords better stabilization effects than computational design [127–129]. The stabilization of a cold shock protein by coulombic interactions also led to the identification of mutations that could not be identified by theoretical analysis. Moreover, the amplitude of the stability changes could not be predicted by electrostatic calculations, which was indicative of the other difficulties in modeling coulombic interactions in a protein [130].

With regards to new protein folds, the strategy was initially applied by the group of Winter to create native-like proteins by fusing DNA encoding a designed polypeptide segment (bait) with random fragments from the *E. coli* genome or from human cDNA [131, 132]. Interestingly, one of these artificial proteins was encoded by an antisense strand of a human gene, and revealed an unexpected potential reservoir for the natural evolution of folded proteins. The structural analysis of a selected 103-residue chimera made from a segment of a cold shock protein (CspA) and the S1 domain of the 30S ribosomal subunit of *E. coli* revealed a segment-swapped, six-stranded beta barrel of unique architecture that assembled to a tetramer [133]. While the CspA segment retained its structural identity, the S1 segment was deformed in the barrel structure. An analysis of the thermodynamic stability and folding kinetics of the tetramer and its monomeric and dimeric intermediates further indicated that segment swapping and oligomerization are both powerful ways of stabilizing proteins, and could be essential driving forces in early protein evolution [134]. More recently, the use of a polypeptide bait with a pre-existing binding activity into a folded domain was shown to provide an evolutionary advantage [135]. When a segment from hen egg lysozyme that, in the native protein, binds to a mAb was combined with random genomic segments, selection by proteolysis in the presence of the antibody led to a folded dimeric protein with an enhanced antibody affinity (at least 2000-fold) compared to the unfolded segment. This supported the idea that functional activities in polypeptide segments may have contributed to domain creation in early evolution. Surprisingly, a stabilizing single heme molecule was also found in the dimer.

24.2.3
Engineering Enzymes by Phage Display

24.2.3.1 Engineering Allosteric Regulation

Phage display has been used to create artificial allosteric sites on an enzyme scaffold (Figure 24.4). In order to create hybrid enzymes endowed with affinity for other molecules, libraries of insertion mutants of β-lactamase TEM-1 were created in which the surface loops were extended by the replacement of one to three residues of the wild-type sequence by random peptide sequences of three to nine residues. *In vivo* preselection was applied to the individual libraries to obtain bacterial clones producing phage-enzymes which were active enough to provide resistance against β-lactam antibiotics. Preselected libraries were then recombined into a large library in which two or three loops have accepted random insertions. The application of this preselection was essential to afford large complex and high-quality libraries [136, 137]. Phage selections were then performed to isolate those insertants endowed with affinities for unrelated mAbs [136], various proteins [138] or, more recently, metallic ions [139] and small organic molecules (J. Fastrez and P. Soumillion, unpublished results). Nanomolar affinities were observed with proteic ligands and submicromolar affinities with ions and small molecules. Moreover, the activities of most of the insertants were modulated by the effectors, with both activation (up to a factor 3) and inhibition (up to a factor 10) being observed. These hybrid

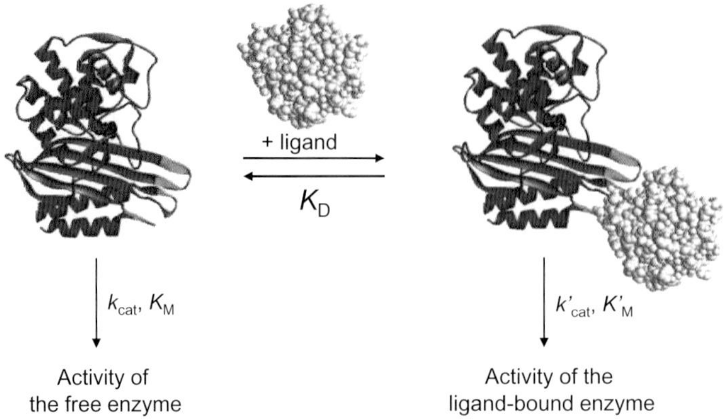

Figure 24.4 Artificial allosteric β-lactamases can be selected from libraries of phage-displayed mutants with random peptides inserted in surface loops at the vicinity of the active site. The mutants selected for binding can be either upregulated or downregulated by the target ligand.

enzymes may find application as diagnostic tools, since the presence of an allosteric effector in a sample may be detected by measuring the specific activity versus that observed in a reference solution. By using one of these insertants, the prostate-specific antigen was detected in a competitive homogeneous immunoassay at concentrations in the nanomolar range [136].

24.2.3.2 **Engineering Catalytic Activity**

Selection for catalytic activity is clearly more difficult than selection for binding, it is necessary to find a way in which the catalytic turnover of an active phage-enzyme can be coupled to its specific capture from a phage library. Several strategies have been developed with this objective in mind.

24.2.3.2.1 **Selection for Binding to Analogues of Substrate, Product or Transition State** The first strategies were applied in projects which aimed at changing the specificity of an enzyme. Here, by using affinity capture with immobilized analogues of substrates or products (Figure 24.5a, upper part), mutants of the staphylococcal nuclease and of the A1-1 glutathione *S*-transferase (GST) featuring altered specificities, could be identified [140, 141].

In a more sophisticated approach which took into account the fact that enzymes must be complementary to the transition state rather than the ground state of the substrate, attempts have been made to select phage-enzymes on immobilized transition-state analogues (TSA)s. In a project aimed at understanding the structural determinants of metal ion affinity in proteins, a small library of phage-displayed carbonic anhydrase II was selected for zinc ion affinity by capture on an immobilized sulfonamide, which was considered to be a TSA. Selected mutants were shown to bind zinc ions with a range of affinities, from a similar level to a

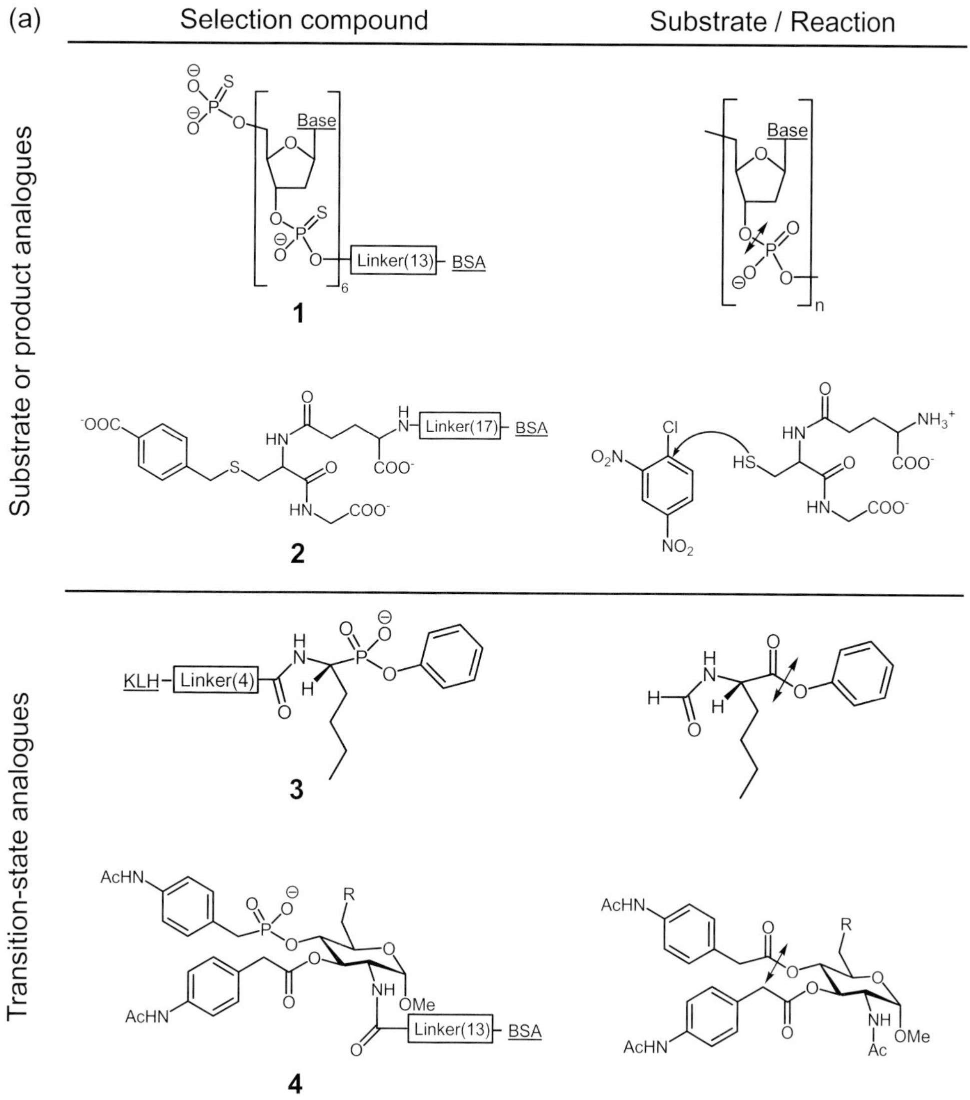

Figure 24.5 Examples of bifunctional molecules designed for covalent or noncovalent capture of phage-enzymes. These molecules usually comprise an immobilization module [BSA, KLH (keyhole limpet hemocyanin) or biotin] and a selection module connected through a linker; the number of atoms separating both modules is indicated in the linker box. (a) **1** and **2** are a nonhydrolyzable substrate analogue and a product analogue for the respective selection of DNase [140] and GST [141]; **3** and **4** are TSAs for the selection of enantioselective [142] or regioselective [143] hydrolytic antibodies, respectively; (b) **5** and **6** are covalent affinity labels used for the selection of thioesterase [144] or aldolase [145] antibodies featuring respectively a reactive cysteine or lysine in their active site; **7**, **8** and **9** are mechanism-based inhibitors for the covalent selection of phage-displayed β-lactamase [146], glycosidase [147] and subtilisin [148], respectively.

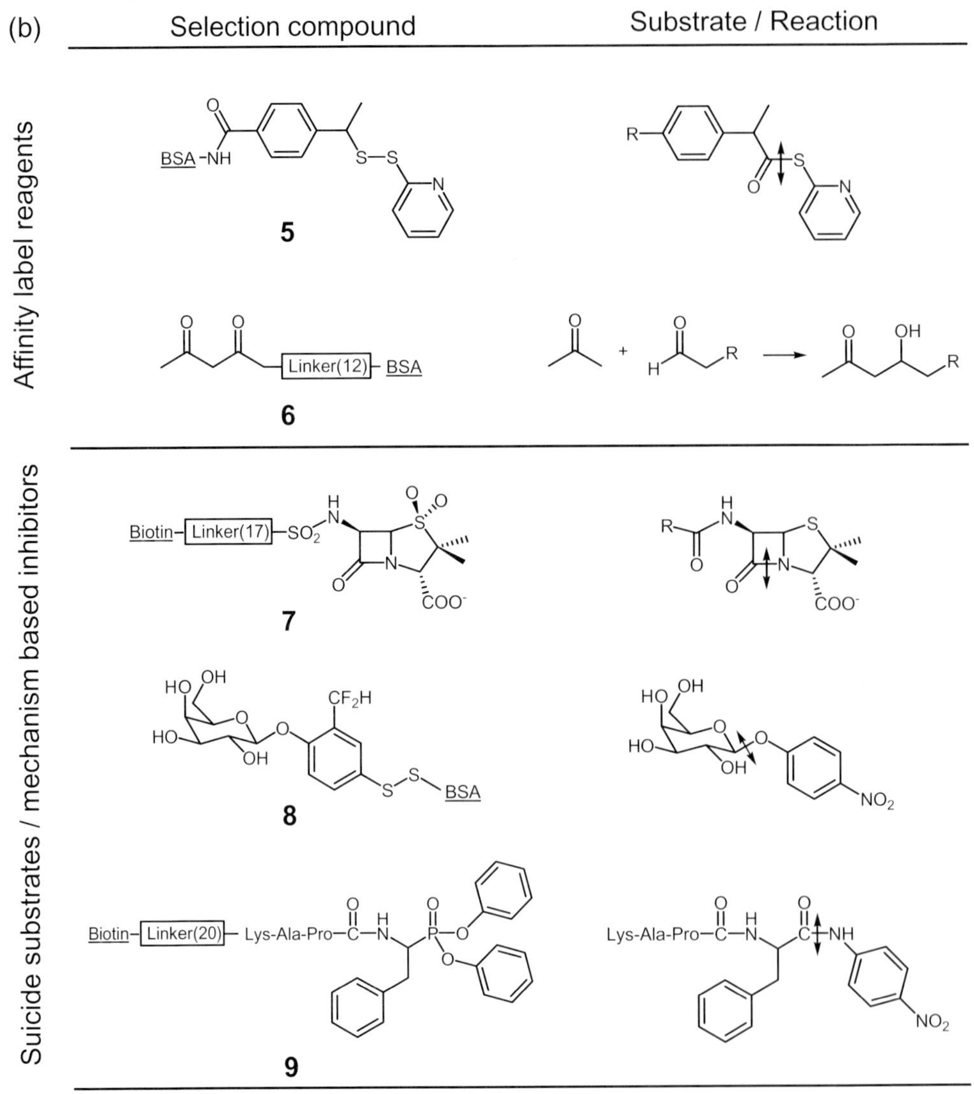

Figure 24.5 *Continued*

100-fold lower level than the wild-type enzyme. Some 80% of the variants, however, had a CO_2 hydrase activity which was close to that of the wild-type enzyme (within a factor of 3) [149]. Selection by capture on a TSA has also been used to extract active enzymes from a library of phage-displayed GST A1-1 mutants. Surprisingly, a mutant with a 20–90-fold lower activity than the wild-type enzyme on a chloronitrobenzene substrate was 13-fold more efficient in catalyzing glutathione (GSH) conjugation with ethacrynic acid (a Michael addition reaction) [150]. In these two examples, the principle of transition-state stabilization as a source of enzymatic catalysis was successfully exploited for the selection of highly active mutants. TSA affinity selection has also been used to extract hydrolytic antibodies from libraries of phage-antibodies (Figure 24.5a, lower part). The TSA affinity maturation of a mAb catalyzing the regioselective deprotection of an acylated carbohydrate was successfully used to improve its activity [143]. The most active mutant had a k_{cat} which was 12-fold higher than that of the wild-type mAb, but a ninefold increase in K_m. Interestingly, the quantitative increase in k_{cat} was comparable to the quantitative increase in TSA affinity. Similarly, the catalytic activity of an hydrolytic antibody was also improved up to 20-fold [151].

24.2.3.2.2 **Covalent Selection with Affinity Labels or Suicide Substrates** Affinity labeling has been used to select active phage-enzymes based on the rationale that enzymatic activities frequently rely on the presence in the active site of certain residues, the reactivity of which being exacerbated by the local environment. The group of Lerner designed a reagent (**5** in Figure 24.5b) that was capable of capturing phage-antibodies which contained a reactive cysteine within their hapten-binding site. A selected mutant was shown to catalyze thioester hydrolysis with the formation of a covalent acyl-intermediate. A 30-fold rate enhancement over background was measured; although this was modest on the acylation step, hydrolysis of the acylated cysteine intermediate appeared to be catalyzed more efficiently (10^4-fold versus the spontaneous hydrolysis of a thioester) [144]. Aldol condensations are catalyzed by amines, and the active sites of many aldolases contain an essential lysine residue. 1,3-Diketones can label highly reactive lysines and have been used in a process called reactive immunization to select an aldolase antibody [145]. By using such a diketone reagent (**6** in Figure 24.5b), enantioselective variants of this antibody endowed with new specificities were selected from a phage-displayed library, one of these being 10-fold more active than the parent antibody [152]. A direct relationship was found between the affinity of the antibody for the diketone and the k_{cat} for the retroaldol reaction, which indicated that the aldolase antibodies operated via an enamine mechanism [153]. Chloromethyl-ketones, in being highly reactive towards serine and cysteine proteases, can also be used to select phage-displayed proteases or to study zymogen activation [154].

In parallel, suicide substrates have been used based on the rationale that they specifically use the enzymatic activity to become activated before irreversibly blocking the active site. A selection strategy based on a biotinylated suicide substrate (**7** in Figure 24.5b) to extract phages displaying β-lactamase from mixtures containing inactive mutants was developed [146]. The method was shown to be suitable for selecting the most active enzymes from a mixture of mutants [155], and was used to select phage displaying β-lactamase antibodies with rate accelerations k_{cat}/k_{uncat} of 5200 and 320 (k_{cat} = 0.29 and 0.018 min^{-1}) [156]. The protocol was also modified to circumvent a difficulty of selections with suicide substrates in mechanisms involving a covalent intermediate: if inhibition arises from a covalent intermediate, then enzymes for which the rate of release of this intermediate is slow will be more efficiently selected. In order to prevent this selection of enzymes that do not turn over efficiently, a counter-selection step was included in the protocol. Thus, the library of mutants was incubated with substrate before adding the biotinylated inhibitor in order to block them as covalent intermediates. In this way the library could be enriched from 6 ppm to 25% active β-lactamases is four rounds of selection [157]. Glycosidase antibodies have also been selected using a suicide substrate (8 in Figure 24.5b). One mutant was shown to catalyze the hydrolysis of *p*-nitrophenyl-β-galactopyranoside with a rate enhancement k_{cat}/k_{uncat} of 7×10^4. For a comparison, the best catalyst obtained from 22 clones using the classical hybridoma technology was 700-fold less active [147].

Phosphonylating agents (e.g. **9** in Figure 24.5b), the status of which is intermediate between covalent transition-state analogues and suicide inhibitors, have been used to select phage displaying active mutants of subtilisin with wild-type-like specificity or with activity on substrates possessing a lysine in the P4 position [148]. These agents have also been used to select antibodies endowed with weak amidase activity [158, 159].

24.2.3.2.3 **Selection with Substrates** Finally, methods in which the direct interaction with substrates is used to select active mutants have been described. In the so-called 'catalytic elution' strategy (Figure 24.6), phages displaying metalloenzymes are immobilized via a substrate and in the absence of the metallic ion [160, 161]. Upon addition of the metallic ion, the enzyme is activated and the substrate turnover leads to elution of the phage. Other strategies use phages displaying both the enzyme and substrate in such a manner that 'intraphage' turnover can take place (Figure 24.7). Phage displaying the reaction product are then selected with an immobilized binder which specifically recognizes that product [162–164]. This self-labeling strategy was used to select improved subtiligases [165], mutants of O^6-alkylguanine-DNA alkyltransferase [166] and RNA polymerases from a DNA polymerase library [167].

24.3 Practical Considerations

A plethora of useful background information and current techniques are already available in several laboratory manuals dedicated to phage display [8–12]. These are excellent guides for those interested in the construction of phage-display libraries and the selection of specific binders. Hence, the following sections will focus on the essential practical aspects of the technology and on the key choices that must be made when starting a phage-display project.

24.3.1 Choosing a Vector

Several vectors have been developed for the display of proteins on filamentous phages. The phage coat is made of several thousands of a small protein (pVIII),

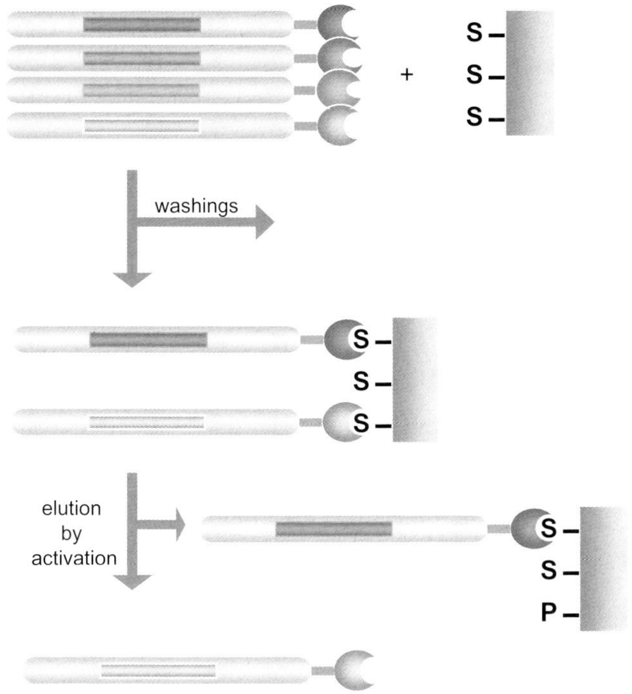

Figure 24.6 The catalytic elution principle for selecting cofactor-dependent enzymes. Phage-displaying apoenzymes are captured using an immobilized substrate and, as the affinity for the product is usually weaker than for the substrate, catalytic turnover upon addition of the required cofactor results in the elution of the active phage-enzymes.

and between three and five copies of each of the four minor proteins at the tips of the filament (pIII and pVI at one tip, pVII and pIX at the other). Although each of these proteins has been used for the display of foreign polypeptides, pIII is generally chosen for protein display. The pIII protein (406 amino acids, 42.5 kDa) is made of three domains, N1 (1–68), N2 (87–217) and CT (257–406), that are connected by glycine-rich sequences. N1 and N2 are necessary for phage infection, while CT (also called the 'anchoring domain') is essential for forming a stable phage particle. The pIII protein is expressed as a precursor with an amino-terminal signal peptide necessary for addressing the protein through the periplasm of *E. coli*. The signal peptide is removed by a specific protease after export, and the pIII ultimately becomes anchored in the bacterial inner membrane. Its assembly in the phage particle will be concomitant with phage extrusion. Besides pIII, it has been shown that the sequence of pVIII can be very efficiently evolved in order to display passenger protein more effectively than authentic pVIII [168–170]. Some engineered pVIIIs have been found to assemble in the phage coat with the C-terminal end outside (e.g. inside-out with respect to wild-type pVIII); this allows the display of protein scaffolds to be anchored to the phage coat by their

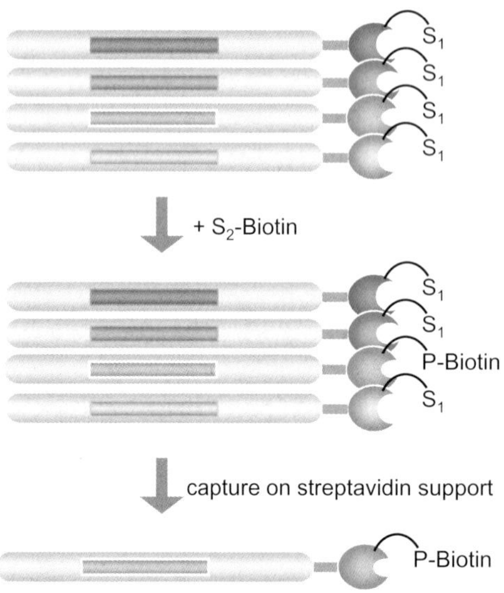

Figure 24.7 Enzyme selection by affinity capture of the reaction product: phage-enzymes are covalently modified by the substrate of the reaction to be selected. After 'intraphage' turnover, the product-labeled phages are selected by specific affinity capture.

N-terminal extremity instead of the C-terminal end, as in other types of construct [170].

Displaying a protein on pIII will necessitate the cloning of its gene between a sequence encoding a signal peptide and a sequence encoding a full or truncated coat protein. Either phage or phagemid vectors can be used, each of which has their own advantages and disadvantages. In the case of phage vectors, the gene of the recombinant protein is included in the phage genome, fused to the full pIII gene and under its native weak and noninducible promoter. Although phage particles should display only fusion proteins, full display remains an exception as *in vivo* proteolysis will often partially remove the fusion protein from the pIII. Typical display levels ranging between 0.2 and 3 proteins per phage particle are generally obtained. Polyvalent phages can afford the selection of weak binders by avidity effects if the density of the immobilized ligand is high. This is important, for example, when capturing enzymes on an immobilized substrate under conditions where the enzyme is inactive [161]. Phage vectors also contain antibiotic resistance markers which allow the detection of infected bacteria as colonies. In terms of disadvantages, all of the cloning and DNA manipulation must be performed using the replicative form (RF) of the phage, which is more difficult to produce in large amounts and with high purity. For some foreign genes, genetic instability is also sometimes evoked. In the case of phagemid vectors, the recombinant protein is encoded as a fusion protein with pIII or truncated pIII. The other proteins required to make a functional phage particle are provided by a 'helper phage'. Phagemids have all the advantages of plasmids with regard to cloning, DNA manipulation, control of expression with promoters and genetic stability. However, common inducible promoters are stronger than the endogenous pIII promoter, and their use may result in toxicity effects. Soluble proteins can be produced without subcloning if an amber codon is inserted between the displayed protein and pIII. Phagemid vectors can then become very efficient expression vectors of selected proteins if a powerful T7-based promoter is included in addition to the classical weaker promoters used for display on the phage surface [92]. This type of vector, although not useful when working with antibodies or poorly folding proteins, are extremely convenient for highly expressed scaffold proteins. One disadvantage of phagemids is the relatively low level of surface display, which results generally in a large proportion of phage particles that do not display any fusion protein. This is due to competition between the fusion protein and the wild-type pIII derived from the helper phage. Selection cycles are also longer because of the need for superinfection with the helper phage, and binders with less than micromolar affinity cannot be selected. Several strategies have been attempted in order to combine the advantages of the phagemid and phage libraries, by creating helper phages mutated or deleted in gene 3 (hyperphage), or packaging bacterial cell lines; in particular; those strategies were described for improving the display level of antibody fragments and the subsequent selection efficiency [171–174].

Phage display absolutely requires that the displayed proteins are efficiently translocated in the periplasm. This was previously a strong limitation as some very promising scaffolds were not efficiently translocated. For example, until recently the extremely stable consensus-based ankyrin protein scaffold [175] could not be used in phage display format, and the classically used phage display vectors relied on Sec B-type periplasmic exportation sequence. A more recently identified family of signal sequence (SRP-based sequences), by promoting cotranslational exportation to the periplasm, led to an efficient solution to this problem and very stable scaffolds can now be efficiently displayed [120]. Similarly, TAT-based signal sequences have been proposed to help the exportation of already folded proteins that are otherwise poorly displayed [176].

With pIII display, the C terminus of the protein of interest is fused to the N terminus of the coat protein via a peptide linker that must be sufficiently long to allow for the proper folding of both proteins. Moreover, in cases where the selection scheme involves capture by covalent trapping of the surface protein, a linker which is cleavable by a specific endoprotease is often of interest. Indeed, phage elution from the selecting support can then be easily performed by treatment with the protease. Finally, the linker should not be too susceptible to *in vivo* proteolysis in order to avoid cleavage of the fusion protein during phage production. AAIEGRAA is a convenient linker which is cleavable by Factor Xa endoprotease or trypsin, and is quite resistant to *in vivo* proteolysis; GGGSGGGS is a noncleavable, flexible and hydrophilic linker which is highly resistant to *in vivo* proteolysis. A monovalent display of N-terminal-anchored proteins using phage vectors has also been achieved through linker engineering of pIII fusion [177].

24.3.2
Phage Production

Although phages are created simply by growing infected bacteria, the quality and quantity produced are not greatly reproducible from one process to another. Infected bacteria can produce phages at various temperatures, typically between 20 and 37 °C. Whilst the rate of phage production will increase with temperature and bacterial density, the level of surface display will usually decrease as the growing temperature is increased. This effect depends on the stability of the displayed protein, since raising the temperature often results in an increased sensitivity of the fusion protein to *in vivo* proteolysis. The culture time is also an important parameter, as it must be long enough for the production of a sufficient amount of phages. On the other hand, the level of display – that is, the number of properly folded proteins displayed per phage particle – will generally decrease with the time of culture. For phagemids, the protocol will be similar but the library of transformed bacteria will be grown under repressing conditions and the phage production will be triggered by helper phage infection during the exponential phase and promoter induction. Phages particles will then be produced over a period ranging from 4 to 24 h.

Two practical protocols for phage production are described below, starting from infected bacteria that are glycerinated (40% glycerol) and stored at –80 °C.

- Protocol A: Inoculate 100 µl of infected bacteria in 250 ml of LB medium containing the appropriate antibiotic (use tetracycline if working with fd-DOG1 phages) in a 1 l flask. Incubate, with agitation, in an orbital shaker at 180 r.p.m. for either 20 h at 37 °C or 72 h at 23 °C. This procedure will produce large amounts of phages, but the level of display will not be highly reproducible.
- Protocol B: Inoculate 100 µl of infected bacteria in 25 ml of LB medium containing the appropriate antibiotic and grow overnight at 37 °C. In the morning, centrifuge the culture to pellet the bacterial cells. These are then resuspended in 250 ml of fresh medium. Incubate, with agitation, in an orbital shaker at 180 r.p.m. for 4 h at 30 or 37 °C. This protocol will generate phages with more reproducible and a generally higher level of display than protocol A, but the amount of phages will be less.

24.3.3
Phage Purification

Phages are typically purified by successive polyethylene glycol (PEG) precipitations. Nevertheless, the purity is not very high as some bacterial products coprecipitate with the phages. The amount of impurities may also vary with the nature of the displayed protein and the time of culture. It is suspected that bacterial lysis or periplasmic release can be provoked by phage extrusion; therefore, when high purity is required a CsCl equilibrium gradient should be performed after PEG precipitation. This will also remove the PEG, which may interfere with the binding of phages to some targets.

24.3.3.1 PEG Precipitation

1. Spin down a bacterial culture of 250 ml at 10 000 r.p.m. (12 000 g) for 10 min.
2. Carefully transfer 200 ml of the supernatant to a tube containing 50 ml of a solution containing 20% PEG (w/v) and 2.5 *M* NaCl.
3. Mix thoroughly and incubate for 1 h on ice.
4. Centrifuge at 10 000 r.p.m. (12 000 g) for 10 min.
5. Carefully discard the supernatant. Re-centrifuge at 10 000 r.p.m. for 1 min and remove the residual liquid.
6. Dissolve the pellet in 20 ml of Tris-buffered saline (TBS) buffer. Filter on a 1 μm Puradisc 25 GD prefilter unit (Whatman), and then on a 0.45 μm Millex-HV unit (Millipore).
7. Add 5 ml of 20% PEG-2.5 *M* NaCl, mix well, and incubate for 30 min on ice.
8. Repeat steps (4) and (5).
9. Dissolve the pellet in 1 ml TBS. Add 0.02% NaN_3 for long-term storage.

24.3.3.2 CsCl Equilibrium Gradient

1. Dissolve 2.5 g of CsCl in 3 ml Tris–EDTA (TE) buffer.
2. Add the phage solution (1 ml) and adjust the volume to 5 ml with TE buffer.
3. Centrifuge at 200 000 g for 17 h and at 15 °C. After centrifugation, the phages appear as a translucent band. PEG appears as a white flocculate below the phage band.
4. Collect the phages by piercing the tube with a needle just below the band and carefully pumping with a syringe. Dialyze twice against 1 liter TBS buffer for ≥6 h at 4 °C. Add 0.02% NaN_3 for long-term storage.

24.3.4 Measuring Phage Titer

1. As a high level of display may impair phage infection, it is recommended that the phages are treated with 10^{-7} *M* trypsin for 30 min before measuring the titer. Note that trypsin will remove the displayed protein only if a cleavable linker is used, or if the protein itself is degraded by trypsin. The stock solution of trypsin (10^{-5} *M*) should be freshly prepared in 20 m*M* acetate buffer, pH 3.0.
2. Prepare serial 10× dilutions of the phage solution.
3. Mix 10 μl of these dilutions with 990 μl of a TG1 culture in exponential phase.
4. Incubate at 37 °C without agitation for 30 min and under agitation for 30 min.
5. Spread 100 μl on Petri dishes containing the appropriate antibiotic (tetracycline for fd-DOG1 phages), and incubate overnight at 37 °C.
6. Count the colonies and calculate the phage titer as colony-forming units (cfu).

24.3.5
Measuring Phage Concentration

Phage concentration is simply obtained by measuring the absorbance at 265 nm and using an appropriate extinction coefficient which is proportional to the genome size. For a 10 kilobase phage, the extinction coefficient is $8.4 \times 10^7\,M^{-1}\,cm^{-1}$. Note that the phage concentration is generally between 20- and 50-fold higher than the phage titer.

24.3.6
Evaluating the Level of Display

The level of display is the average number of proteins displayed per phage particle. For protein display, this level can be evaluated by Western blot. For a wild-type enzyme display, if the activity is not affected by the phage environment, the level of display can be evaluated by dividing the k_{cat} of the phage (see Section 24.3.8) by the k_{cat} of the free enzyme. For some proteins, active-site labeling can also be considered.

24.3.6.1 Western Blot

The protocol is a classical SDS–PAGE (10%) followed by transfer onto a Western blot membrane and immunodetection with an anti-pIII antibody. Nevertheless, special care must be given to the sample preparation as phages are very stable and difficult to denature. The protocol is similar to a typical SDS–PAGE sample preparation, but β-mercaptoethanol should be replaced by fresh dithiothreitol (DTT; 5 mM final) and the samples should be boiled in a water bath for at least 15 min. Moreover, as the pIII-fusion protein is a minor component of the virion, a large amount of phages should be loaded on the gel, typically around 10^{12} per lane.

The level of protein display is roughly evaluated by comparing the relative intensities of the bands corresponding to the protein-pIII fusion protein and the pIII protein alone, and by considering between three and five copies of pIII per phage particle. A quantitative value will be obtained using a fluorimager or a phosphorimager.

24.3.6.2 Active-Site Labeling

Whenever active-site labeling of the displayed protein is amenable with fluorescent or radioactive compounds, the level of display should also be evaluated by this method. The protocol will depend on the displayed protein. It will comprise several steps: labeling the phages; removing the excess label by two PEG precipitations or by dialysis; and comparing the fluorescence or radioactivity of the phages with standards. Note that the phage concentration should not be higher than 10 n*M* in order to avoid insolubility problems.

24.3.7
Measuring the Affinity of a Phage for a Ligand

If phage particles are not polyvalent, dissociation constants can be determined by ELISA, following the procedure described by Friguet *et al.* [178]. Microplates are coated with the ligand and the range of phage concentrations providing a linear signal, directly proportional to the phage concentration, is determined. A concentration giving a good signal, but for which less than 2.5% of the phages are trapped by the coated ligand, is chosen. To measure the dissociation constant, phages at this concentration are incubated with various concentrations of ligand in solution for 1 h at room temperature, with stirring. When the equilibrium is reached the proportion of free phages in each sample is determined on ligand-coated microplates. As the equilibrium in solution is not significantly modified by immobilization, the ELISA signal is proportional to the free phage concentration. If the phage concentration is much lower than the range of ligand concentrations, a sigmoidal titration curve is obtained. The ligand concentration giving 50% of the signal is equal to the dissociation constant.

24.3.8
Measuring the Activity of a Phage-Enzyme

A solution of phage-enzyme can be used as an enzyme solution for measuring kinetic parameters such as k_{cat} and K_M. Usually, phage-displayed enzymes are seen to behave essentially as free enzymes in solution, although an interference is always possible, especially in case of multiple display. It should be noted that the k_{cat} will be the turnover rate of the phage and not of the enzyme, as the level of display is generally just a rough evaluation.

24.3.9
Library Construction

The library will be constructed by error-prone PCR or DNA shuffling or by the incorporation of degenerated oligonucleotides. For error-prone PCR, the average number of mutations per gene is dependent of the number of PCR cycles. The number of different clones present in a library is known as the diversity. Since for any mutagenic method the chances of finding interesting clones in a library will increase as the diversity increases, the major concern when constructing a phage library will be to achieve the highest diversity. Although several methods are available for constructing libraries, the last step is always to ligate a collection of DNA fragments into a phage or a phagemid vector, followed by transformation. The transformation efficiency will generally determine the diversity, and must therefore be optimized. Special attention must be paid to double transformation artifacts [179, 180].

Some handy practical points concerning the final ligation and transformation steps are as follows:

- Use large amounts of a well-purified restricted vector. It is recommended that the vector is prepared using a Qiagen maxipreparation kit, and purified with a CsCl gradient.
- Purify the collection of DNA fragments using acrylamide gel electrophoresis.
- Use a ratio vector : fragment of 1 : 3 for the ligation.
- Purify the ligation mix as thoroughly as possible and concentrate the DNA to approximately 1 μg μl^{-1}.
- Perform a transformation test before creating the library by electroporating 1 μl of ligation mix in 100 μl of competent cells and comparing with a standard such as pUC18. A reasonable objective is between 10^6 and 10^7 transformants per electroporation. Typical libraries are in the order of 10^7 to 10^9 independent clones, although libraries of over 10^{10} independent clones have been described [181].

Combinatorial approaches have also been used to increase the library size up to 10^{12} clones [35, 182, 183].

24.3.10
Library Production

Transformed bacteria must be grown to produce the phage library. At this stage, *in vivo* selection processes and contact-dependent inhibition of growth might favor some clones over others, although any such bias can be reduced by avoiding liquid culture for the first library production. Thus, the library should be grown on large Petri dishes (30 × 30 cm^2) containing agar medium, and under repressing conditions if phagemids are being used. Not more than 10^8 individual transformants should be grown per Petri dish. A typical protocol for preparing the first-generation library is as follows.

1. After electroporation, add 900 μl of LB for 100 μl cells. Incubate for 1 h at 37 °C.
2. Take an aliquot of 10 μl and add 990 μl of LB medium. On small Petri dishes, plate 100 μl of serial 10× dilutions of the cells for measuring the library diversity.
3. Spread the electroporation mix on a large Petri dish containing LB-agar and the appropriate antibiotics (tetracycline for fd-DOG1 phages). Incubate overnight at 37 °C or for 72 h at 23 °C.
4. Recover the bacteria and phages by pouring 30 ml TBS onto the agar and resuspending the bacteria. Repeat this three times for each Petri dish.
5. Centrifuge at 10 000 r.p.m. for 10 min to pellet the bacteria. The bacteria should be resuspended in LB containing 40% glycerol and stored at −80 °C. For phage libraries, the phage particles are also recovered in the supernatant and can be further purified and stored at 4 °C.

Each time that the library must be produced for selection, it should be prepared from the first-generation library of phagemid-transformed bacteria or phages. If starting from phages:

1. Take an aliquot of phages containing at least 100 times more phages than the library diversity.
2. If a trypsin-cleavable linker is present, add 10^{-7} *M* trypsin for removing the displayed proteins and for avoiding infection bias. Incubate for 30 min at room temperature. Remove the trypsin by precipitating the phages with PEG.
3. Dissolve the phages in 1 ml TBS and infect an TG1 culture in exponential phase. The culture should contain at least 10 times more bacteria than the library diversity.
4. Incubate for 1 h at 37 °C, without agitation.
5. Mix the cells and take an aliquot for phage titering. The titer should be at least 10 times more than the library diversity.
6. Grow the bacteria at 37 °C for 4 h or overnight at 23 °C, with agitation. Growing at a lower temperature will improve the display of unstable proteins.
7. Purify the phages as described.

For phagemids, start with at least 10 times more bacteria than the library diversity, and refer to Section 24.3.2.

24.3.11 Selections

24.3.11.1 Affinity-Based Selections

The specific capture of phages on an immobilized target ligand is the most common selection procedure used in phage display. The immobilization step should preserve the structural integrity of the ligand and the accessibility of the interaction site, if defined. The two main immobilization approaches are: (i) the unspecific adsorption on plastic wells or tubes, essentially for proteic ligands; and (ii) the specific capture of a biotinylated ligand on a streptavidin/avidin-coated support such as plastic wells, membranes or magnetic beads. When using biotinylated ligands, the interaction with the phages can be performed in solution, before the immobilization. This provides a better control of the selection pressure by adapting the ligand concentration, and may also prevent a possible accessibility problem of the immobilized ligand. Moreover, when a protein is selected based on its capture on a streptavidin-bound ligand, it is quite common that the protein recognizes the streptavidin/ligand complex but not the free ligand. This is especially problematic when working with small-molecule ligands, but mixing the phages and the ligand in solution should largely prevent this bias. However, any excess of unbound biotinylated ligand should then be removed prior to immobilization, especially when high concentrations are used. Usually, this can be achieved by PEG precipitation or by using a small desalting column. Another way to avoid the selection of phages that are binding to the support is to alternate selection rounds on different supports, for example streptavidin/avidin-coated materials, or plastic wells and magnetic beads.

The phages (typically around 10^{12}) are then incubated with the support for a certain amount of time (typically between 2 and 24 h). The addition of soft detergents (Tween 20 or Triton X-100), albumin or skimmed milk will reduce nonspecific binding. After the capture, the support is washed (usually between five and ten times) with a buffer solution that may contain detergents (e.g. 0.1% Tween 20). The bound phages are then eluted by changing the pH, adding a soluble competitor, or cleaving the linker between the displayed protein and the coat protein. If an acidic pH is used, it should be neutralized immediately after elution. If a soluble competitor is used, then several hours of incubation is recommended because the dissociation of the phage from the support is probably very slow due to its low diffusion in solution. In some cases the phages are bound so strongly that classical elution is inoperative. Disrupting the phage capsid and amplifying its DNA by rolling circle replication has been used [184], but this requires additional restriction, ligation and transformation steps.

Following phage elution comes bacteria infection. When working with possible polyvalent phages, removal of the displayed protein by proteolysis is recommended prior to infection because polyvalent display on pIII may impair phage infectivity. Phage infection titers should also be measured before (in) and after (out) capture and elution, the ratio out/in being an efficiency indicator of the process. Amplified phages can then be injected into a new round of selection. The increase of the out/in ratio from one round to another is an indication of a successful selection. In order to progressively increase the selection pressure from one round to another, reducing the time of incubation between the phages and the ligand (immobilized or not) and increasing the number of washings after the capture can be considered.

24.3.11.2 Activity-Based Selections of Phage-Enzymes

Whenever possible, a model selection should be optimized before starting selections with a library. In this model selection a mixture of active and inactive phage-enzymes is used as a model library. One selection round is then performed and the phage mixture is analyzed before and after selection. The enrichment factor (EF) is then calculated as follows:

$$\mathrm{EF} = \frac{(A_{out}/I_{out})}{(A_{in}/I_{in})}$$

where A and I represent respectively the fractions of active and inactive phages after (out) or before (in) the selection. A selection strategy will be considered as efficient if the enrichment factor is higher than 50.

24.3.11.2.1 Selection with Suicide Substrate The strategy is based on selecting the phages displaying active enzymes by labeling them with a biotinylated suicide substrate and capturing the labeled phages with immobilized streptavidin. The method was initially developed for selecting enzymes that feature a covalent intermediate in their mechanism of action [146]. When the covalent intermediate is formed, the complex can either follow a normal catalytic cycle or go through a suicide event leading to the irreversible labeling which is necessary for the selection. The suicide inhibition efficiency will depend on the ratio between these two rates of reaction, the ratio being in turn dependent upon the nature of the suicide substrate and of the enzyme. Therefore, in these selections it is recommended that a large excess of suicide substrate is used compared to the displayed enzyme.

As the labeling event occurs before – and may even compete with – the last step of the turnover, there is a selection pressure for enzymes that will feature a low last step; that is, enzymes with very low turnover. In order to overcome this problem active sites that do not turn over rapidly can be blocked by an initial incubation with a normal substrate before labeling with the biotinylated suicide substrate [157]. Consequently, special attention must be paid to the kinetic control of the labeling step.

The following protocol is described for the selection of phage-displayed serine β-lactamase with a biotinylated penam-sulfone suicide substrate. For other activities, the concentrations of substrate and suicide substrate, as well as the times of reaction, should most likely be adapted.

- **The Labeling Procedure**
  1. Just before use, prepare 1 m*M* stock solutions of substrate and biotinylated suicide substrate in TBS buffer. If the solubility is too low, dissolve these in pure dimethylsulfoxide (DMSO). Note: a small amount of water in the DMSO may degrade the molecules rapidly as the water activity is high in DMSO; thus, freshly dried DMSO should be used.
  2. In a final volume of 1 ml of phosphate buffer (pH 7, 50 m*M*), mix 10^{12} phages with 10^{-5} *M* substrate and incubate for 10 min at room temperature in order to block all the active sites that are not turning over.
  3. Add 400 µl of PEG 20%-NaCl 2.5 *M*, vortex for a few seconds and centrifuge for 5 min at 14 000 r.p.m. (12 000 g). Discard the supernatant containing excess substrate and dissolve the phage pellet in 1 ml of phosphate buffer.
  4. Add the biotinylated suicide substrate to a final concentration of 10^{-5} *M* and incubate for 20 min at room temperature.
  5. Eliminate the excess of suicide substrate by two PEG precipitations, as described in step (3).
  6. Take an aliquot of 10 µl for measuring the 'input' phage titer and proceed immediately to the capture with the remaining 990 µl.
- **Capture and Elution**
  1. In a microtube, add 1 mg of M-270 streptavidin Dynabeads to 1 ml TBS and place the tube on the magnet to discard the supernatant. Off the magnet, resuspend the beads in 1 ml of MTBS (TBS buffer containing 2% nonfat dried milk) for blocking nonspecific sites. Place the microtube on a rotating wheel for 1 h at room temperature (the wheel is rotated to maintaining bead suspension, and should be slow).
  2. Remove the MTBS on the magnet. Wash the beads with 1 ml TBS and resuspend them in the phage solution supplemented with 1% bovine serum albumin (BSA).

3. Place the microtube on a rotating wheel for 4 h at room temperature or overnight at 4 °C.
4. On the magnet discard the supernatant containing unbound phages. Eventually take an aliquot for measuring the phage titer.
5. Wash the beads five times with 1 ml TTBS (TBS containing 5% v/v Tween 20) and once with TBS (a single washing consists of resuspending the beads off the magnet and discarding the solution on the magnet).
6. If the connecting linker or the displayed enzyme is susceptible to proteolytic cleavage, resuspend the beads in 1 ml TBS containing 10^{-7} *M* trypsin or 5 units of Factor Xa. If the suicide substrate contains a disulfide bridge, elution can also be performed by resuspending the beads in 1 ml TBS containing 10 m*M* DTT. In both cases, incubate on the rotating wheel for 1 h.
7. Recover the phages in the supernatant and add them to 50 ml of a TG1 culture in exponential phase and in LB medium. Mix well and incubate at 37 °C, without agitation for 30 min, and then with agitation for 30 min.
8. Take 300 μl for measuring the 'output' phage titer and transfer the remainder of the culture to a 1 l flask containing 200 ml LB medium with the appropriate antibiotic. Proceed as described in Section 24.3.2.

The selection efficiency is evaluated by the ratio of the titer's 'output' over 'input'. After each selection, the phages are amplified and can be injected into a new selection round. If some clones of the library are effectively selected the ratio should increase from about 10^{-5} (background level) to about 10^{-2}. Typically, the ratio reaches a plateau after between four and eight selection rounds, depending on the starting diversity and on the power of the selection itself.

It may be interesting to increase the selection pressure from one selection round to another by doubling the number of washes and by decreasing by a factor 10 the suicide substrate concentration and the time of incubation. This could lead to the selection of the most efficient catalysts.

At least 20 clones obtained from the last round before the plateau should be sequenced for evaluating diversity after selection. Depending on the ease with which an activity assay is performed, as many clones as possible should also be screened for activity. Monoclonal preparations of phage-enzymes should be assayed first and, if the activity is too low, soluble overexpressed enzymes should be produced for reaching higher concentrations.

24.3.11.2.2 **Other Selection Strategies** Several other strategies have been developed for selecting enzymes on phages. As most of the protocols are very similar, this section will focus only on the major technical and practical differences. Selection with transition-state analogues is not described here as it is simply an affinity selection.

- Selection by suicide leaving group: This strategy is a slight variant of that previously described, as the nature of the selecting substrate is different because it contains a suicide leaving group [147, 185]. In this case the enzymatic turnover will release a reactive species that ultimately will react with a proximal residue and label the phage. Although the advantage is that the selection does not require a covalent intermediate, these materials are very poor suicide substrates due to rapid diffusion of the leaving group. In order to overcome this problem the reaction must be performed with an immobilized substrate on the leaving group side. Also, as the phage-enzyme is not diffusing rapidly in solution the labeling will be efficient enough for selection.

Working with an immobilized substrate can, however, lead to problems that will be difficult to identify. Due to the slow diffusion of the phages, the reaction time should be greatly increased. The control of some parameters, such as interference with the support or the density of immobilization, is difficult; moreover, substrate recognition by the enzyme may be significantly impaired.

- Selection by catalytic elution: In this strategy, phages are affinity-captured on an immobilized substrate under conditions where the enzyme is inactive, for example in the absence of an essential metallic ion or a cofactor [161]. The active phage-enzymes are then eluted by triggering the catalysis upon addition of the metallic ion or cofactor, taking advantage of the enzyme's lower affinity for the product than for the substrate.

  As enzymes do not generally feature a high affinity for their substrates, the initial affinity capture may also be problematic. It is therefore advised that phages be generated with a high level of display (i.e. more than one enzyme per phage) in order to take advantage of possible avidity effects. It is also important to remove the displayed enzymes by proteolysis after elution, as a multivalent display could impair phage infection. In addition to these aspects, the selection essentially resembles a classical affinity selection.
- Selection by product labeling: As shown in Figure 24.1, this protocol starts with the phages being labeled with the substrate, using a variety of approaches [160, 162, 163]. The active enzymes then turn over the substrate into product by intra-phage catalysis. Finally, the product-labeled phages are selected by classical affinity capture.

  In this strategy, as it is very important to avoid inter-phage catalysis, the phage concentration should be kept low (under 10^{-9} *M*) and phages should not be precipitated with PEG. As the removal of excess label will generally be required before affinity capture, it should be completed by either phage dialysis or size-exclusion chromatography.

24.3.12 Troubleshooting

24.3.12.1 Phage Titers are not Reproducible

Phages are 'sticky' towards themselves and towards solid supports and, when concentrated, can form soluble aggregates that will dissociate relatively slowly. Therefore it is recommended that all phage solutions be vortexed thoroughly before infection. As the phage will also stick to micropipette tips, the tips should be changed when performing serial dilutions. Finally, when the phages are highly diluted, a time-dependent loss of infection may result from their binding to the vessel walls. Hence, it is recommended either that silanized microtubes are used, that 1% BSA is added to the solution, or that highly diluted solutions should not be kept for long periods of time.

24.3.12.2 Displayed Protein is Degrading with Time

This effect may be due to the presence of proteases or to an inherently low protein stability. A 'cocktail' of protease inhibitors can be added (Complete tabs, Roche), and freshly prepared phage solutions should always be added when performing selection from libraries.

24.3.12.3 Phages are not Genetically Stable

This may be due to a low toxicity of the fusion protein or to recombination with homologous *E. coli* genes. The solution is to use a *recA* strain such as JM109 to reduce recombination. To avoid problems of toxicity, a phagemid vector such as pHDi.Ex [92] should be used, as this allows control of the fusion protein expression. With this vector, a strong repression will be obtained by adding 1% glucose (catabolic repression).

24.3.12.4 The Ratio 'Out/In' is not Increasing with the Selection Rounds

This may mean that no clones are being selected. For some strategies where low-affinity capture may be necessary for selection, the level of specifically captured phages may always be below the background level. It is therefore worth analyzing the selected phages as an effective enrichment may have occurred.

24.4 Conclusions and Future Challenges

In less than 25 years, phage display has become a robust and extremely powerful technology for creating artificial protein binders that could find applications both in fundamental research and industrial processes. The first therapeutic antibodies evolved by phage display are now available commercially, more are in clinical trials and, most certainly, many more will follow. There is no doubt that the use of antibodies will continue to increase in the near future, although alternative scaffolds are emerging with attractive advantages such as small size, a high level of recombinant expression or intracellular applicability. Now, with a good scaffold and a good library, the selection of strong binders is not exceptional, perhaps even easy, and alternatives to antibodies are changing their status from small outsiders to serious competitors. Even more promising, the engineering of bifunctional proteins that combine a binding site with a fluorescent property, an enzymatic active site or another binding site is probably the next major challenge. These proteins would be capable of both recognizing a target and transmitting a signal, and should find useful applications in diagnostics or immunotherapies.

The phage display technology is much less adapted for enzyme evolution than for engineering binders. Although elegant activity-based selection strategies have been developed, they have been modestly successful. This is most likely because either the methods are not based on the whole catalytic cycle (suicide substrates and affinity labels) or they are based on a single turnover. In that case, a very poor catalyst turning on a minute time scale will generally be selected as well as a very efficient one turning on a millisecond. However, from a fundamental perspective, these studies have been very innovative, especially in the field of catalytic antibodies. In the future, it is unclear whether some of these strategies could evolve towards powerful *in vitro* selection methods for highly efficient catalysts.

From a practical point of view, the key to future success will probably be the capacity to create libraries of both the highest quality and the highest diversity. With regards to the choice of the starting protein, criteria such as thermostability, immunogenicity, size, recombinant expression yield, tolerance to mutations and insertions and multifunctionality may be crucial. With regards to diversification, mutagenesis protocols that are codon-based could afford better quality and better control on the diversity itself. Finally, the combination of the genetic diversity with unnatural amino acid incorporation, or with additional diversification by chemical modifications, could also open the way to innovative and more ambitious protein-engineering projects.

References

1 Smith, G.P. (1985) Filamentous fusion phage – novel expression vectors that display cloned antigens on the virion surface. *Science*, **228**, 1315–17.

2 Leemhuis, H., Stein, V., Griffiths, A.D. and Hollfelder, F. (2005) New genotype-phenotype linkages for directed evolution of functional proteins. *Current Opinion in Structural Biology*, **15**, 472–8.

3 Sergeeva, A., Kolonin, M.G., Molldrem, J.J., Pasqualini, R. and Arap, W. (2006) Display technologies: application for the discovery of drug and gene delivery agents. *Advanced Drug Delivery Reviews*, **58**, 1622–54.

4 Kehoe, J.W. and Kay, B.K. (2005) Filamentous phage display in the new millennium. *Chemical Reviews*, **105**, 4056–72.

5 Smith, G.P. and Petrenko, V.A. (1997) Phage display. *Chemical Reviews*, **97**, 391–410.

6 Katz, B.A. (1997) Structural and mechanistic determinants of affinity and specificity of ligands discovered or engineered by phage display. *Annual Review of Biophysics and Biomolecular Structure*, **26**, 27–45.

7 Winter, G., Griffiths, A.D., Hawkins, R.E. and Hoogenboom, H.R. (1994) Making antibodies by phage display technology. *Annual Review of Immunology*, **12**, 433–55.

8 Sidhu, S.S. (2005) *Phage Display in Biotechnology and Drug Discovery*, CRC Press, Boca Raton.

9 Clackson, T. and Lowman, H.B. (2004) *Phage Display: A Practical Approach*, Oxford University Press, New York.

10 O'Brien, P.M. and Aitken, R. (2002) *Antibody Phage Display – Methods and Protocols*, Humana Press, Totowa.

11 Barbas, C.F., 3rd, Scott, J.K., Silverman, G. and Burton, D.R. (2001) *Phage Display: A Laboratory Manual*, Cold Spring Harbor Laboratory Press, New York.

12 Kay, B.K., Winter, J. and McCafferty, J. (1996) *Phage Display of Peptides and Proteins. A Laboratory Manual*, Academic Press, San Diego.

13 Hust, M., Jostock, T., Menzel, C., Voedisch, B., Mohr, A., Brenneis, M., Kirsch, M.I., Meier, D. *et al.* (2007) Single chain Fab (scFab) fragment. *BMC Biotechnology*, **7**, 14.

14 Todorovska, A., Roovers, R.C., Dolezal, O., Kortt, A.A., Hoogenboom, H.R. and Hudson, P.J. (2001) Design and application of diabodies, triabodies and tetrabodies for cancer targeting. *Journal of Immunological Methods*, **248**, 47–66.

15 Fernández, L. (2004) Prokaryotic expression of antibodies and affibodies. *Current Opinion in Biotechnology*, **15**, 364–73.

16 Benhar, I. (2001) Biotechnological applications of phage and cell display. *Biotechnology Advances*, **19**, 1–33.

17 Muyldermans, S. (2001) Single domain camel antibodies: current status. *Journal of Biotechnology*, **74**, 277–302.

18 De Genst, E., Saerens, D., Muyldermans, S. and Conrath, K. (2006) Antibody repertoire development in camelids. *Developmental and Comparative Immunology*, **30**, 187–98.

19 Davies, J. and Riechmann, L. (1996) Single antibody domains as small recognition units: design and in vitro antigen selection of camelized, human VH domains with improved protein stability. *Protein Engineering*, **9**, 531–7.

20 Aires da Silva, F., Santa-Marta, M., Freitas-Vieira, A., Mascarenhas, P., Barahona, I., Moniz-Pereira, J., Gabuzda, D. and Goncalves, J. (2004) Camelized rabbit-derived VH single-domain intrabodies against Vif strongly neutralize HIV-1 infectivity. *Journal of Molecular Biology*, **340**, 525–42.

21 Tanha, J., Nguyen, T.D., Ng, A., Ryan, S., Ni, F. and Mackenzie, R. (2006) Improving solubility and refolding efficiency of human V(H)s by a novel mutational approach. *Protein Engineering, Design and Selection*, **19**, 503–9.

22 Zarebski, L.M., Urrutia, M. and Goldbaum, F.A. (2005) Llama single domain antibodies as a tool for molecular mimicry. *Journal of Molecular Biology*, **349**, 814–24.

23 Alvarez-Rueda, N., Behar, G., Ferre, V., Pugniere, M., Roquet, F., Gastinel, L., Jacquot, C., Aubry, J. *et al.* (2007) Generation of llama single-domain antibodies against methotrexate, a prototypical hapten. *Molecular Immunology*, **44**, 1680–90.

24 To, R., Hirama, T., Arbabi-Ghahroudi, M., MacKenzie, R., Wang, P., Xu, P., Ni, F. and Tanha, J. (2005) Isolation of monomeric human VHS by a phage selection. *Journal of Biological Chemistry*, **280**, 41395–403.

25 Azzazy, H.M. and Highsmith, W.E., Jr (2002) Phage display technology: clinical applications and recent innovations. *Clinical Biochemistry*, **35**, 425–45.

26 Bradbury, A.R. and Marks, J.D. (2004) Antibodies from phage antibody libraries. *Journal of Immunological Methods*, **290**, 29–49.

27 Hoogenboom, H.R. (2005) Selecting and screening recombinant antibody libraries. *Nature Biotechnology*, **23**, 1105–16.

28 Clackson, T., Hoogenboom, H.R., Griffiths, A.D. and Winter, G. (1991) Making antibody fragments using phage display libraries. *Nature*, **352**, 624–8.

29 Bender, E., Pilkington, G.R. and Burton, D.R. (1994) Human monoclonal Fab fragments from a combinatorial library prepared from an individual with a low serum titer to a virus. *Human Antibodies and Hybridomas*, **5**, 3–8.

30 Williamson, R.A., Burioni, R., Sanna, P.P., Partridge, L.J., Barbas, C.F., 3rd and Burton, D.R. (1993) Human monoclonal antibodies against a plethora of viral pathogens from single combinatorial libraries. *Proceedings of the National Academy of Sciences of the United States of America*, **90**, 4141–5.

31 Griffiths, A.D., Williams, S.C., Hartley, O., Tomlinson, I.M., Waterhouse, P., Crosby, W.L., Kontermann, R.E., Jones, P.T. *et al.* (1994) Isolation of high affinity human antibodies directly from large synthetic repertoires. *EMBO Journal*, **13**, 3245–60.

32 de Haard, H.J., van Neer, N., Reurs, A., Hufton, S.E., Roovers, R.C., Henderikx, P., de Bruine, A.P., Arends, J.W. *et al.* (1999) A large non-immunized human Fab fragment phage library that permits rapid isolation and kinetic analysis of

high affinity antibodies. *Journal of Biological Chemistry*, **274**, 18218–30.

33 Vaughan, T.J., Williams, A.J., Pritchard, K., Osbourn, J.K., Pope, A.R., Earnshaw, J.C., McCafferty, J., Hodits, R.A. *et al.* (1996) Human antibodies with sub-nanomolar affinities isolated from a large non-immunized phage display library. *Nature Biotechnology*, **14**, 309–14.

34 Sheets, M.D., Amersdorfer, P., Finnern, R., Sargent, P., Lindquist, E., Schier, R., Hemingsen, G., Wong, C. *et al.* (1998) Efficient construction of a large nonimmune phage antibody library: the production of high-affinity human single-chain antibodies to protein antigens. *Proceedings of the National Academy of Sciences of the United States of America*, **95**, 6157–62.

35 Sblattero, D. and Bradbury, A. (2000) Exploiting recombination in single bacteria to make large phage antibody libraries. *Nature Biotechnology*, **18**, 75–80.

36 Benhar, I. (2007) Design of synthetic antibody libraries. *Expert Opinion on Biological Therapy*, **7**, 763–79.

37 Barbas, C.F., 3rd, Kang, A.S., Lerner, R.A. and Benkovic, S.J. (1991) Assembly of combinatorial antibody libraries on phage surfaces: the gene III site. *Proceedings of the National Academy of Sciences of the United States of America*, **88**, 7978–82.

38 Soderlind, E., Strandberg, L., Jirholt, P., Kobayashi, N., Alexeiva, V., Aberg, A.M., Nilsson, A., Jansson, B. *et al.* (2000) Recombining germline-derived CDR sequences for creating diverse single-framework antibody libraries. *Nature Biotechnology*, **18**, 852–6.

39 Knappik, A., Ge, L., Honegger, A., Pack, P., Fischer, M., Wellnhofer, G., Hoess, A., Wolle, J. *et al.* (2000) Fully synthetic human combinatorial antibody libraries (HuCAL) based on modular consensus frameworks and CDRs randomized with trinucleotides. *Journal of Molecular Biology*, **296**, 57–86.

40 Hust, M. and Dubel, S. (2004) Mating antibody phage display with proteomics. *Trends in Biotechnology*, **22**, 8–14.

41 Sidhu, S.S. and Kossiakoff, A.A. (2007) Exploring and designing protein function with restricted diversity. *Current Opinion in Chemical Biology*, **11**, 347–54.

42 Deng, S.J., MacKenzie, C.R., Sadowska, J., Michniewicz, J., Young, N.M., Bundle, D.R. and Narang, S.A. (1994) Selection of antibody single-chain variable fragments with improved carbohydrate binding by phage display. *Journal of Biological Chemistry*, **269**, 9533–8.

43 Yang, W.P., Green, K., Pinz-Sweeney, S., Briones, A.T., Burton, D.R. and Barbas, C.F., 3rd (1995) CDR walking mutagenesis for the affinity maturation of a potent human anti-HIV-1 antibody into the picomolar range. *Journal of Molecular Biology*, **254**, 392–403.

44 Yau, K.Y., Dubuc, G., Li, S., Hirama, T., Mackenzie, C.R., Jermutus, L., Hall, J.C. and Tanha, J. (2005) Affinity maturation of a V(H)H by mutational hotspot randomization. *Journal of Immunological Methods*, **297**, 213–24.

45 Schier, R. and Marks, J.D. (1996) Efficient in vitro affinity maturation of phage antibodies using BIAcore guided selections. *Human Antibodies and Hybridomas*, **7**, 97–105.

46 Schier, R., McCall, A., Adams, G.P., Marshall, K.W., Merritt, H., Yim, M., Crawford, R.S., Weiner, L.M. *et al.* (1996) Isolation of picomolar affinity anti-c-erbB-2 single-chain Fv by molecular evolution of the complementarity determining regions in the center of the antibody binding site. *Journal of Molecular Biology*, **263**, 551–67.

47 Rajpal, A., Beyaz, N., Haber, L., Cappuccilli, G., Yee, H., Bhatt, R.R., Takeuchi, T., Lerner, R.A. *et al.* (2005) A general method for greatly improving the affinity of antibodies by using combinatorial libraries. *Proceedings of the National Academy of Sciences of the United States of America*, **102**, 8466–71.

48 Hawkins, R.E., Russell, S.J. and Winter, G. (1992) Selection of phage antibodies by binding affinity. Mimicking affinity maturation. *Journal of Molecular Biology*, **226**, 889–96.

49 Irving, R.A., Kortt, A.A. and Hudson, P.J. (1996) Affinity maturation of recombinant antibodies using *E. coli* mutator cells. *Immunotechnology*, **2**, 127–43.

50 Low, N.M., Holliger, P.H. and Winter, G. (1996) Mimicking somatic hypermutation: affinity maturation of antibodies displayed on bacteriophage using a bacterial mutator strain. *Journal of Molecular Biology*, **260**, 359–68.

51 Crameri, A., Cwirla, S. and Stemmer, W.P. (1996) Construction and evolution of antibody-phage libraries by DNA shuffling. *Nature Medicine*, **2**, 100–2.

52 Marks, J.D., Griffiths, A.D., Malmqvist, M., Clackson, T.P., Bye, J.M. and Winter, G. (1992) By-passing immunization: building high affinity human antibodies by chain shuffling. *Bio/Technology (Nature Publishing Company)*, **10**, 779–83.

53 Daugherty, P.S., Chen, G., Iverson, B.L. and Georgiou, G. (2000) Quantitative analysis of the effect of the mutation frequency on the affinity maturation of single chain Fv antibodies. *Proceedings of the National Academy of Sciences of the United States of America*, **97**, 2029–34.

54 Hoet, R.M., Cohen, E.H., Kent, R.B., Rookey, K., Schoonbroodt, S., Hogan, S., Rem, L., Frans, N. *et al.* (2005) Generation of high-affinity human antibodies by combining donor-derived and synthetic complementarity-determining-region diversity. *Nature Biotechnology*, **23**, 344–8.

55 Zahnd, C., Spinelli, S., Luginbuhl, B., Amstutz, P., Cambillau, C. and Pluckthun, A. (2004) Directed in vitro evolution and crystallographic analysis of a peptide-binding single chain antibody fragment (scFv) with low picomolar affinity. *Journal of Biological Chemistry*, **279**, 18870–7.

56 Schier, R., Bye, J., Apell, G., McCall, A., Adams, G.P., Malmqvist, M., Weiner, L.M. and Marks, J.D. (1996) Isolation of high-affinity monomeric human anti-c-erbB-2 single chain Fv using affinity-driven selection. *Journal of Molecular Biology*, **255**, 28–43.

57 Thompson, J., Pope, T., Tung, J.S., Chan, C., Hollis, G., Mark, G. and Johnson, K.S. (1996) Affinity maturation of a high-affinity human monoclonal antibody against the third hypervariable loop of human immunodeficiency virus: use of phage display to improve affinity and broaden strain reactivity. *Journal of Molecular Biology*, **256**, 77–88.

58 Chowdhury, P.S. and Pastan, I. (1999) Improving antibody affinity by mimicking somatic hypermutation in vitro. *Nature Biotechnology*, **17**, 568–72.

59 Beers, R., Chowdhury, P., Bigner, D. and Pastan, I. (2000) Immunotoxins with increased activity against epidermal growth factor receptor vIII-expressing cells produced by antibody phage display. *Clinical Cancer Research*, **6**, 2835–43.

60 Chames, P., Coulon, S. and Baty, D. (1998) Improving the affinity and the fine specificity of an anti-cortisol antibody by parsimonious mutagenesis and phage display. *Journal of Immunology*, **161**, 5421–9.

61 Iba, Y., Hayashi, N., Sawada, J., Titani, K. and Kurosawa, Y. (1998) Changes in the specificity of antibodies against steroid antigens by introduction of mutations into complementarity-determining regions of the V(H) domain. *Protein Engineering*, **11**, 361–70.

62 Miyazaki, C., Iba, Y., Yamada, Y., Takahashi, H., Sawada, J. and Kurosawa, Y. (1999) Changes in the specificity of antibodies by site-specific mutagenesis followed by random mutagenesis. *Protein Engineering*, **12**, 407–15.

63 Dubreuil, O., Bossus, M., Graille, M., Bilous, M., Savatier, A., Jolivet, M., Menez, A., Stura, E. *et al.* (2005) Fine tuning of the specificity of an anti-progesterone antibody by first and second sphere residue engineering. *Journal of Biological Chemistry*, **280**, 24880–7.

64 Forrer, P., Jung, S. and Pluckthun, A. (1999) Beyond binding: using phage display to select for structure, folding and enzymatic activity in proteins. *Current Opinion in Structural Biology*, **9**, 514–20.

65 Juarez-Gonzalez, V.R., Riano-Umbarila, L., Quintero-Hernandez, V., Olamendi-Portugal, T., Ortiz-Leon, M., Ortiz, E., Possani, L.D. and Becerril, B. (2005) Directed evolution, phage display and combination of evolved mutants: a strategy to recover the neutralization

properties of the scFv version of BCF2 a neutralizing monoclonal antibody specific to scorpion toxin Cn2. *Journal of Molecular Biology*, **346**, 1287–97.

66 De Pascalis, R., Gonzales, N.R., Padlan, E.A., Schuck, P., Batra, S.K., Schlom, J. and Kashmiri, S.V. (2003) In vitro affinity maturation of a specificity-determining region-grafted humanized anticarcinoma antibody: isolation and characterization of minimally immunogenic high-affinity variants. *Clinical Cancer Research*, **9**, 5521–31.

67 Ho, M., Kreitman, R.J., Onda, M. and Pastan, I. (2005) In vitro antibody evolution targeting germline hot spots to increase activity of an anti-CD22 immunotoxin. *Journal of Biological Chemistry*, **280**, 607–17.

68 Ho, M., Nagata, S. and Pastan, I. (2006) Isolation of anti-CD22 Fv with high affinity by Fv display on human cells. *Proceedings of the National Academy of Sciences of the United States of America*, **103**, 9637–42.

69 Mullen, L.M., Nair, S.P., Ward, J.M., Rycroft, A.N. and Henderson, B. (2006) Phage display in the study of infectious diseases. *Trends in Microbiology*, **14**, 141–7.

70 Brekke, O.H. and Loset, G.A. (2003) New technologies in therapeutic antibody development. *Current Opinion in Pharmacology*, **3**, 544–50.

71 Holt, L.J., Herring, C., Jespers, L.S., Woolven, B.P. and Tomlinson, I.M. (2003) Domain antibodies: proteins for therapy. *Trends in Biotechnology*, **21**, 484–90.

72 Jain, M., Kamal, N. and Batra, S.K. (2007) Engineering antibodies for clinical applications. *Trends in Biotechnology*, **25**, 307–16.

73 Rothe, A., Hosse, R.J. and Power, B.E. (2006) In vitro display technologies reveal novel biopharmaceutics. *FASEB Journal*, **20**, 1599–610.

74 Filpula, D. (2007) Antibody engineering and modification technologies. *Biomolecular Engineering*, **24**, 201–15.

75 Pavlou, A.K. and Belsey, M.J. (2005) The therapeutic antibodies market to 2008. *European Journal of Pharmaceutics and Biopharmaceutics*, **59**, 389–96.

76 Andersen, D.C. and Reilly, D.E. (2004) Production technologies for monoclonal antibodies and their fragments. *Current Opinion in Biotechnology*, **15**, 456–62.

77 Sheridan, C. (2007) Pharma consolidates its grip on post-antibody landscape. *Nature Biotechnology*, **25**, 365–6.

78 Gill, D.S. and Damle, N.K. (2006) Biopharmaceutical drug discovery using novel protein scaffolds. *Current Opinion in Biotechnology*, **17**, 653–8.

79 Skerra, A. (2007) Alternative non-antibody scaffolds for molecular recognition. *Current Opinion in Biotechnology*, **18**, 295–304.

80 Smith, G. (1998) Patch engineering: a general approach for creating proteins that have new binding activities. *Trends in Biochemical Sciences*, **23**, 457–60.

81 Lipovsek, D. and Pluckthun, A. (2004) *In-vitro* protein evolution by ribosome display and mRNA display. *Journal of Immunological Methods*, **290**, 51–67.

82 Pluckthun, A., Schaffitzel, C., Hanes, J. and Jermutus, L. (2000) *In vitro* selection and evolution of proteins. *Advances in Protein Chemistry*, **55**, 367–403.

83 Binz, H.K., Amstutz, P. and Pluckthun, A. (2005) Engineering novel binding proteins from nonimmunoglobulin domains. *Nature Biotechnology*, **23**, 1257–68.

84 Binz, H.K. and Pluckthun, A. (2005) Engineered proteins as specific binding reagents. *Current Opinion in Biotechnology*, **16**, 459–69.

85 Hosse, R.J., Rothe, A. and Power, B.E. (2006) A new generation of protein display scaffolds for molecular recognition. *Protein Science*, **15**, 14–27.

86 Mathonet, P. and Fastrez, J. (2004) Engineering of non-natural receptors. *Current Opinion in Structural Biology*, **14**, 505–11.

87 Renberg, B., Nordin, J., Merca, A., Uhlen, M., Feldwisch, J., Nygren, P.A. and Karlstrom, A.E. (2007) Affibody molecules in protein capture microarrays: evaluation of multidomain ligands and different detection formats. *Journal of Proteome Research*, **6**, 171–9.

88 Silverman, J., Liu, Q., Bakker, A., To, W., Duguay, A., Alba, B.M., Smith, R., Rivas, A. *et al.* (2005) Multivalent avimer

proteins evolved by exon shuffling of a family of human receptor domains. *Nature Biotechnology*, **23**, 1556–61.
89 Koide, A., Jordan, M.R., Horner, S.R., Batori, V. and Koide, S. (2001) Stabilization of a fibronectin type III domain by the removal of unfavorable electrostatic interactions on the protein surface. *Biochemistry*, **40**, 10326–33.
90 Koide, A., Bailey, C.W., Huang, X. and Koide, S. (1998) The fibronectin type III domain as a scaffold for novel binding proteins. *Journal of Molecular Biology*, **284**, 1141–51.
91 Skerra, A. (2001) 'Anticalins': a new class of engineered ligand-binding proteins with antibody-like properties. *Journal of Biotechnology*, **74**, 257–75.
92 Heyd, B., Pecorari, F., Collinet, B., Adjadj, E., Desmadril, M. and Minard, P. (2003) In vitro evolution of the binding specificity of neocarzinostatin, an enediyne-binding chromoprotein. *Biochemistry*, **42**, 5674–83.
93 Nord, K., Gunneriusson, E., Ringdahl, J., Stahl, S., Uhlen, M. and Nygren, P.A. (1997) Binding proteins selected from combinatorial libraries of an alpha-helical bacterial receptor domain. *Nature Biotechnology*, **15**, 772–7.
94 Nord, K., Nilsson, J., Nilsson, B., Uhlen, M. and Nygren, P.A. (1995) A combinatorial library of an alpha-helical bacterial receptor domain. *Protein Engineering*, **8**, 601–8.
95 Binz, H.K., Amstutz, P., Kohl, A., Stumpp, M.T., Briand, C., Forrer, P., Grutter, M.G. and Pluckthun, A. (2004) High-affinity binders selected from designed ankyrin repeat protein libraries. *Nature Biotechnology*, **22**, 575–82.
96 Wikman, M., Steffen, A.C., Gunneriusson, E., Tolmachev, V., Adams, G.P., Carlsson, J. and Stahl, S. (2004) Selection and characterization of HER2/neu-binding affibody ligands. *Protein Engineering, Design & Selection*, **17**, 455–62.
97 Gunneriusson, E., Nord, K., Uhlen, M. and Nygren, P. (1999) Affinity maturation of a Taq DNA polymerase specific affibody by helix shuffling. *Protein Engineering*, **12**, 873–8.
98 Hogbom, M., Eklund, M., Nygren, P.A. and Nordlund, P. (2003) Structural basis for recognition by an in vitro evolved affibody. *Proceedings of the National Academy of Sciences of the United States of America*, **100**, 3191–6.
99 Hansson, M., Ringdahl, J., Robert, A., Power, U., Goetsch, L., Nguyen, T.N., Uhlen, M., Stahl, S. *et al.* (1999) An in vitro selected binding protein (affibody) shows conformation-dependent recognition of the respiratory syncytial virus (RSV) G protein. *Immunotechnology*, **4**, 237–52.
100 Wahlberg, E., Lendel, C., Helgstrand, M., Allard, P., Dincbas-Renqvist, V., Hedqvist, A., Berglund, H., Nygren, P.A. *et al.* (2003) An affibody in complex with a target protein: structure and coupled folding. *Proceedings of the National Academy of Sciences of the United States of America*, **100**, 3185–90.
101 Nilsson, F.Y. and Tolmachev, V. (2007) Affibody molecules: new protein domains for molecular imaging and targeted tumor therapy. *Current Opinion in Drug Discovery & Development*, **10**, 167–75.
102 Nord, K., Nord, O., Uhlen, M., Kelley, B., Ljungqvist, C. and Nygren, P.A. (2001) Recombinant human factor VIII-specific affinity ligands selected from phage-displayed combinatorial libraries of protein A. *European Journal of Biochemistry*, **268**, 4269–77.
103 Choo, Y. and Klug, A. (1995) Designing DNA-binding proteins on the surface of filamentous phage. *Current Opinion in Biotechnology*, **6**, 431–6.
104 Segal, D.J., Dreier, B., Beerli, R.R. and Barbas, C.F., 3rd (1999) Toward controlling gene expression at will: selection and design of zinc finger domains recognizing each of the 5′-GNN-3′ DNA target sequences. *Proceedings of the National Academy of Sciences of the United States of America*, **96**, 2758–63.
105 Pabo, C.O., Peisach, E. and Grant, R.A. (2001) Design and selection of novel Cys2His2 zinc finger proteins. *Annual Review of Biochemistry*, **70**, 313–40.
106 Beerli, R.R. and Barbas, C.F., 3rd (2002) Engineering polydactyl zinc-finger transcription factors. *Nature Biotechnology*, **20**, 135–41.

107 Mandell, J.G. and Barbas, C.F., 3rd (2006) Zinc Finger Tools: custom DNA-binding domains for transcription factors and nucleases. *Nucleic Acids Research*, **34**, W516–23.

108 Dhanasekaran, M., Negi, S. and Sugiura, Y. (2006) Designer zinc finger proteins: tools for creating artificial DNA-binding functional proteins. *Accounts of Chemical Research*, **39**, 45–52.

109 Xu, L., Aha, P., Gu, K., Kuimelis, R.G., Kurz, M., Lam, T., Lim, A.C., Liu, H. *et al.* (2002) Directed evolution of high-affinity antibody mimics using mRNA display. *Chemistry & Biology*, **9**, 933–42.

110 Koide, A., Gilbreth, R.N., Esaki, K., Tereshko, V. and Koide, S. (2007) High-affinity single-domain binding proteins with a binary-code interface. *Proceedings of the National Academy of Sciences of the United States of America*, **104**, 6632–7.

111 Beste, G., Schmidt, F.S., Stibora, T. and Skerra, A. (1999) Small antibody-like proteins with prescribed ligand specificities derived from the lipocalin fold. *Proceedings of the National Academy of Sciences of the United States of America*, **96**, 1898–903.

112 Schlehuber, S. and Skerra, A. (2005) Anticalins as an alternative to antibody technology. *Expert Opinion on Biological Therapy*, **5**, 1453–62.

113 Schlehuber, S. and Skerra, A. (2005) Anticalins in drug development. *BioDrugs*, **19**, 279–88.

114 Schlehuber, S. and Skerra, A. (2005) Lipocalins in drug discovery: from natural ligand-binding proteins to 'anticalins'. *Drug Discovery Today*, **10**, 23–33.

115 Forrer, P., Stumpp, M.T., Binz, H.K. and Plückthun, A. (2003) A novel strategy to design binding molecules harnessing the modular nature of repeat proteins. *FEBS Letters*, **539**, 2–6.

116 Forrer, P., Binz, H.K., Stumpp, M.T. and Plückthun, A. (2004) Consensus design of repeat proteins. *Chem Bio Chem*, **5**, 183–9.

117 Zahnd, C., Wyler, E., Schwenk, J.M., Steiner, D., Lawrence, M.C., McKern, N.M., Pecorari, F., Ward, C.W. *et al.* (2007) A designed ankyrin repeat protein evolved to picomolar affinity to Her2. *Journal of Molecular Biology*, **369**, 1015–28.

118 Amstutz, P., Binz, H.K., Parizek, P., Stumpp, M.T., Kohl, A., Grutter, M.G., Forrer, P. and Plückthun, A. (2005) Intracellular kinase inhibitors selected from combinatorial libraries of designed ankyrin repeat proteins. *Journal of Biological Chemistry*, **280**, 24715–22.

119 Sennhauser, G., Amstutz, P., Briand, C., Storchenegger, O. and Grutter, M.G. (2007) Drug export pathway of multidrug exporter AcrB revealed by DARPin inhibitors. *PLoS Biology*, **5**, e7.

120 Steiner, D., Forrer, P., Stumpp, M.T. and Plückthun, A. (2006) Signal sequences directing cotranslational translocation expand the range of proteins amenable to phage display. *Nature Biotechnology*, **24**, 823–31.

121 Stumpp, M.T., Forrer, P., Binz, H.K. and Plückthun, A. (2003) Designing repeat proteins: modular leucine-rich repeat protein libraries based on the mammalian ribonuclease inhibitor family. *Journal of Molecular Biology*, **332**, 471–87.

122 Pancer, Z. and Cooper, M.D. (2006) The evolution of adaptive immunity. *Annual Review of Immunology*, **24**, 497–518.

123 Kristensen, P. and Winter, G. (1998) Proteolytic selection for protein folding using filamentous bacteriophages. *Folding & Design*, **3**, 321–8.

124 Sieber, V., Plückthun, A. and Schmid, F.X. (1998) Selecting proteins with improved stability by a phage-based method. *Nature Biotechnology*, **16**, 955–60.

125 Bai, Y.W. and Feng, H.Q. (2004) Selection of stably folded proteins by phage-display with proteolysis. *European Journal of Biochemistry*, **271**, 1609–14.

126 Pedersen, J.S., Otzen, D.E. and Kristensen, P. (2002) Directed evolution of barnase stability using proteolytic selection. *Journal of Molecular Biology*, **323**, 115–23.

127 Wunderlich, M., Martin, A., Staab, C.A. and Schmid, F.X. (2005) Evolutionary protein stabilization in comparison with computational design. *Journal of Molecular Biology*, **351**, 1160–8.

128 Wunderlich, M. and Schmid, F.X. (2006) In vitro evolution of a hyperstable G beta 1 variant. *Journal of Molecular Biology*, **363**, 545–57.

129 Wunderlich, M., Max, K.E.A., Roske, Y., Mueller, U., Heinemann, U. and Schmid, F.X. (2007) Optimization of the g beta 1 domain by computational design and by in vitro evolution: Structural and energetic basis of stabilization. *Journal of Molecular Biology*, **373**, 775–84.

130 Wunderlich, M., Martin, A. and Schmid, F.X. (2005) Stabilization of the cold shock protein CspB from *Bacillus subtilis* by evolutionary optimization of coulombic interactions. *Journal of Molecular Biology*, **347**, 1063–76.

131 Riechmann, L. and Winter, G. (2000) Novel folded protein domains generated by combinatorial shuffling of polypeptide segments. *Proceedings of the National Academy of Sciences of the United States of America*, **97**, 10068–73.

132 Fischer, N., Riechmann, L. and Winter, G. (2004) A native-like artificial protein from antisense DNA. *Protein Engineering Design & Selection*, **17**, 13–20.

133 de Bono, S., Riechmann, L., Girard, E., Williams, R.L. and Winter, G. (2005) A segment of cold shock protein directs the folding of a combinatorial protein. *Proceedings of the National Academy of Sciences of the United States of America*, **102**, 1396–401.

134 Riechmann, L., Lavenir, I., de Bono, S. and Winter, G. (2005) Folding and stability of a primitive protein. *Journal of Molecular Biology*, **348**, 1261–72.

135 Riechmann, L. and Winter, G. (2006) Early protein evolution: building domains from ligand-binding polypeptide segments. *Journal of Molecular Biology*, **363**, 460–8.

136 Legendre, D., Soumillion, P. and Fastrez, J. (1999) Engineering a regulatable enzyme for homogeneous immunoassays. *Nature Biotechnology*, **17**, 67–72.

137 Mathonet, P., Deherve, J., Soumillion, P. and Fastrez, J. (2006) Active TEM-1 beta-lactamase mutants with random peptides inserted in three contiguous surface loops. *Protein Science*, **15**, 2323–34.

138 Legendre, D., Vucic, B., Hougardy, V., Girboux, A.L., Henrioul, C., Van Haute, J., Soumillion, P. and Fastrez, J. (2002) TEM-1 beta-lactamase as a scaffold for protein recognition and assay. *Protein Science*, **11**, 1506–18.

139 Mathonet, P., Barrios, H., Soumillion, P. and Fastrez, J. (2006) Selection of allosteric beta-lactamase mutants featuring an activity regulation by transition metal ions. *Protein Science*, **15**, 2335–43.

140 Light, J. and Lerner, R.A. (1995) Random mutagenesis of staphylococcal nuclease and phage display selection. *Bioorganic and Medicinal Chemistry*, **3**, 955–67.

141 Widersten, M. and Mannervik, B. (1995) Glutathione transferases with novel active-sites isolated by phage display from a library of random mutants. *Journal of Molecular Biology*, **250**, 115–22.

142 Baca, M., Scanlan, T.S., Stephenson, R.C. and Wells, J.A. (1997) Phage display of a catalytic antibody to optimize affinity for transition-state analog binding. *Proceedings of the National Academy of Sciences of the United States of America*, **94**, 10063–8.

143 Fujii, I., Fukuyama, S., Iwabuchi, Y. and Tanimura, R. (1998) Evolving catalytic antibodies in a phage-displayed combinatorial library. *Nature Biotechnology*, **16**, 463–7.

144 Janda, K.D., Lo, C.H.L., Li, T.Y., Barbas, C.F., Wirsching, P. and Lerner, R.A. (1994) Direct selection for a catalytic mechanism from combinatorial antibody libraries. *Proceedings of the National Academy of Sciences of the United States of America*, **91**, 2532–6.

145 Wagner, J., Lerner, R.A. and Barbas, C.F. (1995) Efficient aldolase catalytic antibodies that use the enamine mechanism of natural enzymes. *Science*, **270**, 1797–800.

146 Soumillion, P., Jespers, L., Bouchet, M., Marchandbrynaert, J., Winter, G. and Fastrez, J. (1994) Selection of beta-lactamase on filamentous bacteriophage by catalytic activity. *Journal of Molecular Biology*, **237**, 415–22.

147 Janda, K.D., Lo, L.C., Lo, C.H.L., Sim, M.M., Wang, R., Wong, C.H. and Lerner, R.A. (1997) Chemical selection for catalysis in combinatorial antibody libraries. *Science*, **275**, 945–8.

148 Legendre, D., Laraki, N., Graslund, T., Bjornvad, M.E., Bouchet, M., Nygren, P.A., Borchert, T.V. and Fastrez, J. (2000) Display of active subtilisin 309 on phage: analysis of parameters influencing the selection of subtilisin variants with changed substrate specificity from libraries using phosphonylating inhibitors. *Journal of Molecular Biology*, **296**, 87–102.

149 Hunt, J.A. and Fierke, C.A. (1997) Selection of carbonic anhydrase variants displayed on phage–aromatic residues in zinc binding site enhance metal affinity and equilibration kinetics. *Journal of Biological Chemistry*, **272**, 20364–72.

150 Hansson, L.O., Widersten, M. and Mannervik, B. (1997) Mechanism-based phage display selection of active-site mutants of human glutathione transferase A1-1 catalyzing SNAr reactions. *Biochemistry*, **36**, 11252–60.

151 Takahashi, N., Kakinuma, H., Liu, L.D., Nishi, Y. and Fujii, I. (2001) In vitro abzyme evolution to optimize antibody recognition for catalysis. *Nature Biotechnology*, **19**, 563–7.

152 Tanaka, F., Lerner, R.A. and Barbas, C.F. (2000) Reconstructing aldolase antibodies to alter their substrate specificity and turnover. *Journal of the American Chemical Society*, **122**, 4835–6.

153 Tanaka, F., Fuller, R., Shim, H., Lerner, R.A. and Barbas, C.F. (2004) Evolution of aldolase antibodies in vitro: correlation of catalytic activity and reaction-based selection. *Journal of Molecular Biology*, **335**, 1007–18.

154 Lasters, I., VanHerzeele, N., Lijnen, H.R., Collen, D. and Jespers, L. (1997) Enzymatic properties of phage-displayed fragments of human plasminogen. *European Journal of Biochemistry*, **244**, 946–52.

155 Vanwetswinkel, S., Avalle, B. and Fastrez, J. (2000) Selection of beta-lactamases and penicillin binding mutants from a library of phage displayed TEM-1 beta-lactamase randomly mutated in the active site Omega-loop. *Journal of Molecular Biology*, **295**, 527–40.

156 Tanaka, F., Almer, H., Lerner, R.A. and Barbas, C.F. (1999) Catalytic single-chain antibodies possessing beta-lactamase activity selected from a phage displayed combinatorial library using a mechanism-based inhibitor. *Tetrahedron Letters*, **40**, 8063–6.

157 Avalle, B., Vanwetswinkel, S. and Fastrez, J. (1997) In vitro selection for catalytic turnover from a library of beta-lactamase mutants and penicillin-binding proteins. *Bioorganic and Medicinal Chemistry Letters*, **7**, 479–84.

158 Paul, S., Tramontano, A., Gololobov, G., Zhou, Y.X., Taguchi, H., Karle, S., Nishiyama, Y., Planque, S. *et al.* (2001) Phosphonate ester probes for proteolytic antibodies. *Journal of Biological Chemistry*, **276**, 28314–20.

159 Reshetnyak, A.V., Armentano, M.F., Ponomarenko, N.A., Vizzuso, D., Durova, O.M., Ziganshin, F., Serebryakova, M., Govorun, V. *et al.* (2007) Routes to covalent catalysis by reactive selection for nascent protein nucleophiles. *Journal of the American Chemical Society*, **129**, 16175–82.

160 Pedersen, H., Holder, S., Sutherlin, D.P., Schwitter, U., King, D.S. and Schultz, P.G. (1998) A method for directed evolution and functional cloning of enzymes. *Proceedings of the National Academy of Sciences of the United States of America*, **95**, 10523–8.

161 Ponsard, I., Galleni, M., Soumillion, P. and Fastrez, J. (2001) Selection of metalloenzymes by catalytic activity using phage display and catalytic elution. *ChemBioChem*, **2**, 253–9.

162 Jestin, J.L., Kristensen, P. and Winter, G. (1999) A method for the selection of catalytic activity using phage display and proximity coupling. *Angewandte Chemie–International Edition*, **38**, 1124–7.

163 Demartis, S., Huber, A., Viti, F., Lozzi, L., Giovannoni, L., Neri, P., Winter, G. and Neri, D. (1999) A strategy for the isolation of catalytic activities from repertoires of enzymes displayed on phage. *Journal of Molecular Biology*, **286**, 617–33.

164 Heinis, C., Huber, A., Demartis, S., Bertschinger, J., Melkko, S., Lozzi, L., Neri, P. and Neri, D. (2001) Selection of catalytically active biotin ligase and trypsin mutants by phage display. *Protein Engineering*, **14**, 1043–52.

165 Atwell, S. and Wells, J.A. (1999) Selection for improved subtiligases by phage display. *Proceedings of the National Academy of Sciences of the United States of America*, **96**, 9497–502.

166 Juillerat, A., Gronemeyer, T., Keppler, A., Gendreizig, S., Pick, H., Vogel, H. and Johnsson, K. (2003) Directed evolution of O-6-alkylguanine-DNA alkyltransferase for efficient labeling of fusion proteins with small molecules in vivo. *Chemistry and Biology*, **10**, 313–17.

167 Xia, G., Chen, L.J., Sera, T., Fa, M., Schultz, P.G. and Romesberg, F.E. (2002) Directed evolution of novel polymerase activities: Mutation of a DNA polymerase into an efficient RNA polymerase. *Proceedings of the National Academy of Sciences of the United States of America*, **99**, 6597–602.

168 Roth, T.A., Weiss, G.A., Eigenbrot, C. and Sidhu, S.S. (2002) A minimized M13 coat protein defines the requirements for assembly into the bacteriophage particle. *Journal of Molecular Biology*, **322**, 357–67.

169 Sidhu, S.S., Weiss, G.A. and Wells, J.A. (2000) High copy display of large proteins on phage for functional selections. *Journal of Molecular Biology*, **296**, 487–95.

170 Sidhu, S.S. (2001) Engineering M13 for phage display. *Biomolecular Engineering*, **18**, 57–63.

171 Rondot, S., Koch, J., Breitling, F. and Dubel, S. (2001) A helper phage to improve single-chain antibody presentation in phage display. *Nature Biotechnology*, **19**, 75–8.

172 Kirsch, M., Zaman, M., Meier, D., Dubel, S. and Hust, M. (2005) Parameters affecting the display of antibodies on phage. *Journal of Immunological Methods*, **301**, 173–85.

173 Soltes, G., Hust, M., Ng, K.K., Bansal, A., Field, J., Stewart, D.I., Dubel, S., Cha, S. *et al.* (2007) On the influence of vector design on antibody phage display. *Journal of Biotechnology*, **127**, 626–37.

174 Chasteen, L., Ayriss, J., Pavlik, P. and Bradbury, A.R. (2006) Eliminating helper phage from phage display. *Nucleic Acids Research*, **34**, e145.

175 Binz, H.K., Stumpp, M.T., Forrer, P., Amstutz, P. and Plückthun, A. (2003) Designing repeat proteins: well-expressed, soluble and stable proteins from combinatorial libraries of consensus ankyrin repeat proteins. *Journal of Molecular Biology*, **332**, 489–503.

176 Thammawong, P., Kasinrerk, W., Turner, R.J. and Tayapiwatana, C. (2006) Twin-arginine signal peptide attributes effective display of CD147 to filamentous phage. *Applied Microbiology and Biotechnology*, **69**, 697–703.

177 Fuh, G. and Sidhu, S.S. (2000) Efficient phage display of polypeptides fused to the carboxy-terminus of the M13 gene-3 minor coat protein. *FEBS Letters*, **480**, 231–4.

178 Friguet, B., Chaffotte, A.F., Djavadiohaniance, L. and Goldberg, M.E. (1985) Measurements of the true affinity constant in solution of antigen-antibody complexes by enzyme-linked immunosorbent-assay. *Journal of Immunological Methods*, **77**, 305–19.

179 Goldsmith, M., Kiss, C., Bradbury, A.R.M. and Tawfik, D.S. (2007) Avoiding and controlling double transformation artifacts. *Protein Engineering Design and Selection*, **20**, 315–18.

180 Velappan, N., Sblattero, D., Chasteen, L., Pavlik, P. and Bradbury, A.R.M. (2007) Plasmid incompatibility: more compatible than previously thought?. *Protein Engineering Design & Selection*, **20**, 309–13.

181 Sidhu, S.S., Lowman, H.B., Cunningham, B.C. and Wells, J.A. (2000) Phage display for selection of novel binding peptides. *Methods in Enzymology*, **328**, 333–63.

182 Waterhouse, P., Griffiths, A.D., Johnson, K.S. and Winter, G. (1993) Combinatorial infection and in vivo recombination – a strategy for making large phage antibody repertoires. *Nucleic Acids Research*, **21**, 2265–6.

183 Cen, X.D., Bi, Q. and Zhu, S.G. (2006) Construction of a large phage display antibody library by in vitro package and

in vivo recombination. *Applied Microbiology and Biotechnology*, **71**, 767–72.

184 Bassindale, A.R., Codina-Barrios, A., Frascione, N. and Taylor, P.G. (2007) An improved phage display methodology for inorganic nanoparticle fabrication. *Chemical Communications*, **28**, 2956–8.

185 Cesaro-Tadic, S., Lagos, D., Honegger, A., Rickard, J.H., Partridge, L.J., Blackburn, G.M. and Plückthun, A. (2003) Turnover-based in vitro selection and evolution of biocatalysts from a fully synthetic antibody library. *Nature Biotechnology*, **21**, 679–85.

25
Screening Methodologies for Glycosidic Bond Formation

Amir Aharoni and Stephen G. Withers

25.1
Introduction

Complex carbohydrates occur in a wide range of contexts in biology, including polysaccharides, proteoglycans, glycoproteins, glycolipids and antibodies. There they play important roles in a number of functions, including: cell growth, cell–cell interactions [1], immune defense [2], inflammation [3] and both viral and parasitic infections [4]. In Nature, the assembly of these complex structures is orchestrated by a series of specific glycosyltransferases (GT)s which sequentially transfer the monosaccharide moieties of their activated sugar donor to the required acceptor, with the correct positional and stereochemical outcome [5]. Consequently, a large number of GTs exists, with widely different specificities. Interestingly, all GTs for which structures are known, to date, fall into one of two structural folds: (i) GTA with a single Rossman fold domain; or (ii) GTB with a twin fold structure [6]. It therefore appears that these frameworks can accommodate whichever specificities may be needed, although in at least one case additional specificity is achieved through an appended lectin domain [7]. The prospects for engineering GTs to generate enzymes of desired specificity are, therefore, very promising.

During the late 1990s, an important alternative to enzymatic synthesis of oligosaccharides by natural GTs emerged. At this time, a new class of glycosidase mutants was introduced which catalyzed the synthesis of new glycosidic linkages but did not hydrolyze the newly formed linkages, therefore driving the reaction in the synthetic direction [8]. These mutant glycosidases, termed 'glycosynthases', were seen to be rendered hydrolytically inactive through the mutation of the nucleophilic residue usually to alanine, glycine or serine residue. When supplied with glycoside fluoride substrates of the opposite anomeric configuration to that of the natural substrate, the enzyme is often able to transfer this activated glycosyl donor to a suitable acceptor [9]. In recent years this class of enzymes has gained popularity for the enzymatic synthesis of complex oligosaccharides. However,

Protein Engineering Handbook. Edited by Stefan Lutz and Uwe T. Bornscheuer

ISBN: 978-3-527-31850-6

in some cases this approach suffers from low catalytic efficiency and limited specificity.

The enzymatic synthesis of biologically active complex oligosaccharides is highly important as it is carried out under mild conditions while maintaining high specificity and regioselectivity without the need for complex protection and deprotection steps. In addition, by contrast with the situation for peptides and oligonucleotides, the chemical synthesis of complex carbohydrates is an extremely challenging and labor-intensive process, and cannot generally be achieved in an automated fashion or on a large scale. New approaches to complex carbohydrate synthesis remain an urgent need in glycobiology in order to further our understanding as well as to facilitate the development of potential therapeutics.

During the past few years directed evolution approaches for protein engineering have proved to especially useful in improving the stability of enzymes [10, 11], and for altering their substrate specificities [12]. One of the most crucial steps in any directed evolution experiment is the development of a screening assay to facilitate the examination of large libraries [13]. However, assaying for transfer activity–and particularly GT activity–is extremely challenging as no obvious change in fluorescence or absorbance is associated with glycosidic bond formation. This is in contrast to many screening strategies that have been developed in recent years for hydrolytic enzymes, including β-lactamases, esterases, lactonases, lipases and phosphotriesterases, which are based on the release of a chromophore or fluorophore [13]. As the screening for a desired phenotype is, in most cases, a random process it is highly desirable to develop high-throughput screening (HTS) methodologies to facilitate the screening of extremely large libraries [14, 15]. These methodologies are particularly valuable for the enrichment and isolation of rare mutants with beneficial activity from large mutant libraries [16]. In this respect, fluorescence-activated cell sorting (FACS) holds great promise for HTS. Modern FACS systems can be used to routinely analyze and sort $>10^7$ events per hour; moreover, fluorescence is a highly sensitive signal which is widely adopted to detect both binding [17] and enzymatic activity [18]. Unfortunately, the use of FACS for enzyme selection is limited mainly due to diffusion of the substrate/product; consequently, in order to overcome this limitation methods of trapping the product in the interior of the cell [19], on the cell surface [18] or by compartmentalization, have been developed [20, 21].

This chapter describes recent advances in the development of screening strategies for glycosidic bond formation, and will also include strategies which have been developed for generating GTs and glycosynthases with improved catalytic efficiencies and new specificity. Attention will be focused on recent studies describing the development of a new screening methodology for GTs, including an overall protocol and practical considerations for screening for GT activity using a FACS-based methodology [22].

25.2 Glycosynthases

Screening methodologies for glycosidic bond formation were developed first for glycosynthases. An on-plate *in vivo* screening assay was developed in order to improve the catalytic efficiency and alter the specificity of the *Agrobacterium* sp. β-glucosidase (Abg). The assay is based on the use of an endoglycosidase as a coupling enzyme *in situ* to release a fluorophore *only* upon product formation [23]. In this system a library of randomly mutated Abg variants, together with the coupling endoglycosidase, is expressed from a single plasmid. A mild cell wall lysis step, involving the addition of D-cycloserine, is included in order to increase the cell permeability and reduce the requirement for high donor sugar concentrations. By using this powerful screening assay, a 50-fold improvement in catalytic efficiency was achieved through saturation mutagenesis of the nucleophile. A further 27-fold increase in catalytic efficiency was achieved by two rounds of random mutagenesis and screening [23], such that the kinetic parameters for glycoside synthesis by the engineered glycosynthase were improved >1300-fold and were remarkably similar to those for glycoside hydrolysis by the wild-type enzyme. The improved glycosynthase containing three mutations (plus the mutated nucleophile) is also able to efficiently use α-xylosyl fluoride as a glycosyl donor, thereby expanding the specificity, given that the wild-type enzyme and the original single-mutant Abg glycosynthase catalyze the transfer of xylosyl moieties only extremely slowly [24]. By using different endoglycosidases as the coupling enzymes, this system may be applicable to many other glycosynthase reactions and could, in principle, be adapted for the screening of GTs.

An alternative approach for the directed evolution of glycosynthase has been reported by Cornish and coworkers [25], who used 'chemical complementation' to select for improved activity of the *Humicola insolens* Cel7B glycosynthase *in vivo* through a yeast three-hybrid assay. Briefly, this methodology involves the chemical conjugation of the glycosyl fluoride donor and glycosyl acceptor with methotrexate and dexamethasone, respectively. A reporter gene (in this case LEU2, required for the production of leucine) is placed under the control of a promoter, requiring a transcription activator (consisting of an activation domain and a DNA-binding domain) in a leucine auxotrophic host. The activation and DNA-binding domains of the transcription activator are separately fused to a glucocorticoid receptor and dihydrofolate reductase (DHFR), respectively. In the presence of an active glycosynthase, the glycosyl donor and acceptor are joined, thereby chemically connecting the methotrexate and dexamethasone through a glycosidic linkage. Interaction of the glucocorticoid receptor and DHFR with these ligands results in the reconstitution of the transcription activator complex, enabling the expression of the LEU2 gene and allowing growth in the absence of leucine. By using this approach, a fivefold increase in activity was achieved for the Cel7B glycosynthase. The three-hybrid approach was shown also to be applicable for the selection of other enzymatic activities such as β-lactamases [26].

25.3 Glycosyltransferases

Engineered GTs with broadened or tailored substrate specificities, have enormous potential for the synthesis of novel, non-natural and biologically relevant carbohydrate structures, either by the synthesis of non-natural linkages or by the incorporation of non-natural monosaccharides. The difficulty of synthesizing the necessary non-natural nucleotide sugars is being overcome by rational design and directed evolution of the anomeric kinases and nucleotidyl transferases. These approaches have produced enzymes that process variously configured D- or L-hexose sugars and tolerate diverse sugar modifications [27, 28]. At this point, two possible strategies exist for the use of GTs in the synthesis of glycosides containing non-natural sugars: (i) promiscuous GTs may be available that catalyze the requisite non-natural sugar transfer; and (ii) a pre-existing GT substrate specificity may be modified to accept the non-natural substrate using protein engineering approaches. At this point we will focus on the recent development of strategies that will enable the engineering of GTs using a directed evolution approach.

Selecting for GT activity remains the most challenging aspect of engineering GTs using directed evolution approaches. Recently, a new phage display selection assay for GT activity has been proposed by Walker and coworkers [29]. This methodology is based on the expression of an active GT on the surface of the M13 phage and acceptor immobilization on the phage surface adjacent to the GT. The phage-linked acceptor is converted to product by incubation of the phages with the appropriate donor sugar. The authors developed a chemically straightforward method for substrate attachment using selenocysteine (Sec) residues fused to the gIII phage coat protein [30]. Because Sec is more nucleophilic than cysteine and reacts at a lower pH, Sec-bearing phage can be derivatized selectively in the presence of other potential side-chain nucleophiles. This system enables the crucial link between the genotype and the phenotype, and thus serves as a platform for the development of a selection assay for large GT mutant libraries. The authors successfully expressed the GT MurG on the surface of the phage in an active form, and the activity of the MurG was detected by incubating the phage with radiolabeled donor sugar and biotinylated lipid I acceptor. Product formation was monitored by spotting the phage on streptavidin-coated membranes at various time points using a radioactive counter. The authors showed clearly that GTs from the GT-B superfamily can be displayed in an active form on the surface of a phage and using an efficient substrate immobilization technique. This is the first step toward the development of a whole phage display selection system that will enable the directed evolution of GTs.

Recently, we have described the development of a new fluorescence-based HTS methodology for the directed evolution of sialyltransferases (ST)s [22]. The methodology is based on the development of a new fluorescence cell-based assay for the detection of the sialyltransfer reaction in intact *E. coli* cells (Figure 25.1). This cell-based assay alleviates the need to lyse the cells and perform many other manipulations otherwise necessary for screening large mutant libraries. The assay

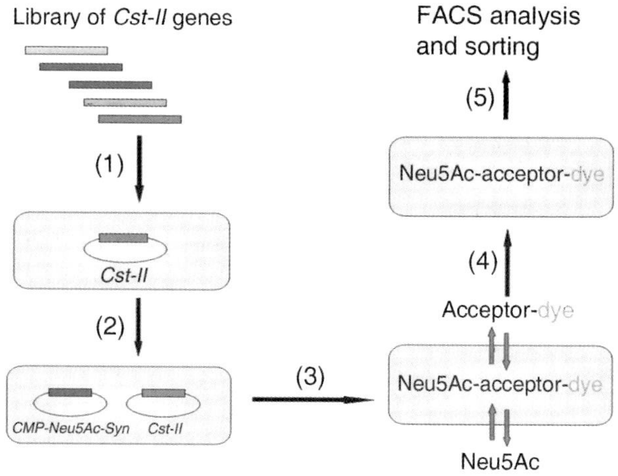

Figure 25.1 A cell-based assay for sialyltransferase (ST) activity. (1) A gene library is transformed and cloned in *E. coli*. (2) The encoded ST protein is expressed in the cytoplasm of the engineered JM107 *NanA*$^-$ strain together with CMP-Neu5Ac-synthetase. (3) Cells are incubated with Neu5Ac donor sugar and fluorescently labeled acceptor sugars (see Figure 25.3). (4) Following incubation, the cells are washed extensively to remove unreacted fluorescent acceptor sugar. (5) The cells are directly analyzed and sorted using fluorescence-activated cell sorting (FACS).

is performed by incubating cells expressing active ST with fluorescently labeled acceptor sugar and the sialic acid (Neu5Ac) donor sugar. The fluorescently labeled acceptor and donor sugar permeate freely into the cell, cells expressing active ST accumulate fluorescently labeled product that, due to its size and charge, is trapped in the cell. Following several washing steps to release the unreacted fluorescently labeled acceptor, the fluorescence of the cell is directly correlated to the amount of product and the enzyme catalytic efficiency (Figures 25.1 and 25.2). The resultant cell population is then analyzed and sorted using FACS to isolate cells expressing highly active ST variants.

We have recently applied this methodology for the directed evolution of the ST *Cst*-II from the human pathogen *Campylobacter jejuni*, the structure of which was recently solved [31]. We described the sensitivity, dynamic range and use of this methodology to detect the transfer reaction to two different fluorescently labeled acceptor sugars labeled with different fluorescent dyes, simultaneously. This methodology enabled the HTS of a library containing ~10^7 different *Cst*-II mutants in a rapid and efficient manner and the isolation of a mutant exhibiting 400-fold increased activity for a fluorescently labeled acceptor sugar. The overall protocol of the methodology and practical considerations for the use of this methodology to screen large GT mutant libraries is described in the following sections. Detailed descriptions, including technical details, are available in Ref. [22], together with supplementary information [22].

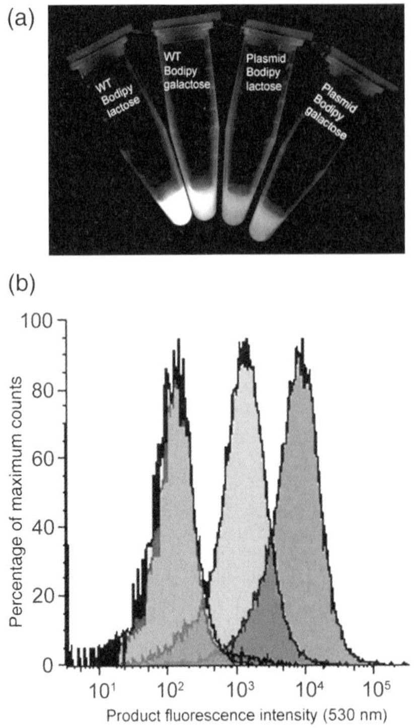

Figure 25.2 A cell-based ST assay for cells expressing wild-type *Cst*-II and pUC18 plasmid (control) (see Section 25.4.5 and Ref. [22]). The different cell samples were incubated with Neu5Ac and bodipy-lactose or bodipy-galactose. Following extensive washing (see text), the cells were analyzed either visually or by FACS. (A) Eppendorf tubes containing the four different cell samples as visualized under a UV lamp; the cell fluorescence intensity is correlated to the *Cst*-II activity (see text); (B) FACS histogram analysis of the different cell samples. The light blue and light green colors (histograms overlap) indicate cells expressing empty pUC18 plasmid incubated with bodipy-galactose or bodipy-lactose; the brown color indicates cells expressing wild-type *Cst*-II incubated with bodipy-galactose; the purple color indicates cells expressing wild-type *Cst*-II incubated with bodipy-lactose.

25.4
Protocol and Practical Considerations for Using HTS Methodology in the Directed Evolution of STs

25.4.1
Cloning of the Target ST and CMP-Neu5Ac-Synthetase

The target ST gene is cloned to pUC18 high-copy number plasmid containing an ampicillin resistance gene to yield pUC18-ST. The CMP-Neu5Ac-syn [32] is cloned to a pACKC18 low-copy number plasmid containing a chloramphenicol resistance gene to yield pACKC18-CMP-syn. The genes in both plasmids are under the control of the *lac* promoter and are induced by the addition of isopropyl β-D-1-thiogalactopyranoside (IPTG) to the growth media (see Section 25.4.4).

- It is crucial to clone the genes to plasmids containing different resistance genes to ensure selection for both plasmids.
- The cloning of the target ST to a high-copy number plasmid increases the expression level upon induction and allows for high-yield plasmid extraction following library selection.
- It is possible to use a polycistronic expression system in which both genes are expressed from the same plasmid.
- All cloning steps can be performed in a standard *E. coli* strain (e.g. DH5α) in rich media containing the appropriate antibiotics.

25.4.2
Synthesis of Fluorescently Labeled Acceptor Sugar

Fluorescent dye is attached to the anomeric center of lactose or galactose through alkyl chains of various lengths in a β (equatorial) configuration (Figure 25.3). Detailed syntheses of the fluorescently labeled acceptor will be described elsewhere.

- It is important to label the acceptor sugar with a bright fluorescent dye that permeates freely into and out of the cells. The Bodipy dye was found to satisfy both criteria and is compatible with the 488 nm laser of the FACS instrument. In contrast, fluorescein-labeled lactose was unable to permeate into the *E. coli* cells, probably due to its size.
- The linker length should be carefully chosen so that the fluorescent dye will be close enough to the acceptor sugar to permit permeation into the cell cytoplasm, and long enough so as not to interfere with binding of the acceptor sugar to the enzyme's active site.
- Different acceptor sugars can be labeled with different dyes such as coumarins and Bodipy so that ST activity could be detected for two different acceptors simultaneously (see Figure 25.2 in supplementary information accompanying Ref. [22]). Dual selection requires a FACS instrument that is also equipped with a UV laser (360 nm or 407 nm) and careful setting up of the instrument.

25.4.3
Cell-Based Assay in JM107 *Nan* A⁻ Strain

Here, the aim is to facilitate the detection of ST activity in living *E. coli* cells.

- To prevent the catabolism of lactose and Neu5Ac, the mutant strain should lack both β-galactosidase (*lac Z*) and Neu5Ac aldolase activities (*NanA*).
- This mutant strain contains specific permeases both for lactose and sialic acid to allow import into the cells.

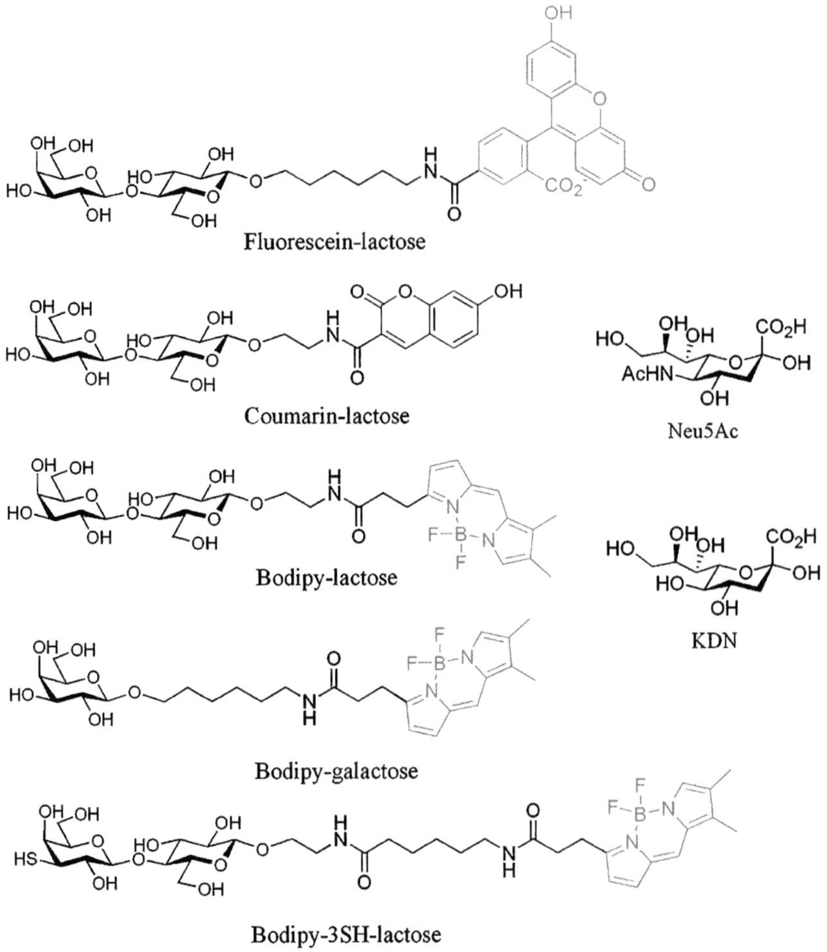

Figure 25.3 Donor sugars and fluorescent acceptor sugars used for the directed evolution of *Cst*-II ST (see Section 25.4.2 and Ref. [22]).

- To allow cell-based synthesis of sialosides, the cell strain must express both CMP-Neu5Ac-synthetase and the target ST genes. CMP-Neu5Ac-synthetase activates the Neu5Ac *in situ* to CMP-Neu5Ac, while the ST uses the latter as its donor sugar for sialyltransfer to β-galactoside acceptors (see Figure 25.1) [33].

25.4.4
Transformation, Growth and Expression of Plasmids Containing ST and CMP-syn Genes in JM107 *Nan* A⁻ Strain

- Transformation of the two plasmids containing ST and CMP-syn genes (as described in Section 25.4.1) to the JM107 *Nan* A⁻ strain is performed sequentially.

First, the pACKC18-CMP-syn plasmid is transformed to the strain and electrocompetent cells are prepared from subsequent clones.

- Plasmid DNA encoding the ST gene or libraries is transformed into these cells and the resulting cell population is used to directly inoculate LB media supplemented with the appropriate antibiotics; the cells are then grown overnight at 37 °C.
- The cells are diluted 1:50 in mineral-cultured media as described elsewhere [33] and then grown at 37 °C until the OD_{600} is 0.6. At this point, 0.5 m*M* IPTG is added and the cells are transferred to a 20 °C environment and grown overnight.

25.4.5 Cell-Based Assay

Here, the cells are incubated with fluorescent acceptor and donor sugars, after which any unreacted fluorescently labeled acceptor sugar is removed by washing.

- A 1 ml-aliquot of the cell suspension is centrifuged at 8000 rpm, using a standard microcentrifuge. The cell pellet is then resuspended in 50 µl of M9 media supplemented with 1 m*M* sialic acid and 0.2–0.5 m*M* concentrations of the different fluorescently labeled acceptor sugars.
- The acceptor concentration and the incubation time are highly dependent on the ST activity. In the case of screening libraries it is possible to increase the selection stringency by reducing the acceptor concentration and incubation time.
- Following an incubation period of 20 min to 1 h, the cells are pelleted (8000 rpm) and the excess acceptor sugar and Neu5Ac are removed by discarding the supernatant. The cells are resuspended in LB media and transferred to a 37 °C environment for 10 min; they are then re-centrifuged and washed twice with phosphate-buffered saline (PBS) before resuspension in 1 ml of PBS. *This step is extremely important to reduce background fluorescence and to facilitate detection of even weak ST activity*. At this stage the cells are ready for visual detection using a UV table and FACS analysis and sorting.

25.4.6 Validation, Sensitivity and Dynamic Range of the Cell-Based Assay

In order to verify that the cell-based assay reflects the target ST activity, and also to characterize the sensitivity and dynamic range of the assay, several control experiments must be performed.

Cells expressing the target ST enzyme and those expressing empty pUC18 plasmid are separately incubated with Neu5Ac and either a highly efficient acceptor or a less-efficient acceptor (e.g. bodipy-lactose and bodipy-galactose respectively for the *Cst*-II ST; Figure 25.3). Following extensive washing, the fluorescence intensity of the cells expressing the target ST should be substantially higher than that of control cells, as can be visualized under a UV lamp (see Figure 25.2a). In order to quantify the difference in fluorescence and test the dynamic range of the cell-based ST assay, the samples are subjected to FACS analysis. The mean fluorescence intensity of the cells expressing the target ST enzyme and incubated with a highly efficient acceptor should be significantly higher than that of the control cells (Figure 25.2b). Larger differences in mean fluorescence between cells expressing active ST and control cells can facilitate the detection of even slow transfer reactions.

25.4.7
Model Selection

Here, cells containing active ST are sorted from a large heterogeneous cell population using FACS.

- Cells expressing the target ST gene and those containing empty plasmid are grown and induced separately, as described above.
- Before incubation with the fluorescently labeled donor sugars, the cells expressing the ST enzymes are mixed with a large excess of cells containing empty plasmid. Several mixtures can be prepared at different ratios, 1:50, 1:100 and 1:200. The cell mixtures are then incubated with the acceptor and donor sugars as described above.
- Following several washing steps, the cells are analyzed by FACS and multiple events within the top 1–2% of the green fluorescence intensity are sorted into growth media in an Eppendorf tube.
- The sorted cells are plated and incubated at 37 °C.
- Randomly picked single colonies (usually 20) are grown and induced for protein expression. The ST activity is tested in crude cell lysates by thin-layer chromatography (TLC) to identify colonies expressing the ST gene.
- An enrichment factor is calculated from the number of colonies expressing the ST divided by the total number of colonies tested. This number is then divided by the initial ratio used for generating the mixture. It is desirable to have an enrichment factor of at least 50-fold from a starting mixture of 1:200.

25.4.8
Generation of Genetic Diversity in the Target ST Gene: Strategies for Constructing Large Mutant Libraries

These procedures involve:

- Inserting random mutations along the gene using error-prone PCR methodologies [34, 35].
- The diversification of several residues in the ST protein simultaneously may be performed by a novel methodology for spiking oligos in the target gene [36].
- In the case of high homology between the target ST gene and STs from other species, family DNA shuffling using the DNaseI method could be performed [37].

For a recent review on library generation methodologies, the reader is referred to Refs. [38, 39].

25.4.9 Library Sorting, Rounds of Enrichment and the Stringency of Selection

- The plasmid DNA library is transformed to JM107 *Nan* A^- cells carrying the CMP synthetase expression plasmid, grown and induced for protein expression (see Figure 25.1, step 2). Care should be taken to ensure a high transformation efficiency ($>10^6$).
- Cells are incubated with Neu5Ac and the fluorescently labeled acceptor (e.g. bodipy lactose), washed, and more than 10^7 cells are analyzed and sorted by FACS (Figure 25.1, steps 3–5).
- The incubation time and concentration of fluorescently labeled acceptor and donor sugar are highly dependent on the specific activity of the target ST. In cases where non-natural acceptor or donor sugars are used, higher concentrations and prolonged incubation times may be needed to detect the transfer activity.
- Three iterative rounds of enrichment are usually performed in order to sort the cells containing active ST and to discard those containing inactive mutants. In each round, multiple 'positive' events ($3–5 \times 10^4$) within the top 1–2% of the green fluorescence intensity are sorted, collected into growth media and plated on agar for a new round of enrichment.
- Following each round of sorting an increase in ST activity of the crude cell lysates should be observed (Figure 25.4a). Accordingly, the mean fluorescence of the library following three rounds of sorting should be substantially higher than that of the wild-type cells (Figure 25.4b).

25.4.10 Identification and Isolation of Improved Mutants

- In order to identify and isolate single clones with improved transfer activity, the plasmid DNA extracted from the third round of sorting is transformed to fresh JM107 *NanA*$^-$ cells.
- At least 20 random clones are picked, individually grown and tested for ST activity. Product formation is analyzed by TLC at different time points and compared to the wild-type ST activity, under identical conditions.
- It is desirable to test the activity with several different acceptor and donor sugars in order to detect changes in substrate specificity. In many cases, improvement toward the target substrate, used in the selection assay, results in a decrease in activity towards other substrates [40].
- Cells from variants showing significantly higher activity than the wild-type ST, are propagated in *E. coli* cells and frozen at –80 °C for further analysis.

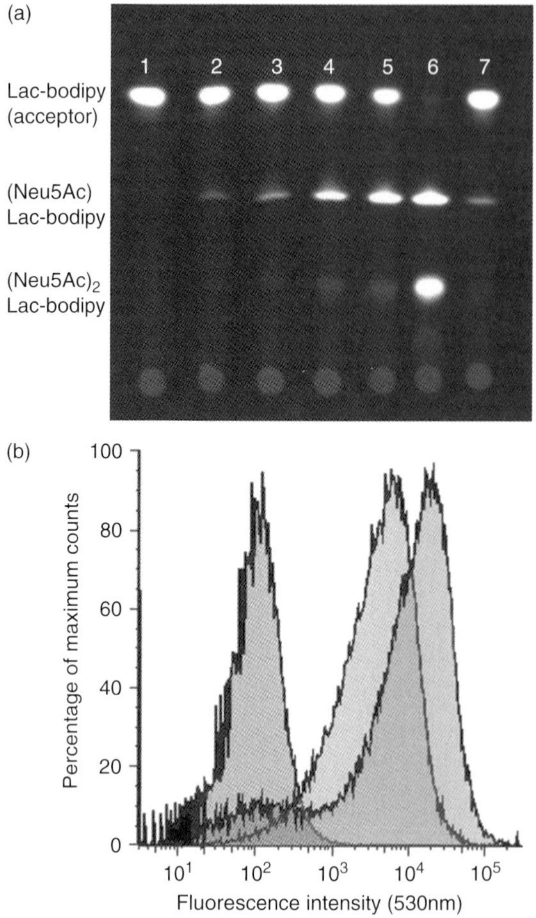

Figure 25.4 Library selection through three iterative rounds of sorting by FACS (see Section 25.4.9 and Ref. [22]). (a) Activity analysis of the various rounds of FACS enrichment. *Cst*-II transfer activity was measured on crude cell lysates prepared from the pool of cells obtained after each round of enrichment and analyzed by TLC. Activity was measured and compared to cells expressing pUC18 plasmid (control), wild-type *Cst*-II and the evolved *Cst*-II F91Y mutant. Lane (1): pUC18 plasmid; Lane (2): Naïve *Cst*-II library; Lanes (3–5): library following one, two and three rounds of enrichment respectively; Lane (6): evolved F91Y *Cst*-II mutant; Lane (7): wild-type *Cst*-II; (b) FACS histogram analysis of cells expressing pUC18 (left), wild-type *Cst*-II (middle) and library following three rounds of enrichment (right) after incubation (1 h) with Neu5Ac (1 m*M*) and bodipy-lactose (0.5 m*M*).

25.4.11
Characterization of Improved ST Mutants

- Improved clones are analyzed by DNA sequencing to identify mutations that affect the transfer activity.

- In most cases improved ST mutants must be subcloned to pET vectors containing a strong T7 promoter for higher overexpression in the BL21 *E. coli* strain.
- Following purification by affinity chromatography, the kinetic parameters of the improved mutants can be determined using a nonradioactive kinetic assay, as described previously [41].

25.5
Challenges and Prospects of GT Engineering

The past few years have seen significant progress in the development and application of HTS systems for enzymatic activity [14, 42, 43]. Despite the difficulties in the development of screening methodologies for GT and other transfer activity, new and innovative methods have been recently developed. The methodology for the directed evolution of GTs described above provides a genuinely HTS approach that facilitates the screening of millions of mutants in a rapid and efficient manner. These investigations open up new avenues for the directed evolution of GTs in the glycosylation of a variety of acceptors. In the case of GTs that form neutral sugar products the ST may serve as a coupling enzyme to trap the reaction product in the cell, thus extending the methodology to other GTs that transfer galactose to fluorescently labeled acceptor sugars. We believe that our methodology, which is based on selectively trapping the transfer product, can be extended to detect other transfer reactions (e.g. phosphorylation or sulfation) in which a charged moiety is transferred to a variety of acceptors. Indeed, preliminary experiments with cells expressing cytosolic sulfotransferases have indicated that the sulfated fluorescent transfer product is trapped in the cells while the unreacted fluorescent substrate is washed away. However, each substrate for the cell-based assay must be carefully designed and examined for penetration into the cells and entrapment following the transfer reaction. In addition, this methodology is also applicable to the detection of transfer reactions by fluorescence resonance energy transfer (FRET) by using acceptors and donors that are both fluorescently labeled. Moreover, the use of FRET will alleviate the need for selective trapping of the product, while extensive washing of the unreacted acceptor sugar will not be necessary in order to reduce the background fluorescence.

References

1 Crocker, P.R. and Feizi, T. (1996) Carbohydrate recognition systems: functional triads in cell-cell interactions. *Current Opinion in Structural Biology*, **6**, 679–91.

2 Rudd, P.M., Elliott, T., Cresswell, P., Wilson, I.A. and Dwek, R.A. (2001) Glycosylation and the immune system. *Science*, **291**, 2370–6.

3 Lowe, J.B. (2003) Glycan-dependent leukocyte adhesion and recruitment in inflammation. *Current Opinion in Cell Biology*, **15**, 531–8.

4 Sacks, D. and Kamhawi, S. (2001) Molecular aspects of parasite-vector and vector-host interactions in leishmaniasis. *Annual Review of Microbiology*, **55**, 453–83.

5 Qasba, P.K., Ramakrishnan, B. and Boeggeman, E. (2005) Substrate-induced conformational changes in glycosyltransferases. *Trends in Biochemical Sciences*, **30**, 53–62.

6 Hu, Y. and Walker, S. (2002) Remarkable structural similarities between diverse glycosyltransferases. *Chemistry and Biology*, **9**, 1287–96.

7 Fritz, T.A., Hurley, J.H., Trinh, L.B., Shiloach, J. and Tabak, L.A. (2004) The beginnings of mucin biosynthesis: the crystal structure of UDP-GalNAc:polypeptide alpha-N-acetylgalactosaminyltransferase-T1. *Proceedings of the National Academy of Sciences of the United States of America*, **101**, 15307–12.

8 MacKenzie, L.F., Wang, Q.P., Warren, R.A.J. and Withers, S.G. (1998) Glycosynthases: mutant glycosidases for oligosaccharide synthesis. *Journal of the American Chemical Society*, **120**, 5583–4.

9 Hancock, S.M., Vaughan, M.D. and Withers, S.G. (2006) Engineering of glycosidases and glycosyltransferases. *Current Opinion in Chemical Biology*, **10**, 509–19.

10 Arnold, F.H., Wintrode, P.L., Miyazaki, K. and Gershenson, A. (2001) How enzymes adapt: lessons from directed evolution. *Trends in Biochemical Sciences*, **26**, 100–6.

11 Tao, H. and Cornish, V.W. (2002) Milestones in directed enzyme evolution. *Current Opinion in Chemical Biology*, **6**, 858–64.

12 Dalby, P.A. (2003) Optimising enzyme function by directed evolution. *Current Opinion in Structural Biology*, **13**, 500–5.

13 Goddard, J.P. and Reymond, J.L. (2004) Enzyme assays for high-throughput screening. *Current Opinion in Biotechnology*, **15**, 314–22.

14 Aharoni, A., Griffiths, A.D. and Tawfik, D.S. (2005) High-throughput screens and selections of enzyme-encoding genes. *Current Opinion in Chemical Biology*, **9**, 210–16.

15 Becker, S., Schmoldt, H.U., Adams, T.M., Wilhelm, S. and Kolmar, H. (2004) Ultra-high-throughput screening based on cell-surface display and fluorescence-activated cell sorting for the identification of novel biocatalysts. *Current Opinion in Biotechnology*, **15**, 323–9.

16 Griffiths, A.D. and Tawfik, D.S. (2003) Directed evolution of an extremely fast phosphotriesterase by in vitro compartmentalization. *The EMBO Journal*, **22**, 24–35.

17 Feldhaus, M.J., Siegel, R.W., Opresko, L.K., Coleman, J.R., Feldhaus, J.M.W., Yeung, Y.A., Cochran, J.R., Heinzelman, P., Colby, D., Swers, J., Graff, C., Wiley, H.S. and Wittrup, K.D. (2003) Flow-cytometric isolation of human antibodies from a nonimmune *Saccharomyces cerevisiae* surface display library. *Nature Biotechnology*, **21**, 163–70.

18 Varadarajan, N., Gam, J., Olsen, M.J., Georgiou, G. and Iverson, B.L. (2005) Engineering of protease variants exhibiting high catalytic activity and exquisite substrate selectivity. *Proceedings of the National Academy of Sciences of the United States of America*, **102**, 6855–60.

19 Griswold, K.E., Kawarasaki, Y., Ghoneim, N., Benkovic, S.J., Iverson, B.L. and Georgiou, G. (2005) Evolution of highly active enzymes by homology-independent recombination. *Proceedings of the National Academy of Sciences of the United States of America*, **102**, 10082–7.

20 Aharoni, A., Amitai, G., Bernath, K., Magdassi, S. and Tawfik, D.S. (2005) High-throughput screening of enzyme libraries: thiolactonases evolved by fluorescence-activated sorting of single cells in emulsion compartments. *Chemistry and Biology*, **12**, 1281–9.

21 Mastrobattista, E., Abecassis, V., Chanudet, E., Treacy, P., Kelly, B.T. and Griffiths, A. (2005) Discovering novel evolutionary pathways to beta-galactosidases using in vitro compartmentalization and fluorescence activated sorting of double emulsions. *Chemistry and Biology*, **12**, 1291–300.

22 Aharoni, A., Thieme, K., Chiu, C.P., Buchini, S., Lairson, L.L., Chen, H., Strynadka, N.C., Wakarchuk, W.W. and Withers, S.G. (2006) High-throughput screening methodology for the directed evolution of glycosyltransferases. *Nature Methods*, **3**, 609–14.

23 Kim, Y.W., Lee, S.S., Warren, R.A. and Withers, S.G. (2004) Directed evolution of a glycosynthase from *Agrobacterium* sp. increases its catalytic activity dramatically and expands its substrate repertoire. *Journal of Biological Chemistry*, **279**, 42787–93.

24 Kempton, J.B. and Withers, S.G. (1992) Mechanism of *Agrobacterium* beta-glucosidase: kinetic studies. *Biochemistry*, **31**, 9961–9.

25 Lin, H., Tao, H. and Cornish, V.W. (2004) Directed evolution of a glycosynthase via chemical complementation. *Journal of the American Chemical Society*, **126**, 15051–9.

26 Carter, B.T., Lin, H., Goldberg, S.D., Althoff, E.A., Raushel, J. and Cornish, V.W. (2005) Investigation of the mechanism of resistance to third-generation cephalosporins by class C beta-lactamases by using chemical complementation. *ChemBioChem*, **6**, 2055–67.

27 Yang, J., Liu, L. and Thorson, J.S. (2004) Structure-based enhancement of the first anomeric glucokinase. *ChemBioChem*, **5**, 992–6.

28 Hoffmeister, D., Yang, J., Liu, L. and Thorson, J.S. (2003) Creation of the first anomeric D/L-sugar kinase by means of directed evolution. *Proceedings of the National Academy of Sciences of the United States of America*, **100**, 13184–9.

29 Love, K.R., Swoboda, J.G., Noren, C.J. and Walker, S. (2006) Enabling glycosyltransferase evolution: a facile substrate-attachment strategy for phage-display enzyme evolution. *ChemBioChem*, **7**, 753–6.

30 Sandman, K.E. and Noren, C.J. (2000) The efficiency of *Escherichia coli* selenocysteine insertion is influenced by the immediate downstream nucleotide. *Nucleic Acids Research*, **28**, 755–61.

31 Chiu, C.P., Watts, A.G., Lairson, L.L., Gilbert, M., Lim, D., Wakarchuk, W.W., Withers, S.G. and Strynadka, N.C. (2004) Structural analysis of the sialyltransferase CstII from *Campylobacter jejuni* in complex with a substrate analog. *Nature Structural and Molecular Biology*, **11**, 163–70.

32 Karwaski, M.F., Wakarchuk, W.W. and Gilbert, M. (2002) High-level expression of recombinant *Neisseria* CMP-sialic acid synthetase in *Escherichia coli*. *Protein Expression and Purification*, **25**, 237–40.

33 Antoine, T., Heyraud, A., Bosso, C. and Samain, E. (2005) Highly efficient biosynthesis of the oligosaccharide moiety of the GD3 ganglioside by using metabolically engineered *Escherichia coli*. *Angewandte Chemie – International Edition in English*, **44**, 1350–2.

34 Vartanian, J.P., Henry, M. and Wain-Hobson, S. (2001) Simulating pseudogene evolution in vitro: determining the true number of mutations in a lineage. *Proceedings of the National Academy of Sciences of the United States of America*, **98**, 13172–6.

35 Zaccolo, M., Williams, D.M., Brown, D.M. and Gherardi, E. (1996) An approach to random mutagenesis of DNA using mixtures of triphosphate derivatives of nucleoside analogues. *Journal of Molecular Biology*, **255**, 589–603.

36 Herman, A. and Tawfik, D.S. (2007) Incorporating synthetic oligonucleotides via gene reassembly (ISOR): a versatile tool for generating targeted libraries. *Protein Engineering, Design and Selection*, **20**, 219–26.

37 Crameri, A., Raillard, S.A., Bermudez, E. and Stemmer, W.P. (1998) DNA shuffling of a family of genes from diverse species accelerates directed evolution. *Nature*, **391**, 288–91.

38 Lutz, S. and Patrick, W.M. (2004) Novel methods for directed evolution of enzymes: quality, not quantity. *Current Opinion in Biotechnology*, **15**, 291–7.

39 Neylon, C. (2004) Chemical and biochemical strategies for the randomization of protein encoding DNA sequences: library construction methods for directed evolution. *Nucleic Acids Research*, **32**, 1448–59.

40 Aharoni, A., Gaidukov, L., Khersonsky, O., Mc, Q.G.S., Roodveldt, C. and Tawfik, D.S. (2005) The 'evolvability' of promiscuous protein functions. *Nature Genetics*, **37**, 73–6.

41 Gosselin, S., Alhussaini, M., Streiff, M.B., Takabayashi, K. and Palcic, M.M. (1994) A continuous spectrophotometric assay for glycosyltransferases. *Analytical Biochemistry*, **220**, 92–7.

42 Boersma, Y.L., Droge, M.J. and Quax, W.J. (2007) Selection strategies for improved biocatalysts. *The FEBS Journal*, **274**, 2181–95.

43 Reymond, J.L. and Babiak, P. (2007) Screening systems. *Advances in Biochemical Engineering/Biotechnology*, **105**, 31–58.

26
Yeast Surface Display in Protein Engineering and Analysis

Benjamin J. Hackel and K. Dane Wittrup

26.1
Review

26.1.1
Introduction

In yeast surface display (YSD), tens of thousands of copies of the protein of interest (POI) are tethered to the exterior of an individual *Saccharomyces cerevisiae* yeast cell while the genetic information for the POI is maintained in the cell interior. The cell–protein linkage begins with the Aga1p subunit of a-agglutinin, which anchors in the cell wall periphery via a β-glucan covalent linkage [1]. The Aga2p subunit, secreted from the yeast cell as a fusion to the POI, attaches to Aga1p via two disulfide bonds. The peptide bond in the fusion protein thus completes the linkage resulting in a 'display' of the POI on the yeast cell. The most common yeast display construct includes a flexible linker (commonly $[Gly_4Ser]_3$) between the N-terminal Aga2p and the C-terminal POI. Epitope tags flank the POI to enable analysis of Aga2p display as well as full-length protein fusion display (Figure 26.1). The reverse orientation of a fusion protein has also been effectively displayed yielding a free N terminus of the POI [2]. Display is achieved through transformation of DNA encoding for the Aga2p–POI fusion, followed by cell growth and induction of both Aga1p and Aga2p–POI protein expression.

The eukaryotic protein production machinery of *S. cerevisiae* has proven to be successful for the high-quality production of a wide variety of proteins. It follows that one of the key advantages of YSD relative to other display formats is the vast array of proteins that can be displayed with post-translational modifications and proper folding. The initial demonstration of YSD yielded display of a single-chain antibody variable fragment (scFv) [3]. Additional molecular recognition formats have been displayed including antigen-binding fragments (Fabs) [4], peptides [5, 6] and the tenth type III domain of fibronectin [7]. The display of T-cell receptors [8], major histocompatability complexes (MHCs), both class I [9] and class II [10], and interleukin-2 (IL-2) [11] have broadened the impact of YSD on

Protein Engineering Handbook. Edited by Stefan Lutz and Uwe T. Bornscheuer

ISBN: 978-3-527-31850-6

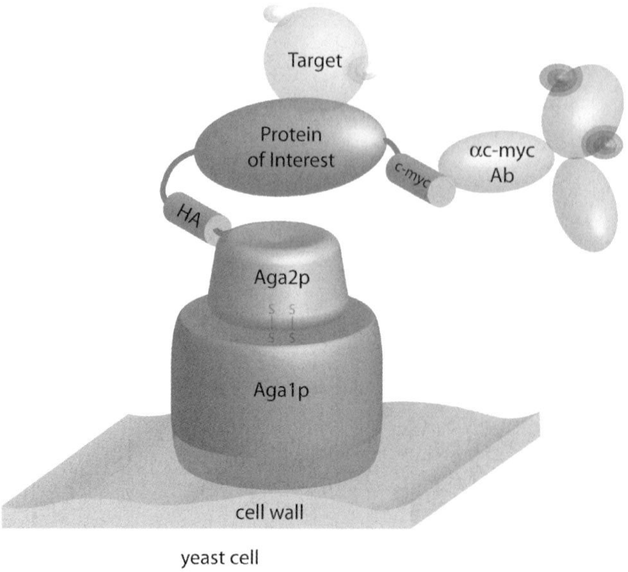

Figure 26.1 Schematic of yeast surface display. The Aga2p subunit is tethered to the yeast cell via two disulfide bonds to Aga1p, which is anchored in the yeast cell wall. The protein of interest is connected to the Aga2p subunit through a flexible peptide linker as a result of genetic fusion. Hemaglutinin and c-myc epitope tags flank the protein of interest on the N and C termini, enabling detection of each displayed molecule using fluorophore-conjugated antibodies. Depending on the protein of interest and induction conditions, tens of thousands of these fusions are displayed per cell. The structural forms of the different elements, aside from the antibody, are not known.

immunological research. E-selectin [12], integrin domains [13, 14], and the ectodomain of epidermal growth factor receptor (EGFR) [15] have been displayed, as have green fluorescent protein (GFP) [16] and epidermal growth factor (EGF) [17]. YSD has also been incorporated into enzyme research, including the display of lignin peroxidase [18], hemolysin [19], glucoamylase [20] and horseradish peroxidase [21].

26.1.2 Protein Engineering

The initial and primary use of YSD, as well as other display formats, is to link genetic information with the selectable phenotype to allow combinatorial library selection. This enables both the isolation of proteins of novel function from naïve libraries as well as the directed evolution of existing clones. Prior to library selection, a library must be created. YSD is amenable to the multitude of techniques available to DNA library construction including synthetic construction, polymerase chain reaction (PCR) amplification of immune or non-immune donor nucleic acids, or randomization of existing sequences [22]. For example, a non-immune library of 10^9 scFvs was created by the PCR of human RNA, and was found to contain multiple nanomolar affinity binders to hapten, peptide and protein targets [23]. In addition, YSD has the distinct option of hybrid library creation through yeast mating [24], which was used effectively to isolate a Fab with 0.8 n*M* affinity for botulinum toxin [25] and to engineer a Fab with a ninefold improvement in affinity [26]. In addition to full-chain shuffling, homologous recombination in yeast permits gene shuffling and facilitates library creation [27].

One important benefit of YSD is the ability to conduct selections via fluorescence-activated cell sorting (FACS). This enables the real-time quantification of protein activity and therefore the ability to design optimized selection protocols [28] and make fine affinity discrimination via protein expression normalization using two-color cytometry [29]. Current cytometers can sort over 10^8 cells per hour, which is sufficient for most situations; for larger libraries magnetic bead sorting has been demonstrated as an effective means of selection [30, 31]. More recently, in order to bind targets in their cellular context, a YSD library was panned against rat brain endothelial cells leading to the isolation of novel antibodies [32], while a density centrifugation protocol was validated for engineering T-cell receptors [33].

26.1.2.1 Affinity Engineering

These techniques have provided significant success in affinity-engineering applications, including the directed evolution of a femtomolar binder to fluorescein, the highest affinity antibody reported to date [34]. A 10 000-fold improvement was engineered for a T-cell receptor against toxic shock syndrome toxin-1 resulting in 180 p*M* affinity [35]. IL-2 variants were engineered for improved binding to the alpha subunit of the IL-2 receptor, leading to a potent stimulation of T-cell proliferation [11]. Isolation from both naïve and immune libraries, followed by directed evolution, has also been demonstrated. A light-chain variable domain against huntingtin exon 1 was isolated, matured to 30 n*M* affinity [36], and demonstrated to rescue toxicity from both neuronal and yeast models of Huntington's disease [37]. A lysozyme-binding fibronectin domain was isolated from a synthetic library and engineered to picomolar affinity [7]. Multiple single-chain antibody variable fragments were isolated for binding to botulinum neurotoxin and evolved to picomolar affinity [38]; in order to generate a broader subtype binding, a desirable attribute of a practical system for toxin detection and neutralization, a 1250-fold improvement in affinity for one toxin serotype was engineered while maintaining an alternative serotype affinity [39].

26.1.2.2 Stability and Expression Engineering

Although the most common use of display technologies is for engineering binding affinity and specificity, YSD is also a valuable tool for engineering stability and protein production. Bovine pancreatic trypsin inhibitor mutants of reduced stability were secreted from yeast at proportionally reduced levels [40]. Similarly, single-chain T-cell receptors of improved stability were secreted at proportionally increased levels [41]; importantly, in the latter case the protein display also correlated with stability, implicating YSD protein level as a readily selectable parameter for stability. This approach has since been employed to improve the stability and expression of a diverse class of proteins. A single-chain T-cell receptor, which typically is insoluble and difficult to express, was engineered to $4\,mg\,ml^{-1}$ solubility and expressed at $7.5\,mg\,l^{-1}$ in yeast culture [42]. The class I MHC molecule L^d was engineered for improved stability, resulting in high-yield bacterial production for structural studies [43]. Directed evolution with YSD of the extracellular domains of the EGFR [44] and p55 tumor necrosis factor receptor (TNFR) [45] increased yeast production levels. A single-chain variable fragment against carcinoembryonic antigen was engineered to maintain 80% activity after a 9-day incubation at 37 °C and was secreted from yeast at $20\,mg\,l^{-1}$, representing a 100-fold improvement in production from the parental clone [46]. The display: secretion correlation has also been used in cellular engineering as YSD selections of a yeast cDNA overexpression library identified cellular proteins capable of increasing soluble heterologous protein expression up to 7.9-fold [47]. However, a recent study indicates that a molten globule three-helix bundle protein can be displayed at high levels and mutant stability did not correlate with display [48], demonstrating a possible limitation to the method.

26.1.2.3 Enzyme Engineering

It should be noted that any selectable activity can be engineered for proteins displayed on the yeast surface. Thus, with clever selection strategies, yeast display can be applied to enzyme engineering. A three-amino acid saturation mutagenesis library of glucoamylase was displayed on the yeast surface and screened using a halo assay; as a result, several clones with 1.4-fold improvement in activity were isolated [20]. Horseradish peroxidase was engineered for altered stereoselectivity [21]. Many other enzymes have been functionally displayed on the yeast surface, but protein engineering results have not been reported [18, 19, 49, 50].

26.1.3 Protein Analysis

While genotype–phenotype linkage enables cell selection, one valuable aspect of YSD is facile protein analysis; rather than make a selection based on phenotype and recover the DNA to identify the genotype, it is possible to display a known genotype and analyze the resultant phenotype. This not only obviates both subcloning into expression vectors and protein purification but also yields functional protein immobilized on a ~5–10 μm sphere coated with a relatively passive cell wall.

26.1.3.1 Clone Characterization

The interaction between displayed protein and soluble ligand can be readily analyzed with flow cytometry, both through association and dissociation kinetic analyses as well as equilibrium affinity titration. Equilibrium dissociation constants determined by YSD are in excellent agreement with those determined via alterna-

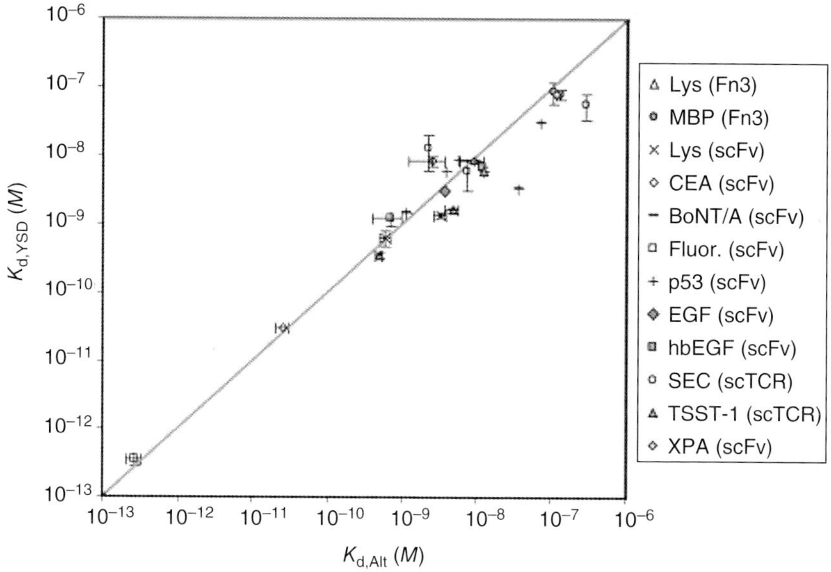

Figure 26.2 Affinity determination by yeast surface display or alternative methods. Equilibrium dissociation constants determined by yeast surface display ($K_{d,YSD}$) are plotted on the vertical axis with values determined by alternative methods ($K_{d,Alt}$) on the horizontal axis. The median difference between $K_{d,YSD}$ and $K_{d,Alt}$ is 41%. Figure adapted from Lipovsek *et al.* [7, 51].

tive methods such as surface plasmon resonance (Figure 26.2). Specificity can also be analyzed either through replicate experiments with multiple targets or directly via multicolor cytometry. Moreover, thermal stability can be determined via YSD and flow cytometry, either through an analysis of the time dependence of denaturation [41] or by equilibrium thermal titrations [52]. The midpoint of thermal denaturation as measured by YSD agreed well with assays using purified proteins for single-chain variable fragments and single-chain T-cell receptors; this agreement also holds for engineered fibronectin type III domains (B.J. Hackel, unpublished results). These techniques enable the rapid characterization of lead clones during library selections.

26.1.3.2 Paratope: Epitope Study

In addition to affinity, specificity and stability determination, YSD facilitates the study of binding interfaces through epitope mapping, rational mutagenesis and shotgun scanning mutagenesis. By displaying various domains of the EGFR on the surface of yeast and assaying binding by soluble antibodies, the binding epitopes of each antibody were ascertained. The status of these epitopes as linear or conformationally specific was resolved by the maintenance or loss of binding with receptor denaturation prior to antibody addition [15]. Epitope mapping was refined to the residue level through FACS selection of randomly mutated clones with loss of binding [53]. Domain-level epitope mapping was also conducted for antibodies to the severe acute respiratory syndrome (SARS) coronavirus nucleocapsid protein [54], and residue-level mapping was conducted for antibodies to West Nile virus [55] and botulinum neurotoxin [56]. In addition to epitope mapping, the paratope can be readily investigated through the display of rational mutants and assay of their binding properties by flow cytometry, as in alanine scanning [57]. For a more descriptive view of the paratope, shotgun scanning mutagenesis could be applied to YSD; in fact, YSD and FACS present an ideal pairing for shotgun scanning mutagenesis because a single expression-normalized selection could be conducted using two-color flow cytometry as opposed to the separate affinity and stability sorts required for phage display [58].

26.1.3.3 YSD in Bioassays

Displaying yeast can also be used as functional beads in bioassays. Serological antibodies can be detected by quantifying (via flow cytometry) the binding of sera to recombinant antigen expressed on the yeast surface (RAYS). The RAYS assay matched the performance of standard ELISA and Western blot procedures for NY-ESO-1 antibody detection, and outperformed the ELISA and phage display detection of SSX2 antibodies [59]. The RAYS technology was also used to study the immune response to a panel of cancer testis antigens [60]. YSD of the Fc-binding ZZ domain enabled protein quantification via a cell-based ELISA [61]. The same ZZ domain-displaying cells were also effective in affinity purification of IgG from serum.

Beyond the quantification and purification of proteins, yeast can function as mimics of alternative cell types through the display of appropriate surface molecules. Cellular rolling kinematics were studied by tracking the motion of E-selectin-displaying yeast on a ligand-coated surface, which was a proxy for ligands on the surface of leucocytes [12]. The impact of binding affinity on rolling adhesion was elucidated by kinematic assays using yeast displaying α-integrin I domains of differing affinities [14]. Additionally, Kranz and colleagues have conducted numerous immunological studies on T-cell receptor and peptide–MHC interactions [62–65].

26.2
Protocols and Practical Considerations

YSD technology can be used to solve a variety of scientific inquiries and engineering challenges, as evidenced by the diversity of applications mentioned in Section 26.1. In the following sections we will address the practical implementation of many techniques, including DNA and cell preparation and culture, library selection and clone characterization. Detailed protocols will be provided for the most common techniques. Although a variety of yeast surface display designs have been implemented, only the most widely used – and formerly commercial – format will be discussed here. Engineering of the fibronectin type III domain molecular recognition scaffold will be used as an example, although the translation to other systems should be readily apparent.

26.2.1 Materials

26.2.1.1 Cells and Plasmids

- EBY100: *S. cerevisiae (a GAL1-AGA1::URA3 ura3-52 trp1 leu2Δ1 his3Δ200 pep4::HIS2 prb1Δ1.6R can1 GAL)*; available from the authors.
- pCT-Fn3: yeast surface display vector; available from authors.

26.2.1.2 Media and Buffers

- YPD: $10\,g\,l^{-1}$ yeast extract, $20\,g\,l^{-1}$ peptone, $20\,g\,l^{-1}$ dextrose; filter sterilize or autoclave.
- SD-CAA: $5.4\,g\,l^{-1}$ Na_2HPO_4, $8.56\,g\,l^{-1}$ $NaH_2PO_4 \cdot H_2O$, $20\,g\,l^{-1}$ dextrose, $6.7\,g\,l^{-1}$ yeast nitrogen base, $5.0\,g\,l^{-1}$ casamino acids; filter-sterilize.
- SG-CAA: $5.4\,g\,l^{-1}$ Na_2HPO_4, $8.56\,g\,l^{-1}$ $NaH_2PO_4 \cdot H_2O$, $20\,g\,l^{-1}$ galactose, $6.7\,g\,l^{-1}$ yeast nitrogen base, $5.0\,g\,l^{-1}$ casamino acids; filter-sterilize

Note: A small amount ($<2\,g\,l^{-1}$) of dextrose can be added to ease the transition into galactose-based induction. But: note that high levels of dextrose suppress the GAL1 promoter.

- SD-CAA plates: 5.4 g Na_2HPO_4, 8.56 g $NaH_2PO_4 \cdot H_2O$, 16 g bacto agar in 0.9 l water; autoclave.
- 20 g dextrose, 6.7 g yeast nitrogen base, 5.0 g casamino acids in 0.1 l water; filter-sterilize into autoclaved buffer once, cool (~50 °C) and pour plates.

Note: Bacterial contamination in yeast media can be inhibited by $100\,kU\,l^{-1}$ penicillin, $0.1\,g\,l^{-1}$ streptomycin, and/or $0.1\,g\,l^{-1}$ kanamycin antibiotics or growth in citrate-buffered (pH 4.5–5.5) media.

26.2.1.3 Buffers

- Buffer E: 10 mM Tris, pH 7.5, 270 m*M* sucrose, 1 m*M* $MgCl_2$; filter-sterilize.
- PBSA (phosphate-buffered saline with bovine serum albumin): 10 m*M* $NaHPO_4$ pH 7.4, 137 m*M* NaCl, 2.7 m*M* KCl, $1\,g\,l^{-1}$ BSA; filter-sterilize.
- PBSM (phosphate-buffered saline for MACS): 10 m*M* $NaHPO_4$ pH 7.4, 137 m*M* NaCl, 2.7 m*M* KCl, $5\,g\,l^{-1}$ BSA, 2 m*M* EDTA; filter-sterilize.

26.2.1.4 Flow Cytometry Reagents

This brief list is provided as a starting point for reagent selection. Many other reagents have been used effectively.

- Primary detection of displayed protein epitope tags (Covance): mouse anti-c-myc IgG clone 9E10, mouse anti-HA clone 16B12.

- Secondary detection of primary mouse antibodies (Invitrogen): *R*-phycoerythrin-conjugated goat anti-mouse IgG (H + L), AlexaFluor488-conjugated goat anti-mouse IgG (H + L).
- Secondary detection of biotinylated protein (Invitrogen): *R*-phycoerythrin-conjugated NeutrAvidin, AlexaFluor488-conjugated streptavidin, AlexaFluor488-conjugated mouse anti-biotin IgG clone 2F5.

Note: Biotin and dye labeling kits from Invitrogen are effective for antigen labeling.

26.2.2 Nucleic Acid and Yeast Preparation

26.2.2.1 DNA Preparation

As with most molecular cloning applications, there is a wide variety of paths that yield effective results. The necessary result is plasmid DNA containing an expression construct encoding for a protein fusion of Aga2p and the protein of interest. A selectable marker to ensure plasmid retention in minimal media is also needed. The pCT-Fn3 vector is detailed in Figure 26.3. An alternative vector has also been created by modifying the $(Gly_4Ser)_3$ encoding region to reduce genetic repetitive-

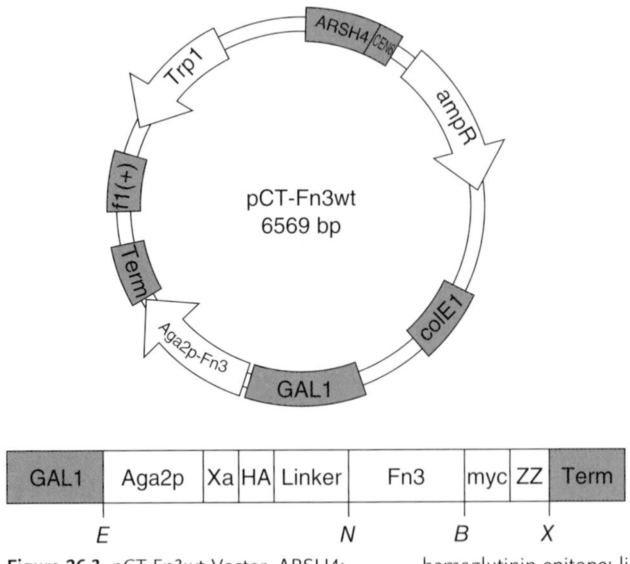

Figure 26.3 pCT-Fn3wt Vector. ARSH4: autonomously replicating sequence H4; CEN6: centromeric sequence 6; ampR: ampicillin resistance gene; colE1: colicin E1 origin of replication; GAL1: GAL1 promoter; E: *Eco*RI site; Aga2p: agglutinin 2p protein subunit; Xa: factor Xa cleavage site; HA: hemaglutinin epitope; linker: $(Gly_4Ser)_3$ linker; N: *Nhe*I site; Fn3: fibronectin tenth type III gene; B: BamHI site; myc: c-myc epitope; ZZ: TAATAG stop codons; X: *Xho*I site; Term: alpha mating factor terminator; f1 (+): f1 origin of replication; Trp1: Trp1 gene. The lower representation is not to scale.

ness to improve the fidelity of homologous recombination. Plasmid DNA can be introduced either in an intact circular form or through homologous recombination of multiple linear segments. Both methods provide good results, although the latter technique facilitates larger library production without the need for ligation and bacterial transformation. A protocol detailing preparation of a linearized vector and gene insert is provided and continued in Section 26.2.2.2 below.

1. Digest pCT-Fn3 with *Nhe*I, *Bam*HI, and *Btg*I (internal to protein of interest).

Note: Although internal digestion is not necessary, it almost eliminates the occurrence of a singly digested vector, which could otherwise enable reclosure without gene insertion during library construction.

2. Purify the vector using agarose gel electrophoresis and gel extraction.
3. Concentrate the DNA by alcohol precipitation (PelletPaint Coprecipitatant (Novagen) is effective). Resuspend in H_2O to 1–5 μg μl^{-1}.
4. Create a genetic construct consisting of the gene of interest flanked on each side by ~50 nucleotides homologous to the DNA flanking the *Nhe*I and *Bam*HI sites.

An error-prone PCR protocol is provided as follows:

- Prepare the PCR mix for mutagenesis:

Amount (μl)	Component	Final concentration
5	10× *Taq* buffer (w/o $MgCl_2$)	1×
2	$MgCl_2$, 50 m*M*	2 m*M*
2.5	5′ primer, 10 μ*M*	0.5 μ*M*
2.5	3′ primer, 10 μ*M*	0.5 μ*M*
1	dNTPs, 10 m*M* each	200 μ*M*
1	template DNA (pCT-Fn3), 0.67 ng μl^{-1}	13.3 pg μl^{-1}
5	8-oxo-dGTP, 20 μ*M*	2 μ*M*
5	dPTP, 20 μ*M*	2 μ*M*
25.5	ddH_2O	–
0.5	*Taq* DNA polymerase	0.05 U μl^{-1}

- Thermally cycle:

Denature:	94 °C for 3 min
15 cycles:	94 °C for 45 s, 60 °C for 30 s, 72 °C for 90 s
Final extension:	72 °C for 10 min

Note: The concentration of DNA and nucleotide analogues, as well as the number of PCR cycles, was optimized to yield one to five amino acid mutations in the Fn3 construct (120 amino acids). A decrease in the nucleotide analogue concentrations or the number of PCR cycles will reduce the mutagenic frequency. For scFv genes, 10 cycles with 2 μ*M* nucleotide analogues is effective.

- Purify gene fragment by agarose gel electrophoresis and gel extraction.
- Prepare the PCR mix for amplification:

Amount (μl)	Component	Final amount/conc.
10	10× *Taq* buffer (w/o $MgCl_2$)	1× *Taq* buffer w/o $MgCl_2$
4	$MgCl_2$, 50 m*M*	2 m*M* $MgCl_2$
5	5′ primer, 10 μ*M*	0.5 μ*M* 5′ primer
5	3′ primer, 10 μ*M*	0.5 μ*M* 3′ primer
2	dNTPs, 10 m*M* each	200 μ*M* each dNTP
4	Extracted PCR product	–
69	ddH_2O	–
1.0	*Taq* DNA polymerase	2.5 units

Note: Multiple tubes may be prepared to yield more DNA for larger library production.

- Thermally cycle

Denature:	94 °C for 3 min.
30 cycles:	94 °C for 45 s, 60 °C for 30 s, 72 °C for 90 s
Final extension:	72 °C for 10 min.

- Purify gene fragment by agarose gel electrophoresis and gel extraction.

Note: This step is optional; it decreases the total yield of gene insert but improves product purity and therefore library accuracy.

- Concentrate DNA by alcohol precipitation. Resuspend in H_2O to 1–5 μg μl^{-1}.

Alternatively, many different variations can be envisioned enabling focused mutagenesis, random mutagenesis, synthetic gene construction, shuffling or other techniques.

The analysis of individual clones can be achieved either through isolation of a cell by combinatorial library selection or by the construction of a clonal plasmid in a manner analogous to library construction.

26.2.2.2 Yeast Transformation

For the transformation of single clones or low-diversity libraries, in which high transformation efficiency is unnecessary, the Frozen-EZ Yeast Transformation Kit II (Zymo Research) provides quick and reliable results. Yeast in logarithmic growth are washed, resuspended in a proprietary solution, and frozen in aliquots at −70 °C; the cells are thawed (at any time within six months), combined with plasmid DNA and poly(ethylene glycol) (PEG) solution, incubated at 30 °C for 1 h and plated. The manufacturer's protocol is effective without modification.

For large library creation, either an optimized lithium acetate method or electroporation can be used. The lithium acetate method has been extensively studied and improved by Gietz and colleagues, and effective protocols have been described [66]. Alternatively, the electroporation of competent cells can yield DNA transformation with high efficiency [67]. The combination of homologous recombination and electroporation routinely yields libraries of 10^6–10^8 transformants. A detailed example protocol is provided as follows (the same protocol may be used with intact plasmid, although the transformation efficiency may be reduced).

1. Inoculate 5 ml of YPD with a colony of EBY100. Incubate overnight at 30 °C, 250 rpm.
2. Inoculate 50 ml of YPD to an $OD_{600\,nm}$ of 0.1 using overnight culture.

Note: 1×10^7 cells ml^{-1} corresponds to $OD_{600} = 1$.

3. Incubate 50 ml culture at 30 °C, 250 rpm until $OD_{600\,nm} = 1.3$–1.5 (~6 h).

Note: It is critical to use cells in the middle of the logarithmic phase of growth. Transformation efficiency is greatly reduced when cells in late logarithmic phase or stationary phase are used.

4. Dissolve 0.195 g of 1,4-dithiothreitol (DTT) in 0.5 ml of 1 *M* Tris, pH 8.0. Filter-sterilize into 50 ml of yeast culture.

Note: the best results are obtained with fresh DTT solution; alternatively, aliquots can be stored at −20 °C for several months.

5. Incubate the culture with DTT at 30 °C for 15 min.

Note: DTT treatment for 10–20 min is acceptable; extended treatment greatly diminishes the transformation efficiency.

6. Centrifuge the cells at 2500 g for 2 min at 4 °C. Discard the supernatant.

Note: From this point until after electroporation, care should be taken to keep the cells chilled. Place the cell-containing tubes on wet ice; also, chill the rotors, pipette tips, tubes and cuvettes.

7. Resuspend the cells in 25 ml of cold buffer E. Centrifuge the cells at 2500 g for 2 min at 4 °C; discard the supernatant.
8. Resuspend the cells in 1 ml of cold buffer E. Centrifuge the cells at 5000 g for 1 min at 4 °C; discard the supernatant.
9. Resuspend the cells in 1 ml of cold buffer E. Centrifuge the cells at 5000 g for 1 min at 4 °C; discard the supernatant.
10. Resuspend the cells to 0.3 ml total volume with cold buffer E.
11. Combine the cells with digested vector DNA (~4 μg) and gene insert DNA (~20 μg).

Note: Less DNA can be used, although the transformation efficiency may decrease; if more DNA is available, it is recommended to increase the scale of the transformation to transform more cells.

12. Aliquot 50–100 μl into each chilled 2 mm electroporation cuvette.
13. Electroporate the cells at 0.54 kV, 25 μF, without pulse control.
14. Note: If alternative cuvettes are used, set the voltage at 0.27 kV per mm of cuvette width.
15. Immediately resuspend the cells in warm (30 °C) YPD. Transfer the cells to a culture tube and incubate at 30 °C, 250 rpm for 1 h.
16. Pellet the cells and remove the supernatant. Resuspend cells in 100–1000 ml SD-CAA. Plate serial dilutions on SD-CAA plates to determine number of transformants. Incubate culture at 30 °C, 250 rpm.

26.2.2.3 **Yeast Culture**

The optimal temperature for EBY100 growth is 30 °C; liquid cultures should be agitated at ~250 rpm with sufficient aeration. Untransformed EBY100 is grown in nonselective YPD medium, whereas EBY100 harboring pCT plasmids are grown in SD-CAA. Expression of both Aga1p and the Aga2p-protein fusion is induced by galactose-containing SG-CAA. Induction at 20 °C can improve the surface display of proteins of reduced stability. Induction timing should be tested for each protein of interest; for example, a 24 h induction is needed for proper display of most scFvs, whereas the tenth type III domain of human fibronectin displays at high levels in about 8 h. Induced cells may be stored at 4 °C for several weeks; the Aga1p anchorage and Aga1p: Aga2p bonds will be preserved, although the folding fidelity of the protein of interest depends on its stability. The number of protein fusions displayed on the yeast cell surface varies depending on the protein with a general range of 10^4–10^5 proteins per cell. For clarity, a protocol for the standard growth and induction of pCT-containing EBY100 is provided as follows.

1. Incubate EBY100 + pCT-based plasmid in SD-CAA for 24–48 h (to middle of logarithmic growth phase).

Note: Optimal display is achieved when cells are induced during the middle of logarithmic growth.

2. Pellet the cells and remove the supernatant. Resuspend cells to a density of $\sim 1 \times 10^7$ cells ml^{-1} in SG-CAA. A concentration of 1×10^7 cells ml^{-1} corresponds to $OD_{600} = 1$.
3. Incubate the cells at 30 °C, 250 rpm for an appropriate time (~8–48 h depending on protein).
4. Use the induced cells or store at 4 °C.

26.2.3 Combinatorial Library Selection

Clones from YSD libraries may be selected based on a variety of phenotypes, such as affinity, specificity and stability, using a variety of methods including FACS, magnetic bead capture and two novel cell–cell interaction techniques.

26.2.4 FACS

A detailed protocol for sorting a scFv library was recently published [68]. A protocol for general FACS selection, as well as a commentary on the experimental design is provided here. Note that the term 'antigen' is used to denote the binding partner of the protein of interest.

1. Grow and induce the yeast library as indicated above.
2. Centrifuge the cells at 12 000*g* for 30 s or 2500*g* for 2 min.

Note: A 10-fold excess of cells, relative to the library diversity, should be labeled and sorted to avoid clone loss, which can result from plasmid loss, cytometer error or incomplete display or labeling.

3. Wash the cells with PBSA. Specifically, resuspend the cells in 1 ml PBSA, then re-pellet and discard the supernatant.
4. Resuspend the cells in PBSA with primary labels: PBSA with biotinylated antigen and mouse anti-c-myc antibody.

Note: Affinity and stoichiometry must both be considered when selecting label concentrations and volume. Sufficient concentrations of antigen and anti-epitope antibody must be used to ensure appropriate labeling. For novel binder selection this is the highest antigen concentration feasible (depending on availability, solubility and lack of nonspecific binding); for affinity maturation, the antigen should be used at approximately 4–10% of the K_d of the parent clone [28]. Anti-epitope antibody should be provided at a saturating concentration. In addition, the volume should be selected such that the total number of molecules of antigen and antibody are in significant excess of the number of displayed proteins in the sample.

5. Incubate the cells in primary labels to sufficiently approach equilibrium.

Note: The time to achieve 95% of the approach to equilibrium is $-\ln(1-0.95)/(k_{on}[Ag]+k_{off})$ where k_{on} is the association rate constant, [Ag] is the antigen concentration, and k_{off} is the dissociation rate constant. Typical k_{on} values for protein–protein interactions are ~10^5–$10^6\,M^{-1}\,s^{-1}$.

6. Wash the cells with PBSA.

Note: To reduce the dissociation of primary labels, use cold PBSA and leave the cells on ice for all of the remaining steps.

7. Resuspend the cells in PBSA with secondary labels: PBSA with fluorophore-conjugated avidin and alternative fluorophore-conjugated anti-mouse antibody. For example, $10\,mg\,l^{-1}$ *R*-phycoerythrin-conjugated streptavidin (33 n*M*) and $20\,mg\,l^{-1}$ AlexaFluor488-conjugated anti-mouse antibody (133 n*M*).
8. Incubate the cells in secondary labels.

Note: the incubation time depends on the status of library selection. For the maturation of a high-affinity binder in which the antigen will not substantially dissociate, ensure extensive secondary labeling; however, for the selection of novel clones of low affinity, it may be desirable to shorten the secondary labeling time to avoid antigen dissociation. Shorter labeling times can be compensated for by increasing the secondary reagent concentration.

9. Wash the cells with PBSA.
10. Resuspend cells in PBSA to 10^7–10^8 cells ml^{-1}. Sort the cells using FACS.

Note: A diagonal sort gate is drawn to collect the cells with the highest binding signal relative to the protein display (Figure 26.4). Only epitope$^+$ cells are selected; moreover, if it is desired to enrich more stable clones, only those cells with the highest display level are selected (see discussion below).

11. Grow the collected cells in SD-CAA at 30 °C, 250 rpm for 24–48 h.
12. Continue selections with collected population as necessary.

Alternative labeling strategies can be used. Two common modifications are the use of fluorophore-conjugated antigens and/or fluorophore-conjugated anti-epitope antibodies. These reagents enable single-step labeling for binding and/or display level. In addition to convenience, this approach can be used to reduce antigen dissociation for the isolation of low-affinity binders because only a single

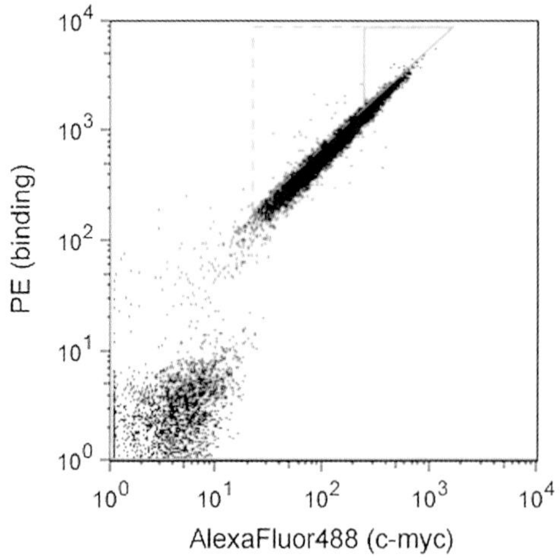

Figure 26.4 Fluorescence-activated cell sorting. Example flow cytometry data and sort gates are shown. Yeast are induced to display a fibronectin type III domain with a C-terminal c-myc epitope tag. Cells are labeled with biotinylated target and mouse anti-c-myc antibody followed by *R*-phycoerythrin (PE)-conjugated streptavidin and AlexaFluor488-conjugated anti-mouse antibody. Cells with the highest binding signal (PE) relative to display (AlexaFluor488) are selected. Either all full-length Fn3 clones are considered (dashed lines) or the highest displaying clones are selected to evolve stability (solid lines).

washing step is required after use, rather than two washes and a secondary label incubation. Unfortunately, however, fluorophore-conjugated antigens are associated with certain drawbacks. For example, fluorophore conjugation can block potential binding epitopes and preclude the signal amplification that is possible with two-step labeling (e.g. an antigen labeled with multiple biotins being bound by avidins with multiple fluorophores each). In the event of very low-affinity binding (such as during the isolation of novel binders from a naïve library), increasing the avidity can be effective for improving the binding signal-to-noise ratio. This can be achieved by multiple methods, including multimerization of biotinylated antigen on fluorophore-conjugated avidin. Similarly, epitope-tagged antigen can be dimerized with a fluorophore-conjugated bivalent antibody against the epitope. The preformed complex is then used in single-step labeling.

If the aim is explicitly to evolve specificity, then slightly modified protocols can be used. For example, in order to preferentially select binders to target A rather than B, unlabeled B may be provided during the primary labeling to block the ability of nonspecific clones to bind target A. Alternatively, targets A and B can be labeled by using different methods (e.g. direct fluorophore labeling with different dyes), while preferential binding can be selected for by multicolor flow cytometry.

Regardless of the individual labeling strategy, care must be taken to avoid non-antigen reagent binders (e.g. avidin binders or fluorophore binders). This is most easily achieved by alternating the secondary reagents or by using different fluorophore–antigen conjugates. The simple rule here is that the antigen should be the only constant element of the labeling approach.

26.2.4.1 **Other Selection Techniques**

Although flow cytometry provides a high-throughput, quantitative means of cell selection and enables fine discrimination between clones, a variety of other selection techniques have been implemented to provide unique advantages.

Two such approaches have been developed to engineer binders to cell-surface receptors; this may sometimes be difficult when using flow cytometry because of the need for moderate quantities of soluble ligand. Yeast can be panned against mammalian cell monolayers, whereupon those yeasts displaying proteins that bind to receptors present on the monolayer are preferentially retained [69]. Alternatively, by using cells in suspension, mammalian cell-bound yeast are separated from unbound yeast by their location after density gradient centrifugation [33]. In addition to eliminating the need for a soluble ligand, these techniques provide the avidity that is conferred by the multivalency of both the yeast-displayed protein and the cell-surface receptor.

Magnetic bead separations, using both paramagnetic Dynal beads [31] and Miltenyi microbeads [30], have been developed to extend screening in laboratories without FACS facilities and to speed up the screening of large libraries (sorting 10^{10} cells by FACS requires 100 h). A typical protocol for AutoMACS separation is as follows:

1. Grow and induce a yeast library as indicated above.
2. Centrifuge the cells at 12 000 *g* for 30 s or 2500 *g* for 2 min.

Note: A 10-fold excess of cells, relative to the library diversity, should be labeled and sorted to avoid clone loss, which can result from plasmid loss or incomplete display or labeling.

3. Wash the cells with PBSM. Specifically, resuspend cells in PBSM, re-pellet and discard the supernatant.
4. Resuspend the cells in PBSM with an antigen at desired concentration.
5. Incubate the cells in the primary label for sufficient time to reach equilibrium. Then incubate on ice for 10 min.
6. Wash cells with PBSM.

Note: To reduce the dissociation of antigen, use cold PBSM and leave the cells on ice for all of the remaining steps.

7. Resuspend the cells to 1×10^9 cells ml^{-1} in PBSM. Add magnetic microbeads; for example, streptavidin or anti-biotin microbeads should be used for biotinylated antigen. For populations with a high frequency of positive clones, the manufacturer recommends 200 μl of beads per 10^8 cells; however, for the isolation of rare clones, 200 μl of beads per 10^{10} cells is sufficient.
8. Incubate the cells on ice for 10 min, with inversion every 2 min.
9. Immediately prior to sorting, add PBSM to dilute the cell suspension to 1×10^8 cells ml^{-1}.
10. Sort cells on the AutoMACS separator; the possel_s program is recommended for the enrichment of rare clones. Place a 50 ml conical flask which is partially filled with SD-CAA under the collection port to collect the cells.

Note: Depending on the library size and column capacity, multiple batches may be needed to avoid column saturation.

11. Grow the collected cells in SD-CAA at 30 °C, 250 rpm for 24–48 h.

26.2.4.2 **Stability**

Stability engineering and activity engineering can be performed simultaneously with flow cytometry. Two primary metrics are available to enable stability evolution: (i) yeast surface display level; and (ii) resistance to thermal denaturation. For the display level, it is important simply to select the cells with the highest epitope-associated signal (see Figure 26.4). It should be noted that the correlation of display level and stability has been verified for single-chain T-cell receptors [41], single-chain class II MHC [70, 71] and the tenth type III domain of fibronectin (B.J. Hackel, unpublished results), but did not hold for the thermostable three-helix bundle α3D [48]. Thus, the existence and strength of the correlation should be tested for the protein of interest if mutants of varying stability exist. The induction of protein expression at elevated temperatures can improve the differences in display among mutants.

An alternative approach to engineer stability for unstable proteins is to thermally stress the displayed proteins prior to selection. Yeast are induced to display the protein library, incubated at elevated temperatures, labeled for activity, and sorted using flow cytometry. Clones with improved thermal stability will undergo less denaturation and maintain activity better than less-stable clones. The protocol is the same as for activity selection methods, but with a brief thermal denaturation step prior to cell labeling. Although yeast cells are relatively thermostable (they maintain their structural integrity in excess of 85 °C and maintain limited viability with brief treatment in excess of 50 °C), the denaturation temperatures required for stable proteins can kill yeast, thus eliminating clonal propagation by growth after sorting. In these cases the selected clones must be identified and/or propagated by DNA isolation and PCR.

26.2.4.3 **Clone Identification**

When the desired phenotype is achieved, or when sequence analysis of the population is otherwise desired, clones can be analyzed by plasmid recovery, bacterial transformation and sequencing, or by whole-cell PCR of plated yeast colonies. Zymoprep kits (Zymo Research) are recommended for plasmid recovery; if kit II is used, additional centrifugation of the neutralized cell materials and additional purification with a DNA spin column are recommended. The population of plasmids is then transformed into standard subcloning efficiency *E. coli*, plated to isolate individual clones, miniprepped, and sequenced. An alternative method of clone isolation is to plate a yeast population on SD-CAA plates, perform whole-cell PCR on individual colonies, and then sequence the PCR product.

26.2.5 Analysis

YSD and flow cytometry enable a diverse array of analytical techniques to be conducted, including investigations of the affinity and kinetics of binding, protein stability and epitope localization.

26.2.5.1 **Binding Measurements**

26.2.5.1.1 **Equilibrium Dissociation Constant** The determination of equilibrium dissociation constants using YSD and flow cytometry is straightforward, and correlates well with other methods (see Figure 26.2). Displaying cells are incubated in at least ten different antigen concentrations extending about 100-fold on both sides of the expected affinity, and the relative binding is quantified using flow cytometry. The volume of each sample must be selected to ensure a significant stoichiometric excess of antigen relative to the displayed protein. An example design is presented in Table 26.1.

In the event of very high-affinity binding, unreasonably large volumes may be required to maintain antigen excess. In moderate cases, this can be remedied by decreasing the number of cells per sample. However, about 1×10^5 cells are

Table 26.1 Affinity titration design example. An example titration for a ~5 n*M* binder is presented. The amounts of buffer, antigen stock and cell suspension are shown in bold type. The resulting antigen concentration [Ag] and antigen: protein ratio is also indicated, assuming 50 000 proteins displayed per cell and a cell density of 1×10^7 cells ml^{-1}.

Tube	[Ag] (n*M*)	PBSA (ml)	Ag (μl)	Stock (n*M*)	Cells (μl)	Ag: Protein
1	0	**0.100**	**0.00**	–	**50**	0
2	0.025	**50.0**	**12.5**	0.1	**50**	30
3	0.072	**50.0**	**36.1**	0.1	**50**	87
4	0.21	**10.0**	**21.0**	0.1	**50**	50
5	0.60	**10.0**	**6.04**	1	**50**	145
6	1.73	**1.50**	**2.69**	1	**50**	65
7	5.0	**1.50**	**7.80**	1	**50**	188
8	14.4	**0.500**	**8.05**	1	**50**	194
9	42	**0.500**	**23.9**	1	**50**	575
10	120	**0.100**	**20.4**	1	**50**	492
11	347	**0.100**	**5.37**	10	**50**	1294
12	1000	**0.100**	**16.7**	10	**50**	4021

needed to form a consistent pellet for washing and labeling steps. For extremely high-affinity binding, a mixture of nondisplaying and displaying cells can be used. Thus, the amount of displayed protein may be decreased to arbitrarily low levels while the nondisplaying cells maintain the ability to form a cell pellet.

1. Starting from a single colony, grow and induce a 5 ml yeast culture as indicated above.
2. Centrifuge 1×10^7 cells at 12 000*g* for 30 s.
3. Wash the cells with PBSA. Specifically, resuspend the cells in 1 ml PBSA, re-pellet and discard the supernatant.
4. Resuspend the cells in 1 ml PBSA.
5. Add an appropriate amount of PBSA, antigen and cell suspension.
6. Incubate the tubes at the desired temperature (often 25 °C or 37 °C) until binding approaches equilibrium.

Note: See the note after primary label incubation in FACS for the commentary on requisite labeling time.

7. Wash the cells with PBSA.

Note: This and all following steps should be conducted on ice with chilled PBSA.

8. Resuspend the cells in secondary label.

Note: For example, a biotinylated antigen is labeled by fluorophore-conjugated avidin. An epitope-tagged antigen is labeled with a fluorophore-conjugated anti-epitope antibody. Directly fluorophore-conjugated antigen does not need a secondary label.

9. Incubate on ice for 10–20 min.
10. Wash the cells with PBSA.
11. Resuspend the cells in 500 μl PBSA. Analyze with flow cytometry.
12. Determine the equilibrium dissociation constant, K_d, using the equation $F = F_{min} + (F_{max} - F_{min})[Ag]/([Ag] + K_d)$, where F is the mean fluorescence per cell (min and max indicate the appropriate extremes without labeling and at saturation), and [Ag] is the antigen concentration, which can be assumed constant at the initial concentration since a significant excess was used. The antigen concentration is known for each tube; F is measured directly; F_{min}, F_{max} and K_d are solved using a nonlinear optimization to minimize the sum of squared errors between experimental and theoretical F.

26.2.5.1.2 Dissociation Rate

1. Starting from a single colony, grow and induce a 5 ml yeast culture as indicated above.
2. Centrifuge 1×10^7 cells at 12 000 g.
3. Wash the cells with PBSA. Specifically, resuspend the cells in 1 ml PBSA, re-pellet and discard the supernatant.
4. Resuspend the cells in PBSA with a saturating amount of antigen.

Note: Saturation is not necessary but increases the signal-to-noise ratio relative to nonsaturating concentrations.

5. Incubate at the temperature of interest until equilibrium is nearly reached.

Note: See the note after primary label incubation in FACS for commentary on requisite labeling time.

6. At various times, remove an aliquot of 1×10^6 labeled cells and wash with PBSA with an excess of unlabeled antigen (relative to number of displayed proteins).

Note: Unlabeled antigen is used to prevent the association of dissociated antigen.

7. Resuspend the cells in PBSA with an excess of unlabeled antigen.

8. Incubate at the temperature of interest to enable dissociation.
9. Repeat steps 6–8 for multiple times.

Note: Select times to yield a range of samples from undissociated to nearly fully dissociated. For example, for a binder with a dissociation half-time of 1 h, use dissociation times of 0, 9, 19, 31, 44, 60, 79, 104, 139 and 199 min. These times are calculated from $-\ln(1-y)/k_{off}$ where y is the fraction of dissociated antigen; the proposed times yield y-values of 0, 0.1, 0.2, . . . , 0.8, 0.9.

10. Simultaneously pellet all samples. Wash with 1 ml PBSA.

Note: This and all following steps should be conducted on ice with chilled buffers.

11. Resuspend the cells in PBSA with secondary label.

Note: See the note after secondary label step in equilibrium dissociation constant for commentary on reagent selection.

12. Incubate on ice for 10–20 min.
13. Wash the cells with PBSA.
14. Resuspend the cells in 500 μl PBSA. Analyze with flow cytometry.
15. Determine the dissociation rate constant, k_{off}, using the equation $F = F_{max}\exp(-k_{off}\ t)$ where F and F_{max} are the mean fluorescence per cell for a sample dissociated for time t and an undissociated sample, respectively. The dissociation time is known for each sample, F, is measured directly, and F_{max} and k_{off} are determined by error minimization.

26.2.5.1.3 **Association Rate**

1. Starting from a single colony, grow and induce a 5 ml yeast culture as indicated above.
2. Centrifuge 1×10^7 cells at 12 000 g.
3. Wash the cells with PBSA. Specifically, resuspend cells in 1 ml PBSA, re-pellet and discard the supernatant.
4. At various times, pellet an aliquot of 1×10^6 cells. Wash with PBSA.
5. Resuspend in PBSA and antigen.

Note: The concentration of antigen must be identical for all samples. The concentration should be above the value of the equilibrium dissociation constant to provide significant signal and to ensure that the observed association rate ($k_{on}[Ag] + k_{off}$) is not dominated by k_{off}; yet the concentration should be low enough to enable multiple samples to be collected prior to saturation.

6. Incubate at temperature of interest to enable association.
7. Repeat steps 4–6 for multiple times.

Note: Select times to yield a range of samples from unassociated to nearly fully associated. For example, for a 1 n*M* binder with a dissociation half-time of 1 h, use association times of 0, 2, 5, 7, 11, 14, 19, 25, 34 and 48 min for 3 n*M* antigen labeling. These times are calculated from $-\ln(1 - y)/k_{obs}$ where y is the fraction of associated antigen; the proposed times yield y-values of 0, 0.1, 0.2, . . . , 0.8, 0.9.

8. Add unlabeled antigen at a concentration in significant excess of labeled antigen and place on ice.

Note: Unlabeled antigen is used to prevent further association during washing and secondary labeling. This and all following steps should be conducted on ice with chilled buffers.

9. Wash cells with PBSA and excess unlabeled antigen.
10. Resuspend cells in PBSA with secondary label.
11. Incubate on ice for 10–20 min.
12. Wash cells with PBSA.
13. Resuspend in 500 μl PBSA. Analyze with flow cytometry.
14. Determine the observed association rate constant, k_{obs}, from the equation $F = F_{max}[1 - \exp(-k_{obs}\,t)]$, where F and F_{max} are the mean fluorescence per cell for a sample associated for time t and a saturated sample.
15. Repeat steps 2–14 with multiple antigen concentrations.
16. Determine the association rate constant, k_{on}, from the equation $k_{obs} = k_{on}[Ag] + k_{off}$. The antigen concentrations [Ag] are known for each experiment, and k_{obs} is determined in step 14. The dissociation rate constant can either be determined independently (see above) or set as a fit parameter. The association rate can then be determined by error minimization.

An alternative approach can be used in which all samples begin association simultaneously. At various times, the association of labeled antigen is effectively terminated by incubation on ice and the addition of excess unlabeled antigen. Although this facilitates more rapid sample handling, it places more importance on the ability of reduced temperature and unlabeled antigen to terminate labeling.

26.2.5.2 **Stability Measurement**

Two metrics of stability – thermal denaturation rate and the midpoint of thermal denaturation – can be readily measured using YSD and flow cytometry. Thermal denaturation rate can be calculated in analogous fashion to binding dissociation.

1. Starting from a single colony, grow and induce a 5 ml yeast culture as indicated above.
2. Centrifuge 1×10^7 cells at 12 000 g.
3. Wash the cells with PBSA. Specifically, resuspend cells in 1 ml PBSA, re-pellet and discard the supernatant.
4. Resuspend the cells in PBSA.
5. At various times, remove an aliquot of 1×10^6 labeled cells and incubate at the temperature of interest to enable unfolding or aggregation.

Note: Yeast cells are structurally tolerant of temperatures in excess of 85 °C.

6. Repeat step 5 for multiple times.

Note: Select times to yield a range of samples from intact to nearly fully denatured.

7. Simultaneously pellet all samples. Wash with 1 ml PBSA.

Note: This and all following steps should be conducted on ice with chilled buffers.

8. Quantify protein integrity of each sample by flow cytometry.

Note: For a binding protein, the activity may be quantified by antigen labeling at an identical concentration (saturating concentration recommended) for all samples. Alternatively, the fidelity of the displayed protein can be quantified by labeling with a conformationally specific antibody if available.

The midpoint of thermal denaturation can be calculated by a temperature titration.

1. Starting from a single colony, grow and induce a 5 ml yeast culture as indicated above.
2. Centrifuge 1×10^7 cells at 12 000 g.
3. Wash the cells with PBSA. Specifically, resuspend cells in 1 ml PBSA, re-pellet and discard the supernatant.
4. Resuspend cells in 1 ml PBSA and make ten 100 μl aliquots.
5. Incubate each sample at a different temperature for 30 min.

Note: Thermal cyclers with the gradient option provide a convenient source of temperature control.

6. Incubate each sample on ice for 15 min.
7. Quantify the protein integrity of each sample with flow cytometry.

Note: For a binding protein, the activity may be quantified by antigen labeling at an identical concentration (saturating concentration recommended) for all samples. Alternatively, the fidelity of the displayed protein can be quantified by labeling with a conformationally specific antibody if available.

26.3
The Future of Yeast Surface Display

Today, research groups have at their disposal multiple display technologies that link genotype and phenotype to enable combinatorial library selection (see also Chapter 24). Whilst each display method can be effective when used in the proper context, unlike prokaryotic and *in vitro* methods, YSD provides eukaryotic expression machinery enabling glycosylation, disulfide bond formation, improved protein folding and a eukaryotic expression bias. Moreover, flow cytometry facilitates high-throughput analysis and selection including fine phenotype discrimination. As a result, YSD has developed into a powerful tool for the evolution of protein engineering and analysis.

Perhaps the predominant criticism of YSD for combinatorial library selection has been that yeast transformation limitations result in an apparently low library diversity relative to other display technologies such as phage, mRNA or ribosome display. However, these comparisons are based on genetic diversity, such as the number of transformed cells or theoretical RNA–protein fusions, rather than on the appropriate metric: functional diversity. The eukaryotic machinery of yeast facilitates proper protein folding which improves the fraction of library clones that are functionally displayed. Moreover, expression bias does not skew the proportional diversity of YSD libraries. In a recent direct comparison of YSD and phage display using the same scFv DNA library and target antigen, YSD identified three times more clones than did phage display, and did not miss a single phage clone [72]. The eukaryotic processing of the scFv was considered the primary cause for the greater success of the YSD selection. Thus, although YSD may be trailing in the number of DNA sequences placed in the test tube, it appears to effectively analyze a higher fraction of these sequences at a protein level. Nevertheless, the vastness of protein sequence space leaves most libraries orders of magnitude smaller than their theoretical diversity, so an improvement in transformation efficiency would yield faster and potentially more successful selections. Thus, the ability to more efficiently create and screen billion-member YSD libraries would be a valuable development for protein engineering. The use of homologous recombination and improvements in yeast transformation protocols have already made a strong impact in this area. Continued development and novel ideas – perhaps including a highly transformation-competent yeast strain – should be realizable.

Another solution by which to overcome the vastness of sequence space is to continue to advance understanding of the sequence–function relationship to focus on the search of combinatorial space. The ability to rapidly analyze clones of known sequence with flow cytometry and YSD permits hypothesis testing to direct library design. Alternatively, as mentioned above, expression-normalized FACS using YSD could enhance shotgun scanning mutagenesis to identify amino acid positions to target for diversification. This semi-rational directed evolution improves the functional distribution of library clones, which can be created by either targeted mutagenesis or nucleic acid synthesis.

Since its initiation almost a decade ago, the primary use of YSD has been to engineer protein–target interactions. Experimental design permits selection on the basis of association and dissociation kinetics, equilibrium affinity or specificity, depending on the desired properties for the application of interest. Moreover, rapid characterization without protein purification aids directed evolution design. In addition to the evolution of binding interactions, ingenious alternative applications have been introduced for YSD including stability engineering, epitope identification, enzyme maturation and immunological study. The demonstration of effective biopanning using YSD [32] opens the door for drug discovery through the coincident identification of novel surface receptors and cognate binding proteins. The continued development of yeast biopanning, and its application to other biological systems, offer an exciting future avenue for YSD. Although already a robust platform for many purposes, YSD should continue to evolve to improve its efficiency and expand its breadth of function.

Abbreviations

Fab	antigen-binding fragment
FACS	fluorescence-activated cell sorting
POI	protein of interest
scFv	single-chain antibody variable fragment
YSD	yeast surface display

Acknowledgments

B.J.H. is supported by a National Science Foundation Graduate Fellowship and an NDSEG Graduate Fellowship. These studies were supported financially by NIH grants CA101830, CA96504 and AI065824.

References

1 Lu, C.-F., Montijn, R.C., Brown, J.L., Klis, F., Kurjan, J., Bussey, H. and Lipke, P.N. (1995) Glycosyl phosphatidylinositol-dependent cross-linking of alpha-agglutinin and beta1,6-glucan in the *Saccharomyces cerevisiae* cell wall. *The Journal of Cell Biology*, **128**, 333–40.

2 Wang, Z., Mathias, A., Stavrou, S. and Neville, D.M. (2005) A new yeast display vector permitting free scFv amino termini can augment ligand binding affinities. *Protein Engineering, Design and Selection*, **18**, 337–43.

3 Boder, E.T. and Wittrup, K.D. (1997) Yeast surface display for screening combinatorial polypeptide libraries. *Nature Biotechnology*, **15**, 553–7.

4 van den Beucken, T., Pieters, H., Steukers, M., van der Vaart, M., Ladner, R.C., Hoogenboom, H.R. and Hufton, S.E. (2003) Affinity maturation of Fab antibody fragments by fluorescent-activated cell sorting of yeast-displayed libraries. *FEBS Letters*, **546**, 288–94.

5 Krauland, E.M., Peelle, B.R., Wittrup, K.D. and Belcher, A.M. (2007) Peptide tags for enhanced cellular and protein adhesion to single-crystalline sapphire. *Biotechnology and Bioengineering*, **97**, 1009–20.

6 Peelle, B.R., Krauland, E.M., Wittrup, K.D. and Belcher, A.M. (2005) Probing the interface between biomolecules and inorganic materials using yeast surface display and genetic engineering. *Acta Biomaterialia*, **1**, 145–54.

7 Lipovsek, D., Lippow, S.M., Hackel, B.J., Gregson, M.W., Cheng, P., Kapila, A. and Wittrup, K.D. (2007) Evolution of an interloop disulfide bond in high-affinity antibody mimics based on fibronectin type III domain and selected by yeast surface display: molecular convergence with single-domain camelid and shark antibodies. *Journal of Molecular Biology*, **368**, 1024–41.

8 Kieke, M.C., Shusta, E.V., Boder, E.T., Teyton, L., Wittrup, K.D. and Kranz, D.M. (1999) Selection of functional T cell receptor mutants from a yeast surface-display library. *Proceedings of the National*

Academy of Sciences of the United States of America, **96**, 5651–6.

9 Brophy, S.E., Holler, P.D. and Kranz, D.M. (2003) A yeast display system for engineering functional peptide-MHC complexes. *Journal of Immunological Methods*, **272**, 235–46.

10 Boder, E.T., Bill, J.R., Nields, A.W., Marrack, P.C. and Kappler, J.W. (2005) Yeast surface display of a noncovalent MHC class II heterodimer complexed with antigenic peptide. *Biotechnology and Bioengineering*, **92**, 485–91.

11 Rao, B.M., Girvin, A.T., Ciardelli, T., Lauffenburger, D.A. and Wittrup, K.D. (2003) Interleukin-2 mutants with enhanced alpha-receptor subunit binding affinity. *Protein Engineering*, **16**, 1081–7.

12 Bhatia, S.K., Swers, J.S., Camphausen, R.T., Wittrup, K.D. and Hammer, D.A. (2003) Rolling adhesion kinematics of yeast engineered to express selectins. *Biotechnology Progress*, **19**, 1033–7.

13 Jin, M., Song, G., Carman, C.V., Kim, Y.S., Astrof, N.S., Shimaoka, M., Wittrup, K.D. and Springer, T.A. (2006) Directed evolution to probe protein allostery and integrin I domains of 200 000-fold higher affinity. *Proceedings of the National Academy of Sciences of the United States of America*, **103**, 5758–63.

14 Pepper, L.R., Hammer, D.A. and Boder, E.T. (2006) Rolling adhesion of alpha(L) I domain mutants decorrelated from binding affinity. *Journal of Molecular Biology*, **360**, 37–44.

15 Cochran, J.R., Kim, Y.S., Olsen, M.J., Bhandari, R. and Wittrup, K.D. (2004) Domain-level antibody epitope mapping through yeast surface display of epidermal growth factor receptor fragments. *Journal of Immunological Methods*, **287**, 147–58.

16 Huang, D. and Shusta, E.V. (2005) Secretion and surface display of green fluorescent protein using the yeast *Saccharomyces cerevisiae*. *Biotechnology Progress*, **21**, 349–57.

17 Cochran, J.R., Kim, Y.S., Lippow, S.M., Rao, B. and Wittrup, K.D. (2006) Improved mutants from directed evolution are biased to orthologous substitutions. *Protein Engineering, Design and Selection*, **19**, 245–53.

18 Ryu, K. and Lee, E.K. (2002) Rapid colorimetric assay and yeast surface display for screening of highly functional fungal lignin peroxidase. *Journal of Chemical Engineering of Japan*, **35**, 527–32.

19 Zhu, K.L., Chi, Z.M., Li, J., Zhang, F.L., Li, M.J., Yasoda, H.N. and Wu, L.F. (2006) The surface display of haemolysin from *Vibrio harveyi* on yeast cells and their potential applications as live vaccine in marine fish. *Vaccine*, **24**, 6046–52.

20 Shiraga, S., Kawakami, M. and Ueda, M. (2004) Construction of combinatorial library of starch-binding domain of *Rhizopus oryzae* glucoamylase and screening of clones with enhanced activity by yeast display method. *Journal of Molecular Catalysis B: Enzymatic*, **28**, 229–34.

21 Lipovsek, D., Antipov, E., Armstrong, K.A., Olsen, M.J., Klibanov, A.M., Tidor, B. and Wittrup, K.D. (2007) Selection of horseradish peroxidase variants with enhanced enantioselectivity by yeast surface display. *Chemistry Biology*, **14**, 1176–85.

22 Neylon, C. (2004) Chemical and biochemical strategies for the randomization of protein encoding DNA sequences: library construction methods for directed evolution. *Nucleic Acids Research*, **32**, 1448–59.

23 Feldhaus, M.J., Siegel, R.W., Opresko, L.K., Coleman, J.R., Feldhaus, J.M.W., Yeung, Y.A., Cochran, J.R., Heinzelman, P., Colby, D., Swers, J., Graff, C., Wiley, H.S. and Wittrup, K.D. (2003) Flow-cytometric isolation of human antibodies from a nonimmune *Saccharomyces cerevisiae* surface display library. *Nature Biotechnology*, **21**, 163–70.

24 Shiomi, N., Murao, K., Koga, H., Kuroda, K., Hosokawa, H. and Katoh, S. (2002) A new method for production of combinatorial libraries by mating of *Saccharomyces cerevisiae* cells. *Journal of Chemical Engineering of Japan*, **35**, 474–8.

25 Weaver-Feldhaus, J.M., Lou, J.L., Coleman, J.R., Siegel, R.W., Marks, J.D. and Feldhaus, M.J. (2004) Yeast mating for combinatorial Fab library generation and surface display. *FEBS Letters*, **564**, 24–34.

26 Blaise, L., Wehnert, A., Steukers, M.P.G., Beucken, T., Hoogenboom, H.R. and Hufton, S.E. (2004) Construction and diversification of yeast cell surface displayed libraries by yeast mating: application to the affinity maturation of Fab antibody fragments. *Gene*, **342**, 211–18.

27 Swers, J.S., Kellogg, B.A. and Wittrup, K.D. (2004) Shuffled antibody libraries created by in vivo homologous recombination and yeast surface display. *Nucleic Acids Research*, **32**, e36.

28 Boder, E.T. and Wittrup, K.D. (1998) Optimal screening of surface-displayed polypeptide libraries. *Biotechnology Progress*, **14**, 55–62.

29 VanAntwerp, J.J. and Wittrup, K.D. (2000) Fine affinity discrimination by yeast surface display and flow cytometry. *Biotechnology Progress*, **16**, 31–7.

30 Siegel, R.W., Coleman, J.R., Miller, K.D. and Feldhaus, M.J. (2004) High efficiency recovery and epitope-specific sorting of an scFv yeast display library. *Journal of Immunological Methods*, **286**, 141–53.

31 Yeung, Y.A. and Wittrup, K.D. (2002) Quantitative screening of yeast surface-displayed polypeptide libraries by magnetic bead capture. *Biotechnology Progress*, **18**, 212–20.

32 Wang, X.X., Cho, Y.K. and Shusta, E.V. (2007) Mining a yeast library for brain endothelial cell-binding antibodies. *Nature Methods*, **4**, 143–5.

33 Richman, S.A., Healan, S.J., Weber, K.S., Donermeyer, D.L., Dossett, M.L., Greenberg, P.D., Allen, P.M. and Kranz, D.M. (2006) Development of a novel strategy for engineering high-affinity proteins by yeast display. *Protein Engineering, Design and Selection*, **19**, 255–64.

34 Boder, E.T., Midelfort, K.S. and Wittrup, K.D. (2000) Directed evolution of antibody fragments with monovalent femtomolar antigen-binding affinity. *Proceedings of the National Academy of Sciences of the United States of America*, **97**, 10701–5.

35 Buonpane, R.A., Moza, B., Sundberg, E.J. and Kranz, D.M. (2005) Characterization of T cell receptors engineered for high affinity against toxic shock syndrome toxin-1. *Journal of Molecular Biology*, **353**, 308–21.

36 Colby, D.W., Garg, P., Holden, T., Chao, G., Webster, J.M., Messer, A., Ingram, V.M. and Wittrup, K.D. (2004) Development of a human light chain variable domain (V-L) intracellular antibody specific for the amino terminus of huntingtin via yeast surface display. *Journal of Molecular Biology*, **342**, 901–12.

37 Colby, D.W., Chu, Y.J., Cassady, J.P., Duennwald, M., Zazulak, H., Webster, J.M., Messer, A., Lindquist, S., Ingram, V.M. and Wittrup, K.D. (2004) Potent inhibition of huntingtin and cytotoxicity by a disulfide bond-free single-domain intracellular antibody. *Proceedings of the National Academy of Sciences of the United States of America*, **101**, 17616–21.

38 Razai, A., Garcia-Rodriguez, C., Lou, J., Geren, I.N., Forsyth, C.M., Robles, Y., Tsai, R., Smith, T.J., Smith, L.A., Siegel, R.W., Feldhaus, M. and Marks, J.D. (2005) Molecular evolution of antibody affinity for sensitive detection of botulinum neurotoxin type A. *Journal of Molecular Biology*, **351**, 158–69.

39 Garcia-Rodriguez, C., Levy, R., Arndt, J.W., Forsyth, C.M., Razai, A., Lou, J.L., Geren, I., Stevens, R.C. and Marks, J.D. (2007) Molecular evolution of antibody cross-reactivity for two subtypes of type A botulinum neurotoxin. *Nature Biotechnology*, **25**, 107–16.

40 Kowalski, J.M., Parekh, R.N., Mao, J. and Wittrup, K.D. (1998) Protein folding stability can determine the efficiency of escape from endoplasmic reticulum quality control. *Journal of Biological Chemistry*, **273**, 19453–8.

41 Shusta, E.V., Kieke, M.C., Parke, E., Kranz, D.M. and Wittrup, K.D. (1999) Yeast polypeptide fusion surface display levels predict thermal stability and soluble secretion efficiency. *Journal of Molecular Biology*, **292**, 949–56.

42 Shusta, E.V., Holler, P.D., Kieke, M.C., Kranz, D.M. and Wittrup, K.D. (2000) Directed evolution of a stable scaffold for T-cell receptor engineering. *Nature Biotechnology*, **18**, 754–9.

43 Jones, L.L., Brophy, S.E., Bankovich, A.J., Colf, L.A., Hanick, N.A., Garcia, K.C. and Kranz, D.M. (2006) Engineering and characterization of a stabilized alpha 1/alpha 2 module of the class I major histocompatibility complex product Ld. *Journal of Biological Chemistry*, **281**, 25734–44.

44 Kim, Y.S., Bhandari, R., Cochran, J.R., Kuriyan, J. and Wittrup, K.D. (2006) Directed evolution of the epidermal growth factor receptor extracellular domain for expression in yeast. *Proteins: Structure, Function, and Bioinformatics*, **62**, 1026–35.

45 Schweickhardt, R.L., Jiang, X.L., Garone, L.M. and Brondyk, W.H. (2003) Structure-expression relationship of tumor necrosis factor receptor mutants that increase expression. *Journal of Biological Chemistry*, **278**, 28961–7.

46 Graff, C.P., Chester, K., Begent, R. and Wittrup, K.D. (2004) Directed evolution of an anti-carcinoembryonic antigen scFv with a 4-day monovalent dissociation half-time at 37 degrees C. *Protein Engineering, Design and Selection*, **17**, 293–304.

47 Wentz, A.E. and Shusta, E.V. (2007) Novel high-throughput screen reveals yeast genes that increase secretion of heterologous proteins. *Applied and Environmental Microbiology*, **73**, 1189–98.

48 Park, S., Xu, Y., Stowell, X.F., Gai, F., Saven, J.G. and Boder, E.T. (2006) Limitations of yeast surface display in engineering proteins of high thermostability. *Protein Engineering, Design and Selection*, **19**, 211–17.

49 Parthasarathy, R., Bajaj, J. and Boder, E.T. (2005) An immobilized biotin ligase: Surface display of *Escherichia coli* BirA on *Saccharomyces cerevisiae*. *Biotechnology Progress*, **21**, 1627–31.

50 Zhang, H.C., Bi, J.Y., Chen, C., Huang, G.L., Qi, Q.S., Xiao, M. and Wang, P.G. (2006) Immobilization of UDP-galactose 4-epimerase from *Escherichia coli* on the yeast cell surface. *Bioscience, Biotechnology, and Biochemistry*, **70**, 2303–6.

51 Koide, A., Gilbreth, R.N., Esaki, K., Tereshko, V. and Koide, S. (2007) High-affinity single-domain binding proteins with a binary-code interface. *Proceedings of the National Academy of Sciences of the United States of America*, **104**, 6632–7.

52 Orr, B.A., Carr, L.M., Wittrup, K.D., Roy, E.J. and Kranz, D.M. (2003) Rapid method for measuring ScFv thermal stability by yeast surface display. *Biotechnology Progress*, **19**, 631–8.

53 Chao, G., Cochran, J.R. and Wittrup, K.D. (2004) Fine epitope mapping anti-epidermal growth factor receptor antibodies through random mutagenesis and yeast surface display. *Journal of Molecular Biology*, **342**, 539–50.

54 Liang, Y.F., Wan, Y., Qiu, L.W., Zhou, J.G., Ni, B., Guo, B., Zou, Q., Zou, L.Y., Zhou, W., Jia, Z.C., Che, X.Y. and Wu, Y.Z. (2005) Comprehensive antibody epitope mapping of the nucleocapsid protein of severe acute respiratory syndrome (SARS) coronavirus: Insight into the humoral immunity of SARS. *Clinical Chemistry*, **51**, 1382–96.

55 Oliphant, T., Nybakken, G.E., Engle, M., Xu, Q., Nelson, C.A., Sukupolvi-Petty, S., Marri, A., Lachmi, B.E., Olshevsky, U., Fremont, D.H., Pierson, T.C. and Diamond, M.S. (2006) Antibody recognition and neutralization determinants on domains I and II of West Nile virus envelope protein. *Journal of Virology*, **80**, 12149–59.

56 Levy, R., Forsyth, C.M., LaPorte, S.L., Geren, I.N., Smith, L.A. and Marks, J.D. (2007) Fine and domain-level epitope mapping of botulinum neurotoxin type A neutralizing antibodies by yeast surface display. *Journal of Molecular Biology*, **365**, 196–210.

57 Cunningham, B.C. and Wells, J.A. (1989) High-resolution epitope mapping of HGH-receptor interactions by alanine-scanning mutagenesis. *Science*, **244**, 1081–5.

58 Pal, G., Kossiakoff, A.A. and Sidhu, S.S. (2003) The functional binding epitope of a high affinity variant of human growth hormone mapped by shotgun alanine-scanning mutagenesis: Insights into the mechanisms responsible for improved affinity. *Journal of Molecular Biology*, **332**, 195–204.

59 Mischo, A., Wadle, A., Watzig, K., Jager, D., Stockert, E., Santiago, D., Ritter, G., Regitz, E., Jager, E., Knuth, A., Old, L.,

Pfreundschuh, M. and Renner, C. (2003) Recombinant antigen expression on yeast surface (RAYS) for the detection of serological immune responses in cancer patients. *Cancer Immunity*, **3**, 5–15.

60 Wadle, A., Kubuschok, B., Imig, J., Wuellner, B., Wittig, C., Zwick, C., Mischo, A., Waetzig, K., Romeike, B.F.M., Lindemann, W., Schilling, M., Pfreundschuh, M. and Renner, C. (2006) Serological immune response to cancer testis antigens in patients with pancreatic cancer. *International Journal of Cancer*, **119**, 117–25.

61 Nakamura, Y., Shibasaki, S., Ueda, M., Tanaka, A., Fukuda, H. and Kondo, A. (2001) Development of novel whole-cell immunoadsorbents by yeast surface display of the IgG-binding domain. *Applied Microbiology and Biotechnology*, **57**, 500–5.

62 Weber, K.S., Donermeyer, D.L., Allen, P.M. and Kranz, D.M. (2005) Class II-restricted T cell receptor engineered in vitro for higher affinity retains peptide specificity and function. *Proceedings of the National Academy of Sciences of the United States of America*, **102**, 19033–8.

63 Kieke, M.C., Sundberg, E., Shusta, E.V., Mariuzza, R.A., Wittrup, K.D. and Kranz, D.M. (2001) High affinity T cell receptors from yeast display libraries block T cell activation by superantigens. *Journal of Molecular Biology*, **307**, 1305–15.

64 Chlewicki, L.K., Holler, P.D., Monti, B.C., Clutter, M.R. and Kranz, D.M. (2005) High-affinity, peptide-specific T cell receptors can be generated by mutations in CDR1, CDR2 or CDR3. *Journal of Molecular Biology*, **346**, 223–39.

65 Holler, P.D., Chlewicki, L.K. and Kranz, D.M. (2003) TCRs with high affinity for foreign pMHC show self-reactivity. *Nature Immunology*, **4**, 55–62.

66 Gietz, R.D. and Woods, R.A. (2002) Transformation of yeast by lithium acetate/single-stranded carrier DNA/polyethylene glycol method. *Methods in Enzymology*, **350**, 87–96.

67 Meilhoc, E., Masson, J.M. and Teissie, J. (1990) High-efficiency transformation of intact yeast cells by electric-field pulses. *Bio-Technology*, **8**, 223–7.

68 Chao, G., Lau, W.L., Hackel, B.J., Sazinsky, S.L., Lippow, S.M. and Wittrup, K.D. (2006) Isolating and engineering human antibodies using yeast surface display. *Nature Protocols*, **1**, 755–68.

69 Wang, X.X. and Shusta, E.V. (2005) The use of scFv-displaying yeast in mammalian cell surface selections. *Journal of Immunological Methods*, **304**, 30–42.

70 Starwalt, S.E., Masteller, E.L., Bluestone, J.A. and Kranz, D.M. (2003) Directed evolution of a single-chain class II MHC product by yeast display. *Protein Engineering*, **16**, 147–56.

71 Esteban, O. and Zhao, H.M. (2004) Directed evolution of soluble single-chain human class II MHC molecules. *Journal of Molecular Biology*, **340**, 81–95.

72 Bowley, D.R., Labrijn, A.F., Zwick, M.B. and Burton, D.R. (2007) Antigen selection from an HIV-1 immune antibody library displayed on yeast yields many novel antibodies compared to selection from the same library displayed on phage. *Protein Engineering, Design and Selection*, **20**, 81–90.

27
In Vitro Compartmentalization (IVC) and Other High-Throughput Screens of Enzyme Libraries

Amir Aharoni and Dan S. Tawfik

27.1
Introduction

Harnessing the potential of available gene repertoires from both natural (genomic, metagenomic, or environmental libraries, and cDNA libraries), and artificial sources (gene libraries), is largely dependent on the ability to screen, or select, these vastly large repertoires for a variety of useful activities. This chapter addresses recent developments in the selection of enzyme-coding genes for directed evolution and functional genomics, with an emphasis on *in vitro* compartmentalization (IVC) in emulsions (for recent reviews, see Refs. [1, 2]).

The basis of all screening and selection methodologies is a linkage between the gene, the enzyme it encodes, and the product of the activity of that enzyme (Figure 27.1). The difference between screening and selection is that, screening is performed on individual genes or clones and requires some spatial organization of the screened variants on agar plates, microtiter plates, arrays, and chips. In contrast, selections act simultaneously on the entire pool of genes. The focus of this chapter is on HTS (high-throughput screening/selection) approaches that enable selection from large libraries ($>10^6$ gene variants) with relatively modest means (i.e. nonrobotic systems), and *in vitro* compartmentalization (IVC), in particular. The examples provided are primarily from the area of directed, or *in vitro*, evolution, although the described methodologies (including phage-display, cell-display and IVC) are also applicable in the area of functional genomics (the screening of natural gene repertoires). Another focus of this chapter is screens and selections for enzymatic functions, although some of the methods described here are applicable to other functions, including binding and regulatory functions.

Directed enzyme evolution has been used during the past two decades as a powerful approach for generating enzymes with desired properties. Enzyme variants were evolved for catalytic activity under extreme conditions such as high temperatures, acidic and alkaline environments and organic solvents [3, 4] and with improved catalytic activities and new substrate specificities [5, 6]. Directed evolution experiments consist of two major steps: (i) the creation of genetic

Protein Engineering Handbook. Edited by Stefan Lutz and Uwe T. Bornscheuer

ISBN: 978-3-527-31850-6

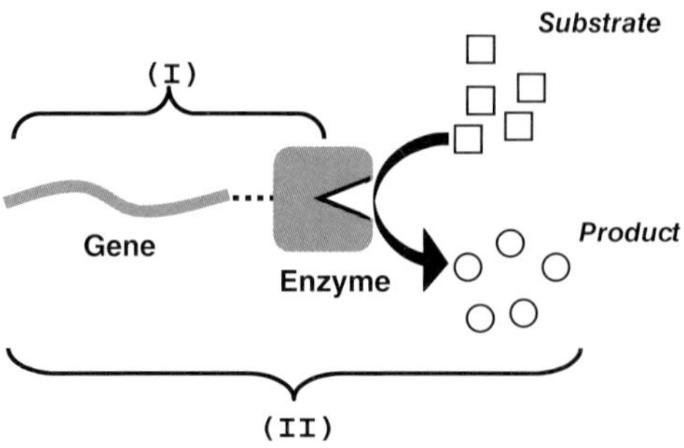

Figure 27.1 The basis of all selection and screens for enzyme-coding genes, be it in a directed evolution or functional genomics context, is the linkage between the gene, the enzyme it encodes, and the products of the activity of that enzyme. The first link, between the enzyme and its coding gene (I) can be achieved in a variety of ways, including the cloning and expression of the library in living cells, phage-display, ribosome and mRNA-peptide display, and by cell-free translation in emulsion droplets (IVC). The second link, between the product of the enzymatic activity and the linking gene (II), is usually harder to obtain. A variety of methods that provide the above link in a genuinely high-throughput (HT) format are discussed in the text.

diversity in the target gene in the form of gene-libraries; and (ii) an effective selection of the library for the desired catalytic activity. A large variety of means for the creation of genetic diversity is currently available (for recent reviews, see Refs. [7, 8]). However, the typical library size is still many orders of magnitude larger than the number of protein variants that can be screened. The same restriction applies to cDNA and genomic libraries derived from natural sources (e.g. environmental libraries), the diversity of which is almost unlimited. Further, while methods for creating gene diversity are generic, the screens for activity need to be tailored for each enzyme and reaction. The 'bottle-neck' for most enzyme isolation endeavors is therefore the availability of a genuinely high-throughput (HT) screen or selection for the target activity.

27.2
The Fundamentals of High-Throughput Screens and Selections

Screening and selection methodologies should meet the following demands:

- They should be, if possible, directly for the property of interest – 'you get what you select for' is the first rule of directed evolution [9]. Thus, the substrate should be identical, or as close as possible to the target substrate, and product detection should be under multiple turnover conditions to ensure the selection of effective catalysts.
- The assay should be sensitive over the desired dynamic range. The first rounds of any evolution experiments demand isolation with high recovery – all improved variants, including those that exhibit only several-fold improvement over the

starting gene, should be recovered. The more advanced rounds must be performed at higher stringency to ensure the isolation of the best variants. A limited dynamic range seems to be the drawback of most selection approaches.
- The procedure should be applicable in a HT format.

Numerous assays enable the detection of enzymatic activities in agar colonies or crude cell lysates by the production of a fluorophore or chromophore (see also Refs. [10–13], Chapter 28). Assays on agar-plated colonies typically enable the screening of $>10^4$ variants in a matter of days, but they are often limited in sensitivity: soluble products diffuse away from the colony and hence only very active variants are detected. Assays based on insoluble products have higher sensitivity, but their scope is rather limited (for example, see Ref. [14]). The range of assays that are applicable for crude cell lysates is obviously much wider, but their throughput is rather restricted. In the absence of sophisticated robotics (that are usually unavailable to academic laboratories), only 10^3–10^4 variants are typically screened [15]. These low- to medium-throughput screens have certainly proved effective for the isolation of enzyme variants with improved properties, as described in a number of recent reviews [6, 16, 17]. However, a far more efficient sampling of sequence space is required for the isolation of rare variants with dramatically altered phenotypes.

Whilst HT selections for binding activity have become abundant [18–20], enzymatic selections remain a challenge. The main obstacle is, that catalytic screens or selections demand a linkage between the gene, the encoded enzyme, and multiple product molecules (Figure 27.1) [21, 22]. This chapter focuses on methodologies that enable the selection of large libraries, typically well over 10^6 variants, for enzymatic activities. This throughput is still beyond the reach of HT technologies that are based on 2-D-arrays and robotics (microplates, chips, etc.).

27.3 Enzyme Selections by Phage-Display

Displaying enzymes (Chapter 24) on the surface of bacteriophages has several advantages: (i) the phage provides the link between the gene and the protein it encodes (Figure 27.1; Step I); (ii) although library size is limited by transformation, 10^7 transformants can be obtained quite easily, and 10^{11}–10^{12} are not beyond reach; and (iii) display on the surface allows unhindered accessibility of the substrate and reaction conditions of choice (e.g. buffer, pH, metals). The main challenge with display systems is to maintain the linkage between the enzyme and the products of its activity (Figure 27.1; Step II). Several selections for enzymatic activities, of either catalytic antibodies or enzymes displayed on phage, were performed indirectly by binding to transition-state analogues or suicide inhibitors (for a review, see Ref. [23]). However, direct selections for product formation under single [24, 25] as well as multiple turnovers have been achieved. A notable example is the selection of a synthetic antibody library for alkaline phosphatase activity [26]. The hydrolysis of a soluble substrate generated a product which acted as an electrophilic reagent and coupled onto the phage particle that displayed the catalytically active antibody variant. These phage particles were then captured by affinity chromatography to the product. Two rounds of selection from an initial library of $>10^9$ antibody variants yielded an antibody with catalytic efficiency (k_{cat}/K_M) that was 1000-fold higher than for other alkaline phosphatase antibodies generated by immunization with a transition-state analogue. This high catalytic efficiency was mainly the result of an increase in k_{cat} value, thus demonstrating the importance of selection for multiple turnovers. Another example involved the selection of catalytic antibodies with peroxidase activity. Selection was based on a biotin-linked tyramine substrate being oxidized and coupled to the antibody's binding site. Phages displaying active antibody catalysts were isolated by binding to streptavidin [27].

Another strategy for selection by phage display is based on linking the substrate to the phage particle expressing the target enzyme. Active enzyme variants transform the phage-linked substrate to product, which remains attached to the phage, and the phage can be isolated by affinity chromatography to the product [24, 25, 28, 29]. The strategy was recently applied for a model selection of phages displaying adenylate cyclase from a large excess of phages that do not display the enzyme. The enzymatic conversion of a chemically linked ATP into cAMP was selected using immobilized anti-cAMP antibodies, and an enrichment factor of about 70-fold was demonstrated [30]. A selection for DNA polymerases was also described [31, 32]. In this case, the substrate (a DNA primer) was also covalently attached via a flexible tether to the phage coat protein. Active variants were selected by virtue of elongation of the linked primer, first by incorporation of the target nucleotides, and finally by a biotinylated nucleotide that was used to capture the phage onto avidin-coated beads. By using this strategy, a DNA polymerase was evolved to efficiently incorporate rNTPs and thereby function as an RNA polymerase [31]. The same selection system was used to evolve DNA polymerases that incorporate 2′-O- methylribonucleosides with high efficiency [32], and even unnatural bases [33, 34]. More recently, thermophilic reverse transcriptases were evolved by the selection of a Taq DNA polymerase library displayed on phage [35].

A phage selection strategy termed 'catalytic elution' has also been described for enzymes, the catalytic activity of which depends on cofactor binding [24, 36]. Following protein expression on the phage surface, the cofactor (e.g. a metal) is removed and the catalytically inactive phages are bound to an immobilized substrate. The cofactor is then added, and phages displaying an active enzyme are eluted by conversion of the substrate into product.

27.4
HTS of Enzymes Using Cell-Display and FACS

The application of cell-surface display for directed evolution has gained much momentum. In particular, screening by fluorescence-activated cell sorting (FACS) has yielded a number of highly potent binding proteins such as antibodies [18, 37]. Cell-surface display can also be used to select for catalysis and has many features in common with phage-display. Thus, sorting for enzymatic activity was thus far achieved only in cases where the product of the enzymatic reaction could be captured within bacterial cells [38] or on their surface [39, 40]. The protease OmpA was displayed on the surface of *E. coli*, and a fluorescence resonance energy transfer (FRET) substrate was added that adheres to the surface of the bacteria. Enzymatic cleavage of the substrate releases a quencher group, and the resulting fluorescent cells were isolated by FACS. This approach allowed the screening of a library of $\sim 10^6$ OmpA proteinase variants and the subsequent isolation of a variant exhibiting a 60-fold improvement in catalytic activity [39]. In a more recent application, the same system was used for a dual selection (a selection for cleavage of a given substrate, and against cleavage of others), and provided a variant of OmpT proteinase with a change in selectivity of three million-fold relative to the wild-type [41].

A complementary approach was developed by Kolmar and coworkers, whereby the substrates of surface-displayed enzymes (esterase or lipase) release a tyramide-biotin moiety that is then covalently linked to the cell surface using peroxidase-activated tyramide conjugation. This labeling enables the isolation of *E. coli* cells presenting active enzyme variants by either FACS, or magnetic-bead selections [42].

27.5
Other FACS-Based Enzyme Screens

The powers of screening by FACS can be harnessed not only for enzymes displayed on cell surfaces, as described above, but also for enzymes expressed within the cytoplasm. The latter, however, depends on the entrapment of the fluorescent product within the cell. This can be achieved primarily when the target reaction involves the modification of a hydrophobic fluorescent substrate with a charged group, such that the unmodified substrate can be washed out of the cells while the product remains within. In some cases this can be easily achieved. For example, modification of the fluorophore 7-amino-4-chloromethyl coumarin with glutathione resulted in its entrapment, and thus large libraries of *E. coli* cells carrying active glutathione-S-transferase (GST) variants could be selected [43, 44].

A more elaborate scheme for the capture of the fluorescent product within *E. coli* cells was devised for the selection of glycosyltransferases [45], and is described in detail in Chapter 25. The capture of fluorescent products can also be achieved by compartmentalization in emulsion droplets, as described in Section 27.8 below.

27.6
In vivo Genetic Screens and Selections

Genetic, or *in vivo*, selections comprised the tool for the earliest directed evolution experiments [46, 47]. However, as the scope of targets for evolution widened the use of this approach became limited. After all, *in vivo* selections are usually based on the evolving activity complementing an auxotroph strain in which an enzyme was knocked-out. Thus, the target for evolution (substrate, reaction, etc.) usually parallels an already existing enzyme. Nevertheless, genetic screens are of much utility, for example in the identification of promiscuous enzyme activities [48, 49], and the directed evolution of many new enzyme variants (for examples see Refs [50, 51]). Recently, a novel selection assay was developed for the production of glycosylated macrolide antibiotics in *E. coli*. Reconstitution of the antibiotic biosynthetic pathway in *E. coli*, coupled to *B. subtilis* growth inhibition assay, allowed the isolation of *E. coli* strains that produce the macrolide 6-deoxyerythromycin D [52].

The scope of genetic selections has been significantly widened by application of the three-hybrid system to link the catalytic activity of an enzyme to the transcription of a reporter gene in yeast cells. This link is achieved via 'chemical complementation', whereby the product of the screened enzymatic reaction triggers translation of a reporter gene [53–55].

27.7 *In vitro* Compartmentalization (IVC)

IVC is based on water-in-oil emulsions, where the water phase is dispersed in the oil phase to form microscopic aqueous compartments. Each droplet contains (on average) a single gene, and serves as an artificial cell in allowing for the transcription, translation and activity of the resulting proteins to take place within the compartment. The oil phase remains largely inert and restricts the diffusion of genes and proteins between compartments (Figure 27.2). The droplet volume (~5 fl) enables a single DNA molecule to be transcribed and translated [56], as well

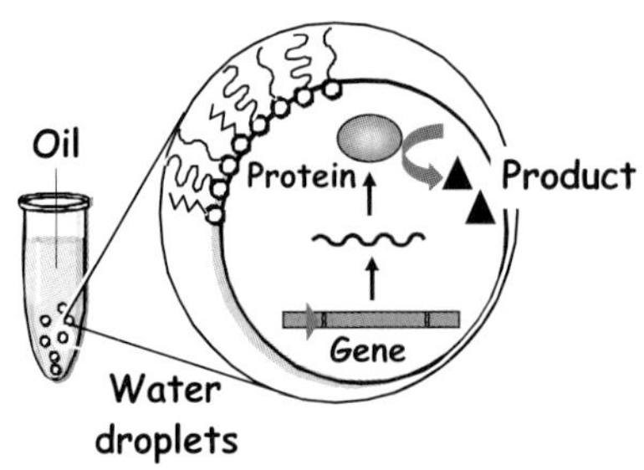

Figure 27.2 *In vitro* compartmentalization (IVC) in water droplets of water-in-oil (w/o) emulsions provides a mean of linking genotype to phenotype (protein activity). Gene libraries are compartmentalized so that, on average, each droplet contains a single gene. The genes are transcribed and translated within their compartments using cell-free translation (single *E. coli* cells expressing gene variants can also be compartmentalized; see Figure 27.4). Gene variants encoding an active protein, for example, an enzyme, generate a product that remains within the same droplet.

as the detection of single enzyme molecules [57]. The high capacity of the system (>10^{10} in 1 ml of emulsion), the ease of preparing emulsions, and their high stability over a broad range of temperatures, render IVC an attractive system for the HTS of enzymes, and of binding and regulatory applications, as well as for many other HT genetic and genomic manipulations (for a recent review, see Ref. [58]).

IVC provides a facile mean for cocompartmentalizing genes and the proteins they encode, but the selection of an enzymatic activity requires a link between the desired reaction product and the gene. One possible selection format is to have the substrate – and subsequently the product – of the desired enzymatic activity physically linked to the gene. Enzyme-encoding genes can then be isolated by virtue of their attachment to the product, while other genes, that encode an inactive protein, carry the unmodified substrate. The simplest applications of this strategy lie in the selection of DNA-modifying enzymes where the gene and substrate comprise the same molecule. Indeed, IVC was first applied for the selection of DNA methyltransferases (MTases) [56]. Selection was performed by extracting the genes from the emulsion and subjecting them to digestion by a cognate restriction enzyme that cleaves the nonmethylated DNA [56, 59, 60]. Other applications include the selection of restriction endonucleases [61] and DNA polymerases [62–66]. The selection of DNA polymerases was based on the fact that inactive variants failed to amplify their own genes and therefore disappeared from the library pool.

The above-described modes of selection are obviously restricted, not only in the scope of the selected enzymatic activities, but also by the stoichiometry – typically, one gene, and hence one substrate molecule (or several substrate sites) is present per droplet, together with 10 to 100 enzyme molecules. Despite these restrictions, several new and interesting enzyme variants were evolved by IVC. A variant of *Hae*III MTase was evolved with up to 670-fold improvement in catalytic efficiency for a nonpalindromic target sequence (AGCC) and ninefold improvement for the original recognition site (GGCC) [60]. Active *Fok*I restriction nuclease variants were also selected by IVC from a large library of mutants with low residual activity [61]. In addition, DNA polymerase variants were evolved with increased thermal stability [62], and the ability to incorporate a diverse range of bases including fluorescent dye-labeled nucleotides, and ribonucleotides [63, 64]. More recently, the same selection system was applied towards the evolution of thermophilic polymerase variants for the amplification of ancient DNA; these variants are capable of extending from primers with as many as four mismatches, and bypassing template lesions [66].

The first application of IVC beyond DNA-modifying enzymes was demonstrated by a selection of bacterial phosphotriesterase variants [57]. The selection strategy was based on two emulsification steps. In the first step, microbeads, each displaying a single gene and multiple copies of the encoded protein variant, were formed by translating genes immobilized to microbeads in emulsion droplets and capturing the resulting protein via an affinity tag. In the second step, the microbeads were isolated and re-emulsified in the presence of a modified phosphotriester substrate. The product and any unreacted substrate were subsequently coupled to the beads. Product-coated beads, displaying active enzymes and the genes that encode them, were detected with fluorescently labeled anti-product antibodies and selected by FACS. Selection from a library of $>10^7$ different variants led to the isolation of a variant with a very high k_{cat} value ($>10^5\,s^{-1}$). Microbead display libraries formed by IVC can also be selected for binding activity [67] (for examples of binding selections by IVC, see Refs. [68–71]).

Some of the IVC selection modes take advantage of the fact that this system is functional purely *in vitro*, and can allow selection for substrates, products and reaction conditions that are incompatible with *in vivo* systems. However, the cell-free translation must be performed under defined pH, buffer, ionic strength and metal ion composition. In the selection for phosphotriesterase using the IVC described above [22], the translation is completely separated from the catalytic selection by using two sequential emulsification steps, allowing selection for catalysis under conditions that are incompatible with translation. However, it is also possible to use a single emulsification step, and to modify the content of the droplets without breaking the emulsion once translation is completed.

There are currently several ways of modulating the emulsion content without affecting its integrity (Figure 27.3). These include the delivery of hydrophobic

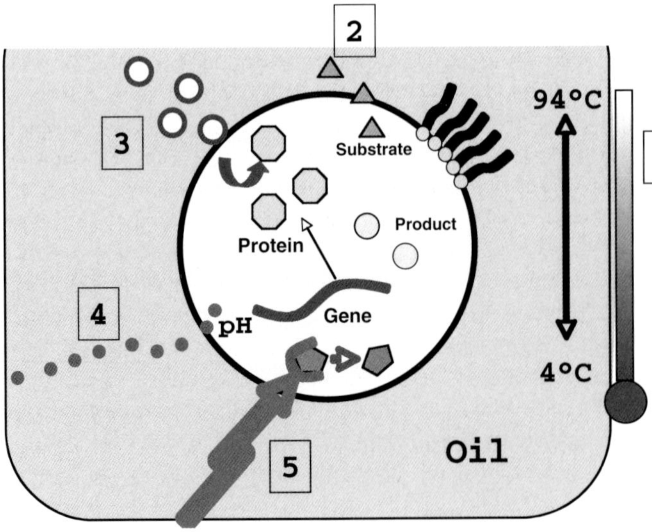

Figure 27.3 Physical and chemical manipulation of the inner content of emulsion droplets. Various means of 'communicating' with droplets, or changing their composition, without compromising their integrity, have been developed. These include: (1) Altering the temperature along a wide range, as demonstrated by the application of PCR in emulsions; (2) the transfer of hydrophobic substrates, or ligands through the oil phase into the water droplets; (3) Delivering water-soluble ligands, such as metal ions, by nanodroplets; (4) Altering the pH by the delivery of acid or base; (5) Substrates or ligands can be photocaged, introduced into the droplets during emulsification, and then activated by irradiation.

substrates through the oil phase, reduction of the droplet's pH by delivery of acid, and photoactivation of a substrate contained within in the aqueous droplets [57]. More recently, a nanodroplet delivery system was developed that allows the transport of various solutes, including metal ions, into the emulsion droplets. This transport mechanism was applied for the selection of DNA-nuclease inhibitors. Inactive DNA-nucleases were cocompartmentalized with a gene library, and once translation has been completed, the nuclease was activated by delivery of nickel or cobalt ions. Genes encoding nuclease inhibitors survived the digestion and were subsequently amplified and isolated. Selection was therefore performed directly for inhibition, and not for binding of the nuclease [72]. Other manipulations included temperature changes, as demonstrated by the application of PCR in emulsions [62, 73] (for more recent references, see [58]), and the fusion of droplets from two water-in-oil (w/o) emulsions to combine reagents contained in the different emulsions [74].

27.8 IVC in Double Emulsions

The need to link the product to the enzyme-coding gene complicates and restricts the scope of selection, especially for non-DNA modifying enzymes. Recently, an alternative strategy has been developed based on compartmentalizing (and sorting) single genes, together with the fluorescent product molecules generated by their encoded enzymes. The technology makes use of double, water-in-oil-in-water (w/o/w) emulsions that are amenable to sorting by FACS (Figure 27.4). This circumvents the need to tailor the selection for each substrate and reaction, and allows the use of a wide variety of existing fluorogenic substrates. The making and sorting of w/o/w emulsion droplets does not disrupt the content of the aqueous droplets of the primary w/o emulsions. Further, sorting by FACS of w/o/w emulsion droplets containing a fluorescent marker and parallel gene enrichments have been demonstrated [75].

W/o/w emulsions were also applied for the directed evolution of two different enzymatic systems: new variants of serum paraoxonase (PON1) with thiolactonase activity [76], and new enzyme variants with β-galactosidase activity [77] were selected from libraries of $>10^7$ mutants. The β-galactosidase variants were translated *in vitro* as with previously described IVC selections [22, 56]. In the case of PON1, intact *E. coli* cells in which the library variants were expressed, were emulsified and FACS-sorted, thus demonstrating the applicability of double emulsions for single-cell phenotyping and directed enzyme evolution [76]. The same strategy has been recently applied to the selection of lactonase activity [78], using an oxo-lactone substrate the enzymatic hydrolysis of which generates a thiol that was subsequently detected with a fluorogenic probe [79]. Detailed protocols for the preparation and sorting of double emulsions are also available [80].

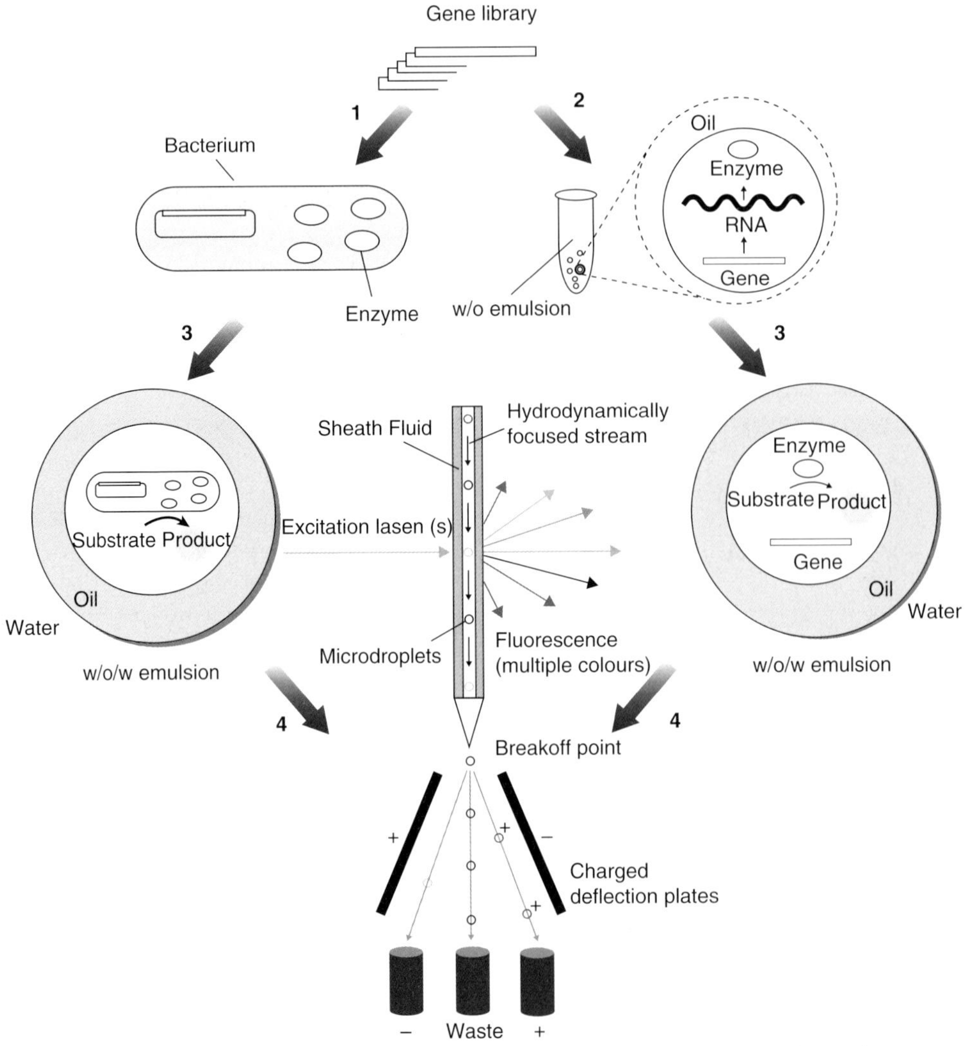

Figure 27.4 Selections by FACS sorting of double emulsion droplets. A gene library is transformed into bacteria, and the encoded proteins expressed in the cytoplasm, periplasm, or the surface of the cells *(Step 1)*. The bacteria are dispersed to form a water-in-oil (w/o) emulsion with typically one cell per aqueous microdroplet. Alternatively, an *in vitro* transcription/translation reaction mixture containing a library of genes is dispersed to form a w/o emulsion with typically one gene per aqueous microdroplet and the genes are transcribed and translated within the microdroplets *(Step 2)*. Proteins with enzymatic activity convert the nonfluorescent substrate into a fluorescent product and the w/o emulsion is converted into a water-in-oil-in-water (w/o/w) emulsion *(Step 3)*. Fluorescent microdroplets are separated from nonfluorescent microdroplets using FACS *(Step 4)*. Bacteria or genes from fluorescent microdroplets, which encode active enzymes, are recovered and the bacteria propagated or the DNA amplified using the polymerase chain reaction. These bacteria or genes can be re-compartmentalized for further rounds of selection.

27.9
What's Next?

An ancient Hebrew proverb claims that, ever since the destruction of the temple, prophecy has been left to fools. That said, there seem to be several obvious applications and potential directions of IVC that are worthy of mention.

One direction regards the means of compartmentalization. To date, all applications of IVC have relied on compartmentalization in emulsions owing to their ease of production and high stability although, following the original protocol [56], various new emulsion formulations were developed [63, 75–77, 81–84]. Large, unilamellar phospholipid vesicles containing a cell-free expression system can also be made by first creating a w/o emulsion [85]. These more cell-like compartments may also have some interesting applications [86, 87].

Other interesting directions follow recent developments in the formation and manipulation of microdroplets in a highly controlled and sophisticated manner, and have been demonstrated with microfluidic systems. Aqueous microdroplets of almost identical size can be created at up to $10\,000\,s^{-1}$, including double emulsion droplets. The combination of microfluidics and emulsion compartments, and of cell-free translation within droplets made in microfluidic systems [88], is of major interest. The high-fidelity manipulation of microdroplets in microfluidic channels, and the ability to amplify DNA, translate genes and perform other chemical and biological experiments in microdroplets, offers the possibility of developing powerful automated instruments. These new instruments can provide an unprecedented level of control and ultra-high-throughput that can be applied for HT enzyme screening and also in other areas (these points are discussed at length in recent reviews [58, 89]).

The ability to compartmentalize intact cells [76] opens the road to single-cell analyses of large populations. In the future, this methodology could be used to directly monitor the actual levels of activities of endogenous cellular enzymes rather than analyzing mRNA or protein expression levels through green fluorescent protein (GFP) fusions. IVC may therefore be applied for genotyping or phenotyping whole cells, and for identifying rare genotypes [73] or phenotypes, in large cell populations.

The use of emulsions to generate DNA–protein microbeads, namely microbeads carrying single genes and multiple copies of the protein that they encode [57, 67], may also become useful in the screening of environmental (metagenome) libraries, functional genomics, medical genetics and proteomics. Microbead display could also develop into an interesting alternative to protein chips. Proteomes displayed on microbead ensembles could be subjected to analysis by mass spectrometry, screened for binding a protein target to identify protein–protein interactions, or compartmentalized and screened for enzymatic as well as regulatory functions. The ability to perform 'microbead cloning' in emulsions, then to transcribe and translate the 'cloned' DNA in emulsion droplets, and finally to sort these emulsion droplets by FACS (Figure 27.4), or to use microfluidic devices, suggests that in the future functional genomics and proteomics could be performed in purely cell-free systems and with ultra-high-throughput. Given that, when overexpressed in *E. coli*, over half of *E. coli*'s own genes are toxic or severely growth-inhibiting [90], then cell-free translation holds great potential. Developments in cell-free translation [91], and in particular the availability of highly purified cell-free systems comprised of only the essential components for transcription and translation [92], make this route increasingly attractive.

27.10
Experimental Details

IVC selections (as almost any other enzyme selection) need to be tailored specifically for each enzyme and reaction; hence, the provision of detailed protocols is rather pointless. In general, the basic emulsion protocol, based on Span/Tween surfactants and homogenization with a magnetic stirrer, is suitable for most enzyme selections [56], although several useful variations on this protocol exist, both in emulsion composition [93] and preparation protocol (e.g. the use of a rotor-stator homogenizer with disposable shafts; [72]). The current literature also describes over a dozen different emulsion protocols, although most of these are for w/o emulsions developed for PCR (for recent examples, see Refs [83, 94, 95]). Detailed protocols for making double (w/o/w) emulsions and setting up FACS-based selection systems have been recently published [80]. Despite the particularity of selection protocols, there are several general points that are worth bearing in mind while establishing an IVC selection system.

(i) Prior to any attempt to compartmentalize, the complete sequence of reactions, including transcription, translation, enzymatic reaction(s), generation of the product-related signal (be it fluorescence, antibody binding or modification of the enzyme coding DNA), and the recovery of the enzyme coding genes, must be successfully reproduced in bulk solution, using conditions that are as close as possible to the compartmentalized reaction. The main difference, and ideally the only one, is DNA concentration. In emulsions, one DNA molecule per compartment is typically applied (yielding ~0.5 n*M* concentration for a single DNA molecule in a ~2 μm droplet). In solution, it may be necessary to use higher concentrations (sometimes an order of magnitude higher) to drive the transcription/translation. DNA stability – and hence the efficiency of recovering the selected genes – may also differ between emulsions and the bulk solution. The cell-free transcription/translation mixtures contain DNases that lead to DNA degradation, and this process is often much faster in bulk solution.

    Other than these factors, the signal-to-noise ratio observed in bulk solution (namely, the selectable signal observed in the presence of DNA coding the selected enzymatic activity, versus control DNA) must be clear and reproducible, or else there is little point in attempting to compartmentalize these reactions.

(ii) The quality of the emulsion (size distribution, and stability) is not a very common problem, but it may occur. Obvious signs are, for example, the external appearance (a good emulsion looks creamy and dense). The Span/Tween, and other emulsion compositions described below, should also exhibit high stability: they should not break (separation of the water and oil phases) or cream (floating of the oil phase) upon long standing, heating to temperatures as high as 94 °C, or centrifugation. (Centrifugation leads to precipitation of the aqueous droplets and a clear oil phase, but not breakage.) A more precise analysis can be performed by examining the emulsions using light microscopy (using ~1 μl undiluted emulsion under a coverslip), or with instruments designed to determine droplet size distributions (e.g. see Refs [56, 72]). These analyses should be performed bearing in mind that the high surface tension induced by the coverslip, or dilutions required for light scattering, may induce changes that include breakage.

(iii) Prior to library selections, it is almost always necessary to apply a set of model selections to ensure that the system is functioning. A model selection would typically involve mixtures (at varying ratios) of DNA coding the selected enzymatic activity (positive genes), versus control DNA. In the absence of an enzyme that performs the selected reaction, it may be possible to use a control gene in conjunction with the relevant product. The model selections (which are described in almost every IVC-based work published to date) help to establish that all steps are performed in an efficient manner, and that a reasonable enrichment and recovery of the 'positive' genes is obtained.

In cases where a high signal-to-noise ratio is observed in bulk solution, but not in emulsion, the translation's yield in the latter may be a problem. This could be further examined by using high DNA concentrations (>>1 gene per compartment), and mixing separately made emulsions with the positive and negative control DNAs. Such an experiment may help to separate the transcription/translation step from product formation and the gene enrichment process.

An equally crucial measure is recovery – namely, the fraction of positive genes recovered after selection. In most systems these parameters are traded-off: high enrichments are observed under very stringent condition, but stringency results in only a small fraction (often well under 1%) of genes surviving the selection (for an example where enrichment and recovery were co-optimized, see Ref. [72]). Such conditions are unsuitable for a library selection. In a model selection which yields high enrichments (e.g. 10^3-fold), and even when starting from a very high ratio of negative-to-positive genes (e.g. 10^4:1), a high copy number of the positive genes is present (a typical emulsion selection begins with $>10^9$ genes per emulsion, i.e. $>10^5$ positive genes). So, even if the recovery is 0.1%, then 10^6 genes will survive overall and these will include 10^5 positive genes. In a library selection the luxury of having 10^5 copies, or a positive gene, is beyond reach, and low recovery may be detrimental even if some degree of redundancy exists in the library.

Acknowledgments

We gratefully acknowledge research grants provided by the Estate of Fannie Sherr, and the Israeli Ministry of Science and Technology.

References

1 Boersma, Y.L., Droge, M.J. and Quax, W.J. (2007) Selection strategies for improved biocatalysts. *The FEBS Journal*, **274**, 2181–95.

2 Reymond, J.L. and Babiak, P. (2007) Screening systems. *Advances in Biochemical Engineering/Biotechnology*, **105**, 31–58.

3 Arnold, F.H., Wintrode, P.L., Miyazaki, K. and Gershenson, A. (2001) How enzymes adapt: lessons from directed evolution. *Trends in Biochemical Sciences*, **26**, 100–6.

4 Tao, H. and Cornish, V.W. (2002) Milestones in directed enzyme evolution. *Current Opinion in Chemical Biology*, **6**, 858–64.

5 Lin, H. and Cornish, V.W. (2002) Screening and selection methods for large-scale analysis of protein function. *Angewandte Chemie – International Edition in English*, **41**, 4402–25.

6 Dalby, P.A. (2003) Optimising enzyme function by directed evolution. *Current Opinion in Structural Biology*, **13**, 500–5.

7 Neylon, C. (2004) Chemical and biochemical strategies for the randomization of protein encoding DNA sequences: library construction methods for directed evolution. *Nucleic Acids Research*, **32**, 1448–59.

8 Lutz, S. and Patrick, W.M. (2004) Novel methods for directed evolution of enzymes: quality, not quantity. *Current Opinion in Biotechnology*, **15**, 291–7.

9 Schmidt-Dannert, C. and Arnold, F.H. (1999) Directed evolution of industrial enzymes. *Trends in Biotechnology*, **17**, 135–6.

10 Bornscheuer, U.T. (2002) Methods to increase enantioselectivity of lipases and esterases. *Current Opinion in Biotechnology*, **13**, 543–7.

11 Goddard, J.P. and Reymond, J.L. (2004) Recent advances in enzyme assays. *Trends in Biotechnology*, **22**, 363–70.

12 Goddard, J.P. and Reymond, J.L. (2004) Enzyme assays for high-throughput screening. *Current Opinion in Biotechnology*, **15**, 314–22.

13 Andexer, J., Guterl, J.K., Pohl, M. and Eggert, T. (2006) A high-throughput screening assay for hydroxynitrile lyase activity. *Chemical Communications (Cambridge, England)*, 4201–3.

14 Khalameyzer, V., Fischer, I., Bornscheuer, U.T. and Altenbuchner, J. (1999) Screening, nucleotide sequence, and biochemical characterization of an esterase from *Pseudomonas fluorescens* with high activity towards lactones. *Applied and Environmental Microbiology*, **65**, 477–82.

15 Geddie, M.L., Rowe, L.A., Alexander, O.B. and Matsumura, I. (2004) High throughput microplate screens for directed protein evolution. *Methods in Enzymology*, **388**, 134–45.

16 Panke, S., Held, M. and Wubbolts, M. (2004) Trends and innovations in industrial biocatalysis for the production of fine chemicals. *Current Opinion in Biotechnology*, **15**, 272–9.

17 Turner, N.J. (2003) Directed evolution of enzymes for applied biocatalysis. *Trends in Biotechnology*, **21**, 474–8.

18 Hayhurst, A. and Georgiou, G. (2001) High-throughput antibody isolation. *Current Opinion in Chemical Biology*, **5**, 683–9.

19 Amstutz, P., Forrer, P., Zahnd, C. and Pluckthun, A. (2001) *In vitro* display technologies: novel developments and applications. *Current Opinion in Biotechnology*, **12**, 400–5.

20 Takahashi, T.T., Austin, R.J. and Roberts, R.W. (2003) mRNA display: ligand discovery, interaction analysis and beyond.

Trends in Biochemical Sciences, **28**, 159–65.
21 Becker, S., Schmoldt, H.U., Adams, T.M., Wilhelm, S. and Kolmar, H. (2004) Ultra-high-throughput screening based on cell-surface display and fluorescence-activated cell sorting for the identification of novel biocatalysts. *Current Opinion in Biotechnology*, **15**, 323–9.
22 Griffiths, A.D. and Tawfik, D.S. (2000) Man-made enzymes – from design to in vitro compartmentalisation. *Current Opinion in Biotechnology*, **11**, 338–53.
23 Fernandez-Gacio, A., Uguen, M. and Fastrez, J. (2003) Phage display as a tool for the directed evolution of enzymes. *Trends in Biotechnology*, **21**, 408–14.
24 Pedersen, H., Holder, S., Sutherlin, D.P., Schwitter, U., King, D.S. and Schultz, P. G. (1998) A method for directed evolution and functional cloning of enzymes. *Proceedings of the National Academy of Sciences of the United States of America*, **95**, 10523–8.
25 Atwell, S. and Wells, J.A. (1999) Selection for improved subtiligases by phage display. *Proceedings of the National Academy of Sciences of the United States of America*, **96**, 9497–502.
26 Cesaro-Tadic, S., Lagos, D., Honegger, A., Rickard, J.H., Partridge, L.J., Blackburn, G.M. and Pluckthun, A. (2003) Turnover-based *in vitro* selection and evolution of biocatalysts from a fully synthetic antibody library. *Nature Biotechnology*, **21**, 679–85.
27 Yin, J., Mills, J.H. and Schultz, P.G. (2004) A catalysis-based selection for peroxidase antibodies with increased activity. *Journal of the American Chemical Society*, **126**, 3006–7.
28 Demartis, S., Huber, A., Viti, F., Lozzi, L., Giovannoni, L., Neri, P., Winter, G. and Neri, D. (1999) A strategy for the isolation of catalytic activities from repertoires of enzymes displayed on phage. *Journal of Molecular Biology*, **286**, 617–33.
29 Jestin, J.L., Kristensen, P. and Winter, G. (1999) A method for the selection of catalytic activity using phage display and proximity coupling. *Angewandte Chemie – International Edition*, **38**, 1124–7.
30 Strobel, H., Ladant, D. and Jestin, J.L. (2003) *In vitro* selection for enzymatic activity: a model study using adenylate cyclase. *Journal of Molecular Biology*, **332**, 1–7.
31 Xia, G., Chen, L., Sera, T., Fa, M., Schultz, P.G. and Romesberg, F.E. (2002) Directed evolution of novel polymerase activities: mutation of a DNA polymerase into an efficient RNA polymerase. *Proceedings of the National Academy of Sciences of the United States of America*, **99**, 6597–602.
32 Fa, M., Radeghieri, A., Henry, A.A. and Romesberg, F.E. (2004) Expanding the substrate repertoire of a DNA polymerase by directed evolution. *Journal of the American Chemical Society*, **126**, 1748–54.
33 Leconte, A.M., Chen, L. and Romesberg, F.E. (2005) Polymerase evolution: efforts toward expansion of the genetic code. *Journal of the American Chemical Society*, **127**, 12470–1.
34 Leconte, A.M., Matsuda, S. and Romesberg, F.E. (2006) An efficiently extended class of unnatural base pairs. *Journal of the American Chemical Society*, **128**, 6780–1.
35 Vichier-Guerre, S., Ferris, S., Auberger, N., Mahiddine, K. and Jestin, J.L. (2006) A population of thermostable reverse transcriptases evolved from *Thermus aquaticus* DNA polymerase I by phage display. *Angewandte Chemie – International Edition in English*, **45**, 6133–7.
36 Ponsard, I., Galleni, M., Soumillion, P. and Fastrez, J. (2001) Selection of metalloenzymes by catalytic activity using phage display and catalytic elution. *Chembiochem*, **2**, 253–9.
37 Feldhaus, M.J., Siegel, R.W., Opresko, L.K., Coleman, J.R., Feldhaus, J.M.W., Yeung, Y.A., Cochran, J.R., Heinzelman, P., Colby, D., Swers, J., Graff, C., Wiley, H.S. and Wittrup, K.D. (2003) Flow-cytometric isolation of human antibodies from a nonimmune *Saccharomyces cerevisiae* surface display library. *Nature Biotechnology*, **21**, 163–70.
38 Kawarasaki, Y., Griswold, K.E., Stevenson, J.D., Selzer, T., Benkovic, S.J., Iverson, B.L. and Georgiou, G. (2003) Enhanced crossover SCRATCHY: construction and high-throughput screening of a combinatorial library containing multiple

non-homologous crossovers. *Nucleic Acids Research*, **31**, e126.

39 Olsen, M.J., Stephens, D., Griffiths, D., Daugherty, P., Georgiou, G. and Iverson, B.L. (2000) Function-based isolation of novel enzymes from a large library. *Nature Biotechnology*, **18**, 1071–4.

40 Olsen, M., Iverson, B. and Georgiou, G. (2000) High-throughput screening of enzyme libraries. *Current Opinion in Biotechnology*, **11**, 331–7.

41 Varadarajan, N., Gam, J., Olsen, M.J., Georgiou, G. and Iverson, B.L. (2005) Engineering of protease variants exhibiting high catalytic activity and exquisite substrate selectivity. *Proceedings of the National Academy of Sciences of the United States of America*, **102**, 6855–60.

42 Becker, S., Michalczyk, A., Wilhelm, S., Jaeger, K.E. and Kolmar, H. (2007) Ultrahigh-throughput screening to identify *E. coli* cells expressing functionally active enzymes on their surface. *Chembiochem*, **8**, 943–9.

43 Griswold, K.E., Aiyappan, N.S., Iverson, B.L. and Georgiou, G. (2006) The evolution of catalytic efficiency and substrate promiscuity in human theta class 1-1 glutathione transferase. *Journal of Molecular Biology*, **364**, 400–10.

44 Griswold, K.E., Kawarasaki, Y., Ghoneim, N., Benkovic, S.J., Iverson, B. L. and Georgiou, G. (2005) Evolution of highly active enzymes by homology-independent recombination. *Proceedings of the National Academy of Sciences of the United States of America*, **102**, 10082–7.

45 Aharoni, A., Thieme, K., Chiu, C.P., Buchini, S., Lairson, L.L., Chen, H., Strynadka, N.C., Wakarchuk, W.W. and Withers, S.G. (2006) High-throughput screening methodology for the directed evolution of glycosyltransferases. *Nature Methods*, **3**, 609–14.

46 Hall, B.G. (1981) Changes in the substrate specificities of an enzyme during directed evolution of new functions. *Biochemistry*, **20**, 4042–9.

47 Brown, J.E., Brown, P.R. and Clarke, P.H. (1969) Butyramide-utilizing mutants of *Pseudomonas aeruginosa* 8602 which produce an amidase with altered substrate specificity. *Journal of General Microbiology*, **57**, 273–85.

48 Yang, K. and Metcalf, W.W. (2004) A new activity for an old enzyme: *Escherichia coli* bacterial alkaline phosphatase is a phosphite-dependent hydrogenase. *Proceedings of the National Academy of Sciences of the United States of America*, **101**, 7919–24.

49 Miller, B.G. and Raines, R.T. (2004) Identifying latent enzyme activities: substrate ambiguity within modern bacterial sugar kinases. *Biochemistry*, **43**, 6387–92.

50 Rothman, S.C. and Kirsch, J.F. (2003) How does an enzyme evolved *in vitro* compare to naturally occurring homologs possessing the targeted function? Tyrosine aminotransferase from aspartate aminotransferase. *Journal of Molecular Biology*, **327**, 593–608.

51 Taylor, S.V., Kast, P. and Hilvert, D. (2001) Investigating and engineering enzymes by genetic selection. *Angewandte Chemie – International Edition in English*, **40**, 3310–35.

52 Lee, H.Y. and Khosla, C. (2007) Bioassay-guided evolution of glycosylated macrolide antibiotics in *Escherichia coli*. *PLoS Biology*, **5**, e45.

53 Baker, K., Bleczinski, C., Lin, H., Salazar-Jimenez, G., Sengupta, D., Krane, S. and Cornish, V.W. (2002) Chemical complementation: a reaction-independent genetic assay for enzyme catalysis. *Proceedings of the National Academy of Sciences of the United States of America*, **99**, 16537–42.

54 Lin, H., Tao, H. and Cornish, V. (2004) Directed evolution of a glycosynthase via chemical complementation. *Journal of the American Chemical Society*, **126**, 15051–9.

55 Carter, B.T., Lin, H., Goldberg, S.D., Althoff, E.A., Raushel, J. and Cornish, V. W. (2005) Investigation of the mechanism of resistance to third-generation cephalosporins by class C beta-lactamases by using chemical complementation. *Chembiochem*, **6**, 2055–67.

56 Tawfik, D.S. and Griffiths, A.D. (1998) Man-made cell-like compartments for molecular evolution. *Nature Biotechnology*, **16**, 652–6.

57 Griffiths, A.D. and Tawfik, D.S. (2003) Directed evolution of an extremely fast phosphotriesterase by *in vitro*

compartmentalization. *The EMBO Journal*, **22**, 24–35.

58 Griffiths, A.D. and Tawfik, D.S. (2006) Miniaturising the laboratory in emulsion droplets. *Trends in Biotechnology*, **24**, 395–402.

59 Lee, Y.F., Tawfik, D.S. and Griffiths, A.D. (2002) Investigating the target recognition of DNA cytosine-5 methyltransferase *Hha*I by library selection using *in vitro* compartmentalisation. *Nucleic Acids Research*, **30**, 4937–44.

60 Cohen, H.M., Tawfik, D.S. and Griffiths, A.D. (2004) Altering the sequence specificity of *Hae*III methyltransferase by directed evolution using *in vitro* compartmentalization. *Protein Engineering, Design and Selection*, **17**, 3–11.

61 Doi, N., Kumadaki, S., Oishi, Y., Matsumura, N. and Yanagawa, H. (2004) In vitro selection of restriction endonucleases by *in vitro* compartmentalization. *Nucleic Acids Research*, **32**, e95.

62 Ghadessy, F.J., Ong, J.L. and Holliger, P. (2001) Directed evolution of polymerase function by compartmentalized self-replication. *Proceedings of the National Academy of Sciences of the United States of America*, **98**, 4552–7.

63 Ghadessy, F.J., Ramsay, N., Boudsocq, F., Loakes, D., Brown, A., Iwai, S., Vaisman, A., Woodgate, R. and Holliger, P. (2004) Generic expansion of the substrate spectrum of a DNA polymerase by directed evolution. *Nature Biotechnology*, **22**, 755–9.

64 Ong, J.L., Loakes, D., Jaroslawski, S., Too, K. and Holliger, P. (2006) Directed evolution of DNA polymerase, RNA polymerase and reverse transcriptase activity in a single polypeptide. *Journal of Molecular Biology*, **361**, 537–50.

65 Ghadessy, F.J. and Holliger, P. (2007) Compartmentalized self-replication: a novel method for the directed evolution of polymerases and other enzymes. *Methods in Molecular Biology*, **352**, 237–48.

66 d'Abbadie, M., Hofreiter, M., Vaisman, A., Loakes, D., Gasparutto, D., Cadet, J., Woodgate, R., Paabo, S. and Holliger, P. (2007) Molecular breeding of polymerases for amplification of ancient DNA. *Nature Biotechnology*, **25**, 939–43.

67 Sepp, A., Tawfik, D.S. and Griffiths, A.D. (2002) Microbead display by *in vitro* compartmentalisation: selection for binding using flow cytometry. *FEBS Letters*, **532**, 455–8.

68 Yonezawa, M., Doi, N., Kawahashi, Y., Higashinakagawa, T. and Yanagawa, H. (2003) DNA display for *in vitro* selection of diverse peptide libraries. *Nucleic Acids Research*, **31**, e118.

69 Doi, N. and Yanagawa, H. (1999) STABLE: protein-DNA fusion system for screening of combinatorial protein libraries *in vitro*. *FEBS Letters*, **457**, 227–30.

70 Bertschinger, J., Grabulovski, D. and Neri, D. (2007) Selection of single domain binding proteins by covalent DNA display. *Protein Engineering, Design and Selection*, **20**, 57–68.

71 Fen, C.X., Coomber, D.W., Lane, D.P. and Ghadessy, F.J. (2007) Directed evolution of p53 variants with altered DNA-binding specificities by *in vitro* compartmentalization. *Journal of Molecular Biology*, **371**, 1238–48.

72 Bernath, K., Magdassi, S. and Tawfik, D.S. (2005) Directed evolution of protein inhibitors of DNA-nucleases by *in vitro* compartmentalization (IVC) and nano-droplet delivery. *Journal of Molecular Biology*, **345**, 1015–26.

73 Dressman, D., Yan, H., Traverso, G., Kinzler, K.W. and Vogelstein, B. (2003) Transforming single DNA molecules into fluorescent magnetic particles for detection and enumeration of genetic variations. *Proceedings of the National Academy of Sciences of the United States of America*, **100**, 8817–22.

74 Pietrini, A.V. and Luisi, P.L. (2004) Cell-free protein synthesis through solubilisate exchange in water/oil emulsion compartments. *Chembiochem*, **5**, 1055–62.

75 Bernath, K., Hai, M., Mastrobattista, E., Griffiths, A.D., Magdassi, S. and Tawfik, D.S. (2004) In vitro compartmentalization by double emulsions: sorting and gene enrichment by fluorescence activated cell sorting. *Analytical Biochemistry*, **325**, 151–7.

76 Aharoni, A., Amitai, G., Bernath, K., Magdassi, S. and Tawfik, D.S. (2005) High-throughput screening of enzyme libraries: thiolactonases evolved by fluorescence-activated sorting of single cells in emulsion compartments. *Chemistry and Biology*, **12**, 1281–9.

77 Mastrobattista, E., Taly, V., Chanudet, E., Treacy, P., Kelly, B.T. and Griffiths, A.D. (2005) High-throughput screening of enzyme libraries: *in vitro* evolution of a beta-galactosidase by fluorescence-activated sorting of double emulsions. *Chemistry and Biology*, **12**, 1291–300.

78 Amitai, G., Devi-Gupta, R. and Tawfik, D.S. (2007) Latent evolutionary potentials under the neutral mutational drift of an enzyme. *HFSP Journal*, **1**, 67–78.

79 Khersonsky, O. and Tawfik, D.S. (2006) Chromogenic and fluorogenic assays for the lactonase activity of serum paraoxonases. *Chembiochem*, **7**, 49–53.

80 Miller, O.J., Bernath, K., Agresti, J.J., Amitai, G., Kelly, B.T., Mastrobattista, E., Taly, V., Magdassi, S., Tawfik, D.S. and Griffiths, A.D. (2006) Directed evolution by *in vitro* compartmentalization. *Nature Methods*, **3**, 561–70.

81 Margulies, M., Egholm, M., Altman, W.E., Attiya, S., Bader, J.S., Bemben, L. A., Berka, J., Braverman, M.S., Chen, Y. J., Chen, Z., Dewell, S.B., Du, L., Fierro, J.M., Gomes, X.V., Godwin, B.C., He, W., Helgesen, S., Ho, C.H., Irzyk, G.P., Jando, S.C., Alenquer, M.L., Jarvie, T.P., Jirage, K.B., Kim, J.B., Knight, J.R., Lanza, J.R., Leamon, J.H., Lefkowitz, S. M., Lei, M., Li, J., Lohman, K.L., Lu, H., Makhijani, V.B., McDade, K.E., McKenna, M.P., Myers, E.W., Nickerson, E., Nobile, J.R., Plant, R., Puc, B.P., Ronan, M.T., Roth, G.T., Sarkis, G.J., Simons, J.F., Simpson, J.W., Srinivasan, M., Tartaro, K.R., Tomasz, A., Vogt, K.A., Volkmer, G.A., Wang, S.H., Wang, Y., Weiner, M.P., Yu, P., Begley, R.F. and Rothberg, J.M. (2005) Genome sequencing in microfabricated high-density picolitre reactors. *Nature*, **437**, 376–80.

82 Diehl, F., Li, M., Dressman, D., He, Y., Shen, D., Szabo, S., Diaz, L.A. Jr, Goodman, S.N., David, K.A., Juhl, H., Kinzler, K.W., and Vogelstein and B. (2005) Detection and quantification of mutations in the plasma of patients with colorectal tumors. *Proceedings of the National Academy of Sciences of the United States of America*, **102**, 16368–73.

83 Kojima, T., Takei, Y., Ohtsuka, M., Kawarasaki, Y., Yamane, T. and Nakano, H. (2005) PCR amplification from single DNA molecules on magnetic beads in emulsion: application for high-throughput screening of transcription factor targets. *Nucleic Acids Research*, **33**, e150.

84 Musyanovych, A., Mailander, V. and Landfester, K. (2005) Miniemulsion droplets as single molecule nanoreactors for polymerase chain reaction. *Biomacromolecules*, **6**, 1824–8.

85 Noireaux, V. and Libchaber, A. (2004) A vesicle bioreactor as a step toward an artificial cell assembly. *Proceedings of the National Academy of Sciences of the United States of America*, **101**, 17669–74.

86 Luisi, P.L., Ferri, F. and Stano, P. (2006) Approaches to semi-synthetic minimal cells: a review. *Naturwissenschaften*, **93**, 1–13.

87 Monnard, P.A. and Deamer, D.W. (2002) Membrane self-assembly processes: steps toward the first cellular life. *The Anatomical Record*, **268**, 196–207.

88 Dittrich, P.S., Jahnz, M. and Schwille, P. (2005) A new embedded process for compartmentalized cell-free protein expression and on-line detection in microfluidic devices. *Chembiochem*, **6**, 811–14.

89 Kelly, B.T., Baret, J.C., Taly, V. and Griffiths, A.D. (2007) Miniaturizing chemistry and biology in microdroplets. *Chemical Communications (Cambridge, England)*, 1773–88.

90 Kitagawa, M., Ara, T., Arifuzzaman, M., Ioka-Nakamichi, T., Inamoto, E., Toyonaga, H. and Mori, H. (2005) Complete set of ORF clones of *Escherichia coli* ASKA library (a complete set of *E. coli* K-12 ORF archive): unique resources for biological research. *DNA Research*, **12**, 291–9.

91 Spirin, A.S. (2004) High-throughput cell-free systems for synthesis of functionally active proteins. *Trends in Biotechnology*, **22**, 538–45.

92 Shimizu, Y., Inoue, A., Tomari, Y., Suzuki, T., Yokogawa, T., Nishikawa, K. and Ueda, T. (2001) Cell-free translation reconstituted with purified components. *Nature Biotechnology*, **19**, 751–5.

93 Ghadessy, F.J. and Holliger, P. (2004) A novel emulsion mixture for *in vitro* compartmentalization of transcription and translation in the rabbit reticulocyte system. *Protein Engineering, Design and Selection*, **17**, 201–4.

94 Williams, R., Peisajovich, S.G., Miller, O.J., Magdassi, S., Tawfik, D.S. and Griffiths, A.D. (2006) Amplification of complex gene libraries by emulsion PCR. *Nature Methods*, **3**, 545–50.

95 Diehl, F., Li, M., He, Y., Kinzler, K.W., Vogelstein, B. and Dressman, D. (2006) BEAMing: single-molecule PCR on microparticles in water-in-oil emulsions. *Nature Methods*, **3**, 551–9.

28
Colorimetric and Fluorescence-Based Screening

Jean-Louis Reymond

28.1
Introduction

Enzyme technology requires repeated measurement of the catalytic activity of enzymes, from their initial discovery during a high-throughput screening (HTS) campaign through their purification, characterization and routine use. A variety of enzyme assays which are capable of indicating enzyme activities reliably, rapidly and selectively, have been developed over the years to address this need. The development of laboratory automation and the systematic search for activities in large microbial collections and mutant libraries has led to a particularly strong demand for HTS enzyme assays suitable for miniaturized formats such as microtiter plates, agar plates and microarrays. Under these conditions, assays producing an ultraviolet/visible (UV/Vis) or fluorescent signal serve as the most convenient to follow enzyme activities, and this format is the most popular for enzyme screening.

This chapter focuses on the different strategies available to screen biocatalytic reactions using HTS assays that produce UV/Vis or fluorescence changes in the solution, highlighting in each case selected time-honored as well as very recent examples of the methods. Such assays are often well-suited for HTS (1000–10 000 assays per day) when used in microtiter plate or microarray format, and can also be used for very high-throughput methods to screen microbial colonies on agar plates (10 000 colonies per plate). With the help of fluorescence-activated cell sorting (FACS), these assays allow the screening of millions of different enzymes either expressed in living cells or inside *in vitro* compartments produced by emulsions. The latter methods reach throughputs which are similar to those attainable in genetic selection experiments where catalyst-producing cells or viruses survive a metabolic selection or affinity-based phage-display selection step. A thorough discussion of these aspects has been recently reviewed in a separate volume covering HTS, genetic selection and fingerprinting, a volume which also includes screening methods based on classical analytical instruments such as high-performance liquid chromatography (HPLC), gas chromatography (GC), mass

Protein Engineering Handbook. Edited by Stefan Lutz and Uwe T. Bornscheuer

ISBN: 978-3-527-31850-6

spectrometry (MS) and nuclear magnetic resonance (NMR) instrumentation. These techniques, although limited in their throughput, find use in the context of industrial projects focusing on specific synthetic targets [1].

In surveying enzyme assays as described in the examples presented here, it should be borne in mind that these methods may have very different levels of performance. The choice of one assay method over another often depends on its most desirable feature. Both, chromogenic and fluorogenic substrates are model substrates that do not represent the true enzyme reaction; rather, they are usually compatible with diverse conditions such as various cosolvent and cell-culture media, and are very simple to use. On the other hand, enzyme-coupled assays and sensor systems involve multiple reagents and typically require well-tuned conditions compatible with the different components involved. An important distinction must also be made between 'real-time' enzyme assays, which facilitate the quantitative determination of kinetic parameters for the enzyme reaction, and 'end-point' assays which give global conversion after a certain time, and are only convenient to use for the qualitative detection of activities. One further parameter of importance in any assay is the extent of signal modulation realized by the assay. This modulation can range from the most desirable off–on switch upon product formation (which is realized whenever the signal is amplified by 20-fold or more upon reaction) to signal modulation of the order of 50% signal intensity observed in certain sensor-based assays.

It should always be remembered that an enzyme assay is just one method for detecting the progress of an enzyme-catalyzed reaction under specific conditions. This is particularly important when screening is undertaken to discover an activity. In such an experiment, a signal which is indicative of enzyme activity should at first always be treated with skepticism, reproduced, and the enzyme activity confirmed by a different analytical method, before concluding on the discovery. This issue is less critical when the identity and activity of the enzyme is already known and the assay is used as a quality control measure. In this case, the assay serves to determine the activity content of the sample in Units, which are defined as micromoles of reference substrate per minute per milligram of protein under a set of reference conditions.

28.2
Enzyme-Coupled Assays

Most enzyme reactions do not produce a visible change in the solution, especially if the enzyme is present only in very small amounts and thus only produces a weak activity – as is often the case during screening. One of the most straightforward methods to render an enzyme-catalyzed reaction detectable consists of further converting the reaction product by a second enzyme to form a second product, and so on, until one of these follow-up reactions produces a detectable spectroscopic change. Such assays are useful for screening if the activity of the follow-up enzymes, which are added as reagents in the assay, is strong and unaffected by the components of the assay. A few examples are discussed below.

28.2.1
Alcohol Dehydrogenase (ADH)-Coupled Assays

The vast majority of enzyme-coupled assays are coupled to an oxido-reductase, usually an alcohol dehydrogenase (ADH), using NAD^+ or $NADP^+$ as cofactors. The reduced and oxidized forms of these cofactors have different UV and fluorescence properties in the near-UV range (300–400 nm) compatible with standard microtiter plate readers. The versatility of NAD-coupled assays is such that in biochemistry the expression 'enzyme assay' is almost synonymous with 'dehydrogenase-coupled assay'. For example, Bornscheuer *et al.* screened lipases and esterases with various enantiomeric acetate esters in microtiter plate format using an acetic acid detection kit which involved a chain of three enzymes eventually producing NADH from NAD^+ (Figure 28.1) [2] (see Chapter 30). In this assay, acetyl CoA synthase (ACS) first activates acetic acid to acetyl CoA with ATP. The acetyl CoA is then condensed with oxaloacetate to form citrate by citrate synthase (CS). Finally, oxaloacetate is produced from L-malate by L-malate dehydrogenase (L-MDH). Each of these three enzymes is highly specific for its substrate and remains unaffected by the enzyme/substrate combination, such that the actual lipase or esterase is reliably indicated by the production of NADH from NAD^+.

A similar enzyme-coupled assay was recently used to screen tagatose 1,6-bisphosphate aldolase mutants in a directed evolution experiment [3] (Figure 28.2). In this case, fructose 1,6-bisphosphate (**1**) was the substrate, and glyceraldehyde

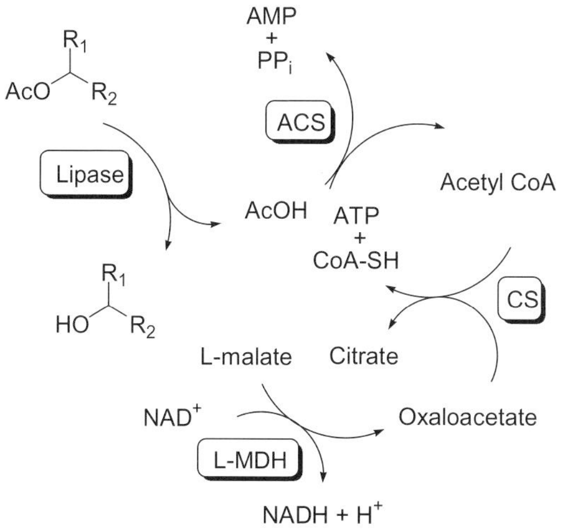

Figure 28.1 Enzyme-coupled detection of acetic acid via NAD/NADH at 340 nm allows lipase and esterase activities to be followed. The lipase may accept a variety of substrates, while the detection enzymes ACS (acetyl CoA synthase), CS (citrate synthase) and L-MDH (L-malate dehydrogenase) are very specific for their substrates.

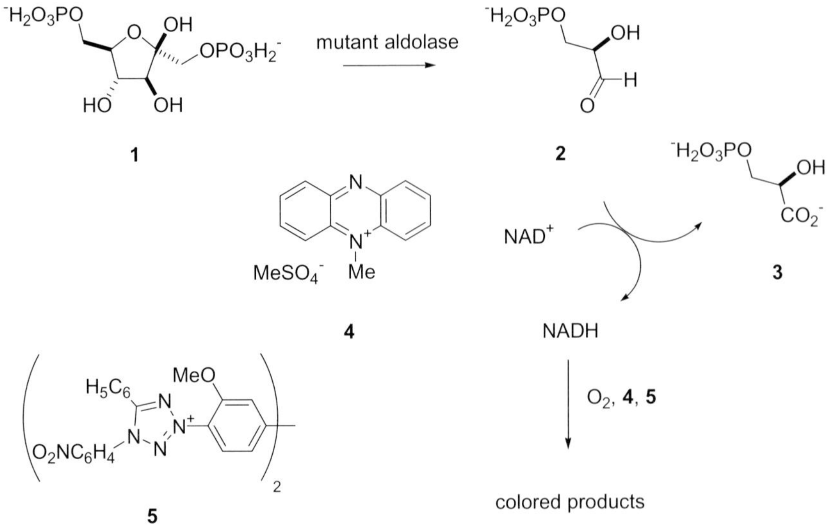

Figure 28.2 Enzyme-coupled assay for screening mutant tagatose-bisphosphate aldolases for altered stereospecificity, using a chromogenic detection of NADH with phenazin methosulfate (**4**) and nitroblue tetrazolium (**5**).

3-phosphate dehydrogenase was the coupled enzyme in conjunction with NAD^+ as cofactor. Oxidation of the glyceraldehyde 3-phosphate product **2** by the dehydrogenase gave 3-phosphoglycerate **3** and NADH. The detection of NADH was not made directly but instead through a specific chromogenic reagent consisting of phenazine methosulfate (**4**) and nitroblue tetrazolium (**5**) and oxygen in a process involving superoxide radicals to produce a colored soluble product. A similar enzyme-coupled assay for sialic acid aldolase relies on the formation of pyruvate, which is detected by reduction with lactate dehydrogenase (LDH) and concomitant oxidation of NADH, and was used to optimize this enzyme for conversion of unnatural sialic acid analogues by site-saturation mutagenesis [4].

The enantioselectivity of ADH enzymes has also been used to determine the enantioselectivity of alcohols to screen enantioselective catalysts. The enantioselective *Thermoanaerobium* sp. AD that converts the (*S*)-enantiomer of 1-phenylpropanol, but is inhibited by the (*R*)-enantiomer, allows one to determine the optical purity of this product formed in a catalyzed addition of diethylzinc to benzaldehyde [5]. In an elegant experiment, a pair of ADHs with opposite enantioselectivities was used to screen the enantiomeric excess (ee) values of chiral alcohols [6]. It was shown that only partially enantioselective enzymes are sufficient to determine ee-values accurately. A related pair of ADHs with complementary enantioselectivities was used to screen a transition metal-catalyzed epoxide opening reaction [7]. The conversion and enantioselectivity of the transition metal-catalyzed conversion of benzaldehyde (**6**) and acetyl cyanide to the corresponding acetylated cyanohydrins **7** was determined in an elegant sequential assay with three enzymes. The addition

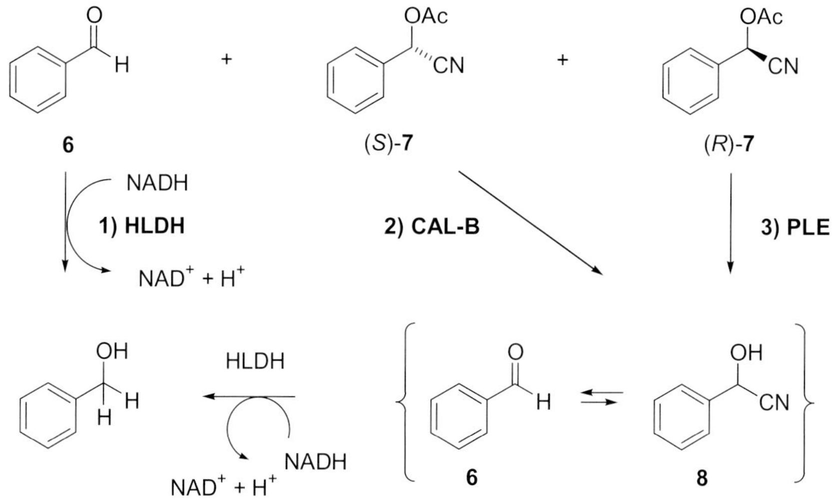

Figure 28.3 Sequential enzyme-coupled assay to determine conversion and enantioselectivity.

of horse liver alcohol dehydrogenase (HLDH) and NADH first allowed determination of the unreacted **6**. Then, addition of the lipase CAL-B hydrolyzed *(S)*-**7** enantioselectively to produce the unstable cyanohydrin **8** and liberate more benzaldehyde for HLDH-catalyzed reduction. Finally, the addition of pig liver esterase (PLE) hydrolyzed the remaining, as-yet unreacted, *(R)*-**7** to produce a third burst of reduction via the same pathway (Figure 28.3) [8, 9].

28.2.2 Peroxidase-Coupled Assays

The reduction of hydrogen peroxide to water, catalyzed by peroxidases, occurs with the oxidation of various chromogenic dyes. This process is one of the most broadly used chromogenic enzyme reactions in analytical biochemistry, in particular for revealing biomolecules in binding assays by using horse radish peroxidase (HRP) conjugates to avidin, antibodies or other labeling proteins. In the framework of enzyme screening, Turner *et al.* have developed a reaction to follow the activity of an enantioselective microbial monoamine oxidase (MAO-N) on amines such as **9** to form achiral imines such as **10**. The MAO-N produces hydrogen peroxide as a byproduct, which is revealed by a peroxidase and 3,3′-diaminobenzidine **11** as a chromogenic reagent (Figure 28.4) [10, 11]. This elegant system has turned out to be very practical for the assay of enzymes in agar plates in the direct detection of enzyme activities in bacteria, and consequently facilitates HTS. The use of HRP as a coupled enzyme to polymerize the naphthol formed by the hydroxylation of naphthalene and used to screened P450 monooxygenases should also be mentioned here [12, 13]. Coupling to a peroxidase was also recently used for a HTS assay of the molecular chaperone Hsp90 (see also Section 28.2.4), exploiting the

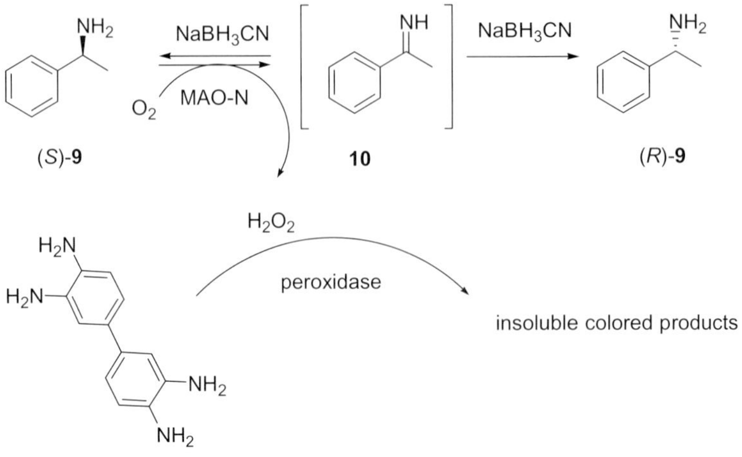

11

Figure 28.4 Detection of hydrogen peroxide by peroxidase and diaminobenzidine **11** as a tool for screening enantioselective monoamine oxidases.

release of inorganic phosphate upon ATP hydrolysis by this enzyme via maltose phosphorylase and glucose oxidase [14].

28.2.3
Hydrolase-Coupled Assays

Glycosidases and proteases have been used as secondary enzymes to follow the production of chromogenic substrates from nonreactive precursors through a primary enzyme. For example, it is possible to follow the cleavage of a glycosidic linkage to galactose using β-galactosidase as secondary enzyme, as for the nitrophenyl disaccharide **12a** and its umbelliferyl analogue **12b** for monitoring α-L-fucosidases acting specifically on L-fucosyl-α(1→2)-galactosyl substrates. While these substrates are unreactive towards β-galactosidase, cleavage by the fucosidase liberates the monogalactosides **13a/b** which undergo fluorogenic cleavage with β-galactosidase (Figure 28.5) [15, 16]. The sialidase substrate **14** operates by the same mechanism [17]. A similar strategy was used for an assay to screen the glucosynthase activity of a transglucosidase using the non-natural glycosyl fluoride donor and β-nitrophenyl-glucoside **15** as glycoside acceptor in the presence of a secondary endocellulase. The endocellulase only cleaved the elongated oligosaccharide **16** and not its shorter precursor [18].

Chymotrypsin has been used as a conformation-sensitive protease to follow the activity of proline-*cis-trans* isomerase enzymes and catalytic antibodies on the chromogenic phenylalanyl-nitroanilide peptide substrate **17a**, which releases the yellow-colored nitroaniline as product (Figure 28.6) [19–21]. Indeed, chymo-

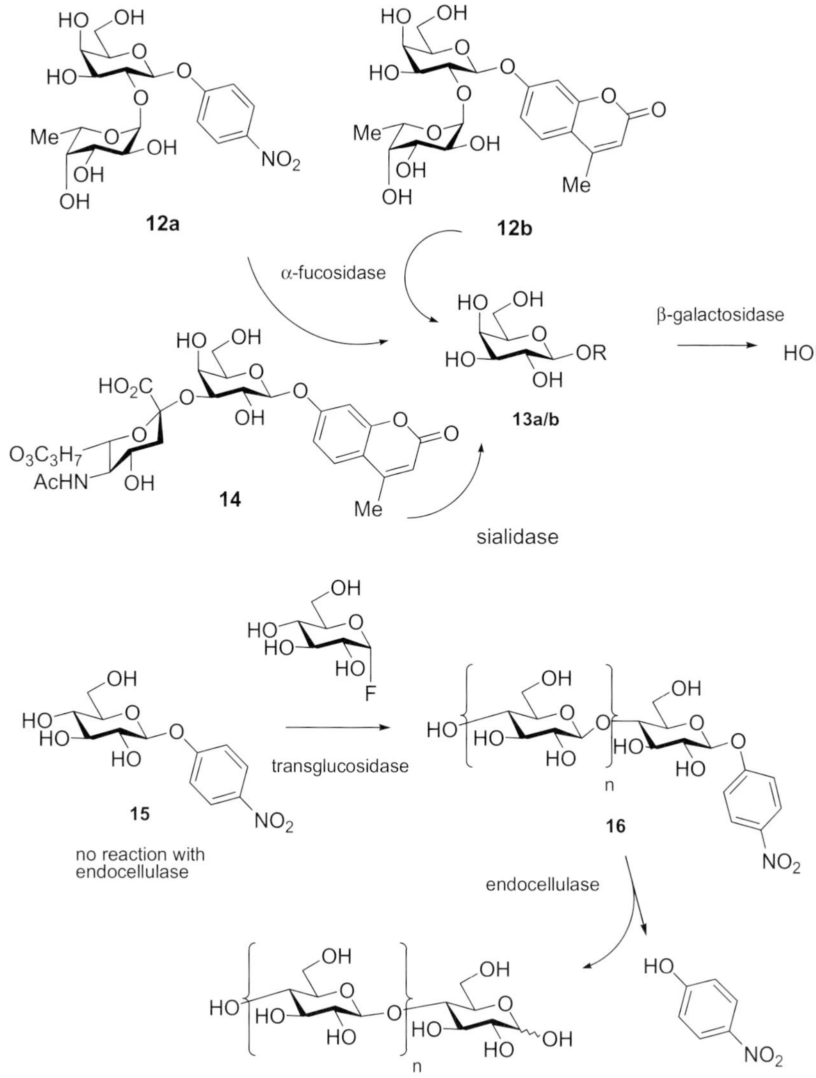

Figure 28.5 Glycosidase-coupled assays.

trypsin does not cleave **17a** with the proline residue in the *cis*-conformation, which is obtained by dissolution in a low-polarity solvent (tetrahydrofuran). The assay is initiated by dilution into aqueous buffer, under which conditions the *trans*-isomer **17b** is thermodynamically favored.

A kinase assay has been reported based on the diminished activity of an aminopeptidase for the fluorogenic release of rhodamine from a phosphorylated labeled peptide in comparison to the nonphosphorylated form [22]. This set-up is similar to a patented assay based on the differential protease sensitivity of fluorescence

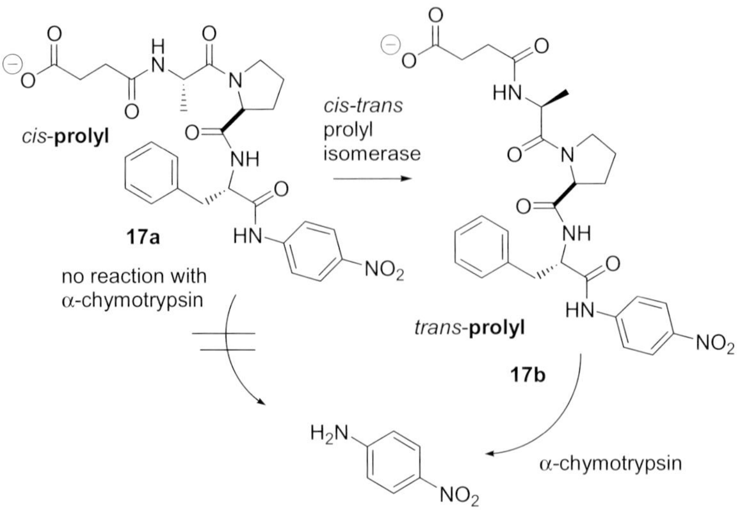

Figure 28.6 A chymotrypsin-coupled assay to follow *cis-trans* proline isomerase.

resonance energy transfer (FRET)-labeled phosphorylatable peptides, commercialized by Invitrogen under the tradename Z′-Lyte [23].

28.2.4
Luciferase-Coupled Assays

Luciferases produce light by oxidation under the consumption of ATP, oxygen and an oxidizable substrate such as luciferin or an aldehyde and reduced flavin. A number of assays have been reported that use a luciferase as the secondary enzyme to screen a reaction producing one of the luciferase substrates as product. The oldest and best method to quantify ATP relies on the use of the firefly luciferase and luciferin [24]. Keinan and coworkers used luminescent bacteria to monitor the activity of aldolase catalytic antibodies [25]. Aldol substrate **18** serves as the test substrate (Figure 28.7). The aldolase catalytic antibodies catalyze a retro-aldol reaction that produces nonanal, which is metabolized by the bacteria to produce light in aldehyde-negative mutants of the bacteria. The assay was used to monitor an antibody-catalyzed retroaldol reaction, and led to new aldolase-antibodies not found by routine techniques. The same phenotypic screen was recently used to discover new oxido-reductase enzymes, whereby the substrate spectrum of the bacterial luciferase was explored and extended [26]. Firefly luciferase was recently used as the substrate undergoing refolding by the chaperone Hsp90 [27]. An assay for monoamine oxidases A and B was recently reported where the phenol-type substrate for the luciferase is released by β-elimination of the primary aldehyde oxidation product of amine **35** (see also below, Figure 28.8).

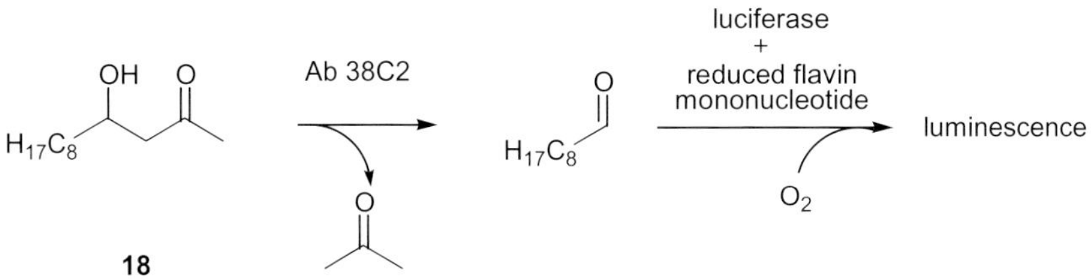

18

Figure 28.7 Luciferase-coupled assay of an aldolase catalytic antibody.

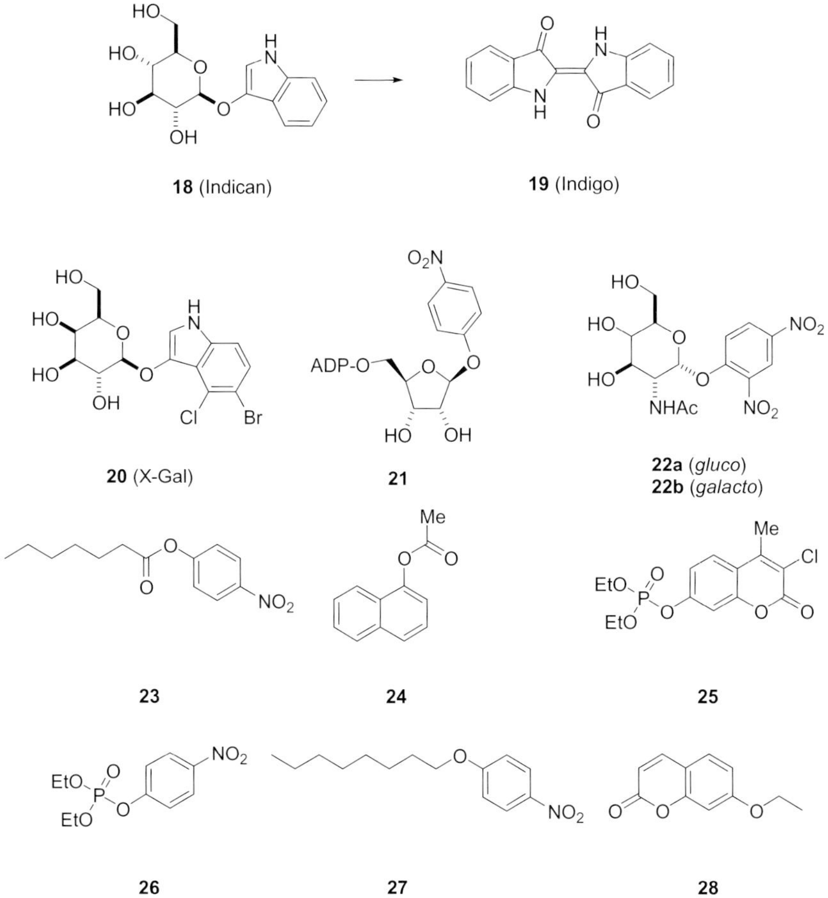

Figure 28.8 Enzyme substrates with indirect release of an aromatic alcohol.

28.3 Fluorogenic and Chromogenic Substrates

Fluorogenic and chromogenic enzyme substrates form the cornerstone of enzyme assay technology. Such substrates incorporate a chromophore, the absorbency or fluorescence properties of which change as a result of the chemical transformation by the enzyme. The key advantage of these substrates is that the signal produced is directly related to the enzyme-catalyzed reaction. The assays are therefore simpler and less prone to artifacts and interferences than are the enzyme-coupled assays discussed above. On the other hand, fluorogenic and chromogenic substrates are usually not identical with the substrate of interest in a biotransformation, and this may be problematic in the context of directed evolution experiments because optimizing enzyme activity towards a chromogenic or fluorogenic substrate may not lead to an improved enzyme with respect to the desired substrate. Generally, chromogenic and fluorogenic substrates are recommended for broad, enzyme-type-directed screens, for kinetic and mechanistic studies, and for the routine quality control of defined enzymes. Furthermore, they are also very useful in the context of *in vivo* imaging studies of enzyme activities and for enzyme inhibitor screening in drug discovery. The examples presented here follow a classification according to the mechanism of chromophore activation.

28.3.1 Release of Aromatic Alcohols

The oldest known chromogenic enzyme substrate is indican (indoxyl-β-D-glucopyranoside, **18**), which belongs to a family of natural product glycosides found in plants such as *Isatis tinctoria* and *Polygonum tinctorum* (Figure 28.9) [28]. These plants, and hence this substrate, have been used since Neolithic times to produce indigo **19** by fermentation. Fermentation results in glycoside hydrolysis to produce indenol, which oxidizes in air to the dimer indigo. The most common such indigo-type substrate today is X-Gal (**20**); this is used to visualize galactosidase activity of the *lacZ* operon used as a marker of plasmid incorporation or as a reporter for gene activation. Glycosides of various fluorescent or colored aromatic alcohols are commercially available for the screening of glycosidases, in particular nitrophenyl and 4-methylumbelliferyl glycosides (e.g. **15**, **13a/b**). The poly(ADP-ribose)polymerase-1 substrate **21** [29], and the readily available dinitrophenyl glycosides such as **22a/b** [30], are interesting recent examples of such glycosidic substrates. Fluorescein bis-glycosides are also available commercially for screening, and an *N*-acetyl-β-glucoside of this type was recently used for screening inhibitors of hexosaminidases and related enzymes [31]. Esters of the same phenols serve as chromogenic and fluorogenic substrates for lipases and esterases, such as nitrophenyl caproate **23**, although such esters tend also to hydrolyze spontaneously without enzyme. Nevertheless, fluorescein monoesters of carbamate-protected amino acids were recently reported as fluorogenic protease substrates [32]. Esterases can be visualized in polyacrylamide gels, the so-called ‘zymograms‘,

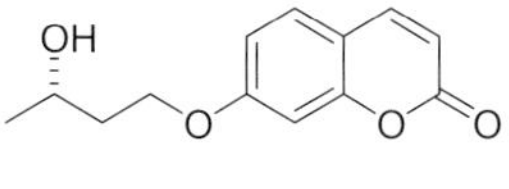

29

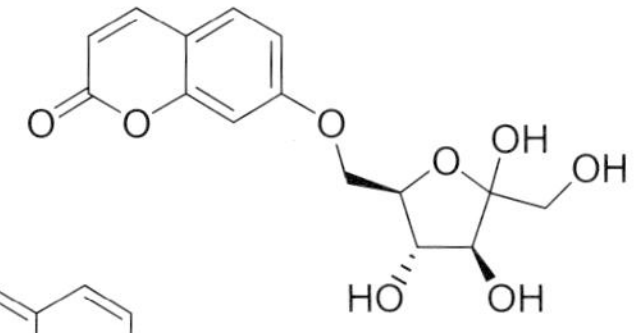

31

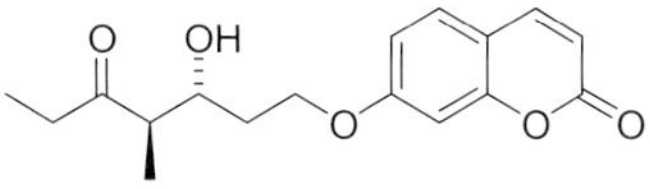

30

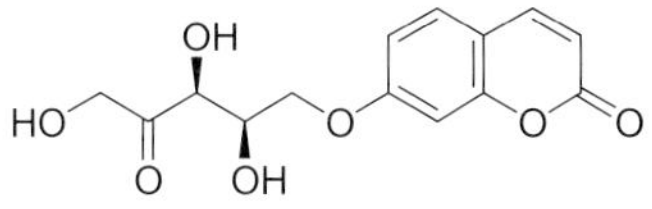

32

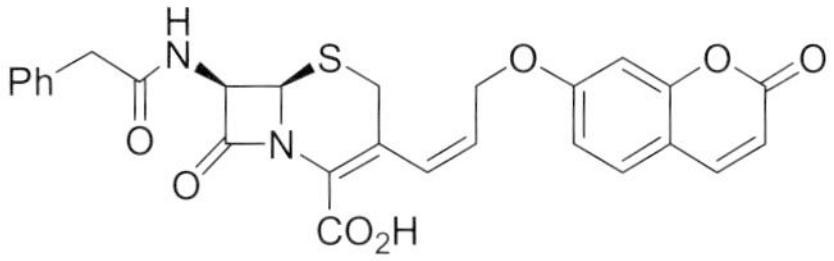

33

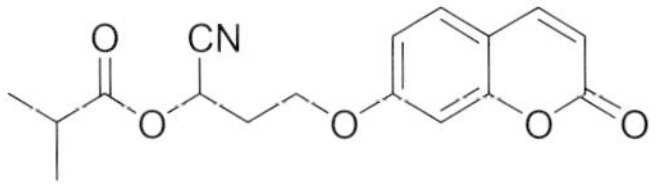

34

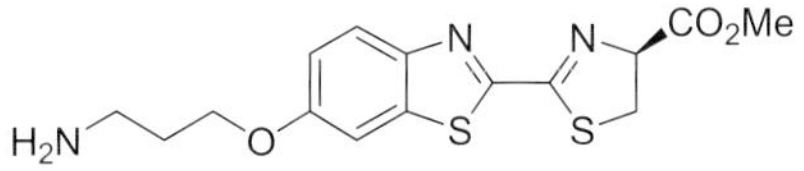

35

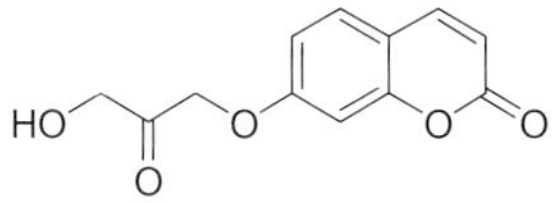

36

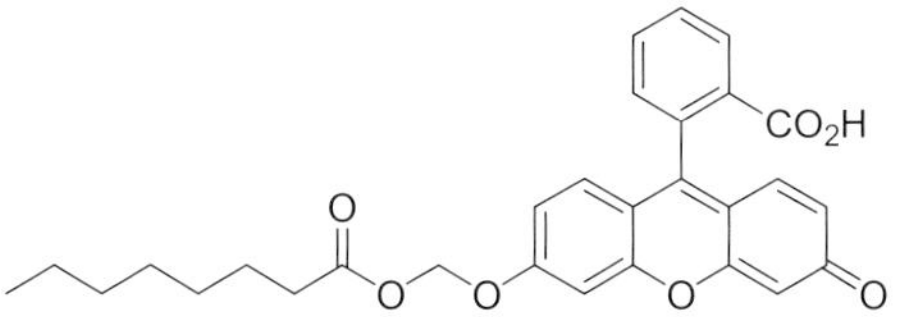

37

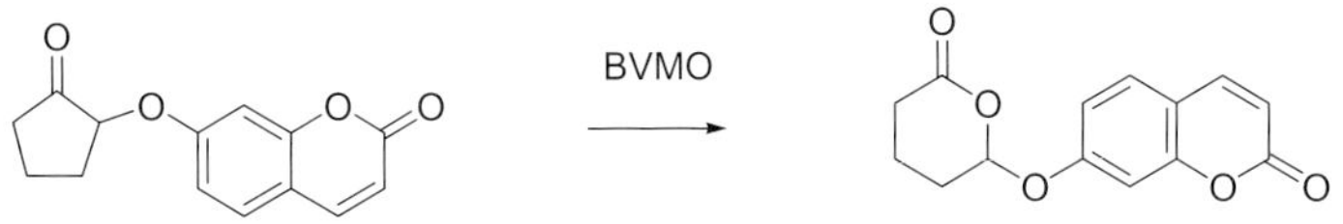

38 **39**

Figure 28.9 Chromogenic and fluorogenic substrates that release an aromatic alcohol at the enzyme-reactive bond.

or in tissue samples, by using naphthyl acetate **24** as a chromogenic substrate. The released naphthol is revealed by the addition of a diazonium salt under basic conditions to form azo-dyes [33]. The phosphotriester coumaphos (**25**) and analogues, which are related to the classical phosphatase substrate 4-methyl-umbelliferyl phosphate and are direct analogues of paraoxon (**26**), were recently reported as tools for the screening of enzymes capable of hydrolyzing chemical warfare agents such as VX [34]. Nitrophenylethers such as **27** and umbelliferyl ethers such as **28** are useful for screening cytochrome P450, which oxidizes the ether methylene group to liberate the colored phenol [35, 36]. The same cytochrome also oxidizes indoles themselves to produce indigo upon secondary oxidation of the hydroxyindole primary product [37].

The release of an aromatic alcohol can also be triggered indirectly in an enzyme substrate, allowing the enzyme-reactive functional group to be separated from the chromogenic or fluorogenic group, and also giving access to a broader diversity of reaction types. For example, the aromatic alcohol may be produced by a β-elimination reaction from a ketone or aldehyde produced by the enzyme reaction, leading to fluorogenic substrates for ADHs (**29**), aldolase antibodies (**30**) [38–40], transaldolases (**31**) [41], transketolases (**32**) [42], β-lactamases (**33**) [43], or, via the decomposition of an intermediate cyanohydrin [44], for lipases (**34**) (Figure 28.8). The monoamine oxidase substrate **35** also relies on β-elimination of an intermediate iminium, but the released aromatic alcohol is revealed as a substrate for luciferase, resulting in a luminescent, enzyme-coupled reaction [45]. The simple mono-ether of dihydroxyacetone **36** serves as a useful chromogenic substrate for aldolase-type catalysts [46, 47]. The alcoholate may also be formed from the decomposition of an hemi-acetal primary product as in the fluorescein acyloxymethyl ether **37**, leading to alternative esterase and lipase substrates with increased stability and specific enzyme reactivity [48–50]. The same mechanism also provides a fluorogenic substrate for Baeyer–Villiger monooxygenases (BVMOs) in the form of the 2-aryloxyketone **38** [51], via the intermediate lactone **39** which may be considered as a lactonase-type probe. Substrate **38** and its analogues are currently the only available fluorogenic substrates for BVMOs, and are readily obtained in one step from the commercially available precursors.

The ketone or aldehyde leading to β-elimination may also be formed by the chemical oxidation of a primary 1,2-diol or 1,2-aminoalcohol reaction product (Figure 28.10) [52]. Examples include the chromogenic lipase substrate **40** [53–55],

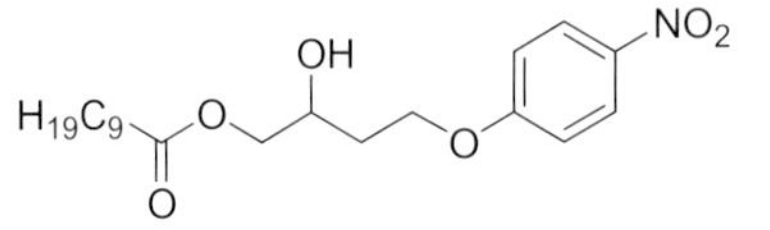

40

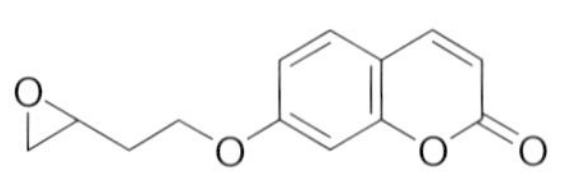

41

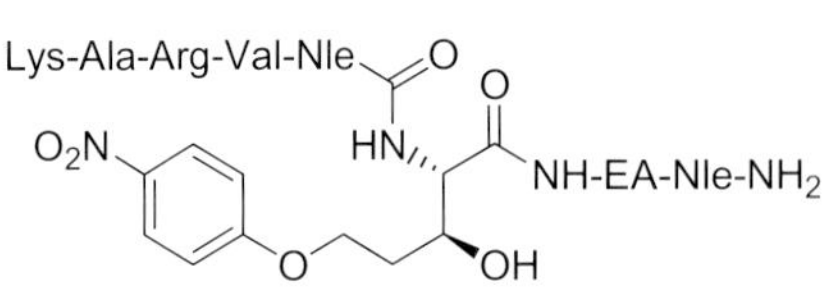

42

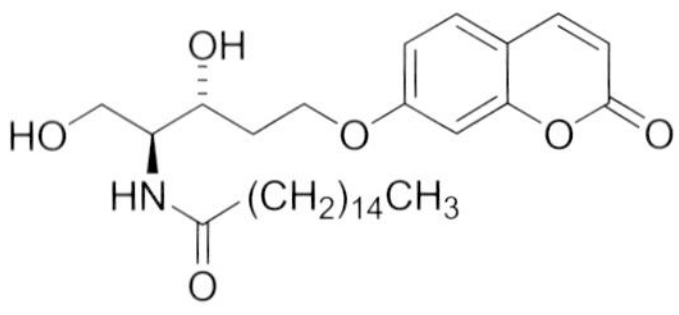

43

Figure 28.10 Periodate-coupled chromogenic and fluorogenic substrates. Nle = Norleucine, a sulfur-free analogue of methionine.

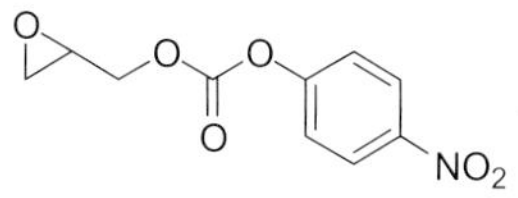

44

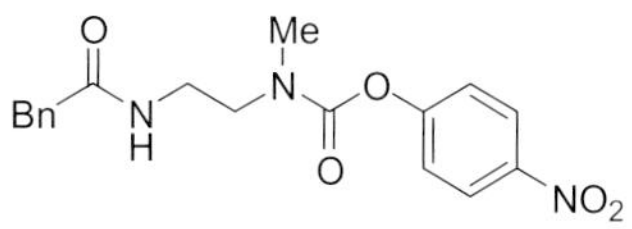

45

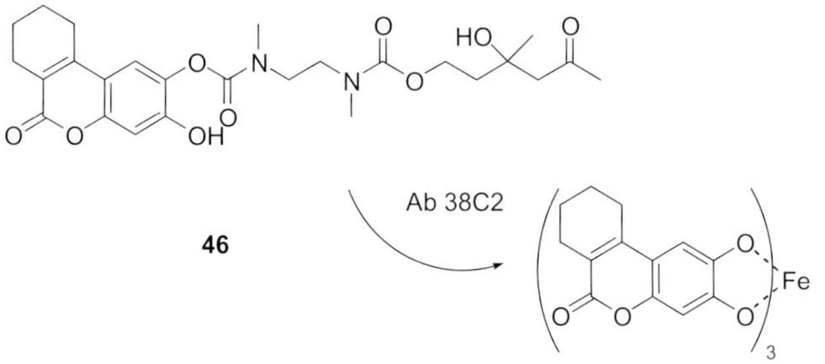

46

Figure 28.11 Indirect release of phenolate via carbonates and carbamates.

for screening thermophilic microorganisms due to its remarkable stability [56], and the fluorogenic epoxide hydrolase substrate **41** [52], which provides a highly reliable probe of epoxide hydrolase activity for screening in microbial cultures [57]. Both substrates are readily synthesized in multigram quantities in two to three steps from simple starting materials. The HIV-protease substrate **42** [58] and the fluorogenic ceramidase substrate **43** [59] are two important examples of substrates where nitrophenol or umbelliferone is released by β-elimination from a carbonyl product formed by the periodate oxidation of a primary 1,2-aminoalcohol product. In both cases, however, the synthesis is much longer, rendering these substrates less accessible for screening.

Further strategies for the indirect release of aromatic alcohols include intramolecular carbamate or carbonate cyclizations, as shown in the epoxide hydrolase substrate **44** [60], the acylase substrate **45** [61], and the related aldolase antibody substrate **46** releasing a catechol which is made visible by the formation of an insoluble black precipitate with iron (Figure 28.11) [62].

28.3.2
Aniline Release

The release of colored or fluorescent anilines from their colorless or nonfluorescent precursors has been used broadly to screen amidases and proteases, either directly as in the recently reported L-alanine amide **47** releasing a yellow-to-red acridine product suitable for agar-plate screening of aminopeptidase-producing microorganisms [63], the protease probe **48** designed for caspases and releasing a rhodamine fluorophore [64], or indirectly via a 'self-immolative' quinone methide mechanism as in the fluorogenic peptide substrate **49** for the prostate-specific

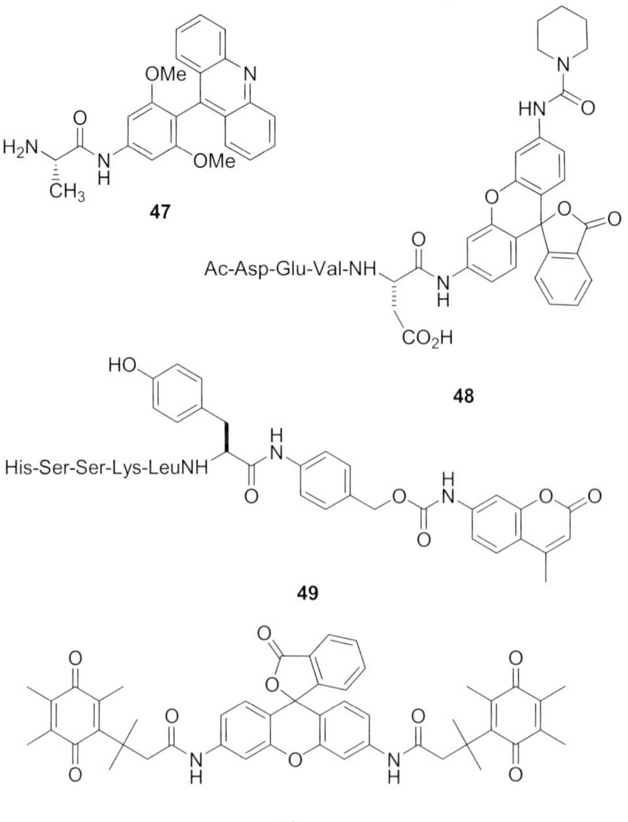

Figure 28.12 Enzyme assays with aniline release.

antigen (PSA), a protease, which releases a fluorescent aminocoumarin product [65]. A similar self-immolative release was used in a substrate for β-galactosidase [66]. The indirect release of anilines also occurs in the so-called 'trimethyl lock' principle, whereby the enzyme-triggered unmasking of a phenol leads to intramolecular cyclization and release of the fluorescent aniline; this principle is used to assay esterases [67] and also diaphorase, using **50** (Figure 28.12) [68].

28.3.3 FRET

The phenomenon of FRET (note that this may also mean Förster resonance energy transfer) relates to the ability of certain chromophores to absorb the fluorescence of a fluorophore present at a short distance within the same molecule (quenching); this may in turn lead to the emission of a photon at a longer wavelength if the quencher is itself a fluorophore (true FRET). Thus, doubly labeled substrates allow a bond-cleavage reaction between the pair of chromophores to be followed. It should be noted that the need for double labeling and the possibility of secondary chromophore–enzyme interactions make FRET substrates relatively expensive to manufacture and quite risky to use in an enzyme discovery program. On the other hand, FRET detection is usually highly selective, sensitive and very useful for *in vivo* investigations.

Numerous examples of FRET have been reported to measure proteases [69–72], cellulases [73], phospholipases (e.g. **51–53** used for *in vivo* imaging) [74–77] and lipases [78, 79] (e.g. **54** for assays at basic pH) [80] (Figure 28.13). Aldehyde **55**, maleimide **56** and fumaramide **57** with the electrophilic carbonyl respectively olefin having an intramolecular quenching effect on the fluorophore, allow a fluorescence assay for the addition of nucleophiles, such as reduction, aldol addition or Diels–Alder reactions [81, 82]. The recently reported 2,3-diazobicyclo[2.2.2]oct-2-ene (DBO) fluorophore with an unusually long excited-state lifetime was used together with tryptophan as a quencher for a FRET assay of carboxypeptidase A with the tripeptide **58** [83]. The quenching efficiency for this fluorophore–quencher pair is particularly sensitive to short distances, an effect which has been exploited to design a FRET-type peptide substrate for protein kinase C based upon a conformational change of the peptide upon reaction [84]. These studies follow numerous previous reports of fluorescence modulation in labeled peptide substrates upon phosphorylation (as recently reviewed [85]), for example by eliminating the quenching interaction of the tyrosine side chain on a pyrene label [86], or by introducing targeted fluorophore–quencher pairs to optimize the dynamic range of the assay [87]. Modulation of the fluorophore microenvironment, which is also involved in the above-mentioned kinase assays, is the principle of an aminonitrobenzofurazane-labeled γ-cyclodextrin reported as an α-amylase fluorogenic substrate, whereby cleavage of the cyclodextrin ring by the amylase exposes the fluorophore to water, leading to a reduction in fluorescence intensity [88]. A complex FRET substrate was recently reported for imaging β-lactamase activities in tissues in the near-infrared spectrum [89]. FRET pairs consisting of fluorescent proteins expressed recombinantly have also been used to detect proteases [90]; the latter were also assayed using a fluorescent polymeric substrate to which quenching groups were attached via a cleavable peptide linker [91].

Fluorescence modulation by aggregation, dilution or phase change can be assimilated to FRET substrates. Thus, lipases can be assayed with 1,3-dioleoyl-2-(4-pyrenylbutanoyl)glycerol in the presence of lipoproteins and albumin [92]. Ester hydrolysis releases the pyrene carboxylate, which then binds to albumin, resulting in an increased fluorescence. The commercially available fluorescein isothiocyanate (FITC)-conjugates of casein are used as protease substrates, whereby proteolysis removes the autoquenching and leads to a stronger fluorescence of the fluorescein chromophores [93]. A fluorescence increase by dilution to remove fluorescein autoquenching also forms the basis of the vesicle-based assay of Matile.

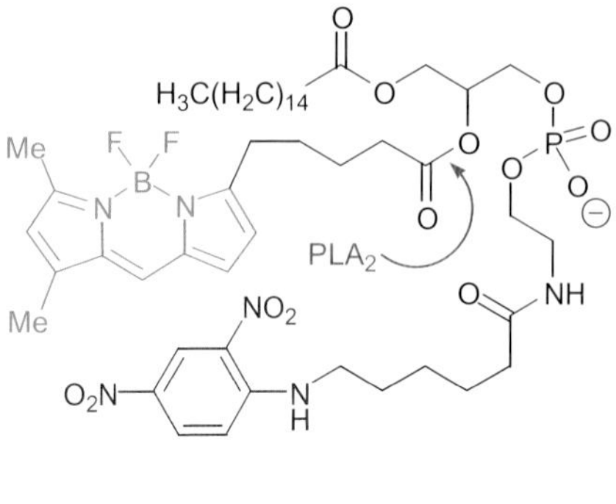

51

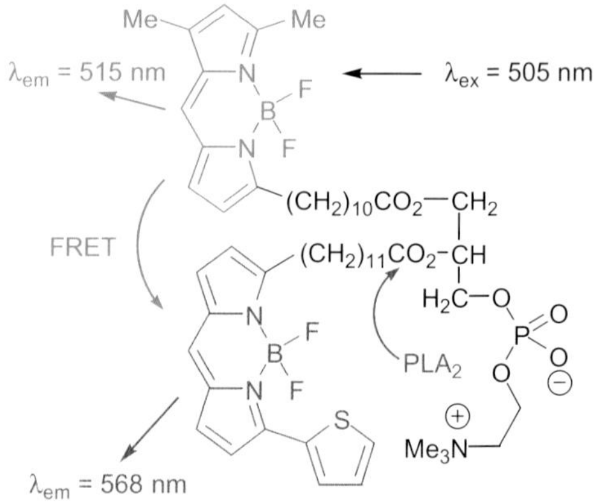

52

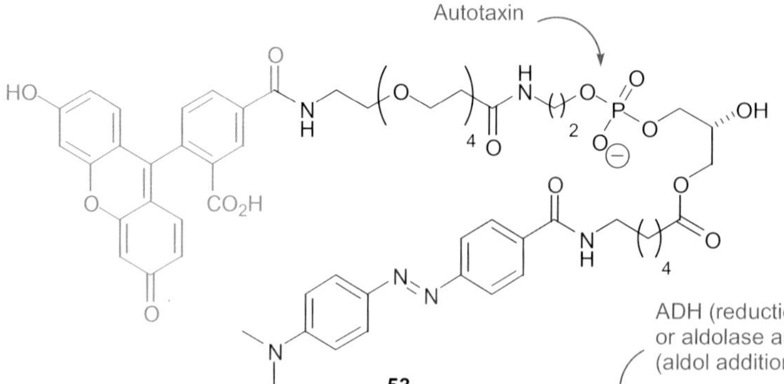

53

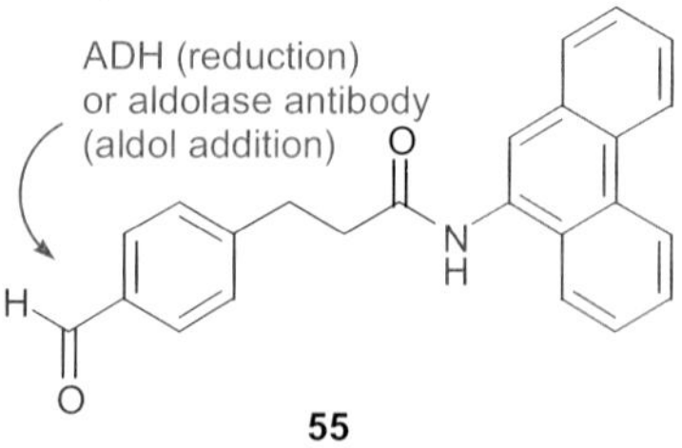

55

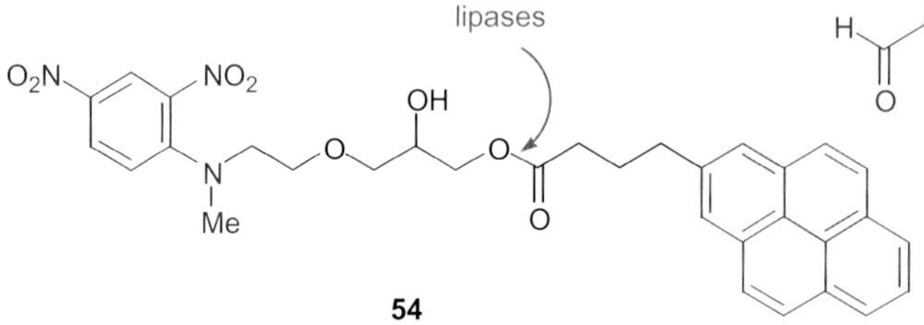

54

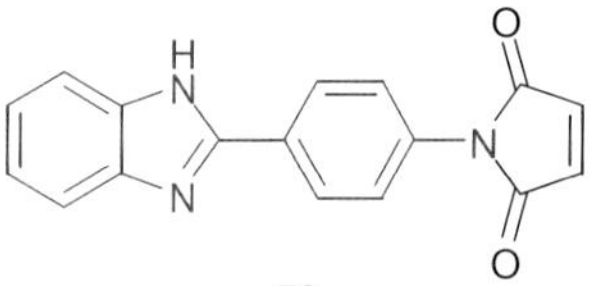

56

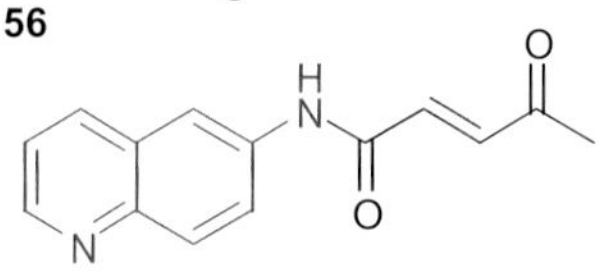

57

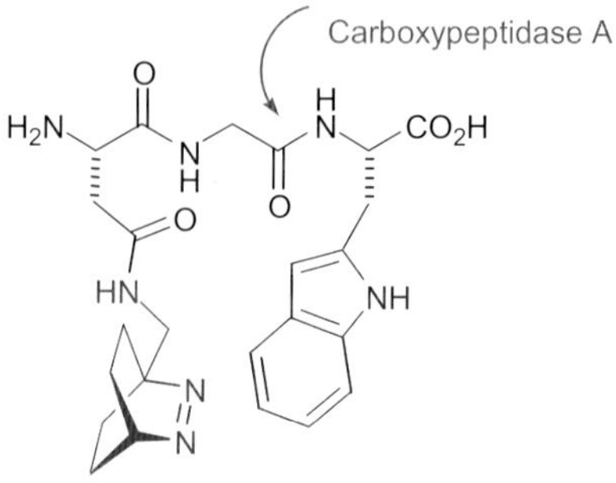

58

Figure 28.13 Examples of FRET enzyme substrates. Color code: Dark blue = quencher; blue, green, brown = blue, green, red fluorophores, respectively.

28.3.4
Reactions that Modify the Chromophore Directly

Many enzymes can be screened using aromatic substrates the chromophore of which is directly affected by the enzyme reaction. For example, microbial growth can be monitored by following the activity of nitroreductases that reduce the nitro group of various 7-nitrocoumarins such as **59** to form the corresponding 7-aminocoumarins as fluorescent products (Figure 28.14) [94]. Peroxidases react with a variety of aromatic systems (e.g. **11**; see Figure 28.4) to form colored products, and these enzymes are indeed often used as secondary enzymes for enzyme-coupled assays (see Section 28.2.2). Alcohol dehydrogenases and aldolases can be screened using 6-methoxynaphthaldehyde-related substrates such as **60** [40]. The elegant substrate **61** recently reported by Sames *et al.* for screening MAO, which oxidizes the primary amine to an aldehyde, ultimately forming the fluorescent indole **62** by condensation with the aniline amino group [95]. The recent report of an assay for fatty acid dehydrogenases using the fluorogenic substrate **63** and analogues by the same group is also of interest [96].

28.3.5
Separation of Labeled Substrates

The presence of a fluorescent or colored label in a substrate may be useful to screen enzyme reactions, even without any change in fluorescence or absorbency,

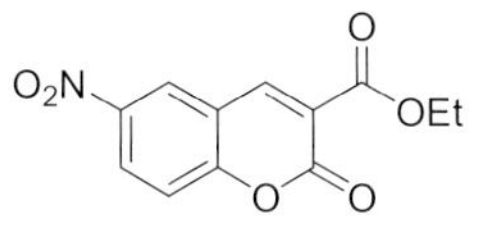

59

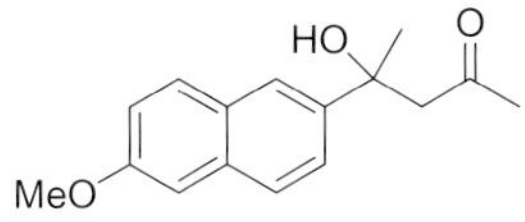

60

61 **62**

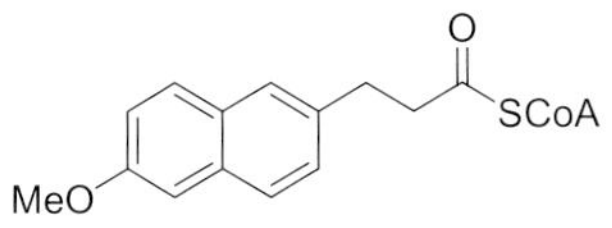

63

Figure 28.14 Fluorogenic substrates with direct modification of the chromophore upon reaction.

by exploiting separation effects. For example, the extremely bright and stable blue fluorescent acridone label can be used to follow reactions in high-throughput by parallel thin-layer chromatography (TLC) analysis, such as the diastereoselective epoxidation of the citronellol derivative **64**, which forms two TLC-separable diastereomeric epoxides **65a** and **65b** [97]. Glycosides labeled with Bodipy dyes such as **66** were used to screen sialyl transferases in living cells, relying on the fact that this substrate – but not its sialylated product – is cell-permeable. Thus, after washing away any unreacted substrate, those cells containing an active enzyme retained the fluorescent product and were separated by FACS (Figure 28.15) [98]. Although rarely used, similar phase-separation or even precipitation effects can be used to follow the reactions of labeled substrates whenever the enzyme reaction induces a strong polarity change in the substrate (see also Section 28.3.6). Landry *et al.* have used derivatives of cocaine bearing a tritiated or ^{14}C-labeled benzoate group to selectively screen for the debenzoylation of cocaine by catalytic antibodies [99, 100]. The samples being assayed were simply acidified and the benzoic acid released selectively extracted with hexane, while the unreacted cocaine substrate and the methyl ester hydrolysis product remained in the aqueous phase due to the positive charge on nitrogen. The benzoic acid formed was then quantitated in high-throughput by counting radioactivity in the organic extract. Active site-directed probes, which are suicide enzyme inhibitors that remain attached covalently to the enzyme after the catalytic turnover, can be labeled by a fluorophore to facilitate the identification of an active enzyme in a proteomic analysis, as reported by several authors for serine hydrolases [101, 102], proteases [103] and lipases and esterases [104].

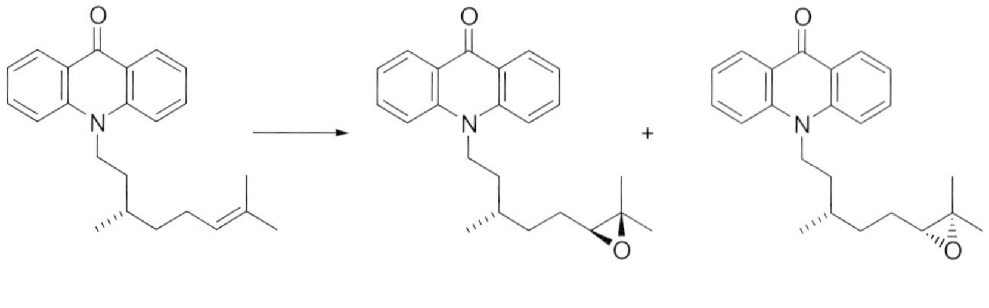

64 **65a** **65b**

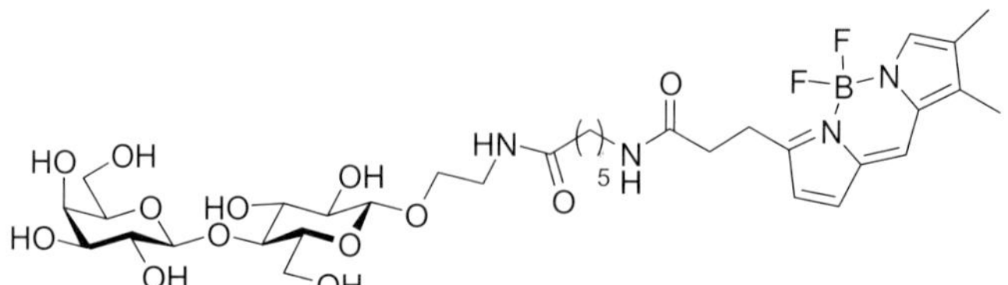

66

Figure 28.15 Fluorescence-labeled enzyme substrates allowing product separation.

28.3.6
Precipitation

A variety of screening assays rely on substrate dissolution or product precipitation or crystallization as the visible signal, without the formation of a colored product in the precipitate, as discussed above with **18, 24** or **46**. Classically, microbial cultures producing active lipases will form a clearing zone on agar plates prepared with tributyrin. Certain polymer-degrading enzymes can be screened by recording the dissolution of an insoluble substrate, in particular cellulases [105]. More sophisticated systems were recently investigated, such as the formation of hydrogels upon dephosphorylation of **67** [106], or upon cleavage of the β-lactamase substrate **68** [107]. The self-immolative substrate **69** leads to the formation of dipeptide-based nanotubes by releasing three Phe-Phe dipeptides through a dendritic cascade amplification upon reaction with penicillin G acylase (Figure 28.16) [108].

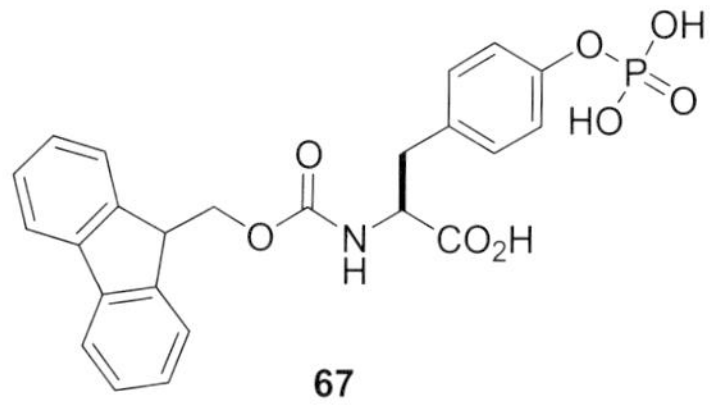

67

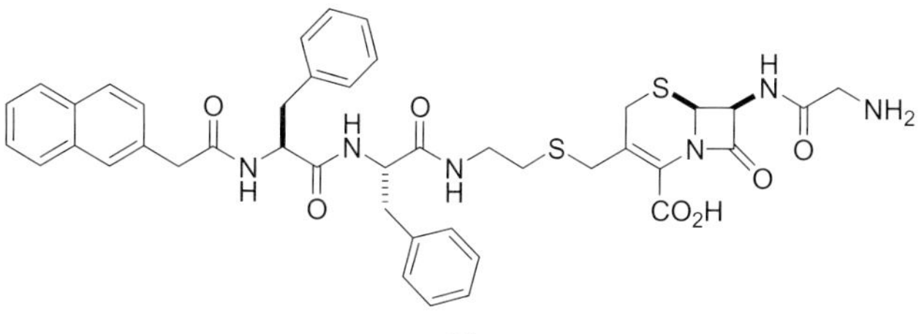

68

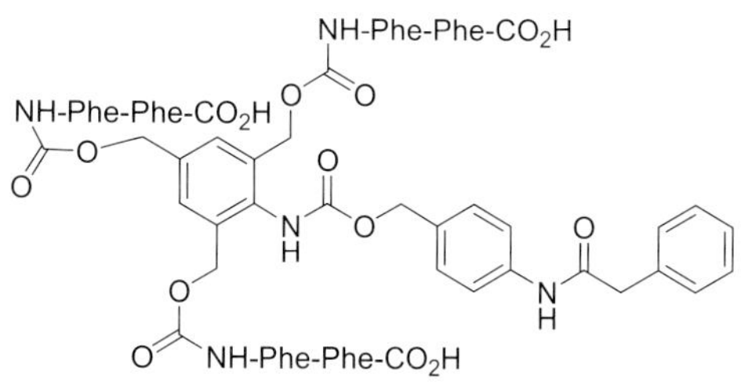

69

Figure 28.16 Enzyme substrates forming hydrogels and nanotubes upon reaction.

28.4 Chemosensors and Biosensors

Today, applied biocatalysis focuses on identifying optimal enzymes that convert specific substrates selectively and efficiently to specific products. Such discovery and optimization programs require screening of the reactions of the authentic substrate to the desired product directly and in high-throughput. As mentioned in Section 28.1, parallelized classical analytical instruments are frequently used. Chemosensors offer a simpler alternative by using chromogenic or fluorogenic reagents that respond to conversion of the authentic substrate in an indirect manner, usually through a functional group-specific reaction. Biosensors binding to either substrate or product may be used similarly. Some available options for such screening assays are discussed below.

28.4.1 Quick-E with pH-Indicators

Kazlauskas *et al.* have exploited the well-known fact that ester hydrolysis lowers the pH of the reaction medium to develop a colorimetric assay for screening lipases and esterases for enantioselectivity [109, 110]. This so-called 'quick-E assay' measures the enzyme-catalyzed esterolysis rates of two enantiomeric esters, revealed by the rate of nitrophenolate protonation at neutral pH, in the presence of a fluorogenic resorufin ester **70** serving as competition substrate (Figure 28.17).

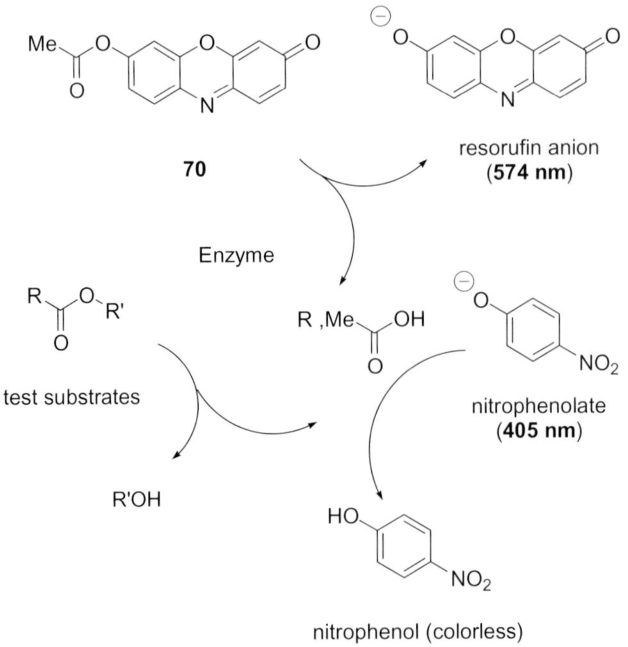

Figure 28.17 The Quick-E assay.

An apparent enantioselectivity E can be deduced from this pair of measurements that accurately predicts the true E-value of the enzyme acting on the racemic ester. pH-indicators have also been used in the context of sol–gel-encapsulated enzymes [111].

28.4.2
Functional Group-Selective Reagents

Functional group-selective reagents can be used to measure product formation or substrate consumption of an enzyme reaction, provided that the key functional group is formed or consumed during the transformation, as in the examples below (Figure 28.18). Amidases can be detected colorimetrically using amine-selective

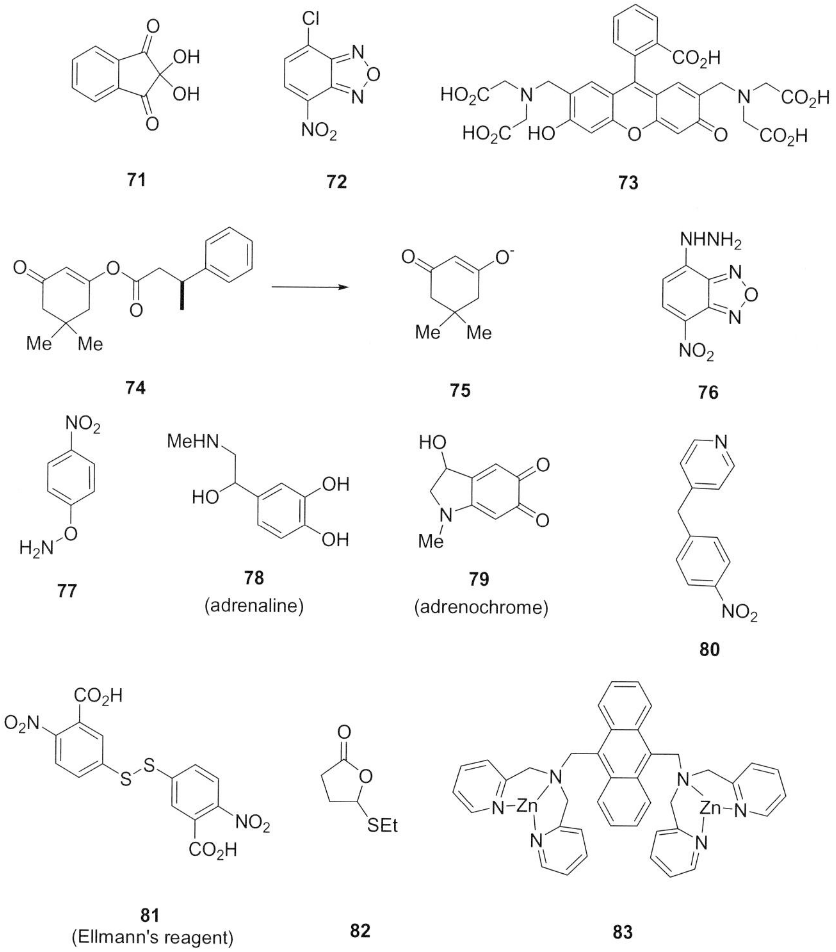

Figure 28.18 Chemosensors useful for enzyme screening.

reagents such as ninhydrin **71** [112] or chloronitrobenzofurazane **72** [113]. Similarly, ammonia released by the hydrolysis of nitriles by nitrilases can be revealed by its reaction with *o*-phthalaldehyde and β-mercaptoethanol to form a fluorescent isoindole derivative [114]. Enzymes producing amino acids from amide precursors – for example, acylases, amidases and proteases – can be screened by fluorescence using the nonfluorescent copper complex of calcein **73** [115, 116]. Indeed, the free amino acids released by the enzyme reaction rapidly and strongly chelate copper, which leads to a fluorescence increase due to the release of free calcein. In a related, simpler – but less sensitive – set-up, amino acid amidases were screened by quantification of the released amino acids by complexation with $CuSO_4$, which forms a colored complex at basic pH [117]. Similarly, dimedone esters such as (*S*)-**74** serve as useful chromogenic substrates for lipases due to the formation of a blue–green complex of the dimedone hydrolysis product **75** with Cu(II) [118]. Aldehydes can be detected by fluorescence using hydrazine **76**, which forms a fluorescent hydrazone, leading to a screen for hydrolases in the synthetic direction using vinyl esters as acyl donors [119]. Formaldehyde can be revealed colorimetrically using oxyamine **77**, which decomposes to nitrophenol after oxime bond formation [120]; this assay was used to visualize hydrolase activities on acyloxymethylether substrates.

In the so-called 'adrenaline test' for enzymes, 1,2-diols or 1,2-aminoalcohols produced or consumed by enzyme reactions such as glyceride or epoxide hydrolysis can be quantified by back-titration of sodium periodate with adrenaline **78**. The adrenaline consumes the unreacted periodate to form the red dye adrenochrome **79**. The method can be used to profile esterases using carbohydrates and polyol acetates [121], and is useful as a screen for epoxide hydrolases [122], although these can also be screened by periodate cleavage of the 1,2-diol product by detecting the aldehyde product either directly [123], or by reaction with a chromogenic Schiff's base reagent [124]. Epoxides can be revealed by alkylation of 4-nitrobenzyl-pyridine **80** in a microtiter plate-based HTS assay, which allows the screening of epoxide hydrolase by quantification of the unreacted substrate [125].

Enzyme substrates that release a thiol group as the product are readily detected using thiol-specific reagents such as Ellman's reagent **81** or a recently reported dinitrobenzensulfonyl fluorescein derivative [126]. The principle was used for a paraoxonase screen using 5-thioalkylbutyrolactones such as **82** as substrates [127]. Chemosensors for phosphorylated peptides such as zinc bipyridyl amines [128, 129] or the commercial reagent 'Pro-Q diamond' [130], allow the kinase reactions of solid-supported peptide substrates to be followed. Similarly, the fluorescent chemosensor **83**, which responds to the diphosphate group of UDP more strongly than to UDP-Gal, was used to screen galactosyl transferase activities [131].

28.4.3
Antibodies, Aptamers and Lectins

Biosensors usually contain a biomacromolecule capable of the selective recognition of an analyte of interest, in particular a reaction substrate or product. This biomacromolecule is often an antibody, an aptamer or an enzyme. Anti-product antibodies have been used to screen a variety of reactions, in particular those of catalytic antibodies. In this method, which is known as 'cat-ELISA', a microtiter plate is coated with a substrate–carrier protein conjugate. The enzyme to be assayed is incubated in the wells, and product formation then detected using the anti-product antibody [132]. Products can also be detected by using a homogeneous assay with soluble substrates and antibodies using an acridone-labeled product analogue in complex with the anti-product antibody as a sensor for product formation [133]. The best implementation of cat-ELISA involves a competitive ELISA for enantioselectivity that allows the product formation and optical purity of a soluble substrate to be quantified [134, 135]. Antibodies have also been used for an elegant 'sandwich assay' to follow metal-catalyzed coupling reactions [136]. Enzyme reactions have also been followed using aptamers that bind selectively to either substrate or product; for example, an AMP-selective aptamer was used to follow the dephosphorylation of ATP by alkaline phosphatase [137]. In a classical set-up (from which the cat-ELISA is inspired) it is possible to follow glycosyl transferases using lectin–enzyme conjugates to detect the product, for example for galactosyl transferases [138, 139]. In a recent example of this method, the peroxidase conjugate of the fucose-selective lectin of *Tetragonolobus purpureas* was used to follow the activity of a fucosyl transferase on disaccharide **84** printed on a microarray by selective binding of this lectin to the product fucoside **85**, a set-up used to screen inhibitors of the enzyme (Figure 28.19) [140, 141].

28.4.4 Gold Nanoparticles

Colorimetric enzyme assays have recently been reported based on gold nanoparticles, utilizing the fact that colloidal suspensions of gold nanoparticles are colored red due to plasmon absorbance. Solutions of Au(III) ($HAuCl_4$) are reduced by NAD(P)H to form gold nanoparticles, thereby allowing a colorimetric assay for redox reactions such as the oxidation of lactate by LDH [142]. Solutions of Au(III) are similarly reduced to nanoparticles by catechols and thiols, enabling colorimetric assays for acetyl choline esterase using thioacetyl choline as substrate [143], and tyrosinase using tyrosine as substrate [144]. Kinases have been assayed using gold nanoparticles coated with a kinase substrate using a biotinylated ATP as substrate, which results in biotinylation of the kinase-reactive substrate, and hence the nanoparticle [145]. Avidin-coated nanoparticles are then added, inducing aggregation and a color change from red to purple-gray which can be used to screen kinase inhibitors [146]. Proteases have been assayed using a synthetic peptide with a protease-specific sequence flanked by a pair of S-acetyl cysteine residues, such as the thrombin-reactive peptide **86** which induced the aggregation of gold nanoparticles by crosslinking. As crosslinking depends on bifunctionality, proteolysis inhibits the color change (Figure 28.20) [147]. Alkaline phosphatase was similarly assayed using a peptide bearing a phosphotyrosine flanked by an arginine and a cysteine. The aggregation of nanoparticles was not induced by the

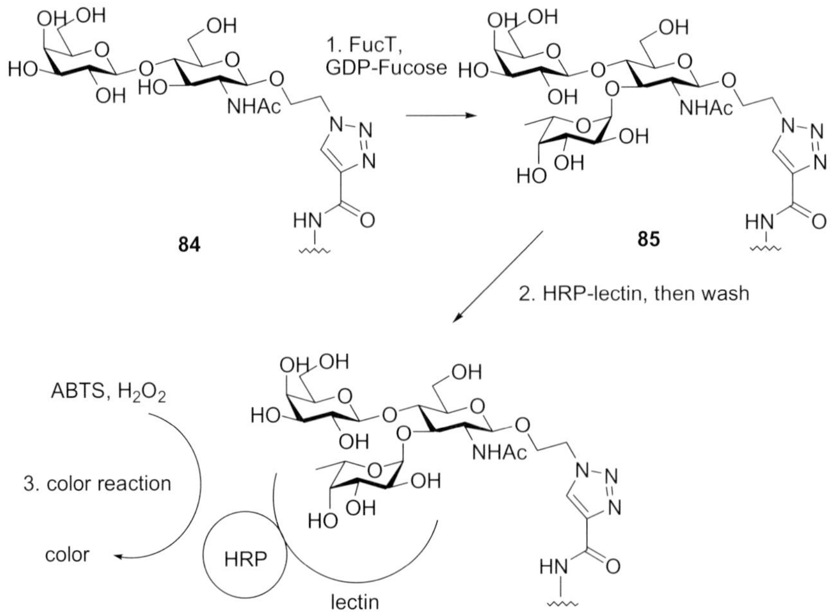

Figure 28.19 Fucosyl-transferase assay on a microarray. FucT = α-(1,3)-fucosyl transferase; HRP = horse radish peroxidase; ABTS = 2,2′-azinobis(3-ethylbenzothiazoline-6-sulfonic acid).

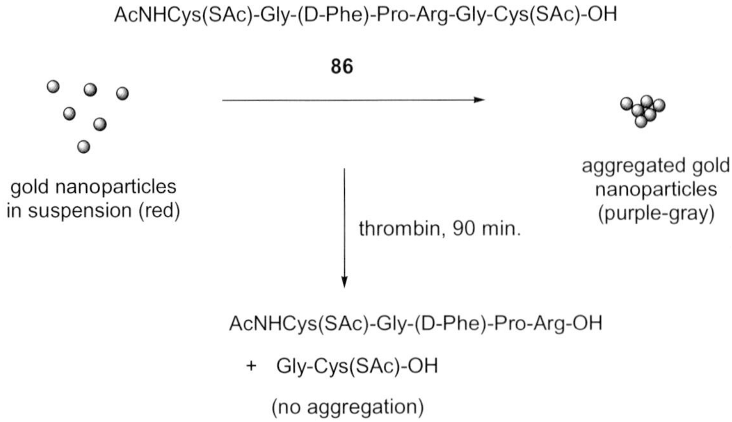

Figure 28.20 A protease assay with gold nanoparticles.

zwitterionic substrate, but took place when induced by the positively charged reaction product due to crosslinking by both the thiol and the positive charge [148]. Endonucleases have been assayed using two sets of gold nanoparticles coated with complementary, single-strand DNA substrates [149]. Here, upon cleavage by the endonuclease the color changed from purple, which was typical of the aggregated nanoparticles, to red, which was typical for dispersed nanoparticles. The assay was used to screen for inhibitors of the enzyme.

28.5 Enzyme Fingerprinting with Multiple Substrates

The information content of screening assays can be increased by analyzing multiple substrates simultaneously, which leads to a reactivity profile – or 'fingerprint' – for the enzyme-containing sample under study [150]. The challenge of enzyme fingerprinting is to provide multiple datapoints that are not only relevant to the enzyme's reactivity but also reproducible and hence can be trusted. The multidimensional datasets produced by multisubstrate assays may be interpreted in terms of the individual reactivities for identifying particularly reactive or unreactive enzyme–substrate pairs, in which case the method essentially represents an advanced version of HTS. Alternatively, the data may be used for functional classification of the enzymes [55]. This classification forms the basis for the identification of microbial strains by the APIZYM array, which is used routinely in hospitals, and has also been reported for proteases on the basis of their substrate specificity, as well as for lipases, as discussed below. Assays with multiple substrates may be carried out simply by using parallel reactions in microtiter plates, as reported for hydrolase profiling [121, 151–153].

28.5.1 APIZYM

The APIZYM system was developed during the 1960s by Bussière *et al.* as a tool to identify microorganisms on the basis of their expressed enzyme activities using 16 different enzyme substrates (Table 28.1) [154]. The assay, named Auxotab, included chromogenic substrates for lipases and esterases, aminopeptidases, chymotrypsin, trypsin, phosphatases, sulfatases and β-galactosidases. In this series, each substrate reacts with its corresponding enzyme. The assay was formulated as a combined set on a filter-paper format, and carried out in parallel on crude microorganism cultures. The APIZYM system was commercialized [155], and became a popular tool in microbiology based on a version with 19 enzyme assays characteristic of 19 enzyme activities and one blank as reference. The system has been used broadly to characterize microorganisms [156]. The apparent activities are classified qualitatively from the enzymic tests using labels such as α for a strong activity and β for a weak activity, and interpreted in terms of the presence/absence of the corresponding enzymes [157, 158]. Newer, more exhaustive variations on this theme have been developed. The current format includes 32 different assays and is available in miniaturized, preformatted plates. Fluorogenic and chromogenic hydrolase substrates are generally in use in microbiology as detection tools for a variety of enzymes [159].

Table 28.1 Enzymes and substrates assayed by the API ZYM system.

N°	Enzyme	Substrate[a]	pH[b]	Color[c]
1	None	None		[d]
2	Alkaline phosphatase	2-Naphthyl phosphate	8.5	Purple
3	Esterase C4	2-Naphthyl butyrate	7.1	Purple
4	Lipase C8	2-Naphthyl caprylate	7.1	Purple
5	Lipase C14	2-Naphthyl myristate	7.1	Purple
6	Leucine aminopeptidase	L-Leucyl 2-naphthylamide	7.5	Orange
7	Valine aminopeptidase	L-Valyl-2-naphthylamide	7.5	Orange
8	Cystine aminopeptidase	L-Cystyl-2-naphthylamide	7.5	Orange
9	Trypsin	*N*-Benzoyl-D,L-arginine 2-naphthylamide	8.5	Orange
10	Chymotrypsin	*N*-Benzoyl-D,L-phenylalanine 2-naphthylamide	7.1	Purple
11	Acid phosphatase	2-Naphthyl phosphate	5.4	Purple
12	Phosphoamidase	Naphthol AS bis-phosphodiamide	5.4	Blue
13	β-Galactosidase	6-Bromo-2-naphthyl-β-D-galactopyranoside	5.4	Purple
14	β-Galactosidase	2-Naphthyl-β-D-galactopyranoside	5.4	Purple
15	β-Glucuronidase	Naphthol AS Bis-β-D-glucuronide	5.4	Blue
16	β-Glucosidase	2-Naphthyl-β-D-glucopyranoside	5.4	Purple
17	β-Glucosidase	6-Bromo-2-naphthyl-β-D-glucopyranoside	5.4	Purple
18	*N*-Acetyl-β-glucosaminidase	1-Naphthyl-*N*-acetyl-β-D-glucosaminide	5.4	Brown
19	β-Mannosidase	6-Bromo-2-naphthyl-β-D-mannopyranoside	5.4	Purple
20	β-Fucosidase	2-Naphthyl-β-L-fucoside	5.4	Purple

Conditions: A fibrous material (filter paper) is impregnated with substrate [a] added as an alcohol solution, then with [b] a pH stabilizer (Tris–HCl > pH 7, or Tris–maleate), and then placed in contact with the biological sample to be analyzed for 2–4 h at 37 °C. [c] Coloration observed after reaction with a solution of Fast Blue BB (*N*-(4-amino-2,5-diethoxyphenyl)-benzamide) in 25% aqueous Tris–HCl containing 10% (w/w) lauryl sulfate. [d] Control with biological sample only.

28.5.2
Protease Profiling

One of the characteristic features of proteases is their sequence specificity, which may be either broad or narrow. Protease profiling consists of testing a given protease with a large number of peptide substrates. The experiment is usually carried out to determine the optimal substrate [160, 161], or with the aim of designing a selective inhibitor [162, 163]. These profiling experiments represent those assays with the largest number of fluorogenic or chromogenic substrates reported to date [164], although protease profiling is not limited to such substrates and can also be conducted using phage-display libraries, for example [165, 166]. Initially, Meldal *et al.* reported the use of synthetic combinatorial libraries of millions of synthetic substrates as FRET substrates on solid supports [167–169], and this highly practical method is still undergoing further development today [170, 171]. The same solid-supported synthesis technique was later used to produce so-called 'positional scanning libraries' of aminocoumarin amides as mixtures of fluorogenic substrates in solution, such as **87–89** (Figure 28.21) [172–178]. Positional scanning libraries of FRET protease substrates were used for profiling cathepsin B [179]. Further methods for protease profiling include peptide nucleic acid (PNA)-tagged libraries [180–183], microarray-displayed substrates [184–187] and active site-directed probes [188–191], all of which are based on colorimetric or fluorescence assays of protease activity.

28.5.3
Cocktail Fingerprinting

Multisubstrate assays can be quite complex when they involve multiple parallel experiments in which each datapoint is acquired in a physically separated location, for example different microtiter plate wells, different microarray positions or different polymer beads in a profiling experiment. In cocktail fingerprinting, a single experiment is carried out by reacting the enzyme-containing sample to be tested with a mixture, or 'cocktail', of different, selected substrates. The reaction is then analyzed in a single step using a separation/quantification method such as GC or HPLC, and the information is extracted from the corresponding chromatogram. This method has the advantages of operational simplicity facilitating multiple fingerprinting measurements, ease of quality control, and the use of GMP-approved instruments with reliable quantification. The principle was recently reported for the fingerprint analysis of lipases and esterases using a cocktail of monoacyl-glycerol analogues [192, 193], and for proteases using a cocktail of five hexapeptides with a domino-type sequence arrangement [194]. A single HPLC-analysis returns the activity fingerprint, which can be used for functional classification of the enzyme. Such characterization tools may prove useful for identifying novel enzymes with unusual selectivities, as well as in the area of diagnostics. In the same way, the APIZYM system (see Section 28.5.1) can be formulated as a cocktail reagent, allowing 16 different enzyme reactivities to be determined in a single

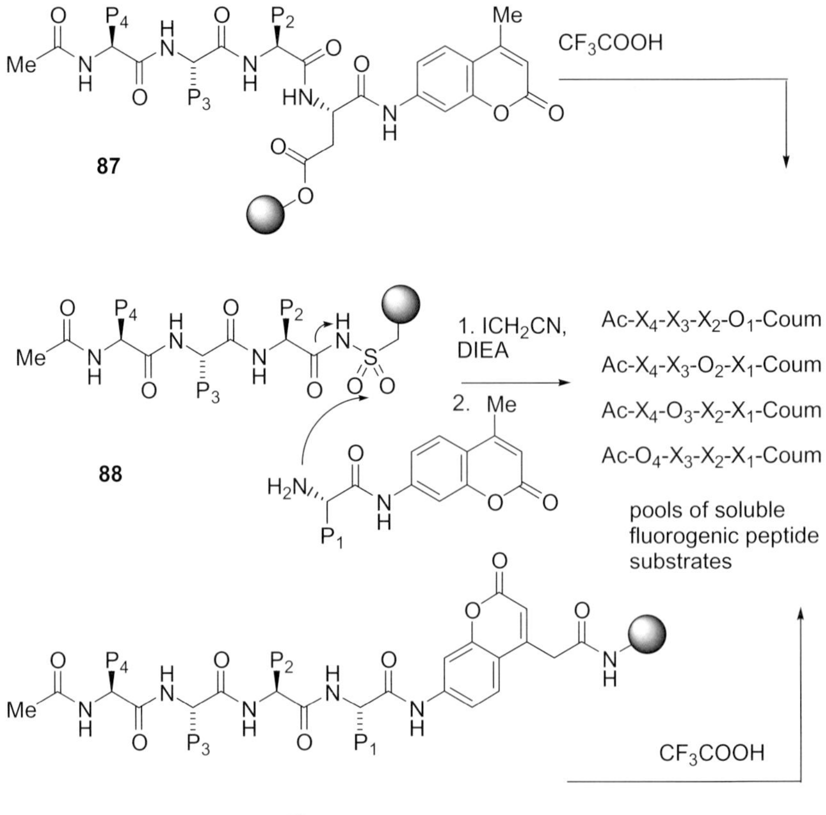

Figure 28.21 Positional scanning libraries for protease profiling [172–178]. The peptide mixtures are prepared by solid-phase peptide synthesis, and then cleaved from the resin for use as substrates. X_n = mixture of proteinogenic amino acids; O_n = defined proteinogenic amino acid (cysteine is omitted, and norleucine replaces methionine). Coum denotes either 4-methyl-7-aminocoumarin or 4-carboxamidomethyl-7-aminocoumarin. P_1–P_4 denote the amino acid side chains occupying the corresponding S_1–S_4 sites on the protease relative to the scissile bond defined by the coumarinamide. DIEA = diisopropyl-ethylamine.

analysis [195]. The method is also suitable for analyzing thermophilic microorganisms and generally extremophiles. Similarly, a cocktail of fluorescent umbelliferyl glycosides was recently used to characterize various glycosidases using an HPLC-based assay [196]. It should be noted here that substrate cocktails can also be analyzed with MS in a variety of settings, including enantioselectivity determination using a mixture of two isotopically labeled pseudoenantiomers [197–203], for the glycosyl transferase reaction using substrate mixtures [204–207], or for protease activity determinations using a mixture of peptide substrates [208].

28.5.4
Substrate Microarrays

Solid-supported assays allow the miniaturization of colorimetric and fluorescence screening below the typical 10 μl volume (which is the lower limit for liquid handling in microtiter plate assays), and can be obtained using specially designed supports and formats. A practical and inexpensive implementation is possible using silica gel plates which have been preimpregnated with a fluorogenic substrate as the reaction medium [209]. A robotic arm is used to dispense the enzyme-containing test solutions in a volume of 1 μl per assay, which results in a homogeneously dispersed spot on the silica gel surface on which the enzyme reacts evenly with the substrate (Figure 28.22). The high-throughput potential of this method was demonstrated by profiling 40 different esterases and lipases across 35 different fluorogenic ester substrates of umbelliferone or 4-methylumbelliferone **91** with various acyl chains (e.g. **90**), using only 50 μl of each enzyme solution, and submilligram quantities of each substrate for a total of over 7000 tests.

Further miniaturization of screening is possible with microarrays printed on glass slides. A nanospray system was used to homogeneously distribute nanodroplets of a solution containing three fluorogenic protease substrates on a microarray on which spots of enzyme-containing solutions had been previously printed, allowing the HTS of enzyme inhibitors [210]. The deposition of fluorogenic substrates

90 **91**

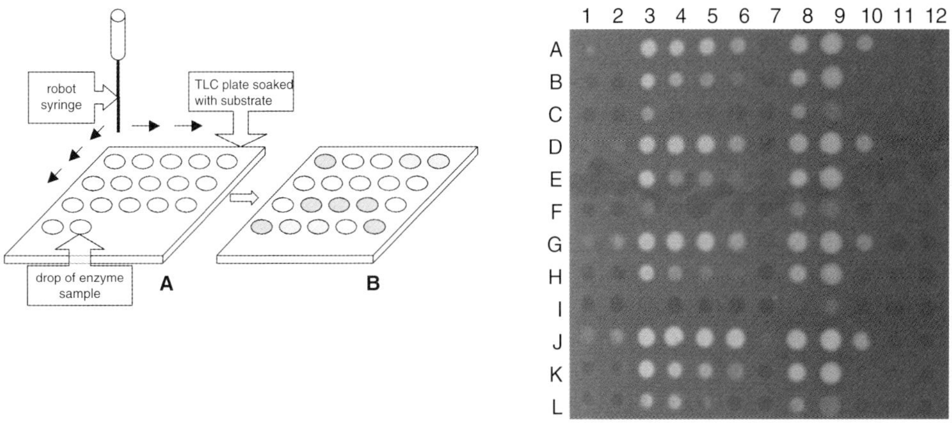

Figure 28.22 High-throughput screening with microliter reaction on silica gel plates.

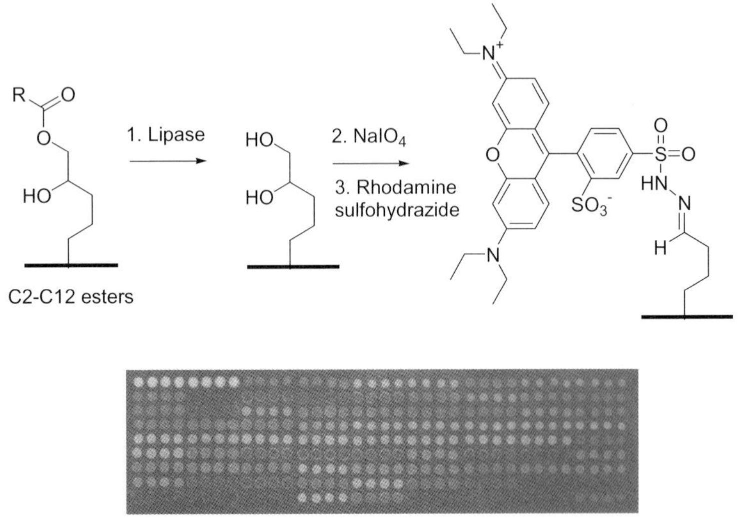

Figure 28.23 Red-fluorescent lipase fingerprinting microarray (Ref. [213]).

on poly-lysine-coated glass slides also allows the efficient assay of various enzymes in nanodroplets [211]. Nanodroplets for enzyme assays can be moved using thermal gradients for mixing [212]. Fluorogenic substrates have also been arrayed with covalent attachment to the surface of glass slides to allow activity profiling experiments with proteases and other hydrolytic enzymes using combinatorial series of fluorogenic, coumarin-derived substrates [184, 185], and with lipases using substrates of varying acyl chain length, relying on the chemoselective oxidation of the 1,2-diol product with sodium periodate followed by reaction with rhodamine sulfohydrazide to detect conversion (Figure 28.23) [213].

Analytical methods using microarray formats with fluorescence and colorimetric readouts have also been reported for screening enantioselectivity, and may potentially be useful for enzyme screening. Microarrays have also been used to estimate the optical purity of amino acids after covalent attachment by reaction with a pseudoenantiomeric pair of labels bearing two different fluorophores [214]. The ratio of fluorophore can be correlated to the optical purity using Horeau's method [215, 216]. The enantioselectivities of reactions with suitably derivatized substrates can also be measured by monitoring color changes induced in a cholesteric-phase liquid crystal by doping with the reaction product in a set-up which is suitable for high-throughput measurements [217].

28.6 Conclusions

Colorimetric and fluorescence-based assays offer a broad diversity of options for screening almost any enzyme of interest. The key advantages of spectrophotometric assays are that they are often very simple, they rely on only a single added reagent, and they allow an enzyme reaction to be followed directly by the naked eye. In this way the instrumental hurdles that render observations indirect and therefore more prone to technical complications are removed. Most of the reagents and methods discussed in this chapter involve the use of commercially available materials, and can be implemented without any major investment in specific instrumentation. In particular, a simple digital camera, combined with image-processing software, can replace a microtiter plate reader, although the quantification of fluorescence or color intensity is less sensitive at low concentrations and is limited in its intensity range due to saturation effects. Today, progress in enzyme assays is an ongoing and highly active area of research where new ideas for converting enzyme reactions into observable signals using simple methods continue to emerge.

Acknowledgments

These studies were supported financially by the University of Berne, the Swiss National Science Foundation, and Protéus SA, Nîmes, France.

References

1 Reymond, J.L. (2005) *Enzyme Assays: High-Throughput Screening, Genetic Selection and Fingerprinting*, Wiley-VCH Verlag GmbH, Weinheim.

2 Baumann, M., Sturmer, R. and Bornscheuer, U.T. (2001) A high-throughput-screening method for the identification of active and enantioselective hydrolases. *Angewandte Chemie – International Edition*, **40**, 4201–4.

3 Williams, G.J., Domann, S., Nelson, A. and Berry, A. (2003) Modifying the stereochemistry of an enzyme-catalyzed reaction by directed evolution. *Proceedings of the National Academy of Sciences of the United States of America*, **100**, 3143–8.

4 Woodhall, T., Williams, G., Berry, A. and Nelson, A. (2005). Creation of a tailored aldolase for the parallel synthesis of sialic acid mimetics. *Angewandte Chemie – International Edition*, **44**, 2109–12.

5 Abato, P. and Seto, C.T. (2001) EMDee: an enzymatic method for determining enantiomeric excess. *Journal of the American Chemical Society*, **123**, 9206–7.

6 Li, Z., Butikofer, L. and Witholt, B. (2004) High-throughput measurement of the enantiomeric excess of chiral alcohols by using two enzymes. *Angewandte Chemie – International Edition*, **43**, 1698–702.

7 Dey, S., Karukurichi, K.R., Shen, W.J. and Berkowitz, D.B. (2005) Double-cuvette ISES: in situ estimation of enantioselectivity and relative rate for catalyst screening. *Journal of the American Chemical Society*, **127**, 8610–11.

8 Hamberg, A., Lundgren, S., Penhoat, M., Moberg, C. and Hult, K. (2006) High-throughput enzymatic method for enantiomeric excess determination of O-acetylated cyanohydrins. *Journal of the American Chemical Society*, **128**, 2234–5.

9 Hamberg, A., Lundgren, S., Wingstrand, E., Moberg, C. and Hult, K. (2007) High-throughput synthesis and analysis of acylated cyanohydrins. *Chemistry – a European Journal*, **13**, 4334–41.

10 Alexeeva, M., Enright, A., Dawson, M.J., Mahmoudian, M. and Turner, N.J. (2002) Deracemization of alpha-methylbenzylamine using an enzyme obtained by in vitro evolution. *Angewandte Chemie – International Edition*, **41**, 3177–80.

11 Carr, R., Alexeeva, M., Enright, A., Eve, T.S.C., Dawson, M.J. and Turner, N.J. (2003) Directed evolution of an amine oxidase possessing both broad substrate specificity and high enantioselectivity. *Angewandte Chemie – International Edition*, **42**, 4807–10.

12 Joo, H., Arisawa, A., Lin, Z.L. and Arnold, F.H. (1999) A high-throughput digital imaging screen for the discovery and directed evolution of oxygenases. *Chemistry and Biology*, **6**, 699–706.

13 Joo, H., Lin, Z.L. and Arnold, F.H. (1999) Laboratory evolution of peroxide-mediated cytochrome P450 hydroxylation. *Nature*, **399**, 670–3.

14 Avila, C., Kornilayev, B.A. and Blagg, B.S.J. (2006) Development and optimization of a useful assay for determining Hsp90's inherent ATPase activity. *Bioorganic and Medicinal Chemistry*, **14**, 1134–42.

15 Dicioccio, R.A., Piskorz, C., Salamida, G., Barlow, J.J. and Matta, K.L. (1981) Synthesis and use of para-nitrophenyl-2-O-(alpha-L-fucopyranosyl)-beta-D-galactopyranoside for the rapid detection of substrate-specific alpha-L-fucosidases. *Analytical Biochemistry*, **111**, 176–83.

16 Vankayalapati, H. and Singh, G. (1999) Synthesis of fucosidase substrates using propane-1,3-diyl phosphate as the anomeric leaving group. *Tetrahedron Letters*, **40**, 3925–8.

17 Indurugalla, D., Watson, J.N. and Bennet, A.J. (2006) Natural sialoside analogues for the determination of enzymatic rate constants. *Organic and Biomolecular Chemistry*, **4**, 4453–9.

18 Mayer, C., Jakeman, D.L., Mah, M., Karjala, G., Gal, L., Warren, R.A.J. and Withers, S.G. (2001) Directed evolution of new glycosynthases from *Agrobacterium* beta-glucosidase: a general screen to detect enzymes for oligosaccharide synthesis. *Chemistry and Biology*, **8**, 437–43.

19 Garciaecheverria, C., Kofron, J.L., Kuzmic, P., Kishore, V. and Rich, D.H. (1992) Continuous fluorometric direct (uncoupled) assay for peptidyl prolyl cis-trans-isomerases. *Journal of the American Chemical Society*, **114**, 2758–9.

20 Kofron, J.L., Kuzmic, P., Kishore, V., Colonbonilla, E. and Rich, D.H. (1991) Determination of kinetic constants for peptidyl prolyl cis trans isomerases by an improved spectrophotometric assay. *Biochemistry*, **30**, 6127–34.

21 Ylikauhaluoma, J.T., Ashley, J.A., Lo, C.H.L., Coakley, J., Wirsching, P. and Janda, K.D. (1996) Catalytic antibodies with peptidyl-prolyl cis-trans isomerase activity. *Journal of the American Chemical Society*, **118**, 5496–7.

22 Kupcho, K., Somberg, R., Bulleit, B. and Goueli, S.A. (2003) A homogeneous, nonradioactive high-throughput fluorogenic protein kinase assay. *Analytical Biochemistry*, **317**, 210–17.

23 Rodems, S.M., Hamman, B.D., Lin, C., Zhao, J., Shah, S., Heidary, D., Makings, L., Stack, J.H. and Pollok, B.A. (2002) A FRET-based assay platform for ultra-high density drug screening of protein kinases and phosphatases. *Assay and Drug Development Technologies*, **1**, 9–19.

24 Molin, O., Nilsson, L. and Ansehn, S. (1983) Rapid detection of bacterial-growth in blood cultures by bioluminescent assay of bacterial ATP. *Journal of Clinical Microbiology*, **18**, 521–5.

25 Shulman, H., Eberhard, A., Eberhard, C., Ulitzur, S. and Keinan, E. (2000) Highly sensitive and rapid detection of antibody catalysis by luminescent bacteria. *Bioorganic and Medicinal Chemistry Letters*, **10**, 2353–6.

26 Krebs, G., Hugonet, L. and Sutherland, J.D. (2006) Substrate ambiguity and catalytic promiscuity within a bacterial proteome probed by an easy phenotypic screen for aldehydes. *Angewandte Chemie – International Edition*, **45**, 301–5.

27 Galam, L., Hadden, M.K., Ma, Z.Q., Ye, Q.Z., Yun, B.G., Blagg, B.S.J. and Matts, R.L. (2007) High-throughput assay for the identification of Hsp90 inhibitors based on Hsp90-dependent refolding of firefly luciferase. *Bioorganic and Medicinal Chemistry*, **15**, 1939–46.

28 Oberthur, C., Graf, H. and Hamburger, M. (2004) The content of indigo precursors in *Isatis tinctoria* leaves – a comparative study of selected accessions and post-harvest treatments. *Phytochemistry*, **65**, 3261–8.

29 Nottbohm, A.C., Dothager, R.S., Putt, K.S., Hoyt, M.T. and Hergenrother, P.J. (2007) A colorimetric substrate for poly(ADP-ribose) polymerase-1, VPARP, and tankyrase-1. *Angewandte Chemie – International Edition*, **46**, 2066–9.

30 Chen, H.M. and Withers, S.G. (2007) Facile synthesis of 2,4-dinitrophenyl alpha-D-glycopyranosides as chromogenic substrates for alpha-glycosidases. *ChemBioChem*, **8**, 719–22.

31 Kim, E.J., Perreira, M., Thomas, C.J. and Hanover, J.A. (2006) An O-GlcNAcase-specific inhibitor and substrate engineered by the extension of the N-acetyl moiety. *Journal of the American Chemical Society*, **128**, 4234–5.

32 Mugherli, L., Burchak, O.N., Chatelain, F. and Balakirev, M.Y. (2006) Fluorogenic ester substrates to assess proteolytic activity. *Bioorganic and Medicinal Chemistry Letters*, **16**, 4488–91.

33 Kass, L. (1979) Cytochemistry of esterases. *CRC Critical Reviews in Clinical Laboratory Sciences*, **10**, 205–23.

34 Briseno-Roa, L., Hill, J., Notman, S., Sellers, D., Smith, A.P., Timperley, C.M., Wetherell, J., Williams, N.H., Williams, G.R., Fersht, A.R. and Griffiths, A.D. (2006) Analogues with fluorescent leaving groups for screening and selection of enzymes that efficiently hydrolyze organophosphorus nerve agents. *Journal of Medicinal Chemistry*, **49**, 246–55.

35 Farinas, E.T., Schwaneberg, U., Glieder, A. and Arnold, F.H. (2001) Directed evolution of a cytochrome P450 monooxygenase for alkane oxidation. *Advanced Synthesis and Catalysis*, **343**, 601–6.

36 Celik, A., Roberts, G.A., White, J.H., Chapman, S.K., Turner, N.J. and Flitsch, S.L. (2006) Probing the substrate specificity of the catalytically self-sufficient cytochrome P450RhF from a *Rhodococcus* sp. *Chemical Communications*, 4492–4.

37 Celik, A., Speight, R.E. and Turner, N.J. (2005) Identification of broad specificity P450(CAM) variants by primary screening against indole as substrate. *Chemical Communications*, 3652–4.

38 Jourdain, N., Carlon, R.P. and Reymond, J.L. (1998) A stereoselective fluorogenic assay for aldolases: detection of an anti-selective aldolase catalytic antibody. *Tetrahedron Letters*, **39**, 9415–18.

39 Perez Carlon, R., Jourdain, N. and Reymond, J.L. (2000) Fluorogenic polypropionate fragments for detecting stereoselective aldolases. *Chemistry*, **6**, 4154–62.

40 List, B., Barbas, C.F. and Lerner, R.A. (1998) Aldol sensors for the rapid generation of tunable fluorescence by antibody catalysis. *Proceedings of the National Academy of Sciences of the United States of America*, **95**, 15351–5.

41 Gonzalez-Garcia, E., Helaine, V., Klein, G., Schuermann, M., Sprenger, G.A., Fessner, W.D. and Reymond, J.L. (2003) Fluorogenic stereochemical probes for transaldolases. *Chemistry – A European Journal*, **9**, 893–9.

42 Sevestre, A., Helaine, V., Guyot, G., Martin, C. and Hecquet, L. (2003) A fluorogenic assay for transketolase from *Saccharomyces cerevisiae*. *Tetrahedron Letters*, **44**, 827–30.

43 Gao, W.Z., Xing, B.G., Tsien, R.Y. and Rao, J.H. (2003) Novel fluorogenic substrates for imaging 6-lactamase gene expression. *Journal of the American Chemical Society*, **125**, 11146–7.

44 Leroy, E., Bensel, N. and Reymond, J.L. (2003) Fluorogenic cyanohydrin esters as chiral probes for esterase and lipase activity. *Advanced Synthesis and Catalysis*, **345**, 859–65.

45 Zhou, W.H., Valley, M.P., Shultz, J., Hawkins, E.M., Bernad, L., Good, T., Good, D., Riss, T.L., Klaubert, D.H. and Wood, K.V. (2006) New bioluminogenic substrates for monoamine oxidase assays. *Journal of the American Chemical Society*, **128**, 3122–3.

46 Kofoed, J., Darbre, T. and Reymond, J.L. (2006) Dual mechanism of zinc-proline catalyzed aldol reactions in water. *Chemical Communications (Cambridge, England)*, 1482–4.

47 Kofoed, J., Darbre, T. and Reymond, J.L. (2006) Artificial aldolases from peptide dendrimer combinatorial libraries. *Organic and Biomolecular Chemistry*, **4**, 3268–81.

48 Bensel, N., Reymond, M.T. and Reymond, J.L. (2001) Pivalase catalytic antibodies: towards abzymatic activation of prodrugs. *Chemistry*, **7**, 4604–12.

49 Leroy, E., Bensel, N. and Reymond, J.L. (2003) A low background high-throughput screening (HTS) fluorescence assay for lipases and esterases using acyloxymethylethers of umbelliferone. *Bioorganic and Medicinal Chemistry Letters*, **13**, 2105–8.

50 Yang, Y.Z., Babiak, P. and Reymond, J.L. (2006) New monofunctionalized fluorescein derivatives for the efficient high-throughput screening of lipases and esterases in aqueous media. *Helvetica Chimica Acta*, **89**, 404–15.

51 Sicard, R., Chen, L.S., Marsaioli, A.J. and Reymond, J.L. (2005) A fluorescence-based assay for Baeyer-Villiger monooxygenases, hydroxylases and lactonases. *Advanced Synthesis and Catalysis*, **347**, 1041–50.

52 Badalassi, F., Wahler, D., Klein, G., Crotti, P. and Reymond, J.L. (2000) A versatile periodate-coupled fluorogenic assay for hydrolytic enzymes. *Angewandte Chemie – International Edition*, **39**, 4067–70.

53 Nyfeler, E., Grognux, J., Wahler, D. and Reymond, J.L. (2003) A sensitive and selective high-throughput screening fluorescence assay for lipases and esterases. *Helvetica Chimica Acta*, **86**, 2919–27.

54 Gonzalez-Garcia, E.M., Grognux, J., Wahler, D. and Reymond, J.L. (2003) Synthesis and evaluation of chromogenic and fluorogenic analogs of glycerol for enzyme assays. *Helvetica Chimica Acta*, **86**, 2458–70.

55 Grognux, J. and Reymond, J.L. (2004) Classifying enzymes from selectivity fingerprints. *ChemBioChem*, **5**, 826–31.

56 Lagarde, D., Nguyen, H.K., Ravot, G., Wahler, D., Reymond, J.L., Hills, G., Veit, T. and Lefevre, F. (2002) High-throughput screening of thermostable esterases for industrial bioconversions. *Organic Process Research and Development*, **6**, 441–5.

57 Bicalho, B., Chen, L.S., Grognux, J., Reymond, J.L. and Marsaioli, A.J. (2004) Studies on whole cell fluorescence-based screening for epoxide hydrolases and Baeyer-Villiger monooxygenases. *Journal of the Brazilian Chemical Society*, **15**, 911–16.

58 Badalassi, F., Nguyen, H.K., Crotti, P. and Reymond, J.L. (2002) A selective HIV-protease assay based on a chromogenic amino acid. *Helvetica Chimica Acta*, **85**, 3090–8.

59 Bedia, C., Casas, J., Garcia, V., Levade, T. and Fabrias, G. (2007) Synthesis of a novel ceramide analogue and its use in a high-throughput fluorogenic assay for ceramidases. *ChemBioChem*, **8**, 642–8.

60 Jones, P.D., Wolf, N.M., Morisseau, C., Whetstone, P., Hock, B. and Hammock, B.D. (2005) Fluorescent substrates for soluble epoxide hydrolase and application to inhibition studies. *Analytical Biochemistry*, **343**, 66–75.

61 Amir, R.J. and Shabat, D. (2004) Self-immolative dendrimer biodegradability by multi-enzymatic triggering. *Chemical Communications*, 1614–15.

62 Shamis, M., Barbas, C.F. and Shabat, D. (2007) A new visual screening assay for catalytic antibodies with retro-aldol retro-Michael activity. *Bioorganic and Medicinal Chemistry Letters*, **17**, 1172–5.

63 James, A.L., Perry, J.D., Rigby, A. and Stanforth, S.P. (2007) Synthesis and evaluation of novel chromogenic aminopeptidase substrates for microorganism detection and identification. *Bioorganic and Medicinal Chemistry Letters*, **17**, 1418–21.

64 Wang, Z.Q., Liao, J.F. and Diwu, Z.J. (2005) N-DEVD-N′-morpholinecarbonyl-rhodamine 110: novel caspase-3 fluorogenic substrates for cell-based apoptosis assay. *Bioorganic and Medicinal Chemistry Letters*, **15**, 2335–8.

65 Jones, G.B., Crasto, C.F., Mathews, J.E., Xie, L.F., Mitchell, M.O., El-Shafey, A., D'Amico, A.V. and Bubley, G.J. (2006) An image contrast agent selectively activated

by prostate specific antigen. *Bioorganic & Medicinal Chemistry*, **14**, 418–25.

66 Ho, N.H., Weissleder, R. and Tung, C.H. (2007) A self-immolative reporter for beta-galactosidase sensing. *ChemBioChem*, **8**, 560–6.

67 Chandran, S.S., Dickson, K.A. and Raines, R.T. (2005) Latent fluorophore based on the trimethyl lock. *Journal of the American Chemical Society*, **127**, 1652–3.

68 Huang, S.T. and Lin, Y.L. (2006) New latent fluorophore for DT diaphorase. *Organic Letters*, **8**, 265–8.

69 Matayoshi, E.D., Wang, G.T., Krafft, G.A. and Erickson, J. (1990) Novel fluorogenic substrates for assaying retroviral proteases by resonance energy-transfer. *Science*, **247**, 954–8.

70 Kainmuller, E.K., Olle, E.P. and Bannwarth, W. (2005). Synthesis of a new pair of fluorescence resonance energy transfer donor and acceptor dyes and its use in a protease assay. *Chemical Communications*, 5459–61.

71 Kainmuller, E.K. and Bannwarth, W. (2006) A new robust and highly sensitive FRET donor-acceptor pair: synthesis, characterization, and application in a thrombin assay. *Helvetica Chimica Acta*, **89**, 3056–70.

72 Warfield, R., Bardelang, P., Saunders, H., Chan, W.C., Penfold, C., James, R. and Thomas, N.R. (2006) Internally quenched peptides for the study of lysostaphin: an antimicrobial protease that kills *Staphylococcus aureus*. *Organic and Biomolecular Chemistry*, **4**, 3626–38.

73 Boyer, V., Fort, S., Frandsen, T.P., Schulein, M., Cottaz, S. and Driguez, H. (2002) Chemoenzymatic synthesis of a bifunctionalized cellohexaoside as a specific substrate for the sensitive assay of cellulase by fluorescence quenching. *Chemistry–A European Journal*, **8**, 1389–94.

74 Farber, S.A., Pack, M., Ho, S.Y., Johnson, L.D., Wagner, D.S., Dosch, R., Mullins, M.C., Hendrickson, H.S., Hendrickson, E.K. and Halpern, M.E. (2001) Genetic analysis of digestive physiology using fluorescent phospholipid reporters. *Science*, **292**, 1385–8.

75 Wichmann, O., Wittbrodt, J. and Schultz, C. (2006) A small-molecule FRET probe to monitor phospholipase A(2) activity in cells and organisms. *Angewandte Chemie–International Edition*, **45**, 508–12.

76 Ferguson, C.G., Bigman, C.S., Richardson, R.D., van Meeteren, L.A., Moolenaar, W.H. and Prestwich, G.D. (2006) Fluorogenic phospholipid substrate to detect lysophospholipase D/autotaxin activity. *Organic Letters*, **8**, 2023–6.

77 Rose, T.M. and Prestwich, G.D. (2006) Synthesis and evaluation of fluorogenic substrates for phospholipase D and phospholipase C. *Organic Letters*, **8**, 2575–8.

78 Zandonella, G., Haalck, L., Spener, F., Faber, K., Paltauf, F. and Hermetter, A. (1996) Enantiomeric perylene-glycerolipids as fluorogenic substrates for a dual wavelength assay of lipase activity and stereoselectivity. *Chirality*, **8**, 481–9.

79 Duque, M., Graupner, M., Stutz, H., Wicher, I., Zechner, R., Paltauf, F. and Hermetter, A. (1996) New fluorogenic triacylglycerol analogs as substrates for the determination and chiral discrimination of lipase activities. *Journal of Lipid Research*, **37**, 868–76.

80 Yang, Y.Z., Babiak, P. and Reymond, J.L. (2006) Low background FRET-substrates for lipases and esterases suitable for high-throughput screening under basic (pH 11) conditions. *Organic and Biomolecular Chemistry*, **4**, 1746–54.

81 Tanaka, F., Thayumanavan, R. and Barbas, C.F. (2003) Fluorescent detection of carbon-carbon bond formation. *Journal of the American Chemical Society*, **125**, 8523–8.

82 Tanaka, F., Mase, N. and Barbas, C.F. (2004) Design and use of fluorogenic aldehydes for monitoring the progress of aldehyde transformations. *Journal of the American Chemical Society*, **126**, 3692–3.

83 Hennig, A., Roth, D., Enderle, T. and Nau, W.M. (2006) Nanosecond time-resolved fluorescence protease assays. *ChemBioChem*, **7**, 733–7.

84 Sahoo, H. and Nau, W.M. (2007) Phosphorylation-induced conformational changes in short peptides probed by

fluorescence resonance energy transfer in the 10 angstrom domain. *ChemBioChem*, **8**, 567–73.
85 Lawrence, D.S. and Wang, Q.Z. (2007) Seeing is believing: peptide-based fluorescent sensors of protein tyrosine kinase activity. *Chembiochem*, **8**, 373–8.
86 Wang, Q.Z., Cahill, S.M., Blumenstein, M. and Lawrence, D.S. (2006) Self-reporting fluorescent substrates of protein tyrosine kinases. *Journal of the American Chemical Society*, **128**, 1808–9.
87 Sharma, V., Agnes, R.S. and Lawrence, D.S. (2007) Deep quench: an expanded dynamic range for protein kinase sensors. *Journal of the American Chemical Society*, **129**, 2742–3.
88 Murayama, T., Tanabe, T., Ikeda, H. and Ueno, A. (2006) Direct assay for alpha-amylase using fluorophore-modified cyclodextrins. *Bioorganic and Medicinal Chemistry*, **14**, 3691–6.
89 Xing, B., Khanamiryan, A. and Rao, J.H. (2005) Cell-permeable near-infrared fluorogenic substrates for imaging beta-lactamase activity. *Journal of the American Chemical Society*, **127**, 4158–9.
90 Kohl, T., Heinze, K.G., Kuhlemann, R., Koltermann, A. and Schwille, P. (2002) A protease assay for two-photon crosscorrelation and FRET analysis based solely on fluorescent proteins. *Proceedings of the National Academy of Sciences of the United States of America*, **99**, 12161–6.
91 Wosnick, J.H., Mello, C.M. and Swager, T.M. (2005) Synthesis and application of poly(phenylene ethynylene)s for bioconjugation: a conjugated polymer-based fluorogenic probe for proteases. *Journal of the American Chemical Society*, **127**, 3400–5.
92 Rosseneu, M., Taveirne, M.J., Caster, H. and Vanbiervliet, J.P. (1985) Hydrolysis of very-low-density lipoproteins labeled with a fluorescent triacylglycerol – 1,3-Dioleoyl-2-(4-Pyrenyl butanoyl)Glycerol. *European Journal of Biochemistry*, **152**, 195–8.
93 Twining, S.S. (1984) Fluorescein isothiocyanate-labeled casein assay for proteolytic-enzymes. *Analytical Biochemistry*, **143**, 30–4.
94 James, A.L., Perry, J.D., Jay, C., Monget, D., Rasburn, J.W. and Gould, F.K. (2001) Fluorogenic substrates for the detection of microbial nitroreductases. *Letters in Applied Microbiology*, **33**, 403–8.
95 Chen, G., Yee, D.J., Gubernator, N.G. and Sames, D. (2005) Design of optical switches as metabolic indicators: new fluorogenic probes for monoamine oxidases (MAO A and B). *Journal of the American Chemical Society*, **127**, 4544–5.
96 Froemming, M.K. and Sames, D. (2006) Fluoromorphic substrates for fatty acid metabolism: Highly sensitive probes for mammalian medium-chain acyl-CoA dehydrogenase. *Angewandte Chemie – International Edition*, **45**, 637–42.
97 Reymond, J.L., Koch, T., Schroer, J. and Tierney, E. (1996) A general assay for antibody catalysis using acridone as a fluorescent tag. *Proceedings of the National Academy of Sciences of the United States of America*, **93**, 4251–6.
98 Aharoni, A., Thieme, K., Chiu, C.P., Buchini, S., Lairson, L.L., Chen, H., Strynadka, N.C., Wakarchuk, W.W. and Withers, S.G. (2006) High-throughput screening methodology for the directed evolution of glycosyltransferases. *Nature Methods*, **3**, 609–14.
99 Landry, D.W., Zhao, K., Yang, G.X.Q., Glickman, M. and Georgiadis, T.M. (1993) Antibody-catalyzed degradation of cocaine. *Science*, **259**, 1899–901.
100 Yang, G., Chun, J., ArakawaUramoto, H., Wang, X., Gawinowicz, M.A., Zhao, K. and Landry, D.W. (1996) Anti-cocaine catalytic antibodies: a synthetic approach to improved antibody diversity. *Journal of the American Chemical Society*, **118**, 5881–90.
101 Liu, Y.S., Patricelli, M.P. and Cravatt, B.F. (1999) Activity-based protein profiling: The serine hydrolases. *Proceedings of the National Academy of Sciences of the United States of America*, **96**, 14694–9.
102 Li, W., Blankman, J.L. and Cravatt, B.F. (2007) A functional proteomic strategy to discover inhibitors for uncharacterized hydrolases. *Journal of the American Chemical Society*, **129**, 9594–5.
103 Greenbaum, D., Medzihradszky, K.F., Burlingame, A. and Bogyo, M. (2000) Epoxide electrophiles as activity-

dependent cysteine protease profiling and discovery tools. *Chemistry and Biology*, **7**, 569–81.

104 Schmidinger, H., Birner-Gruenberger, R., Riesenhuber, G., Saf, R., Susani-Etzerodt, H. and Hermetter, A. (2005) Novel fluorescent phosphonic acid esters for discrimination of lipases and esterases. *ChemBioChem*, **6**, 1776–81.

105 Montenecourt, B.S. and Eveleigh, D.E. (1977) Semiquantitative plate assay for determination of cellulase production by *Trichoderma viride*. *Applied and Environmental Microbiology*, **33**, 178–83.

106 Yang, Z. and Xu, B. (2004) A simple visual assay based on small molecule hydrogels for detecting inhibitors of enzymes. *Chemical Communications (Cambridge, England)*, 2424–5.

107 Yang, Z., Ho, P.L., Liang, G., Chow, K.H., Wang, Q., Cao, Y., Guo, Z. and Xu, B. (2007) Using beta-lactamase to trigger supramolecular hydrogelation. *Journal of the American Chemical Society*, **129**, 266–7.

108 Adler-Abramovich, L., Perry, R., Sagi, A., Gazit, E. and Shabat, D. (2007) Controlled assembly of peptide nanotubes triggered by enzymatic activation of self-immolative dendrimers. *ChemBioChem*, **8**, 859–62.

109 Janes, L.E. and Kazlauskas, R.J. (1997) Quick E. A fast spectrophotometric method to measure the enantioselectivity of hydrolases. *Journal of Organic Chemistry*, **62**, 4560–1.

110 Janes, L.E., Lowendahl, A.C. and Kazlauskas, R.J. (1998) Quantitative screening of hydrolase libraries using pH indicators: identifying active and enantioselective hydrolases. *Chemistry – A European Journal*, **4**, 2324–31.

111 Park, C.B. and Clark, D.S. (2002) Sol-gel encapsulated enzyme arrays for high-throughput screening of biocatalytic activity. *Biotechnology and Bioengineering*, **78**, 229–35.

112 Taylor, S.J.C., Brown, R.C., Keene, P.A. and Taylor, I.N. (1999) Novel screening methods – the key to cloning commercially successful biocatalysts. *Bioorganic and Medicinal Chemistry*, **7**, 2163–8.

113 Henke, E. and Bornscheuer, U.T. (2003) Fluorophoric assay for the high-throughput determination of amidase activity. *Analytical Chemistry*, **75**, 255–60.

114 Banerjee, A., Sharma, R. and Banerjee, U.C. (2003) A rapid and sensitive fluorometric assay method for the determination of nitrilase activity. *Biotechnology and Applied Biochemistry*, **37**, 289–93.

115 Klein, G., Kaufmann, D., Schurch, S. and Reymond, J.L. (2001) A fluorescent metal sensor based on macrocyclic chelation. *Chemical Communications*, 561–2.

116 Dean, K.E.S., Klein, G., Renaudet, O. and Reymond, J.L. (2003) A green fluorescent chemosensor for amino acids provides a versatile high-throughput screening (HTS) assay for proteases. *Bioorganic and Medicinal Chemistry Letters*, **13**, 1653–6.

117 Duchateau, A.L.L., Hillemans-Crombach, M.G., van Duijnhoven, A., Reiss, R. and Sonke, T. (2004) A colorimetric method for determination of amino amidase activity. *Analytical Biochemistry*, **330**, 362–4.

118 Humphrey, C.E., Easson, M.A.M. and Turner, N.J. (2004) Dimedone esters as novel hydrolase substrates and their application in the colorimetric detection of lipase and esterase activity. *ChemBioChem*, **5**, 1144–8.

119 Konarzycka-Bessler, M. and Bornscheuer, U.T. (2003) A high-throughput-screening method for determining the synthetic activity of hydrolases. *Angewandte Chemie – International Edition in English*, **42**, 1418–20.

120 Salahuddin, S., Renaudet, O. and Reymond, J.L. (2004) Aldehyde detection by chromogenic/fluorogenic oxime bond fragmentation. *Organic and Biomolecular Chemistry*, **2**, 1471–5.

121 Wahler, D., Boujard, O., Lefevre, F. and Reymond, J.L. (2004) Adrenaline profiling of lipases and esterases with 1,2-diol and carbohydrate acetates. *Tetrahedron*, **60**, 703–10.

122 Wahler, D. and Reymond, J.L. (2002) The adrenaline test for enzymes. *Angewandte Chemie – International Edition in English*, **41**, 1229–32.

123 Mateo, C., Archelas, A. and Furstoss, R. (2003) A spectrophotometric assay for measuring and detecting an epoxide hydrolase activity. *Analytical Biochemistry*, **314**, 135–41.

124 Doderer, K., Lutz-Wahl, S., Hauer, B. and Schmid, R.D. (2003) Spectrophotometric assay for epoxide hydrolase activity toward any epoxide. *Analytical Biochemistry*, **321**, 131–4.

125 Zocher, F., Enzelberger, M.M., Bornscheuer, U.T., Hauer, B. and Schmid, R.D. (1999) A colorimetric assay suitable for screening epoxide hydrolase activity. *Analytica Chimica Acta*, **391**, 345–51.

126 Maeda, H., Matsuno, H., Ushida, M., Katayama, K., Saeki, K. and Itoh, N. (2005) 2,4-dinitrobenzenesulfonyl fluoresceins as fluorescent alternatives to Ellman's reagent in thiol-quantification enzyme assays. *Angewandte Chemie – International Edition*, **44**, 2922–5.

127 Khersonsky, O. and Tawfik, D.S. (2006) Chromogenic and fluorogenic assays for the lactonase activity of serum paraoxonases. *Chembiochem*, **7**, 49–53.

128 Ojida, A., Mito-oka, Y., Inoue, M. and Hamachi, I. (2002) First artificial receptors and chemosensors toward phosphorylated peptide in aqueous solution. *Journal of the American Chemical Society*, **124**, 6256–8.

129 Yamaguchi, S., Yoshimura, L., Kohira, T., Tamaru, S. and Hamachi, I. (2005) Cooperation between artificial receptors and supramolecular hydrogels for sensing and discriminating phosphate derivatives. *Journal of the American Chemical Society*, **127**, 11835–41.

130 Martin, K., Steinberg, T.H., Cooley, L.A., Gee, K.R., Beechem, J.M. and Patton, W.F. (2003) Quantitative analysis of protein phosphorylation status and protein kinase activity on microarrays using a novel fluorescent phosphorylation sensor dye. *Proteomics*, **3**, 1244–55.

131 Wongkongkatep, J., Miyahara, Y., Ojida, A. and Hamachi, I. (2006) Label-free, real-time glycosyltransferase assay based on a fluorescent artificial chemosensor. *Angewandte Chemie – International Edition*, **45**, 665–8.

132 Tawfik, D.S., Green, B.S., Chap, R., Sela, M. and Eshhar, Z. (1993) catELISA: a facile general route to catalytic antibodies. *Proceedings of the National Academy of Sciences of the United States of America*, **90**, 373–7.

133 Geymayer, P., Bahr, N. and Reymond, J.L. (1999) A general fluorogenic assay for catalysis using antibody sensors. *Chemistry – a European Journal*, **5**, 1006–12.

134 Taran, F., Renard, P.Y., Creminon, C., Valleix, A., Frobert, Y., Pradelles, P., Grassi, J. and Mioskowski, C. (1999) Competitive immunoassay (Cat-EIA), a helpful technique for catalytic antibody detection. Part I. *Tetrahedron Letters*, **40**, 1887–90.

135 Taran, F., Renard, P.Y., Creminon, C., Valleix, A., Frobert, Y., Pradelles, P., Grassi, J. and Mioskowski, C. (1999) Competitive immunoassay (Cat-EIA), a helpful technique for catalytic antibody detection. Part II. *Tetrahedron Letters*, **40**, 1891–4.

136 Vicennati, P., Bensel, N., Wagner, A., Creminon, C. and Taran, F. (2005) Sandwich immunoassay as a high-throughput screening method for cross-coupling reactions. *Angewandte Chemie – International Edition*, **44**, 6863–6.

137 Nutiu, R., Yu, J.M.Y. and Li, Y.F. (2004) Signaling aptamers for monitoring enzymatic activity and for inhibitor screening. *Chembiochem*, **5**, 1139–44.

138 Zatta, P.F., Nyame, K., Cormier, M.J., Mattox, S.A., Prieto, P.A., Smith, D.F. and Cummings, R.D. (1991) A solid-phase assay for Beta-1,4-Galactosyltransferase activity in human serum using recombinant aequorin. *Analytical Biochemistry*, **194**, 185–91.

139 Khraltsova, L.S., Sablina, M.A., Melikhova, T.D., Joziasse, D.H., Kaltner, H., Gabius, H.J. and Bovin, N.V. (2000) An enzyme-linked lectin assay for alpha 1,3-galactosyltransferase. *Analytical Biochemistry*, **280**, 250–7.

140 Bryan, M.C., Lee, L.V. and Wong, C.H. (2004) High-throughput identification of fucosyltransferase inhibitors using carbohydrate microarrays. *Bioorganic and Medicinal Chemistry Letters*, **14**, 3185–8.

141 Bryan, M.C., Plettenburg, O., Sears, P., Rabuka, D., Wacowich-Sgarbi, S. and Wong, C.H. (2002) Saccharide display on microtiter plates. *Chemistry & Biology*, **9**, 713–20.

142 Xiao, Y., Pavlov, V., Levine, S., Niazov, T., Markovitch, G. and Willner, I. (2004) Catalytic growth of Au nanoparticles by NAD(P)H cofactors: optical sensors for $NAD(P)^+$-dependent biocatalyzed transformations. *Angewandte Chemie – International Edition*, **43**, 4519–22.

143 Pavlov, V., Xiao, Y. and Willner, I. (2005) Inhibition of the acetycholine esterase-stimulated growth of Au nanoparticles: nanotechnology-based sensing of nerve gases. *Nano Letters*, **5**, 649–53.

144 Baron, R., Zayats, M. and Willner, I. (2005) Dopamine-, L-DOPA-, adrenaline-, and noradrenaline-induced growth of Au nanoparticles: assays for the detection of neurotransmitters and of tyrosinase activity. *Analytical Chemistry*, **77**, 1566–71.

145 Wang, Z., Lee, J., Cossins, A.R. and Brust, M. (2005) Microarray-based detection of protein binding and functionality by gold nanoparticle probes. *Analytical Chemistry*, **77**, 5770–4.

146 Wang, Z., Levy, R., Fernig, D.G. and Brust, M. (2006) Kinase-catalyzed modification of gold nanoparticles: a new approach to colorimetric kinase activity screening. *Journal of the American Chemical Society*, **128**, 2214–15.

147 Guarise, C., Pasquato, L., De Filippis, V. and Scrimin, P. (2006) Gold nanoparticles-based protease assay. *Proceedings of the National Academy of Sciences of the United States of America*, **103**, 3978–82.

148 Choi, Y., Ho, N.H. and Tung, C.H. (2006) Sensing phosphatase activity by using gold nanoparticles. *Angewandte Chemie – International Edition in English*, **46**, 707–9.

149 Xu, X., Han, M.S. and Mirkin, C.A. (2007) A gold-nanoparticle-based real-time colorimetric screening method for endonuclease activity and inhibition. *Angewandte Chemie – International Edition in English*, **46**, 3468–70.

150 Reymond, J.L. and Wahler, D. (2002) Substrate arrays as enzyme fingerprinting tools. *ChemBioChem*, **3**, 701–8.

151 Liu, A.M.F., Somers, N.A., Kazlauskas, R.J., Brush, T.S., Zocher, F., Enzelberger, M.M., Bornscheuer, U.T., Horsman, G. P., Mezzetti, A., Schmidt-Dannert, C. and Schmid, R.D. (2001) Mapping the substrate selectivity of new hydrolases using colorimetric screening: lipases from *Bacillus thermocatenulatus* and *Ophiostoma piliferum*, esterases from *Pseudomonas fluorescens* and *Streptomyces diastatochromogenes*. *Tetrahedron-Asymmetry*, **12**, 545–56.

152 Wahler, D., Badalassi, F., Crotti, P. and Reymond, J.L. (2001) Enzyme fingerprints by fluorogenic and chromogenic substrate arrays. *Angewandte Chemie – International Edition*, **40**, 4457–60.

153 Wahler, D., Badalassi, F., Crotti, P. and Reymond, J.L. (2002) Enzyme fingerprints of activity, and stereo- and enantioselectivity from fluorogenic and chromogenic substrate arrays. *Chemistry*, **8**, 3211–28.

154 Buissier, J., Fourcard, A. and Colobert, L. (1967). Usage de substrats synthetiques pour l'étude de l'equipement enzymatique de microorganismes. *Comptes Rendus Hebdomadaires des Seances de l'Academie des Sciences Serie D*, **264**, 415.

155 Nardon, P., Monget, D., Didierfichet, M.L. and Dethe, G. (1976) Comparison of zymogram of 3 lymphoblastoid cell lines with a new microtechnique. *Biomedicine*, **24**, 183–90.

156 Humble, M.W., King, A. and Phillips, I. (1977) Api Zym – simple rapid system for detection of bacterial enzymes. *Journal of Clinical Pathology*, **30**, 275–7.

157 Gruner, E., Vongraevenitz, A. and Altwegg, M. (1992) The Api Zym System – a tabulated review from 1977 to date. *Journal of Microbiological Methods*, *16*, 101–18.

158 Garcia-Martos, P., Marin, P., Hernandez-Molina, J.M., Garcia-Agudo, L., Aoufi, S. and Mira, J. (2001) Extracellular

enzymatic activity in 11 Cryptococcus species. *Mycopathologia*, **150**, 1–4.
159 Manafi, M., Kneifel, W. and Bascomb, S. (1991) Fluorogenic and chromogenic substrates used in bacterial diagnostics. *Microbiological Reviews*, **55**, 335–48.
160 Powers, J.C., Asgian, J.L., Ekici, O.D. and James, K.E. (2002) Irreversible inhibitors of serine, cysteine, and threonine proteases. *Chemical Reviews*, **102**, 4639–750.
161 Meldal, M. (2005) Smart combinatorial assays for the determination of protease activity and inhibition. *QSAR and Combinatorial Science*, **24**, 1141–8.
162 Puente, X.S., Sanchez, L.M., Overall, C.M. and Lopez-Otin, C. (2003) Human and mouse proteases: a comparative genomic approach. *Nature Reviews Genetics*, **4**, 544–58.
163 Richardson, P.L. (2002) The determination and use of optimized protease substrates in drug discovery and development. *Current Pharmaceutical Design*, **8**, 2559–81.
164 Maly, D.J., Huang, L. and Ellman, J.A. (2002) Combinatorial strategies for targeting protein families: application to the proteases. *Chembiochem*, **3**, 17–37.
165 Matthews, D.J. and Wells, J.A. (1993) Substrate phage–selection of protease substrates by monovalent phage display. *Science*, **260**, 1113–17.
166 Chaparro-Riggers, J.F., Breves, R., Maurer, K.H. and Bornscheuer, U. (2006) Modulation of infectivity in phage display as a tool to determine the substrate specificity of proteases. *Chembiochem*, **7**, 965–70.
167 Breddam, K. and Meldal, M. (1992) Substrate preferences of glutamic-acid-specific endopeptidases assessed by synthetic peptide-substrates based on intramolecular fluorescence quenching. *European Journal of Biochemistry*, **206**, 103–7.
168 Meldal, M., Svendsen, I., Breddam, K. and Auzanneau, F.I. (1994) Portion-mixing peptide libraries of quenched fluorogenic substrates for complete subsite mapping of endoprotease specificity. *Proceedings of the National Academy of Sciences of the United States of America*, **91**, 3314–18.
169 St Hilaire, P.M., Willert, M., Juliano, M.A., Juliano, L. and Meldal, M. (1999) Fluorescence-quenched solid phase combinatorial libraries in the characterization of cysteine protease substrate specificity. *Journal of Combinatorial Chemistry*, **1**, 509–23.
170 Alves, F.M., Hirata, I.Y., Gouvea, I.E., Alves, M.F.M., Meldal, M., Bromme, D., Juliano, L. and Juliano, M.A. (2007) Controlled peptide solvation in portion-mixing libraries of FRET peptides: improved specificity determination for dengue 2 virus NS2B-NS3 protease and human cathepsin S. *Journal of Combinatorial Chemistry*, **9**, 627–34.
171 Ekici, O.D., Karla, A., Paetzel, M., Lively, M.O., Pei, D.H. and Dalbey, R.E. (2007) Altered-3 substrate specificity of Escherichia coli signal peptidase 1 mutants as revealed by screening a combinatorial peptide library. *Journal of Biological Chemistry*, **282**, 417–25.
172 Thornberry, N.A., Rano, T.A., Peterson, E.P., Rasper, D.M., Timkey, T., Garcia-Calvo, M., Houtzager, V.M., Nordstrom, P.A., Roy, S., Vaillancourt, J.P., Chapman, K.T. and Nicholson, D.W. (1997) A combinatorial approach defines specificities of members of the caspase family and granzyme B. Functional relationships established for key mediators of apoptosis. *Journal of Biological Chemistry*, **272**, 17907–11.
173 Harris, J.L., Backes, B.J., Leonetti, F., Mahrus, S., Ellman, J.A. and Craik, C.S. (2000) Rapid and general profiling of protease specificity by using combinatorial fluorogenic substrate libraries. *Proceedings of the National Academy of Sciences of the United States of America*, **97**, 7754–9.
174 Harris, J.L., Alper, P.B., Li, J., Rechsteiner, M. and Backes, B.J. (2001) Substrate specificity of the human proteasome. *Chemistry and Biology*, **8**, 1131–41.
175 Rano, T.A., Timkey, T., Peterson, E.P., Rotonda, J., Nicholson, D.W., Becker, J. W., Chapman, K.T. and Thornberry, N.A. (1997) A combinatorial approach for determining protease specificities: application to interleukin-1 beta

converting enzyme (ICE). *Chemistry and Biology*, **4**, 149–55.

176 Pinilla, C., Appel, J.R., Blanc, P. and Houghten, R.A. (1992) Rapid identification of high affinity peptide ligands using positional scanning synthetic peptide combinatorial libraries. *Biotechniques*, **13**, 901–5.

177 Kisselev, A.F., Garcia-Calvo, M., Overkleeft, H.S., Peterson, E., Pennington, M.W., Ploegh, H.L., Thornberry, N.A. and Goldberg, A.L. (2003) The caspase-like sites of proteasomes, their substrate specificity, new inhibitors and substrates, and allosteric interactions with the trypsin-like sites. *Journal of Biological Chemistry*, **278**, 35869–77.

178 Choe, Y., Leonetti, F., Greenbaum, D.C., Lecaille, F., Bogyo, M., Bromme, D., Ellman, J.A. and Craik, C.S. (2006) Substrate profiling of cysteine proteases using a combinatorial peptide library identifies functionally unique specificities. *Journal of Biological Chemistry*, **281**, 12824–32.

179 Cotrin, S.S., Puzer, L., Judice, W.A.D., Juliano, L., Carmona, A.K. and Juliano, M.A. (2004) Positional-scanning combinatorial libraries of fluorescence resonance energy transfer peptides to define substrate specificity of carboxydipeptidases: assays with human cathepsin B. *Analytical Biochemistry*, **335**, 244–52.

180 Winssinger, N., Damoiseaux, R., Tully, D.C., Geierstanger, B.H., Burdick, K. and Harris, J.L. (2004) PNA-encoded protease substrate microarrays. *Chemistry and Biology*, **11**, 1351–60.

181 Harris, J.L. and Winssinger, N. (2005) PNA encoding (PNA = peptide nucleic acid): From solution-based libraries to organized microarrays. *Chemistry–A European Journal*, **11**, 6792–801.

182 Debaene, F., Mejias, L., Harris, J.L. and Winssinger, N. (2004) Synthesis of a PNA-encoded cysteine protease inhibitor library. *Tetrahedron*, **60**, 8677–90.

183 Diaz-Mochon, J.J., Bialy, L. and Bradley, M. (2006) Dual colour, microarray-based, analysis of 10 000 protease substrates. *Chemical Communications*, 3984–6.

184 Salisbury, C.M., Maly, D.J. and Ellman, J.A. (2002) Peptide microarrays for the determination of protease substrate specificity. *Journal of the American Chemical Society*, **124**, 14868–70.

185 Zhu, Q., Uttamchandani, M., Li, D.B., Lesaicherre, M.L. and Yao, S.Q. (2003) Enzymatic profiling system in a small-molecule microarray. *Organic Letters*, **5**, 1257–60.

186 Gosalia, D.N., Salisbury, C.M., Maly, D.J., Ellman, J.A. and Diamond, S.L. (2005) Profiling serine protease substrate specificity with solution phase fluorogenic peptide microarrays. *Proteomics*, **5**, 1292–8.

187 Gosalia, D.N., Salisbury, C.M., Ellman, J.A. and Diamond, S.L. (2005) High throughput substrate specificity profiling of serine and cysteine proteases using solution-phase fluorogenic peptide microarrays. *Molecular and Cellular Proteomics*, **4**, 626–36.

188 Borodovsky, A., Ovaa, H., Meester, W.J.N., Venanzi, E.S., Bogyo, M.S., Hekking, B.G., Ploegh, H.L., Kessler, B. M. and Overkleeft, H.S. (2005) Small-molecule inhibitors and probes for ubiquitin- and ubiquitin-like-specific proteases. *Chembiochem*, **6**, 287–91.

189 Greenbaum, D.C., Arnold, W.D., Lu, F., Hayrapetian, L., Baruch, A., Krumrine, J., Toba, S., Chehade, K., Bromme, D., Kuntz, I.D. and Bogyo, M. (2002) Small molecule affinity fingerprinting: a tool for enzyme family subclassification, target identification, and inhibitor design. *Chemistry and Biology*, **9**, 1085–94.

190 Bogyo, M., Verhelst, S., Bellingard-Dubouchaud, V., Toba, S. and Greenbaum, D. (2000) Selective targeting of lysosomal cysteine proteases with radiolabeled electrophilic substrate analogs. *Chemistry and Biology*, **7**, 27–38.

191 Srinivasan, R., Huang, X., Ng, S.L. and Yao, S.Q. (2006) Activity-based fingerprinting of proteases. *Chembiochem*, **7**, 32–6.

192 Goddard, J.P. and Reymond, J.L. (2004) Enzyme activity fingerprinting with substrate cocktails. *Journal of the

American Chemical Society, **126**, 11116–17.

193 Elend, C., Schmeisser, C., Leggewie, C., Babiak, P., Carballeira, J.D., Steele, H. L., Reymond, J.L., Jaeger, K.E. and Streit, W.R. (2006) Isolation and biochemical characterization of two novel metagenome-derived esterases. *Applied and Environmental Microbiology*, **72**, 3637–45.

194 Yongzheng, Y. and Reymond, J.L. (2005) Protease profiling using a fluorescent domino peptide cocktail. *Molecular BioSystems*, **1**, 57–63.

195 Sicard, R., Goddard, J.P., Mazel, M., Audiffrin, C., Fourage, L., Ravot, G., Wahler, D., Lefevre, F. and Reymond, J.L. (2005) Multienzyme profiling of thermophilic microorganisms with a substrate cocktail assay. *Advanced Synthesis and Catalysis*, **347**, 987–96.

196 Park, S. and Shin, I. (2007) Profiling of glycosidase activities using coumarin-conjugated glycoside cocktails. *Organic Letters*, **9**, 619–22.

197 Horeau, A. and Nouaille, A. (1990) Micromethod for determining configuration of secondary alcohols by kinetic reduction – use of mass-spectrography. *Tetrahedron Letters*, **31**, 2707–10.

198 Schoofs, A. and Horeau, A. (1977) New general method for determining enantiomeric purity and absolute-configuration of chiral secondary alcohols. *Tetrahedron Letters*, 3259–62.

199 Guo, J.H., Wu, J.Y., Siuzdak, G. and Finn, M.G. (1999) Measurement of enantiomeric excess by kinetic resolution and mass spectrometry. *Angewandte Chemie – International Edition*, **38**, 1755–8.

200 Reetz, M.T., Becker, M.H., Klein, H.W. and Stockigt, D. (1999) A method for high-throughput screening of enantioselective catalysts. *Angewandte Chemie – International Edition*, **38**, 1758–61.

201 Zha, D.X., Eipper, A. and Reetz, M.T. (2003) Assembly of designed oligonucleotides as an efficient method for gene recombination: a new tool in directed evolution. *ChemBioChem*, **4**, 34–9.

202 Reetz, M.T., Torre, C., Eipper, A., Lohmer, R., Hermes, M., Brunner, B., Maichele, A., Bocola, M., Arand, M., Cronin, A., Genzel, Y., Archelas, A. and Furstoss, R. (2004) Enhancing the enantioselectivity of an epoxide hydrolase by directed evolution. *Organic Letters*, **6**, 177–80.

203 DeSantis, G., Wong, K., Farwell, B., Chatman, K., Zhu, Z.L., Tomlinson, G., Huang, H.J., Tan, X.Q., Bibbs, L., Chen, P., Kretz, K. and Burk, M.J. (2003) Creation of a productive, highly enantioselective nitrilase through gene site saturation mutagenesis (GSSM). *Journal of the American Chemical Society*, **125**, 11476–7.

204 Zea, C.J., MacDonell, S.W. and Pohl, N.L. (2003) Discovery of the archaeal chemical link between glycogen (starch) synthase families using a new mass spectrometry assay. *Journal of the American Chemical Society*, **125**, 13666–7.

205 Yu, Y., Ko, K.S., Zea, C.J. and Pohl, N.L. (2004) Discovery of the chemical function of glycosidases: design, synthesis, and evaluation of mass-differentiated carbohydrate libraries. *Organic Letters*, **6**, 2031–3.

206 Yang, M., Brazier, M., Edwards, R. and Davis, B.G. (2005) High-throughput mass-spectrometry monitoring for multisubstrate enzymes: determining the kinetic parameters and catalytic activities of glycosyltransferases. *Chembiochem*, **6**, 346–57.

207 Nagahori, N. and Nishimura, S.I. (2006) Direct and efficient monitoring of glycosyltransferase reactions on gold colloidal nanoparticles by using mass spectrometry. *Chemistry – A European Journal*, **12**, 6478–85.

208 Basile, F., Ferrer, I., Furlong, E.T. and Voorhees, K.J. (2002) Simultaneous multiple substrate tag detection with ESI-ion trap MS for in vivo bacterial enzyme activity profiling. *Analytical Chemistry*, **74**, 4290–3.

209 Babiak, P. and Reymond, J.L. (2005) A high-throughput, low-volume enzyme assay on solid support. *Analytical Chemistry*, **77**, 373–7.

210 Gosalia, D.N. and Diamond, S.L. (2003) Printing chemical libraries on microarrays for fluid phase nanoliter reactions. *Proceedings of the National Academy of Sciences of the United States of America*, **100**, 8721–6.

211 Uttamchandani, M., Huang, X., Chen, G.Y.J. and Yao, S.Q. (2005) Nanodroplet profiling of enzymatic activities in a microarray. *Bioorganic and Medicinal Chemistry Letters*, **15**, 2135–9.

212 Kotz, K.T., Gu, Y. and Faris, G.W. (2005) Optically addressed droplet-based protein assay. *Journal of the American Chemical Society*, **127**, 5736–7.

213 Grognux, J. and Reymond, J.L. (2006) A red-fluorescent substrate microarray for lipase fingerprinting. *Molecular BioSystems*, **2**, 492–8.

214 Korbel, G.A., Lalic, G. and Shair, M.D. (2001) Reaction microarrays: a method for rapidly determining the enantiomeric excess of thousands of samples. *Journal of the American Chemical Society*, **123**, 361–2.

215 Horeau, A. (1961) Principe et applications d'une nouvelle methode de determination des configurations dite par dedoublement partiel. *Tetrahedron Letters*, 506–12.

216 Horeau, A. (1962) Determination des configurations par dedoublement partiel. 2. Precisions et complements. *Tetrahedron Letters*, 965–9.

217 Eelkema, R., van Delden, R.A. and Feringa, B.L. (2004) Direct visual detection of the stereoselectivity of a catalytic reaction. *Angewandte Chemie – International Edition*, **43**, 5013–16.

29
Confocal and Conventional Fluorescence-Based High Throughput Screening in Protein Engineering

Ulrich Haupts, Oliver Hesse, Michael Strerath, Peter J. Walla and Wayne M. Coco

29.1
General Aspects

29.1.1
HTS and Combinatorial DNA Library Strategies in Protein Engineering

Whether in the field of pharmaceuticals, industrial biotechnology or research, proteins are typically singled out for engineering efforts to make them more suitable for a given purpose. Adapting a protein for a particular application can require changes to the protein's primary amino acid sequence, as well as non-native biological or chemical post-translational modifications. In many cases, protein engineering efforts begin with quantitative goals, such as a desired binding constant anticipated to allow an antibody to reach a certain pharmacodynamic threshold, or the thermostability needed in an industrial enzyme to attain process profitability. Fortunately, in the vast majority of programs, the number of changes required to provide even dramatic improvements to the desired protein phenotype can be quite limited, involving perhaps one to a few dozen amino acid changes in a protein of, say, 50 to 1000 or more amino acids in length. Indeed, successfully completed protein engineering programs requiring fewer than ten mutations are quite common. Nevertheless, the amino acids in proteins typically also have quite complex intra-protein interactions when functionally folded, as well as complex interactions with their environment. A key question in protein engineering is thus often: "out of the overwhelmingly vast sequence space of possible changes, how do I identify the relatively few amino acid positions that should be modified in my protein and what new amino acids should I substitute at each?"

Notwithstanding the ever more efficient high-resolution determination of protein structures, structural information is available only for a small fraction of proteins of potential interest. In addition, despite notable improvements in structurally guided or 'rational' protein design, even in those cases where exquisitely detailed structural or structure–function data are available, it is often not

Protein Engineering Handbook. Edited by Stefan Lutz and Uwe T. Bornscheuer

ISBN: 978-3-527-31850-6

feasible to predict the amino acid changes that will lead to the extent of functional alteration needed to fully satisfy a given protein engineering effort's design criteria [1–3].

The solution to the above dilemma necessarily includes testing changes at more than just the amino acid positions that are ultimately needed to generate the final improved protein. That is, instead of making the needed changes and confirming the final improved activities, in protein engineering one typically generates multiple combinatorial DNA libraries addressing changes at many more than a few dozen positions. The next step is to screen through the created genetic variants using functional assays in order to identify the few improved clones that have some improvement in the desired direction. The number of variants generated in these approaches can range from dozens to over a trillion per combinatorial library, and the amino acid positions modified can range from carefully chosen sites or regions based upon structural modeling or structure function data to completely random mutations by gene-wide random mutagenesis (see, e.g. Ref. [4]). In general, the simultaneous pursuit of a number of different types of combinatorial libraries often leads to the most rapid identification of beneficial mutations, and–all else remaining equal–the higher the screening throughput available, the more chances exist to find the mutations that best improve the protein of interest. Therein lies the main value of high-throughput screening (HTS): it allows the testing of the properties of far more protein variants on an ongoing basis, in a shorter time and at lower cost than is possible by using manual biochemical assays.

The advantages of increasing throughput, however, go beyond the mere efficient identification of improved clones from complex combinatorial libraries. In some cases, mutations that improve a protein in the main optimization goal –for example, thermostability–will also cause undesired changes to the protein, perhaps causing impairments in activity, pH profile, expressibility or other parameters. Among the advantages to be gained from HTS is the more frequent identification of many alternative putatively improved clones (alternative screening 'hits'). The isolation of multiple hits offers the luxury of being able to discard compromise phenotypes, and to move forward only with exclusively positive mutations into subsequent rounds of optimization.

When multiple improved protein variants have been identified by screening, the mutations that lead to the observed improvements in each are typically pooled and recombined. It is generally not the combination of all such mutations, but rather the optimal combination of a subset thereof, that leads to the best ultimate phenotypes. That is, some improved mutations will actually decrease the fitness of a protein containing other improved mutations. The optimal subset is identified by screening a library made by *in vitro* enzymatic, synthetic or *in vivo* genetic recombination (DNA shuffling) of the best improved mutations [4–11]. The recombination of improved mutations often leads to additive or even synergistic increases in the desired protein characteristics. The cycle of variant creation, screening and recombination is typically repeated until the final desired protein properties are obtained. Additional combinatorial library generation methods used to create genetic diversity and to recombine improved mutations are discussed in this Handbook and elsewhere [4].

29.1.2
HTS in Protein Engineering: Coupling Genotype and Phenotype and the Advantages of Clonal Assays

For reasons apparent from the above discussion, the field of protein engineering has grown hand in hand with the field of protein screening and selection. During the 1980s, successes in protein engineering using gene family chimeragenesis relied largely on the manual biochemical assays available at the time [12–14]. Also readily adapted to the protein engineer's toolbox was Martinis Beijerinck's long-before established and ever-refined concept of selective microbial enrichment culturing. A wide variety of clever and/or technically sophisticated selection and screening methods have followed (see, e.g. Ref. [15]). However, common to virtually all methods targeting individual proteins is that, once the protein engineer has identified an improved protein, the gene encoding that protein must be made available for further optimization rounds and/or to express the optimized protein for use in the final application. The coupling of genotype to phenotype is thus an unavoidable additional constraint in every library generation/screening/selection strategy, and often determines their ultimate throughput.

This 'addressing' of a genetic variant (genotype) to the HTS assay readout (phenotype measurement) of the corresponding encoded protein is often accomplished in the wells of a multi-well or 'microtiter' plate. For example, one genotypic variant in an expression host can be deposited into each well of the plate and, after growth of the host and expression of the desired protein, this protein is then characterized. The hosts from wells containing proteins identified as improved are then used to recover the protein's encoding gene. Where the act of characterization will inactivate the host, other approaches are implemented, such as the testing of a sample from the grown well, the rescue and transformation of the inactivated host's DNA into a living host cell, or the generation of a replica plate that preserves each genotype in its living host. This type of screening approach will be discussed in more detail in following sections.

Well-free methods for genotype/phenotype coupling have included protein display, for example on nucleotides, particles, phages or cells, as well as the well-free compartmentation of genes with their encoded proteins, for example in fluid or gel microdrops [16, 17]. The screening methods that accompany the above, well-free coupling strategies include the panning of displayed proteins or fluorescence-activated cell sorting (FACS) screens for either displayed proteins or for proteins in microdrops. These latter methods allow the highest throughputs available to the protein engineer, and have been used in a number of impressive protein optimization successes.

In fact, with the rise of such methods, the temptation might be to predict the eventual obsolescence of well-based HTS. It is worth considering why this has not yet been the case. Reasons for choosing well-based HTS include:

- The ability to ensure analysis of one genotype per well ('clonal' assays) and thus avoid counterproductive population effects, as occur for example in display or growth-selection methods, and which are often difficult to control or even diagnose.
- The ability to produce the unaltered protein variant (e.g. not displayed on a particle, phage or cell).
- The ability to replicate plates with addressing of each genotype, and thus to correlate data from different, mutually incompatible assays for the same variant.
- The ability to submit a replica or sample of each variant to screening conditions that will render unrecoverable the genotype or host.
- Greater flexibility in the amount, timing or type of reagent additions.
- More quantitative determination of the effective number of genotypes screened, and thus of the completeness of screening for a library of a given complexity.
- The ability to gain quantitative or kinetic data on protein function.
- Tremendous flexibility in assay principles and readouts.
- The ability to use conditions that are often more similar to the final application.
- The ability to conduct many types of screens without the need for intellectual property licenses that may be costly or even withheld, and which are common to many well-free technologies.

While some well-free methods can boast one or several of these advantages, none can claim all of them. It should be noted also that many of the above factors contribute to the 'application relevance' and thus to the ultimate predictiveness of the screening process. For these reasons, well-based HTS has continued to be the method of choice in a wide variety of protein engineering applications.

29.1.3
Well-Based HTS Formats

The categorization of a screening method as 'high throughput' generally depends on the context in which it is used. For example, definitions will vary for practitioners referring to the screening of proteins, chemical compounds or mice knockouts. The use of advanced integrated automated systems, other specialized equipment and the ultimate throughput achievable, are determined by the nature of the problem at hand. Some may thus refer to HTS when screening as few as several hundred samples per day in some cases, while fully automated robotic screening campaigns exceeding many tens of thousands of data points per day are sometimes referred to as ultra-high-throughput screening (uHTS). For simplicity, we will refer here to both types as HTS, but with an emphasis on the higher throughput methods among these.

In general, screening capabilities should be matched with the envisioned or required protein engineering strategies, which in turn determine the HTS technology, capital investments and human resources needed. For example, one would generally be ill advised to build and maintain an HTS facility capable of 10^5 data points per day if the library complexity will not exceed 10^3, if library generation or downstream capacities are outpaced by the screening throughput, if only a few libraries will be processed each month, or if protein engineering projects are only sporadically envisioned.

The influence of a well format on assay predictiveness is also an important consideration. The most common well format comprises microtiter plates conforming to SBS (Society for Biomolecular Sciences) standards, with which the vast majority of commercially available, robotics-capable equipment modules are compliant. The choice, however, between 96, 384 and 1536-well plates must be carefully considered. The number of plates processed, the time to process each and the number of wells per plate have a clear relationship to screening throughput. However, even after commitment to state-of-the-art assay development and robotics infrastructure, many assays are still more predictive in the 96- or 384-well formats than in smaller-volume formats. Assay predictiveness is the paramount consideration for any screen, and is far more important than nominal throughput. For this reason, assays should as a rule be compared in different well volumes to determine the best balance between throughput and predictiveness. An additional factor to consider is that many HTS workstations or other equipment are only compatible with either 96/384 or 384/1536-well formats. This is one of several areas where an HTS platform based upon confocal fluorescence measurements has a clear advantage; since only 1 fl of the well is required for data collection in the confocal format, miniaturization to the 1536-well format can be readily achieved while still maintaining high data quality. This topic will be discussed in more depth in Section 29.2.1.5.

Where only lower throughput assays are found to be sufficiently predictive, the protein engineer may compensate for this deficit by using a higher-throughput qualitative or semi-quantitative prescreen to, for example, exclude inactive or low-activity mutants from a subsequent, more resource-intensive, quantitative screen. Agar plate assays are frequently used for this purpose [18]. By including more than one genotype in each well, library clone 'multiplexing' represents a second option for screens of suboptimal throughput. While nominal throughput is thus increased, this method works best for differentiating rare active clones which possess activity from a background of clones which have no activity under the screening conditions. In practice, such conditions can be either difficult or impossible to implement for the improvement of some protein characteristics.

With the prevalence of robotic high-throughput small-molecule library screens in the pharmaceutical industry, it is worth contrasting such screening technologies with the protein screening strategies required for protein engineering. There are several fundamental differences in the approach to HTS in these two disciplines, and the skills, infrastructure and equipment required for each. In compound library screening, the chemical identities of the compounds to be screened are often known, and complicated systems for storing, retrieving and tracking them throughout the screening process ensure the correct correlation between the effect (or phenotype) and the identity (chemical formula) of a 'hit' compound. In addition, each chemical library is generally reused many times across multiple projects to seek, for example, agonists or antagonists to a wide variety of potentially drug-susceptible targets as they become known or available. Finally, when the 10^5 or more compounds are exhausted, they need to be replaced, generally by chemical synthesis.

In protein engineering, in contrast, the libraries of genotypes that enter HTS screens are unique to each protein and are constantly replaced during the progress of each engineering project. This can require the weekly or even daily completion of multiple high-quality combinatorial libraries, as well as constant decision making as to the best types of libraries to be tested at any given point in a project. This means that the exact structures (sequences) of the variants to be tested are rarely known in advance, and must be recovered and identified only after a hit phenotype is identified. In addition, 10^5 or more novel protein variants can be generated and screened each day on an ongoing basis, making a separate, long-term addressable storage of each unscreened protein 'compound' unfeasible. For this reason, the library genotypes often only exist in the individual tested well or a replica well. This, in turn, requires the above-mentioned in-screen genotype–phenotype coupling and in-screen protein expression, which places a unique set of demands on the protein engineer's HTS infrastructure. These demands can include the distribution of individual genotypic variants in single expression host cells to each well on a multiwell plate, as well as growth and protein expression in microbial growth-compatible media, with the concomitant requirement for avoidance of microbial contamination and host cell cross-contamination, and the requirement for assays that can thus tolerate more interferences (e.g. from cells, extraneous cell-derived proteins or cell lysates), all on the HTS platform. It also demands the ability to recover the hit genotype from each hit well (or from a replica well), often within a living organism. As such, microbiology, molecular biology, structural biology and bioinformatics skills are required to run an HTS protein engineering effort, above and beyond the engineering and assay development skills needed to operate the machines and to evaluate each phenotype.

Finally, another critical aspect of HTS in protein engineering is the assay readout technology. A wide variety of assay principles and detection principles are available for measuring a wide range of protein parameters (see Sections 29.2.1.3, 29.2.1.6–29.2.1.9 and 29.4.2). In the section that follows, we will describe in more detail the principles, use and advantages of fluorescence detection in well-based HTS.

29.2 Fluorescence

Due to their relative ease of implementation, high sensitivity, cost effectiveness and broad applicability, fluorescence-based measurement methods continue to outperform other HTS homogeneous assay formats in many protein engineering applications. Protein parameters that may be assayed (and thus engineered) using fluorescence readout technologies include catalytic activity, product yield, binding constants, aggregation, solubility, selectivity, specificity, shelf life, thermostability, temperature optimum, pH optimum, pH range or resistance against proteases, oxidizing agents, inhibitors, detergents, organic solvents as well as other parameters.

In the following sections, we will introduce some principles of fluorescence that can be exploited in protein engineering screening assays, and will discuss the advantages and limitations of fluorescence measurements for HTS. For a more detailed discussion of fluorescence theory and for more example applications, the reader is referred to the excellent textbook by Lakowicz [19].

29.2.1 Overview of Theory and Principles of Fluorescence

The absorption of one or more photons can promote an element or molecule into an electronically excited state. De-excitation to the electronic ground state, when accompanied by the emission of a photon, is referred to as phosphorescence or fluorescence, depending on whether the excitation involves a triplet or singlet state, respectively. Electrons in a fluorophore usually occupy the most stable (or ground) state electronic orbital (S_0) or slightly higher vibrationally excited states. An S_0 electron can be excited to higher energy states (S_1 and above) by the absorption of a photon with energy corresponding to the difference between the final and initial orbital energies. If excited to a slightly higher and very unstable energy state, vibrational relaxation to S_1 is rapid. In fluorescence, the relaxation of an excited S_1 electron to its S_0 ground state is accompanied by the release of a photon of energy S_1–S_0. Two-photon excitation, in contrast, requires the absorption of two photons in rapid succession for an electron to reach the excited state, and thus requires extremely high photon densities, which are achievable using lens-focused, femtosecond-pulsed lasers.

Each fluorophore is characterized by its absorption coefficient and by its quantum yield. The former describes the efficiency of excitation, while the latter the fraction of de-excitation events that lead to photon emission, as opposed to radiation-free de-excitation modes. Accordingly, high absorption coefficient with high quantum yield are generally desirable for a good fluorophore, and will determine the fluorophore's brightness.

Absorption and emission spectra of common organic fluorophores have a half-width of about 50 nm, centered around the absorption and emission maxima, but with significant tailing (Figure 29.1, panel b.). The characteristic difference between the absorption and emission maxima is called the 'Stokes shift', with emission necessarily occurring at longer wavelengths (lower energies) than excitation. This allows spectral separation of excitation and emission light in spectrophotometers, by way of filters or gratings.

A further important physical parameter is the fluorescence lifetime, which describes how quickly the excited state decays. For most organic fluorophores the transition is quite fast, resulting in lifetimes on the scale of 0.1 to 10 ns. Processes

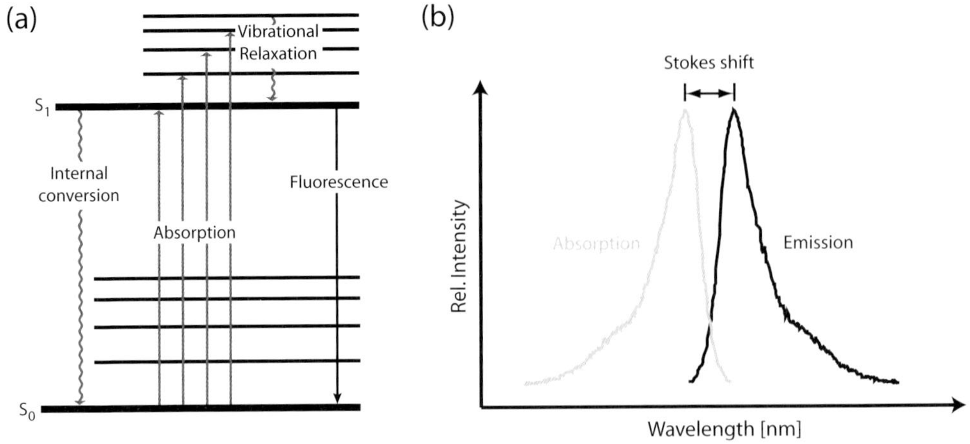

Figure 29.1 (a) The electronic ground state and slightly higher energy states are indicated by the labeled bold horizontal line (S_0) and the thinner lines above it, respectively. Above these are shown the excited state, S_1, and its higher energy states. One or more photons can excite a ground-state electron into an excited state. In fluorescence, relaxation from an excited state to a ground state is accompanied by release of a photon; (b) The Stokes shift describes the energy difference between the absorbed excitation photon and the resulting lower energy fluorescence emission photon. The excitation and emission spectra reflect the photon energies most likely to stimulate excitation or to result from emission, respectively. Rel. Intensity = relative intensity. See text for further explanations.

or environments that enhance radiation-free decay lead to a shortened lifetime. When two fluorophores are in close proximity (tens of Angstroms) and the emission spectrum of the first fluorophore (donor) matches the excitation spectrum of the second fluorophore (acceptor), the excitation energy may be transferred from donor to acceptor in a radiation-free process, called Förster transfer or fluorescence resonance energy transfer (FRET). The transfer efficiency is proportional to $1/r^6$, where r is the distance between the two fluorophores. The distance at which the efficiency drops to 50% is called R_0 (Förster distance), is characteristic for each donor–acceptor pair, and varies between 1.5 and 6.0 nm for typical organic fluorophores.

When polarized light is used as excitation source, the emitted light is also partially polarized. This phenomenon is called fluorescence anisotropy or polarization. The degree of polarization of the emitted light depends on the amount of rotational diffusion that occurs while a fluorophore is in the excited state, and thus reflects the mass of the molecule. Fluorescence polarization can therefore be used as a reporter for mass changes during a reaction of interest. This is widely applicable in assay design, for example, to monitor protein–protein interaction or substrate cleavage.

In order to use fluorescence as a means of following the course of a reaction – be it substrate turnover or a binding event – some property of a fluorophore's emitted light must change. Each of the fluorescence properties introduced above, and also a number to be discussed below, may be exploited to track such reactions and are thus relevant in assay design for protein engineering. The exploitation of fluorescence in assay design will be discussed in more detail later in Section 29.4.2.

29.2.1.1 Choice of Fluorophores in HTS

Several different classes of fluorescent molecules are used in HTS assay design, and each has advantages and limitations. Organic molecules are the most versatile and commonly used fluorophores as they are available for a broad range of wavelengths, can be coupled to different reactive groups for specific labeling, and are small and inexpensive. Highly photostable fluorophores that are sensitive to their environment (hydrophobicity, pH, etc.) can be used for specific applications. However, although a number of companies have emerged offering different families of organic fluorophores, little overall progress has been made in some areas. Common limitations remain hydrophobicity, in particular for the more red-shifted fluorophores, a limited range of fluorescent lifetimes (mostly in the range of 1 to 10 ns), broad spectra and small Stokes shifts (mostly <50 nm), leading to the inefficient separation of emission and excitation light. As a consequence, several excitation wavelengths, and sometimes different light sources, must be used when different fluorophores are employed in multiplexed formats.

Autofluorescent proteins, such as green fluorescent protein (GFP), have revolutionized molecular biological and cellular imaging. GFPs themselves have been subject to extensive protein engineering to modify their spectral properties, stabilities or aggregation states [20]. However, until now there have been relatively few applications in the engineering of nonfluorescent proteins, where GFPs have been used as a reporter. One notable exception is the fusion of GFP with a library of a protein of interest that requires solubility improvement. The misfolding or aggregation of the target protein prevents folding or leads to degradation of the fused GFP. The selection of highly fluorescent cells yields proteins that fold properly and are highly expressed [21].

In the context of assay design, it is important to bear in mind that fluorescence can be emitted by numerous molecules. In particular, many biological materials have a strong 'autofluorescence'. Although their brightness may be quite low, these molecules are often at far higher concentrations compared to the assay fluorophore, and may therefore contribute a significant background signal. This background increases towards the blue side of the spectrum, which explains the trend toward greater use of red-shifted fluorophores.

The use of red-shifted fluorophores, however, is not the only means of reducing autofluorescence interferences. Lanthanide complexes, for example containing Eu(III) or Tb^{3+}, represent another class of complexes with widespread use in HTS. They are characterized by quite long half-lives, in the range of milliseconds; this allows gated detection – that is, fluorescence detection begins tens of microseconds after the excitation light flash, and the fluorescent signal is then integrated for 200–300 μs. During the delay before detection, the intensity of the excitation flash, and any background fluorescence, decays essentially to zero, and the desired assay fluorescence signal is thus detected with strongly reduced background interference. This system is particularly suited for energy transfer to red dyes or fluorescent proteins such as allophycocyanin (APC). Many such systems–for example LANCE (Perkin Elmer) and HTRF (Cisbio)–differ in the type of Eu(III) complexes used, are commercially available and have gained some popularity. Other metal complexes based on ruthenium or rhenium have been described, but their low brightness limits the sensitivity of corresponding assays and they have not found wide acceptance.

Among the most remarkable new developments in recent years is the introduction of reproducibly manufacturable quantum dots. Quantum dot surfaces can be conjugated to biological molecules or substrates, such as by coating them with proteins (e.g. protein A, streptavidin) or by derivatizing them with chemically reactive groups. The emission maxima can be tailored by variation of the size and composition of the quantum dots. Quantum dots have a quite broad excitation spectrum and the extinction coefficient increases towards the blue. Quantum dots are thus ideal for multiplexing as they can be excited at one wavelength while their respective emission spectra can be efficiently separated. However, quantum dots have found only infrequent use in HTS, partly because of their expense.

29.2.1.2 Concentration Requirements for Fluorescent Analytes

Fluorescence measurements are quite sensitive. With state-of-the-art instruments, concentrations of 10 n*M* fluorophore usually yield a robust signal-to-noise ratio for most of the formats introduced below. Although confocal measurements can, in principle, be performed in the picomolar, single-molecule range in buffer systems, in practice the useable concentration of fluorophore depends on the background fluorescence from the biological material in a given assay (see also the discussion of two-photon excitation in Section 29.5).

29.2.1.3 Fluorescence Intensity Measurements with a Precautionary Note on Fluorescent Labeling of Substrates and Binding Partners

As mentioned above, in order to monitor a process via fluorescence assays, at least one property of the emitted light must change during the course of the reaction. That is, the fluorophore must be influenced in one way or another. Changes in total intensity are the most straightforward to measure, and are applied in countless assays. One popular assay modality, for example, exploits the use of fluorogenic substrates to monitor enzyme reactions. These are nonfluorescent or weakly fluorescent as educt but highly fluorescent as product due to a change in chemical structure; an example is the hydrolysis of an esterified hydroxyl group of a chromophore such as *para*-nitrophenol.

In the context of protein engineering, however, a note of caution is warranted. Assay design in general should, whenever possible, avoid placement of the fluorophore where it could influence substrate conformation, or where it may have direct contact with the protein to be optimized. By not heeding this warning, one risks unintentionally optimizing the protein to the fluorophore-labeled substrate and not to the desired native substrate. For example, in the case of antibody affinity or specificity maturation, labeling the antigen at positions that alter its conformation, or that allow contact between the fluorophore and antibody complementarity-determining regions, should be avoided. Analogous problems can occur for fluorescently labeled receptors or ligands in cytokine optimization programs.

Given the note of caution above, it is advisable to use chromogenic substrates, where possible, in indirect, or coupled, reactions (see the example in Section 29.4.4). In this case, total fluorescence intensity is an indirect measure of the reaction of interest.

Note also that total intensity measurements are particularly sensitive to interferences, for example, to particulates from cells or to plate effects. Assay performance may be increased using 'ratiometric' measurements, where a second reference fluorophore (measured at a different wavelength) is added and the ratio of intensities is used rather than absolute intensity.

The above discussion falls under the protein engineer's caution, "... you get what you screen for". As suggested above however, with sufficient experience, a large toolbox of assays, fluorophores and labeling techniques, as well as with careful assay design, predictive and high-quality assays can be developed for a wide range of protein engineering goals.

29.2.1.4 **Confocal Versus Bulk Detection Methods**

Before describing additional fluorescence-based HTS assay principles, we would like to discuss two conceptually different approaches for fluorescence measurements.

Conventional fluorescence detection, as performed on commercially available plate readers and spectrofluorometers, yields signals that are averaged over a vast number of fluorophores and over a large portion of the sample volume to be measured, and may therefore be termed bulk or macroscopic detection methods. In such techniques, the signal quality and amplitude degrades as the sample volume is reduced. This is because the signal amplitude in bulk detection methods is determined by the number of fluorophores in the assay volume, rather than by their concentration and because, as the sample volume shrinks, the relative contributions of noise from light scattering and other sources of background increase.

In contrast to macroscopic fluorescence, confocal or microscopic fluctuation measurements are performed in epi-illuminated confocal formats where the signal-producing volume is defined by the diffraction-limited focal spot of a laser beam and a pinhole in the detection light path [22]. This yields a detection volume of approximately 10^{-15} l (1 fl), which makes these methods intrinsically insensitive to miniaturization. At nanomolar tracer concentrations, few molecules (on average) are present in the detection volume. The actual number of molecules observed at any given time point fluctuates around a mean value, causing fluctuations in the fluorescence intensity signal. As a molecule diffuses through the focal volume it emits a photon burst, the intensity of which is related to the brightness of the molecule, while the duration of the burst is determined by the diffusion time and therefore the mass of the molecule. A number of mathematical algorithms are available that can extract information on specific brightness, diffusion time and concentration from the primary intensity fluctuation signal, even for mixtures of molecules, and are exemplified by the techniques, fluorescence correlation spectroscopy (FCS) and fluorescence intensity distribution analysis (FIDA). These are discussed in greater depth in Sections 29.2.1.8 and 29.2.1.9, respectively.

29.2.1.5 Advantages of the Confocal Fluorescence Detection Format

One obvious advantage of microscopic methods is their insensitivity to assay volume reduction such that the assay volume is limited by pipetting, evaporation and surface effects rather than by signal quality. By comparison, modern conventional fluorescence plate readers have improved significantly and can often yield good data quality down to approximately 10 μl.

As microscopic (confocal) methods extract information exclusively from single-molecule fluctuation measurements, they simultaneously reveal concentration, mass and brightness changes, even in mixtures of different molecular species. Such data are used for multiplexing assays – that is, for the probing of several reactions or molecular species in the same well. On the other hand, the multiple parameters obtained from such microscopic measurements can also be applied for well-based quality control, for example, hit selection based on additional assay quality parameters. FIDA in particular has the power to distinguish background contributed by autofluorescent but dim molecules from the bright fluorophore tracers, even if the background molecules are at much higher concentrations. This represents a powerful advantage in comparison to conventional techniques. For a more detailed discussion of confocal versus bulk methods, see Refs [23–26].

29.2.1.6 Anisotropy

In fluorescence anisotropy (FA), linear polarized light is used to excite fluorescence markers attached to biomolecules. The loss in the linear polarization of the light emitted by the fluorophore then depends on the ratio of the rotational diffusion time of the labeled biomolecule versus the fluorescence lifetime of the fluorescence marker. This loss can be easily measured using two detectors with two perpendicularly aligned linear polarization filters. FA is among the most versatile assay principles, as many protein-driven reactions can be configured to yield a mass change that results in a significant change in the rotational diffusion time. For small molecules (mass <1 kDa), the rotation is fast compared to the lifetime of standard fluorophores (ns timescale) and anisotropy accordingly is low, while the rotation becomes slower and anisotropy increases as mass increases. Beyond masses of approximately 50 kDa, there are only small further increases in polarization possible upon increases in molecular weight. However, below 50 kDa, mass changes are correlated with easily detectable anisotropy changes and have thus served as the assay of choice for a significant proportion of modern, fluorescence-based HTS.

In exploiting such assays, it is important to note that the assay dynamic range for a given analyte mass depends on the lifetime of the fluorophore. For example, long-lived fluorophores ($t_{½} = 20–100\,ns$) such as ruthenium or rhenium complexes have been used to extend anisotropy assays towards higher masses. However, these fluorophores suffer from low quantum yields and therefore confer low assay sensitivities. Independent of the total mass of the labeled molecule/complex, anisotropy may also be reduced by local motions of the fluorophore. Short linkers between protein and fluorophore tend to reduce this effect.

If the reaction of interest leads to a (de)quenching of the fluorophore in addition to the mass change, this must be corrected for quantitatively in the assay data analysis, as the bulk anisotropy is given by the sum of the contributions of the different species weighted by their concentration and brightness [27]. Finally, while FA does not require the confocal format, it benefits from its ease of miniaturization and relative insensitivity to interferences.

29.2.1.7 **FRET/TR-FRET/Lifetime**

The above advantages of ratiometric measurement can also be exploited in energy-transfer assay formats that lead to reduced intensity at the wavelength of the donor emission and increased intensity of the acceptor emission. In order to yield efficient transfer, the two fluorophores must be physically close. This strategy can be used to measure binding reactions, for example, in a complex between antigen and antibody (labeled with donor and acceptor fluorophore, respectively). However, the predictability of transfer efficiency is limited, and finding appropriately labeled binding partners often requires some trial and error with respect to labeling strategy and donor–acceptor compatibility. This lack of predictability is due to the above-mentioned strong distance dependence of the transfer efficiency and because most Förster distance values are in the range of the diameter of typical protein–protein complexes. In establishing such assays, it is also important to bear in mind that ratiometric signals are not directly proportional to fluorophore number – that is, they are not proportional to substrate turnover or to the number of bound molecules.

One special case of energy transfer, homogeneous time-resolved FRET (TR-FRET), involves the use of lanthanide complexes as donors and acceptors in the far red region of the spectrum. Although Eu(III) is excited at 390 nm, where a strong background signal is usually generated, this background signal is avoided using the time-gated detection explained in Section 29.2.1.1. In addition, the Förster distance of such lanthanide complexes (3 to 9 nm) is greater than that of common fluorescence dye pairs, reducing the problem of low energy transfer efficiency in the detection of many protein–protein complexes. Despite the considerably higher reagent costs, TR-FRET has proven valuable for the detection of binding events.

In addition to enabling time-gated data acquisition, modern instrumentation allows the accurate measurement of fluorescence lifetime ($t_{1/2}$) changes, which can arise in a number of ways. Changes in the local fluorophore environment, quenching or dequenching (and in particular FRET) can lead to changes in $t_{1/2}$ that can be exploited in assay design. Although lifetime measurements offer advantages such as sensitivity, concentration independence and high accuracy, this technique is not yet widespread in use. This may be due in part to the limited availability of HTS-compatible instrumentation and a lower predictability of assay design.

29.2.1.8 Fluorescence Correlation Spectroscopy

Fluorescence correlation spectroscopy is based on the detection of fluorescence fluctuations caused by few or even single fluorescently labeled biomolecules randomly diffusing in and out of a femtoliter focal detection volume. FCS analysis of, for example, binding reactions can yield a number of parameters including: diffusion times for bound and unbound species; the total number of molecules in the detection volume; the fraction of bound ligand; and total fluorescence intensity. The assay design for FCS, however, relies on a change in the number of fluorescing particles or a substantial mass change during the reaction. The resolution of translational FCS with respect to mass is relatively low, requiring an eightfold change in mass for a twofold change in diffusion time. This is due to the fact that diffusion is directly proportional to the hydrodynamic radius of the molecule, whereas mass is proportional to the third power of the radius. In addition, acquisition times are comparatively long in order to observe enough molecules for a statistically satisfying result, and the measurement is also sensitive against background fluorescence or particulates in the assay. In addition to the high end instrumentation required, the above points constitute the major factors as to why FCS has not prevailed in HTS. Nonetheless, it represents a valuable tool for certain types of lower throughput assays.

29.2.1.9 FIDA

Somewhat similar to FCS, FIDA analyzes the statistical probability to detect fluorescence bursts of a certain integral brightness, caused by few or even single fluorescently labeled biomolecules randomly diffusing in and out a femtoliter focal detection volume. FIDA allows determination of the concentration and brightness of different molecular species in a mixture. HTS assay design generally requires a change in molecule brightness by a factor of at least two. FIDA has the major advantage of being able to distinguish dim background molecules from bright tracer fluorophores, even if the background molecules are at high concentrations. A further advantage of molecular intensity measurements is the ability to detect binding events that do not lead to a change in total intensity. Examples include ligands bound to multiple receptors on a vesicle.

A further extension of the technique includes two-dimensional (2-D) measurements in which two channels can be differentiated by different wavelengths (2-D-2Color-FIDA) or by fluorescence anisotropy (2-D-FA-FIDA). Overall, FIDA has established itself as a versatile tool for HTS but, because it analyzes only individual molecules it requires the less common, confocal spectroscopic format.

29.3
Hardware and Instrumentation

29.3.1
Confocal and Bulk Concepts

Implementation of the above-mentioned different fluorescence techniques in HTS requires appropriate instrumentation. Standard multi-mode, nonconfocal fluorescence plate readers are available from many suppliers and easily integrated into automated screening tracks. Confocal instruments, on the other hand, are technically far more demanding, although their increased complexity is balanced by high versatility and information-rich data generation. Confocal imaging is well known and used extensively in the field of microscopy. A similar instrumental set-up based on inverted microscopes and laser light sources is also used for confocal, fluorescence fluctuation measurements in homogeneous HTS. In the confocal format, the detection volume is limited in its x, y and z-dimensions by the laser beam profile, microscope objective and pinhole, to yield a 'cigar-shaped' space, which has the approximate volume of an *E. coli* cell [22]. A variation on the confocal format additionally includes the use of multiphoton excitation. As the probability of two-photon excitation is proportional to the square of the light intensity, efficient excitation is limited to the center of the focused beam where photon flux is extremely high. This results in a restricted excited volume element without the need for a pinhole (this is described in more detail in Section 29.5). As highlighted in Sections 29.2.1.4 and 29.2.1.5, the detection volume in the confocal set-up is negligible compared to the total volume of the well, making such measurements independent of the well format and thus particularly easy to miniaturize.

Figure 29.2 depicts the key elements inside confocal and standard fluorescence spectrometers used in HTS. The principal elements common to both systems include automation-friendly plate handling, light source(s), optical elements (mirrors, filters, optical fibers), detectors and data processing electronics and software. Except for plate handling, the requirements for each of these units depend on the chosen measurement format. Robotic integration is a prerequisite for most HTS efforts, and therefore fast and reliable handling of multiwell plates is a must.

The ideal fluorescent spectrometer would have homogeneous performance over a wide wavelength range, but unfortunately none of the main functional parts of the spectrometer is wavelength-independent. The available components, together with their advantages and limitations, are briefly described in Sections 29.3.1.1–29.3.1.4.

29.3.1.1 Light Sources

Three groups of light sources are available, namely lamps, lasers and light-emitting diodes (LEDs).

- Lamps have a broad wavelength spectrum between 250 and 800 nm. The intensity is somewhat dependent on the chosen wavelength. The most common lamp in

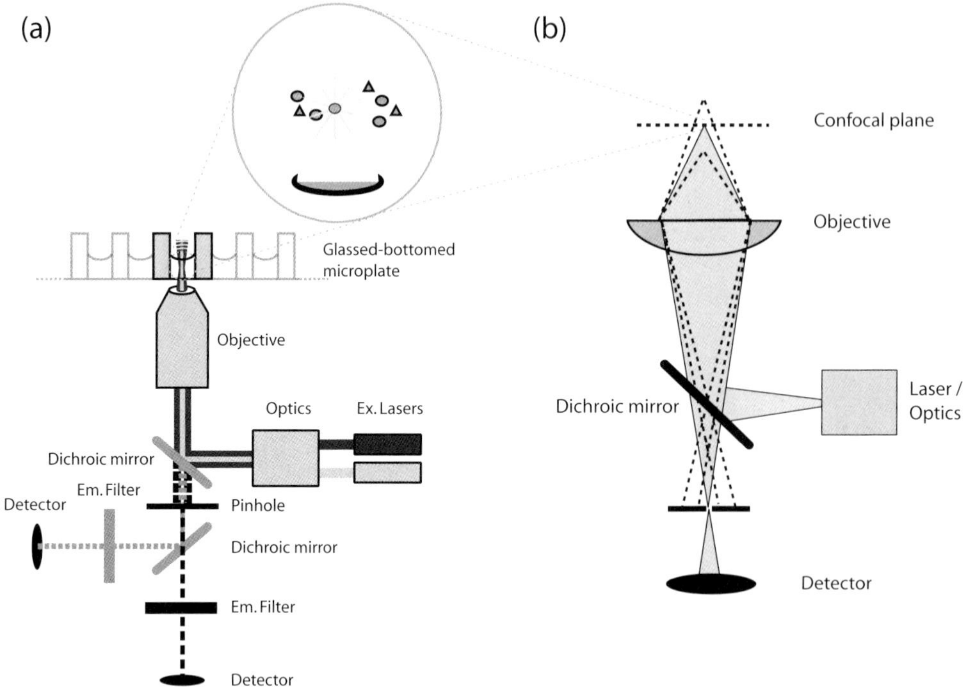

Figure 29.2 (a) Confocal fluorimetry optics. Excitation laser beams are reflected by a dichroic mirror and focused by a microscope lens to illuminate a sample in a glass-bottomed multiwell plate. Emitted fluorescent photons that pass through the lens, dichroic mirror and emission filters (Em. Filter) are converted into an electronic signal by one or more detectors. Emission filters block excitation wavelength photons; (b) Function of the pinhole. The pinhole is required to allow the detection only of photons emitted from fluorophores in the 1 fl volume focus at the confocal plane (solid lines), while blocking photons from fluorophores above or below this plane (dashed lines) in the multiwell plate. Ex. Lasers, excitation lasers.

use is the Xe-flash lamp, which runs typically at 60 Hz. The output is less wavelength-dependent compared to continuous lamps, and offers a stable radiant intensity between 400 and 800 nm. At wavelengths below 400 nm the (ultraviolet; UV) intensity drops continuously to 0 at ~220 nm. Beyond 800 nm the (infrared; IR) intensity depends heavily on the wavelength

- Lasers, in contrast to lamps, emit light with discrete wavelengths rather than a continuous spectrum, thus reducing the versatility of this light source type. On the other hand, pulsed lasers enable other, more sophisticated concepts of measurements such as confocal fluctuation or time-resolved measurements. The pulse width of such lasers is usually in the picosecond range and is therefore small enough for the resolution of fluorescence lifetime, while the repetition rate is high enough to realize short data acquisition times (MHz range). Since

2001, robust Ti:Sa lasers generating 100 fs pulses in the IR spectrum have become available, and are especially well suited for multiphoton excitation. Femtosecond-range fiber lasers are also now available.

- LEDs are comparatively new to fluorescence readers. Although a given LED emits at a single wavelength, different LEDs can be tailored for a wide range of wavelengths and represent a cost-effective alternative light source. Pulsed laser diodes, also available in multiple colors, have a typical pulse width in the picosecond range and repetition rates of 40 MHz, and are thus particularly useful for time-resolved measurements.

29.3.1.2 Wavelength Selection/Filtering

The selection of a particular wavelength from a broad lamp spectrum is usually performed with monochromators or filters. Monochromators contain diffraction gratings that split the light into its spectral components such that only the desired wavelength is allowed to pass through a slit. The optical quality of the gratings and mirrors and the slit width determine the spectral profile and intensity of the transmitted light. Many spectrometers use double monochromators for optimal spectral resolution of excitation or emission light. The clear advantage of this concept is the free choice of any wavelength.

In contrast, filters are made to transmit specific wavelengths. State-of-the-art interference filters consist of vacuum-deposited thin layers on a glass support, and can be customized for specific transmission profiles to match the requirements of particular fluorophores. Most standard plate readers use combinations of filters for excitation and emission. A further important optical element in many plate readers is the dichroic mirror [22] that is configured to reflect the excitation light but to pass the emission light, which is shifted towards longer wavelengths due to the Stokes shift.

29.3.1.3 Detectors

Ultimately, the emitted light must be converted into an electrical signal by a detector. Requirements for detectors are high, and include largely uniform sensitivity across a broad wavelength spectrum, linear response to incident light and, for time-correlated measurements, short and uniform response times. For most applications, photomultiplier tubes (PMT)s are the detectors of choice, and a variety of types with differences in spectral response, geometry and signal processing are commercially available.

For the ultimate sensitivity, which may be required for single-molecule measurements, the detector of choice is the avalanche photo diode (APD). Although unsurpassed in its sensitivity, the APD's drawbacks include a small actuating area, making it more cumbersome to align, and also a limited dynamic range.

Charged coupled devices (CCD)s are becoming increasingly popular for the detection of fluorescence, mainly because of their ease of application and large area of detection. Consequently, they are used with imagers that can measure the signal from a whole plate in a single action.

Whereas, the electronics for steady-state measurements are relatively simple, time-correlated measurements demand more sophisticated solutions. Pulsed signals from the laser must be correlated with the emitted photons, and signals from individual pulses must be processed separately. Because of the high repetition rate of the pulses, the electronics must be extremely fast and accurate.

Overall, the data acquisition, processing and analysis has to keep up with sample processing at high-throughput speed, which can be in the millisecond range per sample. However, given the ever-increasing power of modern electronics, computers and software, this no longer represents a significant limitation.

29.3.1.4 Reader Systems

Currently, a wide range of multimode plate readers is available commercially. At the lower end, there are fluorescent readers with basic functionalities. The most commonly employed low-end devices in HTS are multimode readers, which are generally capable of detecting not only fluorescence but also absorbance or luminescence. High-end devices in this class are equipped with a combination of monochromators and filters, usually Xe-flash lamps but sometimes with additional lasers and LEDs and various PMTs for different readouts. Their optics are capable of reading microtiter plates in top and bottom modes with alignment in all three axes. At the far high end of readers are the confocal instruments with multiple laser sources, which allow assay principles not accessible with even high-end conventional fluorescence plate readers. Although, unfortunately, such confocal systems are no longer commercially available, custom systems can be built that are even more powerful than the previously available commercial readers. For example, a custom-built reader used at DIREVO Biotech AG for protein engineering employs 12 foci per well for rapid, parallel data collection as well as multiphoton, multifluorophore and multiparameter uHTS at speeds in the microsecond per well range. An example of HTS using this custom reader in a two-color fluorescence intensity assay for the engineering of a bacterial phytase is described in Section 29.4.

29.4 Practical Considerations and Screening Protocol

29.4.1 Introduction

In this section, we will provide insights into considerations that are important for setting up a successful HTS effort for protein screening using fluorescence readouts. We will discuss a number of general, practical considerations that are relevant across a wide range of such projects, some practical issues concerning unit operations in HTS, and will further describe specific technical implementations using a successful protein engineering project that we recently completed. The project goals will be described in more detail in Section 29.4.4, but the specific enzyme class from this project will not be revealed for reasons of confidentiality (neither is it relevant to the practical considerations discussed). Because our platform is based upon a number of custom-made equipment modules and software elements, in some cases only the specifications of equipment can be described, and not the equipment vendors. Nevertheless, the issues discussed should have general utility, regardless of the equipment used or enzyme class engineered. Indeed, we chose to describe this particular example because it was based on fluorescence intensity, a fluorescence assay readout that does not require confocal technology. For the described project, however, we again used our custom confocal fluorescence plate reader because it contributed significantly to the quality and throughput of the described assay. Although the main focus here will be on the fluorescence spectroscopy assay design, a number of biological and technical issues that are critical in HTS will also be addressed.

29.4.2
Fluorescence-Based Assay Design: Practical Considerations

29.4.2.1 Choice of Assay Design

For the reasons mentioned above (see Sections 29.1 and 29.2.1.3), a predictive assay is among the most critical parameters in protein engineering. The assay designer must have an appreciation for the advantages and limitations of each assay approach considered. As mentioned in Section 29.2.1.3, direct fluorophore contact with the protein to be optimized, or any modification that may alter conformation of the interaction surface presented to the protein to be optimized, should be avoided. The use of natural substrates or binding partners is thus preferred over non-natural alternatives, even if the latter exhibits a nominally higher assay quality. For example, the activity of a protease can be measured using the natural protein substrate or using a short peptide comprising the cleavage site. Although the assay design for a peptidic substrate is usually much easier, it has been found that the activities of protease variants on respective peptides and protein substrates generally do not correlate well.

Changes in many fluorescent properties can be exploited to monitor the course of a reaction. In order to avoid unnecessary trial and error, highly predictable formats are preferable, as are properties that can be both measured accurately and that facilitate a large assay window. For example, for a small reactant, a significant mass increase during the reaction suggests the use of polarization or translational FCS, whereas binding events between larger entities are often better probed with FRET, TR-FRET or FIDA [25, 28–31]. A system of coupled reactions to measure product generation from engineered enzyme libraries opens up further possibilities in assay design using unmodified substrates. Many of these issues will be discussed in more depth, below.

29.4.2.2 Labeling

Although, in the protocol below (Section 29.4.4) there was no requirement for substrate labeling strategies, we nonetheless offer here some practical considerations on this important topic. Only in the minority of cases are commercially available assays or kits adaptable for HTS, and therefore when coupled reactions are not used most fluorescence-based assay strategies will require the custom labeling of at least one reaction participant. Both, the labeling of substrates as well as their purification and characterization, often require substantial expertise that should not be underestimated. Labeling must be conducted under mild enough conditions so as not to affect the relevant structure or activity of the labeled moiety. Functionality should thus be checked (e.g. specific activity, $k_{cat}/K_{M,}$ affinity) before the use of a labeled protein or substrate.

The specific labeling of proteins, whilst more involved, is generally preferred over statistical, or random, labeling approaches. This provides control over the reaction site, yielding a structurally homogeneous substrate and thus increasing the probability of producing a homogeneously biologically active molecule or an unperturbed biologically relevant conformation. This approach often requires the construction and expression of functionalized protein variants with, for example, a free cysteine at a defined position for labeling. The functional groups to be labeled must be accessible, and this sometimes requires the generation of multiple constructs before a suitable functionalization strategy is identified. Care must also be taken to remove residual free fluorophore after substrate labeling, as this otherwise often leads to a reduction in assay quality. Finally, while genetic fusions with a fluorescent protein can be a useful strategy they are not suitable for every protein or assay type.

In addition to the above points, some assay designs require the *in situ* labeling of protein variants expressed from a combinatorial library. *In situ* covalent labeling of the expressed protein variants via tags is difficult in the background of culture supernatants, and even more so in the presence of cell lysates [32, 33]. *In situ* labeling can often be better accomplished indirectly using a labeled antibody against the protein of interest, or against a suitable tag on the protein of interest. It should be borne in mind, however, that lower-affinity antibodies (e.g. anti-his-tag antibodies) are sometimes unable to yield a sufficient degree of binding at the concentrations of protein desired in the final screening assay (often in the nanomolar range). However, the selection of higher-affinity antibodies can circumvent this problem.

Due to the laborious generation–and sometimes significant cost–of assay components, their consumption should be reduced to a minimum. The reduction of assay components can be accomplished by a reduction of the assay volume as well as by selecting higher sensitivity readout devices. In addition, dead volumes (i.e. the unavailable volume of reagents that remains unused after dispenser priming, in long tubing or in fluid reservoirs) in HTS instrumentation should be minimized, as reagents often either cannot be recovered or have a limited shelf life.

29.4.2.3 Choice of Fluorophore

Another integral part of an assay design is the choice of fluorophore. The characteristics that should be considered include wavelength, photostability, brightness, hydrophobicity, limiting anisotropy and the availability of reactive derivatives. For example, the popular fluorescein and Cy5 photobleach easily, while more photostable fluorophores such as rhodamine green or, in particular, quantum dots are readily available. In general, absorption coefficients of >70 000 $M^{-1}cm^{-1}$ and quantum efficiencies >0.6 should be selected. Fluorophore hydrophobicity is often an additional problem, leading to the aggregation and precipitation of labeled proteins. As a rule of thumb, hydrophobicity tends to increase for more red-shifted fluorophores, as these have more extended conjugated π-electron systems. Sensitivity to the local environment, such as to solvent polarity or pH, may or may not be desired, depending on the application. Some suppliers provide quantum efficiency or lifetime data for both aqueous buffer and methanol (or other solvents). Significant differences in behavior in the two solvents can be indicative of a greater environmental sensitivity. The increased hydrophobicity of a labeled substrate may also lead to problems such as adhesion to tubing, to the surface of microtiter plate wells or association with micelles formed by the detergents that are present in many assay solutions. The latter can lead, for example, to unexpectedly high anisotropy or diffusion times.

Typically, fluorophores are connected to a labeled moiety via organic linkers, which can vary in length and composition. A short linker may be preferable for anisotropy measurements to avoid local motion, whereas long linkers can provide a better signal in some FRET assays by allowing more productive orientations of the donor and acceptor. When labeling oligonucleotides, unexpected effects may also be encountered due to interactions between the fluorophore and the DNA/RNA bases, or quenching due to electron-transfer reactions with the nucleotides.

29.4.3 Assay Quality

29.4.3.1 What Needs to Be Discriminated?

Key to the success of screening campaigns are assay robustness and quality – factors that partly determine the recovered fraction of false-positive and -negative hits. Concepts applying to compound screening in the pharmaceutical industry are often exploited for assay design and HTS by protein engineers. In the vast majority of compound screens, however, inhibitors of enzymatic activity or binding are of interest. In contrast, in protein engineering one often seeks improved protein variants that thus have higher activity compared to the wild-type protein. Accordingly, measures of assay quality often differ (Figure 29.3).

29.4.3.2 Mathematical Description

In compound screening, the Z'-factor is frequently used to characterize assay quality [34]. This factor relates the standard deviation for the measurements of fully active controls and negative controls to the difference between the means (see Equation 29.1); a Z'-factor value >0.5–0.6 is often considered to be the threshold for an acceptable assay. Inhibitor hits are generally found in the area between

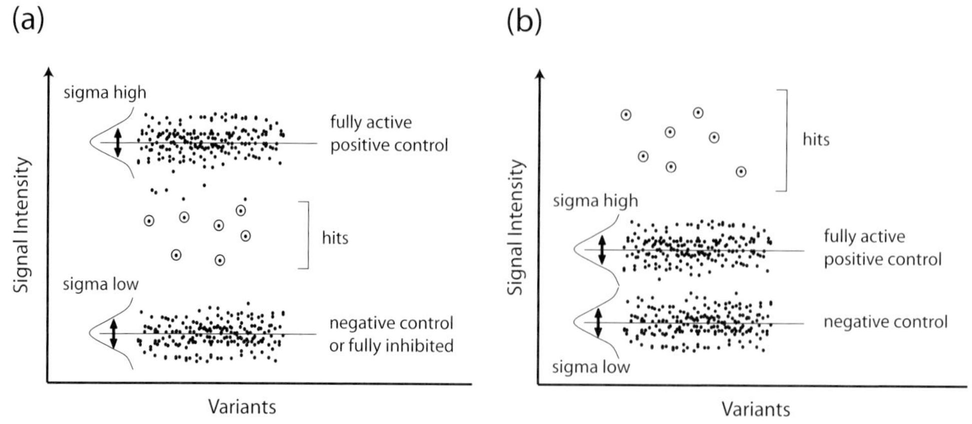

Figure 29.3 Assessing assay quality. (a) The Z'-factor of Zhang *et al.* [34] was devised for compound screens, where hits are often inhibitors that result in assay readouts intermediate in value between the positive and negative controls; (b) In protein engineering, protein improvements often generate assays signals that are greater than the positive control value. In such screens, the Z'-factor has less value as a predictor of assay quality, and the size of the assay window above the positive control value becomes an important factor in addition to the standard deviation of the positive control. The sigma curves depict the spread of data values about the mean. See text for a more complete discussion.

the fully active and negative controls [35]. For a more detailed discussion, see Ref. [34].

$$Z' = -\frac{3(\sigma_p + \sigma_n)}{|\mu_p - \mu_n|} \tag{29.1}$$

Here, σ is the standard deviation, μ is the mean value, p is the positive control, and n the negative control [34].

When the protein engineer seeks to enhance activity, assay quality is often not directly related to the ideal separation of the negative and positive control values. The wild-type (fully active) activity, or that of the most recent improved variant, becomes the low activity control, and there is often no satisfactory control for how improved variants will perform at the upper limits of the assay as the sought-after improved enzymes do not yet exist. This deficiency can sometimes be addressed indirectly by measuring the activity of a more concentrated control protein, or by using longer reaction times. Such measurements can then be used to determine the size of the assay window, but often do not reflect the standard deviation that would arise from an improved biological control grown and assayed under conditions identical to the combinatorial library screen. The more relevant issue in such cases is thus the probability that an improved protein can be reliably differentiated from the wild-type, or the probability that the best variants can be discriminated from their also-improved peers. The tools available are often limited to the standard deviation of the wild-type control and the size of the assay window in the direction of higher values. Optimizing resolution of activities in this region is often purchased at the price of less resolution between the negative control activity and the wild-type activity. However, as no hits are to be found in this region, this is often the preferred course of action. The conclusion, however, is that maximizing the *Z*-factor can actually result in a poorer screen for many protein engineering applications.

A retrospective assessment of the quality of a running screen can be less challenging. The top (perhaps 50 or few hundred) hits from a given screening day can be analyzed in highly accurate confirmatory assays, for example using purified protein from each variant. If the top few screening hits remain in the top ranking after the more intensive characterization steps, then the predictiveness of the screen is confirmed.

29.4.4 A Specific HTS Protein Engineering Program Using a Fluorescence-Based Screen

In animal husbandry, enzymes are commonly added to feedstuffs to support the bioavailability of nutrients. During the industrial processing to form the final feed pellet product, temperatures may rise to above 80 °C, at which a wide variety of useful enzymes would be irreversibly unfolded. As thermostability is not inherent in all enzymes that can carry out the desired reactions, protein engineering is used to increase this parameter. Besides thermostability at feed-pelleting temperatures, the enzyme must also remain sufficiently active in the digestive tract of, for example, monogastric livestock such as swine (i.e. at pH $\leq$2.5) and in the presence of the broad-specificity digestive protease, pepsin.

The goal of this recently completed project was thus the optimization of a feed enzyme for four different optimization parameters (thermostability, pH, pepsin stability and substrate preference). Although the project resulted in the successful optimization of all four parameters, for simplicity only thermostability and activity under application pH will be discussed here.

29.4.5 The Assay

In the described application, we were able to use the natural substrate. The assay was based on the detection of a reaction product via a system of coupled enzyme reactions, which ultimately led to the oxidation of fluorogenic Amplex Red to fluorescent Resorufin (supplier: Molecular Probes). In this case, the readout of fluorescence intensity of resorufin is directly proportional to release of the reaction product, and therefore accurately reflects differences in enzymatic activity of the protein variants screened.

Resorufin is pH-sensitive and yields its highest intensity at pH 7–8. Accordingly, buffering of the final assay solution is essential for optimal results. While commonly excited at 543 nm using green He-Ne lasers, resorufin is also amenable to two-photon excitation at 800 nm, providing the advantages discussed in Section 29.5. The use of two-photon excitation allowed the simultaneous readout of a second fluorophore with a longer emission wavelength, which in turn enabled an additional quality control measure with respect to liquid handling, evaporation or other effects during the screening process.

In this example of screening for thermostability, the screen exploited a heat challenge step. The heat challenge temperature was chosen to lead to a desired fraction of still-active variants from each combinatorial library. For libraries that contained large gains in activity (data not shown), the wild-type or previous best improved clone had no residual activity at the heat challenge temperature that was needed to distinguish between the best hits (again highlighting the deficiency of the *Z*-factor calculation for some screening assays). Positive, negative and library controls were then included with a more permissive heat challenge solely to verify the performance of the assay reagents and screening mechanics. For libraries delivering more modest improvements, heat inactivation was adjusted to best resolve the thermostability levels at or beyond the upper end of the positive control distribution. By using this approach of adjusting the assay window to match the improvement level inherent in each combinatorial library, we were able to routinely isolate improved clonal variants with improvements in inactivating temperature as small as 0.5 °C, and resolved the best hits from variants improved over 10 °C (Figure 29.4).

29.4.5.1 **Expression Host**

Most HTS protein engineering efforts rely on microbial expression systems. In this section, we attempt to provide an appreciation for some of the issues that guide the choice of an appropriate HTS expression host, and will introduce the expression system used in the below protocol.

Since on many HTS platforms, the organisms used for expression are not fully contained, the expression host must be compatible with applicable safety regulations. The lack of containment also generally makes useful a means for selecting the used host in order to reduce contamination, as well as stably maintaining the genes of interest within the host. An inducible promoter also contributes to plasmid stability and comparable levels of protein production from well to well. A further requirement is the availability of vectors and protocols that allow easy genetic manipulation of the genes of interest and the host. In particular, a high transformation rate is desired for the generation of combinatorial DNA libraries of high complexity. Where transformation protocols for a particular host are not sufficiently efficient, they can often be optimized, albeit at the expense of an investment of time and resources. In addition, a sufficient growth rate is desirable to reduce growth times in the screening process (which affects the staggering of multiple screens) and cells should be stable enough to avoid autolysis or death during the screen. Finally, the medium requirements of the strain must be considered. The complex media required by some strains can limit the assay options of some screens, although two-photon-based assays can often yield excellent results, even in the presence of complex media. Cell lysis may also be integrated

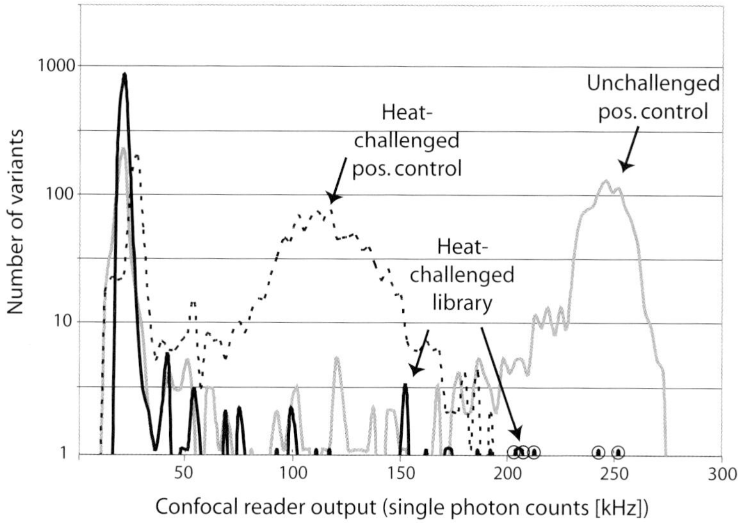

Figure 29.4 Assay quality in the example screen. Representative data histograms of three 1536-well plates from an HTS screen. One positive control plate contained clones expressing the wild-type enzyme without any heat inactivation (gray line). The same positive control is also shown after the expressed enzyme was subjected to a heat challenge (dashed black line). The solid black line represents one 1536-well plate of a screened library, in this case a highly mutagenized library with a low proportion of active variants. Due to limiting dilution library dispensing in this screen, 30% of the wells remained empty and appear as the peak at the low signal side along with the completely inactive clones. Hits from the heat-challenged library plate are circled. Note that the poor separation between negative positive controls would yield a nominally poor *Z*′-factor, yet the hits are well resolved from the heat-challenged positive control, reflecting a well-functioning assay.

in the screening process, but protein secretion is often preferable due to its simplicity and the purity of the final protein solution to be measured. Protein secretion requires the availability of efficient secretion signal sequences for the host. Finally, an often paramount consideration in host choice is the production of a sufficient and from well-to-well consistent amount of the recombinant protein of choice. For more information concerning these and further aspects of the biology of expression hosts for protein screening, the reader is referred to recent review articles [4].

In the below protocol, *Bacillus subtilis* was used for secreted protein expression. The transformation rates were optimized to achieve 10^6 to 10^7 colony-forming units (CFU) per microgram of plasmid DNA, and growth periods of 29 h at 37 °C yielded stable and homogeneous expression levels. Hit recovery from the original expression culture was possible for more than 5 days, although hits were routinely picked within 24 h of culture growth.

29.4.6
Multiwell Format and Unit Operations in the HTS Protocol

Typically, we target over 50 000 samples per day throughput for our screening campaigns. In order to accomplish this, we use sample carriers with 1536 wells and total working volumes in each well limited to about 0.5–4 µl. The unit operations include liquid handling, incubation for growth, expression and assays, centrifugation, fluorescence-based confocal assay readout and data processing. The efficient scheduling of the processes should also be mentioned as an important step. We will provide a short overview of these operations and then highlight their use in the process. As with all activities in protein engineering, experienced staff, well-maintained equipment and permanent quality control is critical for the successful execution of HTS.

29.4.6.1 Liquid Handling

In general, two basic types of instrument are used for liquid handling, namely pipettors and dispensers. Among the former are the massively parallel 384-channel devices using positive air displacement; examples are the 'CyBios' and 'CyBiWell' instruments. By pipetting up and down, pipettors may also be used for mixing which otherwise may be difficult to achieve in small-volume, multiwell plates. However, one drawback of this device is that pipetting channels cannot be addressed individually. Other suppliers also offer parallel pipettors, and for these details the reader is referred to the respective websites (e.g. www.tecan.com; www.matrixtechcorp.com; www.beckman.com).

Dispensers, on the other hand, often use solenoid valve technology, as exemplified by Genomic Solutions' 'PreSys'. This instrument can address a maximum of eight channels individually, and allows 'on-the-fly' dispensing (dispensing into the wells while moving continuously across the plate). Although the specifications allow dispensed volumes as low as 50 nl, in our hands reproducible results with this instrument were obtained only for volumes greater than 150 nl. In general, robust and reproducible results from both types of liquid handler can, unfortunately, require extensive testing and adjustments. The specifications of the liquid handlers used in the described protocol will be described upon their introduction.

29.4.6.2 Incubation

Available incubators are similar in concept. The selection criteria include storage size, temperature range, CO_2 compatibility, robustness and integration with the robotic and software environment. Long incubation times generally require either a humidified environment to reduce evaporation or, alternatively, the sealing of plates. Differences can include technical features such as cooling with external cryostats or built-in compressors. The selection of an appropriate device will depend on the specific context and requirements. In the present protocol, Liconic STX incubators (Liconic AG, Liechtenstein) were used. Humidification was also chosen over plate-sealing, based on considerations of timing and cost. The circumvention of downstream unsealing or seal-piercing steps further contributed to this decision.

29.4.6.3 **Centrifugation**

Centrifugation is an important step in HTS protocols, not only to ensure that the dispensed fluid is collected at the bottom of the plate but also to eliminate air bubbles from the solutions. For these purposes, only minimal *g*-forces are required. In screening protocols, centrifugation at high *g*-forces may be used to reduce or remove cells or cell debris from analyte or assay supernatants, although problems may occur with respect to handling of the plates and the precise positioning of the rotor when the plate is loaded, if less precise robots are used. However, the vendors of these systems have addressed these issues and presented elegant solutions (e.g. Vspin by Velocity11, Menlo Park, USA or Hettich GmbH and Co KG, Tuttlingen, Germany).

29.4.6.4 **Scheduling**

The coordination of different unit operations and multiple instruments requires sophisticated scheduling software. This application provides tools for setting up and pre-calculating all processes, organizing communication between the various devices, and ultimately scheduling the complete processing of each sample carrier.

While most vendors of HTS hardware also provide scheduling software, other independent companies also offer platform-independent solutions (e.g. Ovcrlord by PAA; www.paa.co.uk; Genera by ReTiSoft; www.retisoft.ca).

Available scheduling software is based on two fundamentally different concepts for the algorithms used, namely static schedulers and dynamic or event-driven schedulers. Static schedules are set up prior to the actual process, using predetermined durations for each step, whereas event-driven schedulers decide on events in real time, in an attempt to use the currently available devices in optimal fashion. Both solutions have advantages and drawbacks, and some schedulers thus combine both concepts. In many screening campaigns for protein engineering, the reproducibility of unit operation times has been found to be quite important, and in such cases static scheduling is preferred. Unintentional deviations from preset timings during the process, however, can lead to the subsequent suboptimal use of resources and prolong the affected incubation times. Ultimately, this can result in timeout errors, with the worst-case scenario being the failure of a screening run. Approaches that combine static scheduling with real-time analysis of the system status improve the system's overall performance as they can better react to delays or external interventions. In the present protocol, the Tecan FACTS scheduler (Tecan AG, Maennedorf, Switzerland) was used.

29.4.6.5 **Screening Protocol**

A schematic overview of the process is provided in Figure 29.5. The entire process was run in custom-made, 1536-well, glass-bottomed plates which were suitable for confocal readout and high-temperature incubation [36]. The entire process

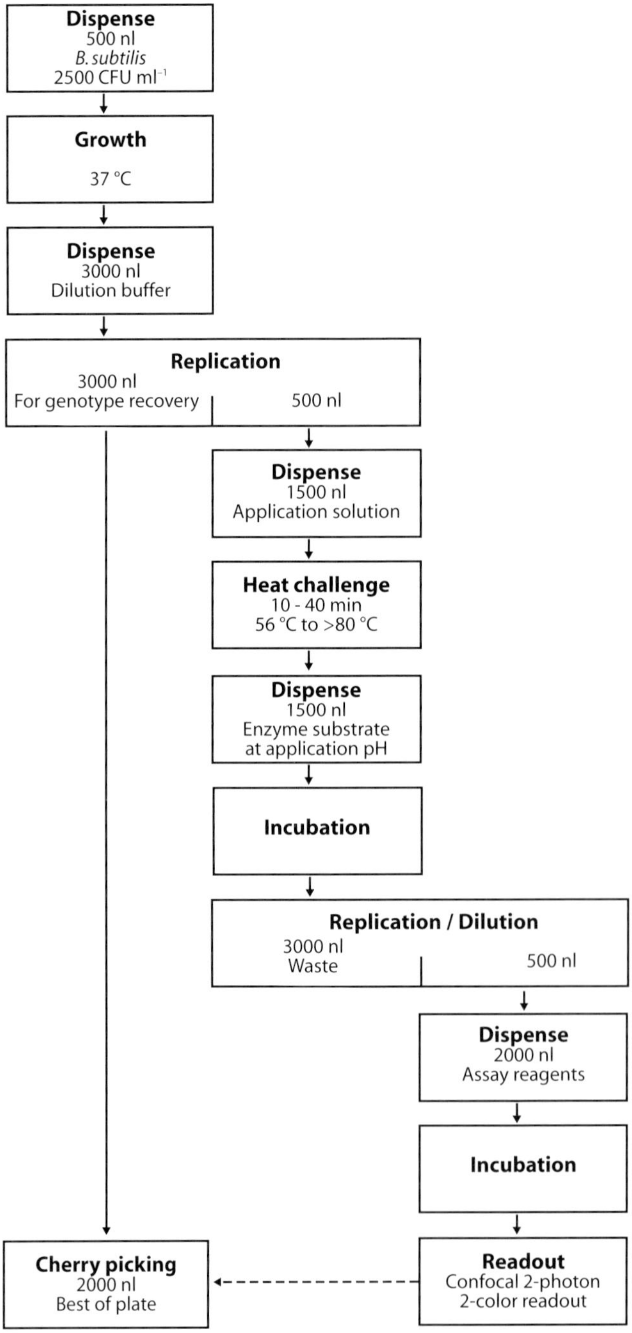

Figure 29.5 An overview of steps in the described high-throughput screen.

was fully automated on a customized robotics platform, comprising all steps from library dispensing to hit-picking. The typical throughput for this complex process was approximately 60 000 data points in 36 h, excluding bacterial growth.

In the first step, host cells transformed with the combinatorial library were compartmentalized with one organism per compartment of the 1536-well plates. Although we generally distribute one cell into each well by FACS, an unusual constellation of specific medium requirements and cell physiology made a limiting dilution approach using a solenoid valve dispenser the method of choice in this case. Growth at 37 °C was followed by an initial dilution step, realized by adding 3000 nl of a NaCl solution to each well. This was necessary due to the high expression levels of the employed expression system.

As the plates containing the proteins must be heat-challenged to test protein variants for thermostability, a replica of the growth plate (master plate) was generated to enable recovery of the encoding hit gene from live host cells at the end of the screening process. A 500 nl sample was transferred from the master plate to a replica plate that was later subjected to the heat step. The master plate was stored at room temperature for the final 'cherry picking' of hits.

After replication, 1500 nl of solution was added, followed by a short centrifugation step to collect the fluid at the bottom of the plate and to remove air bubbles. Heat-inactivation was performed by a high-temperature incubation, followed by cooling to room temperature. Over the course of the optimization campaign, the heat-challenge temperature was gradually increased from the initial 56 °C to over 80 °C, and the time of heat inactivation was increased from 10 min to 40 min, reflecting the progressive increase in thermostability of the target enzyme.

Residual enzyme activity was assayed by adding a substrate solution with the solenoid valve dispensing device, followed by a reaction period of 1 h at 37 °C in a humidity-controlled incubator (Liconic AG). An additional component added with the application solution was the mentioned inert dye for quality control (Dy-647; Dyomics GmbH, Jena, Germany). The enzymatic reaction was performed at low pH, reflecting the application in the animal gut. The reaction product was transformed into a fluorescent signal by coupled enzymatic reactions. To neutralize the solution pH before the assay readout step, another replication/dilution step was conducted, using a similar device and pipetting script as used for the above replication step, and this was followed by dispensing of the neutralization solution, which also stopped the reaction. This was followed by another centrifugation step and an incubation period of 100 min at 37 °C.

The readout was performed using the above-described two-photon excitation reader in the 12-focus data collection mode. The ability to excite and detect two different dyes simultaneously was exploited to determine both the resorufin concentration, which reflects the protein performance, as well as to simultaneously measure the concentration of the inert, assay quality control dye. The best-performing enzyme variants were identified directly by analyzing the generated data in real time using custom data analysis algorithms. The output included a work list for a single-channel cherry-picking device, which transferred the hit-containing expression hosts from the original 1536-well master plate to 96-well destination plates for further processing.

By using the above approach, with a variety of combinatorial DNA library strategies, we were able to successfully engineer an increase of over 20 °C in thermostability, with simultaneous improvement of pH stability, pepsin stability and specific activity.

29.5 Challenges and Future Directions

Confocal fluorescence spectroscopy provides unique access to a variety of novel parameters that can be used to track the progress of a range of enzymatic reactions and biomolecular interactions. Informative parameters, in addition to the above-mentioned diffusion times, include the direct monitoring of the number of fluorescing particles in the ~1 fl focal volume, which enables, for example, the determination of concentration changes in the picomolar to nanomolar range. It also enables measurement of the molecular brightness of single-molecular species, which allows the monitoring of, for example, aggregation processes. As described in the preceding sections, the total number of protein parameters and processes that can be monitored is quite large, and thus of great potential value to the protein engineer. Nevertheless, the hopes that such techniques would gain widespread use in the protein engineering community have not yet been realized, in part due to the challenges in implementing automated confocal HTS screens. Recent and anticipated technological developments will continue to shape the application of confocal as well as conventional fluorescence assays for the HTS protein engineer. There are also, however, several challenges to be met in these areas. Both, the challenges and future potential – including more speculative directions – will be discussed in this section.

The two main reasons for the lack of wider use of confocal HTS detection systems are surely the capital equipment costs and the specialized skills required to operate and maintain such systems. In addition, some techniques based on single-molecule fluorescence fluctuations can require long acquisition times to obtain sufficient statistics, and react poorly to artifacts such as cells diffusing through the focal volume.

For certain applications, two-photon excitation (TPE) fluorimetry ameliorates or eliminates these problems [37–39]. TPE is well suited to confocal detection applications in complex biological samples including, for example, industrial samples containing particulate plant fibers in the case of cellulase/hemicellulase engineering, or containing human serum, in the case of human therapeutic protease optimization. (DIREVO Biotech AG, unpublished observations). As will be described in more detail below, for fluorescence detection in HTS applications in general, TPE has many advantages to offer.

In TPE, two photons of approximately half a fluorophore's absorption energy are required for electron excitation. This generally places the excitation wavelength in the IR range (~700–1100 nm), which is far from the typical emission wavelengths of most commonly used fluorophores, and thus significantly reduces the background from the laser excitation. In addition, TPE requires the use of only one variable filter-set for fluorophore emission, while the dichroic mirror and IR-blocking filter are fixed in the HTS reader to separate the excitation light (IR) from the fluorescence (visible light) for all fluorophores in the sample to be analyzed (Figure 29.6). For confocal HTS applications, TPE has the additional important advantage that no alignment-demanding pinhole is necessary, because the excitation volume is intrinsically restricted to a focal diameter of about 200–500 nm [39]. The reason for this is the quadratic light intensity dependence of the excitation probability of this nonlinear optical process [38]. That is, only at the focus is the light intensity high enough to excite the analyte fluorophore, thus obviating the need to use a pinhole to restrict the focal plane from which fluorescent events are detected. Consequently, the practitioner can quickly switch between different assays and fluorescent dyes without time-consuming optical alignments. Finally, a single TPE wavelength can often be found to excite several fluorophores of

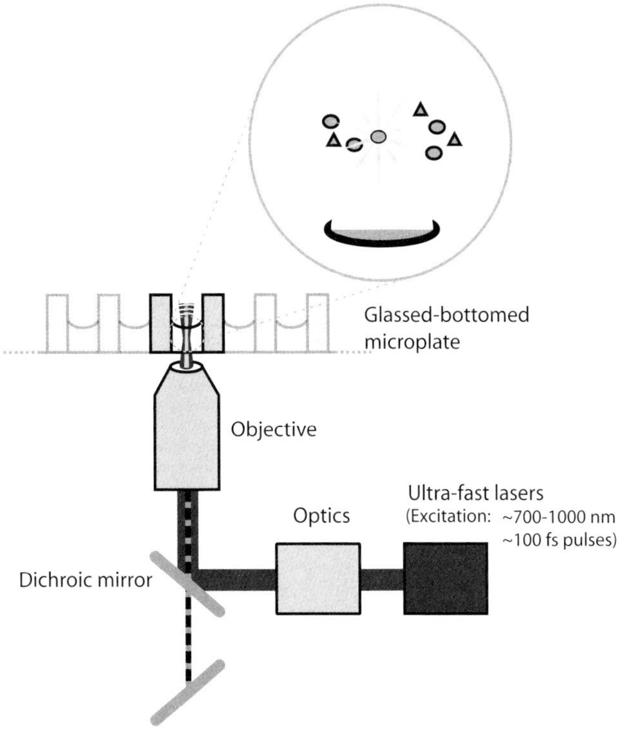

Figure 29.6 Robust ultrafast lasers enable two-photon excitation in a fully automated HTS environment. The fluorophore excitation region of about 200 × 500 nm per axis is intrinsically restricted to the focal volume (inset). Only one alignment-free set of optical components is necessary for the excitation and detection of several different fluorescence dyes.

different emission wavelength simultaneously in one well. If different detectors are used for each dye, then multiple assays can be performed in one measurement, and this can help not only to improve the predictive value of the assay but also to avoid false-positive hits. Although TPE is characterized by low absorption coefficients, and thus requires pulsed lasers with extraordinarily high photon fluxes, in a wide variety of applications, its advantages outweigh its drawbacks.

The above pinhole-free configuration has further enabled the exploitation of multiple femtoliter-sized excitation volumes in a single sample by using microlens arrays, beam splitter arrays or Wollaston prisms (Figure 29.7) [40, 41]. In these applications, not one but many laser beams are focused within an individual well for independent data collection. Such a multifocal detection scheme significantly decreases the necessary measuring time per well and also improves the precision of the assay.

For example, with constant photon flux per focus, N simultaneously measured foci reduce the required measuring time by a factor of N, which can allow single

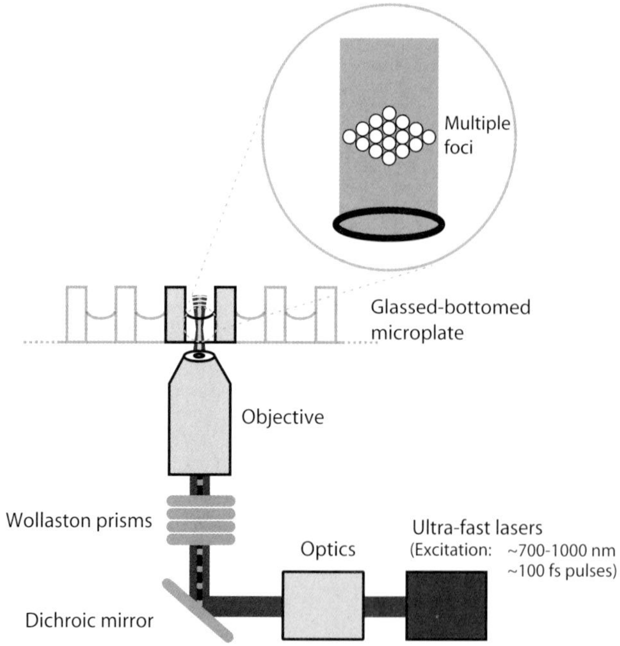

Figure 29.7 Creation of 2^n single-femtoliter excitation volumes by a series of n Wollaston prisms. N simultaneously measured foci reduce the required measuring time N-fold, and/or increase the assay predictability or information content by averaging the information from all volumes or by performing replicate assays in each volume simultaneously. The use of Wollaston prisms allows the creation of near-identical intensities and focal profiles (point-spread functions) of the individual foci which is especially advantageous for single-molecule fluorescence techniques. We have combined this advantage with two-photon excitation fluorescence assays to exploit the benefits of both advances. For details, see Ref. [40].

molecule and molecular fluctuation-based data collection times within an acceptable timeframe for HTS (generally not more than a few seconds per well, and usually much less). Alternatively, it also enables additional measurements of different assay parameters in each focus simultaneously in a total sample size smaller than a single microliter. Finally, the measurement in several confocal volumes simultaneously allows one also to sort out artifacts measured in individual confocal detection volumes on a statistical basis. Even FIDA, which has proven to be a quite practical single-molecule assay technique in current HTS applications, also profits from significantly reduced measuring times and improved assay performance in the multifocus data collection mode [42].

The skills needed to run and maintain HTS platforms exploiting TPE with multifocal confocal fluorimetric detection systems include knowledge in the areas of ultrafast laser optics and devices. For example, current implementations of these advances require a painstaking alignment of prisms and mirrors. We predict, however, that these technologies may in the future be implemented in fully automated industrial environments by the HTS engineer with no specialized training. Modern developments, especially in the field of semiconductor fabrication, are poised to eliminate many of the interventions and adjustments needed in confocal HTS detectors based on femtosecond laser-excitation.

In addition to the above, new camera technologies such as highly sensitive and fast on-chip amplification CCD cameras may soon represent an alternative technology for the simultaneous measurement of correlational information from multiple foci. This CCD camera technology appears to be poised to also allow HTS FCS imaging applications [43].

In the context of methods based on confocally detected single-molecule fluctuations, one powerful method which has not yet been mentioned is that of cross-correlation analysis, which provides direct information about the percentage of binding or cleavage events among molecules labeled with different fluorophores. In comparison to fluorescence polarization, FCS or FRET techniques, cross-correlation analysis methods allow a far wider range of masses or label distances of the interacting partners or molecules to be cleaved [39, 44]. Cross-correlation analysis is based on a statistical coincidence analysis of, for example, two differently labeled binding partners in the confocal detection volume. While cross-correlation analysis is more sensitive to particulates and to residual free dyes than other confocal techniques, it can provide a valuable alternative if other assay principles prove unsatisfactory.

When using conventional one-photon excitation, cross-correlation or coincidence analysis requires the separate alignment in a single femtoliter focal volume of the excitation and detection volumes for the two different dyes. The spectral separation of the two lasers' excitation and the two fluorophores' emission wavelengths is often technically demanding. However, these obstacles can again be circumvented by exploiting TPE, as a single IR laser can excite multiple fluorophores in a given focal region, and the resulting fluorescence emissions in the visible region from two dyes can easily be chosen for spectral separation from the IR laser excitation and from each other (Figures 29.8 and 29.9).

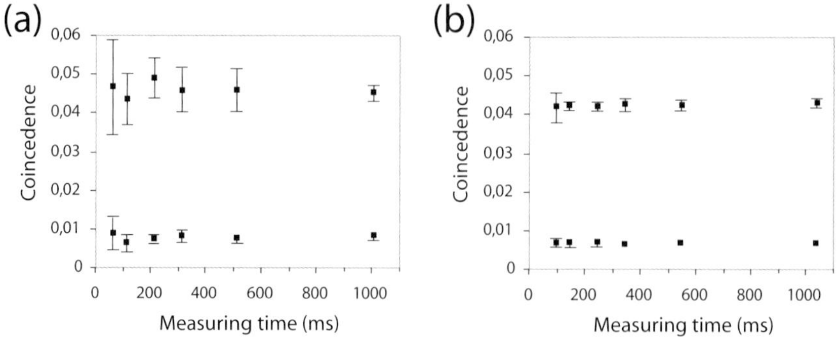

Figure 29.8 Comparison of fluorescence coincidence analysis of (a) single-focus detection (b) with 12-focus detection. For the coincidence analysis, a solution of fluorophores were used to represent fully coupled binding partners (upper data set) or fully separated binding partners (lower data set). Note the better assay quality in the 12-focus approach, even at 1/10 the data collection time per sample. For details, see Ref. [40].

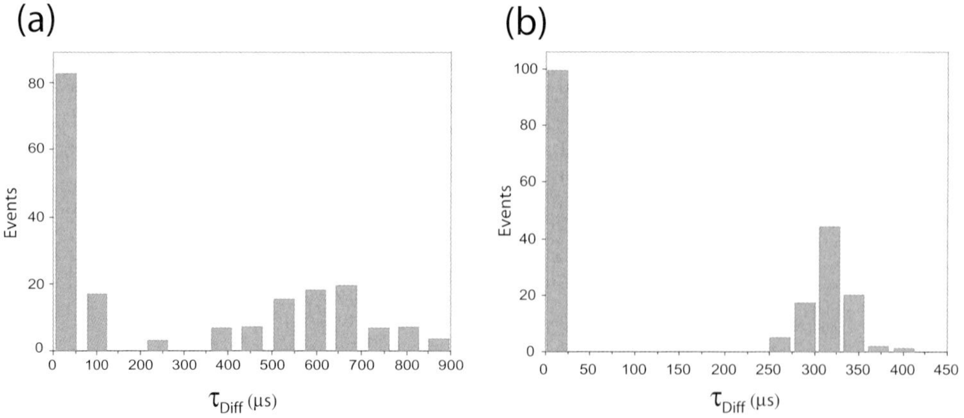

Figure 29.9 Fluorescence correlation analysis comparison for diffusion of the fluorescence dye, TMR (tetra-methyl rhodamine), and TMR-labeled BSA (bovine serum albumin), measured with (a) one-photon, one-focus detection, versus with (b) two-photon, 12-focus detection. The measuring time was 1 s per sample. Note the more precise determination of diffusion time in the latter approach. For details, see Ref. [40].

Another advantage of using ultrafast laser excitation for TPE is the intrinsically pulsed laser excitation. Since ultrafast detection electronics enable time-correlated single-photon counting, the pulsed laser allows fluorescence lifetime analysis for a variety of fluorescent samples [45, 46]. This application also takes advantage of gated detection on the nanosecond timescale for further reduced excitation background from particulates or other sources.

Finally, it may appear that the cost of ultrafast laser systems is a factor in deciding on its use for HTS applications. The cost of the laser alone typically is of the order of several hundred thousand Euros, although in comparison to the non-reader capital costs of a complete, fully automated HTS platform, the expenditure for such equipment is well within the budget of many well-funded efforts in automated HTS protein engineering.

As shown in Section 29.3, many of the above-described approaches based on ultrafast, nonlinear and multiphoton, multifocus excitation of fluorophores have already been routinely implemented in the HTS protein engineering environment. However, to our knowledge this has only been established in our laboratories.

It is expected that the implementation of the above-mentioned recent advances in the fields of laser technology, solid-state detection devices and fluctuation-based data analysis will continue to drive the ever-expanding range of applications of the confocal assay format. We anticipate further developments in this field based upon a number of recently published developments. Fluorescence fluctuation data gathered by confocal methods, for example, is expected to continue to increase in accuracy. At present, a number of dynamic processes can cause fluorescence intensity fluctuations which, at present, are not typically compensated. However, new spatial brightness distribution calculations have, for example, recently been created to allow a better reflection of the 'true' spatial brightness distribution [47].

In another development, solutions are being developed for interferences in two-color and FRET HTS assays. These involve two-color global fluorescence correlation spectroscopy (2CG-FCS), which can be used to perform a global analysis of the simultaneously recorded auto- and cross-correlation data from two-photon detectors that monitor the fluorescence emission of different colors. This advance appears better than traditional auto- and cross-correlation analysis at resolving and quantifying fluorescent species which differ in their diffusional characteristics and/or their molecular brightness. It is thus able to better compensate for photobleaching, crosstalk or concentration variations in sample preparation, while still allowing data collection times of less than 2 s, which makes it suitable for HTS applications. This method has recently been combined with, and significantly improves, FRET assay applications [48].

Beyond the above, however, is the possibility to similarly exploit other nonlinear and ultrafast optical detection technologies. A multitude of these techniques have either been demonstrated or are currently being established in basic research applications. Examples include the label-free detection and identification of biomolecular species and nanoparticles via the nonlinear CARS (coherent anti-stokes Raman scattering) technology [49]. Recently, it has also been shown that classical femtosecond-pump-probe spectroscopy can be used for HTS analysis of biomolecular reactions [50]. Although this study was based on changes in the ultrafast response of a marker molecule, in principle the complex ultrafast response of the biomolecules themselves may also be suitable for label-free, HTS analysis in analytes containing low background interferences. It is difficult to predict with certainty those advances which will become the widely used tools of the protein engineer, those which will find only specialized use, or which of the more speculative among these may never be used in the screening of combinatorial protein libraries. It is certain, however, that a myriad of advances in a range of technical disciplines will continue to drive the rapid progress in fluorescence-based assays for HTS protein engineering.

Abbreviations

APC	Allophycocyanin
APD	Avalanche photo diode
CCD	Charge-coupled device
DNA	Deoxyribonucleic acid
FA	Fluorescence anisotropy
FACS	Fluorescence-activated cell sorting
FCS	Fluorescence correlation spectroscopy
FIDA	Fluorescence intensity distribution analysis
FRET	Förster or Fluorescence resonance energy transfer
GFP	Green fluorescent protein
He-Ne	Helium-Neon
HTS	High-throughput screening
IR	Infrared
LED	Light-emitting diode
PMT	Photomultiplier tube
RNA	Ribonucleic acid
SBS	Society for Biomolecular Sciences
TPE	Two-photon excitation
TR-FRET	Time-resolved fluorescence resonance energy transfer

Acknowledgments

The authors thank Dr. rer. nat. Michael Lammers for critically reading this manuscript, and Dipl. Inform. Otto Birr for assistance in its preparation. The authors also thank Dr. Ing. Christian Votsmeier for critically reading the manuscript and for providing information about the TR-FRET assay that he established with H. Plittersdorf for monoclonal antibody binding at DIREVO Biotech AG.

References

1 Leisola, M. and Turunen, O. (2007) Protein engineering: opportunities and challenges. *Applied Microbiology and Biotechnology*, **75**, 1225–32.

2 Chica, R.A., Doucet, N. and Pelletier, J.N. (2005) Semi-rational approaches to engineering enzyme activity: combining the benefits of directed evolution and

rational design. *Current Opinion in Biotechnology*, **16**, 378–84.

3 Chaparro-Riggers, J.F., Polizzi, K.M. and Bommarius, A.S. (2007) Better library design: data-driven protein engineering. *Biotechnology Journal*, **2**, 180–91.

4 Arnold, F.H. and Georgiou, G. (2003) *Directed Evolution Library Creation Methods and Protocols in Methods in Molecular Biology*, Vol. **230**, Humana Press, Inc., Totowa.

5 Pompon, D. and Nicolas, A. (1989) Protein engineering by cDNA recombination in yeasts: shuffling of mammalian cytochrome P-450 functions. *Gene*, **83**, 15–24.

6 Stemmer, W.P. (1994) DNA shuffling by random fragmentation and reassembly: in vitro recombination for molecular evolution. *Proceedings of the National Academy of Sciences of the United States of America*, **91**, 10747–51.

7 Volkov, A.A., Shao, Z. and Arnold, F.H. (1999) Recombination and chimeragenesis by in vitro heteroduplex formation and in vivo repair. *Nucleic Acids Research*, **27**, e18.

8 Coco, W.M., Levinson, W.E., Crist, M.J., Hektor, H.J., Darzins, A., Pienkos, P.T., Squires, C.H. and Monticello, D.J. (2001) DNA shuffling method for generating highly recombined genes and evolved enzymes. *Nature Biotechnology*, **19**, 354–9.

9 Zhao, H.M., Giver, L., Shao, Z.X., Affholter, J.A. and Arnold, F.H. (1998) Molecular evolution by staggered extension process (step) in vitro recombination. *Nature Biotechnology*, **16**, 258–61.

10 Coco, W.M., Encell, L.P., Levinson, W.E., Crist, M.J., Loomis, A.K., Licato, L.L., Arensdorf, J.J., Sica, N., Pienkos, P.T. and Monticello, D.J. (2002) Growth factor engineering by degenerate homoduplex gene family recombination. *Nature Biotechnology*, **20**, 1246–50.

11 Ness, J.E., Kim, S., Gottman, A., Pak, R., Krebber, A., Borchert, T.V., Govindarajan, S., Mundorff, E.C. and Minshull, J. (2002) Synthetic shuffling expands functional protein diversity by allowing amino acids to recombine independently. *Nature Biotechnology*, **20**, 1251–5.

12 Ortaldo, J.R., Mason, A., Rehberg, E., Moschera, J., Kelder, B., Pestka, S. and Herberman, R.B. (1983) Effects of recombinant and hybrid recombinant human leukocyte interferons on cytotoxic activity of natural killer cells. *The Journal of Biological Chemistry*, **258**, 15011–15.

13 Meister, A., Uze, G., Mogensen, K.E., Gresser, I., Tovey, M.G., Grutter, M. and Meyer, F. (1986) Biological activities and receptor binding of two human recombinant interferons and their hybrids. *The Journal of General Virology*, **67**, 1633–43.

14 Streuli, M., Hall, A., Boll, W., Stewart, W.E., Nagata, S. and Weissmann, C. (1981) Target cell specificity of two species of human interferon-alpha produced in *Escherichia coli* and of hybrid molecules derived from them. *Proceedings of the National Academy of Sciences of the United States of America*, **78**, 2848–52.

15 Arnold, F.H. and Georgiou, G. (2003) *Directed enzyme Evolution Screening and Selection Methods in Methods in Molecular Biology*, Vol. **231**, Humana Press, Inc., Totowa.

16 Tawfik, D.S. and Griffiths, A.D. (1998) Man-made cell-like compartments for molecular evolution. *Nature Biotechnology*, **16**, 652–6.

17 Powell, K.T. and Weaver, J.C. (1990) Gel microdroplets and flow cytometry: rapid determination of antibody secretion by individual cells within a cell population. *Biotechnology (N. Y.)*, **8**, 333–7.

18 Fox, R.J., Davis, S.C., Mundorff, E.C., Newman, L.M., Gavrilovic, V., Ma, S.K., Chung, L.M., Ching, C., Tam, S., Muley, S., Grate, J., Gruber, J., Whitman, J.C., Sheldon, R.A. and Huisman, G.W. (2007) Improving catalytic function by ProSAR-driven enzyme evolution. *Nature Biotechnology*, **25**, 338–44.

19 Lakowicz, J.R. (2006) *Principles of Fluorescence Spectroscopy*, Springer, Berlin.

20 Schuster, S., Enzelberger, M., Trauthwein, H., Schmid, R.D. and Urlacher, V.B. (2005) pHluorin-based in vivo assay for hydrolase screening. *Analytical Chemistry*, **77**, 2727–32.

21 Waldo, G.S. (2003) Genetic screens and directed evolution for protein solubility. *Current Opinion in Chemical Biology*, **7**, 33–8.

22 Eigen, M. (1971) Self-organization of matter and the evolution of biological macromolecules. *Naturwissenschaften*, **58**, 465–523.

23 Eggeling, C., Fries, J.R., Brand, L., Günther, R. and Seidel, C.A.M. (1998) Monitoring conformational dynamics of a single molecule by selective fluorescence spectroscopy. *Proceedings of the National Academy of Sciences of the United States of America*, **95**, 1556–61.

24 Jager, S., Brand, L. and Eggeling, C. (2003) New fluorescence techniques for high-throughput drug discovery. *Current Pharmaceutical Biotechnology*, **4**, 463–76.

25 Haupts, U., Rudiger, M. and Pope, A.J. (2001) Macroscopic versus microscopic fluorescence techniques in (ultra)-high-throughput screening. *Drug Discovery*, **6**, 3–9.

26 Moore, K.J., Turconi, S., Ashman, S., Ruediger, M., Haupts, U., Emerick, V. and Pope, A.J. (1999) Single molecule detection technologies in miniaturized high throughput screening: Fluorescence correlation spectroscopy. *Journal of Biomolecular Screening*, **4**, 335–53.

27 Jameson, D.M. and Seifried, S.E. (1999) Quantification of protein-protein interactions using fluorescence polarization. *Methods*, **19**, 222–33.

28 Haupts, U., Maiti, S., Schwille, P. and Webb, W.W. (1998) Dynamics of fluorescence fluctuations in green fluorescent protein observed by fluorescence correlation spectroscopy. *Proceedings of the National Academy of Sciences of the United States of America*, **95**, 13573–8.

29 Maiti, S., Haupts, U. and Webb, W.W. (1997) Fluorescence correlation spectroscopy–diagnostics for sparse molecules. *Proceedings of the National Academy of Sciences of the United States of America*, **94**, 11753–7.

30 Comley, J. (2006) TR-FRET based assays–getting better with age. *Drug Discovery World*, **7**, 22–38.

31 Glickman, J.F., Wu, X., Mercuri, R., Illy, C., Bowen, B.R., He, Y. and Sills, M. (2002) A comparison of ALPHAScreen, TR-FRET, and TRF as assay methods for FXR nuclear receptors. *Journal of Biomolecular Screening*, **7**, 3–10.

32 Keppler, A., Pick, H., Arrivoli, C., Vogel, H. and Johnsson, K. (2004) Labeling of fusion proteins with synthetic fluorophores in live cells. *Proceedings of the National Academy of Sciences*, **101**, 9955–9.

33 Tirat, A., Freuler, F., Stettler, T., Mayr, L.M. and Leder, L. (2006) Evaluation of two novel tag-based labelling technologies for site-specific modification of proteins. *International Journal of Biological Macromolecules*, **39**, 66–76.

34 Zhang, J.H., Chung, T.D.Y. and Oldenburg, K.R. (1999) A simple statistical parameter for use in evaluation and validation of high throughput screening assays. *Journal of Biomolecular Screening*, **4**, 67–73.

35 Li, J., Cook, R., Dede, K. and Chaiken, I. (1996) Single chain human interleukin 5 and its asymmetric mutagenesis for mapping receptor binding sites. *The Journal of Biological Chemistry*, **271**, 1817–20.

36 Brakmann, S., Peuker, H., Simm, W., Kettling,, U., Koltermann, A., Stephan, J., Winkler, T., Doerre, K and Manfred, E. (2001) Structured reaction substrate, Patent WO0124933.

37 Denk, W., Strickler, J.H. and Webb, W.W. (1990) Two-photon laser scanning fluorescence microscopy. *Science*, **248**, 73–6.

38 Friedrich, D.M. (1982) Two-photon molecular spectroscopy. *Journal of Chemical Education*, **59**, 472–81.

39 Schwille, P. (2001) *Fluorescence Correlation Spectroscopy–Theory and Applications* (eds R. Rigler and E. Elson), Springer-Verlag, Heidelberg, pp. 360–78.

40 Walla, P.J., Kettling, U., Koltermann, A. and Scharte, M. (2004) Multi-Parameter fluorimetric analysis in a massively parallel multi-focal arrangement and the use thereof, Patent EP 1411345.

41 Straub, M. and Hell, S.W. (1998) Multifocal multiphoton microscopy: a fast and efficient tool for 3-D fluorescence imaging. *Bioimaging*, **6**, 177–85.

42 Kask, P., Palo, K., Ullmann, D. and Gall, K. (1999) Fluorescence-intensity

distribution analysis and its application in biomolecular detection technology. *Proceedings of the National Academy of Sciences of the United States of America*, **96**, 13756–61.

43 Burkhart, M. and Schwille, P. (2006) Electron multiplying CCD based detection for spatially resolved fluorescence correlation spectroscopy. *Optics Express*, **14**, 5013–20.

44 Kettling Ulrich, D.R., Koltermann Andre, D.R., Schwille, P. and Eigen, M. (1998) Real-time enzyme kinetics monitored by dual-color fluorescence cross-correlation spectroscopy. *Biochemistry*, **95**, 1416–20.

45 Fries, J.R., Brand, L., Eggeling, C., Köllner, M. and Seidel, C.A.M. (1998) Quantitative identification of different single molecules by selective time-resolved. *Journal of Physical Chemistry*, **102**, 6601–13.

46 Brand, L., Eggeling, C., Zander, C., Drexhage, K.H. and Seidel, C.A.M. (1997) Single-molecule identification of coumarin-120 by time-resolved fluorescence detection: comparison of one- and two-photon excitation in solution. *Journal of Physical Chemistry A*, **101**, 4313–21.

47 Palo, K., Mets, U., Loorits, V. and Kask, P. (2006) Calculation of photon-count number distributions via master equations. *Biophysical Journal*, **90**, 2179–91.

48 Eggeling, C., Kask, P., Winkler, D. and Jager, S. (2005) Rapid analysis of Forster resonance energy transfer by two-color global fluorescence correlation spectroscopy: trypsin proteinase reaction. *Biophysical Journal*, **89**, 605–18.

49 Volkmer, A., Cheng, J.-X. and Xie, X.S. (2001) Vibrational imaging with high sensitivity via epidetected coherent anti-stokes raman scattering microscopy. *Physical Review Letters*, **87**, 023901.

50 Quentmeier, C.C., Wehling, A. and Walla, P.J. (2007) A bioassay based on the ultrafast response of a reporter molecule. *Journal of Biomolecular Screening*, **12**, 341–50.

30
Alteration of Substrate Specificity and Stereoselectivity of Lipases and Esterases

Dominique Böttcher, Marlen Schmidt and Uwe T. Bornscheuer

30.1
Introduction

Lipases (EC 3.1.1.1, triacylglycerol hydrolases) and esterases (EC 3.1.1.3, carboxyl esterases) belong to the enzyme class of hydrolases. These enzymes have been the subject of extensive academic research for many decades, and are very important biocatalysts for the industrial production of bulk chemicals and pharmaceuticals [1–4]. The wide application of these enzymes can be mostly attributed to their high stability, their activity in a range of solvent systems (and especially in pure organic solvents), and their extremely broad substrate specificity combined with usually very good to excellent enantioselectivity. In addition, they do not require cofactors, which makes their use rather simple and versatile. Furthermore, a considerable number of lipases – and to a smaller extent also esterases – are commercially available on a bulk scale at reasonable prices. The most important applications are in organic synthesis, where lipases and esterases can be regarded as the 'work-horses' of biocatalysis and are used not only in kinetic resolutions or asymmetric synthesis, but also for the deprotection or synthesis of simple esters.

Independent of the reaction studied and enzyme investigated for a given biocatalytic process, very often the enzyme does not meet the requirements for a large-scale application, and its properties must therefore be optimized. Hence, not only the chemoselectivity, regioselectivity and stereoselectivity of the biocatalyst must be determined, but also details of any process-related aspects such as long-term stability at certain temperatures or pH-values, as well as enzymatic activity in the presence of high substrate concentrations to achieve the highest productivity. In addition to the rather classical strategies such as immobilization, additives or process engineering, molecular biology techniques today represent the most important methodologies for tailoring the design of an enzyme for a given process. Here, two different – but increasingly complementary – strategies are the methods of choice: (i) rational protein design; or (ii) directed (molecular) evolution (Figure 30.1). Consequently, this chapter focuses on the major achievements made during the past decade in the protein engineering of lipases and esterases using these methodologies.

Protein Engineering Handbook. Edited by Stefan Lutz and Uwe T. Bornscheuer

ISBN: 978-3-527-31850-6

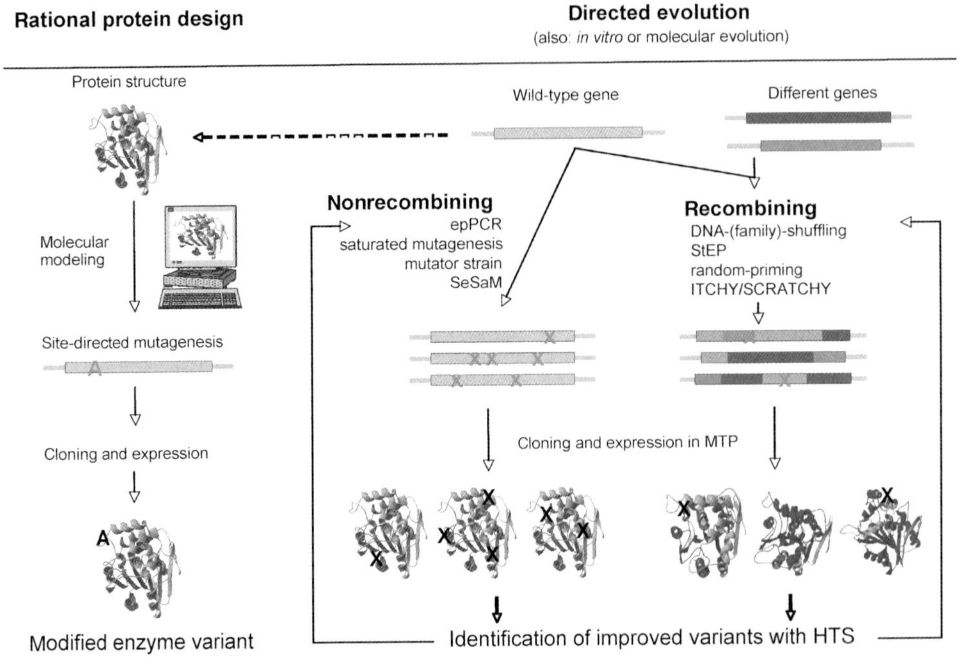

Figure 30.1 Schematic comparison of rational protein design and directed evolution. Rational design starts from a protein structure (or a homology model), from which key amino acid residues are identified. These are then introduced on the gene level and the resulting mutant is produced and verified for desired properties. Directed evolution starts from one or several (homologous) genes, which are subjected to a range of random mutagenesis methods. From the resulting libraries of mutants or chimeras, desired variants are then, after production in a microtiter plate (MTP) format, identified by screening or selection.

Interestingly, the majority of the techniques for directed evolution and the examples described elsewhere and in this chapter have been developed to improve or alter the properties of these hydrolases. This is related not only to the great importance of these biocatalysts but also to their ease of application as they do not require cofactors (in contrast to e.g. NAD(P)H-dependent dehydrogenases) or further enzymes (e.g. P450 monooxygenases need to reductase) to maintain their function.

30.2 Background of Protein Engineering Methods

30.2.1 Directed Evolution

In principle, directed evolution is comprised of two steps: (i) the random generation of mutant libraries; and (ii) the identification of desired variants within these libraries using a suitable screening or selection system. Two different strategies for the generation of mutant libraries have been described (Figure 30.1):

- Asexual (nonrecombining) evolution, in which a parent gene is subjected to random mutagenesis to yield variants with point mutations.
- Sexual (recombining) evolution, in which several parental genes are randomly fragmented, shuffled and reconstructed to create a pool of recombined chimera.

During the past 15 years a plethora of methods have been developed, the most widely used of which are described briefly in the following sections. Here, emphasis is placed on the tools applied to lipases or esterases; for a deeper and more detailed coverage, the reader is referred to a review [5], and also to Chapters 9, 16, 17, 20, 21, 24, 27, 28, 29).

The most widely used nonrecombining method is the error-prone polymerase chain reaction (epPCR). Here, nonoptimal reaction conditions are used to create a mutant library [6, 7]. For example, increasing the Mg^{2+} concentration, adding Mn^{2+} and the use of unbalanced dNTP concentrations, can substantially increase the error rate of the commonly used polymerase from *Thermus aquaticus* (*Taq*) from 0.001 to ~1%. The error rate should kept low enough to generate adaptive mutations, as a higher error rate will lead to deleterious mutations or inactive variants. Nevertheless, a homogeneous mutational spectrum in an unbiased library can hardly be achieved using the *Taq* DNA polymerase. One option is to use other polymerases; that is, the Mutazyme kit from Stratagene.

Another very simple way to obtain mutant libraries is to use mutator strains, such as the commercially available *Epicurian coli* XL1-Red strain (Stratagene). This strain has three defects in its DNA repair pathways (*mut*D, interferences in 3′5′-exonuclease activity; *mut*S, deficient mismatch repairing; and *mut*T, no hydrolysis of 8-oxodGTP), and exhibits a mutation rate which is approximately 5000 times higher than that of the wild-type *E. coli* strain; moreover, the mutations are maintained during replication. The only disadvantages compared to the epPCR are that the mutation rate cannot be adjusted, and that the entire plasmid is mutated and not only the gene of interest.

The first recombining method, named DNA shuffling or gene shuffling, was developed by Stemmer, and also found application in the generation of lipase libraries (see Section 30.4). The procedure consists of a DNase-catalyzed degradation followed by subsequent recombination of the fragments without primers (self-priming PCR) and finally a PCR with primers. This method was improved during the past decade and is now also applied as DNA family shuffling or molecular breeding. A related method is MUltiplex-PCR-based RECombination (MUPREC), which already has been applied to improve lipase evolution [8]. MUPREC was claimed to be more useful in the combination of epPCR-derived libraries containing various point mutations. Within the primers, which anneal on regions of the mutation sites to be recombined, two different vectors bearing the same gene and outside primers (one binds forward at one vector and another reverse at the second vector) are needed, and the procedure can take place in one or in two separated steps. The advantages here are that no amplification of the template occurs and no undesired base substitutions take place.

Other mutagenesis methods, which were used especially for the generation of lipase and esterase libraries, include circular permutation and SIMPLEX; details of these, together with examples of their application, are described in Section 30.4.

30.2.2 **Rational Design**

The engineering of a protein by rational design requires the enzyme's tertiary structure to be available, or at least a homology model of sufficient quality. Furthermore, detailed information regarding the structure–function relationship (and usually also the reaction mechanism) is required to allow for the prediction of amino acid residues to be mutated. Rational protein design is, therefore, a highly information-intensive method. During the past few decades, the number of protein structures deposited in the Brookhaven protein database (pdb) and sequence information in various databases has substantially facilitated the rational design of proteins. Furthermore, a plethora of modeling software has been developed, which not only makes this methodology more easy to use but also enhances the success rate of modeling predictions. Usually, the information derived from computer modeling identifies certain amino acids ('hot spots') which should be altered to lead to a change in the enzyme's properties; examples include a broadened or restricted substrate range or an altered enantioselectivity. Site-directed mutagenesis (SDM) is then carried out at these positions using, for example, the QuikChange SDM method from Stratagene. In many cases, it may be more advantageous to perform a saturation mutagenesis directly at the selected position(s), as this will introduce all 19 proteinogenic amino acids and hence increase the chances of finding desired variants [9–11]. A combination of rational protein design with directed evolution has been described by Reetz and co-workers (Iterative Saturation Mutagenesis (ISM) and subcategory CASTing (Combinatorial Active-Site Saturation Test)) [9, 10, Chapter 16] and researchers from Codexis (ProSAR) [12].

For example, ISM was used to increase the thermostability and enantioselectivity of an esterase [9–11]. First, the enzyme's crystal structure was studied to identify all of the amino acid positions, which appeared to be important for improvement. Next, all sites were subjected to saturation mutagenesis and the libraries screened for the property of interest. Subsequently, the best hit in each library was used as a template for further rounds of saturation mutagenesis, which was then continued in an iterative manner. CASTing focuses rather on the active site and its surrounding region in a radius of ~10 Å. Next, those amino acid residues pointing towards the substrate were subjected to *simultaneous* saturation mutagenesis, followed by a screening of the library for improved properties. A computer program (CASTER), which facilitates the design of CASTing libraries is available (at no cost) at http://www.kofo.mpg.de/kofo/institut/arbeitsbereiche/reetz/deutsch/reetz_forschung1.html.

30.3 Assay Systems

Rational protein design usually leads to the prediction of only a handful of variants, which can be analyzed using standard analytical methods such as gas chromatography (GC) or high-performance liquid chromatography (HPLC). In sharp contrast, directed evolution generates huge mutant libraries, and the key to a successful directed evolution experiment is therefore the availability of rapid and highly reliable assay systems enabling high-throughput as the variants in a library (typically in the order of 10^4–10^6) can hardly be identified after the individual expression of variants using classical analytical tools. This section provides an overview concerning suitable screening or selection systems to determine the activity and enantioselectivity of esterases and lipases in a high-throughput fashion. Another very important requirement here is that the assay system should be as close as possible to the problem of interest, because "... you get what you screen for". A broad overview of recently described assays can be found in Ref. [13] and in a number of reviews [14–18, Chapter 28].

30.3.1 Selection

Selection is a very powerful tool which is used for the screening of very large libraries and hence the discovery of protein mutants. Unfortunately, its application is rather limited as these biological selections are based on the complementation of key enzymes in pathways, auxotrophy or resistance to cytotoxic agents such as antibiotics. In the context of lipase or esterase screening, selection is usually difficult to implement; however, selections can be carried out *in vitro* or *in vivo* and either in solid-phase or microtiter plates.

30.3.1.1 Display Techniques

Phage display is one of the most commonly used techniques for the *in vitro* selection of the 'fittest' mutants in a large library. The procedure consists of cloning the gene of interest (in this case each individual in a library of mutants) in fusion with a gene encoding a coat protein of the virion. A physical linkage between the gene and expression product is achieved by means of a phage particle. The phages are then captured by affinity interaction of the displayed enzyme with an immobilized ligand. The nature of this binder depends on the enzyme; for example, the tag can be a substrate, or a suicide-substrate coupled to biotin (which is subsequently captured on streptavidin beads), or an immobilized transition-state analogue. The selected phages are eluted, replicated and amplified by simple infection [19, Chapter 24].

Protein libraries can be displayed not only on the surface of bacteriophages but also on bacteria and yeast. Bacterial display has certain advantages over the much more widely used phage display, which is applied primarily in antibody research:

- Only one host is needed to propagate the library, compared to two (the bacteriophage and bacterium) in phage display.
- The selected variants can be directly amplified without further transfer of the genetic material to another host.
- The risk of affinity artifacts due to avidity effects may be less pronounced.

The bacterial display of an esterase from *Burkholderia gladioli* was achieved by Schultheiss *et al.* by using an artificial gene composed of the esterase gene and the essential autotransporter domains in *E. coli*. The esterase activity was successfully directed to the outer membrane fraction, as confirmed by different techniques [20].

Bacterial cell surface display coupled with flow cytometric screening has also been adapted to the screening of enzyme libraries [21]. In fact, this technology currently represents the only general approach for the quantitative examination of enzyme catalytic activity at the single cell level and in very large populations of mutants. Furthermore, the display of enzymes on the bacterial surface provides a free access of synthetic substrates to the enzyme.

The ability to form a physical link between a fluorescent product of a reaction and the cell that expresses the respective enzyme on its surface proved to be the key for the quantitative determination of catalytic activity at the single cell level. Today, several routes are available to display enzymes on the microbial cell surface, most of which have been developed for *E. coli* [22]. Very recently, Kolmar and coworkers showed that *E. coli* bacteria which display esterases or lipases on their cell surface together with horseradish peroxidase (HRP) are capable of hydrolyzing carboxylic acid esters of biotin tyramide. The tyramide radicals generated by the coupled lipase–peroxidase reaction were short-lived, and therefore became covalently attached to reactive tyrosine residues located in close vicinity to the surface of a bacterial cell that displayed hydrolase activity. Differences in cellular esterase activity were found to correlate well with the amount of biotin tyramide deposited on the cell surface. This selective biotin tyramide labeling of cells with lipase activity allowed their subsequent isolation by magnetic cell sorting [23].

30.3.1.2 *In vivo* Selection

In vivo selection can be performed when the target activity is essential for viability and growth, for example, overcoming increasing concentrations of antibiotics or providing an essential nutrient [24]. The principle was demonstrated by Reetz and Rüggeberg, using an example in which survival was coupled to the hydrolysis of a certain enantiomer that releases a growth-inhibiting compound [25]. Thus, microorganisms expressing the lipase variant with the adequate enantioselectivity cannot cleave this compound and therefore survive, promoting an effective enrichment of the culture in the enzyme variant with the desired enantioselectivity.

Often, selection is performed as a complementation approach, whereby only a mutated enzyme variant produces an essential metabolite [26, 27].

30.3.2
Screening

High-throughput screening (HTS) methods are used to substitute the elaborate techniques of GC, HPLC or mass spectrometry (MS), where usually only a hundred samples can be analyzed each day. In order to screen mutant libraries, it is advantageous to analyze 10^3–10^4 samples per day, preferably using either the true substrate or its surrogate derivative to determine the activity both rapidly and also very accurately. There should also be no background reaction, the use of standard equipment should be sufficient, and expensive reagents (i.e. isotopically labeled compounds) should be avoided. In addition, the detection limit should be as low as possible and the assay materials should be either commercially available or easily synthesized.

For libraries expressed in microorganisms, HTS may occasionally be performed directly on colonies growing in a solid culture, such as an agar plate. Assays conducted on agar-plated colonies typically enable the screening of $>10^4$ variants in a matter of days, but they are often limited in their sensitivity. This is because soluble products diffuse away from the colony and hence only very active variants are detected, or false-positive reactions may occur. Assays based on insoluble products have a higher sensitivity, but their scope is rather limited. Solid-phase screening relies on product solubilization following an enzymatic reaction that gives rise to a zone of clearance, a fluorescent product, a pH-shift visualized by a pH-indicator, or a strongly absorbing (chromogenic) product such as X-gal or α-naphthyl acetate and Fast Blue/Fast Red, as an example of esterase activity detection [28, 29]. Lipolytic activity can still be screened in a high-throughput format, on-plate, with triolein- or tributyrin-agar through halo formation. Alternatively, a high-throughput assay in solid phase was recently developed by Babiak and Reymond using esters of coumarin [30].

Unfortunately, however, many assays cannot be applied in a solid-phase format, and the individual clones must be grown and assayed in microtiter plates (MTP). These assays are significantly more time-consuming than solid-phase assays, but by using robot automation and colony-picking technology the throughput can be substantially increased. On the other hand, MTP-assays have the major advantage that screening provides significantly more information compared to a selection approach, as the activity can be directly and quantitatively measured and even allow the kinetics to be determined. Furthermore, screening enables a direct determination of the enantioselectivity of an enzyme, which is very often the key property that needs to be improved for industrial biocatalysis.

The hydrolytic activity of esters can be determined using a wide variety of substrates, preferentially using the 'true' compound of interest compared to surrogates – that is, non-natural substrates designed to provide an intense, detectable signal when they are converted by the enzyme. Nevertheless, not all activity assays are susceptible to implementation in a high-throughput format required for screening the vast libraries created by the mutagenesis protocols used in directed evolution. An example is a simple pH-stat assay using tributyrin or triolein emulsions as substrates. Colorimetric and fluorometric assays are undoubtedly the most widespread screens for determining hydrolytic activities, and involve the cleavage of an ester to yield a chromophore/fluorophore that can be measured photometrically. The most commonly used chromophores/fluorophores are *p*-nitrophenol, fluorescein, resorufin or coumarin (Table 30.1). The major disadvantages when using these artificial substrates are that they differ from the true substrate and hence can lead to false-positive hits; they are also often not available commercially.

An assay in which acetates can be directly used as 'true' substrates (i.e. an acetate of a secondary alcohol resembles the compound used in chemical syntheses of chiral alcohols) is based on a commercially available 'acetic acid' test (R-Biopharm GmbH, Darmstadt, Germany). This couples the hydrolysis of acetates with an acetate-dependent enzymatic cascade, leading to the stoichiometric formation of NADH [44]. If enantiomerically pure chiral (*R*)- and (*S*)-acetates are used in separate experiments assaying the same enzyme variant, this method also allows for determination of the apparent[1)] enantioselectivity (E_{app}).

One important disadvantage of the hydrolytic activity assays of lipases and esterases is the rather poor solubility of most substrates in aqueous media, and the risk of strong autohydrolysis at extreme pH or elevated temperature using chromogenic (i.e. *p*-nitrophenyl esters) or fluorogenic (i.e. umbelliferyl esters) substrates.

To circumvent this problem, two strategies have been described. First, the esters of *p*-nitrophenol or coumarin were replaced by the corresponding acyloxymethylethers, or diacylglycerol analogues. This renders the substrate much more stable, as the ester which is susceptible to enzymatic cleavage is separated from the chromophore (or fluorophore), avoiding autohydrolysis as the alcohol moiety is now a worse leaving group compared to coumarin or the *p*-nitrophenoxide ion. Depending on its particular structure, the cleaved alcohol is then directly decarboxylated or is first oxidized with periodate and then subjected to bovine serum albumin (BSA)-catalyzed β-elimination in order to release the chromophore/fluorophore [33, 46, 47]. This methodology is also applicable to the screening and characterization of enantioselective enzymes [48]. The disadvantages are the need to synthesize specifically designed substrates, and that only end-point measurements are possible rather than a quantification of the enzyme kinetics. A variation of the above-described method uses the back-titration with adrenaline of the sodium periodate consumed in the oxidation of the diol generated by enzymatic cleavage [41, 49].

1) The apparent enantioselectivity (E_{app}) is measured using the enantiomers in individual reactions by calculating the ratio of the reaction rates, whereas the true enantioselectivity (E_{true}) is based on a kinetic resolution using a racemic substrate. Calculation of E_{true} can then be performed using, for example, the equations developed by Chen et al. [45].

Table 30.1 Selected examples of chromogenic assay substrates for screening of esterases and lipases libraries.

Substrate	Detected product	Coupled reaction	Detection wavelength [nm]	Reaction type	Assay format	Reference(s)
Resorufin-(esters)	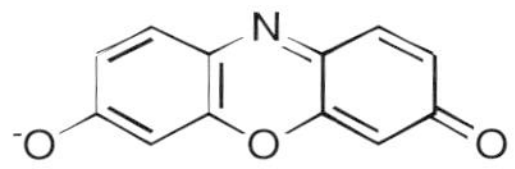	–	$570_{ex}/585_{em}$	Hydrolysis	MTP	[31, 32]
p-Nitrophenol-(esters)	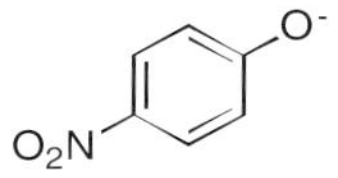	–	410	Hydrolysis, Synthesis	MTP	[33, 34, 35]
α-Naphthol-(esters)	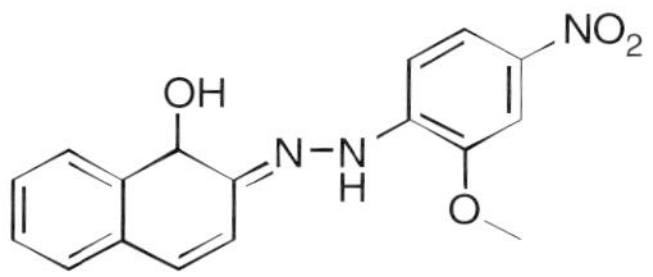 Diazo complex	Fast Red	–	Hydrolysis	Agar plate	[36]
Rhodamine B	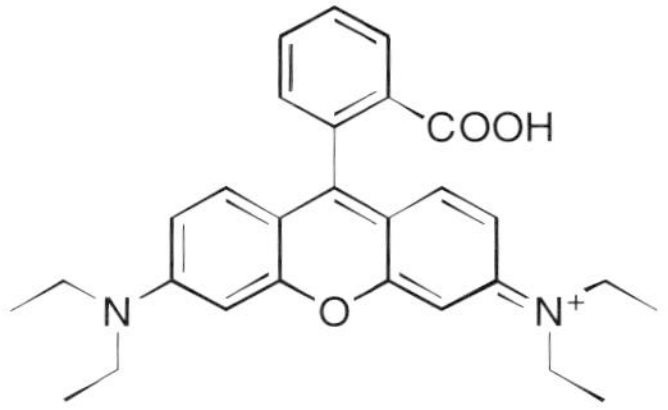	–	$485_{ex}/535_{em}$	Hydrolysis	Agar plate or MTP	[37, 38]

Table 30.1 *Continued*

Substrate	Detected product	Coupled reaction	Detection wavelength [nm]	Reaction type	Assay format	Reference(s)
Coumarins (e.g. umbelliferone)	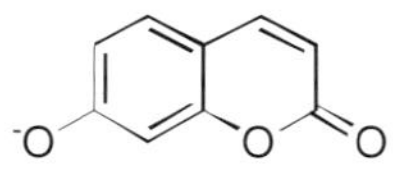	Sodium periodate/BSA	$360_{ex}/460_{em}$	Hydrolysis	MTP	[39, 40]
Adrenaline	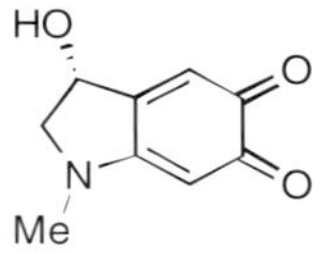 Adrenochrome	Sodium periodate	490	Hydrolysis	MTP	[41]
4-Hydrazino-7-nitro-2,1,3-benzoxadiazole (NBD-H)	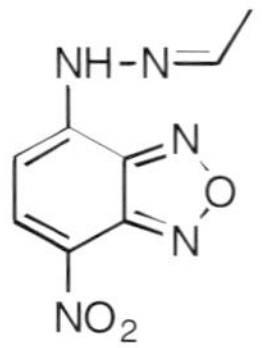	Acetaldehyde	$485_{ex}/520_{em}$	Synthesis	MTP	[42]
4-Amino-3-hydrazino-5-mercapto-1,2,4-triazole (Purpald)	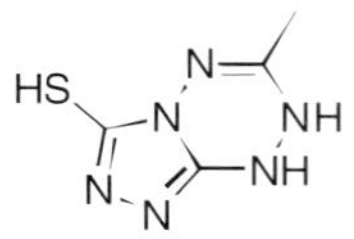	Acetaldehyde	–	Synthesis	Filter paper	[43]

In 1997, Janes and Kazlauskas described the first HTS method for enantioselectivity based on the separate hydrolysis of the enantiomers of *p*-nitrophenyl esters of a chiral acid, named 'Quick E'. In this method, E_{app} was determined by measuring initial rates of hydrolysis of pure enantiomers of 4-nitrophenyl-2-phenylpropanoate and a reference compound (resorufin tetradecanoate). The advantages were the short measurement time, the need for a much smaller amount of hydrolase, and the easy measurement of high enantioselectivities. However, some disadvantages were apparent, such as the need for pure enantiomers, and that the test could only be used with chromogenic substrates [50]. The Quick E method has been applied successfully in order to test the substrate selectivity of a series of hydrolases [51–53].

In order to use a racemate as substrate, a separation technique is needed prior to the quantification of each enantiomer. This separation can be made according to the chirality or, in the case of isotopically labeled substrates, according to the molecular mass. HPLC and GC have each been adapted to HTS in order to be used [54], and with this set-up approximately 700 measurements were carried out each day when screening a mutant library from *Pseudomonas aeruginosa* lipase for the enantioselective esterification of 2-phenylpropanol. MS is used with one isotopically labeled compound in an enantiomer pair in kinetic resolutions (a pseudo-racemate) [55] or in the biotransformation of a *meso*-compound. Capillary electrophoresis using chiral selectors (e.g. cyclodextrins) as a pseudo-stationary phase in the electrolyte has also been adapted to process as many as 96 samples in parallel, allowing the determination of 7000 samples of derivatized chiral amines per day [56].

Recently, a HTS method based on the coexpression of pHluorin, a pH-sensitive mutant of green fluorescent protein (GFP) and an esterase was reported. As the recombinant esterase catalyzed the hydrolysis of the substrate, an acid is released that causes an intracellular pH shift, leading to a change in the emission spectrum of pHluorin; the pH change can be quantified as its extent depends on the hydrolase activity. However, this technique is limited to substrates that are not converted by the *E. coli* host cell, and which can enter the cell either alone or with the help of a cosolvent such as dimethylsulfoxide (DMSO). The assay can be performed either in a MTP or using a flow cytometer, in order to separate cells expressing esterase activity from control cells expressing an inactive esterase [57]. Modern flow cytometers can be used for the routine analysis and sorting of $>10^6$ events per hour; moreover, fluorescence is a sensitive signal and can be widely adopted to detect both binding and enzymatic reactions. Indeed, flow cytometry has achieved much success in the screening of gene libraries to yield highly potent binding proteins such as antibodies [58–60]. The potential and utility of flow cytometry for enzyme selections are discussed in detail in Chapter 27.

The majority of assays allow only the determination of hydrolytic activity in aqueous media, despite the fact that the application of lipases or esterases is very often in (pure) organic solvents. Thus, only a limited number of assays for determining synthetic activity (i.e. esterification and transesterification) in HTS have been described [61–63]. A fluorometric method to determine transesterification by

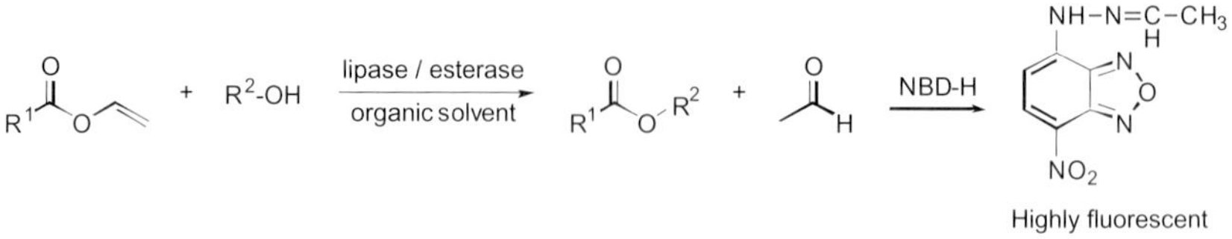

Scheme 30.1 High-throughput assay for the determination of transesterification activity of lipases and esterases.

lipases and esterases in organic solvents was described by Konarzycka-Bessler and Bornscheuer [42]. The assay is based on the transesterification between an alcohol and a vinyl ester of a carboxylic acid. Acetaldehyde generated from the vinyl alcohol by keto–enol tautomerization is reacted with a (nonfluorescent) hydrazine (NBD-H) to produce the corresponding highly fluorescent hydrazone, which is then quantified by fluorimetric measurement (Scheme 30.1). This principle allows the rapid identification of active enzymes, which could be demonstrated for a range of hydrolase enzymes in a MTP format and for a broad range of solvents (e.g. toluene, hexane, ether).

A recent publication described a screening approach for measuring esterase activity in an organic solvent using a modified *p*-nitrophenyl ester assay [64]. The transesterification activity of immobilized esterases was determined by sampling the *p*-nitrophenol released and its quantification by spectrophotometric measurement in a MTP format. A similar method was reported very recently for the determination of the synthesis activity of a lipase [65].

30.4 Examples

One of the first example for a successful directed evolution experiment was an esterase variant from *Bacillus subtilis* (BsubpNBE) with a 150-fold higher activity in 15% DMF compared to the wild-type, created by combing epPCR and shuffling. The enzyme is therefore applicable for the deprotection of a precursor in the production of the antibiotic Loracarbef in the presence of DMF as cosolvent [66].

The thermostability of the same enzyme was also increased by 14 °C in eight rounds of epPCR and recombination [67]. A similar esterase, BS2, which differs only by 11 amino acids from BsubpNBE, was evolved by rational design in our group, and the enantioselectivity towards an ester of the tertiary alcohol 2-phenyl-3-butin-2-yl acetate could be increased sixfold to E = 19, and towards linalyl acetate inverted from (*R*) to (*S*) preference with E = 6 [68]. In a later study, this mutant (Gly105Ala) showed a good enantioselectivity towards 2-phenyl-3-butin-2-yl acetate (E = 54) in 20% (v/v) DMSO, and an E-value of >100 towards the trifluoromethyl analogue [69]. Another point mutation (Glu188Asp) gave similar high enantioselectivity towards both substrates, as well as a series of other tertiary alcohol acetates [70]. By using a focused directed evolution approach, we recently created a library

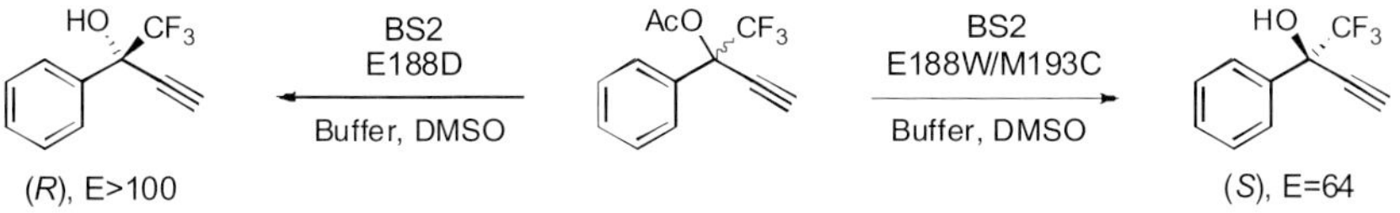

Scheme 30.2 Rational protein design identified a variant of *Bacillus subtilis* esterase (BS2) having excellent enantioselectivity towards the *(R)*-enantiomer, whereas a focused directed evolution approach led to an *(S)*-selective double mutant.

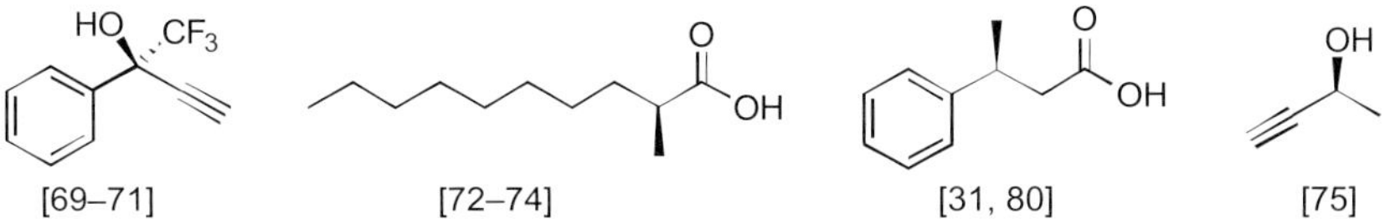

Scheme 30.3 Selected examples of chiral compounds obtained using designed esterases or lipases.

covering three residues by saturation mutagenesis ($20^3 = 8000$ theoretical variants) and screening with the acetate assay described above identified a double mutant (Glu188Trp/Met193Cys) with an *(S)*-preference and an E-value of ~70 towards 1,1,1-trifluoro-2-phenylbut-1-yn-3-ol, thus leading to a variant with completely inverted enantioselectivity [71] (Scheme 30.2). Notably, neither of the single mutants was found to be useful: Glu188Trp showed only E = 26 for the (*S*)-enantiomer, while Met193Cys was still (*R*)-selective (E = 16). Thus, only the combination (which is unlikely be obtained by 'normal' random mutagenesis) resulted in the desired and significant inversion of enantioselectivity. This synergistic manner is thus an excellent argument for using focused directed evolution.

Shortly after the first reports of an evolved esterase from the Arnold group, the directed evolution of a lipase from *Pseudomonas aeruginosa* was noted by Reetz and coworkers (see Chapter 16 for details) (Scheme 30.3). The initial enantioselectivity in the kinetic resolution of 2-methyl-decanoic acid *p*-nitrophenyl ester (MDA) was E = 1.1 (in favor of the (*S*)-acid), and after four rounds of epPCR an E = 11 was obtained. Further mutants were created by combining mutations on the positions identified to be critical in the generation of the best variants for every round, which led to the identification of a more enantioselective variant (E = 21) [72]. For this same reaction, a DNA-shuffling approach proved effective, yielding a variant that exhibited E = 32. Furthermore, a modified version of Stemmer's combinatorial multiple-cassette mutagenesis was applied to two of the obtained mutants and a mutagenic oligocassette, which allowed simultaneous randomization at previously determined 'hot spots'. This resulted in the most enantioselective variant (X, with six mutations), displaying a selectivity factor of E = 51. In addition, variants with good (*R*)-selectivity (E = 30) were also identified [73]. Interestingly, only one mutation was seen to be located next to the active site, whereas all other substitutions were remote. After a theoretical study, ten new variants were prepared and the double-mutant M8 (S53P/L162G), which was considered to show high enantioselectivity, gave an E-value of 64 at a conversion of 20% [74].

Attempts to resolve racemic mixtures of esters of secondary alcohols with mutants of an esterase from *Pseudomonas fluorescens* (PFE-I) have also been reported. The present authors' group used the mutator strain *E. coli* XL-1 Red to develop an enantiopreference in the hydrolysis of methyl 3-phenylbutyrate [31]. The same enzyme was also evolved by epPCR and the acetic acid assay for its application in the enantioselective hydrolysis of the very important building block 3-butyn-2-ol with an enantiomeric excess (ee) value >99% [75]. Here, only a single round of epPCR was necessary to obtain a variant containing three point mutations that showed increased enantioselectivity and was therefore useful for the production of chiral building block in optically pure form. Subsequently, the role of each mutation on the enantioselectivity, the reaction rate and solubility of the mutant were studied. Based on the results of these studies, it was concluded that mutations close to the active site may also have a substantial (negative) effect on protein folding, and that this point should receive more attention in protein engineering investigations. In contrast, Kazlauskas and coworkers showed, in a rational protein design study of PFE-I, that closer mutations are significantly better than random mutagenesis of the entire enzyme, as exemplified for the kinetic resolution of 3-bromo-2-methylpropionate. Here, mutagenesis of the substrate binding site resulted in up to fivefold better E-values (best: E = 61) compared to the wild-type esterase [76].

Candida antarctica lipase B (CAL-B) is probably the most useful lipase for organic synthesis, and numerous examples have been described [77]. Nevertheless, it was not possible to obtain more thermostable variants by rational design. When the directed evolution approach was used, variants were found after two rounds of epPCR, that were 20-fold more stable at 70 °C than the wild-type. Positions 221 and 281 were found to be critical for preventing irreversible inactivation and protein aggregation, and the variants were also found to be more active against *p*-nitrophenyl butyrate and 6,8-difluoro-4-methylumbelliferyl octanoate [78].

CAL-B was also engineered by shuffling its gene with those of lipases from *Hyphozyma* sp. CBS 648.91 and *Crytococcus tsukubaensis* ATCC 24555 in order to create a lipase B variant with increased activity in the hydrolysis of diethyl 3-(3′,4′-dichlorophenyl)glutarate, that yields a chiral synthon for the preparation of an NK1/NK2 dual antagonist [79].

The enantioselectivity of *Burkholderia cepacia* KWI-56 lipase was evolved towards (*R*)-enantioselectivity in the hydrolysis of 3-phenylbutyric acid *p*-nitrophenyl ester using a novel technique for the construction and screening of a protein library by single-molecule DNA amplification by PCR followed by an *in vitro*-coupled transcription/translation system (SIMPLEX). The library was generated by saturating separately four positions (L17, F119, L167 and L266), and then diluted until only five molecules of DNA were present per well of a MTP. These molecules were amplified using a single-molecule PCR product and expressed *in vitro*, as each gene fragment already carried a T7 promoter, a ribosome-binding site and a T7 terminator. The DNA corresponding to active wells showing the desired enantioselectivity was once again diluted to give one molecule per well, which was re-amplified and re-checked. The best mutant exhibited a selectivity factor of E = 38 towards the *(R)*-enantiomer, whereas the wild-type exhibited E = 33 for the (*S*) enantiomer [80].

A completely different strategy to evolve enzymes without introducing mutations was developed by the group of Lutz, namely circular permutation (CP) [81]. These authors linked the native N and C termini of the gene encoding CAL-B and subsequently linearized it by random digestion to yield variants bearing alternative N and C termini (see Chapters 18 and 21). Surprisingly, this not only led to an active lipase, but some variants also showed a higher catalytic efficiency in comparison to the wild-type (up to 11-fold against *p*-nitrophenol butyrate and 75-fold against 6,8-difluoro-4-methylumbelliferyl octanoate, while K_M values were almost the same). For the most active variant (cp283), kinetic experiments demonstrated that CP of this enzyme did not compromise the enantioselectivity in the resolution of some chiral secondary alcohols [82].

The chain-length selectivity of lipases was altered by rational protein design and SDM, as shown by Joerger and Haas for the *Rhizopus oryzae* (formerly *Rhizopus delemar*) lipase (RDL) [83, 84]. First, the authors used the crystal structure of *Rh. miehei* lipase to identify important residues in the substrate-binding region representing the best target for mutagenesis. Three years later, after the crystal structure of *Rh. oryzae* had been elucidated, they applied molecular modeling to identify the molecular determinants of acyl chain length specificity of this enzyme. In another directed evolution approach–while trying to isolate new enzyme variants of the extracellular lipase from *Thermomyces lanuginosa* with enhanced activity in the presence of detergent–Danielsen *et al.* randomized nine amino acids in two regions flanking the flexible α-helical lid. A S83T mutation was found in six of the seven most active variants, which in the homologous RDL had been proven to determine the chain-length preference [85].

Directed evolution and rational protein design can be also used to alter the regiospecificity of an enzyme (see e.g. the explanation of the regioselectivity of a lipase from *Ps. cepacia* in the acylation of 2′-deoxynucleosides [86]) and to have it accept other functional groups than in its natural substrate. Some enzymes already catalyze reactions on alternative functional groups, but at a very slow rate compared to their main catalytic function. This 'catalytic promiscuity' is currently a major research theme, and forms the subject of several recent reviews [87–89]. Already, Fujii *et al.* have reported an enhancement of the amidase activity of a *Pseudomonas aeruginosa* lipase after a single round of random mutagenesis. Mutant libraries were screened for hydrolytic activity against oleyl-naphthylamide and compared to the hydrolysis of the corresponding ester. Three mutational sites were identified as enhancing amidase activity, and the double-mutant Phe207Ser/Ala213Asp was found to have the highest amidase activity–twice that of the wild-type. These mutations were located near the calcium binding site, far from the active site [90].

Selected examples of improved lipases and esterases with applied mutagenesis methods and screening or selection assays are summarized in Table 30.2.

Table 30.2 Selected examples of esterases and lipases improved by directed evolution methods.

Enzyme (origin)	Target	Mutagenesis method	Assay	Improved property	Reference(s)
P. fluorescens esterase	Enantioselectivity towards tertiary alcohols	Mutator strain	MTP assay with resorufin esters	Increased enantioselectivity	[31]
P. fluorescens esterase	Enantioselectivity towards 3-butyn-2-ol	epPCR	Acetate assay	Increased enantioselectivity	[75]
P. fluorescens esterase	Enantioselectivity towards 3-bromo-2-methylpropionate	Saturation mutagenesis near the active site	*p*-NPA assay, QuickE	Increased enantioselectivity	[53]
P. fluorescens esterase	Enhanced stability	epPCR and StEP	MTP assay with heat treatment	Increase in thermostability and decrease in substrate inhibition	[91]
P. aeruginosa lipase	Enantioselectivity	epPCR, Saturation mutagenesis, Cassette mutagenesis	MTP assay with chiral *p*-nitrophenyl esters	Increased enantioselectivity	[72–74]
P. aeruginosa lipase	Improve amidase activity	epPCR, Saturation mutagenesis	Activity staining on agar plate	Twofold higher amidase activity	[90]
P. aeruginosa lipase	Substrate specificity	CASTing	MTP assay with *p*-nitrophenyl esters	Expanded substrate acceptance for different carboxylic acid esters	[92]
B. subtilis esterase	High activity in DMF towards Loracarbef	epPCR, Saturation mutagenesis Shuffling	MTP assay with *p*-nitrophenyl esters	Activity increased 150-fold in 15% DMF, increased temperature stability	[66]
B. subtilis esterase	Enantioselectivity towards tertiary alcohols	Rational design	–	Increased enantioselectivity	[70]

B. subtilis esterase	Enantioselectivity towards tertiary alcohols	Focused saturation mutagenesis (CASTing)	Acetate assay	Inversed enantioselectivity	[71]
B. cepacia lipase	Enantioselectivity towards 3-phenylbutyrate	SIMPLEX	MTP assay with *p*-nitrophenyl esters	Inversed enantioselectivity	[80]
R. oryzae lipase	Substrate specificity	Saturation mutagenesis	Agar-plates containing rhodamine B	Altered chain-length preference	[83, 84]
C. antarctica lipase B	Enantioselectivity towards halohydrins	Rational design	–	Increased enantioselectivity	[93]
C. antarctica lipase B	Improve activity towards 3-(3′,4′-dichlorophenyl) glutarate	Shuffling	pH indicator	20-fold higher activity	[79]
C. antarctica lipase B	Thermostability	epPCR, Saturation mutagenesis	In MTP with *p*-nitrophenyl esters and heat treatment	>20-fold improvement in half-life at 70 °C	[78]
B. gladioli esterase	Enantioselectivity towards hydroxyisobutyrate	epPCR, Saturation mutagenesis	MTP assay with *p*-nitrophenyl esters	Inverted enantioselectivity	[94]
B. gladioli esterase	Stability in organic solvents	epPCR	pH indicator	100-fold improvement of activity in 35% DMF	[95]
B. subtilis lipase A	Enantioselectivity towards 1,2-O-isopropylidene-glycerol	Saturation mutagenesis near the active site	Phage display using suicide inhibitor substrates	Inverted enantioselectivity	[96]
R. arrhizus lipase	Thermostability	epPCR, Shuffling	Agar-plates containing rhodamine B	Improved thermostability and higher temperature optimum	[97]

DMF = dimethylformamide.

30.5 Conclusions

Since the time when directed evolution was first described for protein engineering a few years ago, this technology has emerged as a very powerful tool for designing and altering the properties of enzymes. The method has quickly found application for a broad range of proteins, the vast majority of which are of interest to biocatalysis. Consequently, a diverse set of molecular biology tools to create not only well-balanced mutant libraries but also suitable HTS methods has been developed to make the application of directed evolution both simpler and feasible. Within only a decade, directed evolution has emerged as a standard methodology in protein design, and can be used either as a complement to, or in combination with, rational protein design to meet the demand for industrially applicable biocatalysts that demonstrate desirable chemoselectivity, regioselectivity and stereoselectivity. Moreover, such biocatalysts are capable not only of withstanding extreme conditions such as high substrate concentrations and exposure to solvents and high temperatures, but also of displaying long-term stability.

References

1 Bornscheuer, U.T. and Kazlauskas, R.J. (2006) *Hydrolases in Organic Synthesis: Regio- and Stereoselective Biotransformations*, 2nd edn, Wiley-VCH Verlag GmbH, Weinheim.
2 Drauz, K. and Waldmann, H. (eds) (2002) *Enzyme Catalysis in Organic Synthesis*, 2nd edn, Vol. **1–3**, Wiley-VCH Verlag GmbH, Weinheim.
3 Faber, K. (2004) *Biotransformations in Organic Chemistry. A Textbook*, 5th edn, Springer, Berlin.
4 Patel, R.N. (ed.) (2006) *Biocatalysis in the Pharmaceutical and Biotechnological Industries*, 1st edn, CRC Press/Taylor & Francis Group, New York.
5 Bornscheuer, U.T. and Pohl, M. (2001) Improved biocatalysts by directed evolution and rational protein design. *Current Opinion in Chemical Biology*, **5**, 137–43.
6 Caldwell, R.C. and Joyce, G.F. (1992) Randomization of genes by PCR mutagenesis. *PCR Methods and Applications*, **2**, 28–33.
7 Leung, D.W., Chen, E. and Goeddel, D.V. (1989) A method for random mutagenesis of a defined DNA segment using a modified polymerase chain reaction. *Technique*, **1**, 11–15.
8 Eggert, T., Funke, S.A., Rao, N.M., Acharya, P., Krumm, H., Reetz, M.T. and Jaeger, K.E. (2005) Multiplex-PCR-based recombination as a novel high-fidelity method for directed evolution. *ChemBioChem*, **6**, 1062–7.
9 Reetz, M.T. and Carballeira, J.D. (2007) Iterative saturation mutagenesis (ISM) for rapid directed evolution of functional enzymes. *Nature Biotechnology*, **2**, 891–903.
10 Reetz, M.T., Carballeira, J.D. and Vogel, A. (2006) Iterative saturation mutagenesis on the basis of B factors as a strategy for increasing protein thermostability. *Angewandte Chemie – International Edition*, **45**, 7745–51.
11 Reetz, M.T., Wang, L.W. and Bocola, M. (2006) Directed evolution of enantioselective enzymes: Iterative cycles of CASTing for probing protein-sequence space. *Angewandte Chemie – International Edition*, **45**, 1236–41; erratum 2494.
12 Fox, R.J., Davis, S.C., Mundorff, E.C., Newman, L.M., Gavrilovic, V., Ma, S.K.,

Chung, L.M., Ching, C., Tam, S., Muley, S., Grate, J., Gruber, J., Whitman, J.C., Sheldon, R.A. and Huisman, G.W. (2007) Improving catalytic function by ProSAR-driven enzyme evolution. *Nature Biotechnology*, **25**, 338–44.

13 Reymond, J.L. (ed.) (2005) *Enzyme Assays*, Wiley-VCH Verlag GmbH, Weinheim.

14 Bornscheuer, U.T. (2004) High-throughput-screening systems for hydrolases. *Engineering in Life Sciences*, **4**, 539–42.

15 Schmidt, M. and Bornscheuer, U.T. (2005) High-throughput assays for lipases and esterases. *Biomolecular Engineering*, **22**, 51–6.

16 Wahler, D. and Reymond, J.L. (2001) High-throughput screening for biocatalysts. *Current Opinion in Biotechnology*, **12**, 535–44.

17 Wahler, D. and Reymond, J.-L. (2001) Novel methods for biocatalyst screening. *Current Opinion in Biotechnology*, **5**, 152–8.

18 Goddard, J.P. and Reymond, J.-L. (2004) Recent advances in enzyme assays. *Trends in Biotechnology*, **22**, 363–70.

19 Soumillon, P. (2004) Selection of phage-displayed enzymes, in *Evolutionary Methods in Biotechnology. Clever Tricks for Directed Evolution* (ed. A. Schwienhorst), Wiley-VCH Verlag GmbH, Weinheim, pp. 47–64.

20 Schultheiss, E., Paar, C., Schwab, H. and Jose, J. (2002) Functional esterase surface display by the autotransporter pathway in *Escherichia coli*. *Journal of Molecular Catalysis B, Enzymatic*, **18**, 89–97.

21 Olsen, M.J., Stephens, D., Griffiths, D., Daugherty, P., Georgiou, G. and Iverson, B.L. (2000) Function-based isolation of novel enzymes from a large library. *Nature Biotechnology*, **18**, 1071–4.

22 Becker, S., Schmoldt, H.U., Adams, T.M., Wilhelm, S. and Kolmar, H. (2004) Ultra-high-throughput screening based on cell-surface display and fluorescence-activated cell sorting for the identification of novel biocatalysts. *Current Opinion in Biotechnology*, **15**, 323–9.

23 Becker, S., Michalczyk, A., Wilhelm, S., Jaeger, K.E. and Kolmar, H. (2007) Ultrahigh-throughput screening to identify *E. coli* cells expressing functionally active enzymes on their surface. *ChemBioChem*, **8**, 943–9.

24 Lorenz, P. and Eck, J. (2004) Screening for novel industrial biocatalysts. *Engineering in Life Sciences*, **4**, 501–4.

25 Reetz, M.T. and Rüggeberg, C.J. (2002) A screening system for enantioselective enzymes based on differential cell growth. *Chemical Communications*, 1428–9.

26 Jürgens, C., Strom, A., Wegener, D., Hettwer, S., Wilmanns, M. and Sterner, R. (2000) Directed evolution of a (beta alpha)(8)-barrel enzyme to catalyze related reactions in two different metabolic pathways. *Proceedings of the National Academy of Sciences of the United States of America*, **97**, 9925–30.

27 MacBeath, G., Kast, P. and Hilvert, D. (1998) Redesigning enzyme topology by directed evolution. *Science*, **279**, 1958–61.

28 Bornscheuer, U.T., Altenbuchner, J. and Meyer, H.H. (1999) Directed evolution of an esterase: Screening of enzyme libraries based on pH-indicators and a growth assay. *Bioorganic and Medicinal Chemistry*, **7**, 2169–73.

29 Bornscheuer, U.T., Altenbuchner, J. and Meyer, H.H. (1998) Directed evolution of an esterase for the stereoselective resolution of a key intermediate in the synthesis of epothilones. *Biotechnology and Bioengineering*, **58**, 554–9.

30 Babiak, P. and Reymond, J.L. (2005) A high-throughput, low-volume enzyme assay on solid support. *Analytical Chemistry*, **77**, 373–7.

31 Henke, E. and Bornscheuer, U.T. (1999) Directed evolution of an esterase from *Pseudomonas fluorescens*. Random mutagenesis by error-prone PCR or a mutator strain and identification of mutants showing enhanced enantioselectivity by a resorufin-based fluorescence assay. *Biological Chemistry*, **380**, 1029–33.

32 Kramer, D.N. and Guilbault, G.G. (1964) Resorufin acetate as substrate for determination of hydrolytic enzymes at low enzyme and substrate concentrations. *Analytical Chemistry*, **36**, 1662–3.

33 Grognux, J., Wahler, D., Nyfeler, E. and Reymond, J.L. (2004) Universal chromogenic substrates for lipases and esterases. *Tetrahedron: Asymmetry*, **15**, 2981–9.

34 Krebsfänger, N., Schierholz, K. and Bornscheuer, U.T. (1998) Enantioselectivity of a recombinant esterase from *Pseudomonas fluorescens* towards alcohols and carboxylic acids. *Journal of Biotechnology*, **60**, 105–11.

35 Kurioka, S. and Matsuda, M. (1976) Phospholipase-C assay using para-nitrophenylphosphorylcholine together with sorbitol and its application to studying metal and detergent requirement of enzyme. *Analytical Biochemistry*, **75**, 281–9.

36 Bornscheuer, U., Reif, O.W., Lausch, R., Freitag, R., Scheper, T., Kolisis, F.N. and Menge, U. (1994) Lipase of *Pseudomonas cepacia* for biotechnological purposes – purification, crystallization and characterization. *Biochimica et Biophysica Acta*, **1201**, 55–60.

37 Jette, J.F. and Ziomek, E. (1994) Determination of lipase activity by a rhodamine-triglyceride-agarose assay. *Analytical Biochemistry*, **219**, 256–60.

38 Kouker, G. and Jaeger, K.E. (1987) Specific and sensitive plate assay for bacterial lipases. *Applied and Environmental Microbiology*, **53**, 211–13.

39 Klein, G. and Reymond, J.L. (1999) Enantioselective fluorogenic assay of acetate hydrolysis for detecting lipase catalytic antibodies. *Helvetica Chimica Acta*, **82**, 400–7.

40 Wahler, D. and Reymond, J.L. (2001) High-throughput screening for biocatalysts. *Current Opinion in Biotechnology*, **12**, 535–44.

41 Wahler, D. and Reymond, J.L. (2002) The adrenaline test for enzymes. *Angewandte Chemie – International Edition*, **41**, 1229–32.

42 Konarzycka-Bessler, M. and Bornscheuer, U.T. (2003) A high-throughput-screening method for determining the synthetic activity of hydrolases. *Angewandte Chemie – International Edition*, **42**, 1418–20.

43 Wong, T.S., Schwaneberg, U., Stürmer, R., Hauer, B. and Breuer, M. (2006) A filter paper-based assay for laboratory evolution of hydrolases and dehydrogenases. *Combinatorial Chemistry and High Throughput Screening*, **9**, 289–93.

44 Baumann, M., Stürmer, R. and Bornscheuer, U.T. (2001) A high-throughput-screening method for the identification of active and enantioselective hydrolases. *Angewandte Chemie – International Edition*, **40**, 4201–4.

45 Chen, C.S., Fujimoto, Y., Girdaukas, G. and Sih, C.J. (1982) Quantitative analyses of biochemical kinetic resolutions of enantiomers. *Journal of the American Chemical Society*, **104**, 7294–9.

46 Klein, G. and Reymond, J.L. (1998) An enantioselective fluorimetric assay for alcohol dehydrogenases using albumin-catalyzed beta-elimination of umbelliferone. *Bioorganic and Medicinal Chemistry Letters*, **8**, 1113–16.

47 Leroy, E., Bensel, N. and Reymond, J.L. (2003) A low background high-throughput screening (HTS) fluorescence assay for lipases and esterases using acyloxymethylethers of umbelliferone. *Bioorganic and Medicinal Chemistry Letters*, **13**, 2105–8.

48 Grognux, J. and Reymond, J.L. (2004) Classifying enzymes from selectivity fingerprints. *ChemBioChem*, **5**, 826–31.

49 Wahler, D., Boujard, O., Lefevre, F. and Reymond, J.L. (2004) Adrenaline profiling of lipases and esterases with 1,2-diol and carbohydrate acetates. *Tetrahedron*, **60**, 703–10.

50 Janes, L.E. and Kazlauskas, R.J. (1997) Quick E. A fast spectroscopic method to measure the enantioselectivity of hydrolases. *The Journal of Organic Chemistry*, **62**, 4560–1.

51 Somers, N.A. and Kazlauskas, R.J. (2004) Mapping the substrate selectivity and enantioselectivity of esterases from thermophiles. *Tetrahedron Asymmetry*, **15**, 2991–3004.

52 Liu, A.M.F., Somers, N.A., Kazlauskas, R.J., Brush, T.S., Zocher, F., Enzelberger, M.M., Bornscheuer, U.T., Horsman, G.P., Mezzetti, A., Schmidt-Dannert, C. and Schmid, R.D. (2001) Mapping the substrate selectivity of new hydrolases using colorimetric screening: lipases from *Bacillus thermocatenulatus* and *Phiostoma piliferum*, esterases from *Pseudomonas fluorescens* and *Streptomyces* *diastatochromogenes*. *Tetrahedron Asymmetry*, **12**, 545–56.

53 Horsman, G.P., Liu, A.M.F., Henke, E., Bornscheuer, U.T. and Kazlauskas, R.J. (2003) Mutations in distant residues moderately increase the enantioselectivity of *Pseudomonas fluorescens* esterase toward methyl 3-bromo-2-methylpropanoate and ethyl 3-phenylbutyrate. *Chemistry–A European Journal*, **9**, 1933–9.

54 Reetz, M.T., Kühling, K.M., Wilensek, S., Husmann, H., Häusig, U.W. and Hermes, M. (2001) A GC-based method for high-throughput screening of enantioselective catalysts. *Catalysis Today*, **67**, 389–96.

55 Reetz, M.T., Becker, M.H., Klein, H.W. and Stöckigt, D. (1999) A method for high-throughput screening of enantioselective catalysts. *Angewandte Chemie–International Edition*, **38**, 1758–61.

56 Reetz, M.T., Kühling, K.M., Deege, A., Hinrichs, H. and Belder, D. (2000) Super-high-throughput screening of enantioselective catalysts by using capillary array electrophoresis. *Angewandte Chemie–International Edition*, **39**, 3891–3.

57 Schuster, S., Enzelberger, M., Trauthwein, H., Schmid, R.D. and Urlacher, V.B. (2005) pHluorin-based *in vivo* assay for hydrolase screening. *Analytical Biochemistry*, **77**, 2727–32.

58 Harvey, B.R., Georgiou, G., Hayhurst, A., Jeong, K.J. and Iverson, B.L. (2004) Anchored periplasmic expression, a versatile technology for the isolation of high-affinity antibodies from *Escherichia coli*-expressed libraries. *Proceedings of the National Academy of Sciences of the United States of America*, **101**, 9193–8.

59 Hayhurst, A. and Georgiou, G. (2001) High-throughput antibody isolation. *Current Opinion in Biotechnology*, **5**, 683–9.

60 Wittrup, K.D. (2001) Protein engineering by cell-surface display. *Current Opinion in Biotechnology*, **12**, 395–9.

61 de Maria, P.D., Martinez-Alzamora, F., Moreno, S.P., Valero, F., Rua, M.L., Sanchez-Montero, J.M., Sinisterra, J.V. and Alcantara, A.R. (2002) Heptyl oleate synthesis as useful tool to discriminate between lipases, proteases and other hydrolases in crude preparations. *Enzyme and Microbial Technology*, **31**, 283–8.

62 Furutani, T., Su, R.H., Ooshima, H. and Kato, J. (1995) Simple screening method for lipase for transesterification in organic solvent. *Enzyme and Microbial Technology*, **17**, 1067–72.

63 Kiran, K.R., Krishna, S.H., Babu, C.V.S., Karanth, N.G. and Divakar, S. (2000) An esterification method for determination of lipase activity. *Biotechnology Letters*, **22**, 1511–14.

64 Brandt, B., Hidalgo, A. and Bornscheuer, U.T. (2006) Immobilization of enzymes in microtiter plate scale. *Biotechnology Journal*, **1**, 582–7.

65 Teng, Y. and Xu, Y. (2007) A modified para-nitrophenyl palmitate assay for lipase synthetic activity determination in organic solvent. *Analytical Biochemistry*, **363**, 297–9.

66 Moore, J.C. and Arnold, F.H. (1996) Directed evolution of a *para*-nitrobenzyl esterase for aqueous-organic solvents. *Nature Biotechnology*, **14**, 458–67.

67 Giver, L., Gershenson, A., Freskgard, P.O. and Arnold, F.H. (1998) Directed evolution of a thermostable esterase. *Proceedings of the National Academy of Sciences of the United States of America*, **95**, 12809–13.

68 Henke, E., Bornscheuer, U.T., Schmid, R.D. and Pleiss, J. (2003) A molecular mechanism of enantiorecognition of tertiary alcohols by carboxylesterases. *ChemBioChem*, **4**, 485–93.

69 Heinze, B., Kourist, R., Fransson, L., Hult, K. and Bornscheuer, U.T. (2007) Highly enantioselective kinetic resolution of two tertiary alcohols using mutants of an esterase from *Bacillus subtilis*. *Protein Engineering, Design and Selection*, **20**, 125–31.

70 Kourist, R., Bartsch, S. and Bornscheuer, U.T. (2007) Highly enantioselective synthesis of arylaliphatic tertiary alcohols using mutants of an esterase from *Bacillus subtilis*. *Advanced Synthesis and Catalysis*, **349**, 1393–8.

71 Bartsch, S., Kourist, R. and Bornscheuer, U.T. (2008) Complete inversion of enantioselectivity towards acetylated tertiary alcohols by a double mutant of a

Bacillus subtilis esterase. *Angewandte Chemie – International Edition*, **47**, 1508–11.

72 Liebeton, K., Zonta, A., Schimossek, K., Nardini, M., Lang, D., Dijkstra, B.W., Reetz, M.T. and Jaeger, K.E. (2000) Directed evolution of an enantioselective lipase. *Chemistry and Biology*, **7**, 709–18.

73 Reetz, M.T., Wilensek, S., Zha, D. and Jaeger, K.E. (2001) Directed evolution of an enantioselective enzyme through combinatorial multiple-cassette mutagenesis. *Angewandte Chemie – International Edition*, **40**, 3589–91.

74 Reetz, M.T., Puls, M., Carballeira, J.D., Vogel, A., Jaeger, K.E., Eggert, T., Thiel, W., Bocola, M. and Otte, N. (2007) Learning from directed evolution: further lessons from theoretical investigations into cooperative mutations in lipase enantioselectivity. *ChemBioChem*, **8**, 106–12.

75 Schmidt, M., Hasenpusch, D., Kähler, M., Kirchner, U., Wiggenhorn, K., Langel, W. and Bornscheuer, U.T. (2006) Directed evolution of an esterase from *Pseudomonas fluorescens* yields a mutant with excellent enantioselectivity and activity for the kinetic resolution of a chiral building block. *ChemBioChem*, **7**, 805–9.

76 Park, S., Morley, K.L., Horsman, G.P., Holmquist, M., Hult, K. and Kazlauskas, R.J. (2005) Focusing mutations into the *P. fluorescens* esterase binding site increases enantioselectivity more effectively than distant mutations. *Chemistry and Biology*, **12**, 45–54.

77 Anderson, E.M., Larsson, K.M. and Kirk, O. (1998) One biocatalyst – many applications: The use of *Candida antarctica* B-lipase in organic synthesis. *Biocatalysis and Biotransformation*, **16**, 181–204.

78 Zhang, N., Suen, W.C., Windsor, W., Xiao, L., Madison, V. and Zaks, A. (2003) Improving tolerance of *Candida antarctica* lipase B towards irreversible thermal inactivation through directed evolution. *Protein Engineering*, **16**, 599–605.

79 Suen, W.C., Zhang, N., Xiao, L., Madison, V. and Zaks, A. (2004) Improved activity and thermostability of *Candida antarctica* lipase B by DNA family shuffling. *Protein Engineering, Design and Selection*, **17**, 133–40.

80 Koga, Y., Kato, K., Nakano, H. and Yamane, T. (2003) Inverting enantioselectivity of *Burkholderia cepacia* KWI-56 Lipase by combinatorial mutation and high-throughput screening using single-molecule PCR and *in vitro* expression. *Journal of Molecular Biology*, **331**, 585–92.

81 Qian, Z. and Lutz, S. (2005) Improving the catalytic activity of *Candida antarctica* lipase B by circular permutation. *Journal of the American Chemical Society*, **127**, 13466–7.

82 Qian, Z., Fields, C.J. and Lutz, S. (2007) Investigating the structural and functional consequences of circular permutation on lipase B from *Candida antarctica*. *ChemBioChem*, **8**, 1989–96.

83 Joerger, R.D. and Haas, M.J. (1994) Alteration of chain length selectivity of a *Rhizopus delemar* lipase through site-directed mutagenesis. *Lipids*, **29**, 377–84.

84 Klein, R.R., King, G., Moreau, R.A. and Haas, M.J. (1997) Altered acyl chain length specificity of *Rhizopus delemar* lipase through mutagenesis and molecular modeling. *Lipids*, **32**, 123–30.

85 Danielsen, S., Eklund, M., Deussen, H.J., Graslund, T., Nygren, P.A. and Borchert, T.V. (2001) *In vitro* selection of enzymatically active lipase variants from phage libraries using a mechanism-based inhibitor. *Gene*, **272**, 267–74.

86 Lavandera, I., Fernandez, S., Magdalena, J., Ferrero, M., Grewal, H., Savile, C.K., Kazlauskas, R.J. and Gotor, V. (2006) Remote interactions explain the unusual regioselectivity of lipase from *Pseudomonas cepacia* toward the secondary hydroxyl of 2′-deoxynucleosides. *ChemBioChem*, **7**, 693–8.

87 Bornscheuer, U.T. and Kazlauskas, R.J. (2004) Catalytic promiscuity in biocatalysis: using old enzymes to form new bonds and follow new pathways. *Angewandte Chemie – International Edition*, **43**, 6032–40.

88 Aharoni, A., Gaidukov, L., Khersonsky, O., McQ Gould, S., Roodveldt, C. and Tawfik, D.S. (2004) The "evolvability" of promiscuous protein functions. *Nature Genetics*, **37**, 73–6.

89 Hult, K. and Berglund, P. (2007) Enzyme promiscuity: mechanism and applications. *Trends in Biotechnology*, **25**, 231–8.

90 Fujii, R., Nakagawa, Y., Hiratake, J., Sogabe, A. and Sakata, K. (2005) Directed evolution of *Pseudomonas aeruginosa* lipase for improved amide-hydrolyzing activity. *Protein Engineering, Design and Selection*, **18**, 93–101.

91 Kim, J.H., Choi, G.S., Kim, S.B., Kim, W.H., Lee, J.Y., Ryu, Y.W. and Kim, G.J. (2004) Enhanced thermostability and tolerance of high substrate concentration of an esterase by directed evolution. *Journal of Molecular Catalysis B, Enzymatic*, **27**, 169–75.

92 Reetz, M.T., Bocola, M., Carballeira, J.D., Zha, D. and Vogel, A. (2005) Expanding the range of substrate acceptance of enzymes: combinatorial active-site saturation test. *Angewandte Chemie – International Edition*, **44**, 4192–6.

93 Rotticci, D., Rotticci-Mulder, J.C., Denman, S., Norin, T. and Hult, K. (2001) Improved enantioselectivity of a lipase by rational protein engineering. *ChemBioChem*, **2**, 766–70.

94 Ivancic, M., Valinger, G., Gruber, K. and Schwab, H. (2007) Inverting enantioselectivity of *Burkholderia gladioli* esterase EstB by directed and designed evolution. *Journal of Biotechnology*, **129**, 109–22.

95 Valinger, G., Hermann, M., Wagner, U.G. and Schwab, H. (2007) Stability and activity improvement of cephalosporin esterase EstB from *Burkholderia gladioli* by directed evolution and structural interpretation of muteins. *Journal of Biotechnology*, **129**, 98–108.

96 Dröge, M.J., Boersma, Y.L., van Pouderoyen, G., Vrenken, T.E., Rüggeberg, C.J., Reetz, M.T., Dijkstra, B. W. and Quax, W.J. (2006) Directed evolution of *Bacillus subtilis* lipase A by use of enantiomeric phosphonate inhibitors: Crystal structures and phage display selection. *ChemBioChem*, **7**, 149–57.

97 Niu, W.N., Li, Z.P., Zhang, D.W., Yu, M.R. and Tan, T.W. (2006) Improved thermostability and the optimum temperature of *Rhizopus arrhizus* lipase by directed evolution. *Journal of Molecular Catalysis B, Enzymatic*, **43**, 33–9.

31
Altering Enzyme Substrate and Cofactor Specificity via Protein Engineering

Matthew DeSieno, Jing Du and Huimin Zhao

31.1
Introduction

31.1.1
Overview

Enzymes are useful tools in synthetic chemistry, medicine, and industrial biocatalysis. On the basis of their high selectivity and specificity, enzymes typically enable regioselective, chemoselective and stereoselective reactions to be conducted, and can also greatly simplify downstream separation processes. In addition, they function under mild conditions such as ambient temperatures and pressures, thus reducing energy costs. Unfortunately, however, the rigid and narrow substrate and cofactor specificities confine the number of compounds that are acceptable for the enzyme-catalyzed reaction, thus limiting the widespread application of this powerful tool.

One example of such rigid and narrow enzyme substrate specificity is the large family of oxygenases which can perform difficult oxygenation reactions with high selectivity and specificity. These enzymes, which catalyze two types of reaction that introduce either one or two oxygen atoms into substrates, are referred to as monooxygenases and dioxygenases, respectively. In the past, the cytochrome P450 monooxygenases have undergone extensive study because of their application in the oxygenation of a diverse array of aromatic and aliphatic compounds, including alkanes and other complex endogenous molecules [1]. Wild-type cytochrome P450cam performs only a monooxygenation reaction towards camphor, but not towards substrates such as short-chain alkanes and polycyclic aromatic hydrocarbons (PAHs) [2]. Wild-type flavocytochrome P450 BM-3 only hydroxylates saturated and unsaturated fatty acids of chain length C_{12}—C_{16} at the ω-1 to ω-3 positions, but not alkylammonium compounds [3]. The Baeyer–Villiger monooxygenases represent another important group of enzymes in this class, and are known to differ both structurally and functionally from P450 monooxygenases; these enzymes efficiently convert aliphatic and cyclic ketones into esters and lactones,

Protein Engineering Handbook. Edited by Stefan Lutz and Uwe T. Bornscheuer

ISBN: 978-3-527-31850-6

respectively [4]. The Baeyer–Villiger monooxygenases catalyze a wide range of reactions, including alkane degradation (potentially useful for the clean-up of hydrocarbon pollution) and the production of secondary metabolites [5]. Dioxygenases are also widely studied because of their potential application in environmental bioremediation. Although these enzymes are capable of oxidizing aromatic compounds, their relatively narrow substrate specificity confines their wider application. For example, biphenyl dioxygenase can oxidize polychlorinated biphenyls, albeit with a limited range [6], while toluene dioxygenase can oxidize toluene but not heterocyclic substrates such as 4-methylpyridine [7].

Another example of rigid and narrow enzyme substrate specificity is the acyltransferase domain within polyketide synthases (PKSs), the enzyme complexes responsible for polyketide biosynthesis. Polyketides are a class of natural products that are of great importance as they possess antibiotic, anticancer, immunosuppressive or other useful biological properties for the development of therapeutic and agricultural chemicals [8]. During the past few decades, PKSs have attracted a great deal of interest, and many investigations have been conducted to generate polyketide synthases that are capable of biosynthesizing novel polyketides (see also Chapter 32). The acyltransferase domain in PKSs catalyzes the transacylation of the monomer unit Coenzyme A (CoA) to the phosphopantetheine arm of the acyl carrier protein, and thus determines the extender unit, as well as the starter unit for each module in modular PKS systems [9]. However, the natural acyltransferase domain in PKSs only accepts a limited number of substrates. In order to develop PKSs capable of biosynthesizing novel important polyketides, researchers have had to alter the substrate specificity of the acyltransferase domains.

The cofactor specificity of an enzyme is also an important issue in biocatalysis and metabolic engineering. Many enzyme-catalyzed reactions require cofactors such as NAD(H) or NADP(H), and cofactor imbalance problems in metabolic pathways often hamper the overproduction of certain compounds of interest or the metabolic flux towards favorable directions. One such example is the use of D-xylose in recombinant *Saccharomyces cerevisiae*, where an imbalance between NADP(H) and NAD(H) is created, ultimately hampering the utilization process of D-xylose in recombinant strains [10].

Fortunately, Nature has evolved divergent evolution that makes structurally similar enzymes catalyze different reactions, as well as convergent evolution that makes structurally different enzymes catalyze the same reactions. These facts may suggest that only a small structural change can shift the substrate specificity completely, and that more than one solution may exist for a chemical problem [11, 12]. Many enzymes also catalyze side reactions to a small extent, and such promiscuous activities may provide the starting point for protein engineering in the evolution of novel enzymes with new activity and substrate/cofactor specificity [13].

31.1.2 Approaches

A number of different approaches have been developed for altering the cofactor and substrate specificities of enzymes. These are divisible into three main categories, namely rational design, directed evolution and semi-rational design (Figure 31.1).

31.1.2.1 Rational Design

Rational design approaches rely on an understanding of enzyme structure and catalytic mechanism, and change enzyme specificity by replacing amino acids in certain sites. Due to a limited understanding of protein folding and the relationship between protein structure and function, the application of a rational design approach is limited to several well-studied systems. Recently, some noticeable progress has been made in understanding and predicting protein cofactor and substrate specificities with the availability of more crystal structures, an increased knowledge in relation to the relationship between enzyme structure and function, and also more powerful computational tools.

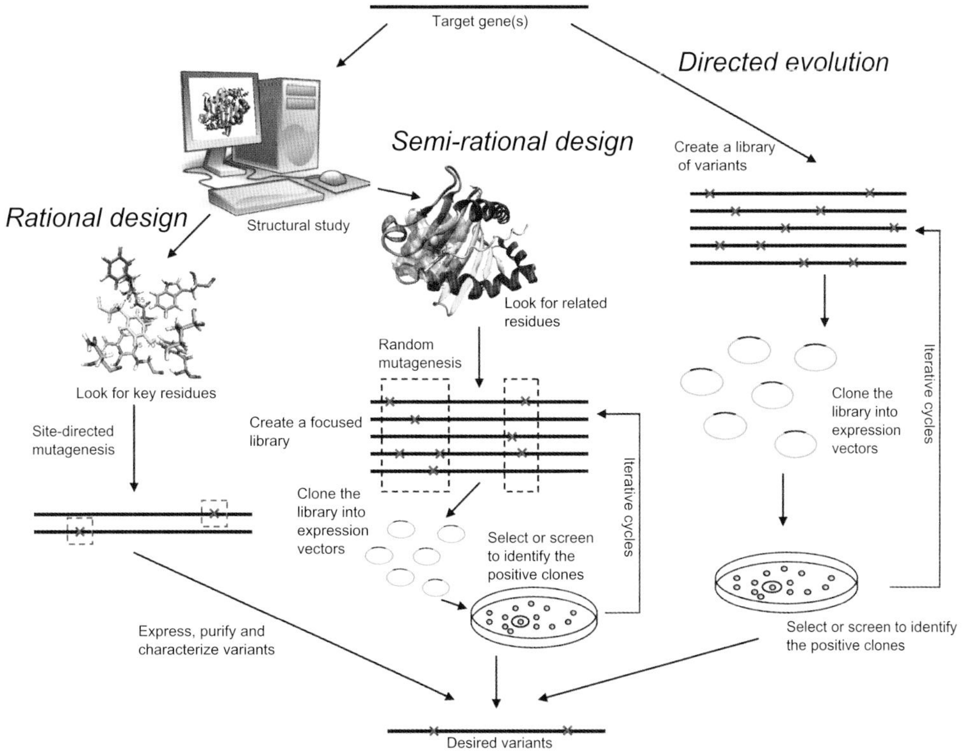

Figure 31.1 General scheme for rational design, semi-rational design and directed evolution.

Wilson and coworkers [14] developed an algorithm to evaluate the enzyme–substrate interaction by analyzing a library of side-chain rotamers to sample conformation space within the binding of the complex. By using this algorithm, they designed a protease that is both highly active and specific for an unnatural substrate solely on the basis of empirical energy calculations. Yadav and coworkers [15] have recently developed a computational approach to predict the domain organization and substrate specificity of modular PKSs. Based on an extensive sequence analysis of experimentally characterized PKS clusters, this team developed an automated computational protocol for the unambiguous identification of various PKS domains in a polypeptide sequence. Using a structure-based computational approach, they also identified the putative active-site residues of acyltransferase (AT) domains and provided an explanation for the experimentally observed chiral preference for the AT domain. The results of these studies, and the computational tools that were developed for the prediction of substrate specificity, have been organized into a computerized database and may serve as a valuable tool for further rational design studies of PKSs.

Following the identification of those amino acids related to substrate specificity, and determination of the amino acids used to replace them, the most widely used technique for altering substrate specificity by rational design is that of site-directed mutagenesis (SDM). For example, Dang and coworkers [16] shifted the balance between procoagulant and anticoagulant activities of serine protease by rational SDM of those residues which control Na^+ binding. The resultant mutants showed a reduced specificity toward fibrinogen, but an enhanced specificity towards protein C. It was found that the replacement of residue 225 on a serine protease could be used to alter the functional properties or to engineer the Na^+-binding capabilities over a wide range of proteases.

In another example, Wang and coworkers [17] altered the substrate specificity of methylamine dehydrogenase by indirect SDM. It had been reported previously that residue αPhe55 controls the substrate preference of methylamine dehydrogenase towards longer or shorter carbon chain amines. In the studies conducted by Wang's group, SDM was applied to residue βIle107, which makes close contact with αPhe55. The subsequent mutation on βIle107 caused a repositioning of αPhe55 and thus affected the substrate preference of the enzyme. The results of this study suggested that SDM on one residue could be used to reposition another residue and thus to alter enzyme specificity [18].

Although many successful examples of altering substrate specificity by rational design have been reported, the possibility of shifting substrate specificity is usually limited, even for well-characterized enzymes. For example, Lanio and coworkers [19] performed a structure-guided approach to engineer variants of the well-characterized restriction enzyme *Eco*RV. However, the mutants which were created according to the structural analysis provided slightly or even no improvement in cleavage of the unnatural nucleotide sequence compared with wild-type *Eco*RV.

Another limitation of rational design rests on the high requirements for knowledge of the protein structure, although this may be overcome as more protein structures and information regarding protein structure–function relationships become available. In a recent study conducted by Geddie and coworkers [20], site-directed insertion mutagenesis and heterologous expression were used to design transcription factor p53 to be specifically activated by designated effectors. In this study, insertions were made to the region within the N and C termini of p53, which are thought to be intrinsically unstructured, and the resulting p53 mutant was activated up to 100-fold by novel effectors. These results suggested that, for certain classes of proteins, the absence of a protein crystal structure may sometimes be overcome for rational engineering to proceed.

31.1.2.2 **Directed Evolution**

By mimicking the process of natural evolution in a test tube, directed evolution has provided the possibility of engineering enzyme activities and specificities, without any adequate structural and mechanistic understanding [21]. Directed evolution typically consists of repeated cycles of creating genetic diversity using random mutagenesis and/or gene recombination, followed by library screening or selection. Numerous methods for creating genetic diversity have been developed, such as error-prone PCR, DNA shuffling, family shuffling and non-homologous DNA recombination (see Chapters 16–21). During the past decade, directed evolution has been successfully used to alter enzyme substrate and cofactor specificities. For example, in comparison to the limited improvement on the substrate specificity of *Eco*RV achieved by rational design, three rounds of random mutagenesis on *Eco*RV led to the generation of variants which differed in their substrate specificity by two orders of magnitude [22].

The studies of Collins and coworkers [23] have provided an excellent example for altering substrate specificity using a two-step approach. In the first step a positive selection is used to improve the substrate affinity towards unnatural substrates; in the second step a negative selection is used to decrease the substrate specificity towards natural substrates. By applying this approach, Collins and colleagues enhanced the signaling specificity of a variant of transcriptional activator LuxR, and shifted the response of LuxR from 3-oxo-hexanoyl-homoserine lactone (3OC6HSL) to straight-chain acyl-HSLs.

The main issue in the application of directed evolution rests on the creation of large libraries and the efficient screening or selection to identify positive mutants. In order to overcome these problems, many clever approaches have been developed for screening large libraries for individual applications. At the same time, an analysis on protein structure based on a limited knowledge of protein structure was performed before random mutagenesis to reduce the number of residues subjected to mutation; thus, the semi-rational design approach was introduced.

31.1.2.3 **Semi-Rational Design**

Although current knowledge on protein structure–function relationships is insufficient for an entirely rational design of enzyme substrate and cofactor specificity, structural analysis can facilitate the evolutionary manipulation of enzymes in a more targeted manner. Several strategies have been developed for semi-rational design. As more protein structural data have become available, comparative studies between enzymes sharing similar catalytic activity and sequence homology can provide certain information as to which residues are related with substrate binding and cofactor affinity. Subsequently, random mutagenesis or saturation mutagenesis can be applied to certain regions associated with substrate/cofactor binding, after which they can be subjected to rounds of directed evolution. This focused mutagenesis approach takes advantage of the large library and evolvability of directed evolution, and also overcomes the disadvantage of screening unnecessarily large libraries by confining the region for random mutagenesis. For example, Santoro and coworkers [24] performed a targeted random mutagenesis to shift the substrate specificity of a Cre DNA recombinase to recognize a new *loxP* site not recognized by wild-type Cre recombinase. Two regions of the Cre protein were suspected of being important for recombination activity, and so were subjected to random mutagenesis to create two libraries. When the libraries had been screened using fluorescence-activated cell sorting (FACS), a mutant with novel substrate specificity was identified.

Reetz and coworkers [25] developed a method for expanding the scope of substrate acceptance of enzymes, including the design and generation of relatively small focused libraries produced by mutagenesis at several sets of two spatially close amino acid positions around the active site. For a given pair of residues in which the side chains point in the direction of the binding site, according to the geometric relationship, if one member of the pair is at residue n, the other one should be at $(n + 1)$ in a loop, $(n + 2)$ in a β-sheet, $(n + 3)$ in a 3_{10} helix, and $(n + 4)$ in an α-helix. The identified pairs were then subjected to a process known as combinatorial active-site saturation testing (CAST), which enables the discovery of positive hits within a small fraction of the library instead of the whole region for random mutagenesis. The method was then applied to broaden the substrate acceptance of a lipase from *Pseudomonas aeruginosa* as a catalyst in the hydrolysis of a carboxylic acid ester, and positive mutants were identified after screening a small library of only 3000 colonies (see also Chapter 16).

31.2
Specific Examples

Directed evolution and rational design – and hybrids of these two approaches – have been used successfully to create enzymes with altered or novel cofactor and substrate specificities. These powerful methods, which have been performed on a wide range of enzymes, can be used either for specific industrial processes or to provide insights into some fundamental biological questions.

31.2.1
Cofactor Specificity

Many enzymes require cofactors for their catalyzed reactions to proceed. These small molecules typically contain a specific functional group, which makes them interchangeable in some cases as long as the functional group remains. Protein engineering has been used to switch the cofactor specificity between natural cofactors and also to introduce unnatural counterparts to the enzymatic reactions.

31.2.1.1 NAD(P)(H)

Oxidoreductases often require the cofactors nicotinamide adenine dinucleotide (NAD(H)) and nicotinamide adenine dinucleotide phosphate (NADP(H)), which become the driving force for enzymatic oxidations and reductions through the reversible reduction of the nicotinamide ring. The high cost of these cofactors necessitates either an *in situ* regeneration method or an adjustment of the cofactor specificity, because addition of the cofactor in stoichiometric amounts becomes impractical on an industrial scale [26].

Petscacher and coworkers used the crystal structure and SDM in order to switch the cofactor specificity of xylose reductase from *Candida tenius* (CtXR) [27]. The wild-type CtXR is able to utilize both NADPH and NADH as cofactors in the reduction of xylose to xylitol, but with a 33-fold preference for NADPH. The crystal structure for CtXR bound to NADPH and NADH was used to identify which residues may interact with the 2′-phosphate on the NADPH cofactor, and SDM was performed to remove these interactions to favor NADH activity. Six single mutants, a double-mutant and the wild-type enzyme were purified and characterized for their catalytic activity with NADPH and NADH. The double-mutant, Lys274Arg/Asn276Asp, showed a fivefold preference for NADH over NADPH with limited alteration to the original NADH and overall catalytic activity.

Woodyer and coworkers were able to relax the cofactor specificity of *Pseudomonas stutzeri* phosphite dehydrogenase (PTDH) using homology modeling and SDM [28]. Two key residues, Glu175 and Ala176, were identified as the determinants to cofactor specificity based on sequence alignment with three other D-hydroxy acid dehydrogenases and their resulting homology model. These two residues were mutated separately and in combination to create three mutants with relaxed cofactor specificity and improved catalytic activity for both cofactors. The Ala175 and Arg176 double-mutant performed best, with a 3.6-fold higher catalytic efficiency for NAD^+, a 1000-fold higher catalytic efficiency for $NADP^+$, and a threefold cofactor specificity for $NADP^+$ compared to 100-fold in favor of NAD^+ for the wild-type PTDH.

31.2.1.2 ATP

Protein kinases play a pivotal role in the signal transduction pathway inside eukaryotic cells, including regulation of the cell cycle, stress responses and ion-channel regulation. These enzymes catalyze the phosphorylation of tyrosine, serine, threonine or histidine residues of cellular proteins. Liu and coworkers used SDM in engineering v-Src, a tyrosine kinase, to preferentially accept an ATP analogue, N^6-(benzyl)-ATP, as its substrate instead of the natural substrate, ATP [29]. Based on structural and sequence alignments with other protein kinases, two residues were identified as potential targets for adjusting the substrate specificity of v-Src, Val323 and Ile338. After several rounds of iterative mutagenesis on these two residues, the authors determined that Ile338 alone controlled the substrate specificity. Upon mutating the residue to glycine, a mutant protein kinase was formed that was highly specific for N^6-(benzyl)-ATP, even in the presence of high concentrations of the natural ATP. The same protein engineering approach can be taken for many other protein kinases except for a few, one of which is c-Abl [30]. A new strategy for conferring the ATP variant specificity on the c-Abl had to be determined, as the single-residue mutation approach was unsuccessful. The highly homologous SH1 catalytic domain of v-Src was used as a blueprint in the protein engineering of c-Abl, as the domain is primarily responsible for ATP specificity in kinases. Consequently, the N-terminal lobe of the SH1 domain of v-Src was switched with that of c-Abl, thus creating a chimeric kinase that could accept the N^6-(benzyl)-ATP variant but still retain the same peptide specificity of the wild-type c-Abl.

31.2.1.3 **Summary and Comments for Cofactor Specificity**

A majority of the reported cases involving the relaxing or switching of cofactor specificity have utilized structural information, typically from either X-ray crystallography or homology modeling, in identifying those amino acid residues likely to affect the specificity. In the examples described above, the residues which were believed to interact with either the 2′-phosphate of NADPH or the N^6 amine of ATP were then rationally mutated. The changes that are required to alter the cofactor specificity of enzymes appear to be localized around the active site, which makes this approach far more effective than a totally randomized method such as directed evolution.

When altering cofactor specificity, generalizations cannot be made even over small ranges of enzymes, which means that the protein engineering of any enzyme should be analyzed on a case-by-case basis. For example, the cofactor specificity of *Haloferax volcanii* Ds-*threo*-isocitrate dehydrogenase (ICDH) and *Shewanella* sp. Ac10 alanine dehydrogenase were both converted from NADPH-dependent to NADH-dependent [31, 32]. Despite both enzymes being dehydrogenases, the cofactor switch in ICDH occurred after five mutations had been made, while the alanine dehydrogenase preferred NADH after only a single mutation. Together, these reports provided evidence as to the complexity and diversity of enzymes, even within the same family.

31.2.2 Substrate Specificity

The level of substrate specificity that enzymes display varies a great deal. For example, some have a high preference for a single substrate, while others are more promiscuous and can accept a range of similar molecules. Accordingly, protein engineering is capable of broadening the substrate specificity of a highly specific enzyme, of narrowing the substrate specificity of a less-specific enzyme, or even creating new activity and substrate specificity where there was none to begin with.

31.2.2.1 **P450s**

The cytochrome P450 superfamily is a very diverse group of heme-containing enzymes found in virtually all organisms, including bacteria, yeast and eukaryotes. They catalyze the monooxygenation of a large range of substrates with high regioselectivity and stereoselectivity, which makes them promising candidates in both the chemical and pharmaceutical industries [33].

Lindeberg and coworkers were able to completely switch the substrate specificity between two similar cytochrome P450 monooxygenases using SDM [34]. $P450_{15\alpha}$ and $P450_{coh}$ are both 494 amino acids long, differ only in 11 of the residues, and catalyze the 15α-hydroxylation of Δ^4 3-ketone steroids (e.g. testosterone) and the 7-hydroxylation of coumarin, respectively. Eleven different mutants were made based on the wild-type $P450_{coh}$, where each of the differing amino acids was switched to the corresponding one from $P450_{15\alpha}$. Three of the mutants displayed activity that was 35–45% of the wild-type $P450_{coh}$ coumarin hydroxylation, which indicated that Val117, Phe209 and Met365 are significant to the substrate specificity and overall activity of the enzyme. In particular, the single substitution of Phe209 to Leu converts $P450_{coh}$ towards a preference of steroid hydroxylation, and the triple mutant combining all the amino acid changes was able to totally remove any coumarin activity.

Glieder and coworkers used directed evolution to broaden the substrate specificity of P450 BM-3 from *Bacillus megaterium*, a medium-chain (C_{12}—C_{18}) fatty acid monooxygenase, into a catalyst for the conversion of small- to medium-chain (C_3—C_8) alkanes to alcohols [35]. Five rounds of random mutagenesis on the enzyme yielded P450 BM-3 mutant 139-3, which had a total of 11 residue substitutions. The mutant enzyme had high activity for octane, hexane, cyclopentane and pentane. Interestingly, it was still active on the fatty acid substrates of the wild-type. The evolved enzyme was approximately twofold more active on palmitic acid, which was already a good substrate for the P450 BM-3. Meinhold and coworkers extended these studies to smaller alkanes, as the P450 BM-3 139-3 was unable to produce appreciable amounts of 1-propanol from propane [36]. Successive rounds of random mutagenesis and screening focused on the heme domain of the P450 BM-3 139-3 resulted in a mutant which was capable of producing significantly more 1-propanol compared to the earlier mutant. Using the wild-type enzyme crystal structure, 11 residues in the active site were within 5 Å of the bound substrate, making them good targets for saturation mutagenesis. The newest mutant contained three active-site substitutions, all of which led to a closing of the binding pocket by substituting alanine with a larger residue, and it was able to hydroxylate ethane to produce ethanol. The P450 BM-3 mutants represent promising candidates for alkane hydroxylases over naturally occurring hydroxylases due to their increased solubility, high substrate selectivity and ability to be heterologously expressed.

31.2.2.2 **Aldolases**

Aldolases have attracted increasing attention recently due to their ability to efficiently catalyze a carbon–carbon bond with high stereoselectivity at this position.

Protein engineering of this group of enzymes has been focused primarily on an improvement of its instability and substrate specificity. Hsu and coworkers used directed evolution in order to relax the substrate specificity and enantioselectivity of D-sialic acid aldolase to form an efficient L-3-deoxy-manno-2-octulosonic acid (L-KDO) aldolase [37]. Five rounds of error-prone PCR and iterative screening using a coupled reaction with the oxidation of NADH to NAD^+ were performed. The final mutant displayed a >1000-fold improvement in the specificity constant for the unnatural substrate, L-KDO, which was comparable to the wild-type D-sialic acid aldolase and its natural substrate.

DeSantis and coworkers used SDM in order to broaden the substrate specificity of 2-deoxyribose-5-phosphate-aldolase (DERA) [38]. This enzyme catalyzes the reversible reaction between acetaldehyde and D-glyceraldehyde-3-phosphate, and demonstrates a strong preference for phosphorylated substrates. The crystal structure and proposed catalytic mechanism of DERA were used to identify residues within close proximity of the phosphate group of the substrate. Five different variants were created, one for each residue thought to be in the phosphate-binding pocket, where the two basic and three neutral side chains were mutated to acidic forms in order to create a net repulsion on the phosphate. The S238D mutant was found to be most active, with a 2.5-fold improvement over the wild-type in the reaction with the unnatural, unphosphorylated D-2-deoxyribose.

31.2.2.3 Transfer-RNA Synthetases

Transfer-RNA synthetases are the bridge between the DNA genetic code and the 20 amino acids composing all life. Except for the rare selenocysteine and pyrrolysine, the same 20 comprise all naturally produced proteins [39]. Recently, efforts have been made to expand the number of encoded amino acids to include unnatural ones. For example, Wang and colleagues used a combined directed evolution and rational design approach to generate an orthogonal transfer RNA/aminoacyl-tRNA synthetase which was able to effectively expand the number of genetically encoded amino acids in *E. coli* [40]. Based on the crystal structure of an homologous synthetase, a library of synthetase variants was created by mutating five residues in the active site. Rounds of alternating positive and negative selections were used in order to ensure that the orthogonal synthetase only recognized the target unnatural amino acid, *O*-methyl-L-tyrosine and no other endogenous *E. coli* tRNAs. Upon introduction into *E. coli*, the mutant pair selectively inserted *O*-methyl-L-tyrosine into proteins in response to an amber nonsense codon, with translational fidelity >99%. The same methodology was used to create orthogonal transfer RNA/aminoacyl-tRNA synthetase pairs that extended the genetic code and allowed the incorporation of more than 30 unnatural amino acids in *E. coli*, yeast or mammalian cells [41] (Figure 31.2).

31.2.2.4 Restriction Endonucleases

Type II restriction endonucleases have made a major contribution to recombinant DNA technology and the subsequent biotechnology revolution. These enzymes

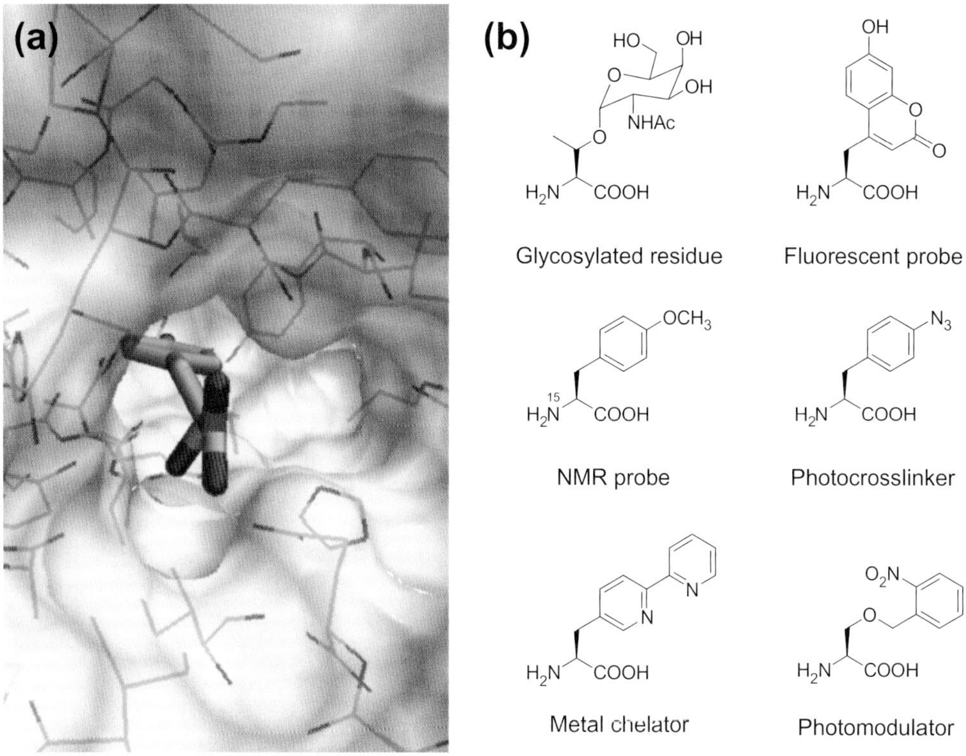

Figure 31.2 Protein engineering of an orthogonal transfer RNA/aminoacyl-tRNA synthetase which successfully expands the number of genetically encoded amino acids. (a) Crystal structure of binding pocket used to create initial *O*-methyl-L-tyrosine orthogonal pair. Residues forming the pocket are targets for the further protein engineering of other pairs; (b) Several representative unnatural amino acids that have been added to the genetic codes of *E. coli*, yeast or mammalian cells, including their potential experimental applications.

cleave highly specific restriction sites on DNA, and this property is used by bacteria as a natural defense mechanism against foreign DNA from invading bacteriophages. As phage genomes are typically short, the bacteria would logically prefer frequent cutters, which would result in only a few naturally occurring endonucleases having recognition sites of greater than six base pairs [19].

Lanio and coworkers used directed evolution in order to change *Eco*RV (restriction site: 5′-GAT-ATC-3′) into a rare-cutter that could recognize eight or ten base pairs [22]. By using the crystal structure of *Eco*RV, the authors identified three regions (a total of 22 amino acid residues) on the endonuclease that were within 5 Å of the base pairs flanking the restriction site. After three rounds of random mutagenesis, two classes of mutant were obtained that could effectively discriminate between AT- and GC-flanked cleavage sites, without losing much overall catalytic activity. A triple-mutant was identified that had a 25-fold higher rate of cleaving *Eco*RV restriction sites flanked by AT-rich regions compared to GC-rich regions. Samuelson and colleagues used directed evolution to increase the substrate specificity of *Bst*YI (restriction site: 5′-R-GATCY-3′, where R = A or G and Y = T or C), so that it would only cleave AGATCT rather than the degenerate wild-type RGATCY [42]. Random mutagenesis was performed on the gene encoding the endonuclease, followed by three rounds of *in vivo* screening of the clones, which was able to eliminate approximately 10^7 clones. The 45 remaining clones were then screened for the altered specificity and overall activity. Two variants were identified that had higher specificity for AGATCT over the three other possible cleavage sites. These two mutations were combined to form a final variant that displayed a 12-fold preference of the desired AGATCT over GGATCT or AGATCC, and had no cleavage of GGATCC.

31.2.2.5 **Homing Endonucleases**

Homing endonucleases recognize long DNA sequences, typically 15–40 base pairs in length, and efficiently cleave at these homing sites to promote the transfer of introns. The redesign of homing endonucleases is widely considered much easier than for type II restriction endonucleases. The cleavage specificity and interactions involved in the DNA recognition of I-CreI, a homodimer with a 22-base pair recognition sequence, were investigated by Seligman and coworkers, using SDM [43]. The recognition domain of the homing endonuclease consisted of nine residues, each of which (along with a tenth amino acid that was predicted to participate in water-mediated interactions) was mutated to alanine. These ten mutants were screened using two *in vivo* methods, where active I-CreI was seen to cleave at a homing site; this made the strain kanamycin-resistant in one assay, with conversion from LacZ+ to LacZ− in the other assay. The homing recognition site for I-CreI was seen to be semi-palindromic, with seven of the 11 base pairs being identical. These palindromic sites were found to be the most crucial to successful site recognition, as any single mutation in five of the seven led to a resistance in recognition and cleavage. The authors were able to see cleavage of novel targets and loss of cleavage at the wild-type I-CreI homing site by some of the mutants.

Chevalier and coworkers used a computational approach to create a highly specific, artificial homing endonuclease by fusing domains from two other endonucleases, I-DmoI and I-CreI [44]. The chimeric endonuclease, E-DreI (Engineered I-DmoI/I-CreI), was generated by fusing the N-terminal domain of I-DmoI to an I-CreI monomer and inserting a short peptide linker in between the two domains. A computational program was used to redesign the endonuclease interface in order to minimize the steric hindrance caused by the residue side chains. The redesign was focused on 14 residues along the interface by simulating saturation mutagenesis on each. Three of the best redesigned variants were screened *in vivo* for folding and solubility. The resulting E-DreI was a novel endonuclease with altered DNA target specificity.

31.2.2.6 Polymerases

DNA polymerases, including DNA pol I from *Thermus aquaticus* (*Taq*), are exceptional enzymes as they only misincorporate the incorrect base pair once approximately every 10 000 bases, despite the fact they recognize four different substrates with extraordinarily high specificity. Ong and coworkers devised a novel technique for expanding the substrate specificity of polymerases called short-patch compartmentalized self-replication (spCSR) [45]. The engineering of DNA polymerases requires an adjustment of the active site to accommodate the unnatural nucleotide, which may include bulky substituents on the nucleobase or ribofuranose scaffold, whilst retaining the ability to pass all of the fidelity checkpoints. The spCSR technique described by these authors was an extension of their previous studies with compartmentalized self-replication (CSR), where the polymerases replicate its own encoding gene. Through a positive feedback loop, any positive gains by the polymerase are maintained by more efficient translation. The difference between the two methods was that CSR required the polymerase to translate the entire gene, while spCSR only required translation of a short, defined portion of the gene. Two short segments of the active site of the *Taq* polymerase were targeted for the spCSR. One polymerase variant was able to incorporate RNA (NTPs) whilst retaining activity for DNA (dNTPs), thus creating a single polypeptide with DNA polymerase, RNA polymerase and reverse-transcriptase capabilities.

31.2.2.7 Summary and Comments for Substrate Specificity

When modifying the substrate specificity of an enzyme, the interactions become far more difficult – if not impossible – to predict in comparison to cofactor specificity. Mutations away from the active site with no direct contact to the substrate are also important determinants of the catalytic activity. Ideally, a particular region or domain on the enzyme can be selected for directed evolution, although the latter does not require any prior structural and mechanistic information relating to the enzyme of interest.

As shown by the examples described here, a wide range of enzymes have undergone substrate specificity modifications. The methods for creating the variant libraries were nearly as diverse, including rational design, directed evolution, chimeric proteins and even the novel spCSR. Although a majority of the above-discussed studies have utilized directed evolution, the major problem – regardless of the method – is the ability to efficiently identify the best variants. The screening method must be high-throughput in nature so as to allow large libraries to be assayed, but sensitive enough to allow small changes in activity to be measured. Some of the screening methods used in these examples have proven to be difficult, because the target substrate had such a low activity in the initial rounds of mutants. In the case of the sialic acid aldolase (see Section 31.2.2.2), an analogue substrate had to be used initially when switching the enantioselectivity of the enzyme. However, once a viable diversity generation and screening methods had been designed, it became possible to create enzymes with enhanced substrate specificities.

31.3
Challenges and Future Prospects

Despite previous major successes in enzyme cofactor and substrate specificity engineering, many challenges remain. Whilst the creation of novel cofactor or substrate specificities is difficult, due to the effect of distant mutations on cofactor and substrate specificities and the lack of powerful analytical tools to study protein conformation dynamics, deciphering the molecular basis for altered cofactor or substrate specificity also represents a major challenge. It follows that new protein engineering tools and strategies will continue to be developed to tackle these problems, and the use of these engineered enzymes will continue to expand in areas of metabolic engineering, combinatorial biosynthesis and gene therapy during the coming years.

31.3.1
New Strategies for Engineering Cofactor/Substrate Specificity

As noted in the above-described examples, the number of introduced mutations required to alter substrate and/or cofactor specificities is small. Those mutations introduced by rational design are in close proximity to the enzyme active site, while many of those identified by directed evolution are distal. Morley and Kazlauskas have suggested, that although both types of mutation may lead to improvements, the closer mutations have led to greater improvements in specificity [46]. Hence, a targeted directed evolution approach would be most effective as it focuses on mutations in close proximity to the enzyme active site. Although this requires knowledge of the protein structure, limiting the sequence space searched by random mutagenesis greatly improves the efficiency of the entire protein engineering process. In the absence of a protein structure, homology modeling may be used to build a working model for the enzyme of interest.

The introduction of unnatural amino acids into a protein, using engineered orthogonal transfer RNA/aminoacyl-tRNA synthetases, opens new avenues for engineering enzyme cofactor and substrate specificities. As such, many unnatural amino acids with interesting chemical and physical properties have been introduced into proteins to create new properties [41]. As yet, unnatural amino acids have not been introduced into the active site of an enzyme, although if this were to be achieved then a variant with novel catalytic activity and substrate specificities may result.

Another drastic means of engineering enzymes with new cofactor and substrate specificities would be to use *de novo* protein design, with enzyme function being determined solely on the protein primary sequence [47]. Currently, our ability to predict the backbone structure of a protein based on amino acid sequence is limited [48], and thus the *de novo* design of enzymes with desired cofactor or substrate specificities represents a major challenge. However, with recent advances in structural genomics, bioinformatics and computational biology having led to an ability to design proteins *de novo*, it is likely that novel substrate/cofactor specificities and new active sites with completely new chemistry may soon be created simply by adjusting protein primary sequences.

31.3.2
Cofactor/Substrate Specificity Engineering for Combinatorial Biosynthesis

While combinatorial biosynthesis represents a powerful method for the creation of novel products through the engineering of biosynthetic pathways, this approach has so far been underexploited, due mainly to problems of altering enzyme substrate specificity. In principle, the broadening of substrate specificity of enzymes should lead to the creation of new and diversified libraries of compounds, and the number of possible products would increase exponentially with the number of mutated proteins [49]. However, successful combinatorial biosynthesis requires a certain level of promiscuity from those enzymes that comprise the pathway. Moreover, without a broadened substrate specificity downstream there would most likely be a build-up of intermediate, potentially toxic, compounds within the cell.

One such example is the creation of long-chain carotenoids; these are natural pigments which act as precursors for several other biosynthetic pathways [50] (see Chapter 34). Whilst the carotenoid pathway studied was incapable of producing pigments with backbones longer than C_{40}, mutagenesis of the carotenoid synthase led to the production of carotenoids with backbone lengths of C_{45} and C_{50}. The downstream enzymes (which include desaturases and cyclases) that modify the carotenoids are extremely promiscuous, and efficiently produce novel carotenoid derivatives (Figure 31.3). A further example is the generation of large PKS libraries

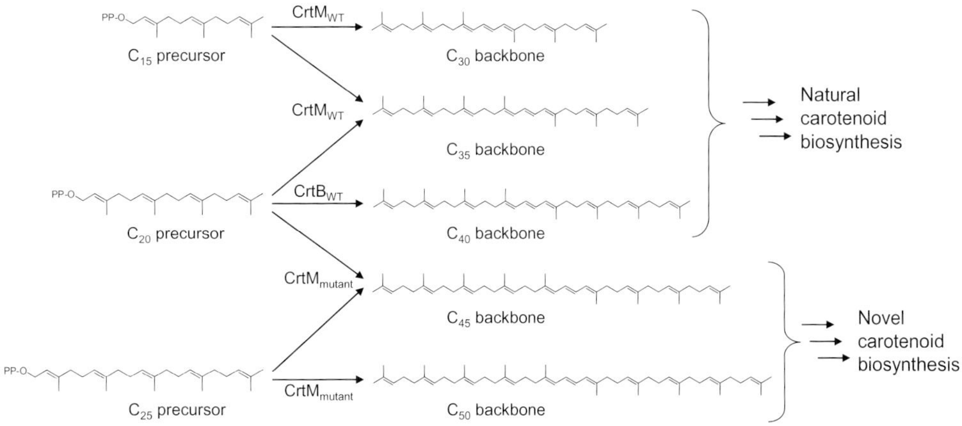

Figure 31.3 Pathways for creating natural and novel carotenoids. Mutant synthases allowed the formation of previously unknown C_{45} and C_{50} carotenoid backbones. Adapted from Ref. [50].

through combinatorial biosynthesis [51] (see also Chapter 32). The products of PKSs can be engineered by altering the accepted primer and extender units or the number of enzymatic modules present. The combinatorial biosynthesis of PKSs is an emerging field that has shown great potential in the field of drug discovery.

31.3.3
Cofactor/Substrate Specificity Engineering for Metabolic Engineering

The aim of metabolic engineering is to create enzymatic pathways capable of either producing value-added molecules or degrading harmful compounds, in simple, effective manner. The process involves the directed improvement of product formation or cellular properties by modifying specific biochemical reaction(s) or introducing new reactions via recombinant DNA technology [52]. This area is especially important in biotechnology as it offers ways of improving existing bioprocesses, of designing new bioprocesses, and of producing novel chemicals and pharmaceuticals. Not surprisingly, therefore, the ability to produce enzymes with altered or novel substrate/cofactor specificities plays a critical role in metabolic engineering.

As an example, protein engineering can be used to balance the intracellular cofactor pool so as to increase the yield and productivity of a target product. As mentioned in Section 31.2.1.1, the first enzyme in the xylose-utilizing pathway, xylose reductase, has a preference for NADPH, while the second enzyme in the xylose-utilizing pathway, xylitol dehydrogenase, requires NAD^+ as it oxidizes xylitol to xylulose [53]. Without modifying the cofactor specificity, there would be a cofactor imbalance with an accumulation of $NADP^+$ and NADH. Hence, if the cell were unable to compensate, not only would xylose metabolism be slowed but also many other cellular metabolic pathways that rely on these cofactors. Recently, when an engineered $NADP^+$-specific xylitol dehydrogenase was combined with a NADPH-specific xylose reductase in *S. cerevisiae*, the result was an increased production of ethanol and a significant reduction in formation of the side product, xylitol [54].

Another possible exploitation of substrate specificity engineering would be the better use of an available substrate pool. For example, if the main substrate for an enzyme were to be mixed with other, potentially useful, compounds, then it may be advantageous for an enzyme's substrate specificity to be relaxed such that it could accept these alternative compounds. As an example, hemicellulosic waste contains mostly D-xylose and L-arabinose; both compounds may be used to synthesize xylitol, but the conversion of L-arabinose would require several extra enzymes [55]. Hence, a xylose reductase with a relaxed specificity could co-utilize both substrates to produce xylitol. Similarly, catechol-2,3-dioxygenase from *Pseudomonas stutzeri* OX1 is a key enzyme in the catabolic degradation of aromatic molecules, but has a limited specificity [56]. An expansion in the number and type of catalyzed substrates for this enzyme would create an excellent target for bioremediation processes.

31.3.4 Cofactor/Substrate Specificity Engineering for Gene Therapy

Gene therapy is defined as the introduction of genetic modifications into target cells in order to cure a disease or to treat its symptoms. Potentially, such as procedure should have a significant influence on human health, and indeed new treatments for many inherited and acquired diseases are promised [57]. Redesigned homing endonucleases with defined DNA sequence specificity represent potential agents for gene therapy [58]. The most promising means of creating a highly specific endonuclease would be to use homing endonucleases as scaffolds and then to alter their specificity, either by introducing mutations at the active site or through a chimeric protein approach [59]. An alternative here is zinc-finger nucleases; these are constructed by linking a zinc-finger protein, which is capable of recognizing a target DNA sequence, and a nonspecific DNA cleavage domain of *FokI*, a Type IIS restriction enzyme [60]. The use of engineered zinc-finger proteins with novel sequence specificity may permit the creation of a wide variety of custom-made zinc finger nucleases to target different diseases.

Small-molecule-regulated gene expression systems (so-called 'gene switches') may also be useful for gene therapy, as they may be used to control target gene expression by regulating when, how, and to what extent, a gene will become expressed [61]. Recently, an orthogonal ligand–receptor pair was created for use in gene switches by combining rational design and directed evolution [62]. A similar method could be used to create other receptors that are inducible by small-molecule drugs.

Further *in vivo* testing would be required in order to monitor the efficacy and toxicity associated with homing endonucleases and gene switches. However, both methods may be effective in treating single-gene diseases such as sickle cell anemia, cystic fibrosis, hemophilia and muscular dystrophy.

Acknowledgments

We gratefully acknowledge financial support from Biotechnology Research and Development Consortium (BRDC) (Project 2-4-121), Office of Naval Research (N00014-02-1-0725), National Science Foundation (BES-0348107), National Institutes of Health (GM 077596 and HL 089418), Department of Energy, and the DuPont Company. M.D. acknowledges support from the National Institutes of Health Chemistry-Biology Interface Training Grant Program.

References

1 Miles, C.S., Ost, T.W., Noble, M.A., Munro, A.W. and Chapman, S.K. (2000) Protein engineering of cytochromes p-450. *Biochimica et Biophysica Acta*, **1543**, 383–407.

2 Shimoji, M., Yin, H., Higgins, L. and Jones, J.P. (1998) Design of a novel P450: a functional bacterial-human cytochrome P450 chimera. *Biochemistry*, **37**, 8848–52.

3 Oliver, C.F., Modi, S., Primrose, W.U., Lian, L.Y. and Roberts, G.C. (1997) Engineering the substrate specificity of *Bacillus megaterium* cytochrome P-450 BM3: hydroxylation of alkyl trimethylammonium compounds. *Journal of Biochemistry*, **327**, 537–44.

4 Mihovilovic, M.D., Rudroff, F., Winninger, A., Schneider, T., Schulz, F. and Reetz, M.T. (2006) Microbial Baeyer-Villiger oxidation: stereopreference and substrate acceptance of cyclohexanone monooxygenase mutants prepared by directed evolution. *Organic Letters*, **8**, 1221–4.

5 Kirschner, A., Altenbuchner, J. and Bornscheuer, U.T. (2007) Design of a secondary alcohol degradation pathway from *Pseudomonas fluorescens* DSM 50106 in an engineered *Escherichia coli*. *Applied Microbiology and Biotechnology*, **75**, 1095–101.

6 Suenaga, H., Goto, M. and Furukawa, K. (2001) Emergence of multifunctional oxygenase activities by random priming recombination. *The Journal of Biological Chemistry*, **276**, 22500–6.

7 Sakamoto, T., Joern, J.M., Arisawa, A. and Arnold, F.H. (2001) Laboratory evolution of toluene dioxygenase to accept 4-picoline as a substrate. *Applied and Environmental Microbiology*, **67**, 3882–7.

8 O'Hagan, D. (1992) Biosynthesis of polyketide metabolites. *Natural Product Reports*, **9**, 447–79.

9 Wu, K., Chung, L., Revill, W.P., Katz, L. and Reeves, C.D. (2000) The FK520 gene cluster of *Streptomyces hygroscopicus* var. *ascomyceticus* (ATCC 14891) contains genes for biosynthesis of unusual polyketide extender units. *Gene*, **251**, 81–90.

10 Rizzi, M., Harwart, K., Erlemann, P., Buithanh, N.A. and Dellweg, H. (1989) Purification and properties of the NAD^+-xylitol dehydrogenase from the yeast *Pichia stipitis*. *Journal of Fermentation and Bioengineering*, **67**, 20–4.

11 Gerlt, J.A. and Babbitt, P.C. (1998) Mechanistically diverse enzyme superfamilies: the importance of chemistry in the evolution of catalysis. *Current Opinion in Chemical Biology*, **2**, 607–12.

12 Todd, A.E., Orengo, C.A. and Thornton, J.M. (2002) Plasticity of enzyme active sites. *Trends in Biochemical Sciences*, **27**, 419–26.

13 Yarnell, A. (2003) The power of promiscuity. *Chemical and Engineering News*, **81**, 33–5.

14 Wilson, C., Mace, J.E. and Agard, D.A. (1991) Computational method for the design of enzymes with altered substrate specificity. *Journal of Molecular Biology*, **220**, 495–506.

15 Yadav, G., Gokhale, R.S. and Mohanty, B. (2003) Computational approach for prediction of domain organization and substrate specificity of modular polyketide synthases. *Journal of Molecular Biology*, **328**, 335–63.

16 Dang, Q.D., Guinto, E.R. and di Cera, E. (1997) Rational engineering of activity and specificity in a serine protease. *Journal of Molecular Biology*, **15**, 146–9.

17 Zhu, Z., Sun, D. and Davidson, V.L. (2000) Conversion of methylamine dehydrogenase to a long-chain amine dehydrogenase by mutagenesis of a single residue. *Biochemistry*, **39**, 11184–6.

18 Wang, Y.T., Sun, D.P. and Davidson, V.L. (2002) Use of indirect site-directed mutagenesis to alter the substrate specificity of methylamine dehydrogenase. *The Journal of Biological Chemistry*, **277**, 4119–22.

19 Lanio, T., Jeltsch, A. and Pingoud, A. (2000) On the possibilities and limitations of rational protein design to expand the specificity of restriction enzymes: a case study employing EcoRV as the target. *Protein Engineering*, **13**, 275–81.

20 Geddie, M.L., O'Loughlin, T.L., Woods, K.K. and Matsumura, I. (2005) Rational design of p53, an intrinsically unstructured protein, for the fabrication of novel molecular sensors. *The Journal of Biological Chemistry*, **280**, 35641–6.

21 Arnold, F.H. (1998) Design by directed evolution. *Accounts of Chemical Research*, **31**, 125–31.

22 Lanio, T., Jeltsch, A. and Pingoud, A. (1998) Towards the design of rare cutting restriction endonucleases: Using directed evolution to generate variants of EcoRV differing in their substrate specificity by two orders of magnitude. *Journal of Molecular Biology*, **283**, 59–69.

23 Collins, C.H., Leadbetter, J.R. and Arnold, F.H. (2006) Dual selection enhances the signaling specificity of a variant of the quorum-sensing transcriptional activator LuxR. *Nature Biotechnology*, **24**, 708–12.

24 Santoro, S.W. and Schultz, P.G. (2002) Directed evolution of the site specificity of Cre recombinase. *Proceedings of the National Academy of Sciences of the United States of America*, **99**, 4185–90.

25 Reetz, M.T., Bocola, M., Carballeira, J.D., Zha, D.X. and Vogel, A. (2005) Expanding the range of substrate acceptance of enzymes: combinatorial active-site saturation test. *Angewandte Chemie – International Edition*, **44**, 4192–6.

26 Rubin-Pitel, S.B. and Zhao, H.M. (2006) Recent advances in biocatalysis by directed enzyme evolution. *Combinatorial Chemistry and High Throughput Screening*, **9**, 247–57.

27 Petschacher, B., Leitgeb, S., Kavanagh, K.L., Wilson, D.K. and Nidetzky, B. (2005) The coenzyme specificity of *Candida tenuis* xylose reductase (AKR2BIX5) explored by site-directed mutagenesis and X-ray crystallography. *Journal of Biochemistry*, **385**, 75–83.

28 Woodyer, R., van der Donk, W.A. and Zhao, H.M. (2003) Relaxing the nicotinamide cofactor specificity of phosphite dehydrogenase by rational design. *Biochemistry*, **42**, 11604–14.

29 Liu, Y., Shah, K., Yang, F., Witucki, L. and Shokat, K.M. (1998) Engineering Src family protein kinases with unnatural nucleotide specificity. *Chemistry and Biology*, **5**, 91–101.

30 Liu, Y., Witucki, L.A., Shah, K., Bishop, A.C. and Shokat, K.M. (2000) Src-Abl tyrosine kinase chimeras: Replacement of the adenine binding pocket of c-Abl with v-Src to swap nucleotide and inhibitor specificities. *Biochemistry*, **39**, 14400–8.

31 Rodriguez-Arnedo, A., Camacho, M., Llorca, F. and Bonete, M.J. (2005) Complete reversal of coenzyme specificity of isocitrate dehydrogenase from *Haloferax volcanii*. *The Protein Journal*, **24**, 259–66.

32 Flores, H. and Ellington, A.D. (2005) A modified consensus approach to mutagenesis inverts the cofactor specificity of *Bacillus stearothermophilus* lactate dehydrogenase. *Protein Engineering, Design and Selection*, **18**, 369–77.

33 Wong, L.L. (1998) Cytochrome P450 monooxygenases. *Current Opinion in Chemical Biology*, **2**, 263–8.

34 Lindberg, R.L.P. and Negishi, M. (1989) Alteration of mouse cytochrome $P450_{coh}$ substrate specificity by mutation of a single amino acid residue. *Nature*, **339**, 632–4.

35 Glieder, A., Farinas, E.T. and Arnold, F.H. (2002) Laboratory evolution of a soluble, self-sufficient, highly active alkane hydroxylase. *Nature Biotechnology*, **20**, 1135–9.

36 Meinhold, P., Peters, M.W., Chen, M.M.Y., Takahashi, K. and Arnold, F.H. (2005) Direct conversion of ethane to ethanol by engineered cytochrome P450 BM-3. *ChemBioChem*, **6**, 1765–8.

37 Hsu, C.C., Hong, Z.Y., Wada, M., Franke, D. and Wong, C.H. (2005) Directed evolution of D-sialic acid aldolase to L-3-deoxy-manno-2-octulosonic acid (L-KDO) aldolase. *Proceedings of the National Academy of Sciences of the United States of America*, **102**, 9122–6.

38 DeSantis, G., Liu, J.J., Clark, D.P., Heine, A., Wilson, I.A. and Wong, C.H. (2003) Structure-based mutagenesis approaches toward expanding the substrate specificity of D-2-deoxyribose-5-phosphate aldolase. *Bioorganic and Medicinal Chemistry*, **11**, 43–52.

39 Wang, L., Brock, A., Herberich, B. and Schultz, P.G. (2001) Expanding the genetic code of *Escherichia coli*. *Science*, **292**, 498–500.

40 Wang, L. (2003) Expanding the genetic code. *Science*, **302**, 584–5.

41 Wang, L., Xie, J. and Schultz, P.G. (2006) Expanding the genetic code. *Annual Review of Biophysics and Biomolecular Structure*, **35**, 225–49.

42 Samuelson, J.C. and Xu, S.Y. (2002) Directed evolution of restriction endonuclease BstYI to achieve increased substrate specificity. *Journal of Molecular Biology*, **319**, 673–83.

43 Seligman, L.M., Chevalier, B.S., Chadsey, M.S., Edwards, S.T., Savage, J.H. and Veillet, A.L. (2002) Mutations altering the cleavage specificity of a homing endonuclease. *Nucleic Acids Research*, **30**, 3870–9.

44 Chevalier, B.S., Kortemme, T., Chadsey, M.S., Baker, D., Monnat, R.J. and Stoddard, B.L. (2002) Design, activity, and structure of a highly specific artificial endonuclease. *Molecular Cell*, **10**, 895–905.

45 Ong, J.L., Loakes, D., Jaroslawski, S., Too, K. and Holliger, P. (2006) Directed evolution of DNA polymerase, RNA polymerase and reverse transcriptase activity in a single polypeptide. *Journal of Molecular Biology*, **361**, 537–50.

46 Morley, K.L. and Kazlauskas, R.J. (2005) Improving enzyme properties: when are closer mutations better? *Trends in Biotechnology*, **23**, 231–7.

47 Arnold, F.H. (2001) Combinatorial and computational challenges for biocatalyst design. *Nature*, **409**, 253–7.

48 Dahiyat, B.I. and Mayo, S.L. (1997) De novo protein design: Fully automated sequence selection. *Science*, **278**, 82–7.

49 Chartrain, M., Salmon, P.M., Robinson, D.K. and Buckland, B.C. (2000) Metabolic engineering and directed evolution for the production of pharmaceuticals. *Current Opinion in Biotechnology*, **11**, 209–14.

50 Umeno, D. and Arnold, F.H. (2004) Evolution of a pathway to novel long-chain carotenoids. *Journal of Bacteriology*, **186**, 1531–6.

51 Khosla, C. and Zawada, R.J.X. (1996) Generation of polyketide libraries via combinatorial biosynthesis. *Trends in Biotechnology*, **14**, 335–41.

52 Stephanopoulos, G.N., Aristidou, A.A. and Nielsen, J. (1998) *Metabolic Engineering: Principles and Methodologies*, Academic Press, London, UK.

53 Pitkanen, J.P., Aristidou, A., Salusjarvi, L., Ruohonen, L. and Penttila, M. (2003) Metabolic flux analysis of xylose metabolism in recombinant *Saccharomyces cerevisiae* using continuous culture. *Metabolic Engineering*, **5**, 16–31.

54 Watanabe, S., Saleh, A.A., Pack, S.P., Annaluru, N., Kodaki, T. and Makino, K. (2007) Ethanol production from xylose by recombinant *Saccharomyces cerevisiae* expression protein engineered $NADP^+$-dependent xylitol dehydrogenase. *Journal of Bacteriology*, **130**, 316–19.

55 Nair, N. and Zhao, H.M. (2007) Biochemical characterization of an L-xylulose reductase from *Neurospora crassa*. *Applied and Environmental Microbiology*, **73**, 2001–4.

56 Siani, L., Viggiani, A., Notomista, E., Pezzella, A. and Di Donato, A. (2006) The role of residue Thr249 in modulating the catalytic efficiency and substrate specificity of catechol-2,3-dioxygenase from *Pseudomonas stutzeri* OX1. *The FEBS Journal*, **273**, 2963–76.

57 Friedman, T. (1999) *The Development of Human Gene Therapy*, Cold Spring Harbor Laboratory Press, Cold Spring Harbor, NY.

58 Chevalier, B.S. and Stoddard, B.L. (2001) Homing endonucleases: structural and functional insight into the catalysts of intron/intein mobility. *Nucleic Acids Research*, **29**, 3757–74.

59 Paques, F. and Duchateau, P. (2007) Meganucleases and DNA double-strand break-induced recombination: perspectives for gene therapy. *Current Gene Therapy*, **7**, 49–66.

60 Kim, Y.G., Cha, J. and Chandrasegaran, S. (1996) Hybrid restriction enzymes: zinc finger fusions to FokI cleavage domain. *Proceedings of the National Academy of Sciences of the United States of America*, **93**, 1156–60.

61 Harvey, D.M. and Caskey, C.T. (1998) Inducible control of gene expression: prospects for gene therapy. *Current Opinion in Chemical Biology*, **2**, 512–18.

62 Chockalingam, K., Chen, Z.L., Katzenellenbogen, J.A. and Zhao, H.M. (2005) Directed evolution of specific receptor-ligand pairs for use in the creation of gene switches. *Proceedings of the National Academy of Sciences of the United States of America*, **102**, 5691–6.

32
Protein Engineering of Modular Polyketide Synthases

Alice Y. Chen and Chaitan Khosla

32.1
Introduction

Polyketide-derived compounds are a family of secondary metabolites that display antibiotic (erythromycin, tetracycline, rifamycin), antifungal (amphotericin), immunosuppressant (FK506 and rapamycin), antitumor (doxorubicin, geldanamycin, epothilone) and other important biological activities (Figure 32.1). Modular polyketide synthases (PKS)s are the enormous multidomain proteins that produce these structurally diverse and therapeutically useful compounds [1–6]. In a modular PKS such as the 6-deoxyerythronolide B synthase (DEBS; Figure 32.2), sets of catalytic domains are organized into modules. These highly conserved domains are covalently linked by relatively unconserved intervening linker sequences. The catalytic domains from a given module cooperate to synthesize polyketides via a thiotemplate mechanism similar to that of fatty acid synthesis. The nascent polyketide intermediates are then passed down to the next module in the protein assembly line.

The modular architecture of polyketide synthases has motivated much effort towards the combinatorial biosynthesis of novel polyketides. The mixing and matching of catalytic domains or modules from various PKSs has generated polyketide analogues with varying degrees of success [1, 7–10]. The first part of this chapter will discuss the structural insights and current engineering approaches on polyketide biosynthesis, while the second part will review the common genetic manipulation and characterization techniques for the engineered PKS systems. Through this chapter, we hope to summarize the current progress in PKS engineering and demonstrate the potential of combinatorial polyketide biosynthesis.

Protein Engineering Handbook. Edited by Stefan Lutz and Uwe T. Bornscheuer

ISBN: 978-3-527-31850-6

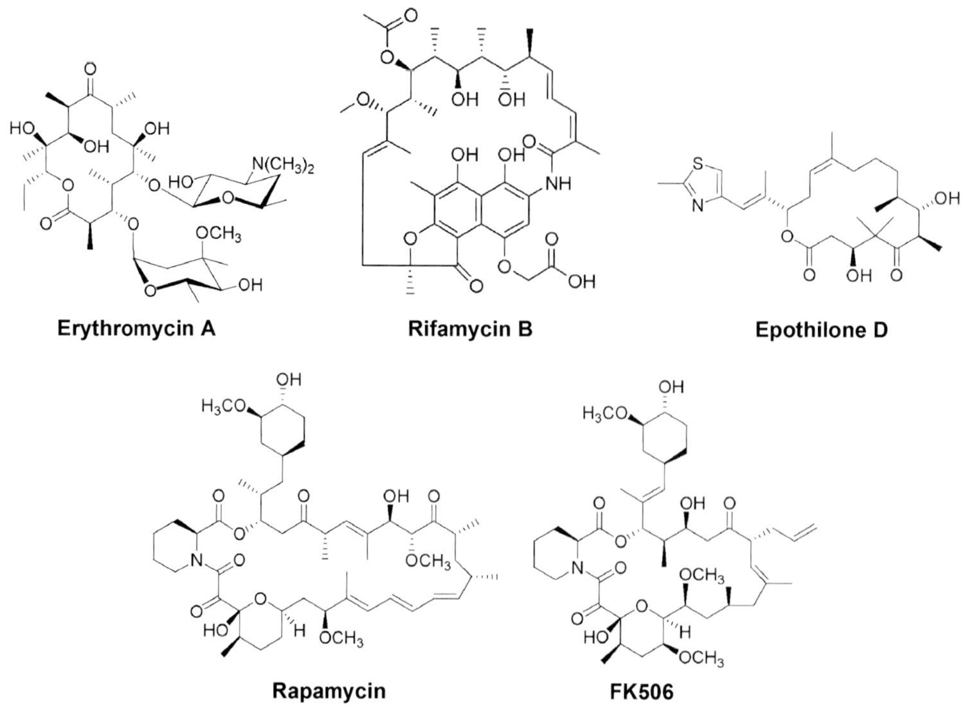

Figure 32.1 Examples of medicinally relevant polyketide-derived compounds. Antibiotics erythromycin and rifamycin, antitumor agent epothilone, and immunosuppressant agents rapamycin and FK506.

32.2
Polyketide Biosynthesis and Engineering

The colinearity between the organization of a PKS and the structure of the corresponding product has made PKS genetic engineering an emerging strategy for polyketide analogue generation. It circumvents the total synthesis or semi-synthetic chemical modification of these structurally complex molecules. This rational recombinant approach is prompted – and is continued to be encouraged – by the discovery of PKS gene clusters with a diverse array of catalytic domains. By properly recombining these domains through insertion, deletion and substitution, it is possible to reprogram a polyketide biosynthetic machinery to produce compounds with exotic building blocks and desired functionalities.

Despite the complex structures of polyketides, the chemical logic of a PKS system is rather straightforward. A set of chemical building blocks in the form of activated carboxylic acids (acyl-CoA thioesters such as acetyl-CoA, malonyl-CoA and methylmalonyl-CoA) are successively assembled into polyketide oligomers through Claisen-like decarboxylative condensation. The condensed polyketide

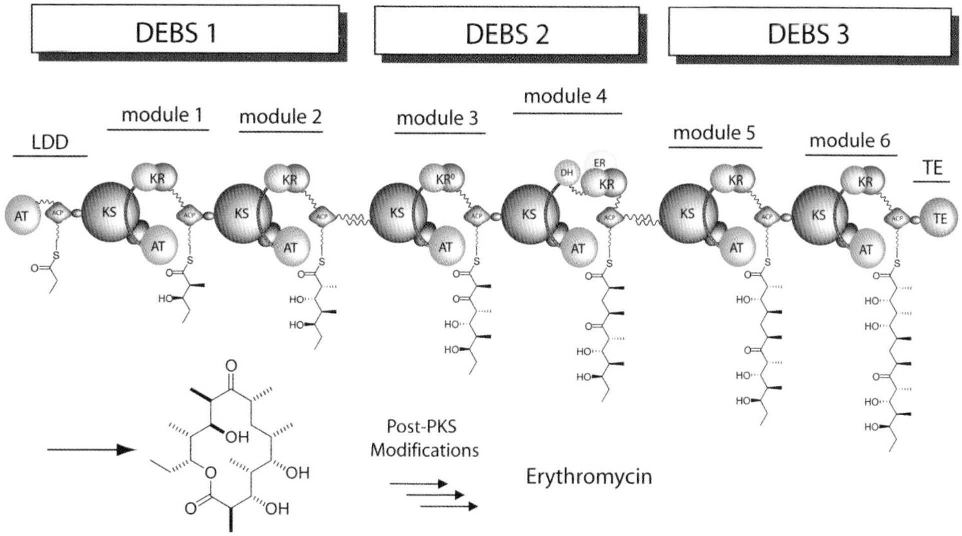

Figure 32.2 Modular organization of 6-deoxyerythronolide B synthase (DEBS) from *Saccharopolyspora erythraea*. Organized into three homodimeric polypeptides (DEBS1-3), DEBS consists of six catalytic modules, each containing a unique set of covalently linked catalytic domains. Together, the six modules of this megasynthase produce 6-deoxyerythronolide B (6-dEB), the polyketide portion of the antibiotic erythromycin. The pathway is primed by a propionyl unit, which is extended with 6 methylmalonyl-extender units through successive decarboxylative condensations. KS = β-keto synthase; AT = acyltransferase; ACP = acyl carrier protein; KR = ketoreductase; DH = dehydratase; ER = enoylreductase; TE = thioesterase; LDD = loading di-domain. KR° in module 3 is inactive.

chain is then subjected to optional β-carbon processing and other modifications to increase its structural diversity. The specific chemical transformations performed by a polyketide synthase are detailed in Figure 32.3.

While numerous PKS systems have been discovered and characterized, most of our structural and engineering insights in polyketide biosynthesis are derived from DEBS (see Figure 32.2). With its 28 catalytic domains that are organized into six homodimeric modules, DEBS is a relatively simple polyketide synthase (though still with a molecular weight of approximately 2 MDa). The enzyme produces 6-deoxyerythronolide B (6-dEB), the 14-member macrolactone precursor to the antibiotic erythromycin. The advancements in PKS engineering, many of which were achieved in the context of DEBS, are highlighted in the following sections.

32.2.1
Active Sites and Domain Boundaries in Multimodular PKSs

In order to identify predictive strategies for engineering multimodular PKSs, active-site motifs and putative domain boundaries have been defined through

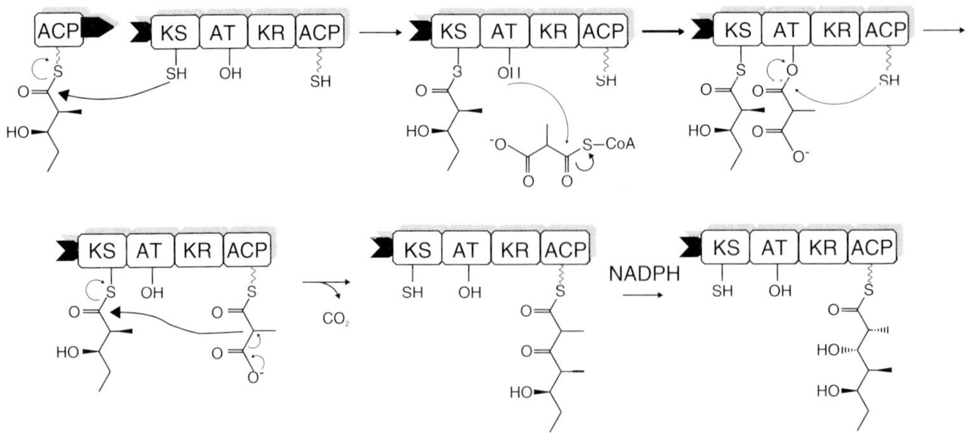

Figure 32.3 Chain elongation cycle and β-carbon processing catalyzed by a typical PKS module. Chain elongation occurs minimally through the combined action of the β-keto synthase (KS), acyltransferase (AT) and acyl carrier protein (ACP) domains. The electrophile (either the primer unit for the KS of the first module or the growing polyketide chain for the KS of subsequent modules) is first tethered onto the KS via a thioester linkage to the active-site cysteine. Meanwhile, the AT captures a nucleophilic β-carboxyacyl-CoA extender unit and transfers it to the phosphopantetheine arm of the ACP domain. Once both the growing chain and the extender units are covalently loaded onto the module, a KS-mediated, Claisen-like decarboxylative condensation occurs to form a β-ketothioester-ACP intermediate. With the release of carbon dioxide, this elongation step extends the polyketide intermediate by two carbon atoms. In addition to these essential condensing domains, optional reductive domains including ketoreductase (KR), dehydratase (DH) and enoylreductase (ER) domains control the ultimate oxidation state of the β-carbon. The aforementioned reductive domains, along with other tailoring domains such as methyltransferase (MT), significantly increased the structural diversity of type I polyketides. Once processed, the polyketide chain is either passed to the KS domain of the downstream module or cyclized and released by the thioesterase (TE) domain at the C terminus of the polyketide synthase.

sequence alignment and limited proteolysis. The active-site motifs of the common PKS domains are listed in Table 32.1 [1, 11]. Sequence alignment between vertebrate fatty acid synthases (FAS) and PKSs facilitated the early assignment of PKS domain boundaries [12]. These boundaries have since been adjusted or verified by alignment among PKS modules and limited proteolysis experiments. For instance, through limited proteolysis Leadlay and coworkers have observed various DEBS1 fragments with partial activities, while Khosla and colleagues have identified conserved junction sequences that led to the expression of a catalytic β-keto synthase- acyltransferase (KS-AT) didomain and standalone KS and AT recombinant proteins [13–16]. Some predicted PKS domain boundaries are shown in Figure 32.4.

Although sequence homology and proteolyzed enzymatic fragments have validated the modular architecture of PKS, many hybrid PKS attempts have been suboptimal due partly to inaccurate choices of domain boundaries (in addition to

Table 32.1 Signature motifs of PKS domain active sites. For KS and AT, the active sites are presented with the nucleophilic residues in boldface and underlined. For KR and ER, the NADPH-binding motif is presented. For DH, the conserved motif that carries part of the His/Glu catalytic diad is presented. For MT, the SAM-binding motif is presented. For ACP, the phosphopantetheine arm attachment site is presented with the nucleophilic residues in boldface and underlined. For TE, the catalytic triad is presented.

PKS domain	Active-site motif
KS	GPxxxxxTA**C**SS
AT	GH**S**xG
KR, ER	GxGxxAxxxA
DH	HxxxGxxxxP
MT	LExGxGxG
ACP	LGxD**S**LxxVE
TE	Ser142, His259, Asp169

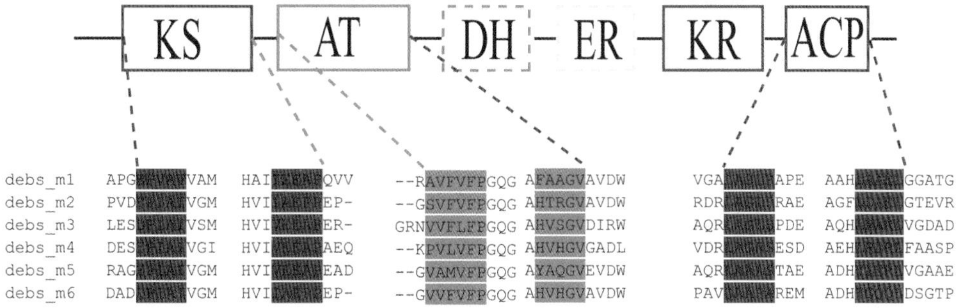

Figure 32.4 Sequence alignment of the six 6-deoxyerythronolide B synthase (DEBS) modules. Highly conserved sequences are present at the termini of PKS domains. Figure reproduced from Ref. [1].

the intrinsic incompatibility of the domain components). Fortunately, high-resolution structures of many PKS fragments, along with several FAS, have recently been reported [17–28]. The homodimeric KS-AT didomain structures from DEBS modules 3 and 5, for example, revealed the extensive interactions between the two domains and their relative positioning [17, 18]. These studies also reported the structural roles of interdomain linkers that were previously predicted to be merely connective in nature. Similarly, the ketoreductase structure from DEBS module 1 refined the boundaries of the three PKS reductive domains, namely ketoreductase (KR), dehydratase (DH) and enoylreductase (ER), and helped in predicting their spatial arrangement within a module [19]. The NMR structure of the acyl carrier protein (ACP) from DEBS module 2 presented its steric and electrostatic surface features that may contribute to domain–domain recognition between ACP and other domains [20]. Finally, the thioesterase (TE) domain structures from DEBS and pikromycin synthase elucidated the potential substrate channeling and release mechanism [21–23].

In addition to the catalytic domains, the structure of the 'docking domain' peptides between DEBS modules 4 and 5 (i.e. the C terminus peptide from DEBS2 and the N terminus peptide from DEBS3) has also been solved, and observed again in the KS-AT structure of DEBS module 5 [17, 24]. From the structures we can deduce that the docking domains contribute to both the noncovalent association between the PKS polypeptides and the homodimerization of each polypeptide. The mammalian, fungal and yeast FAS structures, though with features different from those of PKSs, can shed light on the overall organization of PKSs, which remains to be determined [25–28]. These structural studies will continue to aid future engineering endeavors by offering insights into domain boundary, active site environment, three-dimensional domain organization and domain–domain interaction.

32.2.2 Past Achievements in Genetic Reprogramming of Polyketide Biosynthesis

With our understanding of PKS organization and reaction mechanism, the engineering opportunities that have been independently explored include the incorporation of unnatural starter and extender units, the extent and stereochemistry of reduction of the condensed polyketide chain, and the chain length of the polyketide product.

32.2.2.1 Starter Unit Incorporation

In a PKS assembly line, a loading module containing an acyltransferase (AT_L) and an acyl carrier protein (ACP_L) precedes the condensing modules and primes the assembly line with a starter, or primer, unit. In order to incorporate unnatural starter units in a PKS system, two major approaches have been taken: (i) loading module (AT_L-ACP_L) swapping; and (ii) precursor-directed biosynthesis. In the first approach, various loading modules were covalently substituted with their counterparts in other PKS systems [29–32]. By swapping the propionyl-specific DEBS loading module with that of avermectin or oleandomycin PKS, for example, the resultant fusion systems were capable of producing polyketides with diverse primers such as isobutyrate, 2-methylbutyrate or acetate (Figure 32.5a) [29, 30]. A variety of starter units other than the most common acetate or propionate have been observed in numerous type I PKS systems, such as cyclohexanecarboxylate in asukamycin, *p*-aminobenzoate (PABA) in candicidin, 3-amino-5-hydroxyben-

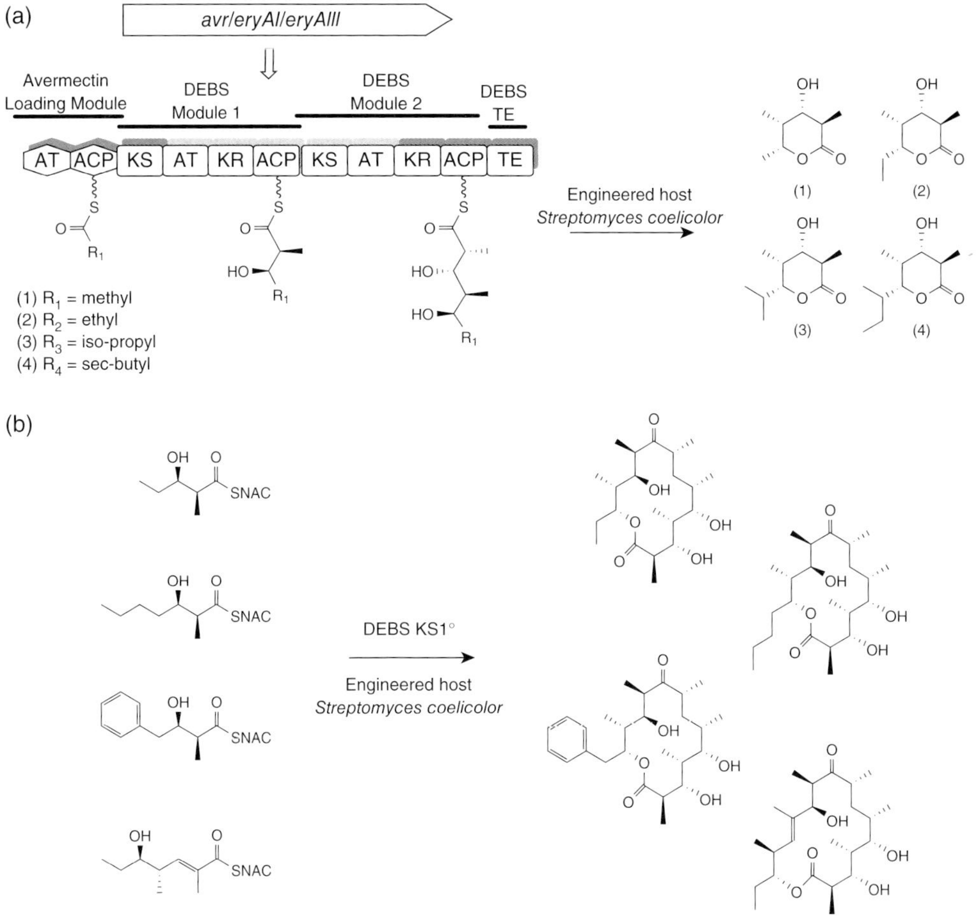

Figure 32.5 Polyketide analogues with unnatural starter units. (a) Triketide lactone analogues produced by loading module swapping. In this study, propionyl-specific DEBS loading module was substituted with avermectin loading module with a broader specificity. Figure adapted from Ref. [29]; (b) 6-Deoxyerythronolide B (6-dEB) analogues produced by precursor-directed biosynthesis. In this study, unnatural ketide-SNAC substrates were fed to KS1-inactivated DEBS mutant. Figure adapted from Ref. [34].

zoic acid (AHBA) in geldanamycin and rifamycin, and the aforementioned isobutyrate and (*S*)-2methylbutyrate in avermectins [33]. Given the appropriate engineering tools, this remarkable diversity can expand our engineering vocabularies in starter unit incorporation.

In the second approach, instead of priming the assembly line with the loading module, synthetic ketide intermediates were introduced as *N*-acetyl cysteamine (SNAC) thioesters to the system [34–39]. The *N*-acetyl cysteamine entity that carries the ketide intermediate behaves as a chemical surrogate for an electrophile-donating ACP domain. This precursor-directed strategy was often coupled with the inactivation of the loading module and/or the first ketosynthase (KS1) to relax starter unit selectivity. By inactivating DEBS KS1, Jacobsen and coworkers incorporated phenyl groups and other unnatural starter units as ketide SNAC substrates, and demonstrated the remarkable substrate tolerance of the downstream modules (Figure 32.5b) [34]. Alternatively, the loading module of DEBS has been eliminated to accommodate 'mono-ketide' substrates such as butyryl-SNAC, which are easier to synthesize than diketide SNACs [38]. In addition to modifying the PKS directly, disturbing the regulation of natural starter unit production can also promote unnatural unit incorporation [39]. Precursor-directed biosynthesis allows the incorporation of synthetic moieties of various lengths and functional groups into the final polyketide products. Advanced intermediates (such as triketides and tetraketides) can also be chemically synthesized and incorporated into the biosynthetic pathway to circumvent potential substrate selectivity of the PKS enzymes [40].

32.2.2.2 Extender Unit Incorporation

In a PKS system, the starter unit, once loaded, is elongated with extender units via decarboxylative condensation as it progresses through the protein assembly line. These extender units are selected by the acyltransferase (AT) domain, dubbed the 'gatekeeper' domain, in each of the PKS module. To introduce diversity through unnatural extender units, AT domain swapping, site-directed mutagenesis (SDM) and *in trans* complementation have been performed on various PKS systems. By replacing the methylmalonyl-specific DEBS AT domains with those that are malonyl- (from rapamycin and pikromycin PKSs) [41–44], ethylmalonyl- (from niddamycin PKS) [45], or methoxylmalonyl-specific (from ansamitocin) [46], different chemical moieties were incorporated into 6-dEB analogues. Similarly, analogues of FK520 and geldanamycin were produced via AT domain swapping [47, 48]. It is worth noting that the yield of the analogue compounds was often reduced relative to their corresponding 'wild-type' compounds. Such a decrease in hybrid protein catalytic efficiency may be attributed to inaccurate domain boundary definition and possible incompatibility between the domains.

Instead of substituting the complete domain, an alternative approach is to identify the peptide sequences that distinguish, for example, methylmalonyl-specific ATs from malonyl-specific ATs. Indeed, sequence alignments of PKS modules revealed three AT sequence motifs that are well correlated with AT substrate specificity [49, 50]. For instance, mutations on short-sequence motifs from **YA**S**H** to **HA**F**H** or from GHS**Q**G to GHS**L**G (adjacent to the active site Ser) were sufficient to relax methylmalonyl-specific ATs to accept malonyl-CoA as their substrate [50, 51]. In addition to domain swapping and mutagenesis, Kumar and coworkers demonstrated the feasibility of utilizing exogenous type II transacylase *in trans* [52]. In this study, the endogenous AT domain was first inactivated via mutagenesis, and a type II malonyl-CoA:ACP transacylase (MAT) was coexpressed *in trans* to rescue the AT function. This approach minimizes structural perturbation to the overall PKS, but relies on productive interactions between noncognate, noncovalently connected protein partners. The availability of AT domains from modular PKSs as standalone proteins should facilitate its further exploitation [16].

The extender unit collection, though not as impressive as that of the starter units, still provides considerable diversity with malonyl-, methylmalonyl-, ethylmalonyl- and glycerol-derived methoxymalonyl-units [33, 53]. Recent studies have also reported the existence of hydroxymalonyl-ACP and aminomalonyl-ACP, adding free hydroxyl and amino functional groups to the extender unit toolbox [54]. The incorporation of substituted malonates into polyketide backbones also contributes to the stereochemistry of these natural products. The precise mechanism of control of methyl (or other functional groups) branched stereocenters remains a mystery. A list of acyltransferase specificity from the major type I PKS systems has been compiled by Yadav and coworkers [55].

32.2.2.3 **β-Carbon Processing**

Unlike fatty acid synthesis in which the growing chain is fully reduced, PKS systems offer a varying degree and stereochemistry of β-ketoreduction with the optional presence of KR, DH and ER. The condensed polyketide chain is processed by these tailoring domains before progressing to the next module. The reactions performed by the three reductive domains are detailed in Figure 32.6.

The various oxidation states offer different functional groups at the β-carbon. A hydroxyl group generated by KR activity, for example, can provide post-PKS modification sites and change the physical properties of the polyketide compound, such as its solubility [56]. To achieve such diversity, domain deletion or inactivation [43, 57], domain insertion [58, 59] and reductive cassette shuffling [43, 60] have successfully manipulated the extent of reduction. Specifically, with combinations of AT swapping, KR deletion and reductive domain shuffling in the context of DEBS, McDaniel and coworkers were able to biosynthesize over 50 6-dEB analogues, demonstrating the viability of combinatorial biosynthesis of type I polyketides (Figure 32.7) [43].

In addition to the final oxidation state, the stereocontrol of β-ketoreduction is also under intense scrutiny. It has been shown that the KR domain determines

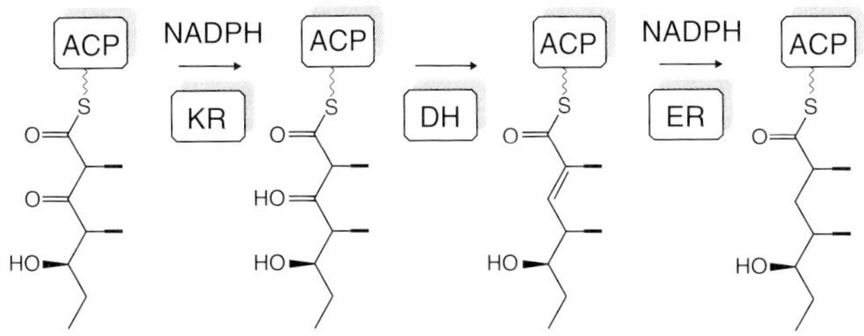

Figure 32.6 Reductive actions of ketoreductase (KR), dehydratase (DH) and enoylreductase (ER). After chain elongation, KR is responsible for the reduction of the condensed β-ketoacyl-S-ACP to a β-hydroxyacyl-S-ACP. DH, if present, then dehydrates the hydroxyl intermediate to a α,β-enoyl-S-ACP. If the module also contains an ER domain, the enoyl intermediate is further reduced to a saturated acyl-S-ACP before the nascent polyketide is channeled to the next module.

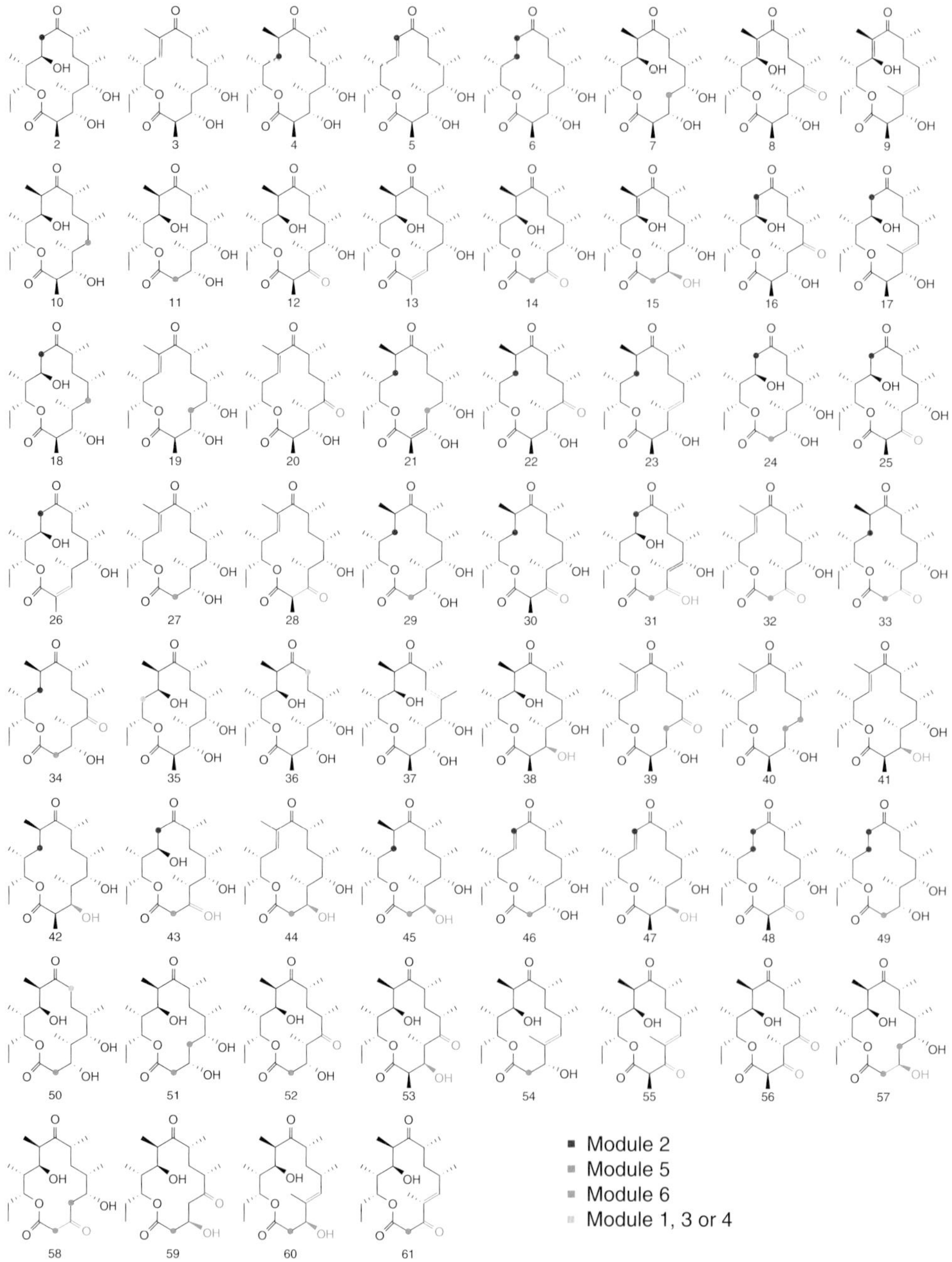

Figure 32.7 6-Deoxyerythronolide B (6-dEB) combinatorial library. 6-dEB analogues produced via combinations of AT swapping, KR deletion and reductive domain shuffling in the context of DEBS. Figure reproduced Ref. [43].

the stereospecificity of ketide reduction, though recent studies reported that different KRs possess varying degrees of stereocontrol [61, 62]. As in AT specificity alteration, signature amino acid motifs have been correlated with the stereochemistry of KR activity (an LDD motif in one case and an FxxPxxxG motif in the other) [63, 64]. The subsequent motif mutagenesis switched the reduction stereospecificity of DEBS KR domains with mixed success [64, 65]. SDM of these motifs was also shown to broaden the KR substrate specificity toward nonpolyketide substrates such as cyclo- and linear alkyl-ketone moieties, further establishing the significance of these motifs in reduction substrate control [66]. In order to accelerate our understanding in this front, robust analytical tools (e.g. standard molecule synthesis and GC/LC-MS methods) and experimentally convenient PKS systems (e.g. the dissociated modular systems; see Section 32.2.2.6) have been developed [67].

32.2.2.4 **Chain Length Control**

In type I PKS, the final chain length is generally controlled by the number of modules within the system in addition to the chain length of the starter unit. Module deletion and addition of DEBS indeed produced the truncated and extended polyketide products, respectively [68, 69]. An alternative approach to chain-length control is the repositioning of the TE domain. The latter, which is located at the C terminus of a PKS, catalyzes the release of the full-length polyketide product through lactonization. In DEBS, when the TE domain was inserted after module 2, 3 and 5, the expected truncated products–triketide, tetraketide and a 12-membered macrolactone, respectively–were indeed observed [68, 70, 71].

32.2.2.5 **Additional Modifications**

Much of the chemical diversity of polyketides arises from PKS tailoring domains. In addition to the reductive domains presented above, PKSs that produce compounds such as epothilone, myxothiazol, lovastatin and yersiniabactin synthases further modify the polyketide chains through their methyltransferase (MT) activities [72–75]. Methyl transferases transfer methyl groups from *S*-adenosylmethonines to the acyl intermediates, thereby increasing the hydrophobicity and steric impact of the final products while reducing their hydrogen-bonding capacity. An aminotransferase (AMT) domain from mycosubtilin synthase has recently been identified and characterized [76, 77]. This AMT domain transfers an amine to the β-carbon of the acyl intermediate, adding another modification capacity to PKS engineering.

32.2.2.6 **Other PKS Engineering Opportunities**

In addition to domain insertion, deletion, substitution, mutagenesis and *in trans* complementation, other PKS engineering opportunities include the constructions of multimolular hybrid PKSs, efficient intermodular and interdomain linkers, and dissociated module systems. Non-PKS systems with analogous chemical logic, such as non-ribosomal peptide synthases (NRPSs), can also be integrated to expand the existing PKS engineering repertoire.

32.2.2.6.1 **Intact Modules as Interchangeable Cassette** Numerous hybrid PKSs have been constructed via module shuffling to examine PKS modularity, substrate specificity and the ability to biosynthesize unnatural polyketides [69, 78–83]. By rearranging intact modules, this approach minimizes structural perturbations that accompany individual domain swaps. It also preserves the domain combinations that presumably render the highest catalytic efficiency. Gokhale and coworkers established that modules from different PKS systems (DEBS and rifamycin PKS in this case) could communicate efficiently when appropriate linker peptides were in place [78]. Subsequent studies further probed into hybrid PKS features such as chain-length modification, intermodular interaction and substrate tolerance [69, 79–82]. A recent study by Menzella and colleagues demonstrated the power of module rearrangement in biosynthesizing polyketide analogues [83]. By shuffling 14 elongation modules from eight PKS clusters into 154 bimodular combinations, Menzella and coworkers were able to generate 72 productive hybrid systems. This bimodular library subsequently facilitated the rational assembly of trimodular PKSs with the addition of a recombinant linker interface [84].

The major limitations of this module shuffling approach are: (i) the substrate tolerance of PKS domains and downstream enzymes that need to process noncognate substrates; and (ii) the communication between noncognate modules in substrate channeling. While the former concern is often addressed with domain evolution (e.g. via mutagenesis, as described previously), the latter has been tackled by docking domain studies.

32.2.2.6.2 **Docking Domain Studies** Substrate channeling is largely mediated by the short but structured peptide fragments at the N and C termini of a PKS subunit such as DEBS1. These short peptides, which often are termed 'docking domains' or interpolypeptide linkers, are organized into α-helix bundles [24]. Through hydrophobic interactions, these bundles bridge the consecutive but noncovalently connected polypeptides, such as DEBS1 and 2 and DEBS2 and 3. It has been well established that proper linker pairing can yield efficient substrate channeling between noncognate module partners, while improper linker pairing can abolish chain transfer between native partners [85–88]. Furthermore, these linker pairs are found to be entirely modular. For example, the linker pair between DEBS1 and 2 can be replaced by the linker pair between DEBS2 and 3 to render comparable chain-transfer efficiency [85, 86]. The results of recent studies have shown that the docking domain specificity can be altered simply by exchanging the docking domain helices–rather than the entire docking domain–that are involved in docking recognition [89]. The portability of docking domains presents favorable prospect in noncovalently linking heterologous modules.

32.2.2.6.3 **Dissociated Module Engineering** To improve hybrid PKS constructions, an alternative approach is to separate the domains into standalone enzymes. The isolation of functional domains validates domain and linker boundaries and allows the concentration ratio of domains to be varied (as opposed to the fixed 1:1 ratio in covalently linked modules). By treating domains as individual reagents, this approach facilitates the interrogation of domain substrate specificity, product stereocontrol and protein–protein interaction, all of which are essential in advancing PKS engineering. Benefiting from sequence homology and structural information, KS-AT didomains and standalone KS, AT, KR and ACP domains from various DEBS modules have been expressed as soluble proteins [14–16, 61]. In addition to these catalytic domains, we can now also visualize the well-defined interdomain structures of KS-AT linker from DEBS modules 3 and 5 [17, 18], AT-KR linker from DEBS module 1 [19, 90], and DH-ER linker from DEBS module 4 [90]. Despite the valuable insights they offer, the dissociated systems tend to suffer kinetic penalties. The ideal approach is, then, to apply the information of individual domains from dissociated systems in the context of covalently connected modules, thereby garnering both the detailed domain knowledge and the preferred kinetics of intramodular chemistry.

32.2.2.6.4 **PKS/NRPS Hybrid Systems** Analogous to a PKS system, a NRPS system consists minimally of three domains organized in a modular fashion: the adenylation domain (A, analogous to AT in PKS); the peptidyl carrier protein (PCP, analogous to ACP in PKS); and the condensation domain (C, analogous to KS in PKS) [91]. The A-domain selects and activates the cognate amino acid before transferring the amino moiety to a PCP, and the C-domain then catalyzes the formation of the peptide bond. In addition to the large amino acid substrate repertoire, modifications such as N-methylations, epimerizations and cyclizations further expand the chemical diversity offered by NRPS systems [92, 93]. By harvesting the power and flexibility of PKS and NRPS, complex compounds such as antitumor agents epothilones and bleomycin, and immunosuppressant agents rapamycin and FK506, have been synthesized by PKS/NRPS hybrid megasynthetases (see Figure 32.1) [2]. By manipulating the number and class of building blocks incorporated, the order and configuration of their incorporation, and the combination of modifications, PKS/NRPS hybrid systems provide a complicated yet promising route for the biosynthesis of novel, intricate and biologically active compounds.

32.2.3 Pre-/Post-PKS Pathway Engineering

Pathways supplying the starter and extender units are often necessary in polyketide biosynthesis, especially when the precursor units are unusual metabolites in the host organisms. Moreover, many polyketide compounds require post-PKS modifications to render bioactive molecules. Consequently, *in vivo* precursor productions and post-PKS modifications must be considered in order to engineer systems that produce the fully active polyketide-derived compounds.

32.2.3.1 Precursor Production

While many PKSs are primed with acetate- or propionate-derived moieties, others require specialized starter units such as 3,4-dihydroxycyclohexanecarboxylic acid

(DHCHC, for rapamycin and FK506 syntheses) [94] and AHBA for rifamycin synthesis) [95, 96]. Several specialized starter unit production pathways have been identified from their associated PKS clusters, permitting the transfer of such pathways into heterologous hosts [2]. For example, Watanabe and coworkers constructed an AHBA biosynthetic pathway in *Escherichia coli* by introducing seven genes from rifamycin and ansamitocin producers [97]. Similarly, to maintain sufficient intracellular supplies of extender units, biosynthetic pathways for acyl-CoA species such as methylmalonyl-CoA [98–100] and ethylmalonyl-CoA [45] have been constructed in various organisms. Precursor pathway engineering has unquestionably facilitated PKS characterization in readily controlled host organisms such as *E. coli*.

32.2.3.2 **Post-PKS Modification**

Nascent polyketides synthesized by a PKS are typically bioinactive until they are further processed by post-PKS modification enzymes. Modifications such as glycosylation, oxidation and alkylation alter the physico-chemical features of the polyketides and thereby confer their biological activities [101]. The aminosugar desosamine on the antibiotic erythromycin, for example, is essential in the ribosomal binding of the molecule [102]. As such, unnatural sugars have been introduced to heterologous hosts through gene exchange between different sugar-producing organisms, demonstrating once again the power of combinatorial biosynthesis in novel polyketide production [103, 104]. In addition to glycosylation, hydroxylation, epoxidation and other transformations performed by oxygenases (OX) such as cytochrome P-450, monooxygenases can drastically change the structure and properties of the polyketide substrate [101]. Mixing and matching these modification pathways with their counterparts from different organisms can hold tremendous promise in introducing further diversity and complexity to existing PKS systems.

32.3
Engineering and Characterization Techniques

PKS-encoding genes are known for their large cluster sizes and high GC contents. In order to accommodate these unusual characteristics, various genetic manipulations strategies have been adapted and tailored toward PKS engineering. Similarly, the complexity of the encoded PKS enzymes, in both protein structure and reaction mechanism, has motivated the developments of PKS expression and characterization methods. Some of the common methods used in PKS genetic engineering and protein chemistry studies are highlighted in the following sections.

32.3.1
Common Genetic Techniques for PKS Engineering

Special challenges in molecular biology often accompany the large cluster sizes of PKS genes (>30 kb). For instance, restriction site-based manipulations are limited by the rare occurrence of unique restriction sites in large DNA fragments. Furthermore, the efficiency of standard procedures such as polymerase chain reaction (PCR), mutagenesis and ligation decreases dramatically with increasing DNA sizes. Here, we summarize the common genetic techniques that have been evolved to address these issues.

32.3.1.1 Restriction Site Engineering

To exploit the modular architecture of PKS and construct hybrid PKS systems accordingly, one obvious approach is to fashion PKS domains into genetic 'cassettes' by introducing unique restriction cleavage sites at the domain junctions [43, 60, 69, 79, 81, 82]. The resulting domain cassettes can then be easily inserted, deleted and substituted by their counterparts from other modules. McDaniel and coworkers, for example engineered unique restriction sites between: (i) KS and AT (*Bam*HI); (ii) AT and the reductive domains (*Pst*I); and (iii) the reductive domains and ACP (*Xba*I) of various PKS modules [43]. These sites allowed rapid domain shuffling, and consequently led to the seminal study that generated >50 6-dEB analogues (see Figure 32.7). Likewise, Watanabe and coworkers framed various PKS modules with *Bsa*BI sites prior to their KS domains and *Spe*I sites following their ACP domains, making intact modules into portable cassettes for module shuffling [81]. By utilizing standard molecular biology, this method allowed the construction of PKS 'genetic Lego blocks' that could be conveniently mixed and matched to render desired hybrid constructs. However, each gene must be precisely engineered before it can enter the shuffling repertoire, and the introduction of nonsilent mutations is sometimes inevitable.

32.3.1.2 Gene SOEing

When it is impractical to engineer unique restriction sites at domain junctions, gene splicing by overlap extension (gene SOEing) provides an alternative to *in vitro* PKS recombination [105, 106]. Gene SOEing is a PCR-based technique that joins two DNA sequences without relying on restriction site placement. As shown in Figure 32.8, by designing the 5' end of PCR primers, we can promote hybridization of two DNA sequences derived from different genes, thereby 'fusing' two DNA fragments via PCR. The PCR product consequently encodes the desired fusion protein. This method, unlike restriction site engineering, does not require mutagenesis on potentially large genes, and can practically recombine any two fragments of DNA. Nonetheless, because each construct is individualized, this technique is still laborious and the cost of primers may be high. Also, additional cloning steps are required if the DNA fragments to be joined are too large to obtain directly through PCR [15].

32.3.1.3 Red/ET Homology Recombination

To circumvent the limitations imposed by *in vitro* recombination, Red/ET technologies have been introduced to PKS engineering [48, 100, 107–109]. Red/ET recombination employs phage protein pair Redα/Redβ or RecE/RecT to recombine large DNA fragments with engineered homology regions in place of unique restriction sites [110, 111]. As shown in Figure 32.9, applications of Red/ET

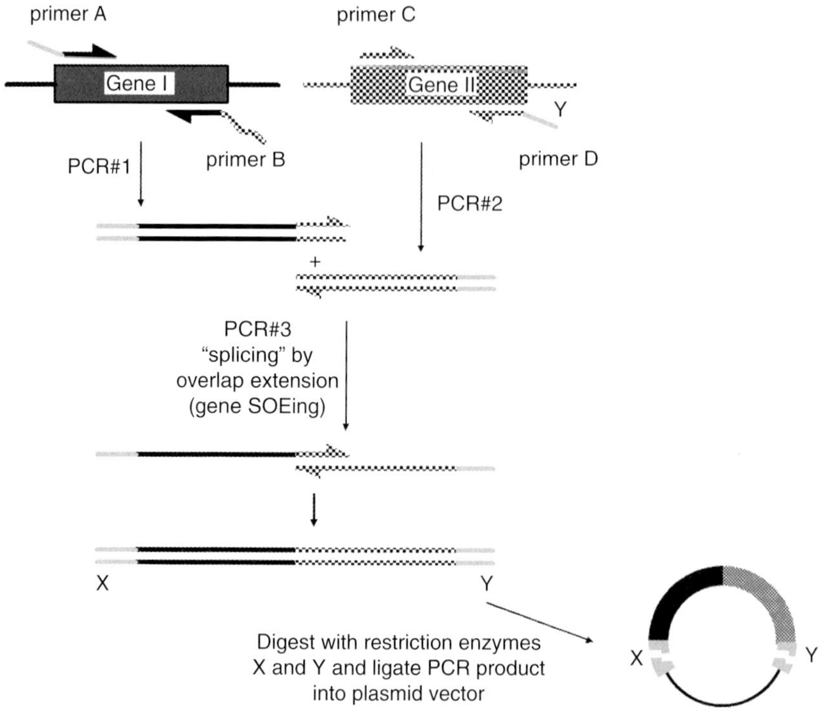

Figure 32.8 Schematics for Gene SOEing. Fragments of two genes can be fused together via gene splicing by overlap extension (gene SOEing). The 5′ end of SOEing primer B is designed to partially complement primer C. This allows the fragments from PCR#1 and PCR#2 to prime each other in PCR#3, thereby producing the recombinant product. The product can then be ligated into a plasmid vector with restriction sites inserted via primers A and D.

include DNA insertion, deletion, substitution, subcloning and point mutation. In the realm of PKS, Red/ET recombination has been used for domain substitution and mutation [48, 107], metabolic pathway engineering [100] and the heterologous expression of various gene clusters [108, 109]. A newly developed method ALFIRE (Assisted Large Fragment Insertion with Red/ET) further improved the existing Red/ET approach with strategic placements of homology arms and restriction sites for downstream cloning steps [112]. While Red/ET is ideal for large DNA fragment recombination, manipulations within the recombined DNA fragments (such as domain mutation [107] and gene shuffling [113]), if needed, would still need to be performed using *in vitro* techniques.

32.3.1.4 **Gene Synthesis**

Total gene synthesis has become increasingly popular due to the opportunity for codon optimization and rational placement of restriction sites. Kodumal and coworkers, for example, semi-synthesized the complete DEBS gene cluster from

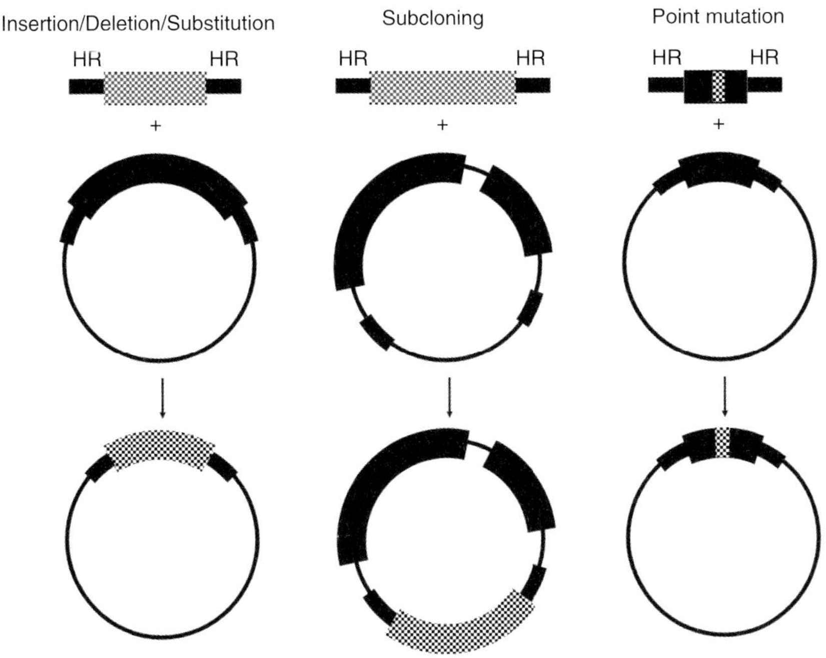

Figure 32.9 Schematics for Red/ET homology recombination. Red/ET recombination employs phage protein pair Redα/Redβ or RecE/RecT to recombine large DNA fragments with engineered homology regions (HR). Its applications include DNA insertion, deletion, substitution, subcloning and point mutation.

500–800 bp of synthetic DNA 'synthons' [114]. These precisely designed synthons were assembled into ~5 kb 'multisynthons' and, eventually, the 31.7 kb gene cluster containing codons optimized for heterologous host *E. coli*. By combining this technique with the conventional restriction site engineering approach, Menzella and coworkers synthesized 14 codon-optimized PKS modules with standardized unique restriction sites flanking their domains and linkers [83]. Derived from eight PKS clusters, these 14 synthesized module cassettes were then shuffled into 154 bimodular combinations, with almost half (72) of these combinations deemed productive. By using properly designed synthetic building blocks, this study demonstrated both the advantages of total gene synthesis and the modularity and flexibility of PKS systems.

32.3.1.5 **Gene Shuffling**

Directed evolution methods such as gene shuffling followed by Red/ET recombination have also been applied to hybrid PKS constructions. The family shuffling of a pikromycin PKS loading module with its DEBS counterpart by Kim and coworkers successfully generated 4000+ variants of the loading module [113, 115]. Through functional complementation in *Streptomyces venezuelae* and *in vivo* screening bioassays, three hybrid PKS capable of producing pikromycin were identified.

Although this study demonstrated the power of gene shuffling in rapid constructions of PKS hybrids, it also hinted at the low hit rate that this strategy is likely to suffer due to the intrinsic complexity of PKS enzymes. Nonetheless, with the advancements in gene shuffling and other directed evolution techniques, PKS hybrid junction sites will hopefully be better predicted and thereby increase the feasibility of random PKS recombination.

32.3.2
In vitro Characterization

32.3.2.1 Protein Expression

To facilitate *in vitro* protein characterizations, the recombinant proteins need to be expressible in practical quantities. In our laboratory, genes are usually inserted into expression vectors such as pET21 and pET28 and transformed into *E. coli* BL21 (DE3) or its modified strains for expression. Cells containing the plasmid(s) are grown at 37 °C in LB media with the proper resistance(s) to an OD_{600} of 0.6. At this point the medium is cooled to the expression temperature and induced with inducing agents such as isopropyl β-D-1-thiogalactopyranoside (IPTG) for 4–48 h. Typically, higher expression temperatures promote cell growth and require shorter expression time, whereas lower expression temperatures promote proper protein folding and require a longer expression time. While 13–25 °C has been optimal for PKS proteins expressed in our laboratory, temperatures ranging from 4 to 37 °C have been used. After expression, the cells are harvested via centrifugation and resuspended in lysis buffer (typically with 100 m*M* phosphate at pH 7.2–7.6, with optional protease inhibitors). The resuspended cells are disrupted by sonication, and cellular debris is removed by centrifugation prior to protein purification [9, 14, 15, 17].

32.3.2.2 Protein Purification

Protein purification methods such as ammonium sulfate precipitation have yielded satisfying results [9]. However, chromatographic methods such as affinity column purification [e.g. with nickel-nitroloacetic acid (NTA) or glutathione *S*-transferase (GST) resins] and ion-exchange column purification (e.g. with cation- or anion-exchange columns) have rapidly replaced the traditional methods. In our laboratory, proteins are usually expressed with His_6-tags and are therefore purified with nickel-NTA resins. The proteins are first incubated with the resins for 2 h at 4 °C, after which the resin-bound proteins are washed and eluted with an imidazole concentration gradient. Because Ni-NTA-purified proteins are generally 90–95% pure, fast protein liquid chromatography (FPLC) with ion-exchange columns are normally used to further purify the eluted proteins. The FPLC elution gradient is usually set up with a NaCl concentration ranging from 0 to 1 *M* [9, 14, 15, 17]. If additional purification is required, size-exclusion columns can be employed thereafter.

When multiple tags are needed, FLAG tag (peptide sequence DYKDDDDK) and AviTag (peptide sequence LNDIFEAQKIEWH, a biotinylation recognition sequence) can also be engineered into the protein termini [9, 57]. The FLAG-tagged proteins bind to FLAG agarose and can be eluted with FLAG peptide. Similarly, the AviTag-linked proteins, which are biotinylated, can bind to a Neutravidin bead column prior to bioassays and be eluted with SDS–PAGE loading buffer containing dithiothreitol (DTT). All purification steps are carried out at 4 °C. Regardless of the purification steps involved, the purified proteins are usually dialyzed into desired storage buffer (with relatively low salt concentration) and concentrated before flash-freezing with liquid nitrogen and stored at –80 °C.

32.3.2.3 **Protein Characterization**

32.3.2.3.1 **Radiolabeling and Elongation Assays** Radiolabeled substrates are often used to interrogate the activities of PKS enzymes [9, 14–16]. To examine the AT acylation and AT→ACP transacylation activities, for example, ^{14}C-labeled acyl-CoA substrates can be incubated with the proteins of interest. Due to the hydrolytic activity of AT domains, AT-associated acylation assays are often performed on ice for 5–20 min. The acylation assays are followed by SDS–PAGE separation, and the labeling intensities are quantified via autoradiographic analysis. Similarly, KS acylation can be tested through protein labeling and SDS–PAGE if radioactive ketide substrates, usually in the form of SNAC thioesters, are available.

When the acylation activities of individual domains have been established, elongation activity (if of interest) can be evaluated by incubating the KS, AT and ACP domains with the appropriate acyl-CoA and ketide-SNAC substrates. Elongation assays are carried out at 20–30 °C for 5–120 min, and the acyl-CoA is often the substrate carrying the ^{14}C-label (due to its commercial availability). The radiolabeled elongation products are then analyzed with thin-layer chromatography (TLC) and quantified via autoradiographic analysis. If reductive domains such as KR are present, the cofactor NADPH can be added to the elongation reaction to assess the reductive activities of these domains. While SDS–PAGE assays provide qualitative protein labeling data, kinetic information can be extracted from TLC time-course analyses. The main limitations of radioactive assays are the signal detection level and the diversity of available radiolabeled substrates. Furthermore, radioactive SDS–PAGE or TLC assays do not monitor individual chemical transformations in reactions involving two or more steps.

32.3.2.3.2 **Reaction Intermediate Analyses** The direct detection of protein-bound intermediates along the PKS assembly line can greatly enhance our understanding of PKS reaction mechanisms. Toward this end, high-resolution mass spectrometry (MS) techniques have been developed to monitor PKS-mediate chemical transformations through mass information [40, 116, 117]. To visualize protein-bound intermediates, reactions such as acylation, elongation and reduction are first performed with the proteins and substrates of interest. The reaction mixtures then undergo limited trypsin proteolysis prior to high-resolution MS analyses. Unlike radioactive assays, MS-related methods do not rely on radiolabeled or otherwise labeled molecules, and can therefore be used to screen far more substrates than are commercially available or synthetically accessible.

Pioneered by Kelleher and coworkers, Fourier-transform mass spectrometry (FTMS) technology has provided remarkable insights into various features of polyketide and nonribosomal peptide biosyntheses [116]. Panels of compounds with minor differences in structure can be simultaneously evaluated to derive PKS substrate selectivity. A real-time visualization of protein-bound intermediates further reveals the timing of the tailoring reaction, covalent linkage formation and intermodular transfer of intermediates. Despite the wealth of information that high-resolution MS provides, FTMS instrumentation is costly and cannot distinguish between stereoisomers in competition assays. It is also more time-consuming to perform enzyme kinetic evaluations compared to assays using radiolabels.

32.3.2.3.3 **Protein Chemical Assays** In addition to tandem proteolysis-MS experiments, enzyme crosslinking represents another protein chemical technique which is used in PKS characterization. By using covalent crosslinking it is possible to probe the oligomeric state of proteins and their relative orientations within the homo- or hetero-oligomeric complexes [9]. The common crosslinking chemicals are bifunctional reagents that can react with sulfhydryl groups of cysteine residues and/or the amino groups of lysine residues. For example, ethylene glycol bis(succinimidylsuccinate) (EGS) are homobifunctional towards amino groups, bismaleimidohexane (BMH) are homobifunctional towards sulfhydryl groups, while sulfosuccinimidyl-4-(*N*-maleimidomethyl)cyclohexane-1-carboxylate (Sulfo-SMCC) contains one amine-reactive end and one sulfhydryl-reactive end. The spacing between the two functional groups can be customized as needed.

In a recent study, Worthington and coworkers extended the usage of crosslinking to investigate protein–protein interactions in fatty acid biosynthesis [118]. By designing substrate-mimicking crosslinkers, fatty acid ACP was crosslinked to its condensing partner ketosynthases (fatty acid KSAI and KASII). With appropriate modifications in spacer length and terminal moieties, this technique has been successfully transferred to modular PKS systems in our laboratory. Coupled with structural elucidation of the crosslinked adducts, the aim is to capture the interaction between PKS ketosynthase and ACPs in chain elongation.

32.3.3
In vivo Characterization

32.3.3.1 Host Engineering

For the *in vivo* production of polyketide compounds, it is essential that the host organisms are well characterized, readily controlled and easy to engineer for the necessary pathways. Although native hosts are capable of producing and modifying polyketides, they are often resistant to genetic manipulations. Moreover, the doubling times of these organisms are often too long to make them ideal for laboratory experiments. Model actinomycetes such as *Streptomyces coelicolor*, *S. lividans* and *S. collinus* have been adapted for PKS heterogeneous expression [9]. Stassi and coworkers, for example, inserted the gene encoding crotonyl-CoA reductase (*ccr*) into *S. collinus* to promote the production of butyryl-CoA, which could then be converted to ethylmalonyl-CoA *in vivo* [45]. This modification increased the limiting supply of ethylmalonyl-CoA and consequently allowed the engineered PKS to produce the expected analogue product.

To further simplify *in vivo* PKS characterizations, the widely used, well-characterized *E. coli* stain BL21(DE3) has been engineered for polyketide production [119]. Propionyl-CoA carboxylase (*pcc*), for example, was expressed in BL21(DE3) to promote the conversion of propionyl-CoA into methylmalonyl-CoA. Coupled with the engineered accumulation of propionyl-CoA, the recombinant *E. coli* strain was able to enrich intracellular methylmalonyl-CoA concentrations, supplying this essential building block to the biosynthesis of macrolides such as 6-dEB. Similarly, the *sfp* phosphopantetheinyl transferase gene was inserted via homologous recombination to activate the *apo* form of ACP domains with phosphopantetheine moieties. Post-PKS glycosylation pathway has also recently been reconstituted in *E. coli* to enable the production of fully processed, bioactive compounds [120, 121]. These modifications allowed us to harness the benefits of *E. coli* engineering, such as rapid and characterized growth, straightforward genetic manipulation procedures and the absence of PKS-related pathways that affords a clean background for PKS characterization.

32.3.3.2 **High-Throughput Screening Assay**

Host engineering that led to the production of bioactive products has made it possible to devise *in vivo* bioactivity-based screening methods. Kim and coworkers, for example, used modified *Bacillus subtilis*–which is susceptible to antibiotic agents–to screen for pikromycin-producing colonies generated by gene shuffling [113]. *E. coli* and other organisms have likewise been used as the indicating strains in overlay inhibition assays [122]. By overlaying the engineered test strains onto LB agar seeded with *B. subtilis* or other indicating strains, only those engineered colonies capable of producing the target antibiotics will inhibit the growth of the indicating strains, thereby forming a clear zone ('halo') on the agar plate. By extending the application of this 'halo test', Lee and colleagues were able to evolve erythromycin-producing *E. coli* through repeated *B. subtilis* screening assays (Figure 32.10) [121].

As the above examples show, this assay allows the screening of antibiotic-producing colonies in high-throughput fashion. Furthermore, by linking the survival of a colony with its ability to utilize a heterologous biosynthetic machinery, the assay promotes the directed evolution of an antibiotic pathway. The versatility of this screening assay can be expanded by altering the screening organism to tailor toward different bioactivities, and by introducing additional stimuli to accelerate pathway evolution. Despite the high sensitivity, easy visualization and high-throughput condition of these *in vivo* bioactivity-based assays, they still rely on the production of bioactive compounds. As mentioned above, polyketide compounds are typically bioinactive until they are further decorated by post-PKS modifying

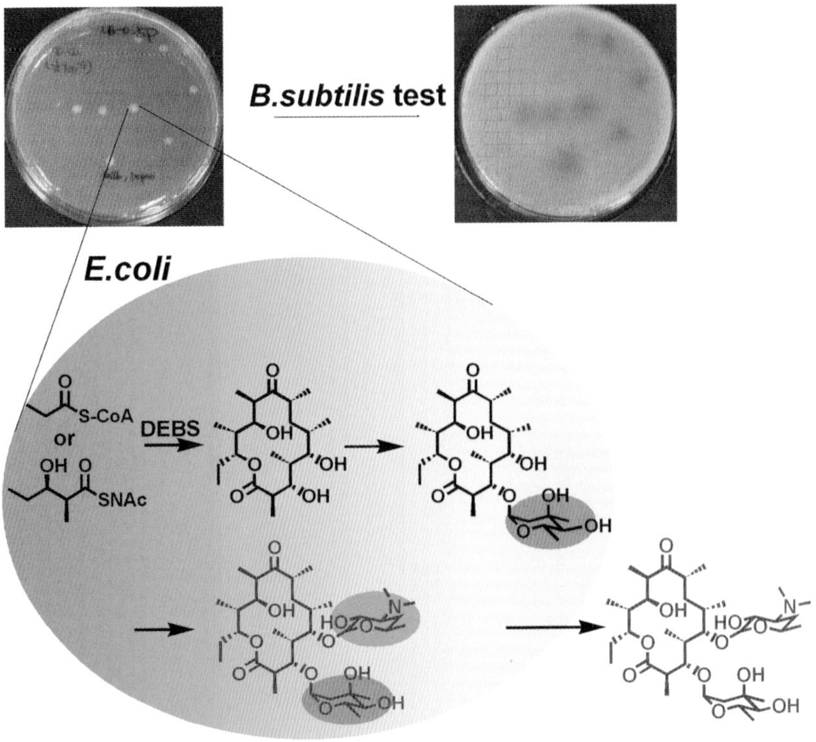

Figure 32.10 Sample agar plate for single-colony bioactivity assay. Transformants of the engineered *E. coli* strains are overlaid onto LB agar seeded with indicating strain *B. subtilis*. The *E. coli* colonies capable of producing the target antibiotics (erythromycin in this case) will inhibit the growth of *B. subtilis*, thereby forming a clear zone on the agar plate. Figure adapted from Ref. [121].

enzymes. When such modifying capacity is unavailable – as in the case of many heterologous expression systems – HPLC and other analytical techniques have been employed to analyze polyketide compounds produced *in vivo* after the compounds have been properly isolated.

32.4
The Path Forward

The modular nature of polyketide synthases presents a wealth of engineering opportunities for the combinatorial biosynthesis of medicinally relevant compounds. It is hoped that the studies highlighted in this chapter have familiarized the reader with the current progress in PKS engineering. Despite their combinatorial potential, modular PKSs are well known for their complexity in both protein structures and reaction mechanisms; hence, a lack of knowledge of each subunit along these protein assembly lines has hampered our ability to fully harness the advantages of an engineering-friendly PKS architecture.

Fortunately, recent structural studies have provided high-resolution descriptions of PKS domains and facilitated the rational engineering of hybrid polyketide pathways. The continuing discovery and characterization of new PKS gene clusters will further fuel the genetic reprogramming of these biosynthetic machineries by providing building blocks with interesting structural and chemical features. With regards to the engineering of these systems, gene synthesis technology will greatly benefit the redesign of PKS genes to include features such as straightforward heterologous expression and readily interchangeable units, while steady advances in directed evolution will promote the rapid recombination of PKS genes. The resultant engineered PKS systems can then undergo analysis by high-throughput screening assays and/or precise analytical techniques such as FTMS.

By combining our expanding knowledge on PKS organization and reaction mechanisms with the technological advancements in gene manipulation and analytical techniques, predictive engineering strategies and rapid feedback systems are anticipated to emerge. These will undoubtedly accelerate the scanning of therapeutic polyketide-derived compounds and help us to understand the productive versus nonproductive recombinations, pushing us closer to truly combinatorial polyketide biosynthesis.

Abbreviations

PKS	Polyketide synthase
FAS	Fatty acid synthase
NRPS	Nonribosomal peptide synthase
DEBS	6-Deoxyerythronolide B synthase
6-dEB	6-Deoxyerythronolide B
KS	β-Keto synthase
AT	Acyltransferase
ACP	Acyl carrier protein
KR	Ketoreductase
DH	Dehydratase
ER	Enoylreductase
TE	Thioesterase
MT	Methyltransferase
AMT	Aminotransferase

References

1 Khosla, C., Tang, Y., Chen, A.Y., Schnarr, N.A. and Cane, D.E. (2007) Structure and mechanism of the 6-deoxyerythronolide B synthase. *Annual Review of Biochemistry*, **76**, 195–221.

2 Fischbach, M.A. and Walsh, C.T. (2006) Assembly-line enzymology for polyketide and nonribosomal peptide antibiotics: Logic, machinery, and mechanisms. *Chemical Reviews*, **106**, 3468–96.
3 Hill, A.M. (2006) The biosynthesis, molecular genetics and enzymology of the polyketide-derived metabolites. *Natural Product Reports*, **23**, 256–320.
4 Weissman, K.J. and Leadlay, P.F. (2005) Combinatorial biosynthesis of reduced polyketides. *Nature Reviews Microbiology*, **3**, 925–36.
5 Wenzel, S.C. and Muller, R. (2005) Formation of novel secondary metabolites by bacterial multimodular assembly lines: deviations from textbook biosynthetic logic. *Current Opinion in Chemical Biology*, **9**, 447–58.
6 McDaniel, R., Welch, M. and Hutchinson, C.R. (2005) Genetic approaches to polyketide antibiotics. 1. *Chemical Reviews*, **105**, 543–58.
7 Kittendorf, J.D. and Sherman, D.H. (2006) Developing tools for engineering hybrid polyketide synthetic pathways. *Current Opinion in Biotechnology*, **17**, 597–605.
8 Wenzel, S.C. and Muller, R. (2005) Recent developments towards the heterologous expression of complex bacterial natural product biosynthetic pathways. *Current Opinion in Biotechnology*, **16**, 594–606.
9 Kumar, P., Khosla, C. and Tang, Y. (2004) Manipulation and analysis of polyketide synthases. *Methods in Enzymology*, **388**, 269–93.
10 Donadio, S. and Sosio, M. (2003) Strategies for combinatorial biosynthesis with modular polyketide synthases. *Combinatorial Chemistry and High Throughput Screening*, **6**, 489–500.
11 El-Sayed, A.K., Hothersall, J., Cooper, S.M., Stephens, E., Simpson, T.J. and Thomas, C.M. (2003) Characterization of the mupirocin biosynthesis gene cluster from *Pseudomonas fluorescens* NCIMB 10586. *Chemistry and Biology*, **10**, 419–30.
12 Donadio, S. and Katz, L. (1992) Organization of the enzymatic domains in the multifunctional polyketide synthase involved in erythromycin formation in *Saccharopolyspora erythraea*. *Gene*, **111**, 51–60.
13 Aparicio, J.F., Caffrey, P., Marsden, A.F., Staunton, J. and Leadlay, P.F. (1994) Limited proteolysis and active-site studies of the first multienzyme component of the erythromycin-producing polyketide synthase. *The Journal of Biological Chemistry*, **269**, 8524–8.
14 Kim, C.Y., Alekseyev, V.Y., Chen, A.Y., Tang, Y., Cane, D.E. and Khosla, C. (2004) Reconstituting modular activity from separated domains of 6-deoxyerythronolide B synthase. *Biochemistry*, **43**, 13892–8.
15 Chen, A.Y., Schnarr, N.A., Kim, C.Y., Cane, D.E. and Khosla, C. (2006) Extender unit and acyl carrier protein specificity of ketosynthase domains of the 6-deoxyerythronolide B synthase. *Journal of the American Chemical Society*, **128**, 3067–74.
16 Chen, A.Y., Cane, D.E. and Khosla, C. (2007) Structure-based dissociation of a type I polyketide synthase module. *Chemistry and Biology*, **14**, 784–92.
17 Tang, Y., Kim, C.Y., Mathews, I.I., Cane, D.E. and Khosla, C. (2006) The 2.7-angstrom crystal structure of a 194-kDa homodimeric fragment of the 6-deoxyerythronolide B synthase. *Proceedings of the National Academy of Sciences of the United States of America*, **103**, 11124–9.
18 Tang, Y., Chen, A.Y., Kim, C.Y., Cane, D.E. and Khosla, C. (2007) Structural and mechanistic analysis of protein interactions in module 3 of the 6-deoxyerythronolide B synthase. *Chemistry and Biology*, **14**, 931–43.
19 Keatinge-Clay, A.T. and Stroud, R.M. (2006) The structure of a ketoreductase determines the organization of the beta-carbon processing enzymes of modular polyketide synthases. *Structure*, **14**, 737–48.
20 Alekseyev, V.Y., Liu, C.W., Cane, D.E., Puglisi, J.D. and Khosla, C. (2007) Solution structure and proposed domain-domain recognition interface of an acyl carrier protein domain from a modular polyketide synthase. *Protein Science*, **16**, 2093–107.

21 Tsai, S.C., Miercke, L.J., Krucinski, J., Gokhale, R., Chen, J.C., Foster, P.G., Cane, D.E. et al. (2001) Crystal structure of the macrocycle-forming thioesterase domain of the erythromycin polyketide synthase: versatility from a unique substrate channel. *Proceedings of the National Academy of Sciences of the United States of America*, **98**, 14808–13.

22 Tsai, S.C., Lu, H., Cane, D.E., Khosla, C. and Stroud, R.M. (2002) Insights into channel architecture and substrate specificity from crystal structures of two macrocycle-forming thioesterases of modular polyketide synthases. *Biochemistry*, **41**, 12598–606.

23 Giraldes, J.W., Akey, D.L., Kittendorf, J.D., Sherman, D.H., Smith, J.L. and Fecik, R.A. (2006) Structural and mechanistic insights into polyketide macrolactonization from polyketide-based affinity labels. *Nature Chemical Biology*, **2**, 531–6.

24 Broadhurst, R.W., Nietlispach, D., Wheatcroft, M.P., Leadlay, P.F. and Weissman, K.J. (2003) The structure of docking domains in modular polyketide synthases. *Chemistry and Biology*, **10**, 723–31.

25 Maier, T., Jenni, S. and Ban, N. (2006) Architecture of mammalian fatty acid synthase at 4.5 Å resolution. *Science*, **311**, 1258–62.

26 Jenni, S., Leibundgut, M., Maier, T. and Ban, N. (2006) Architecture of a fungal fatty acid synthase at 5 Å resolution. *Science*, **311**, 1263–7.

27 Leibundgut, M., Jenni, S., Frick, C. and Ban, N. (2007) Structural basis for substrate delivery by acyl carrier protein in the yeast fatty acid synthase. *Science*, **316**, 288–90.

28 Lomakin, I.B., Xiong, Y. and Steitz, T.A. (2007) The crystal structure of yeast fatty acid synthase, a cellular machine with eight active sites working together. *Cell*, **129**, 319–32.

29 Marsden, A.F., Wilkinson, B., Cortes, J., Dunster, N.J., Staunton, J. and Leadlay, P.F. (1998) Engineering broader specificity into an antibiotic-producing polyketide synthase. *Science*, **279**, 199–202.

30 Long, P.F., Wilkinson, C.J., Bisang, C.P., Cortes, J., Dunster, N., Oliynyk, M., McCormick, E. *et al.* (2002) Engineering specificity of starter unit selection by the erythromycin-producing polyketide synthase. *Molecular Microbiology*, **43**, 1215–25.

31 Kuhstoss, S., Huber, M., Turner, J.R., Paschal, J.W. and Rao, R.N. (1996) Production of a novel polyketide through the construction of a hybrid polyketide synthase. *Gene*, **183**, 231–6.

32 Sheehan, L.S., Lill, R.E., Wilkinson, B., Sheridan, R.M., Vousden, W.A., Kaja, A. L., Crouse, G.D. *et al.* (2006) Engineering of the spinosyn PKS: directing starter unit incorporation. *Journal of Natural Products*, **69**, 1702–10.

33 Moore, B.S. and Hertweck, C. (2002) Biosynthesis and attachment of novel bacterial polyketide synthase starter units. *Natural Product Reports*, **19**, 70–99.

34 Jacobsen, J.R., Hutchinson, C.R., Cane, D.E. and Khosla, C. (1997) Precursor-directed biosynthesis of erythromycin analogs by an engineered polyketide synthase. *Science*, **277**, 367–9.

35 Cane, D.E., Kudo, F., Kinoshita, K. and Khosla, C. (2002) Precursor-directed biosynthesis: biochemical basis of the remarkable selectivity of the erythromycin polyketide synthase toward unsaturated triketides. *Chemistry and Biology*, **9**, 131–42.

36 Kinoshita, K., Pfeifer, B.A., Khosla, C. and Cane, D.E. (2003) Precursor-directed polyketide biosynthesis in *Escherichia coli*. *Bioorganic and Medicinal Chemistry Letters*, **13**, 3701–4.

37 Hartung, I.V., Rude, M.A., Schnarr, N.A., Hunziker, D. and Khosla, C. (2005) Stereochemical assignment of intermediates in the rifamycin biosynthetic pathway by precursor-directed biosynthesis. *Journal of the American Chemical Society*, **127**, 11202–3.

38 Murli, S., MacMillan, K.S., Hu, Z., Ashley, G.W., Dong, S.D., Kealey, J.T., Reeves, C.D. and Kennedy, J. (2005) Chemobiosynthesis of novel 6-deoxyerythronolide B analogues by mutation of the loading module of 6-deoxyerythronolide B synthase 1. *Applied*

and Environmental Microbiology, **71**, 4503–9.

39 Gregory, M.A., Petkovic, H., Lill, R.E., Moss, S.J., Wilkinson, B., Gaisser, S., Leadlay, P.F. and Sheridan, R.M. (2005) Mutasynthesis of rapamycin analogues through the manipulation of a gene governing starter unit biosynthesis. *Angewandte Chemie - International Edition in English*, **44**, 4757–60.

40 Schnarr, N.A., Chen, A.Y., Cane, D.E. and Khosla, C. (2005) Analysis of covalently bound polyketide intermediates on 6-deoxyerythronolide B synthase by tandem proteolysis-mass spectrometry. *Biochemistry*, **44**, 11836–42.

41 Oliynyk, M., Brown, M.J., Cortes, J., Staunton, J. and Leadlay, P.F. (1996) A hybrid modular polyketide synthase obtained by domain swapping. *Chemistry and Biology*, **3**, 833–9.

42 Ruan, X., Pereda, A., Stassi, D.L., Zeidner, D., Summers, R.G., Jackson, M., Shivakumar, A. *et al.* (1997) Acyltransferase domain substitutions in erythromycin polyketide synthase yield novel erythromycin derivatives. *Journal of Bacteriology*, **179**, 6416–25.

43 McDaniel, R., Thamchaipenet, A., Gustafsson, C., Fu, H., Betlach, M. and Ashley, G. (1999) Multiple genetic modifications of the erythromycin polyketide synthase to produce a library of novel 'unnatural' natural products. *Proceedings of the National Academy of Sciences of the United States of America*, **96**, 1846–51.

44 Lau, J., Fu, H., Cane, D.E. and Khosla, C. (1999) Dissecting the role of acyltransferase domains of modular polyketide synthases in the choice and stereochemical fate of extender units. *Biochemistry*, **38**, 1643–51.

45 Stassi, D.L., Kakavas, S.J., Reynolds, K.A., Gunawardana, G., Swanson, S., Zeidner, D., Jackson, M. *et al.* (1998) Ethyl-substituted erythromycin derivatives produced by directed metabolic engineering. *Proceedings of the National Academy of Sciences of the United States of America*, **95**, 7305–9.

46 Kato, Y., Bai, L., Xue, Q., Revill, W.P., Yu, T.W. and Floss, H.G. (2002) Functional expression of genes involved in the biosynthesis of the novel polyketide chain extension unit, methoxymalonyl-acyl carrier protein, and engineered biosynthesis of 2-desmethyl-2-methoxy-6-deoxyerythronolide B. *Journal of the American Chemical Society*, **124**, 5268–9.

47 Reeves, C.D., Chung, L.M., Liu, Y., Xue, Q., Carney, J.R., Revill, W.P. and Katz, L. (2002) A new substrate specificity for acyl transferase domains of the ascomycin polyketide synthase in *Streptomyces hygroscopicus*. *The Journal of Biological Chemistry*, **277**, 9155–9.

48 Patel, K., Piagentini, M., Rascher, A., Tian, Z.Q., Buchanan, G.O., Regentin, R., Hu, Z. *et al.* (2004) Engineered biosynthesis of geldanamycin analogs for hsp90 inhibition. *Chemistry and Biology*, **11**, 1625–33.

49 Haydock, S.F., Aparicio, J.F., Molnar, I., Schwecke, T., Khaw, L.E., Konig, A., Marsden, A.F. *et al.* (1995) Divergent sequence motifs correlated with the substrate specificity of (methyl)malonyl-CoA:Acyl carrier protein transacylase domains in modular polyketide synthases. *FEBS Letters*, **374**, 246–8.

50 Reeves, C.D., Murli, S., Ashley, G.W., Piagentini, M., Hutchinson, C.R. and McDaniel, R. (2001) Alteration of the substrate specificity of a modular polyketide synthase acyltransferase domain through site-specific mutations. *Biochemistry*, **40**, 15464–70.

51 Del Vecchio, F., Petkovic, H., Kendrew, S.G., Low, L., Wilkinson, B., Lill, R., Cortes, J. *et al.* (2003) Active-site residue, domain and module swaps in modular polyketide synthases. *Journal of Industrial Microbiology and Biotechnology*, **30**, 489–94.

52 Kumar, P., Koppisch, A.T., Cane, D.E. and Khosla, C. (2003) Enhancing the modularity of the modular polyketide synthases: Transacylation in modular polyketide synthases catalyzed by malonyl-CoA:ACP transacylase. *Journal of the American Chemical Society*, **125**, 14307–12.

53 Walton, L.J., Corre, C. and Challis, G.L. (2006) Mechanisms for incorporation of glycerol-derived precursors into

polyketide metabolites. *Journal of Industrial Microbiology and Biotechnology*, **33**, 105–20.
54 Chan, Y.A., Boyne, M.T. 2nd, Podevels, A.M., Klimowicz, A.K., Handelsman, J., Kelleher, N.L. and Thomas, M.G. (2006) Hydroxymalonyl-acyl carrier protein (ACP) and aminomalonyl-ACP are two additional type I polyketide synthase extender units. *Proceedings of the National Academy of Sciences of the United States of America*, **103**, 14349–54.
55 Yadav, G., Gokhale, R.S. and Mohanty, D. (2003) Computational approach for prediction of domain organization and substrate specificity of modular polyketide synthases. *Journal of Molecular Biology*, **328**, 335–63.
56 Borgos, S.E., Tsan, P., Sletta, H., Ellingsen, T.E., Lancelin, J.M. and Zotchev, S.B. (2006) Probing the structure-function relationship of polyene macrolides: Engineered biosynthesis of soluble nystatin analogues. *Journal of Medicinal Chemistry*, **49**, 2431–9.
57 Reid, R., Piagentini, M., Rodriguez, E., Ashley, G., Viswanathan, N., Carney, J., Santi, D.V. *et al.* (2003) A model of structure and catalysis for ketoreductase domains in modular polyketide synthases. *Biochemistry*, **42**, 72–9.
58 Xue, Q., Ashley, G., Hutchinson, C.R. and Santi, D.V. (1999) A multiplasmid approach to preparing large libraries of polyketides. *Proceedings of the National Academy of Sciences of the United States of America*, **96**, 11740–5.
59 Zhang, X., Chen, Z., Li, M., Wen, Y., Song, Y. and Li, J. (2006) Construction of ivermectin producer by domain swaps of avermectin polyketide synthase in *Streptomyces avermitilis*. *Applied Microbiology and Biotechnology*, **72**, 986–94.
60 McDaniel, R., Kao, C.M., Hwang, S.J. and Khosla, C. (1997) Engineered intermodular and intramodular polyketide synthase fusions. *Chemistry and Biology*, **4**, 667–74.
61 Siskos, A.P., Baerga-Ortiz, A., Bali, S., Stein, V., Mamdani, H., Spiteller, D., Popovic, B. *et al.* (2005) Molecular basis of Celmer's rules: Stereochemistry of catalysis by isolated ketoreductase domains from modular polyketide synthases. *Chemistry and Biology*, **12**, 1145–53.
62 Kao, C.M., McPherson, M., McDaniel, R., Fu, H., Cane, D.E. and Khosla, C. (1998) Alcohol stereochemistry in polyketide backbones is controlled by the beta-ketoreductase domains of modular polyketide synthases. *Journal of the American Chemical Society*, **120**, 2478–9.
63 Caffrey, P. (2003) Conserved amino acid residues correlating with ketoreductase stereospecificity in modular polyketide synthases. *ChemBioChem*, **4**, 654–7.
64 O'Hare, H.M., Baerga-Ortiz, A., Popovic, B., Spencer, J.B. and Leadlay, P.F. (2006) High-throughput mutagenesis to evaluate models of stereochemical control in ketoreductase domains from the erythromycin polyketide synthase. *Chemistry and Biology*, **13**, 287–96.
65 Baerga-Ortiz, A., Popovic, B., Siskos, A.P., O'Hare, H.M., Spiteller, D., Williams, M.G., Campillo, N. *et al.* (2006) Directed mutagenesis alters the stereochemistry of catalysis by isolated ketoreductase domains from the erythromycin polyketide synthase. *Chemistry and Biology*, **13**, 277–85.
66 Bali, S., O'Hare, H.M. and Weissman, K.J. (2006) Broad substrate specificity of ketoreductases derived from modular polyketide synthases. *ChemBioChem*, **7**, 478–84.
67 Castonguay, R., He, W., Chen, A.Y., Khosla, C. and Cane, D.E. (2007) Stereospecificity of ketoreductase domains of the 6-deoxyerythronolide B synthase. *Journal of the American Chemical Society*, **129**, 13758–69.
68 Kao, C.M., Luo, G., Katz, L., Cane, D.E. and Khosla, C. (1995) Manipulation of macrolide ring size by directed mutagenesis of a modular polyketide synthase. *Journal of the American Chemical Society*, **117**, 9105–6.
69 Rowe, C.J., Bohm, I.U., Thomas, I.P., Wilkinson, B., Rudd, B.A., Foster, G., Blackaby, A.P. *et al.* (2001) Engineering a polyketide with a longer chain by insertion of an extra module into the

erythromycin-producing polyketide synthase. *Chemistry and Biology*, **8**, 475–85.

70 Cortes, J., Wiesmann, K.E., Roberts, G.A., Brown, M.J., Staunton, J. and Leadlay, P.F. (1995) Repositioning of a domain in a modular polyketide synthase to promote specific chain cleavage. *Science*, **268**, 1487–9.

71 Gokhale, R.S., Hunziker, D., Cane, D.E. and Khosla, C. (1999) Mechanism and specificity of the terminal thioesterase domain from the erythromycin polyketide synthase. *Chemistry and Biology*, **6**, 117–25.

72 Tang, L., Shah, S., Chung, L., Carney, J., Katz, L., Khosla, C. and Julien, B. (2000) Cloning and heterologous expression of the epothilone gene cluster. *Science*, **287**, 640–2.

73 Silakowski, B., Schairer, H.U., Ehret, H., Kunze, B., Weinig, S., Nordsiek, G., Brandt, P. *et al.* (1999) New lessons for combinatorial biosynthesis from myxobacteria. The myxothiazol biosynthetic gene cluster of *Stigmatella aurantiaca* DW4/3-1. *The Journal of Biological Chemistry*, **274**, 37391–9.

74 Kennedy, J., Auclair, K., Kendrew, S.G., Park, C., Vederas, J.C. and Hutchinson, C.R. (1999) Modulation of polyketide synthase activity by accessory proteins during lovastatin biosynthesis. *Science*, **284**, 1368–72.

75 Gehring, A.M., DeMoll, E., Fetherston, J.D., Mori, I., Mayhew, G.F., Blattner, F.R., Walsh, C.T. and Perry, R.D. (1998) Iron acquisition in plague: modular logic in enzymatic biogenesis of yersiniabactin by *Yersinia pestis*. *Chemistry and Biology*, **5**, 573–86.

76 Duitman, E.H., Hamoen, L.W., Rembold, M., Venema, G., Seitz, H., Saenger, W., Bernhard, F. *et al.* (1999) The mycosubtilin synthetase of *Bacillus subtilis* ATCC6633: a multifunctional hybrid between a peptide synthetase, an amino transferase, and a fatty acid synthase. *Proceedings of the National Academy of Sciences of the United States of America*, **96**, 13294–9.

77 Aron, Z.D., Dorrestein, P.C., Blackhall, J.R., Kelleher, N.L. and Walsh, C.T. (2005) Characterization of a new tailoring domain in polyketide biogenesis: the amine transferase domain of MYCA in the mycosubtilin gene cluster. *Journal of the American Chemical Society*, **127**, 14986–7.

78 Gokhale, R.S., Tsuji, S.Y., Cane, D.E. and Khosla, C. (1999) Dissecting and exploiting intermodular communication in polyketide synthases. *Science*, **284**, 482–5.

79 Yoon, Y.J., Beck, B.J., Kim, B.S., Kang, H.Y., Reynolds, K.A. and Sherman, D.H. (2002) Generation of multiple bioactive macrolides by hybrid modular polyketide synthases in *Streptomyces venezuelae*. *Chemistry and Biology*, **9**, 203–14.

80 Kim, B.S., Cropp, T.A., Florova, G., Lindsay, Y., Sherman, D.H. and Reynolds, K.A. (2002) An unexpected interaction between the modular polyketide synthases, erythromycin DEBS1 and pikromycin PikAIV, leads to efficient triketide lactone synthesis. *Biochemistry*, **41**, 10827–33.

81 Watanabe, K., Wang, C.C., Boddy, C.N., Cane, D.E. and Khosla, C. (2003) Understanding substrate specificity of polyketide synthase modules by generating hybrid multimodular synthases. *The Journal of Biological Chemistry*, **278**, 42020–6.

82 Murli, S., Piagentini, M., McDaniel, R. and Hutchinson, C.R. (2004) Identification of domains within megalomicin and erythromycin polyketide synthase modules responsible for differences in polyketide production levels in *Escherichia coli*. *Biochemistry*, **43**, 15884–90.

83 Menzella, H.G., Reid, R., Carney, J.R., Chandran, S.S., Reisinger, S.J., Patel, K.G., Hopwood, D.A. and Santi, D.V. (2005) Combinatorial polyketide biosynthesis by de novo design and rearrangement of modular polyketide synthase genes. *Nature Biotechnology*, **23**, 1171–6.

84 Menzella, H.G., Carney, J.R. and Santi, D.V. (2007) Rational design and assembly of synthetic trimodular polyketide synthases. *Chemistry and Biology*, **14**, 143–51.

85 Tsuji, S.Y., Cane, D.E. and Khosla, C. (2001) Selective protein-protein interactions direct channeling of intermediates between polyketide synthase modules. *Biochemistry*, **40**, 2326–31.

86 Tsuji, S.Y., Wu, N. and Khosla, C. (2001) Intermodular communication in polyketide synthases: comparing the role of protein-protein interactions to those in other multidomain proteins. *Biochemistry*, **40**, 2317–25.

87 Wu, N., Tsuji, S.Y., Cane, D.E. and Khosla, C. (2001) Assessing the balance between protein-protein interactions and enzyme-substrate interactions in the channeling of intermediates between polyketide synthase modules. *Journal of the American Chemical Society*, **123**, 6465–74.

88 Wu, N., Cane, D.E. and Khosla, C. (2002) Quantitative analysis of the relative contributions of donor acyl carrier proteins, acceptor ketosynthases, and linker regions to intermodular transfer of intermediates in hybrid polyketide synthases. *Biochemistry*, **41**, 5056–66.

89 Weissman, K.J. (2006) Single amino acid substitutions alter the efficiency of docking in modular polyketide biosynthesis. *ChemBioChem*, **7**, 1334–42.

90 Richter, C.D., Stanmore, D.A., Miguel, R.N., Moncrieffe, M.C., Tran, L., Brewerton, S., Meersman, F. *et al.* (2007) Autonomous folding of interdomain regions of a modular polyketide synthase. *The FEBS Journal*, **274**, 2196–209.

91 Finking, R. and Marahiel, M.A. (2004) Biosynthesis of nonribosomal peptides. *Annual Review of Microbiology*, **58**, 453–88.

92 Marahiel, M.A., Stachelhaus, T. and Mootz, H.D. (1997) Modular peptide synthetases involved in nonribosomal peptide synthesis. *Chemical Reviews*, **97**, 2651–74.

93 Cane, D.E., Walsh, C.T. and Khosla, C. (1998) Harnessing the biosynthetic code: combinations, permutations, and mutations. *Science*, **282**, 63–8.

94 Lowden, P.A., Wilkinson, B., Bohm, G.A., Handa, S., Floss, H.G., Leadlay, P.F. and Staunton, J. (2001) Origin and true nature of the starter unit for the rapamycin polyketide synthase. *Angewandte Chemie - International Edition in English*, **40**, 777–9.

95 Admiraal, S.J., Khosla, C. and Walsh, C.T. (2002) The loading and initial elongation modules of rifamycin synthetase collaborate to produce mixed aryl ketide products. *Biochemistry*, **41**, 5313–24.

96 Yu, T.W., Muller, R., Muller, M., Zhang, X., Draeger, G., Kim, C.G., Leistner, E. and Floss, H.G. (2001) Mutational analysis and reconstituted expression of the biosynthetic genes involved in the formation of 3-amino-5-hydroxybenzoic acid, the starter unit of rifamycin biosynthesis in *Amycolatopsis mediterranei* S699. *The Journal of Biological Chemistry*, **276**, 12546–55.

97 Watanabe, K., Rude, M.A., Walsh, C.T. and Khosla, C. (2003) Engineered biosynthesis of an ansamycin polyketide precursor in *Escherichia coli*. *Proceedings of the National Academy of Sciences of the United States of America*, **100**, 9774–8.

98 Dayem, L.C., Carney, J.R., Santi, D.V., Pfeifer, B.A., Khosla, C. and Kealey, J.T. (2002) Metabolic engineering of a methylmalonyl-CoA mutase-epimerase pathway for complex polyketide biosynthesis in *Escherichia coli*. *Biochemistry*, **41**, 5193–201.

99 Mutka, S.C., Bondi, S.M., Carney, J.R., Da Silva, N.A. and Kealey, J.T. (2006) Metabolic pathway engineering for complex polyketide biosynthesis in *Saccharomyces cerevisiae*. *FEMS Yeast Research*, **6**, 40–7.

100 Gross, F., Ring, M.W., Perlova, O., Fu, J., Schneider, S., Gerth, K., Kuhlmann, S. *et al.* (2006) Metabolic engineering of *Pseudomonas putida* for methylmalonyl-CoA biosynthesis to enable complex heterologous secondary metabolite formation. *Chemistry and Biology*, **13**, 1253–64.

101 Rix, U., Fischer, C., Remsing, L.L. and Rohr, J. (2002) Modification of post-PKS tailoring steps through combinatorial biosynthesis. *Natural Product Reports*, **19**, 542–80.

102 Schlunzen, F., Zarivach, R., Harms, J., Bashan, A., Tocilj, A., Albrecht, R.,

Yonath, A. and Franceschi, F. (2001) Structural basis for the interaction of antibiotics with the peptidyl transferase centre in eubacteria. *Nature*, **413**, 814–21.

103 Butler, A.R., Bate, N., Kiehl, D.E., Kirst, H.A. and Cundliffe, E. (2002) Genetic engineering of aminodeoxyhexose biosynthesis in *Streptomyces fradiae*. *Nature Biotechnology*, **20**, 713–16.

104 Rodriguez, L., Aguirrezabalaga, I., Allende, N., Brana, A.F., Mendez, C. and Salas, J.A. (2002) Engineering deoxysugar biosynthetic pathways from antibiotic-producing microorganisms. A tool to produce novel glycosylated bioactive compounds. *Chemistry and Biology*, **9**, 721–9.

105 Horton, R.M., Cai, Z.L., Ho, S.N. and Pease, L.R. (1990) Gene splicing by overlap extension: tailor-made genes using the polymerase chain reaction. *Biotechniques*, **8**, 528–35.

106 Horton, R.M. (1995) PCR-mediated recombination and mutagenesis. SOEing together tailor-made genes. *Molecular Biotechnology*, **3**, 93–9.

107 Vetcher, L., Tian, Z.Q., McDaniel, R., Rascher, A., Revill, W.P., Hutchinson, C.R. and Hu, Z. (2005) Rapid engineering of the geldanamycin biosynthesis pathway by Red/ET recombination and gene complementation. *Applied and Environmental Microbiology*, **71**, 1829–35.

108 Wenzel, S.C., Gross, F., Zhang, Y., Fu, J., Stewart, A.F. and Muller, R. (2005) Heterologous expression of a myxobacterial natural products assembly line in pseudomonads via Red/ET recombineering. *Chemistry and Biology*, **12**, 349–56.

109 Perlova, O., Fu, J., Kuhlmann, S., Krug, D., Stewart, A.F., Zhang, Y. and Muller, R. (2006) Reconstitution of the myxothiazol biosynthetic gene cluster by Red/ET recombination and heterologous expression in *Myxococcus xanthus*. *Applied and Environmental Microbiology*, **72**, 7485–94.

110 Zhang, Y., Buchholz, F., Muyrers, J.P. and Stewart, A.F. (1998) A new logic for DNA engineering using recombination in *Escherichia coli*. *Nature Genetics*, **20**, 123–8.

111 Zhang, Y., Muyrers, J.P., Testa, G. and Stewart, A.F. (2000) DNA cloning by homologous recombination in *Escherichia coli*. *Nature Biotechnology*, **18**, 1314–17.

112 Rivero-Muller, A., Lajic, S. and Huhtaniemi, I. (2007) Assisted large fragment insertion by Red/ET-recombination (ALFIRE)–an alternative and enhanced method for large fragment recombineering. *Nucleic Acids Research*, **35**, e78.

113 Kim, B.S., Sherman, D.H. and Reynolds, K.A. (2004) An efficient method for creation and functional analysis of libraries of hybrid type I polyketide synthases. *Protein Engineering, Design and Selection*, **17**, 277–84.

114 Kodumal, S.J., Patel, K.G., Reid, R., Menzella, H.G., Welch, M. and Santi, D.V. (2004) Total synthesis of long DNA sequences: synthesis of a contiguous 32-kb polyketide synthase gene cluster. *Proceedings of the National Academy of Sciences of the United States of America*, **101**, 15573–8.

115 Stemmer, W.P. (1994) Rapid evolution of a protein in vitro by DNA shuffling. *Nature*, **370**, 389–91.

116 Dorrestein, P.C., Bumpus, S.B., Calderone, C.T., Garneau-Tsodikova, S., Aron, Z.D., Straight, P.D., Kolter, R. *et al.* (2006) Facile detection of acyl and peptidyl intermediates on thiotemplate carrier domains via phosphopantetheinyl elimination reactions during tandem mass spectrometry. *Biochemistry*, **45**, 12756–66.

117 Schnarr, N.A. and Khosla, C. (2006) Trapping transient protein-protein interactions in polyketide biosynthesis. *ACS Chemical Biology*, **1**, 679–80.

118 Worthington, A.S., Rivera, H., Torpey, J.W., Alexander, M.D. and Burkart, M.D. (2006) Mechanism-based protein cross-linking probes to investigate carrier protein-mediated biosynthesis. *ACS Chemical Biology*, **1**, 687–91.

119 Pfeifer, B.A. and Khosla, C. (2001) Biosynthesis of polyketides in heterologous hosts. *Microbiology and Molecular Biology Reviews*, **65**, 106–18.

120 Peiru, S., Menzella, H.G., Rodriguez, E., Carney, J. and Gramajo, H. (2005) Production of the potent antibacterial polyketide erythromycin C in *Escherichia coli*. *Applied and Environmental Microbiology*, **71**, 2539–47.

121 Lee, H.Y. and Khosla, C. (2007) Bioassay-guided evolution of glycosylated macrolide antibiotics in *Escherichia coli*. *PLoS Biology*, **5**, e45.

122 Jin, M., Fischbach, M.A. and Clardy, J. (2006) A biosynthetic gene cluster for the acetyl-CoA carboxylase inhibitor andrimid. *Journal of the American Chemical Society*, **128**, 10660–1.

33
Cyanophycin Synthetases

Anna Steinle and Alexander Steinbüchel

33.1
Introduction

Cyanophycin, also referred to as multi-L-arginyl-poly-[L-aspartic acid] or cyanophycin granule peptide (CGP), is a nonribosomally synthesized polypeptide consisting of a poly(aspartic acid) backbone with arginine residues linked to the β-carboxyl group of each aspartate by the α-amino group [1]. CGP serves as temporary reserve material for nitrogen, energy and possibly also carbon [2–4]. The polymer has special solubility properties: it is insoluble at physiological ionic strength and neutral pH, but soluble in diluted acids and bases [5]. In cells, CGP is deposited in membraneless granules [6]. Although inclusions containing this polymer were first observed by Borzi in 1887 during microscopic studies of cyanobacteria [7], it was not until 1971 that Simon isolated the material from cells of *Anabaena cylindrica* by taking advantage of its solubility behavior [5].

Besides CGP, two other poly(amino acids) have been described, namely poly(glutamic acid) and poly(lysine) [8, 9]. One exceptional feature of these compounds is, in contrast to proteins, their ribosome-independent synthesis [10]. As a consequence, these polymers possess several characteristics which distinguish them from proteins synthesized by ribosomes [11]:

- Proteins consist of a mixture of 22 amino acids, whereas poly(amino acids) contain – at least in their backbone – only one type of amino acid.
- The sizes of proteins are exactly defined, whereas the length of poly(amino acids) vary; they also show a polydisperse rather than a monodisperse size range.
- The amide bonds formed between the itemized monomers are linked differently. In proteins, the bonds occur only between α-amino groups and α-carboxyl groups, whereas in poly(amino acids) the monomers may be linked through β- and γ-carboxyl groups or ε-amino groups.
- Poly(amino acid) synthesis cannot be affected by translational inhibitors such as chloramphenicol [12–14].

Protein Engineering Handbook. Edited by Stefan Lutz and Uwe T. Bornscheuer

ISBN: 978-3-527-31850-6

For the synthesis of CGP only one enzyme is essential, and this is referred to as cyanophycin synthetase (CphA) [15]. CphA is encoded by the cyanophycin synthetase gene (*cphA*). The interruption of *cphA* has been performed in cells of *Synechocystis* sp. strain PCC 6803 [3] and *Anabaena variabilis* strain ATCC 29413 [16], after which neither deletion strain was capable of accumulating CGP.

33.2
Occurrence of Cyanophycin Synthetases

For a long time the occurrence of CphA was restricted to oxygenic phototrophic bacteria [1, 5, 6, 17, 18]. However, Krehenbrink *et al.* showed in 2002 that heterotrophic bacteria, such as strains of *Acinetobacter baylyi, Desulfitobacterium hafniense, Bordetella bronchiseptica, B. pertussis, B. parapertussis* and *Nitrosomonas europaea,* also possessed *cphA* homologous genes. The functionality of CphA from *A. baylyi* strain ADP1 [19] and *D. hafniense* strain DSM 10664 was approved [20]. The occurrence of different *cph* genes involved in CGP metabolism in several Eubacteria has been investigated *in silico* by Füser and Steinbüchel (2007), whereas in the genomes from Archaea and Eukarya genes with sequence similarities towards *cphA* or other *cph* genes could not be detected [21].

In addition, *cphA*s from different phototrophic and chemotrophic bacteria have been expressed heterologously in other organisms. As a host organism the model prokaryote *E. coli* was mostly used, but other industrially relevant microorganisms such as *Corynebacterium glutamicum, Ralstonia eutropha* and *Pseudomonas putida* were also applied (Table 33.1) [23, 24, 29]. Recently, the ability to produce CGP was also established in eukaryotes. For example, Neumann *et al.* (2005) detected 1.1% (w/w) of CGP from recombinant tobacco and potato plants expressing *cphA* from *Thermosynechococcus elongatus* strain BP-1 (Table 33.1) [34]. In addition, *cphA* from *Synechocystis* sp. strain PCC 6308 was heterologously expressed in different strains of the yeasts *Pichia pastoris* and *Saccharomyces cerevisiae* (A. Steinle and A. Steinbüchel, unpublished data) [37].

33.3
General Features

CGP isolated from cyanobacteria is highly polydisperse, with a molecular weight range of 25 to 100 kDa, as determined by sodium dodecyl sulfate–polyacrylamide gel electrophoresis (SDS–PAGE) [1, 4, 5, 27]. The degree of polymerization was found to be between 90 and 400. The heterologous expression of CphA results in the formation of CGP with a lower molecular weight and less polydispersity. The molecular weight of CGP produced in recombinant *E. coli* ranges from 25 to 35 kDa [28, 30, 32], while CGP isolated from *A. baylyi* strain ADP1 shows a molecular weight of 21–28 kDa [19]. This phenomenon was also shown to occur in CGP isolated from recombinant cells of *C. glutamicum, P. putida* and *R. eutropha* [29].

Table 33.1 Heterologous expression of different CGP synthetase genes.

Origin of *cphA*	Expression organism	Reference(s)
Acinetobacter baylyi strain ADP1	*E. coli*	[19], [22]
Anabaena sp. strain PCC 7120	*E. coli, P. putida, R. eutropha*	[23], [24]
Anabaena variabilis strain ATCC 29413	*E. coli*	[15], [25]
Desulfitobacterium hafniense strain DSM 10664	*E. coli*	[20]
Nostoc ellipsosporum strain NE1	*E. coli*	[26]
Synechococcus sp. strain MA19	*E. coli, P. putida, R. eutropha*	[23], [27], [28]
Synechocystis sp. strain PCC 6308	*E. coli, C. glutamicum, P. putida, R. eutropha, P. pastoris, S. cerevisiae*	[23], [29], [30], [31], [24], A. Steinle and A. Steinbüchel (unpublished results)
Synechocystis sp. strain PCC 6803	*E. coli, R. eutropha, P. putida*	[15], [23], [32], [33]
Thermosynechococcus elongatus strain BP-1	Tobacco and potato plants	[34]

Polydisperse CGP isolated from recombinant tobacco and potato plants exhibited a maximum size of 35 kDa, with a degree of polymerization of approximately 125 [34]. These observed differences in molecular weight, depending on the host organism, can be explained by insufficiencies and differences in the physiological background of the producing organisms. For example, there may be an insufficiency of inevitable substrates, an absence of essential catalytic factors, or an unfavorable enzyme : substrate ratio [30].

The molecular weight of denatured CphA, as ascertained by SDS–PAGE, is 90–130 kDa [15, 27, 28], and this correlates well with the molecular weight calculated for CphA. As the native enzyme shows an apparent molecular mass of approximately 230 kDa, as determined by size-exclusion chromatography (SEC), the active form of CphA is most probably a homodimer [15].

33.4 Reaction Mechanism

Several components are required for the *in vitro* synthesis of CGP. Typically, for each catalysis cycle the incorporated substrates L-aspartic acid and L-arginine are required, together with ATP as an energy source and a CGP-primer [35], which must consist of at least three dipeptide units [(β-Asp-Arg)$_3$]. The incorporated amino acid substrates are added to the C terminus of the primer during catalysis [25]. Moreover, Mg^{2+}, K^+ and a sulfur compound such as β-mercaptoethanol at low concentrations are essential for the catalytic activity of CphA [35]. The optimum pH value for CphA activity is 8.5, although the binding of CGP remains relatively constant between pH 9.0 and 6.3, and falls significantly only at pH values <6 [22, 31, 35]. The maximum activity of CphA is obtained at 50 °C, at which temperature prolonged incubation leads to the inactivation of CphA after 30 min; however, the enzyme stays active at 28 °C [31].

By conducting experiments employing [α-^{32}P]ATP, Ziegler *et al.* were able to show that during catalysis one molecule of ATP was converted to ADP and phosphate, but AMP was not detected [15]. This led to the conclusion that amino acids in CGP synthesis are activated by phosphorylation of the amino acid's carboxyl groups [15]. In 2000, Berg *et al.* postulated that one catalytic cycle may be divided into two successive condensation steps during polymerization. Subsequent experiments using matrix-assisted laser desorption ionization mass spectrometry (MALDI-MS) to detect reaction products and using synthetic primers showed that the constituent amino acids of CGP were incorporated stepwise and consecutively, in the order of aspartic acid followed by arginine, into the growing polymer [25]. The postulated reaction cycle as described by Berg *et al.* is shown in Figure 33.1 [25]. During the first step the carboxylic acid group of the polyaspartic acid backbone is activated through transfer of the γ-phosphoryl group of ATP (Figure 33.1, Step A), and the amino group of aspartic acid is bound through a peptide bond at the C terminus of the CGP primer (Step B). During the second step, a second molecule of ATP is hydrolyzed (Figure 33.1, Step C), and arginine is bound through an iso-peptide bond at the crescive primer, which is being elongated due to the addition of amino acids (Step D). Presumably, CphA exhibits two different active centers, each of which is involved in one catalytic step [25].

33.5 Substrate Specificity

The first characterizations of the CphA of *Anabaena cylindrica* were made by Simon, in 1976, after the enzyme had been 92-fold enriched from the soluble cell fraction [35]. Meanwhile, CphAs from *A. variabilis* [15], from the thermophilic strain *Synechococcus* sp. strain MA19 [27] and from recombinant *E. coli* harboring *cphA* genes from various cyanobacterial strains such as *Synechocystis* sp. strain PCC 6803 [15], *A. variabilis* strain ATCC 29413 [25], *Synechocystis* sp. strain PCC 6308 [31] or *A. baylyi* strain ADP1 [22], were purified by using several chromatographic steps and characterized on the basis of their substrate specificity and affinity.

The K_m-values for the purified CphA from *Synechocystis* sp. strain PCC 6308 were 450 μ*M*, 49 μ*M*, 200 μ*M* and 35 μg ml^{-1} for L-aspartic acid, L-arginine, ATP

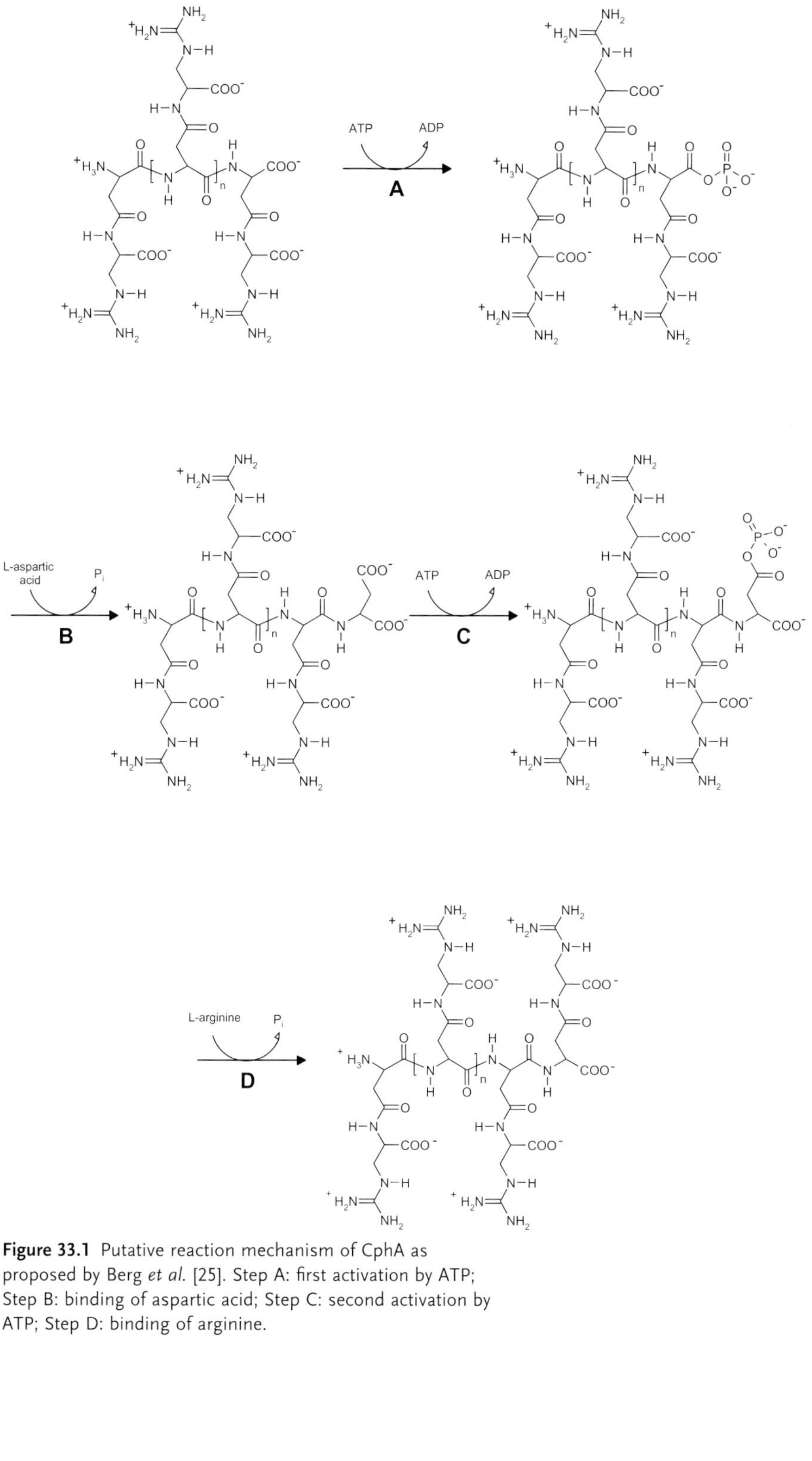

Figure 33.1 Putative reaction mechanism of CphA as proposed by Berg *et al.* [25]. Step A: first activation by ATP; Step B: binding of aspartic acid; Step C: second activation by ATP; Step D: binding of arginine.

and cyanophycin, respectively; this indicated that the enzyme has a high affinity towards these reactants, and especially to arginine [31]. The K_m-values for L-arginine (47 μ*M*) and L-aspartic acid (240 μ*M*) determined for CphA from *A. baylyi* strain ADP1 ($CphA_{ADP1}$) were similar to those of known CphAs from cyanobacteria [22]. For $CphA_{ADP1}$, the two different ATP-binding sites of the enzyme were characterized independently of each other with respect to their affinities for ATP. To determine the K_m-value of the ATP-binding site involved in the incorporation of arginine, CGP-Asp was used as primer in the reaction mixture. It was observed that the ATP-binding site responsible for the addition of arginine exhibited a much higher affinity for ATP (38 μ*M*) than those responsible for the addition of aspartic acid (210 m*M*). Further enzyme activity experiments were carried out regarding the essentiality of Mg^{2+} [22]. Binding of $CphA_{ADP1}$ to CGP-Arg was shown to be independent of Mg^{2+}, whereas binding to CGP-Asp required the presence of Mg^{2+} in order to be effective.

CphAs expressed in cyanobacteria synthesize a polymer that consists of aspartic acid and arginine in approximately equimolar amounts [1, 36]. This amino acid composition is also found in CGP isolated from the chemotrophic bacterium *A. baylyi* strain ADP1 [19] and in a recombinant strain of *E. coli* expressing *cphA* from the cyanobacterium *Nostoc ellipsosporum* [26]. CGPs with divergent amino acid composition have also been observed [15, 18, 19, 34, 37, 38]. Upon heterologous expression of cyanobacterial CphAs from *Synechocystis* sp. strain PCC 6308, *Synechocystis* sp. strain PCC 6803, *T. elongatus* strain BP-1, and of the chemotrophic bacterium *D. hafniense* in *E. coli*, in transgenic tobacco and potato plants, or yeasts such as in *P. pastoris* and *S. cerevisiae*, the incorporation of up to 14 mol% lysine instead of arginine has been determined [19, 20, 34] (also A. Steinle and A. Steinbüchel, unpublished data).

A thorough substitution of arginine by glutamic acid has been also reported. After cultivation of *Synechocystis* sp. strain PCC 6308 under nitrogen-limited conditions, an incorporation of glutamate instead of arginine was observed [38]. Glutamate incorporation was also observed by Wingard *et al.* in CGP produced by nitrogen-starved cells of *Synechococcus* sp. strain G2.1 after the addition of chloramphenicol and nitrate [18]. If the glutamic acid is replacing arginine or aspartic acid could not be detected, the investigated composition was seen to be 29 mol% arginine, 35 mol% aspartic acid and 15 mol% glutamic acid; the remaining 21 mol% was explained by background amounts of a wide variety of other amino acids [18].

The results of several *in vitro* experiments have shown that different structural analogues of arginine and aspartic acid are capable of: (i) substituting the natural CGP constituents; or (ii) affecting the activity of CphA [25, 31]. Aboulmagd *et al.* carried out different experiments concerning substrate specificity using CphA from *Synechocystis* sp. strain PCC 6308 [31], while Berg *et al.* purified CphA from *A. variabilis* strain ATCC 29413 to study the mechanism of the CphA reaction [25]. The chemical structures of a selection of investigated compounds are shown in Figure 33.2. By using L-[4,5-^{3}H]lysine and L-[U-^{3}H]canavanine (15 and 13% enzyme activity, respectively), the incorporation of lysine and canavanine instead

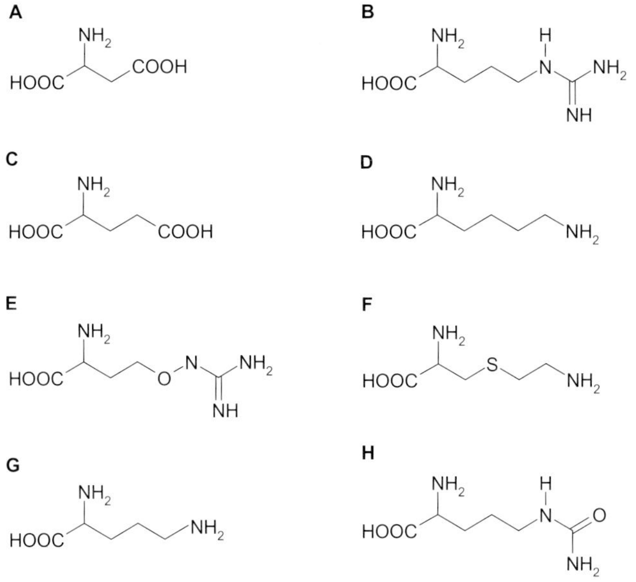

Figure 33.2 Aspartic acid (A), arginine (B) and analogues thereof which can be incorporated into CGP *in vivo* (glutamate (C), lysine (D)) [15, 18, 19, 38] or *in vitro* by purified CGP synthetases (canavanine (E), *S*-(2-aminoethyl) cysteine (F), ornithine (G), citrulline (H)) [25, 28, 31].

of arginine by the investigated enzymes was confirmed [31]. The study conducted by Berg and colleagues, using synthetic primers, showed that arginine could be replaced by lysine, citrulline or ornithine, with all three amino acids being incorporated at the C terminus of the CGP-primer (β-Asp-Arg)$_3$-Asp but not at those of (β-Asp-Arg)$_3$ [25]. In contrast, the incorporation of glutamic acid instead of aspartic acid [38] or arginine as described previously *in vivo* [18, 38] could not be confirmed *in vitro* by the use of L-[U-^{14}C]glutamic acid, which yielded only 0.3 and 0.1% enzyme activity, respectively [25]. The same observation was made when characterizing CphA from *Synechococcus* sp. strain MA19 [28] and CphA from *A. variabilis* strain ATCC 29413 using synthetic primers [25]. Glutamic acid was not ligated to the used primers (β-Asp-Arg)$_3$ or (β-Asp-Arg)$_3$-Asp, which showed that this enzyme could not incorporate glutamic acid instead of arginine or aspartic acid [25].

Further experiments revealed several compounds which inhibited the incorporation of amino acids into CGP [31]. Two such groups could be distinguished. Compounds belonging to the first group, such as arginine methyl ester, argininamide, *S*-(2-aminoethyl) cysteine, β-hydroxy aspartic acid, aspartic acid β-methyl ester, norvaline, citrulline and asparagine, exhibited an equal inhibitory effect on the incorporation of either substrate (aspartic acid and arginine). However, compounds of the second group, such as canavanine, lysine, agmatine, D-aspartic acid,

L-glutamic acid and ornithine, inhibited the incorporation of arginine to a greater extent than the incorporation of aspartic acid. No effect on the incorporation of arginine could be detected in presence of other proteinogenic amino acids such as alanine, histidine, leucine, proline, tryptophan or glycine [31]. Experiments with purified CphA from *A. baylyi* strain ADP1 confirmed that this enzyme was unable to incorporate lysine instead of arginine, nor could arginine or aspartic acid be replaced by glutamate [22]. Moreover, experiments with CphA from *Synechococcus* sp. strain MA19 revealed that α-arginyl aspartic acid dipeptide, citrulline, ornithine, arginine amide, agmatine or norvaline could not replace arginine [28]. On the other hand, this enzyme was able to incorporate canavanine instead of arginine and β-hydroxyaspartic acid instead of L-aspartic acid, at significant rates [28]. Furthermore, this thermostable enzyme accepted several compounds as primers, including α-arginyl aspartic acid dipeptide, modified cyanophycin containing less arginine, and poly-α,β-D,L-aspartic acid [28].

33.6 Primary Structure Analysis

As the three-dimensional structure of CphA has not yet been determined, the detailed mechanism of catalysis has not been assessed. However, an analysis of the primary structure revealed different binding motifs or domains in CphA [15, 19, 25]. A comparison of the amino acid sequences of CphA and homologous proteins showed that CphA could be divided into N-terminal and C-terminal regions [15, 25]. The result of a NCBI BLAST search using the CDD (Conserved Domain Database) tool after submission of the amino acid sequence of CphA from *Synechocystis* sp. strain PCC 6308 is shown in Figure 33.3 [39]. Several proteins show sequence similarities to the N or C termini of CphA. In 2003, Berg postulated that arginine was bound in the C-terminal region, whereas aspartic acid is bound in the N-terminal region of CphA [40]. Both regions were known to possess an ATP-binding motif, and hence the enzyme may contain two active sites [25]. From this it was concluded that each of the incorporated amino acids would be bound in one active site [25].

The N-terminal regions of CphA show high similarities to a superfamily of ATP-dependent ligases [41]. Enzymes belonging to this family activate carboxylates for nucleophilic attack by phosphorylation with Mg^{2+} ATP. Two characteristics of these enzymes are the ATP-binding motif, referred to as ATP-grasp or B-loop, and the so-called J-loop (Figure 33.4). Both loops are flexible in an ATP-free state but become rigid in the ATP-bound state [42]. Sequences beyond these domains show rather low identities to CphA sequences. Members of the ATP-grasp enzyme superfamily include, among others, D-alanine-D-alanine ligase [43], biotin carboxylase α-chain [44], glutathione synthetase [45, 46], succinyl-CoA synthetase [47] and carbamoylphosphate synthetase [48].

The C-terminal regions of CphA, which start approximately at amino acid residue 370 [15, 25], show high sequence similarities to the well-characterized

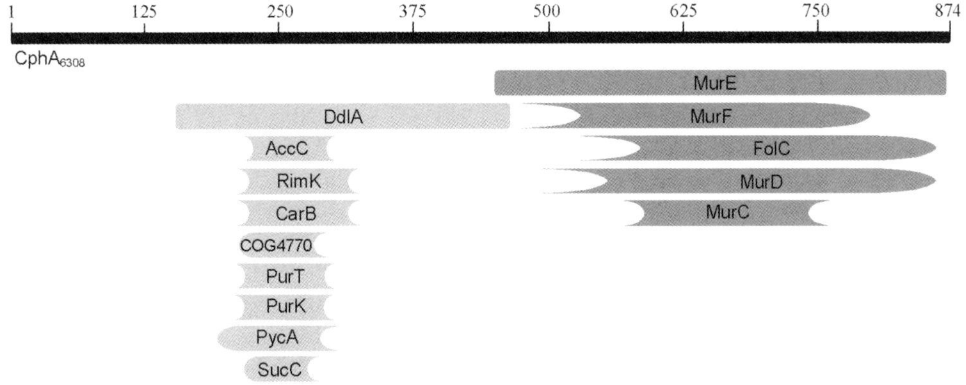

Figure 33.3 Proteins or protein domains homologous to CphA from *Synechocystis* sp. strain PCC 6308 ($CphA_{6308}$). Values in brackets indicate the e-value as determined by the NCBI BLAST CDD (Conserved Domain Database) tool [39]. Straight ends symbolize the end or start of a protein, convex or concave ends indicate that proteins still continue but the sequence does not show any similarity to CphA. DdlA = D-alanine-D-alanine ligase ($2e^{-4}$), AccC = biotin carboxylase alpha-chain ($6e^{-10}$), RimK = glutathione synthetase ($2e^{-4}$), CarB = carbamoylphosphate synthetase large subunit ($5e^{-7}$), COG4770 = acetyl/propionyl-CoA carboxylase ($2e^{-4}$), PurT = formate-dependent phosphoribosylglycinamide formyltransferase ($3e^{-4}$), PurK = phosphoribosylaminoimidazole carboxylase ($7e^{-4}$), PycA = pyruvate carboxylase (0.001), SucC = succinyl-CoA synthetase, beta subunit (0.002), MurE = UDP-*N*-acetylmuramoyl-L-alanyl-D-glutamate: *meso*-diaminopimelate ligase ($1e^{-48}$), MurF = UDP-*N*-acetylmuramoyl-tripeptide: D-alanyl-D-alanine ligase ($9e^{-13}$), FolC = folyl-poly-glutamate synthetase-dihydrofolate synthetase ($8e^{-10}$), MurD = UDP-*N*-acetylmuramoyl-L-alanine: D-glutamate ligase ($6e^{-8}$), MurC = UDP-*N*-acetylmuramate: L-alanine ligase ($2e^{-5}$).

superfamily of murein ligases (MurC, MurD, MurE, MurF) and folyl-poly-γ-glutamate synthetase-dihydrofolate synthetase (FolC) [49–52]. All enzymes belonging to this superfamily share the following enzymatic features [52]: (i) they catalyze the formation of a peptide or amide bond through the hydrolysis of ATP to ADP and P_i; and (ii) they function via a similar mechanism involving the formation of acyl phosphate [53] and tetrahedral intermediates [54, 55]. One characteristic motif of all these enzymes is the so-called 'P-loop motif', which is involved in the binding of ATP [56]. This shows a Rossmann fold which is also referred to as a dinucleotide-binding fold [57–59]. Besides the above-mentioned motifs, the enzyme shows several other conserved residues [15, 50, 51]. In the further amino acid analysis of CphA, the murein ligases in particular play an important role due to the similar substrates that are bound during catalysis. As with CphA the murein ligases add one amino acid to an amino acid primer, with concomitant degradation of ATP to ADP and P_i. The murein ligases, MurC, MurD, MurE and MurF, are responsible for the successive addition of L-alanine, D-glutamate, *meso*-diaminopimelate or L-lysine, and D-alanyl-D-alanine to UDP-*N*-acetylmuramic acid. As their three-dimensional structures have been determined [57, 60–65], the functions of several amino acid residues and regions during the binding of substrates to these enzymes have been assessed [65]. Regions involved in the binding of the primer, ATP, and the amino acid substrates are indicated in Figure 33.4. Due to the primary structure similarities between these enzymes and CphA, it may be assumed that not only ATP is bound in the corresponding region of CphA but also the amino acid substrate and the primer. An alignment of all functional CphAs and a selection of proteins which show sequence similarities to the C or N terminus to CphA is shown in Figure 33.4. Here, the conserved motifs and substrate-binding regions are indicated.

33.7
Enzyme Engineering

An initial series of site-directed mutagenesis (SDM) studies have been performed with CphAs from *N. ellipsosporum, A. variabilis* strain ATCC 29413 and *Synechocystis* sp. strain PCC 6308 [26, 40] (also A. Steinle and A. Steinbüchel, unpublished data). In addition, truncated *cphAs* from *N. ellipsosporum* strain NE1 ($CphA_{NE1}$) and *Synechocystis* sp. strain PCC 6308 ($CphA_{6308}$) have been constructed [26]) (A. Steinle and A. Steinbüchel, unpublished data). The truncation of a stretch of 31 amino acids at the C terminus of $CphA_{NE1}$ resulted in an 2.1-fold increase in enzyme activity compared to the wild-type $CphA_{NE1}$. In contrast, the truncation of a stretch of 59 amino acids at the C terminus of $CphA_{NE1}$ led to a complete loss of enzyme activity. As shown in Figure 33.4, $CphA_{NE1}$ is one of the longest CphA proteins (901 amino acids) when compared with other investigated CphAs (873–914 amino acids). The enhanced activity of truncated $CphA_{NE1}$, which had a length of 870 amino acids, showed that amino acid residues at the far C terminus of CphA are clearly not essential for catalysis [26]. On the other hand, a length of about 870

Figure 33.4 Sequence comparison of different CphAs and proteins with high sequence similarities to CphA. Amino acid residues are specified by standard, one-letter abbreviations. Gaps (-) were introduced to improve the alignment. The areas shaded in black indicate positions where an amino acid is conserved in 14 of the 15 sequences. Areas shaded in dark gray indicate positions where an amino acid is conserved in at least 11 of the 15 sequences. Areas shaded in light gray indicate positions where an amino acid is conserved in at least seven of the 15 sequences. The overlined areas indicate conserved ATP-binding motifs. Substrate-binding regions of Mur-ligases as determined by Mol *et al.* (2003) are underlined. The aligned CphAs (1–9) are from: 1 *Anabaena variabilis* strain ATCC 29413; 2 *Nostoc* sp. strain PCC 7120; 3 *Nostoc ellipsosporum*; 4 *Synechococcus* sp. strain MA19; 5 *Synechocystis* sp. strain PCC 6308; 6 *Synechocystis* sp. strain PCC 6803; 7 *Thermosynechococcus elongatus* strain BP-1; 8 *Acinetobacter baylyi* strain ADP1; and 9 *Desulfitobacterium hafniense* strain DCB-2. The aligned enzymes with high sequence similarities to CphA (10–15) are from *E. coli*: 10 D-alanine-D-alanine ligase; 11 UDP-*N*-acetylmuramoyl-L-alanyl-D-glutamate: *meso*-diaminopimelate ligase (MurE); 12 UDP-*N*-acetylmuramoyl-tripeptide: D-alanyl-D-alanine ligase (MurF); 13 folyl-poly-γ-glutamate synthetase-dihydrofolate synthetase (FolC); 14 UDP-*N*-acetylmuramoyl-L-alanine: D-glutamate ligase (MurD); 15 UDP-*N*-acetylmuramate: L-alanine ligase (MurC).

```
                   *        20         *        40         *        60         *        80
 1 : MRILKIQTLRGPNYWSIRRHKLIVMRLDLETLAETPSNEIPGFYEGLVEALPSLEGHYCSPGCHGGFLMRVREGTMMGHIVEHVAL :  86
 2 : MRILKIQTLRGPNYWSIRRHKLIVMRLDLETLAETPSNEIPGFYEGLVEALPSLEGHYCSPGCHGGFLMRVREGTMMGHIVEHVAL :  86
 3 : MRILKIQTLRGRNYWSIRRHKLIVMRLDLETLAETPSNEIPGFYEGLVEALPSLDGHYCSPGCHGGFLMRVREGTMMRHIVEHVAL :  86
 4 : MRILKIQTLRGPNYWSIRRHKLIVMRLDLENLAETPSNEIPGFYEGLVEALPSLESHYCSPGCRGGFLKRVREGTMMGHIVEHVAL :  86
 5 : MKILKTQTLRGPNYWSIRRQKLIQMRLDLEDVAEKPSNLIPGFYEGLVKILPSLVEHFCSRDHRGGFLERVQEGTYMGHIVEHIAL :  86
 6 : MKILKTLTLRGPNYWSIRRKKLIVMRLDLEDLAERPSNSIPGFYEGLIRVLPSLVEHFCSPGHRGGFLARVREGTYMGHIVEHVAL :  86
 7 : MKILKLQTLRGPNYWSIRRHKLIVMRLDLEEVANTPSNQISGFVDGLVRVLPSLYNHFCSLGHEGGFLTRLREGTYLGHVVEHVAL :  86
 8 : MNIISTSVYVGPNVYASIPLIRLVIDLNPHYITQLAS-MGSEVLENLEKVIPTLKTEQDAKLQHKLEELRQAPQQQIGELVAILAL :  85
 9 : MEILKIQAIPGANVYSYR--PVIRAVVDLQEWTERTSDTFGDFNTRLVQCLPSLYEHFCSRGKPGGFVERLKEGTLVGHIIEHVTI :  84
10 : -------------------------------------------------------------------------------------- :   -
11 : -------------------------------------------------------------------------------------- :   -
12 : -------------------------------------------------------------------------------------- :   -
13 : -------------------------------------------------------------------------------------- :   -
14 : -------------------------------------------------------------------------------------- :   -
15 : -------------------------------------------------------------------------------------- :   -

        *       100         *       120         *       140         *       160         *
 1 : ELQELAGMHVGFGRTR-ETATPGIYQVVIEYLNEEAGRYAGRAAVRLCQSIVDRGRY-PKAELEQDIQDLKDLWRDA-SLGPSTEAI : 169
 2 : ELQELAGMHVGFGRTR-ETATPGIYQVVIEYLNEEAGRYAGRAAVRLCQSIVDRGRY-PKAELEQDIQDLKDLWRDA-SLGPSTEAI : 169
 3 : ELQELAGMHVGFGRTR-ETATPGIYQVVIEYLNEEAGRYAGRAAVRLCQSIVDRGRY-PKAELEQDIQDLKDLWRDA-SLGPSTEAI : 169
 4 : ELQELAGMHVGFGRTR-QTSTPSVYQVVFEYQNEEAGRYAGRAAVRLCQSIVDRGRY-HKARAEQDLQDLKDLWRDA-ALGPSTESI : 169
 5 : ELQELAGMPVGFGRTR-ETSTPGIYNVVFEYVYEEAGRYAGRVAVRLCNSIITTGAY-GLDELAQDLSDLKDLRANS-ALGPSTETI : 169
 6 : ELQELVGMTAGFGRTR-ETSTPGIYNVVYEYVDEQAGRYAGRAAVRLCRSLVDTGDY-SLTELEKDLEDLRDLGANS-ALGPSTETI : 169
 7 : ELQELAGMPVGFGRTR-ETSTPGVYQVVYEYQVEEAGRYAGRAAVRLCQSIIDTGTY-PQQELDQDLADLRELKAKA-SLGPSTEAI : 169
 8 : HLQRLAGQKG--GAAFSAYCHEDETEILYSYESEEIGIEAGEVVCDMLVALAKAHEAGDQIDLNRDVKGFLRYADRF-ALGPSALAL : 168
 9 : ELLTRAGQNIPYGKTLCLPEHPGHYEIIFNYDSLEGGLEGFKQGYALVQELLAG----QKPNVTNRIERIREVIQRF-ELGASTRAI : 165
10 : -------------------------------------MTDKIAVLLGGTSAER-EVSLNSGAAVLAGLREGGIDAYPVDPK-EVD :  46
11 : -------------------------------------------------------------------------------------- :   -
12 : -------------------------------------------------------------------------------------- :   -
13 : -------------------------------------------------------------------------------------- :   -
14 : -------------------------------------------------------------------------------------- :   -
15 : -------------------------------------------------------------------------------------- :   -

          180         *       200         *       220         *       240         *       2
 1 : IVKEAEKRGIP--WMQLSARFLIQLGYGVNHKRMQATMTDKTGILGVELACDKEATKRILAASGVPV------PRGTVINFLDDLE : 249
 2 : IVKEAEKRGIP--WMQLSARFLIQLGYGVNHKRMQATMTDKTGILGVELACDKEATKRILAASGVPV------PRGTVINFLDDLE : 249
 3 : IVKEAEKRGIP--WMQLSARFLIQLGYGVNHKRMQATMTDKTGILGVELACDKEATKRILAASGVPV------PRGTVINFLDDLE : 249
 4 : IVKEAEKRGIP--WMQLGARFLIQLGYGVNQKRIQATMTDQTGILGVELACDKEATKRILANAGIPV------PKGTVINFLDDLE : 249
 5 : IIKEAEARQIP--WMLLSARAMVQLGYGANQQRIQATLSNKTGILGVELACDKEGTKTTLAEAGIPV------PRGTVIYYADELA : 249
 6 : IVTEADARKIP--WMLLSARAMVQLGYGVHQQRIQATLSSHSGILGVELACDKEGTKTILQDAGIPV------PRGTTIQYFDDLE : 249
 7 : LVREAEARNIP--WFELSSRSIIQLGYGARSHRMQATLSDRSSILAVELASDKEGAKRLLQDAGIPV------PKGTVIRYIEDLP : 249
 8 : IVQAAEERNIP--WYRLNDASLIQVGQGKYQKRIEAALTSGTSHIAVEIAGDKNVCNQLLQDLGLPV------PKQRVVYDIDDAV : 247
 9 : IIEAAEGRGIP--VIRLNDSSLLQLGYGRNQKRVQAAMSDQTSCIGVDIACDKGLTKKLLYEGGIPV------PDGVVTRNEDEAV : 244
10 : VTQLKSMGFQKVFIALHGRG-GEDGTLQGMLELMGLPYTGSGVMASALSMDKLRSKLLWQGA-GLPVAPWVALTRAEFEKGLSDKQ : 130
11 : -------------------------------------------------------------------------------------- :   -
12 : -------------------------------------------------------------------------------------- :   -
13 : -------------------------------------------------------------------------------------- :   -
14 : -------------------------------------------------------------------------------------- :   -
15 : -------------------------------------------------------------------------------------- :   -

                  B-Loop
             ________________
     60       *       280         *       300         *       320         *       340
 1 : EAIEYVGGYPIVIKPLDGNHGRGITIDIRSWEEAEAAYEAARQVS--RSIIVERYYVGRDHRVLVVDGKVVAVAERVPAHVIGNGR : 333
 2 : EAIEYVGGYPIVIKPLDGNHGRGITIDIRSWEEAEAAYEAARQVS--RSIIVERYYVGRDHRVLVVDGKVVAVAERVPAHVIGNGR : 333
 3 : EAIEYVGGYPIVIKPLDGNHGRGITIDIRSWEDAEAAYEAARQVS--RSIIVERYYVGRDDRVLVVDGKVVAVAERVPAHVIGNGR : 333
 4 : EAIEYVGGYPIVIKPLDGNHGRGITINIQNWEEAEAAYDAARQIS--RSIIVERYYQGRDHRVLVVD-ASSAVAERVPAHVVGDGR : 332
 5 : DAIADVGGYPIVLKPLDGNHGRGITIDINSQQEAEEAYDLASAASKTRSVIVERYYKGNDHRVLVINGKLVAVSERIPAHVTGNGS : 335
 6 : EAINDVGGYPVVIKPLDGNHGRGITINVRHWEEAIAAYDLAAEESKSRSIIVERYYEGSDHRVLVVNGKLVAVAERIPAHVTGDGT : 335
 7 : EAIEEIGGYPIVIKPLNGNHGRGITIDINSLEAAEEAFEIASSIS--KSVIVERYHAGRDFRVLVVNGKVVAVAERVPAHVIGDGH : 333
 8 : RAARRVG-FPVVLKPLDGNHGRGVSVNLTTDEAVEAAFDIAMSEG--SAVIVESMLYGDDHRLLVVNGELVAAARRVPGHIVGDGK : 330
 9 : EVFRQLD-RLVVVKPYNGNQGKGVTLKLGTEAEVRAAFRVAQTYE--EQVVVEEYIEGKNYRLLVVDGKMAAAAERIPAHVIGDGV : 327
10 : LAEISALGLPVIVKPSREGSSVGMS-KVVAENALQDALRLAFQHDEEVLIEKWLSGPEFTVAIL-GEEILPSIRIQPSGTFYDYEA : 214
11 : -------------------------------------------------------------------------------------- :   -
12 : -------------------------------------------------------------------------------------- :   -
13 : -------------------------------------------------------------------------------------- :   -
14 : -------------------------------------------------------------------------------------- :   -
15 : -------------------------------------------------------------------------------------- :   -
```

Figure 33.4 *Continued*

Figure 33.4 *Continued*

amino acids seems to be the minimum, as can be seen from the truncated $CphA_{NE1}$ that consists of 842 amino acids [26], as well as from truncated $CphA_{6308}$ with a length of 854 amino acids, both of which are completely inactive (A. Steinle and A. Steinbüchel, unpublished data).

Moreover, several point mutations have been carried out with $CphA_{6308}$ and CphA from *A. variabilis* strain ATCC 29413 ($CphA_{29413}$) [40] (A. Steinle and A. Steinbüchel, unpublished data). All performed point mutations are listed in Table 33.2 with respect to their formation of cell inclusions, the possibility of CGP isolation by the acid extraction method as described previously [32], and to the

Table 33.2 Performed point mutations in CphA from *Synechocystis* sp. strain PCC 6308 and *Anabaena variabilis* strain ATCC 29413.[a] All genes were expressed in *E. coli*, and expression of CphA was investigated with respect to the occurrence of cytoplasmic inclusions, the possibility for CGP-isolation [32], measurement of the enzyme activity [31] and the determined CGP content. *E. coli* harboring the wild-type *cphA* from *Synechocystis* sp. strain PCC 6308 was used as a positive control; *E. coli* without *cphA* was used as a negative control.

CphA Genotype	Occurrence of cell inclusions	Possibility of CGP isolation	Enzyme activity (%)	CGP content (%, w/w)
Wild-type $cphA_{6308}$	+	+	100	10.09
None	–	–	0.0	–
C59A	+	–	2.0	–
C133A	+	+	2.0	4.32
C218A	+	+	1.9	3.80
K261G	–	–	0.0	–
K261G[a]	–	–	0.0	–
K497A	–	–	0.0	–
K497A[a]	–	–	0.0	–
C595S	+	+	120	12.50
F692H	–	–	0.0	–
R731E	–	–	0.0	–
H748F	+	+	34.0	2.11
R777K	+	+	3.0	2.47
E800R	–	–	2.0	–
D802E	–	–	0.0	–
R805K	+	+	38.0	0.49
E835Q	+	+	5.0	3.42

a *Anabaena variabilis* strain ATCC 29413.

CGP content in the cells. For the analysis, all mutated *cphA* genes were heterologously expressed in *E. coli*. In both genes the amino acid residues Lys261 and Lys497, which belong to the putative C-terminal and N-terminal ATP-binding domains, have been mutated to glycine and alanine, resulting in K261G or K497A, respectively. As expected, neither of the mutated *cphA* genes led to active CphAs. In 2003, Berg showed that mutant K261G of $CphA_{29413}$ still incorporated arginine into the polymer but not aspartic acid, whereas mutant K497A of $CphA_{29413}$ still incorporated aspartic acid but not arginine. These results led to the conclusion that arginine is bound in the C terminus of CphA and aspartic acid in the N terminus [40]. All of the remaining mutants, except for C595S, showed a reduced rate of CGP synthesis or were completely inactive. Mutant C595S showed an increase in enzyme activity and of CGP content of approximately 20% (A. Steinle and A. Steinbüchel, unpublished data).

Additionally, Berg expressed the N- and C-terminal regions of $CphA_{29413}$ independently and constructed CphAs with deleted areas [40]. The resultant enzyme expressing the N-terminal region consisted of amino acids 1–447, while the enzyme expressing the C-terminal region consisted of amino acids 429–901. The deletion mutant ΔN lacked amino acid residues 233–307, while in deletion mutant ΔC amino acid residues 482–758 were missing. None of these mutant genes was expressing an active enzyme. It seemed most likely that CphA could not be folded correctly in the mutants, or that the catalysis performed at the N terminus of CphA was also dependent on amino acid residues occurring in the C terminus, and *vice versa* [40].

33.8
Biotechnical Applications

During the past few years, research into CGP has steadily increased due to its potential for technical applications. However, as the demand for CGP has risen, it has become clear that biotechnological processes would be required for its production in large quantities. Consequently, a variety of fermentation schemes for CGP production via industrially relevant microorganisms have been established [23, 24, 32, 66–68], as has production in transgenic eukaryotic systems [34, 37] (A. Steinle and A. Steinbüchel, unpublished data).

Currently, the possible applications for cyanophycin-derived biopolymers are numerous. For example, poly(aspartic acid) derived from CGP by the hydrolytic removal of arginine could serve as an environmentally friendly substitute for nonbiodegradable poly(acrylic acid) [69–71]. Bulk chemicals such as acrylonitrile and urea may in future also be derived from CGP [72, 73]. Finally, CGP derived via the action of intracellular cyanophycinases in cyanobacteria [74] may represent a source of Asp-Arg or Asp-Lys dipeptides for use in high-value pharmaceutical products. Alternatively, these materials may be produced by extracellular cyanophycinases secreted from heterotrophic bacteria growing on CGP as the sole carbon and nitrogen source [75–77].

Acknowledgments

The support for these investigations was provided by the Fachagentur für Nachwachsende Rohstoffe (Gülzow), Bayer AG (Leverkusen) and SenterNovem (The Netherlands), and is gratefully acknowledged by the authors.

References

1 Simon, R.D. and Weathers, P. (1976) Determination of the structure of the novel polypeptide containing aspartic acid and arginine which is found in cyanobacteria. *Biochimica et Biophysica Acta*, **420**, 165–76.

2 Mackerras, A.H., De Cazal, N.M. and Smith, G.D. (1990) Transient

accumulation of cyanophycin in *Anabaena cylindrica* and *Synechocystis* 6308. *Journal of General Microbiology*, **136**, 2057–65.

3 Li, H., Sherman, D.M., Bao, S. and Sherman, L.A. (2001) Pattern of cyanophycin accumulation in nitrogen-fixing and non-nitrogen-fixing cyanobacteria. *Archives of Microbiology*, **176**, 9–18.

4 Simon, R.D. (1973a) Measurement of the cyanophycin granule polypeptide contained in the blue-green alga *Anabaena cylindrica. Journal of Bacteriology*, **114**, 1213–16.

5 Simon, R.D. (1971) Cyanophycin granules from the blue-green alga *Anabaena cylindrica*: a reserve material consisting of copolymers of aspartic acid and arginine. *Proceedings of the National Academy of Sciences of the United States of America*, **68**, 265–7.

6 Lawry, N.H. and Simon, R.D. (1982) The normal and induced occurrence of cyanophycin inclusion bodies in several blue green algae. *Journal of Phycology*, **18**, 391–9.

7 Borzi, A. (1887) Le comunicazioni intracellulari delle Nostochinee. *Malpighia*, **1**, 28–74.

8 Ashiuchi, M. and Misono, H. (2003) Poly-γ-glutamic acid, in *Biopolymers. Volume 7: Polyamides and Complex Proteinaceous Materials I.* (eds S.R. Fahnenstock and A. Steinbüchel), Wiley-VCH Verlag GmbH, Weinheim, pp. 123–73.

9 Yoshida, T., Hiraki, J. and Nagasawa, T. (2003) ε-Poly-L-lysine, in *Biopolymers. Volume 7: Polyamides and Complex Proteinaceous Materials I* (eds S.R. Fahnenstock and A. Steinbüchel), Wiley-VCH Verlag GmbH, Weinheim, pp. 107–21.

10 Von Döhren, H. (2003) Non-ribosomal biosynthesis of linear and cyclic oligopeptides, in *Biopolymers. Volume 7: Polyamides and Complex Proteinaceous Materials I* (eds S.R. Fahnenstock and A. Steinbüchel), Wiley-VCH Verlag GmbH, Weinheim, pp. 51–81.

11 Oppermann-Sanio, F.B. and Steinbüchel, A. (2002) Occurrence, functions and biosynthesis of polyamides in microorganisms and biotechnological production. *Naturwissenschaften*, **89**, 11–22.

12 Simon, R.D. (1973) The effect of chloramphenicol on the production of cyanophycin granule polypeptide in the blue-green alga *Anabaena cylindrica. Archiv für Mikrobiologie*, **92**, 115–22.

13 Allen, M.M., Yuen, C., Medeiros, L., Zizlsperger, N., Farooq, M. and Kolodny, N.H. (2005) Effects of light and chloramphenicol stress on incorporation of nitrogen into cyanophycin in *Synechocystis* sp. strain PCC 6308. *Biochimica et Biophysica Acta*, **1725**, 241–6.

14 Allen, M.M. and Hawley, M.A. (1983) Protein degradation and synthesis of cyanophycin granule polypeptide in *Aphanocapsa* sp. *Journal of Bacteriology*, **154**, 1480–4.

15 Ziegler, K., Diener, A., Herpin, C., Richter, R., Deutzmann, R. and Lockau, W. (1998) Molecular characterization of cyanophycin synthetase, the enzyme catalyzing the biosynthesis of the cyanobacterial reserve material multi-L-arginyl-poly-L-aspartate (cyanophycin). *European Journal of Biochemistry*, **254**, 154–9.

16 Ziegler, K., Stephan, D.P., Pistorius, E.K., Ruppel, H.G. and Lockau, W. (2001) A mutant of the cyanobacterium *Anabaena variabilis* ATCC 29413 lacking cyanophycin synthetase: growth properties and ultrastructural aspects. *FEMS Microbiology Letters*, **196**, 13–18.

17 Allen, M.M. (1984) Cyanobacterial cell inclusions. *Annual Review of Microbiology*, **38**, 1–25.

18 Wingard, L.L., Miller, S.R., Sellker, J.M., Stenn, E., Allen, M.M. and Wood, A.M. (2002) Cyanophycin production in a phycoerythrin-containing marine *Synechococcus* strain of unusual phylogenetic affinity. *Applied and Environmental Microbiology*, **68**, 1772–7.

19 Krehenbrink, M., Oppermann-Sanio, F.B. and Steinbüchel, A. (2002) Evaluation of non-cyanobacterial genome sequences for occurrence of genes encoding proteins homologous to cyanophycin synthetase and cloning of an active cyanophycin synthetase from *Acinetobacter* sp. strain DSM 587. *Archives of Microbiology*, **177**, 371–80.

20 Ziegler, K., Deutzmann, R. and Lockau, W. (2002) Cyanophycin synthetase-like enzymes of non-cyanobacterial Eubacteria: characterization of the polymer produced by a recombinant synthetase of *Desulfitobacterium hafniense*. *Zeitschrift fur Naturforschung*, **57**, 522–9.

21 Füser, G. and Steinbüchel, A. (2007) Analysis of genome sequences for genes of cyanophycin metabolism: identifying putative cyanophycin metabolizing prokaryotes. *Macromolecular Bioscience*, **7**, 278–96.

22 Krehenbrink, M. and Steinbüchel, A. (2004) Partial purification and characterization of a non-cyanobacterial cyanophycin synthetase from *Acinetobacter calcoaceticus* strain ADP1 with regard to substrate specificity, substrate affinity and binding to cyanophycin. *Microbiology*, **150**, 2599–608.

23 Voss, I., Cardoso Diniz, S., Aboulmagd, E. and Steinbüchel, A. (2004) Identification of the *Anabaena* sp. strain PCC7120 cyanophycin synthetase as suitable enzyme for production of cyanophycin in Gram-negative bacteria like *Pseudomonas putida* and *Ralstonia eutropha*. *Biomacromolecules*, **5**, 1588–95.

24 Cardoso Diniz, S., Voss, I. and Steinbüchel, A. (2006) Optimization of cyanophycin production in recombinant strains of *Pseudomonas putida* and *Ralstonia eutropha* employing elementary mode analysis and statistical experimental design. *Biotechnology and Bioengineering*, **93**, 698–717.

25 Berg, H., Ziegler, K., Piotukh, K., Baier, K., Lockau, W. and Volkmer-Engert, R. (2000) Biosynthesis of the cyanobacterial reserve polymer multi-L-arginyl-poly-L-aspartic acid (cyanophycin). Mechanism of the cyanophycin synthetase reaction studied with synthetic primers. *European Journal of Biochemistry*, **267**, 5561–70.

26 Hai, T., Frey, K.M. and Steinbüchel, A. (2006) Activation of cyanophycin synthetase of *Nostoc ellipsosporum* strain NE1 by truncation at the carboxy-terminal region. *Applied Microbiology and Biotechnology*, **72**, 7652–60.

27 Hai, T., Oppermann-Sanio, F.B. and Steinbüchel, A. (1999) Purification and characterization of cyanophycin and cyanophycin synthetase from the thermophilic *Synechococcus* sp. MA19. *FEMS Microbiology Letters*, **181**, 229–36.

28 Hai, T., Oppermann-Sanio, F.B. and Steinbüchel, A. (2002) Molecular characterization of a thermostable cyanophycin synthetase from the thermophilic cyanobacterium *Synechococcus* sp. MA19 and in vitro synthesis of cyanophycin and related polyamides. *Applied and Environmental Microbiology*, **68**, 93–101.

29 Aboulmagd, E., Voss, I., Oppermann-Sanio, F.B. and Steinbüchel, A. (2001b) Heterologous expression of cyanophycin synthetase and cyanophycin synthesis in the industrial relevant bacteria *Corynebacterium glutamicum* and *Ralstonia eutropha* and in *Pseudomonas putida*. *Biomacromolecules*, **2**, 1338–42.

30 Aboulmagd, E., Oppermann-Sanio, F.B. and Steinbüchel, A. (2000) Molecular characterization of the cyanophycin synthetase from *Synechocystis* sp. strain PCC6308. *Archives of Microbiology*, **174**, 297–306.

31 Aboulmagd, E., Oppermann-Sanio, F.B. and Steinbüchel, A. (2001a) Purification of *Synechocystis* sp. strain PCC6308 cyanophycin synthetase and its characterization with respect to substrate and primer specificity. *Applied and Environmental Microbiology*, **67**, 2176–82.

32 Frey, K.M., Oppermann-Sanio, F.B., Schmidt, H. and Steinbüchel, A. (2002) Technical-scale production of cyanophycin with recombinant strains of *Escherichia coli*. *Applied and Environmental Microbiology*, **68**, 3377–84.

33 Hai, T., Ahlers, H., Gorenflo, H. and Steinbüchel, A. (2000) Axenic cultivation of anoxygenic phototrophic bacteria, cyanobacteria, and microalgae in a new closed tubular glass photobioreactor. *Applied Microbiology and Biotechnology*, **53**, 383–9.

34 Neumann, K., Stephan, D.P., Ziegler, K., Hühns, M., Broer, I., Lockau, W. and Pistorius, E.K. (2005) Production of cyanophycin, a suitable source for the biodegradable polymer polyaspartate, in transgenic plants. *Plant Biotechnology Journal*, **3**, 249–58.

35 Simon, R.D. (1976) The biosynthesis of multi-L-arginyl-poly(L-aspartic acid) in the filamentous cyanobacterium *Anabaena cylindrica*. *Biochimica et Biophysica Acta*, **422**, 407–18.

36 Allen, M.M. and Weathers, P.J. (1980) Structure and composition of cyanophycin granules in the cyanobacterium *Aphanocapsa* 6308. *Journal of Bacteriology*, **141**, 959–62.

37 Steinle, A., Oppermann-Sanio, F.B., Reichelt, R. and Steinbüchel, A. (2008) Synthesis and accumulation of cyanophycin in transgenic strains of *Saccharomyces cerevisiae*. *Applied and Environmental Microbiology*, **74**, 3410–18.

38 Merritt, M.V., Sid, S.S., Mesh, L. and Allen, M.M. (1994) Variations in the amino acid composition of cyanophycin in the cyanobacterium *Synechocystis* sp. PCC6308 as a function of growth conditions. *Archives of Microbiology*, **162**, 158–66.

39 Altschul, S.F., Madden, T.L., Schäffer, A.A., Zhang, J., Zhang, Z., Miller, W. and Lipman, D.J. (1997) Gapped BLAST and PSI-BLAST: a new generation of protein database search programs. *Nucleic Acids Research*, **25**, 3389–402.

40 Berg, H. (2003) Untersuchungen zu Funktion und Struktur der Cyanophycin-Synthetase von Anabaena variabilis ATCC 29413, Humboldt-Universität zu Berlin (Dissertation).

41 Galperin, M.Y. and Koonin, E.V. (1997) A diverse superfamily of enzymes with ATP-dependent carboxylate-amine/thiol ligase activity. *Protein Science*, **6**, 2639–43.

42 Hibi, T., Nishioka, T., Kato, H., Tanizawa, K., Fukui, T., Katsube, Y. and Oda, J. (1996) Structure of the multifunctional loops in the nonclassical ATP-binding fold of glutathione synthetase. *Nature Structural Biology*, **3**, 16–18.

43 Fan, C., Moews, P.C., Shi, Y., Walsh, C.T. and Knox, J.R. (1995) A common fold for peptide synthetases cleaving ATP to ADP: Glutathione synthetase and D-alanine:D-alanine ligase of *Escherichia coli*. *Proceedings of the National Academy of Sciences of the United States of America*, **92**, 1172–6.

44 Kondo, S., Nakajima, Y., Sugio, S., Yong-Biao, J., Sueda, S. and Kondo, H. (2004) Structure of the biotin carboxylase subunit of pyruvate carboxylase from *Aquifex aeolicus* at 2.2 Å resolution. *Acta Crystallographica. Section D, Biological Crystallography*, **60**, 486–92.

45 Matsuda, K., Mizuguchi, K., Nishioka, T., Kato, H., Go, N. and Oda, J. (1996) Crystal structure of glutathione synthetase at optimal pH: domain architecture and structural similarity with other proteins. *Protein Engineering*, **9**, 1083–92.

46 Vergauwen, B., De Vos, D. and van Beeumen, J.J. (2006) Characterization of the bifunctional gamma-glutamate-cysteine ligase/glutathione synthetase (GshF) of *Pasteurella multocida*. *The Journal of Biological Chemistry*, **281**, 4380–94.

47 Joyce, M.A., Fraser, M.E., James, M.N., Bridger, W.A. and Wolodko, W.T. (2000) ADP-binding site of *Escherichia coli* succinyl-CoA synthetase revealed by X-ray crystallography. *Biochemistry*, **39**, 17–25.

48 Thoden, J.B., Holden, H.M., Wesenberg, G., Raushel, F.M. and Rayment, I. (1997) Structure of carbamoyl phosphate synthetase: a journey of 96 Å from substrate to product. *Biochemistry*, **36**, 6305–16.

49 Sun, X., Bognar, A.L., Baker, E.N. and Smith, C.A. (1998) Structural homologies with ATP and folate-binding enzymes in the crystal structure of folylpolyglutamate synthetase. *Proceedings of the National Academy of Sciences of the United States of America*, **95**, 6647–52.

50 Bouhss, A., Mengin-Lecreulx, D., Blanot, D., van Heijenoort, J. and Parquet, C. (1997) Invariant amino acids in the Mur peptide synthetases of bacterial peptidoglycan synthesis and their modification by site-directed mutagenesis in the UDP-MurNAc:L-alanine ligase from *Escherichia coli*. *Biochemistry*, **36**, 11556–63.

51 Eveland, S.S., Pompliano, D.L. and Anderson, M.S. (1997) Conditionally lethal *Escherichia coli* murein mutants contain point defects that map to regions conserved among murein and folyl poly-(-glutamate ligases: identification of a ligase superfamily. *Biochemistry*, **36**, 6223–9.

52 Dementin, S., Bouhss, A., Auger, G., Parquet, C., Mengin-Lecreulx, D.,

Dideberg, O., van Heijenoort, J. and Blanot, D. (2001) Evidence of a functional requirement for a carbamoylated lysine residue in MurD, MurE and MurF synthetases as established by chemical rescue experiments. *European Journal of Biochemistry*, **268**, 5800–7.

53 Bouhss, A., Dementin, S., Parquet, C., Mengin-Lecreulx, D., Bertrand, J.A., Le Beller, D., Dideberg, O., van Heijenoort, J. and Blanot, D. (1999) Role of the ortholog and paralog amino acid invariants in the active site of the UDP-MurNAc-L-alanine:D-glutamate ligase (MurD). *Biochemistry*, **38**, 12240–7.

54 Gegnas, L.D., Waddell, S.T., Chabin, R.M., Reddy, S. and Wong, K.K. (1998) Inhibitors of the bacterial cell wall biosynthesis enzyme MurD. *Bioorganic and Medicinal Chemistry Letters*, **8**, 1643–8.

55 Tanner, M.E., Vaganay, S., van Heijenoort, J. and Blanot, D. (1996) Phosphinate inhibitors of the D-glutamic acid-adding enzyme of peptidoglycan biosynthesis. *The Journal of Organic Chemistry*, **61**, 1756–60.

56 Saraste, M., Sibbald, P.R. and Wittinghofer, A. (1990) The P-Loop – a common motif in ATP- and GTP-binding proteins. *Trends in Biochemical Sciences*, **15**, 430–4.

57 Bertrand, J.A., Auger, G., Martin, L., Fanchon, E., Blanot, D., van Heijenoort, J. and Dideberg, O. (1997) Crystal structure of UDP-N-acetylmuramoyl-L-alanine:D-glutamate ligase from *Escherichia coli*. *The EMBO Journal*, **16**, 3416–25.

58 Schulz, G.E. (1992) Binding of nucleotides by proteins. *Current Opinion in Structural Biology*, **2**, 61–7.

59 Rossmann, M.G., Liljas, A., Branden, C.I. and Banaszak, L.J. (1975) Evolutionary and structural relationships among dehydrogenases, in *The Enzymes* (ed. P.D. Boyer), Academic Press, New York, pp. 61–102.

60 Bertrand, J.A., Auger, G., Fanchon, E., Martin, L., Blanot, D., Le Beller, D., van Heijenoort, J. and Dideberg, O. (1999) Determination of the MurD mechanism through crystallographic analysis of enzyme complexes. *Journal of Molecular Biology*, **289**, 579–90.

61 Bertrand, J.A., Fanchon, E., Martin, L., Chantalat, L., Auger, G., Blanot, D., van Heijenoort, J. and Dideberg, O. (2000) "Open" structures of MurD: domain movements and structural similarities with folylpolyglutamate synthetase. *Journal of Molecular Biology*, **301**, 1257–66.

62 Yan, Y., Munshi, S., Li, Y., Pryor, K.A.D., Marsilio, F. and Leiting, B. (1999) Crystallization and preliminary X-ray analysis of the *Escherichia coli* UDP-MurNAc-tripeptide D-alanyl-D-alanine-adding enzyme (MurF). *Acta Crystallographica. Section D, Biological Crystallography*, **55**, 2033–4.

63 Yan, Y., Munshi, S., Leiting, B., Anderson, M.S., Chrzas, J. and Chen, Z. (2000) Crystal structure of *Escherichia coli* UDP-MurNAc-tripeptide D-alanyl-D-alanine-adding enzyme (MurF) at 2.3 Å resolution. *Journal of Molecular Biology*, **304**, 435–45.

64 Gordon, E., Flouret, B., Chantalat, L., van Heijenoort, J., Mengin-Lecreulx, D., and Dideberg, O. (2001) Crystal structure of UDP-N-acetylmuramoyl-L-alanyl-D-glutamate: meso-diaminopimelate ligase from *Escherichia coli*. *The Journal of Biological Chemistry*, **276**, 10999–1006.

65 Mol, C.D., Brooun, A., Dougan, D.R., Hilgers, M.T., Tari, L.W., Wijnands, R.A., Knuth, M.W., McRee, D.E. and Swanson, R.V. (2003) Crystal structures of active fully assembled substrate- and product-bound complexes of UDP-N-acetyl-muramic acid:L-alanine-ligase (MurC) from *Haemophilus influenzae*. *Journal of Bacteriology*, **185**, 4152–62.

66 Elbahloul, Y., Krehenbrink, M., Reichelt, R. and Steinbüchel, A. (2005) Physiological conditions conducive to high cyanophycin content in biomass of *Acinetobacter calcoaceticus* strain ADP1. *Applied and Environmental Microbiology*, **71**, 858–66.

67 Elbahloul, Y. and Steinbüchel, A. (2006) Engineering the genotype of *Acinetobacter* sp. strain ADP1 to enhance biosynthesis of cyanophycin. *Applied and Environmental Microbiology*, **72**, 1410–19.

68 Voss, I. and Steinbüchel, A. (2006) Application of a KDPG-aldolase gene-dependent addiction system for enhanced production of cyanophycin in *Ralstonia eutropha* strain H16. *Metabolic Engineering*, **8**, 66–78.

69 Schwamborn, M. (1998) Chemical synthesis of polyaspartates: a biodegradable alternative to currently used polycarboxylate homo- and copolymers. *Polymer Degradation and Stability*, **59**, 39–45.

70 Joentgen, W., Groth, T., Steinbüchel, A., Hai, T. and Oppermann, F.B. (1998) Polyasparaginic acid homopolymers and copolymers, biotechnical production and use thereof. International Patent Application WO 98/39090.

71 Conrad, U. (2005) Polymers from plants to develop biodegradable plastics. *Trends in Plant Science*, **10**, 511–12.

72 Scott, E., Peter, F. and Sanders, J. (2007) Biomass in the manufacture of industrial products–the use of proteins and amino acids. *Applied Microbiology and Biotechnology*, **75**, 751–62.

73 Sanders, J., Scott, E., Weusthuis, R. and Mooibroek, H. (2007) Bio-refinery as the bio-inspired process to bulk chemicals. *Macromolecular Bioscience*, **7**, 105–17.

74 Richter, R., Hejazi, M., Kraft, R., Ziegler, K. and Lockau, W. (1999) Cyanophycinase, a peptidase degrading the cyanobacterial reserve material multi-L-arginyl-poly-L-aspartic acid (cyanophycin): molecular cloning of the gene of *Synechocystis* sp. PCC 6803, expression in *Escherichia coli*, and biochemical characterization of the purified enzyme. *European Journal of Biochemistry*, **263**, 163–9.

75 Obst, M., Oppermann-Sanio, F.B., Luftmann, H. and Steinbüchel, A. (2002) Isolation of cyanophycin-degrading bacteria, cloning and characterization of an extracellular cyanophycinase gene (*cphE*) from *Pseudomonas anguilliseptica* strain BI. *The Journal of Biological Chemistry*, **277**, 25096–105.

76 Obst, M., Sallam, A., Luftmann, H. and Steinbüchel, A. (2004) Isolation and characterization of Gram-positive cyanophycin-degrading bacteria–kinetic studies on cyanophycin depolymerase activity in aerobic bacteria. *Biomacromolecules*, **5**, 153–61.

77 Obst, M., Krug, A., Luftmann, H. and Steinbüchel, A. (2005) Degradation of cyanophycin by *Sedimentibacter hongkongensis* strain KI and *Citrobacter amalonaticus* strain G isolated from an anaerobic bacterium consortium. *Applied and Environmental Microbiology*, **71**, 3642–52.

34
Biosynthetic Pathway Engineering Strategies

Claudia Schmidt-Dannert and Alexander Pisarchik

34.1
Introduction

Biosynthetic pathways are composed of a series of enzyme-catalyzed chemical reactions that lead to the synthesis of one or several small molecules. In turn, a network of hundreds of biological reaction sequences lead to the synthesis of primary and secondary metabolites of cells. The discovery of microbial fermentation and the subsequent development of modern industrial biotechnologies has enabled these metabolic machineries to be exploited in the large-scale production of small molecules for chemical, agricultural and pharmaceutical applications. Initially, microbial biotechnologies relied on the metabolic repertoire identified in native microorganisms, but subsequently screening and strain selection was used to improve product yields and to produce new compounds. However, with the development of molecular methods it became feasible to engineer desired metabolic traits, and the term "metabolic engineering" was coined during the 1980s [1]. Advances in genetic engineering and increasing knowledge of these metabolic processes enabled not only the manipulation of individual metabolic steps for increased production yields, but also the design of new biosynthetic reaction sequences for the creation of new compounds [2].

Genes from different sources can be combined into new multienzyme pathways, and the catalytic functions of individual enzymes manipulated to optimize pathway function and/or diversify the metabolic output of an assembled pathway [3–6]. Because structural information is only available for very few well-characterized metabolic enzymes, which limits the use of rational protein design for pathway engineering, directed or *in vitro* evolution methods (see Chapters 16, 17, 30) are increasingly used to alter the activities of metabolic enzymes. Efforts in genomics and metagenomics continue to provide increasing resources for the identification of new metabolic activities that enable the design of new biosynthetic reaction sequences. Genome data, together with other high-throughput "omics" technologies (e.g. metabolomics, proteomics, DNA microarray analysis) provide a better understanding on how metabolic networks operate and how these

Protein Engineering Handbook. Edited by Stefan Lutz and Uwe T. Bornscheuer

ISBN: 978-3-527-31850-6

networks can be manipulated for the integration of heterologous biosynthetic pathways [7].

In this chapter we will discuss current strategies in the design, optimization and diversification of multistep biosynthetic pathways. In general, we will use the biosynthesis of different isoprenoid-derived compounds in metabolically engineered microbial hosts as examples with which to illustrate the current design principles of complex biosynthetic pathways, and of frequently encountered problems in heterologous pathway engineering.

The biosynthesis of isoprenoid (or terpenoid) compounds, along with polyketides (PK)s and non-ribosomal peptides (NRP)s (see Chapters 32 and 36), currently represent the most explored biosynthetic pathways in terms of engineering strategies applied to pathway optimization and the production of novel and unnatural compounds. Isoprenoids form a remarkable diverse group of compounds, with more than 50 000 structures derived from five-carbon building blocks, isopentenyl diphosphate (IPP) (also known as pyrophosphate, PP) and dimethylallyl diphosphate (DMAPP) that are step-wise condensed by prenyl-transferases [8–10]. Variations in chain length, cyclization and modification of the isoprene chain are responsible for the enormous structural diversity of this class of natural products (Figure 34.1). Many essential oils are isoprenoids, including myrcene ($C_{10}H_{16}$) from bay leaves, limonene ($C_{10}H_{15}$) from lemon oil and zingiberene ($C_{15}H_{24}$) from ginger. In addition to their role as fragrant compounds, the terpenes, composed of two (C10, monoterpenes), three (C15, sesquiterpenes) or four (C20, diterpenes) isoprene units, often have antimicrobial and pharmaceutical properties. Other important isoprenoid classes are those composed of six (C30, triterpenes) and eight (C40, tetraterpenes) isoprene units. Cholesterol and ergosterol are examples of biologically important triterpenoids, while most carotenoids are tetraterpenes. Carotenoids are orange- to purple-colored pigments that, for humans, have important biological functions in signaling and protection against reactive oxygen species (ROS) [11]. These pigments are also used commercially as food colorants, feed supplements, nutraceuticals, and also for cosmetic and pharmaceutical purposes. Isoprene side chains are found in many metabolites such as coenzymes Q6–Q10, vitamin E and chlorophylls. Because of the complexity of these molecules, their chemical synthesis is often impractical and expensive; consequently, numerous isoprenoid pathways have been engineered into microbial hosts for the production and generation of new isoprenoid compounds [12, 13].

34.2
Initial Pathway Design

34.2.1
Functional Pathway Assembly

The engineering of biosynthetic pathways into heterologous hosts becomes possible only when most of the enzymatic steps of the pathway are known and, equally

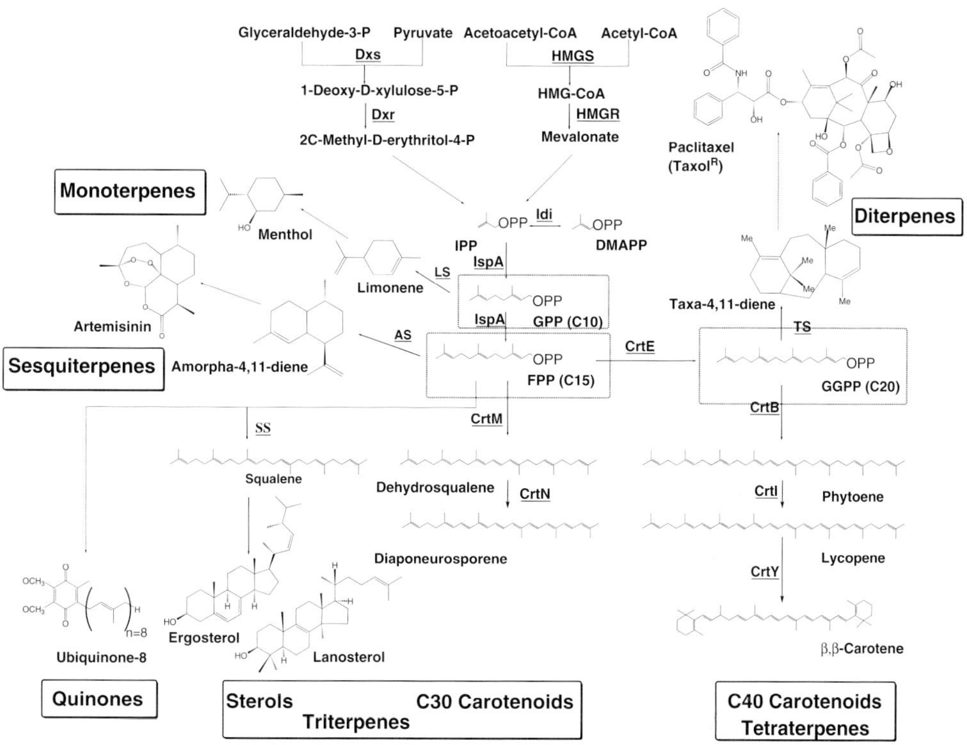

Figure 34.1 Biosynthetic pathways leading to select isoprenoid-derived compounds. Major biosynthetic routes and biosynthetic enzymes discussed in this chapter are shown. Key isoprene diphosphate intermediates, from which pathways leading to the different terpenoid classes branch off, are highlighted as the major terpenoid classes. Enzyme abbreviations as follows: Dxs = 1-Deoxy-D-xylulose synthase; Dxr = 1-Deoxy-D-xylulose reductase; HMGS = hydroxy-3-methylglutaryl-CoA synthase; HMGR = hydroxy-3-methylglutaryl-CoA reductase; Idi = Isoptenyl diphosphate isomerase; IspA = FPP synthase; LS = limonene synthase; AS = amorphadiene synthase; CrtE = GGPP synthase; TS = taxadiene synthase; CrtM = dehydrosqualene/diapophytoene synthase; CrtN = diapophytene desaturase; CrtB = phytoene synthase; CrtI = phytoene desaturase; CrtY = lycopene cyclase; SS = squalene synthase.

important, have been shown to be functional in a heterologous system. In order for an engineered pathway to be functional, the host must have the capability to synthesize necessary precursor molecules and enzyme cofactors. In addition, when assembled into a heterologous pathway the enzymes need to interact functionally, which is not always the case when the corresponding genes were cloned from very different sources. If these basic conditions are met, then a core pathway can be assembled that will take precursors supplied by the host organisms to synthesize major metabolites for future engineering experiments. New enzymes can be added to this pathway to generate novel products which can be analyzed and used to extend the pathway even further. Genes are added to the pathway in a stepwise manner to ensure that every individual enzyme is expressed and active. Additional new products can be created by changing the catalytic activities of assembled enzymes, using rational and evolutionary design methods (Figure 34.2).

Experimentally uncharacterized gene functions from, for example, genome or metagenome sequences can be added to a recombinant pathway. In this case the analysis of metabolites provides valuable information about the activity and product profile of the enzyme of interest, and pathway assembly becomes not only a means of producing compounds or creating novel molecules but also an important tool for studying unknown or putative enzymes. Following the initial pathway assembly, it is typically necessary to improve product yields by optimizing the precursor supply and flux through the assembled pathway. This can be achieved by balancing gene expression levels, improving the activities of rate-limiting enzymes, and by eliminating any undesired side reactions and competing pathways.

The manipulation of isoprenoid pathways has been so successful because their biosynthesis is reasonably well understood, and the majority of isoprenoid biosynthetic genes can be functionally expressed in a variety of heterologous systems [14]. Moreover, isoprenoid genes isolated from very different sources (e.g. plants, bacteria, fungi) can be assembled into functional pathways, which allows for the mixing and matching of enzyme function to increase production levels and the diversity of products. Carotenoid (crt) biosynthesis has proved to be

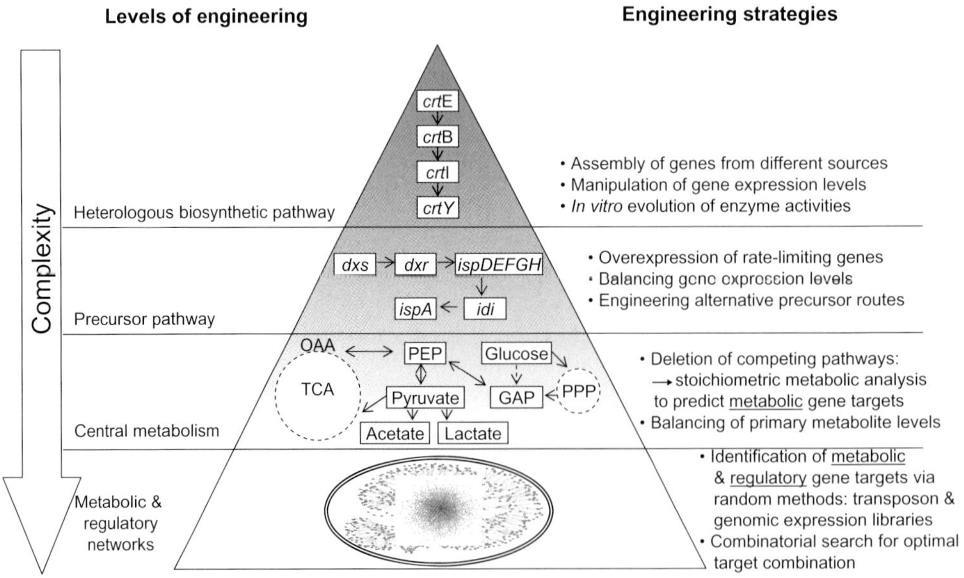

Figure 34.2 Summary of pathway engineering strategies organized into levels of increasing complexity. Metabolic engineering of isoprenoid biosynthesis is used to exemplify different pathway engineering strategies.

especially attractive for pathway engineering because the pathways can be easily assembled with *crt* genes from different sources, while pigment production in noncarotenogenic hosts can be semi-qualitatively followed by the visible inspection of cell color [15]. The latter fact has made these pigment pathways a convenient system for the development of new metabolic engineering strategies (see Section 34.6) [13].

The recombinant expression of carotenoid pathways has initially been used to elucidate the functions of carotenoid enzymes [16–21], and later explored for the production of diverse carotenoid structures [15, 22, 23]. The results of these studies showed that carotenoid biosynthesis requires a minimum of three genes to produce a colored carotenoid structure from the universally synthesized C15 isoprenoid precursor farnesyl diphosphate (FPP). In order to create the linear C40 carotenoid lycopene in a noncarotenogenic host such as *E. coli*, FPP must first be extended by GGPP synthase (CrtB) to produce the C20 isoprenoid molecule, geranyl geranyl diphosphate. Two molecules of GGPP are then condensed by phytoene synthase (CrtE) to form the linear carotenoid backbone of the colorless carotenoid phytoene. The subsequent introduction of four double bonds into this backbone by a desaturase (CrtI) forms the orange-red chromophore lycopene, the cyclization of which then yields cyclic carotenoids (e.g. β-carotene) that may be further modified to yield various oxygen-containing structures. *Crt* genes from bacteria [24–26], fungi [27–29] and plants [30–33] were expressed in *E. coli* for the *de novo* biosynthesis of lycopene, β-carotene and zeaxanthin. Progress in molecular biology–and more recently in genome sequencing–has led to a rapid increase in the discovery of new *crt* enzymes over the past decade. At present, pathways for more than 100 carotenoids have been described on a molecular level and functionally assembled in a heterologous host (for reviews, see Refs [34–38]). Currently, up to eight *crt* genes have been assembled into various biosynthetic routes for carotenoid synthesis in *E. coli* (see Figure 34.1) [39–41].

The introduction of terpene cyclases (or terpene synthases) [42] isolated from plants or fungi into heterologous hosts has enabled the biosynthesis of a range of medically important terpenoid compounds. For example, the heterologous expression of a sesquiterpene cyclase isolated from *Artemisia annua* leads to the production of amorphadiene, the terpene precursor of the potent antimalarial drug artemisinin, in *E. coli* or yeast from the endogenous C15 isoprenoid precursor FPP [43–45] (see Figure 34.1). In another example, the diterpene taxadiene, as the precursor of the potent chemotherapeutic compound paclitaxel (Taxol®) was produced in engineered microbial cells by the overexpression of a diterpene cyclase isolated from the pacific yew tree (*Taxus brevifolia*) [46, 47] (Figure 34.1). This enzyme catalyzes cyclization of the C20 isoprenoid precursor GGPP into a tricyclic terpene structure. In *E. coli* and yeast, the formation of GGPP requires an extension of their endogenous FPP pathways with a heterologous enzyme (GGPP synthase) that adds one C5 isoprene unit to FPP. More recently, a series of labdane diterpenoids was synthesized in *E. coli* by expressing plant-derived diterpene cyclases in this host [48]. These compounds provide the scaffolds for further modification to numerous biologically active plant terpene natural products.

34.2.2 Selection of the Heterologous Host

Complex pathways are frequently first assembled into recombinant hosts such as *E. coli* or *Saccharomyces cerevisiae*, for which a range of genetic tools and methods are available. Unfortunately, production yields obtained in these strains are often low, but may be improved by the use of metabolic strain engineering. Occasionally, the selection of a different production host may be necessary, as illustrated by current efforts to find alternative hosts for recombinant carotenoid production. Carotenoid production levels in *E. coli* are relative low (e.g. 1–2 $mg\,g^{-1}$ dry cell weight; DCW) [49] compared to the up to several hundred-fold higher levels produced by carotenogenic algae and microbial strains [50]. The production levels of these hydrophobic, membrane-bound compounds are influenced not only by the available precursor pools and pathway flux (which can be optimized by additional metabolic engineering efforts) but also by the availability of sufficient membrane storage capabilities (available membrane storage may be most easily increased by selecting a different host).

Yeasts are capable of accumulating large quantities of the isoprenoid-derived ergosterol in their membranes, and the subsequent redirection of their efficient isoprenoid metabolism to carotenoid biosynthesis should facilitate the accumulation of hydrophobic carotenoids in their membranes. Many yeast strains are also generally recognized as safe (GRAS), and can be used for pharmaceutical, nutritional and feed applications. To date, the metabolic engineering of different yeast strains for carotenoid production has not been as widely explored as *E. coli*, although many reported examples show great promise for carotenoid production [51–56].

Alternatively, photosynthetic bacteria that naturally synthesize significant amounts of carotenoids in their intracellular membranes have been investigated for recombinant carotenoid production [20, 57–59]. Other nonphotosynthetic carotenoid-producing bacteria have also been engineered for the production of astaxanthin and the aryl carotenoid chlorobactene [60, 61]. So far, the yields have not been significantly improved over those obtained with recombinant *E. coli*, but most carotenoid pathway engineering attempts in these organisms have not proceeded much beyond the simple overexpression *crt* genes. It is possible that additional engineering of, for example, precursor pools may lead to significant increases in product yields.

The heterologous production of the steroid hydrocortisone represents another example of the construction of a complex isoprenoid-derived pathway [62], and also the importance of selecting an appropriate production host. A total of 13 genes from yeast, mammalian, mitochondrial and plant pathways was assembled into a functional pathway in the yeast *S. cerevisiae*. This organism provided the appropriate conditions for functional expression of the mostly membrane-bound mammalian enzymes and also for an accumulation of the hydrophobic steroid in its membrane.

Many of the downstream modifying enzymes involved in terpene biosynthesis are of the cytochrome P450 family, and catalyze the oxygenation steps that are important for the synthesis of biologically active terpenoid molecules. These enzymes are often difficult to express functionally as they may be membrane-associated and require electron-transfer proteins in order to function. An eukaryotic host such as *S. cerevisiae* is often necessary for the functional expression of mammalian and plant P450 enzymes. Recently, several P450 enzymes from the taxol biosynthetic pathway were functionally expressed in a heterologous system [47] (see Figure 34.1). The random sequencing of cDNAs from suppression subtractive hybridization libraries generated from induced (taxol-producing) and uninduced *Taxus* cell cultures resulted in the identification of 19 enzymes that were thought to catalyze the complete synthesis, and eight of these were expected to be cytochrome P450-oxygenases [63]. Five of the enzymes were coexpressed in yeast, resulting mainly in the taxadiene product of the terpene cyclase, whereas the subsequent P450 reaction generated only minimal amounts of the next intermediate [47]. Because yeasts are able to express this P450 enzyme as a soluble and active protein, the hope remains that coexpressing the proper *Taxus* NADPH: cytochrome P450 reductase with the P450 monooxygenase will improve the activity of this reaction.

Although the above example illustrates the difficulties of P450 expression, artemisinin production by both *E. coli* and yeast shows that it is possible to obtain active P450 enzymes in heterologous expression hosts. Artemesinin, a sesquiterpene from *Artemisia annua*, is used to combat malarial infection and currently is extracted from the native plant. Pathway engineering efforts have developed both yeast and *E. coli* strains for the production of amorphodiene, the cyclized terpene scaffold that is further modified by subsequent oxidation and reduction reactions to artemisinin [43–45] (see Figure 34.1). The production levels of amorphodiene by recombinant *E. coli* have been improved by the engineering of an alternative isoprene precursor pathway (the mevalonate pathway; see Section 34.3.1) [64] and fermentation strategies to collect the volatile product [65]. A breakthrough in the biosynthesis of amorphadiene to artemisinin occurred with the cloning and functional expression of the P450 enzyme that catalyzes the three-step oxidation of amorphodiene to artemisinic acid [44]. Consequently, the coexpression of amorphadiene cyclase and oxidase produced up to $100\,mg\,l^{-1}$ terpenoid in yeast.

34.3 Optimization of the Precursor Supply

After having established the functional expression of a heterologous biosynthetic pathway in a host, the next step towards improving production levels typically involves the maximization of precursor supply for the engineered recombinant pathway. The metabolic engineering of recombinant terpenoid biosynthesis shows different strategies that can be used for increasing the precursor pools (see Figure 34.2). In some cases it may be sufficient to overexpress individual enzymes of a precursor pathway and to identify one or several enzymatic steps that limit flux through this pathway. However, in another strategy it may be necessary to complement the native precursor pathway in a host with an alternate heterologous pathway that uses different primary metabolites and is no longer under the direct regulation of the host's metabolism. The engineering of new control loops that regulate flux through a precursor pathway is yet another approach used to increase production levels.

34.3.1 Identification and Overexpression of Rate-Limiting Enzymes

Depending on the organism involved, the synthesis of the first isoprene unit, IPP, occurs either via the mevalonate or the DXP (1-deoxy-D-xylulose-5-phosphate or nonmevalonate) pathway (for a review, see Ref. [66]) (see Figure 34.1). The mevalonate pathway is present in archeae, fungi, animals, some bacteria and plants (except plastids), and utilizes acetyl-CoA and mevalonate as the starting precursors for IPP synthesis. Most eubacteria and plant plastids, however, synthesize IPP from the glycolytic intermediates pyruvate and glyceraldehyde-3-phosphate (GAP) via the DXP pathway [67]. Individual enzymes in both pathways have been targeted for overexpression to increase isoprenoid flux in both *E. coli* (DXP pathway) and *S. cerevisiae* (mevalonate pathway).

As *E. coli* produces only small quantities of isoprenoid-derived metabolites such as dolichols and quinones, flux through its isoprenoid precursor pathway (DXP pathway) is small, which limits the production of terpenoid compounds in this organism. Consequently, among the first approaches to increase isoprenoid yields in *E. coli* was an overexpression of the first DXP pathway enzyme, 1-deoxy-D-xylulose 5-phosphate synthase (DXP-synthase, DXS). This enzyme condenses two precursors from the central metabolism, D-GAP and pyruvate, to form DXP. The overexpression of heterologous *dxs* genes in *E. coli* increased lycopene levels two- to threefold [68–70], and also benefited the recombinant production of taxadiene, the key intermediate of Taxol biosynthesis [46].

Isomerization of the DXP pathway product IPP by isopentenyl diphosphate isomerase (Idi) to the reactive DMAPP required for isoprene chain elongation has been identified as another key rate-limiting step for isoprenoid biosynthesis in *E. coli*. This enzyme was found not to be essential for *E. coli* growth [71] because spontaneous IPP isomerization appeared to be sufficient for the synthesis of the low levels of isoprenoid metabolites required in *E. coli*. However, the overexpression of Idi in engineered *E. coli* had a major effect on carotenoid (β-carotene) production levels [72], illustrating how recombinant pathways pose new demands on a host's metabolism.

The DXP pathway studies also showed that rate-limiting enzymes identified in precursor pathways may act synergistically on production levels when coexpressed. For example, the coexpression of DXS with the next enzyme in the pathway,

1-deoxy-D-xylose 5-phosphate reducto-isomerase (DXR) had an additive effect on lycopene levels in *E. coli* [73]. Similarly, the coexpression of all three as rate-limiting DXP enzymes (DXR, DXS and Idi) increased production levels of zeaxanthin [74] and astaxanthin [75] in *E. coli*.

Isoprene biosynthesis in the host *S. cerevisiae* utilizes the mevalonate pathway; this begins with the condensation of acetyl-CoA and acetoacetyl-CoA to 3-hydroxy-3-methylglutaryl-CoA (HMG-CoA), catalyzed by HMG-CoA synthase. Acetoacetyl-CoA is synthesized from two molecules of acetyl-CoA by the enzyme acetoacetyl-CoA thiolase. Instead of overexpressing the mevalonate pathway enzymes, studies aimed at increasing isoprenoid levels in *S. cerevisiae* increased carbon flux into the mevalonate pathway by engineering the pyruvate dehydrogenase bypass pathway. This pathway converts pyruvate into the isoprenoid precursor acetyl-CoA by the action of pyruvate decarboxylase, acetaldehyde dehydrogenase and acetyl-CoA synthetase [76]. Recently, the overexpression of acetaldehyde dehydrogenase and the introduction of a *Salmonella enterica* acetyl-CoA synthetase in *S. cerevisiae* increased production yields of the sesquiterpenoid amorphadiene, the precursor to the anti-malarial drug artemisinin [64] (see Figure 34.1).

The overexpression of a precursor pathway enzyme not only increases the activity of an enzymatic step, but the use of a heterologous promoter allows it to be removed from transcriptional regulation by the host. This is particularly important in the case of the mevalonate pathway, where several key enzymatic steps are regulated by intracellular steroid levels via a complex mechanism. In *S. cerevisiae*, sterol levels regulate the expression of acetoacetyl-CoA thiolase, HMG-CoA synthase and the next enzyme in the pathway, HMG-reductase, which converts HMG-CoA into mevalonate [77–85]. The reduction of HMG is the major rate-limiting step in isoprenoid and ergosterol biosynthesis in *S. cerevisiae*. The reaction is catalyzed by two HMG-CoA reductase isoenzymes (*HMG1* and *HMG2*) that are anchored via an N-terminal domain to the endoplasmic reticulum (ER) of *S. cerevisiae*. However, while *HMG1* is predominantly controlled at the transcriptional level, *HMG2* is also regulated by protein degradation via FPP and sterol-derived small-molecule signals. The overexpression of a truncated, cytosolic form of HMG-CoA reductase (HMG1) is widely used to increase flux through the mevalonate pathway [86, 87]. Expression of HMG-CoA reductase in yeast cells expressing *crt* genes for lycopene biosynthesis resulted in a sevenfold increase in the yield of this carotenoid [88]. Yeast were also metabolically engineered for an increased production of the terpenoid epicedrol by overexpressing a truncated HMG-CoA reductase [89].

When IPP has been synthesized through the DXP or mevalonate pathways, the first committed step in the synthesis of the various terpenoids in all organisms is chain elongation by successive head-to-tail condensation of DMAPP, the reactive IPP isomer, initially to IPP and subsequently to the growing polyprenyl diphosphate chain. The isomerization of IPP to DMAPP is catalyzed by Idi, while chain elongation is catalyzed by chain-length-specific prenyl transferases (known as prenyl synthases) that produce for example, geranyl diphosphate (GPP, C10), FPP (C15) or GGPP (C20). FPP is the direct precursor for sterols, dolichols, hopanoids, sesqui- and triterpenoids, and the isoprenoid chain of quinones, while GGPP is the precursor for diterpenoids and most carotenoids (except for C30 carotenoids). Because *E. coli* and yeasts do not produce GGPP, carotenoid and diterpenoid biosynthesis in these hosts require a recombinant GGPP synthase that either catalyzes the chain elongation of FPP (CrtE) or IPP to GGPP (Gps, multi-functional synthase). Several studies have shown that FPP and GGPP synthesis are rate-limiting steps in the conversion of isoprene precursor to terpenoid products in engineered hosts. The overexpression of FPP synthase (ispA) and/or GGPP synthase (crtE) alleviated this bottleneck and significantly improved the production rates of sterols in yeast [90] as well as of carotenoids [26, 75] and the diterpenoid taxadiene [46] in *E. coli*.

34.4
Engineering of Control Loops

Instead of increasing carbon flux into or through a precursor pathway by gene overexpression, yet another strategy is to engineer a dynamic control loop that couples carbon flux into a precursor pathway in response to the availability of glycolytic intermediates. Farmer and Liao [91] placed the precursor supply of the DXP pathway in *E. coli* under dynamic control to increase lycopene production levels. Because the DXP pathway precursor pyruvate and GAP are not available at equal concentrations in the cell (GAP is less available compared to pyruvate), flux from pyruvate to GAP was increased by the overexpression of phosphoenolpyruvate synthase (*pps*) or phosphoenolpyruvate carboxykinase (*pck*), or by the inactivation of pyruvate kinase (*pyk*). This first engineering step improved lycopene production in *E. coli* up to fivefold. In a subsequent step, *idi* (IPP isomerase, see above) and *pps* expression were placed under the dynamic control of an engineered intracellular control loop (Ntr regulon) that senses excess glycolytic flux and in turn upregulates *idi* and *pps* expression. The dynamic control of *idi* and *pps* expression led to a significant further increase of lycopene production due to an effective utilization of excess pyruvate [91].

34.5
Engineering of Alternative Precursor Routes

Precursor pathways are part of a tightly controlled metabolic network, and the overexpression of rate-limiting steps may not be sufficient to improve flux into or through a precursor pathway for increased product yields of an engineered biosynthetic pathway. Problems with optimizing a tightly controlled endogenous pathway, however, may be overcome by the engineering of a heterologous precursor pathway. Installation of the mevalonate pathway as an alternative isoprenoid precursor route in *E. coli* resulted in a significantly increased terpenoid production in *E. coli*.

Previous efforts aimed at manipulating the gene expression levels of *E. coli*'s endogenous DXP pathway generally afforded only moderate improvements of isoprenoid production levels. The DXP pathway appears to be strictly regulated, and perturbation of expression levels of several pathway enzymes (especially when combined with each other) proved to be detrimental for *E. coli* [49]. Keasling's group [45] reconstructed the complete mevalonate pathway from *S. cerevisiae* in *E. coli* to increase production of the sesquiterpenoid amorpha-4,11-diene. The supply of isoprenoid precursor through the alternative pathway improved the production of this antimalarial drug precursor 11-fold. However, expression of the mevalonate pathway alone in *E. coli* led to the accumulation of IPP or FPP, which proved to be toxic for the cells and needed to be removed by conversion to downstream metabolites (e.g. sesquiterpenoid amorpha-4,11-diene). More recently, this group has further optimized the mevalonate pathway in *E. coli* by optimizing rate-limiting steps previously identified in yeast. As in the native yeast pathway, HMG-CoA reductase activity was found to be rate-limiting and to cause a build-up of excessive amounts of HMG-CoA in *E. coli*. However, this problem was solved by increasing the expression of the enzyme [92], which in turn increased mevalonate production.

The construction of alternative supply routes for isoprenoid precursor in *E. coli* was also used to increase carotenoid (lycopene) production in this host [93, 94]. Yoon *et al.* [93] expressed the lower part of the mevalonate pathway from *Streptococcus pneumonia* (note that some Gram-positive cocci utilize the mevalonate pathway) in *E. coli* strains overexpressing *idi*, which enabled the production of lycopene from exogenously fed mevalonate. Production levels obtained in this study may well be close to the maximum for lycopene in *E. coli*, as cell clump formation was observed which suggested that the accumulation of high levels of lycopene in the cell membrane was causing a significant level of stress to the cells. The clump formation was subsequently reduced by adding surfactants to the medium.

34.6 Balancing Gene Expression Levels and Activities of Metabolic Enzymes

An engineered pathway poses a considerable metabolic burden on the host cell, as precursor molecules are diverted from their native metabolic pathways and energy resources are tapped into for reactions such as protein synthesis and plasmid replication. A minimization of metabolic burden and optimization of flux through an engineered pathway requires the fine-tuning of all involved enzyme activities (Figure 34.2). Metabolic flux analysis can be used to determine metabolite concentrations and identify rate-limiting steps in a pathway that requires optimization. The optimization of metabolic activities typically occurs at the protein expression level, although it sometimes also involves manipulation at the protein level to improve catalytic activities. Because it is difficult to predict mutations in coding or regulatory regions that would yield optimal expression or enzyme activity levels, approaches that involve the generation and screening of gene or regulatory variant libraries are typically used for the optimization of pathway activities.

The expression levels of a metabolic enzyme can be optimized by varying the copy number of the plasmid, regulation and strength of the promoter, and the efficiency of translation of the mRNA product into protein. For example, the effects on plasmid copy number on heterologous protein expression were explored by generating a variant library of a regulatory protein that controls plasmid replication [95]. The quantification of product formation and real-time PCR confirmed that the plasmid copy number and product formation were positively correlated. In a different example, a continuous variation of gene expression levels was obtained by creating libraries of expression cassettes with variations in the sequences for promoter, ribosome-binding site and mRNA-stabilizing regions [96].

Different strategies have been explored to balance the carotenoid and isoprenoid gene expression levels in order to relieve growth inhibition caused by overexpression of isoprenoid enzymes, and to increase flux through the carotenoid pathway [70, 73]. For example, Kim *et al.* [73] optimized expression levels of the first enzyme of the DXP isoprenoid precursor pathway, DXP synthase, in *E. coli* to optimize carotenoid production levels. Variation of the plasmid copy number and promoter strength showed, that optimal lycopene production was obtained with medium-copy number plasmids and weaker promoters. Further studies conducted by Jones *et al.* [70] with a variety of plasmids and promoters showed similar results.

In another study, a series of promoters with different strengths [97] was generated using error-prone PCR and green fluorescent protein (GFP) fluorescence to isolate variants with different promoter strengths. The series of promoters was then used to drive the expression of DXP synthase and the lycopene levels were measured. Lycopene production peaked at an optimal expression level for *dxs*, while further increases in the promoter strength decreased the lycopene production; this suggested that intermediate products had accumulated and presumably had toxic effects. However, when additional downstream enzymes of the DXP pathway were overexpressed, lycopene production followed the *dxs* promoter strength, which suggested that *dxs* was rate-limiting.

Instead of limiting mutations to control the elements of transcription (promoter), random mutation can also be introduced into the coding region of biosynthetic genes in order to optimize enzyme activities in an assembled pathway. Such a directed evolution strategy was used to optimize GGPP synthase (Gps) activity in astaxanthin-producing *E. coli* cells. Because the multifunctional *gps* has been cloned from the hyperthermophile *Archaeoglobus fulgidus*, expression and activity were suboptimal in *E. coli*. A variant was obtained that increased lycopene production about twofold, allowing the production of 45 $mg\,g^{-1}$ DCW of lycopene when *dxs* was coexpressed [98].

The modulation of enzyme expression levels has also been used as a strategy to achieve the accumulation of different ratios of carotenoid intermediates by controlling metabolic flux through the carotenoid pathway. Engineering the mRNA stability of two genes encoding phytoene desaturase (CrtI) and lycopene cyclase (CrtY) (see Figure 34.1) varied the production levels of β-carotene relative to lycopene

300-fold, and intermediates such as phytoene accumulated that were not observed in *E. coli* containing the recombinant *Erwinia* carotenoid operon [99]. The importance of intracellular biosynthetic enzyme activities for intermediate accumulation and terminal product formation is also reflected by the different synthetic routes observed during *in vitro* and *in vivo* astaxanthin formation leading to the accumulation of different intermediates [100, 101].

Another common problem of pathway optimization is poor protein expression because of suboptimal codon usage of a heterologous gene. Poor expression of plant enzymes was identified as a major limiting factor for sesquiterpene production in *E. coli* [102]. However, the problem was overcome by expressing a codon-optimized synthetic terpene cyclase gene in *E. coli*, which significantly improved amorphadiene production levels in this host [45].

34.7 Metabolic Network Integration and Optimization

In order for an engineered, heterologous pathway to operate optimally, it is not only necessary to link the pathway to its immediate precursor routes in a host, but also to ensure its full integration into the host's metabolic network (see Figure 34.2). Metabolic network integration may require the simultaneous optimization and rewiring of several metabolic pathways, including an optimization of the host's regulatory network. One strategy which is commonly used to identify optimization targets is metabolic modeling. Unfortunately, current metabolic models do not capture complex regulatory interactions between metabolic enzymes, genes and metabolites that result in nonlinear kinetics; rather, they are impacted by cultivation conditions and slight changes in genetic background. As a consequence, random approaches have been used to identify targets that cannot be deduced from metabolic models and biochemical knowledge on the function of a biosynthetic pathway. In addition, random strategies allow the identification of multiple optimization targets that can be explored in a combinatorial manner to enhance the performance of an engineered pathway even further.

The incorporation of multiple changes into genomes by whole genome shuffling has been used successfully to significantly improve microbial cells as hosts for engineered pathways. Whole-genome shuffling has been derived from strategies used in classical strain improvement. However, instead of subjecting a cell population to a mutagen followed by screening for mutants with improved phenotypes – as in classical strain improvement – whole-genome shuffling goes further by incorporating successive rounds of genomic recombination and screening following an initial treatment of a cell population with a chemical mutagen to introduce random mutations. High-throughput screening (HTS) methods are then used to identify strains with improved properties in the initial random library. The genomes of these selected strains are then shuffled by protoplast fusion to create strains that have combinations of the beneficial mutations. In this manner, after only a few rounds of selection the final strains have incorporated many beneficial changes and are significantly fitter than strains developed using classical strain improvement.

Genome shuffling was originally developed to increase the production of a polyketide antibiotic in an industrial *Streptomyces* producer strain [103]. After only two rounds of genome shuffling, strains were obtained with production levels comparable to those of strains which had been developed over two decades and via 20 rounds of classical improvement methods. In a second study, genome shuffling was used to obtain lactate-producing *Lactobacillus* strains capable of growth at low pH [104]. Five rounds of genome shuffling led to the isolation of strains which thrived at pH 4 and produced lactate at elevated levels. Genome shuffling has also been used to improve degradation of the pesticide pentachlorophenol (PCP) by *Sphingobium chlorophenolicum* [105]. Genome shuffling requires an efficient protoplast fusion in order for genomic recombination to take place. The efficiency of the protoplast fusion and recombination is much lower in Gram-negative bacteria such as *E. coli* than in the Gram-positive bacteria (*Streptomyces, Lactobacillus, Sphingobium*) used in genome shuffling experiments. Until recently, the efficiency of protoplast fusion of *E. coli* was not sufficient for strain engineering application. However, Dai *et al.* [106] developed a protoplast fusion method for *E. coli* with an improvement by several orders of magnitude in efficiency, and which now makes it feasible to apply genome shuffling to pathway engineering in this common host.

Both, metabolic model-based and random strategies, either alone or in combination, have been applied to carotenoid overproduction in *E. coli*. A stoichiometric metabolic model of *E. coli* was used to identify knockout targets for increased carotenoid production [107]. Of the several metabolic genes identified, all but one were directly involved in the production of isoprenoid precursors, and the others increased the concentration of NADPH. A series of *E. coli* strains containing a combination of predicted gene deletion were created and the lycopene production analyzed. The best combination of gene deletions increased lycopene production levels up to 40% in an *E. coli* strain that chromosomally overexpressed four genes of the DXP pathway [107, 108].

Random transposon mutagenesis was then used to identify additional gene-knockout targets not predicted by the metabolic model. The resultant *E. coli* library was screened for clones with darker-colored phenotypes as the result of higher lycopene production levels [109]. Interestingly, the three gene-knockout targets identified by this approach were distinct from those predicted by the stoichiometric model. Moreover, two of the three disrupted genes encoded hypothetical proteins without known functions, which underlines how much is still to be learned about the regulation and function of metabolic networks in even a well-studied model organism such as *E. coli*. With these three additional gene-knockouts, the same study group explored 64 different combinations of stoichiometric and random gene-knockouts to identify the maximum lycopene-overproducing strains. A comparison of lycopene production levels by the generated deletion strains showed that metabolic gene-knockouts identified by stoichiometric analysis had an additive effect on lycopene production, while the effects of random gene-knockouts were not additive because they may have involved incompatible regulatory functions. Moreover, combinations of random gene-knockouts and metabolic gene-knockouts impacted on the lycopene levels very differently. Fermentation studies with lycopene-overproducing strains containing either only stoichiometric-predicted gene deletions or a combination of stoichiometric and random gene also exhibited very different levels of production [108].

The random search for gene targets that improve lycopene production in *E. coli* has more recently been expanded to include the overexpression of genes [110, 111]. For example, a genomic library of *E. coli* was expressed in lycopene-producing *E. coli* strains and screened for genomic inserts that increased lycopene production [111]. Targets identified by this strategy included genes previously identified as rate-limiting in the nonmevalonate pathway (*dxs, idi*), regulatory genes (*rpoS, appY*) and several experimentally uncharacterized and putative metabolic and regulatory genes. Combinations of previously identified gene-knockouts and the new overexpression targets resulted in a strain with similar production levels created previously by gene-knockouts alone, indicating that many parallel routes exist for strain improvement. *Rpo*S, encoding the main sigma factor σ^{70}, identified in this screen has subsequently been subjected to directed evolution to identify mutants that would have a global effect on the transcription machinery of *E. coli* [112]. Sigma-factor variants were identified that indeed increased lycopene production levels in previously created *E. coli* knockout strains, stressing again the importance of the regulatory machinery for the improvement of cellular phenotypes.

34.8
Engineering Pathways for the Production of Diverse Compounds

New biosynthetic pathways can be created by combining enzymes (or catalytic domains in the case of the modular PK and NRP synthases) from different sources and pathways into new biosynthetic reaction sequences. This strategy – originally demonstrated for polyketide synthesis [113, 114] – has been termed 'combinatorial biosynthesis' in analogy to combinatorial chemical synthetic methods developed for the creation of molecular diversity for drug discovery. New products can also be created by altering the product and substrate spectrum of individual enzymes in an assembled pathway. Both approaches can be combined to synthesize an even greater diversity of compounds. The catalytic promiscuity of the enzymes in an assembled reaction sequence determines the ease and efficiency with which new substrate and product spectra can be generated and synthesized in sufficient yields [115, 116]. The conversion of precursors into new products may therefore require additional adaptation of the substrate and product spectra of enzymes located next to the key diversity-generating enzyme(s). *In vitro* evolution methods can be applied to increase the conversion of selected substrates from a range of related substrates to a desired new product, or to decrease catalytic promiscuity such that only one particular product is formed.

Many biosynthetic enzymes, and especially those from secondary metabolic pathways, exhibit some flexibility in their substrate and product ranges, as this allows an organism to produce a series of related bioactive compounds which may provide a selective advantage under altered environmental conditions. The promiscuity of tailoring enzymes is particularly well documented for carotenoid pathways. At a very early stage several research groups found that not only was it possible to combine carotenoid genes from very different sources into functional pathways, but that many tailoring carotenoid enzymes appear to recognize the end-group of a carotenoid structure rather than the entire structure [15]. The relaxed substrate specificity of carotenoid enzymes has been extensively explored for the combinatorial biosynthesis of carotenoid via gene combination [34, 36, 38–41].

The spectrum of chemical compounds produced by a pathway is often controlled by the specificity of so-called 'gatekeeping enzymes' that generate the initial substrate scaffold for downstream enzymes in the pathway. Scaffold-modifying enzymes may frequently accept a range of related compounds sent down a biosynthetic pathway by gatekeeping enzymes. Terpene and carotenoid pathways illustrate how these 'gatekeepers' control the product diversity of a pathway and how a relaxation of their specificity or replacement with homologues with different activities is used to engineering novel biosynthetic routes.

The first gatekeeping enzymes for isoprenoid biosynthetic pathways are prenyl transferases. These provide prenyldiphosphates with different chain lengths to pathways that synthesize the different types of terpenoid compounds. Several groups have investigated and manipulated the substrate and product specificity of these enzymes using rational protein engineering and directed evolution methods [117–123]. These studies results showed that the chain-length specificity of these enzymes could be readily manipulated by the introduction of often only one mutation that alters the size of the binding tunnel for the growing prenyl chain. For example, the chemical mutagenesis of a *Bacillus* FPP synthase led to the selection of variants that produced GGPP in *E. coli* [117]. GGPP synthesis was conveniently detected in *E. coli* strains that contained two genes (phytoene synthase (CrtB) and desaturase (CrtI)) necessary for the conversion of GGPP into the pink carotenoid lycopene. A similar *in vivo* screen for GGPP synthesis was used to examine mutant libraries of *E. coli* FPP synthase (IspA) and an archeael FGPP (C25, farnesylgeranyl diphosphate) synthase created by error-prone PCR [122, 123]. For each enzyme several mutations were identified that led either to an increase (FPP synthase) or decrease (FGPP synthase) of the chain length of the prenyl diphosphate product. Prenyltransferase with tailored chain length specificities can therefore be easily created or are already available to supply prenyl diphosphate precursors for the engineered production of compounds with novel isoprenoid backbones.

Recently, Arnold's group demonstrated the biosynthesis of carotenoid compounds with unnatural isoprenoid backbones such as asymmetric C40 and C35, C45 and C50 chains [34, 124–126]. The formation of these unnatural carotenoid backbones from prenyl diphosphate precursors with different chain lengths requires that the first carotenoid pathway enzyme (diapophytoene or phytonene synthase) can catalyze the head-to-head condensation of prenyl diphosphate precursor with unnatural chain lengths. Normally, these enzymes catalyze the condensation of two GGPP or FPP molecules to synthesize the first C40 or C30 carotenoids phytoene or diapophytoene, respectively. However, when the C30 diapophytoene synthase CrtM has access to both C30 (FPP) and C40 (GGPP) isoprenoid precursor molecules, it produces an asymmetric C35 carotenoid [124]. Interestingly, when this pathway was extended with enzymes that catalyze modifications of C40 carotenoid structures, these modifying C40 carotenoid enzymes continued to act on the GGPP-derived half of the molecule, generating ten new compounds [124]. The directed evolution of the C30 diapophytoene synthase CrtM resulted in a large number of variants that synthesized the C40 carotenoid phytoene from two GGPP molecules [126]. Mutant enzymes obtained in this directed evolution experiment also retained some of their original C30 activity. Additional rounds of directed evolution led to CrtM mutants that produced C45 and C50 carotenoids with previously unknown chain lengths when expressed in *E. coli* strains that expressed a mutant FPP synthase from *B. stearothermophilus* capable of producing C25 FGPP prenyl precursors [125]. When these C45 and C50 pathways were extended with a phytoene desaturase gene *crtI* that normally introduces four double bonds into the backbone of the colorless C40 carotenoid phytoene, unnatural desaturated carotenoids were produced in *E. coli* [34].

The above examples show that altering the activities of upstream located pathway enzymes – the gatekeepers of a pathway – can give rise to new compound diversity. Other examples of such upstream-located enzymes that impact the structural diversity generated by a pathway are carotenoid desaturases [127] and terpenoid cyclases [42]. Carotenoid desaturases introduce double bonds into the initial colorless carotenoid backbone until a colored carotenoid is formed; this is then further modified via various tailoring enzymes. Terpenoid cyclases convert prenyl diphosphate chains into cyclic terpene scaffolds that are the substrate of additional tailoring enzymes that produce biologically active molecules.

Rational protein engineering and directed evolution of the terpene cyclases has been used to optimize and diversify terpene biosynthesis in recombinant hosts [43]. A single terpene cyclase often has a broad product spectrum, producing a mixture of terpenes in different amounts from a single linear prenyl substrate. This heterogeneity is the result of a carbocation-initiated cyclization mechanism of these enzymes, where the cyclase controls the migration of a reactive carbocation through the isoprene chain to produce a terminal carbocation that is finally quenched by a base [128]. The 'sloppiness' of a cyclase in controlling carbocation migration and quenching is reflected by its product spectrum. Most terpene cyclases have evolved to guide the cyclization preferentially to a limited number of products; however, only minimal changes are required to alter fundamentally the product spectrum of the reaction and relatively few amino acid residues may control the relative amounts of each terpene [129]. For example, error-prone PCR and directed evolution schemes have altered cadinene synthase to produce larger amounts of the related molecule germacrene [130], while rational engineering of the γ-humulene synthase yielded seven families of terpene sequences each able to produce specifically unique terpene scaffolds [131].

Carotenoid desaturases typically introduce four double bonds into phytoene to produce lycopene. DNA shuffling of two *Erwinia* phytoene desaturases and transformation of the resulting library into phytoene producing *E. coli* cells led to the identification of a phytoene desaturase variant that introduced six double bonds into phytoene to produce the fully conjugated linear carotenoid, tetradehydrolycopene [132]. The biosynthesis of cylic carotenoids requires the cyclization of carotenoid end-groups catalyzed by the enzyme lycopene cyclase. When the evolved tetradehydrolycopene pathway was extended with a library of DNA-shuffled lycopene cyclase mutants, a variant was isolated that produced the monocyclic carotenoid torulene in *E. coli* [132].

The catalytic flexibility of end-group-modifying carotenoid enzymes has made it possible to extend the *in vitro*-evolved torulene and tetradehydrolycopene pathways with tailoring enzymes to synthesize a number of structurally novel carotenoid compounds in *E. coli* [39]. For example, extension of the tetradehydrolycopene pathway with a *Rhodobacter* spheroidene monooxygenase *crt*A generated novel oxygenated carotenoids, including the violet carotenoid phillipsiaxanthin. Extension of the evolved tetradehydrolycopene pathway with a novel carotenoid oxygenase gene (*crt*Ox) isolated from *Staphylococcus aureus* resulted in the biosynthesis of the fully conjugated deep-purple carotenoid tetradehydrolycopene-dial [40]. The coexpression of β-carotene desaturase (*crt*U) from *Streptomyces griseus*, which is known to aromatize β-carotene with the genes encoding the evolved torulene pathway, resulted in an accumulation of aromatic torulene in *E. coli*. Similarly, the *Erwinia uredevora* enzymes β-carotene hydroxylase (CrtZ) and zeaxanthin glucosylase (CrtX) involved in zeaxanthin glucoside biosynthesis were capable of modifying torulene to the corresponding hydroxylated and glucosylated structures. A β-carotene C(4) oxygenase (CrtO) from *Synechocystis* sp. allowed the production of ketolated torulene in *E. coli* [39].

34.9
Future Perspectives

The past two decades have witnessed significant improvements in our ability to assemble and manipulate complex heterologous pathways. In addition, sophisticated analytical and modeling tools have been developed for metabolic flux analysis, and genome-sequencing efforts have provided an enormous resource of new gene functions that can be explored for biosynthetic purposes. The combination of genes from different sources (and even different pathways) into new biosynthetic reaction sequences, as well as the engineering of biosynthetic activities via rational protein design or *in vitro* evolution methods, has enabled the biosynthesis of novel structures in engineered production hosts. However, despite these successes in the creation of complex heterologous pathways, the production levels of compounds are typically too low for industrial-scale production and are not competitive with traditional synthetic methods.

It has become clear that the manipulation of gene expression levels of a heterologous pathway and of precursor pathways alone is, in most cases, not sufficient to maximize production yields. Instead, it is necessary to integrate an engineered pathway into the metabolic network of the host. Data from metabolic modeling, together with information obtained from genome data, proteome, metabolome and transcriptome analysis, have provided increasing insights into the design principles and complexity of cellular metabolic networks [133, 134]. An appreciation for the complexity of cellular metabolic networks consequently led to the development of new approaches that use evolutionary principles and systems-engineering methods to achieve pathway optimization and metabolic integration of recombinant pathways. Of particular importance is the development of dynamic models that describe, on a systems level, the complex interactions among metabolites, proteins and genes. This will enable the engineering of specific regulatory circuits that will allow a recombinant pathway to become part of a host's metabolic network. Such regulatory circuits that self-regulate gene expression and enzyme activity in response to the metabolic state of the host cell will be an important next step in metabolic engineering. Recent advances in synthetic biology hold promise for the development of such new tools and methods for metabolic network design (for reviews, see Refs [135–138]).

Further advances in metabolic engineering may come from the development of alternative heterologous hosts (other than the model hosts *E. coli* and *S. cerevisiae*) and the simplification of host genomes. Fewer metabolic and regulatory systems may decrease the number of unintended interactions and provide efficient metabolic pathways for natural and novel compounds. To this end, a fully synthetic genome with a characterized list of genes may in future be possible for biosynthetic applications. Alternatively, a reduction in the size of the genome to its essential features may provide a route to the ideal microbial host [139–141].

Abbreviations

AS	amorphadiene synthase
Crt	carotenoid
CrtB	phytoene synthase
CrtE	GGPP synthase
CrtI	phytoene desaturase
CrtM	dehydrosqualene/diapophytoene synthase
CrtN	diapophytene desaturase
CrtY	lycopene cyclase
DCW	dry cell weight
DMAPP	dimethylallyl diphosphate
DXP	1-deoxy-D-xylulose-5-phosphate

Dxr	1-Deoxy-D-xylulose reductase
Dxs	1-Deoxy-D-xylulose synthase
FPP	farnesyldiphosphate
GAP	glyceraldehyde-3-phosphate
GGPP	geranyl geranyldiphosphate
GPP	geranyl diphosphate
HMG	hydroxy-3-methylglutarate
HMGR	hydroxy-3-methylglutaryl-CoA reductase
HMGS	hydroxy-3-methylglutaryl-CoA synthase
Idi	isoptenyl diphosphate isomerase
IPP	isopentenyl diphosphate
IspA	FPP synthase
LS	limonene synthase
OAA	oxaloacetate pathway
PCK	phosphoenolpyruvate carboxykinase
PEP	phosphoenol pyruvate phosphatase
PK	pyruvate kinase
PPP	pentose phosphate pathway
PPS	phosphoenolpyruvate synthase
SS	squalene synthase
TCA	tricarboxylic acid cycle
TS	taxadiene synthase

References

1 Bailey, J.E. (1991) Toward a science of metabolic engineering. *Science*, **252**, 1668–75.

2 Nielsen, J. (2001) Metabolic engineering. *Applied Microbiology and Biotechnology*, **55**, 263–83.

3 Hibbert, E.G., Baganz, F., Hailes, H.C., Ward, J.M., Lye, G.J., Woodley, J.M. and Dalby, P.A. (2005) Directed evolution of biocatalytic processes. *Biomolecular Engineering*, **22**, 11–19.

4 Koffas, M. (2005) Evolutionary metabolic engineering. *Metabolic Engineering*, **7**, 1–3.

5 McDaniel, R. and Weiss, R. (2005) Advances in synthetic biology: on the path from prototypes to applications. *Current Opinion in Biotechnology*, **16**, 476–83.

6 Watts, K.T., Mijts, B.N. and Schmidt-Dannert, C. (2005) Current and emerging approaches for natural product biosynthesis in microbial cells. *Advanced Synthesis and Catalysis*, **347**, 927–40.

7 Tyo, K.E., Alper, H.S. and Stephanopoulos, G.N. (2007) Expanding the metabolic engineering toolbox: more options to engineer cells. *Trends in Biotechnology*, **25**, 132–7.

8 Sacchettini, J.C. and Poulter, C.D. (1997) Creating isoprenoid diversity. *Science*, **277**, 1788–9.

9 Fraga, B.M. (2006) Natural sesquiterpenoids. *Natural Product Reports*, **23**, 943–72.

10 Hanson, J.R. (2006) Diterpenoids. *Natural Product Reports*, **23**, 875–85.

11 Vershinin, A. (1999) Biological functions of carotenoids – diversity and evolution. *Biofactors*, **10**, 99–104.

12 Lee, P.C. and Schmidt-Dannert, C. (2002) Metabolic engineering towards biotechnological production of

carotenoids in microorganisms. *Applied Microbiology and Biotechnology*, **60**, 1–11.
13 Klein-Marcuschamer, D., Ajikumar, P.K. and Stephanopoulos, G. (2007) Engineering microbial cell factories for biosynthesis of isoprenoid molecules: beyond lycopene. *Trends in Biotechnology*, **25**, 417–24.
14 Maury, J., Asadollahi, M.A., Moller, K., Clark, A. and Nielsen, J. (2005) Microbial isoprenoid production: an example of green chemistry through metabolic engineering. *Advances in Biochemical Engineering/Biotechnology*, **100**, 19–51.
15 Sandmann, G. (2002) Combinatorial biosynthesis of carotenoids in a heterologous host: a powerful approach for the biosynthesis of novel structures. *ChemBioChem*, **3**, 629–35.
16 Fraser, P.D., Misawa, N., Linden, M., Yamano, S., Kobayashi, K. and Sandmann, G. (1992) Expression in *Escherichia coli*, purification, and reactivation of the recombinant *Erwinia uredovora* phytoene desaturase. *The Journal of Biological Chemistry*, **267**, 19891–5.
17 Cunningham, F., Chamovitz, D., Misawa, N., Gantt, E. and Hirschberg, J. (1993) Cloning and functional expression in *Escherichia coli* of a cyanobacterial gene for lycopene cyclase, the enzyme that catalyzes the biosynthesis of beta-carotene. *FEBS Letters*, **328**, 130–8.
18 Bauer, C., Bollivar, D. and Suzuki, J. (1993) Genetic analysis of photopigment biosynthesis in eubacteria – a guiding light for algae and plants. *Journal of Bacteriology*, **175**, 3919–25.
19 Armstrong, G. (1994) Eubacteria show their true colors: genetics of carotenoid pigment biosynthesis from microbes to plants. *Journal of Bacteriology*, **176**, 4795–802.
20 Ausich, R. (1994) Production of carotenoids by recombinant DNA technology. *Pure and Applied Chemistry*, **66**, 1057–62.
21 Hundle, B. Alberti, M., Nievelstein, V., Beyer, P., Kleinig, H., Armstrong, G., Burke, D. and Hearst, J. (1994) Functional assignment of *Erwinia herbicola* Eho10 carotenoid genes expressed in *Escherichia coli*. *Molecular and General Genetics*, **245**, 406–16.
22 Sandmann, G. (2003) Novel carotenoids genetically engineered in a heterologous host. *Chemistry and Biology*, **10**, 478–9.
23 Sandmann, G., Albrecht, M., Schnurr, G., Knorzer, O. and Boger, P. (1999) The biotechnological potential and design of novel carotenoids by gene combination in *Escherichia coli*. *Trends in Biotechnology*, **17**, 233–7.
24 Armstrong, G.A., Alberti, M., Leach, F. and Hearst, J.E. (1989) Nucleotide sequence, organization, and nature of the protein products of the carotenoid biosynthesis gene cluster of *Rhodobacter capsulatus*. *Molecular and General Genetics*, **216**, 254–68.
25 Misawa, N., Nakagawa, M., Kobayashi, K., Yamano, S., Izawa, Y., Nakamura, K. and Harashima, K. (1990) Elucidation of the *Erwinia uredovora* carotenoid biosynthetic pathway by functional analysis of gene products expressed in *Escherichia coli*. *Journal of Bacteriology*, **172**, 6704–12.
26 Ruther, A., Misawa, N., Boger, P. and Sandmann, G. (1997) Production of zeaxanthin in *Escherichia coli* transformed with different carotenogenic plasmids. *Applied Microbiology and Biotechnology*, **48**, 162–7.
27 Hausmann, A. and Sandmann, G. (2000) A single five-step desaturase is involved in the carotenoid biosynthesis pathway to beta-carotene and torulene in *Neurospora crassa*. *Fungal Genetics and Biology*, **30**, 147–53.
28 Ruiz-Hidalgo, M.Y., Benito, E., Sandmann, G. and Eslava, A. (1997) The phytoene dehydrogenase gene of *Phycomyces*: regulation of its expression by blue light and vitamin A. *Molecular and General Genetics*, **253**, 734–44.
29 Sandmann, G., Misawa, N., Wiedemann, M., Vittorioso, P., Carattoli, A., Morelli, G. and Macino, G. (1993) Functional identification of al-3 from *Neurospora crassa* as the gene for geranylgeranyl pyrophosphate synthase by complementation with crt genes, in vitro characterization of the gene product and mutant analysis. *Journal of Photochemistry and Photobiology. B, Biology*, **18**, 245–51.

30 Quinlan, R.F., Jaradat, T.T. and Wurtzel, E.T. (2007) *Escherichia coli* as a platform for functional expression of plant P450 carotene hydroxylases. *Archives of Biochemistry and Biophysics*, **458**, 146–57.

31 Pecker, I., Chamovitz, D., Linden, H., Sandmann, G. and Hirschberg, J. (1992) A single polypeptide catalyzing the conversion of phytoene to zeta-carotene is transcriptionally regulated during tomato fruit ripening. *Proceedings of the National Academy of Sciences of the United States of America*, **89**, 4962–6.

32 Pecker, I., Gabbay, R., Cunningham, F. and Hirschberg, J. (1996) Cloning and characterization of the cDNA for lycopene beta-cyclase from tomato reveals decrease in its expression during fruit ripening. *Plant Molecular Biology*, **30**, 807–19.

33 Cunningham, F., Pogson, B., Sun, Z., McDonald, K., DellaPenna, D. and Gantt, E. (1996) Functional analysis of the beta and epsilon lycopene cyclase enzymes of *Arabidopsis* reveals a mechanism for control of cyclic carotenoid formation. *Plant Cell*, **8**, 1613–26.

34 Tobias, A.V. and Arnold, F.H. (2006) Biosynthesis of novel carotenoid families based on unnatural carbon backbones: a model for diversification of natural product pathways. *Biochimica et Biophysica Acta*, **1761**, 235–46.

35 Cheng, Q. (2006) Structural diversity and functional novelty of new carotenoid biosynthesis genes. *Journal of Industrial Microbiology and Biotechnology*, **33**, 552–9.

36 Sandmann, G. (2003) Combinatorial biosynthesis of novel carotenoids in *E. coli*. *Methods in Molecular Biology*, **205**, 303–14.

37 Schmidt-Dannert, C. (2000) Engineering novel carotenoids in microorganisms. *Current Opinion in Biotechnology*, **11**, 255–61.

38 Schmidt-Dannert, C., Lee, P. and Mitjs, B. (2006) Creating carotenoid diversity in *E. coli* cells using combinatorial and directed evolution strategies. *Phytochemistry Reviews*, **5**, 67–74.

39 Lee, P.C., Momen, A.Z., Mijts, B.N. and Schmidt-Dannert, C. (2003) Biosynthesis of structurally novel carotenoids in *Escherichia coli*. *Chemistry and Biology*, **10**, 453–62.

40 Mijts, B.N., Lee, P.C. and Schmidt-Dannert, C. (2005) Identification of a carotenoid oxygenase synthesizing acyclic xanthophylls: Combinatorial biosynthesis and directed evolution. *Chemistry and Biology*, **12**, 453–60.

41 Albrecht, M., Takaichi, S., Steiger, S., Wang, Z.Y. and Sandmann, G. (2000) Novel hydroxycarotenoids with improved antioxidative properties produced by gene combination in *Escherichia coli*. *Nature Biotechnology*, **18**, 843–6.

42 Christianson, D.W. (2006) Structural biology and chemistry of the terpenoid cyclases. *Chemical Reviews*, **106**, 3412–42.

43 Chang, M.C. and Keasling, J.D. (2006) Production of isoprenoid pharmaceuticals by engineered microbes. *Nature Chemical Biology*, **2**, 674–81.

44 Ro, D.K., Paradise, E.M., Ouellet, M., Fisher, K.J., Newman, K.L., Ndungu, J. M., Ho, K.A., Eachus, R.A., Ham, T.S., Kirby, J. *et al.* (2006) Production of the antimalarial drug precursor artemisinic acid in engineered yeast. *Nature*, **440**, 940–3.

45 Martin, V.J., Pitera, D.J., Withers, S.T., Newman, J.D. and Keasling, J.D. (2003) Engineering a mevalonate pathway in *Escherichia coli* for production of terpenoids. *Nature Biotechnology*, **21**, 796–802.

46 Huang, Q., Roessner, C.A., Croteau, R. and Scott, A.I. (2001) Engineering *Escherichia coli* for the synthesis of taxadiene, a key intermediate in the biosynthesis of taxol. *Bioorganic and Medicinal Chemistry*, **9**, 2237–42.

47 DeJong, J.M., Liu, Y.L., Bollon, A.P., Long, R.M., Jennewein, S., Williams, D. and Croteau, R.B. (2006) Genetic engineering of Taxol biosynthetic genes in *Saccharomyces cerevisiae*. *Biotechnology and Bioengineering*, **93**, 212–24.

48 Cyr, A., Wilderman, P.R., Determan, M. and Peters, R.J. (2007) A modular approach for facile biosynthesis of labdane-related diterpenes. *Journal of the American Chemical Society*, **129**, 6684–6.

49 Lee, P.C., Mijts, B.N. and Schmidt-Dannert, C. (2004) Investigation of factors influencing production of the monocyclic carotenoid torulene in metabolically engineered *Escherichia coli*. *Applied Microbiology and Biotechnology*, **65**, 538–46.

50 Johnson, E. and Schroeder, W. (1995) Microbial carotenoids. *Advances in Biochemical Engineering/Biotechnology*, **53**, 119–78.

51 Shimada, H., Kondo, K., Fraser, P., Miura, Y., Saito, T. and Misawa, N. (1998) Increased carotenoid production by the food yeast *Candida utilis* through metabolic engineering of the isoprenoid pathway. *Applied and Environmental Microbiology*, **64**, 2676–80.

52 Verdoes, J.C., Sandmann, G., Visser, H., Diaz, M., van Mossel, M. and van Ooyen, A.J. (2003) Metabolic engineering of the carotenoid biosynthetic pathway in the yeast *Xanthophyllomyces dendrorhous* (*Phaffia rhodozyma*). *Applied and Environmental Microbiology*, **69**, 3728–38.

53 Yamano, S., Ishii, T., Nakagawa, M., Ikenaga, H. and Misawa, N. (1994) Metabolic engineering for production of beta-carotene and lycopene in *Saccharomyces cerevisiae*. *Bioscience, Biotechnology and Biochemistry*, **58**, 1112–14.

54 Miura, Y., Kondo, K., Saito, T., Shimada, H., Fraser, P. and Misawa, N. (1998) Production of the carotenoid lycopene, beta-carotene, and astaxanthin in the food yeast *Candida utilis*. *Applied and Environmental Microbiology*, **64**, 1226–9.

55 Miura, Y., Kondo, K., Shimada, H., Saito, T., Nakamura, K. and Misawa, N. (1998) Production of lycopene by the food yeast, *Candida utilis* that does not naturally synthesize carotenoid. *Biotechnology and Bioengineering*, **58**, 306–8.

56 Verwaal, R., Wang, J., Meijnen, J.P., Visser, H., Sandmann, G., van den Berg, J.A. and van Ooyen, A.J. (2007) High-level production of beta-carotene in *Saccharomyces cerevisiae* by successive transformation with carotenogenic genes from *Xanthophyllomyces dendrorhous*. *Applied and Environmental Microbiology*, **73**, 4342–50.

57 Harker, M. and Hirschberg, J. (1997) Biosynthesis of ketocarotenoids in transgenic cyanobacteria expressing the algal gene for beta-C-4-oxygenase, crtO. *FEBS Letters*, **404**, 129–34.

58 Garcia-Asua, G., Cogdell, R.J. and Hunter, C.N. (2002) Functional assembly of the foreign carotenoid lycopene into the photosynthetic apparatus of *Rhodobacter sphaeroides*, achieved by replacement of the native 3-step phytoene desaturase with its 4-step counterpart from Erwinia herbicola. *Molecular Microbiology*, **44**, 233–44.

59 Mukoyama, D., Takeyama, H., Kondo, Y. and Matsunaga, T. (2006) Astaxanthin formation in the marine photosynthetic bacterium *Rhodovulum sulfidophilum* expressing crtI, crtY, crtW and crtZ. *FEMS Microbiology Letters*, **265**, 69–75.

60 Ye, R.W., Yao, H., Stead, K., Wang, T., Tao, L., Cheng, Q., Sharpe, P.L., Suh, W., Nagel, E., Arcilla, D. *et al.* (2007) Construction of the astaxanthin biosynthetic pathway in a methanotrophic bacterium *Methylomonas* sp. strain 16a. *Journal of Industrial Microbiology and Biotechnology*, **34**, 289–99.

61 Tao, L., Wagner, L.W., Rouviere, P.E. and Cheng, Q. (2006) Metabolic engineering for synthesis of aryl carotenoids in *Rhodococcus*. *Applied Microbiology and Biotechnology*, **70**, 222–8.

62 Szczebara, F.M., Chandelier, C., Villeret, C., Masurel, A., Bourot, S., Duport, C., Blanchard, S., Groisillier, A., Testet, E., Costaglioli, P. *et al.* (2003) Total biosynthesis of hydrocortisone from a simple carbon source in yeast. *Nature Biotechnology*, **21**, 143–9.

63 Jennewein, S., Wildung, M.R., Chau, M., Walker, K. and Croteau, R. (2004) Random sequencing of an induced *Taxus* cell cDNA library for identification of clones involved in Taxol biosynthesis. *Proceedings of the National Academy of Sciences of the United States of America*, **101**, 9149–54.

64 Shiba, Y., Paradise, E.M., Kirby, J., Ro, D.K. and Keasling, J.D. (2007) Engineering of the pyruvate dehydrogenase bypass in *Saccharomyces* *cerevisiae* for high-level production of isoprenoids. *Metabolic Engineering*, **9**, 160–8.

65 Newman, J.D., Marshall, J., Chang, M., Nowroozi, F., Paradise, E., Pitera, D., Newman, K.L. and Keasling, J.D. (2006) High-level production of amorpha-4,11-diene in a two-phase partitioning bioreactor of metabolically engineered *Escherichia coli*. *Biotechnology and Bioengineering*, **95**, 684–91.

66 Eisenreich, W., Bacher, A., Arigoni, D. and Rohdich, F. (2004) Biosynthesis of isoprenoids via the non-mevalonate pathway. *Cellular and Molecular Life Sciences*, **61**, 1401–26.

67 Rohmer, M. (2007) Diversity in isoprene unit biosynthesis: The methylerythritol phosphate pathway in bacteria and plastids. *Pure and Applied Chemistry*, **79**, 739–51.

68 Harker, M. and Bramley, P.M. (1999) Expression of prokaryotic 1-deoxy-D-xylulose-5-phosphatases in *Escherichia coli* increases carotenoid and ubiquinone biosynthesis. *FEBS Letters*, **448**, 115–19.

69 Matthews, P.D. and Wurtzel, E.T. (2000) Metabolic engineering of carotenoid accumulation in *Escherichia coli* by modulation of the isoprenoid precursor pool with expression of deoxyxylulose phosphate synthase. *Applied Microbiology and Biotechnology*, **53**, 396–400.

70 Jones, K.L., Kim, S.W. and Keasling, J.D. (2000) Low-copy plasmids can perform as well as or better than high-copy plasmids for metabolic engineering of bacteria. *Metabolic Engineering*, **2**, 328–38.

71 Hahn, F.M., Hurlburt, A.P. and Poulter, C.D. (1999) *Escherichia coli* open reading frame 696 is idi, a nonessential gene encoding isopentenyl diphosphate isomerase. *Journal of Bacteriology*, **181**, 4499–504.

72 Kajiwara, S., Fraser, P.D., Kondo, K. and Misawa, N. (1997) Expression of an exogenous isopentenyl diphosphate isomerase gene enhances isoprenoid biosynthesis in *Escherichia coli*. *The Biochemical Journal*, **324** (*Pt 2*), 421–6.

73 Kim, S.W. and Keasling, J.D. (2001) Metabolic engineering of the nonmevalonate isopentenyl diphosphate synthesis pathway in *Escherichia coli* enhances lycopene production. *Biotechnology and Bioengineering*, **72**, 408–15.

74 Albrecht, M., Misawa, N. and Sandmann, G. (1999) Metabolic engineering of the terpenoid biosynthetic pathway of *Escherichia coli* for production of the carotenoids beta-carotene and zeaxanthin. *Biotechnology Letters*, **21**, 791–5.

75 Wang, C.W., Oh, M.K. and Liao, J.C. (1999) Engineered isoprenoid pathway enhances astaxanthin production in *Escherichia coli*. *Biotechnology and Bioengineering*, **62**, 235–41.

76 Pronk, J.T., Yde Steensma, H. and Van Dijken, J.P. (1996) Pyruvate metabolism in *Saccharomyces cerevisiae*. *Yeast*, **12**, 1607–33.

77 Servouse, M. and Karst, F. (1986) Regulation of early enzymes of ergosterol biosynthesis in *Saccharomyces cerevisiae*. *The Biochemical Journal*, **240**, 541–7.

78 Dimster-Denk, D. and Rine, J. (1996) Transcriptional regulation of a sterol-biosynthetic enzyme by sterol levels in *Saccharomyces cerevisiae*. *Molecular and Cellular Biology*, **16**, 3981–9.

79 Dimster-Denk, D. *et al.* (1999) Comprehensive evaluation of isoprenoid biosynthesis regulation in *Saccharomyces cerevisiae* utilizing the Genome Reporter Matrix. *Journal of Lipid Research*, **40**, 850–60.

80 Osborne, T.F., Gil, G., Goldstein, J.L. and Brown, M.S. (1988) Operator constitutive mutation of 3-hydroxy-3-methylglutaryl coenzyme A reductase promoter abolishes protein binding to sterol regulatory element. *The Journal of Biological Chemistry*, **263**, 3380–7.

81 Smith, J.R., Osborne, T.F., Brown, M.S., Goldstein, J.L. and Gil, G. (1988) Multiple sterol regulatory elements in promoter for hamster 3-hydroxy-3-methylglutaryl-coenzyme A synthase. *The Journal of Biological Chemistry*, **263**, 18480–7.

82 Goldstein, J.L. and Brown, M.S. (1990) Regulation of the mevalonate pathway. *Nature*, **343**, 425–30.

83 Wang, X., Sato, R., Brown, M.S., Hua, X. and Goldstein, J.L. (1994) SREBP-1, a membrane-bound transcription factor released by sterol-regulated proteolysis. *Cell*, **77**, 53–62.

84 Hua, X., Yokoyama, C., Wu, J., Briggs, M.R., Brown, M.S., Goldstein, J.L. and Wang, X. (1993) SREBP-2, a second basic-helix-loop-helix-leucine zipper protein that stimulates transcription by binding to a sterol regulatory element. *Proceedings of the National Academy of Sciences of the United States of America*, **90**, 11603–7.

85 Veen, M. and Lang, C. (2004) Production of lipid compounds in the yeast *Saccharomyces cerevisiae*. *Applied Microbiology and Biotechnology*, **63**, 635–46.

86 Donald, K.A.G., Hampton, R.Y. and Fritz, I.B. (1997) Effects of overproduction of the catalytic domain of 3-hydroxy-3-methylglutaryl coenzyme A reductase on squalene synthesis in *Saccharomyces cerevisiae*. *Applied and Environmental Microbiology*, **63**, 3341–4.

87 Polakowski, T., Stahl, U. and Lang, C. (1998) Overexpression of a cytosolic hydroxymethylglutaryl-CoA reductase leads to squalene accumulation in yeast. *Applied Microbiology and Biotechnology*, **49**, 66–71.

88 Shimada, H., Kondo, K., Fraser, P.D., Miura, Y., Saito, T. and Misawa, N. (1998) Increased carotenoid production by the food yeast *Candida utilis* through metabolic engineering of the isoprenoid pathway. *Applied and Environmental Microbiology*, **64**, 2676–80.

89 Jackson, B.E., Hart-Wells, E.A. and Matsuda, S.P. (2003) Metabolic engineering to produce sesquiterpenes in yeast. *Organic Letters*, **5**, 1629–32.

90 Szkopinska, A., Swiezewska, E. and Karst, F. (2000) The regulation of activity of main mevalonic acid pathway enzymes: farnesyl diphosphate synthase, 3-hydroxy-3-methylglutaryl-CoA reductase, and squalene synthase in yeast *Saccharomyces cerevisiae*. *Biochemical and Biophysical Research Communications*, **267**, 473–7.

91 Farmer, W.R. and Liao, J.C. (2001) Precursor balancing for metabolic engineering of lycopene production in *Escherichia coli*. *Biotechnology Progress*, **17**, 57–61.

92 Pitera, D.J., Paddon, C.J., Newman, J.D. and Keasling, J.D. (2007) Balancing a heterologous mevalonate pathway for improved isoprenoid production in *Escherichia coli*. *Metabolic Engineering*, **9**, 193–207.

93 Yoon, S.H. *et al.* (2006) Enhanced lycopene production in *Escherichia coli* engineered to synthesize isopentenyl diphosphate and dimethylallyl diphosphate from mevalonate. *Biotechnology and Bioengineering*, **94**, 1025–32.

94 Vadali, R.V., Fu, Y., Bennett, G.N. and San, K.Y. (2005) Enhanced lycopene productivity by manipulation of carbon flow to isopentenyl diphosphate in *Escherichia coli*. *Biotechnology Progress*, **21**, 1558–61.

95 Tao, L., Jackson, R.E. and Cheng, Q. (2005) Directed evolution of copy number of a broad host range plasmid for metabolic engineering. *Metabolic Engineering*, **7**, 10–17.

96 Meynial-Salles, I., Cervin, M.A. and Soucaille, P. (2005) New tool for metabolic pathway engineering in *Escherichia coli*: one-step method to modulate expression of chromosomal genes. *Applied and Environmental Microbiology*, **71**, 2140–4.

97 Alper, H., Fischer, C., Nevoigt, E. and Stephanopoulos, G. (2005) Tuning genetic control through promoter engineering. *Proceedings of the National Academy of Sciences of the United States of America*, **102**, 12678–83.

98 Wang, C., Oh, M.K. and Liao, J.C. (2000) Directed evolution of metabolically engineered *Escherichia coli* for carotenoid production. *Biotechnology Progress*, **16**, 922–6.

99 Smolke, C.D., Martin, V.J. and Keasling, J.D. (2001) Controlling the metabolic flux through the carotenoid pathway using directed mRNA processing and stabilization. *Metabolic Engineering*, **3**, 313–21.

100 Kajiwara, S., Kakizono, T., Saito, T., Kondo, K., Ohtani, T., Nishio, N., Nagai, S. and Misawa, N. (1995) Isolation and functional identification of a novel cDNA for astaxanthin biosynthesis from *Haematococcus pluvialis*, and astaxanthin synthesis in *Escherichia coli*. *Plant Molecular Biology*, **29**, 343–52.

101 Misawa, N., Kajiwara, S., Kondo, K., Yokoyama, A., Satomi, Y., Saito, T., Miki, W. and Ohtani, T. (1995) Canthaxanthin biosynthesis by the conversion of methylene to keto groups in a hydrocarbon beta-carotene by a single-gene. *Biochemical and Biophysical Research Communications*, **209**, 867–76.

102 Martin, V.J., Yoshikuni, Y. and Keasling, J.D. (2001) The in vivo synthesis of plant sesquiterpenes by *Escherichia coli*. *Biotechnology and Bioengineering*, **75**, 497–503.

103 Zhang, Y., Perry, K., Vinci, V., Powell, K., Stemmer, W. and del Cardayre, S. (2002) Genome shuffling leads to rapid phenotypic improvement in bacteria. *Nature*, **415**, 644–6.

104 Patnaik, R., Louie, S., Gavrilovic, V., Perry, K., Stemmer, W.P.C., Ryan, C.M. and del Cardayre, S. (2002) Genome shuffling of *Lactobacillus* for improved acid tolerance. *Nature Biotechnology*, **20**, 707–12.

105 Dai, M.H. and Copley, S.D. (2004) Genome shuffling improves degradation of the anthropogenic pesticide pentachlorophenol by *Sphingobium chlorophenolicum* ATCC 39723. *Applied and Environmental Microbiology*, **70**, 2391–7.

106 Dai, M.H., Ziesman, S., Ratcliffe, T., Gill, R.T. and Copley, S.D. (2005) Visualization of protoplast fusion and quantitation of recombination in fused protoplasts of auxotrophic strains of *Escherichia coli*. *Metabolic Engineering*, **7**, 45–52.

107 Alper, H., Jin, Y.S., Moxley, J.F. and Stephanopoulos, G. (2005) Identifying gene targets for the metabolic engineering of lycopene biosynthesis in *Escherichia coli*. *Metabolic Engineering*, **7**, 155–64.

108 Alper, H., Miyaoku, K. and Stephanopoulos, G. (2006) Characterization of lycopene-overproducing *E. coli* strains in high cell density fermentations. *Applied Microbiology and Biotechnology*, **72**, 968–74.

109 Alper, H., Miyaoku, K. and Stephanopoulos, G. (2005) Construction of lycopene-overproducing *E. coli* strains by combining systematic and combinatorial gene knockout targets. *Nature Biotechnology*, **23**, 612–16.

110 Kang, M.J., Yoon, S.H., Lee, Y.M., Lee, S.H., Kim, J.E., Jung, K.H., Shin, Y.C. and Kim, S.W. (2005) Enhancement of lycopene production in *Escherichia coli* by optimization of the lycopene synthetic pathway. *Journal of Microbiology and Biotechnology*, **15**, 880–6.

111 Jin, Y.S. and Stephanopoulos, G. (2007) Multi-dimensional gene target search for improving lycopene biosynthesis in *Escherichia coli*. *Metabolic Engineering*, **9**, 337–47.

112 Alper, H. and Stephanopoulos, G. (2007) Global transcription machinery engineering: a new approach for improving cellular phenotype. *Metabolic Engineering*, **9**, 258–67.

113 Tsoi, C.J. and Khosla, C. (1995) Combinatorial biosynthesis of "unnatural" natural products: the polyketide example. *Chemistry and Biology*, **2**, 355–62.

114 Cane, D.E., Walsh, C.T. and Khosla, C. (1998) Harnessing the biosynthetic code: combinations, permutations, and mutations. *Science*, **282**, 63–8.

115 Khersonsky, O., Roodveldt, C. and Tawfik, D.S. (2006) Enzyme promiscuity: evolutionary and mechanistic aspects. *Current Opinion in Chemical Biology*, **10**, 498–508.

116 Fischbach, M.A. and Clardy, J. (2007) One pathway, many products. *Nature Chemical Biology*, **3**, 353–5.

117 Ohnuma, S., Nakazawa, T., Hemmi, H., Hallberg, A.M., Koyama, T., Ogura, K. and Nishino, T. (1996) Conversion from farnesyl diphosphate synthase to geranylgeranyl diphosphate synthase by random chemical mutagenesis. *The Journal of Biological Chemistry*, **271**, 10087–95.

118 Ohnuma, S., Hirooka, K., Hemmi, H., Ishida, C., Ohto, C. and Nishino, T. (1996) Conversion of product specificity of archaebacterial geranylgeranyl-diphosphate synthase. Identification of essential amino acid residues for chain length determination of prenyltransferase reaction. *The Journal of Biological Chemistry*, **271**, 18831–7.

119 Ohnuma, S., Hirooka, K., Tsuruoka, N., Yano, M., Ohto, C., Nakane, H. and Nishino, T. (1998) A pathway where polyprenyl diphosphate elongates in prenyltransferase. Insight into a common mechanism of chain length determination of prenyltransferases. *The Journal of Biological Chemistry*, **273**, 26705–13.

120 Ohnuma, S., Narita, K., Nakazawa, T., Ishida, C., Takeuchi, Y., Ohto, C. and Nishino, T. (1996) A role of the amino acid residue located on the fifth position before the first aspartate-rich motif of farnesyl diphosphate synthase on determination of the final product. *The Journal of Biological Chemistry*, **271**, 30748–54.

121 Tarshis, L.C., Proteau, P.J., Kellogg, B.A., Sacchettini, J.C. and Poulter, C.D. (1996) Regulation of product chain length by isoprenyl diphosphate synthases. *Proceedings of the National Academy of Sciences of the United States of America*, **93**, 15018–23.

122 Lee, P.C., Mijts, B.N., Petri, R., Watts, K.T. and Schmidt-Dannert, C. (2004) Alteration of product specificity of *Aeropyrum pernix* farnesylgeranyl diphosphate synthase (Fgs) by directed evolution. *Protein Engineering, Design and Selection*, **17**, 771–7.

123 Lee, P.C., Petri, R., Mijts, B.N., Watts, K.T. and Schmidt-Dannert, C. (2005) Directed evolution of *Escherichia coli* farnesyl diphosphate synthase (IspA) reveals novel structural determinants of chain length specificity. *Metabolic Engineering*, **7**, 18–26.

124 Umeno, D. and Arnold, F.H. (2003) A C35 carotenoid biosynthetic pathway. *Applied and Environmental Microbiology*, **69**, 3573–9.

125 Umeno, D. and Arnold, F.H. (2004) Evolution of a pathway to novel long-chain carotenoids. *Journal of Bacteriology*, **186**, 1531–6.

126 Umeno, D., Tobias, A.V. and Arnold, F.H. (2002) Evolution of the C30 carotenoid synthase CrtM for function in a C40 pathway. *Journal of Bacteriology*, **184**, 6690–9.

127 Raisig, A. and Sandmann, G. (2001) Functional properties of diapophytoene and related desaturases of C(30) and C(40) carotenoid biosynthetic pathways. *Plant Molecular Biology*, **1533**, 164–70.

128 Segura, M.J.R., Jackson, B.E. and Matsuda, S.P.T. (2003) Mutagenesis approaches to deduce structure-function relationships in terpene synthases. *Natural Product Reports*, **20**, 304–17.

129 Greenhagen, B.T., O'Maille, P.E., Noel, J.P. and Chappell, J. (2006) Identifying and manipulating structural determinates linking catalytic specificities in terpene synthases. *Proceedings of the National Academy of Sciences of the United States of America*, **103**, 9826–31.

130 Yoshikuni, Y., Martin, V.J.J., Ferrin, T.E. and Keasling, J.D. (2006) Engineering cotton (+)-delta-cadinene synthase to an altered function: Germacrene D-4-ol synthase. *Chemistry and Biology*, **13**, 91–8.

131 Yoshikuni, Y., Ferrin, T.E. and Keasling, J.D. (2006) Designed divergent evolution of enzyme function. *Nature*, **440**, 1078–82.

132 Schmidt-Dannert, C., Umeno, D. and Arnold, F.H. (2000) Molecular breeding of carotenoid biosynthetic pathways. *Nature Biotechnology*, **18**, 750–3.

133 Patil, K.R., Rocha, I., Forster, J. and Nielsen, J. (2005) Evolutionary programming as a platform for in silico metabolic engineering. *BMC Bioinformatics*, **6**, 308.

134 Vemuri, G.N. and Aristidou, A.A. (2005) Metabolic engineering in the -omics era: elucidating and modulating regulatory networks. *Microbiology and Molecular Biology Reviews*, **69**, 197–216.

135 Chin, J.W. (2006) Programming and engineering biological networks. *Current Opinion in Structural Biology*, **16**, 551–6.

136 Endy, D. (2005) Foundations for engineering biology. *Nature*, **438**, 449–53.

137 Sprinzak, D. and Elowitz, M.B. (2005) Reconstruction of genetic circuits. *Nature*, **438**, 443–8.

138 McDaniel, R. and Weiss, R. (2005) Advances in synthetic biology: on the path from prototypes to applications.

Current Opinion in Biotechnology, **16**, 476–83.

139 Smith, H.O., Hutchison, C.A.3rd , Pfannkoch, C. and Venter, J.C. (2003) Generating a synthetic genome by whole genome assembly: phiX174 bacteriophage from synthetic oligonucleotides. *Proceedings of the National Academy of Sciences of the United States of America*, **100**, 15440–5.

140 Posfai, G., Plunkett, G.3rd, , Feher, T., Frisch, D., Keil, G.M., Umenhoffer, K., Kolisnychenko, V., Stahl, B., Sharma, S. S., de Arruda, M. *et al.* (2006) Emergent properties of reduced-genome *Escherichia coli*. *Science*, **312**, 1044–6.

141 Glass, J.I., Assad-Garcia, N., Alperovich, N., Yooseph, S., Lewis, M.R., Maruf, M., Hutchison, C.A.3rd, , Smith, H.O. and Venter, J.C. (2006) Essential genes of a minimal bacterium. *Proceedings of the National Academy of Sciences of the United States of America*, **103**, 425–30.

35
Natural Polyester-Related Proteins: Structure, Function, Evolution and Engineering

Seiichi Taguchi and Takeharu Tsuge

35.1
Introduction

Polyhydroxyalkanoates (PHAs) form a category of natural polyesters that many microorganisms accumulate in the form of intracellular granules in order to store carbon and reducing equivalents [1, 2]. PHAs that have been identified to date are primarily linear, head-to-tail polyesters consisting of 3-hydroxy fatty acid monomers. As shown in Figure 35.1, the carboxyl group of one monomer forms an ester bond with the hydroxyl group of the neighboring monomer. There are over 140 possible constituent monomers, which have been traditionally classified as short-chain-length (scl: C4 and C5) and medium-chain-length (mcl: C6 to C14) 3-hydroxyalkanoates (3HAs) [3]. The hydroxyl-substituted carbon atom is of the (*R*)-configuration in all PHAs characterized so far. In addition to the length and alkyl substituent of the side chains, the position of the hydroxyl group is also variable, and 4-, 5- and 6-hydroxy acids have been incorporated into PHAs.

PHAs are unique among eco-friendly polymers because the polymerization process occurs in living cell systems. In contrast, polylactic acid – a representative biobased polyester – is chemically polymerized using a monomeric constituent (lactic acid) produced by microbial fermentation. In addition, PHAs are biodegraded completely in various environments or biosystems. The monomeric composition of PHA, which is an important factor affecting polymer material properties, basically depends on the polymerizing enzyme, PHA synthase (or polymerase) [4] and on the hydroxyacyl (HA)-CoA thioester precursors supplied to the enzyme, which in turn depend on the metabolic pathways operating in the cell and on the external carbon source [5]. The primary aims of PHA metabolic engineering include the management of different factors that determine physical properties, such as monomeric composition, molecular weight and copolymer microstructure, as well as optimizing yield. Modification of the PHA polymer can also be achieved to architect much more high-performance materials by using various PHA-related proteins as affinitive supporters or additional functionalizing items.

Protein Engineering Handbook. Edited by Stefan Lutz and Uwe T. Bornscheuer

ISBN: 978-3-527-31850-6

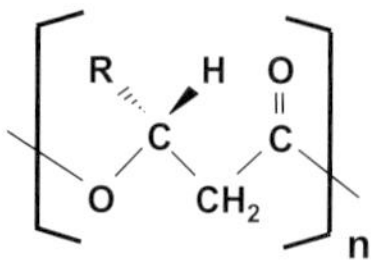

Figure 35.1 Chemical structure of PHAs. PHAs generally consist of (*R*)-3-hydroxy fatty acids with various side chains. Other fatty acids have the hydroxy group at the 4, 5 or 6 position. The pendant group (*R*) varies from C1 to C13, and is saturated or unsaturated or contains substituents. The well-known PHA is P(3HB) (R = methyl).

Protein engineering via structure-based mutagenesis and function-based molecular evolution allows metabolic engineers to optimize enzyme performance to obtain desirable polymer properties and yield. This chapter focuses on the engineering of proteins (enzymes) relevant for PHAs based on their structures, functions and evolutionary insights.

35.2
Enzymes Related to the Synthesis and Degradation of PHA

Three main naturally occurring PHA biosynthesis pathways (I, II and III) are shown schematically in Figure 35.2. Pathway I, which generates (*R*)-3-hydroxybutyrate ((*R*)-3HB) monomers from two acetyl-CoA molecules, is probably the most common, and has been found in a wide range of bacteria accumulating scl-PHA. This pathway has been extensively investigated for *Ralstonia eutropha* [6, 7]. In *R. eutropha*, two acetyl-CoA molecules are condensed to yield acetoacetyl-CoA by 3-ketothiolase (PhaA). The resultant acetoacetyl-CoA is subsequently reduced to (*R*)-3HB-CoA by an NADH- or NADPH-dependent acetoacetyl-CoA reductase (PhaB). Only (*R*)-isomers are accepted as the substrates of the polymerizing enzyme, PHA synthase (PhaC) [6, 7].

Pathways II and III, which generate mainly mcl-(*R*)-3HA monomers from fatty acid β-oxidation and fatty acid biosynthesis intermediates, respectively, have been found in various fluorescent pseudomonads such as *Pseudomonas putida, P. oleovorans* and *P. aeruginosa* [6, 7]. The intermediates in these pathways are effectively converted by some specialized enzymes to generate (*R*)-3HA-CoA monomers for PHA synthases. As shown in Figure 35.2, (*R*)-specific enoyl-CoA hydratase (PhaJ) and (*R*)-3HA-ACP-CoA transferase (PhaG) are capable of supplying (*R*)-3HA-CoA from *trans*-2-enoyl-CoA and (*R*)-3HA-ACP, respectively. It has also been shown that 3-ketoacyl-ACP reductase FabG can accept not only acyl-ACP but also acyl-CoA as a substrate and is capable of supplying mcl-(*R*)-3HA-CoA from fatty acid β-oxidation in *Escherichia coli*.

As in the case of the biodegradation of PHA, environmental degradation and intracellular degradation occur by the action of different types of microbial depolymerase (PhaZ). Environmental degradation is catalyzed by extracellular PHA depolymerases [8]; this enzyme is known to be secreted by various natural microbes to hydrolyze the crystalline PHA into water-soluble monomers and oligomers. These degradation products are assimilated by microbes as carbon sources and finally mineralized into CO_2 and H_2O under aerobic conditions. Meanwhile, intracellular PHA depolymerases are involved in the mobilization of PHA by hydrolyz-

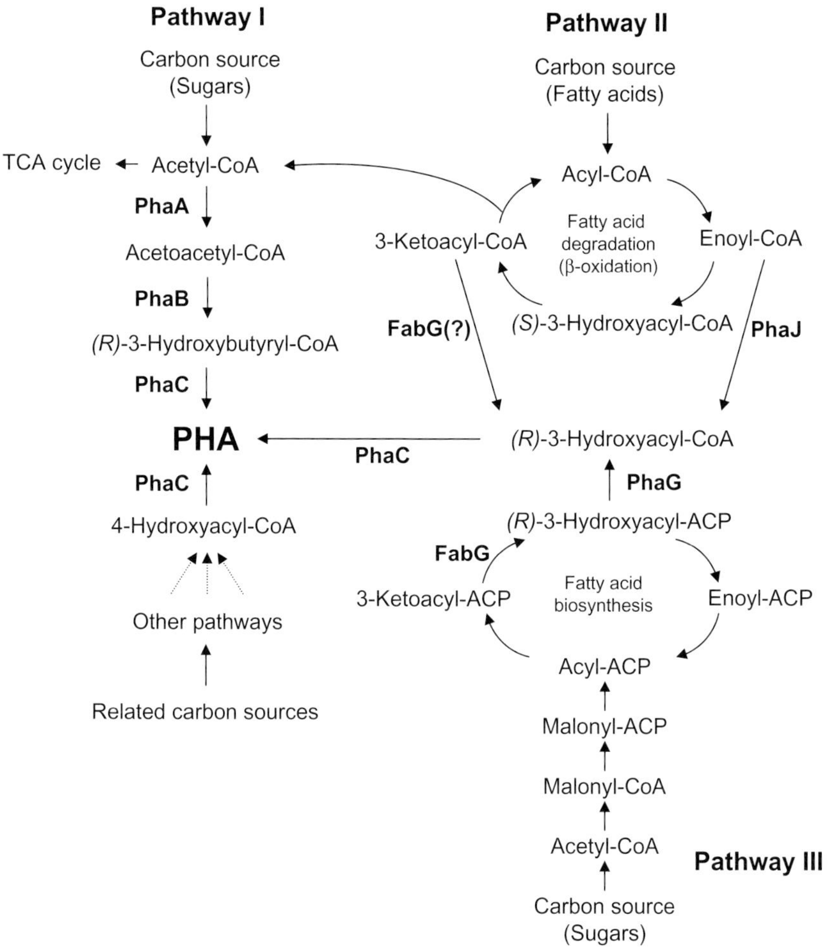

Figure 35.2 Metabolic pathways that supply various hydroxyalkanoate monomers for PHA biosynthesis.

ing amorphous PHA into PHA oligomers in the cells. The oligomers are then converted into monomeric 3HAs by oligomer hydrolase, and metabolized [9].

35.3 Structure-Based Engineering of PHA Synthase and Monomer-Supplying Enzymes

In this section, details of sequence homology- or structure-based protein engineering for PHA synthases and monomer-supplying enzymes are presented, together with their basic enzymatic properties (see Table 35.1). Here, structure-based protein engineering includes not only site-specific point/deletion/insertion mutagenesis but also protein fusion with tagged partners of interest.

Table 35.1 Enzymes and proteins involved in PHA metabolism.

Abbreviation	Enzyme/Protein	Function and role in PHA metabolism
PhaA	3-Ketothiolase	Monomer-supplying enzyme
PhaB	Acetoacetyl-CoA reductase	Monomer-supplying enzyme
PhaC	PHA synthase	Subunit of type I, II, III, IV PHA synthase
PhaD	Putative regulator protein	Regulator involved in mcl-PHA synthesis
PhaE	PHA synthase subunit	Type III PHA synthase subunit
PhaF	MCL-PHA GAP1	MCL -PHA granule associate protein
PhaG	*R*-3-hydroxyacyl-ACP-CoA transferase	Monomer-supplying enzyme
PhaI	MCL-PHA GAP2	MCL-PHA granule associate protein
PhaJ	*R*-specific enoyl-CoA hydratase	Monomer-supplying enzyme
PhaP	SCL-PHA GAP	SCL-PHA granule associate protein
PhaR	Regulator protein or PHA synthase	Transcriptional regulator or type IV PHA synthase subunits
PhaY	PHA oligomer hydrolase	Degradation of PHA oligomer
PhaZ	PHA depolymerase	Extracellular or intracellular PHA depolymerase

35.3.1 PHA Synthase (PhaC, PhaEC, PhaRC)

PHA synthases are the key enzymes of PHA biosynthesis, and catalyze the polymerization reaction of HA to PHA. The immediate substrates of PHA synthase are mainly 3HA-CoAs with various side-chain lengths, and only *R*-enantiomer HA-CoAs are accepted for polymerization by synthase. With the concomitant release of CoA, PHA synthases polymerize the HA moiety of HA-CoA and generate high-molecular-weight polyester molecules that are stored in the form of water-insoluble inclusions of PHA [10, 11].

At present, approximately 60 different PHA synthases have been isolated and characterized [10]. Based on their substrate specificities and subunit compositions of enzymes, PHA synthases have been classified into four types (Table 35.2). Types

Table 35.2 Types of PHA synthase and varieties of PHAs.

Substrate specificity[a]	Type	Subunit(s)	Microorganism[b]	Polymers produced[c]
SCL-3HA-CoA (C3–C5)	I	PhaC	*Ralstonia eutropha*	P(3HB), P(3HB-*co*-3HV)
	III	PhaC, PhaE	*Allochromatium vinosum*	
	IV	PhaC, PhaR	*Bacillus megaterium*	
MCL-3HA-CoA (C6–C14)	III	PhaC	*Pseudomonas oleovorans* *Pseudomonas putida*	PHA
SCL-MCL-3HA-CoA (C3–C14)	I	PhaC	*Aeromonas caviae*	P(3HB-*co*-3HA)
	II	PhaC	*Pseudomonas* sp. 61-3	

a Substrate preferred by the PHA synthase.
SCL-3HA-CoA (C3-C5), short-chain-length-3-hydroxyacyl-coenzyme A (3–5 carbons in length);
MCL-3HA-CoA (C6-C14), medium-chain-length-3-hydroxyacyl-coenzyme A (6–14 carbons in length);
SCL-MCL-3HA-CoA (C3-C14), short-medium-chain-length-3-hydroxyacyl-coenzyme A (3–14 carbons in length).

b Native microorganism where the PHA synthase and polymer are found.

c Polymers produced. P(3HB), poly(3-hydroxybutyrate); P(3HB-*co*-3HV), poly(3-hydroxybutyrate-*co*-3-hydroxyvalerate); PHA, poly(3-hydroxyalkanoate); P(3HB-*co*-3HA), poly(3-hydroxybutyrate-*co*-3-hydroxyalkanoate).

I and II PHA synthases consist of single subunits (PhaC) with molecular masses between 61 and 73 kDa [10]. Type I PHA synthases, represented by the *R. eutropha* enzyme, mainly polymerize scl-monomers (C3–C5), whereas type II PHA synthases, represented by the *P. oleovorans* enzyme, polymerize mcl-monomers (C6–C14). Type III PHA synthases, represented by *Allochromatium vinosum* enzyme, consist of two heterosubunits (PhaC and PhaE). PhaC subunits of type III synthase (approx. 40 kDa) are smaller than those of type I and II synthases, but possess catalytic residues. The amino acid sequence similarities of PhaC subunits between type III and type I/II are 21–28% [10], however, PhaE subunits (approx. 40 kDa) have no significant similarities to the other PHA synthases. Like the type I synthases, these PHA synthases prefer to polymerize scl-monomers (C3–C5). Type IV PHA synthases, represented by *Bacillus megaterium*, are similar to the type III PHA synthases with respect to possessing two subunits. However, unlike the PhaE of type III PHA synthases a small protein of approximately 20 kDa designated as PhaR is required for full activity expression of type IV PhaC.

Alignment analysis of 59 PHA synthases identified eight conserved amino acid residues and six conserved amino acid sequence regions with strong similarity in all synthases [10]. These conserved amino acid residues and sequence regions are expected to play an important role in the enzyme function. The N-terminal region (approx. 100 amino acids) of PhaC such as type I synthase is highly variable and dispensable for a functionally active enzyme, as revealed by the analysis of truncated *R. eutropha* PHA synthases that lacked 60 or even 78 amino acids [12] (see Table 35.3). In contrast to the N-terminal region, the more conserved C-terminal region is required for enzyme activity. Thus, the insertion and deletion of amino acid sequences in the C-terminal region is not permissive; even a five-residue deletion at the C-terminal end led to activity loss of synthase [13]. However, a remarkable exception is shown in the type I synthase of *Delftia acidovorans* [14]; this enzyme has an extra-large-insertion which consists of approximately 40 amino acids in the C-terminal region, and deletion of the insertion was permissive.

In all PHA synthases, two glycine residues adjacent to the active-site cysteine (G-X-C-X-G) are conserved [10]. This sequence pattern is similar to a lipase box (G-X-S-X-G), but the essential active-site serine of lipase is replaced with a cysteine in the PHA synthase. Besides cysteine, aspartic acid and histidine have been proposed to be directly involved in covalent catalysis. In *R. eutropha* PHA synthase, active-site Cys319, Asp480 and His508 are proposed to form a catalytic triad based on site-directed mutagenesis (SDM) studies [15–17]. Other protein-engineering studies (as summarized in Table 35.3) have contributed to the present understanding of the catalytic mechanism of synthase. PHA synthases are classified into the α/β-hydrolase superfamily and are structurally similar to lipases, which are the typical members of this superfamily. On the basis of an homology search, all PHA synthases contain the α/β-hydrolase domain at the C-terminal region of PhaC. Unfortunately, the tertiary structure of PHA synthases has not yet been resolved by X-ray diffraction analysis due to difficulties in the crystallization of PHA synthase. However, several tertiary structure models of PHA synthases have been proposed (Figure 35.3), based on the homology with crystallographically resolved lipases [13, 18–20]. One significant difference in higher-order architecture between PHA synthase and lipase is in the quaternary structure of the active form of the enzymes [11]. Unlike PHA synthases, which are active only in their multimeric form, lipases function in their monomeric form.

35.3.2
3-Ketoacyl-CoA Thiolase (PhaA)

As shown in Figure 35.2, the most well-known P(3HB) biosynthetic pathway starting from acetyl-CoA consists of three enzymatic reactions catalyzed by 3-ketoacyl-CoA thiolase (PhaA), acetoacetyl-CoA reductase (PhaB) and PHA synthase [6, 7]. First, 3-ketothiolase (PhaA) catalyzes the condensation reaction of two acetyl-CoA molecules to form acetoacetyl-CoA, which is then reduced to *(R)*-3HB-CoA by acetoacetyl-CoA reductase (PhaB, see below). P(3HB) polymers are then synthesized from *(R)*-3HB-CoA by the action of PHA synthase. A bacterium having type I PHA synthase most likely possesses this three-step PHA biosynthesis pathway, and the genes encoding PhaA and PhaB are found in the same operon of the PHA synthase gene. (There are exceptions, such as in the case of *D. acidovorans*, whereby

Table 35.3 Sequence- and structure-based protein engineering of PHA synthases.

Method	Enzyme	Position	Purpose or changed property	Active or inactive[a]	Reference(s)
Site-specific mutagenesis	*R. eutropha* PhaC	Y75E/F/G/P	Conformational change at N-terminal region	A	[12]
		A81E/G/M/P	Conformational change at N-terminal region	A	[12]
		Y75F/A81M, Y75P/A81P, Y75G/A81G, Y75E/A81E	Double mutation: Conformational change at N-terminal region	A	[12]
		S260A	Identification of phosphopantetheinylation position	I	[16]
		E267K	Conserved amino acid: substrate specificity change	A	[13]
		C319S/A	Active-site residue identification	I	[15], [17]
		T323S/I	Conserved amino acid	I	[13]
		C438G	Conserved amino acid	I	[13]
		Y445F	Conserved amino acid: substrate specificity change	A	[13]
		L446K	Conserved amino acid: substrate specificity change	A	[13]

Table 35.3 *Continued*

Method	Enzyme	Position	Purpose or changed property	Active or inactive[a]	Reference(s)
		C459S	Active-site residue identification	A	[15]
		D430N	Active-site residue identification	I	[17]
		H481Q	Active-site residue identification	A	[17]
		H508Q	Active-site residue identification	I	[17]
		S546I	Identification of phosphopantetheinylation position	I	[16]
	P. aeruginosa PhaC1	C296A	Active-site residue identification	I	[18]
		C296S	Substrate specificity change	A	[18]
		W398A/F	Dimerization property	I	[18]
		D452N	Active-site residue identification	I	[18]
		H453Q	Substrate specificity change	A	[18]
		H480Q	Substrate specificity change	A	[18]

	A. vinosum PhaEC	C130A/S	Active-site residue identification	A	[19, 21]
		C149A	Active-site residue identification	I	[21]
		C149S	Active-site residue identification	A	[19]
		C292A	Active-site residue identification	A	[21]
		D302A/N	Active-site residue identification	A	[19, 22]
		H303Q	Active-site residue identification	A	[19]
		H331Q	Active-site residue identification	A	[19]
		C149S/H331Q	Active-site residue identification	I	[19]
Site-specific deletion/insertion	*R. eutropha* PhaC	N terminus	60–78 aa deletion	A	[12]
		N terminus	88–98 aa deletion	I	[12]
		C terminus and in-frame	5–12 aa deletion	I	[13]
	D. acidovorans PhaC	in-frame	40 aa deletion	A	[14]

a A: active mutant, I: mutant which shows significantly reduced activity.

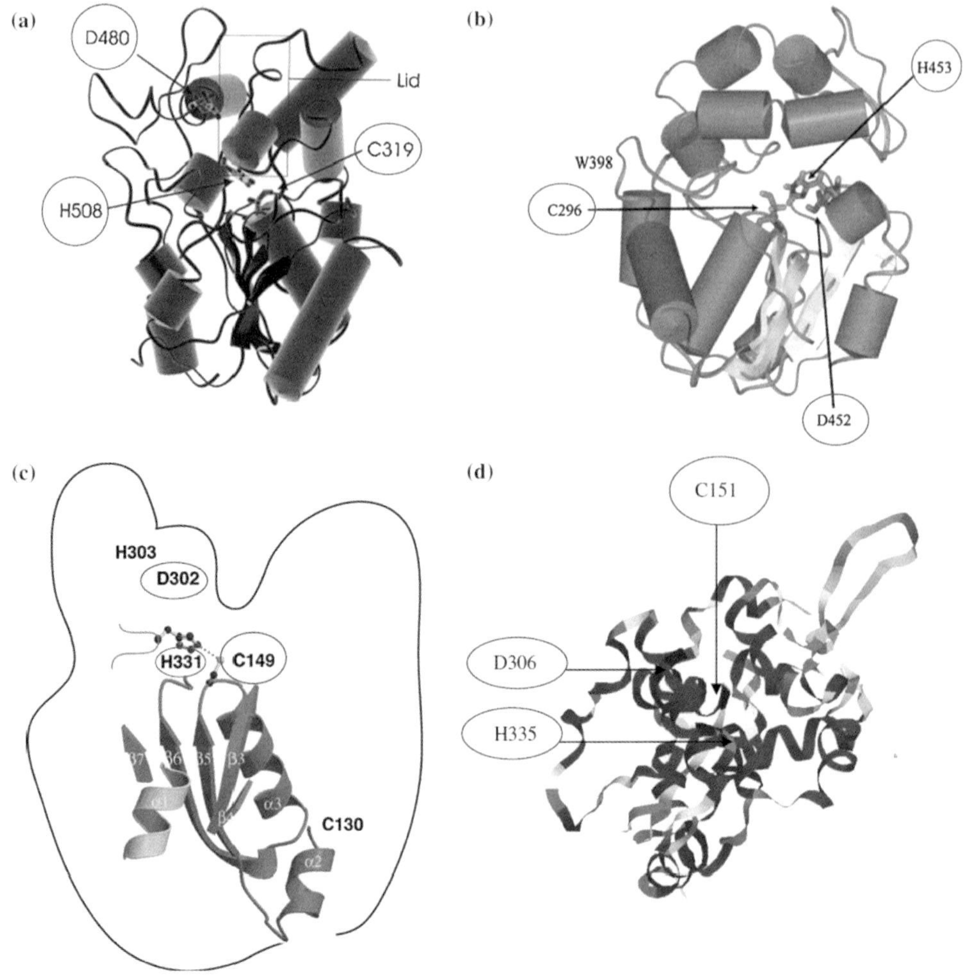

Figure 35.3 Schematic models of PHA synthases. (a) Type I PhaC from *R. eutropha* [13]; (b) type II PhaC from *P. aeruginosa* [18]; (c) type III PhaC from *A. vinosum* [19]; (d) type IV PhaC from *B. cereus* [20]. The catalytic triad residues are encircled. (Adopted from Valappil *et al.* [20].)

only *phaC* and *phaA* are in a single operon.) Meanwhile, in the case of a bacterium having type III or IV PHA synthase, the likely situation is that the same three-step PHA biosynthesis pathway is present, but the gene encoding PhaA is located in a different operon compared to the genes encoding PhaC and PhaB.

3-Ketoacyl-CoA thiolase (acetyl-CoA:acetyl-CoA-acetyl transferase; EC 2.3.1.9) is known as an enzyme that catalyzes the thiolytic cleavage of acetoacetyl-CoA into acyl-CoA plus acetyl-CoA, this reaction is the final step of fatty acid β-oxidation. However, in P(3HB) biosynthesis, the reverse reaction of thiolytic cleavage in β-oxidation is catalyzed by PhaA. In addition, the substrate specificity of PhaA is narrow, in the range of chain length of C3–C5 monomers [23]; hence, PhaA is specialized for scl-PHA biosynthesis. *R. eutropha* contains two 3-ketothiolases (PhaA and BktB) that are able to act in the biosynthetic pathway for PHA synthesis. The significant difference between these two enzymes is in their substrate specificity [24]: BktB shows substrate specificity toward longer monomers (C4 to C10) than PhaA, and also contributes to the 3-hydroxyvalerate (3HV) monomer supply for P(3HB-*co*-3HV) biosynthesis. The molecular weights of PhaA and BktB are 44 kDa and 46 kDa, respectively, and both thiolases exist as a homotetramer in the native state [23, 24].

Several protein-engineering studies (see Table 35.4) were conducted to provide an understanding of the catalytic mechanism of PhaA from *Zoogloea ramigera* [25–27]. Two cysteines in the active site of PhaA were identified by site-specific mutagenesis. The crystal structure of *Z. ramigera* PHA-specific 3-ketothiolase has been solved (see Figure 35.4) [28, 29], and the following functions of the amino acids have been proposed: the catalytic residues His348 (activation of Cys89), Cys89 (covalent acyl-CoA intermediate formation), and Cys378 (substrate activation).

35.3.3
Acetoacetyl-CoA Reductase (PhaB)

Acetoacetyl-CoA reductase is an *(R)*-3HA-CoA dehydrogenase (EC 1.1.1.36) and catalyzes the second step in the three-step P(3HB) biosynthetic pathway. This enzyme converts acetoacetyl-CoA into (*R*)-3-hydroxybutyryl-CoA with the simultaneous oxidation of NADPH or NADH as the cofactor. The acetoacetyl-CoA reductases from *Z. ramigera* and *R. eutropha* require NADPH for the reduction of acetoacetyl-CoA, and most PhaB are considered to be NADPH-dependent enzymes [6, 7, 30]. On the other hand, NADH-dependent acetoacetyl-CoA reductase involved in P(3HB) formation was found in *A. vinosum* and *Azotobacter viinerandi* [6, 7]. The PhaB from *Z. ramigera* is a homotetramer with identical 25 kDa subunits [31], but its crystal structure has not yet been solved.

PhaB plays an important role in PHA biosynthesis because the availability of reducing equivalents in the form of NADPH is considered to be the driving force for P(3HB) formation. Thus, the reaction catalyzed by PhaB is considered to be a bottleneck step, and PHA synthesis is probably influenced by the NADPH concentration in the cells. Indeed, gene dosage of the PHA biosynthetic pathway in recombinant bacteria did not enhance PHA accumulation markedly due to a limitation of the NADPH supply [7]. The substrate specificity of PhaB is broader than that of PhaA, as revealed by *in vivo* studies, in that PhaB can provide longer than C6 monomers based on a combination of substrate range analysis in recombinant bacteria [32]. To date, however, no protein-engineering studies have been conducted on this enzyme.

3-Ketoacyl-ACP reductase (FabG), the constituent of fatty acid biosynthesis pathway, is known as a PhaB homologous protein. The amino acid sequence of the PhaB of *B. megaterium* shows high homology (50% identity, 66%

Table 35.4 Sequence- and structure-based protein engineering of monomer-supplying enzymes.

Enzyme	Position	Purpose or changed property	Active or inactive[a]	Reference(s)
Z. ramigera PhaA	C89S	Active-site residue identification	I	[25]
	C378G/S	Active-site residue identification	I	[26, 27]
A. caviae PhaJ	D31A	Active-site residue identification	I	[42]
	H36A/N	Active-site residue identification	I	[42]
	S62A	Active-site residue identification	A	[42]
	L65A/G/I/V	Substrate specificity change	A	[43]
	V130A/G	Substrate specificity change	A	[43]
	L65A/V130G, L65G/V130G	Double mutation: substrate specificity change	I	[43]
A. hydrophila PhaJ	L65A/G	Substrate specificity change	A	[44]
	V130A/G	Substrate specificity change	A	[44]
	L65A/V130A, L65A/V130G	Double mutation: substrate specificity change	I	[44]
P. putida PhaG	D60A/E	Active-site residue identification	I	[47]
	S102A/T	Active-site residue identification	I	[47]
	H177A/R	Role of H-X(4)-D motif	I	[47]
	D182A/E	Role of H-X(4)-D motif	A	[47]
	H192A/R	Active-site residue identification	I	[47]
	D223A/E	Active-site residue identification	I	[47]
	H251A/R	Active-site residue identification	I	[47]

Table 35.4 *Continued*

Enzyme	Position	Purpose or changed property	Active or inactive[a]	Reference(s)
Pseudomonas sp. 61-3 PhaG	H177A	Role of H-X(4)-D motif	I	[46]
	A182D	Role of H-X(4)-D motif	A	[46]
	E183D	Role of H-X(4)-D motif	A	[46]
	E183Q	Role of H-X(4)-D motif	A	[46]
E. coli FabH	F89X	Saturation mutagenesis: substrate specificity change	A	[49]
R. eutropha PhaA-PhaB fusion	–	Simplifying the expression of P(3HB) biosynthetic pathway in plants	A	[51]

a A: active mutant, I: (inactive) mutant which shows significantly reduced activity.

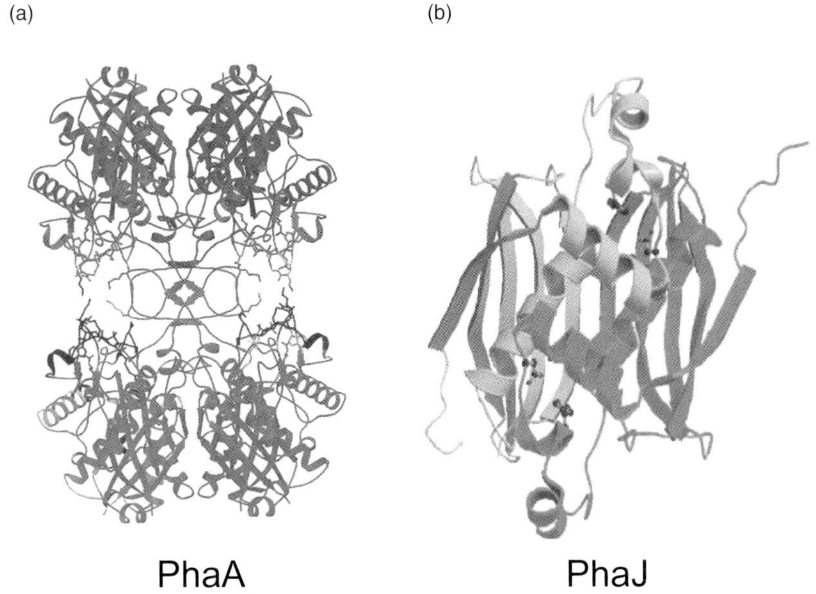

Figure 35.4 Crystal structure of monomer supplying enzymes for PHA biosynthesis. (a) PhaA from *Z. ramigera* (pdb: 1dlu, 1dlv, 1qfl, 1dm3) [28, 29]; (b) PhaJ from *A. caviae* (pdb: 1IQ6) [42].

similarity) to that of FabG of *Synechocystis* PCC 6803 [33], which suggests that PhaB and FabG may have evolved from the same origin. FabG has been shown to function as a monomer supplier for PHA biosynthesis in recombinant *E. coli*, as well as PhaB [33, 34]. In addition, *P. aeruginosa* RhlG involved in rhamnolipid biosynthesis has been identified as a PhaB and FabG homologous protein, and although the RhlG plays a role of bypass, it is not essential for PHA synthesis [35].

35.3.4
(*R*)-Specific Enoyl-CoA Hydratase (PhaJ)

This enzyme catalyzes the stereoselective ((*R*)-specific) hydration of *trans*-2-enoyl-CoA and generates (*R*)-3HA-CoA for PHA biosynthesis. Generally, (*S*)-specific enoyl-CoA hydratase is a constituent of β-oxidation, and therefore PhaJ and (*S*)-hydratase catalyze the competitive reaction towards β-oxidation intermediates. The $PhaJ_{Ac}$ from *Aeromonas caviae* is the first (*R*)-hydratase shown to be involved in PHA biosynthesis, and has been well studied [36]. The active $PhaJ_{Ac}$ forms a homodimer, and shows (*R*)-specific hydration activity towards *trans*-2-enoyl-CoA with four to six carbon atoms [37]. The size of the $PhaJ_{Ac}$ subunit is 14 kDa, and is the smallest among known (*R*)-hydratases. The recombinant product of a gene encoding the hydratase homologue ($PhaJ_{Rr}$) found in *R. rubrum* shows similar substrate specificity to $PhaJ_{Ac}$, but its characteristic in that $PhaJ_{Rr}$ possesses a C-terminal extension of amino acid residues and exists as a homotetramer in the native state [38]. Four *P. aeruginosa* genes homologous to the $phaJ_{Ac}$, referred to as $phaJ1_{Pa}$ to $phaJ4_{Pa}$, were also cloned and investigated as to whether their recombinant gene products conferred the ability to supply monomer units for PHA [39, 40]. Different substrate specificities were found for these hydratases; that is, PhaJ1 was specific for scl-enoyl-CoAs similar to $PhaJ_{Ac}$, while $PhaJ2_{Pa}$, $PhaJ3_{Pa}$, $PhaJ4_{Pa}$ preferred mcl- enoyl-CoAs. Qin *et al.* [41] defined the hydratase 2 motif as [YF]-X(1,2)-[LVIG]-[STGC]-G-D-X-N-P-[LIV]-H-X(5)-[AS] by analyzing the amino acid sequence of eukaryotic (*R*)-hydratases. A sequence similar to the hydratase 2 motif is also found in bacterial (*R*)-hydratases.

The crystal structure of $PhaJ_{Ac}$ at a resolution of 1.5 Å (Protein Data Bank accession code 1IQ6) was determined by Hisano *et al.* [42]. As shown in Figure 35.4, the monomer structure of the enzyme consists of a five-stranded antiparallel β-sheet and five α-helices. Two of the monomers are associated with one another to form a functional homodimer with an extended 10-strand β-sheet; this structure is generally referred to as a 'hotdog fold'. The catalytic residues, Asp31 and His36, are located deep in the substrate-binding pocket; these amino acids define the hydratase 2 motifs and are highly conserved among (*R*)-hydratases. They also play critical roles in catalysis, as revealed by a mutational study [42]. A structural docking model of this enzyme with substrate provides information on its substrate recognition. The binding site for the acyl moiety of the substrate is surrounded by side chains consisting of Ser62, Leu65, Pro70, Ser74, Tyr76 and Val130. The side chains consisting of these residues define the depth and width of the acyl-chain-binding pocket, which specifically accepts C4 to C6 substrates but does not permit the entrance of substrates longer than C6.

Site-specific mutagenesis studies of $PhaJ_{Ac}$ were carried out to broaden the substrate specificity [43]. Based on structural information, amino acid substitutions were introduced into $PhaJ_{Ac}$ by site-specific mutagenesis to Ser62, Leu65 and Val130, which are expected to define the width and depth of the acyl-chain-binding pocket. Of the mutants generated, Leu65Ala, Leu65Gly and Val130Gly exhibited significantly higher activities toward octenoyl-CoA than the wild-type enzyme; hence, the acyl chain length substrate specificity of $PhaJ_{Ac}$ can be varied. This study was the first protein-engineering to report successfully an achieved enzyme modification based on rational design. Recently, a similar study was performed for the (*R*)-hydratase from *Aeromonas hydrophila* [44].

35.3.5
(*R*)-3-Hydroxyacyl-ACP-CoA Transferase (PhaG)

Some pseudomonads produce mcl-PHA from sugars through fatty acid *de novo* biosynthesis. In this pathway, a new enzymatic activity is required to convert from (*R*)-3-hydroxyacyl-ACP to (*R*)-3-hydroxyacyl-CoA. Rehm *et al.* [45] determined that the gene product of *phaG* from *P. putida* is responsible for this conversion. Later, other genes encoding PhaG were cloned from some pseudomonads. The molecular weight of PhaG is 34 kDa, as deduced from the nucleotide sequences, but information on the quaternary structure of this protein has not been reported.

A H-X(4)-D or similar motif-encoding sequence is found in all PhaG, and this sequence appears to be commonly shared with a variety of glycerolipid acyltransferases. Matsumoto *et al.* [46] studied the small sequence difference of H-X(4)-D-like motif, H-X(4)-A–E, found in $PhaG_{Ps}$ from *Pseudomonas* sp. These authors showed the importance of His177 in the H^{177}-X(4)-A–E on PhaG activity by substitution with Ala. They also studied the role of Asp in the H-X(4)-D motif. If H-X(4)-D were the optimal sequence motif for PhaG catalysis, then an activity elevation of $PhaG_{Ps}$ could be achieved by a substitution of Ala182 by Asp in the H-X(4)-A^{182}-E sequence. To this end, the mutant A182D was generated and examined, but no significant change observed. Furthermore, it was also demonstrated, by generating two mutants (E183D/Q), that Glu183 is not essential for PhaG activity. In contrast, Hoffmann *et al.* [47] showed Ser102 and His251 to be essential for the activity of *P. putida* PhaG and to form part of a catalytic triad. Thus, these results suggest that the acyltransfer mechanism of PhaG may be different from that of glycerolipid acyltransferases.

35.3.6
3-Ketoacyl-ACP Synthase III (FabH)

3-Ketoacyl-acyl carrier protein (ACP) synthase III (FabH) is a constituent of fatty acid biosynthesis, and it is not a specialized enzyme for PHA synthesis. It was shown previously that coexpression of the PHA synthase gene and the 3-ketoacyl-ACP synthase III gene (*fabH*) from *E. coli* led to the production of a P(3HB) homopolymer in recombinant *E. coli* grown in the presence of glucose [48].

Recently, Nomura *et al.* [49, 50] demonstrated PHA copolymer production in recombinant *E. coli* by using engineered FabH mutants. The FabH proteins from various bacterial species have been shown to have very different substrate specificities. For example, the *E. coli* FabH protein has specificity for carbon substrates of two to four carbons in length, while the *Mycobacterium tuberculosis* FabH protein displays specificity for carbon substrates of 10 to 16 carbons in length *in vitro*. In addition, the crystal structures of both the *E. coli* FabH protein and the *M. tuberculosis* FabH protein have been determined; a comparison of the primary amino acid sequences and the two crystal structures revealed a potential explanation for the difference in substrate specificity observed between the *E. coli* FabH and the *M. tuberculosis* FabH. In the *E. coli* FabH protein, there is a Phe residue at position 87; this amino acid occupies the end of the predicted substrate-binding pocket and obstructs the binding of straight-chain fatty acids longer than four carbons. Thus, the FabH from *E. coli* was modified by saturation point mutagenesis at the codon encoding amino acid 87 of the FabH protein sequence, and the abilities of these strains to accumulate PHA from glucose were assessed. The overexpression of several of the FabH mutants enabled recombinant *E. coli* to induce the production of C4 to C10 monomers and subsequently to produce unusual P(3HB-*co*-3HA) copolymers containing scl- and mcl-units. This was another successful case of structure-based enzyme modification and control of copolymer composition following the modification of PhaJ.

35.4
Directed Evolution of PHA Synthases

As described above, PHA synthase (PhaC) catalyzes the polymerization of 3-HA-CoA monomers to PHAs, and has been subjected to various forms of protein engineering to improve the enzyme activity or substrate specificity. This section describes the recent studies on PHA synthase engineering and also summarizes the investigations that have utilized engineered PHA synthases. Applications of genetic engineering to improve PHA synthases are summarized in Table 35.3. Unfortunately, for the 'rational' design of the enzyme, the lack of a suitable structural model for any PHA synthase has limited attempts to improve the activity and stability and to alter the substrate specificity of these enzymes to 'irrational' approaches, such as random mutagenesis and gene shuffling. Once function-related residues have been identified by these 'irrational' approaches, site-specific saturation mutagenesis of residues known to affect the activity of the enzyme, and the recombination of beneficial mutations can be used to improve the enzymes [52–54]. Generally, natural diversity provides us with attractive starting materials for artificial evolution as it represents functionalized sequence spaces to some extent. A vast population (over 60 species) of randomly screened PHA-producing bacteria suggests that attractive prototype enzymes for molecular breeding should exist. Among these, an enzyme evolution approach has been applied to the following type I and type II PHA synthases derived from bacteria such as *R. eutropha* (type I), *Aeromonas caviae* (type I) and *Pseudomonas* sp. 61-3 (type II) (Table 35.5).

35.4.1
Engineering of the Type I Synthases

The first study to establish methods for genetically engineering PHA synthase were conducted in 2001, using the best biochemically studied enzyme, the *R. eutropha* PHA synthase [55]. This pioneering study involved an establishment of the '*in vitro* evolutionary technique', which consisted of an error-prone PCR method coupled to two convenient screening methods to generate a 'fitness landscape' representative of mutations with varied functions [55]. A mutant library of the *R. eutropha phaC* gene was prepared by colony formation of transformant cells of *E. coli* JM109. The change in P(3HB) accumulation resulting from the introduction of mutations into the *R. eutropha phaC* gene was judged by using the convenient and sensitive colony viable staining plate assay (primary screening). For the precise quantification of many primary screens, P(3HB) content was determined with high-performance liquid chromatography (HPLC) on a smaller scale by chemically converting P(3HB) into crotonic acid. The synthase activity correlated well with the level of P(3HB) accumulation in *E. coli*, which implied that the synthase activity could be readily predicted by monitoring the level of P(3HB) accumulation based on the *in vivo* assay system established.

In the past it has been difficult for research groups to obtain *R. eutropha* $PhaC_{Re}$ mutant enzymes with higher activities. Therefore, a more reliable evolutionary program was required to explore the beneficial mutations for activity improvement, because it is not easily possible to distinguish subtle differences in activity among many activity-positive mutants and wild-type enzymes when using the existing screening conditions. In order to search for beneficial mutations for activity improvement of this enzyme, multistep mutations [56], including an activity loss and an intragenic suppression-type activity reversion, were attempted. When the mutant enzymes had been identified by a primary mutation analysis, a secondary round of mutation was used to evolve these enzymes to proteins with better characteristics than the wild-type enzyme. This approach was technically advantageous, and allowed (in positive-selection manner) improved mutant PHA synthases to be obtained from the primary mutant enzyme with reduced activity, compared to a screening from the library including the wild-type molecules and molecules with apparently increased activity.

Initially, by using this 'intragenic suppression-type mutagenesis', a 2.4-fold increase in specific activity towards 3HB-CoA compared to the wild-type enzyme was acquired by a mutation of Phe420Ser in a type I PHA synthase [57]. The next screened beneficial mutation, Gly4Asp, exhibited higher levels of protein accumulation and P(3HB) production compared to the recombinant *E. coli*

Table 35.5 Summary of evolutionary engineering studies of PHA synthases.

Year	Enzyme source[a]	Method	Polyester[c]	Changed enzyme property	Reference
2001	*Ralstonia eutropha*	Random mutagenesis	P(3HB)	Activity	[55]
2002	*Ralstonia eutropha*	Random mutagenesis	P(3HB)	Activity, thermostability	[57]
	Aeromonas caviae (*punctata*)	Random mutagenesis	P(3HB-*co*-3HHx)	Activity, substrate specificity	[65]
	Ralstonia eutropha	Intragenic suppression mutagenesis	PHA	Activity, substrate specificity	[57]
	Ralstonia eutropha and *Pseudomonas aeruginosa*	Gene shuffling	PHA	Activity, substrate specificity	[13]
	Aeromonas caviae (*punctata*)	Random mutagenesis (*in vivo*)	P(3HB-*co*-3HHx)	Activity, substrate specificity	[66]
2003	*Pseudomonas* sp. 61-3	Random mutagenesis, Site-specific saturation mutagenesis, Recombination	P(3HB-*co*-3HA)	Activity, substrate specificity	[67]
	Pseudomonas resinovorans	Site-specific chimeragenesis[b]	PHA	Substrate specificity	[77]
2004	*Pseudomonas* sp. 61-3	Random mutagenesis, Site-specific saturation mutagenesis, Recombination	P(3HB-*co*-3HA)	Activity, substrate specificity	[68]
	Pseudomonas putida GPo1	Localized semi-random mutagenesis	PHA	Activity, substrate specificity	[78]
	Pseudomonas oleovorans	PCR-mediated random chimeragenesis[b]	PHA	Activity, substrate specificity	[79]
	Ralstonia eutropha	Site-specific saturation mutagenesis	P(3HB), P(3HB-*co*-3HA)	Activity, substrate specificity	[69]

Table 35.5 *Continued*

Year	Enzyme source[a]	Method	Polyester[c]	Changed enzyme property	Reference
2005	*Ralstonia eutropha*	Intragenic suppression mutagenesis, Site-specific saturation mutagenesis	P(3HB)	Protein expression	[58]
	Ralstonia eutropha	Recombination	P(3HB)	Activity	[61]
	Pseudomonas sp. 61-3	Site-specific saturation mutagenesis, Recombination	P(3HB-*co*-3HA)	Activity, substrate specificity	[70]
2006	*Pseudomonas* sp. 61-3	Site-specific saturation mutagenesis Recombination	P(3HB), P(3HB-*co*-3HA)	Activity, substrate specificity	[71]
2007	*Aeromonas caviae* (*punctata*)	Site-specific saturation mutagenesis	P(3HB-*co*-3HHx)	Activity, substrate specificity	[74]

a Strain of bacteria from which the PHA synthase was derived.
b Method for generation of chimeric genes used in this study (see text).
c Polymer produced by the study. P(3HB), poly(3-hydroxybutyrate); P(3HB-*co*-3HHx), poly(3-hydroxybutyrate-*co*-3-hydroxyhexanoate); PHA, poly(3-hydroxyalkanoate); P(3HB-*co*-3HA), poly(3-hydroxybutyrate-*co*-3-hydroxyalkanoate).

strain harboring the wild-type PHA synthase [58]. As with intragenic suppression-type mutagenesis, the second-site reversion was seen to be dependent or independent of primary mutation in the activity. Secondary mutations of Phe420Ser and Gly4Asp, as the latter cases, were independent of primary mutation. A similar case was demonstrated for the thermostabilization of yeast iso-1-cytochrome *c* [59]. In contrast, a directed evolution study on the cold-adaptation of an industrial protease, subtilisin BPN′ [56], showed that the second-site reversion was dependent on the primary mutation in the activity. Thus, it is feasible that diverse pathways for enzyme improvement might be demonstrated by using primary mutants of PhaC as a set for secondary mutagenesis; this would allow an efficient expansion of the possibility to explore beneficial mutations.

Site-specific saturation mutagenesis represents a powerful tool for genetic engineering and further examination of potentially beneficial sequences in PHA synthases, as shown for other types of enzyme [60]. Site-specific saturation mutagenesis was performed on the codon encoding the Gly4 residue of the *R. eutropha* PHA synthase, and many substitutions resulted in a much higher P(3HB) content as well as higher molecular weights of the polymers. These substitutions were subsequently combined with the Phe420Ser mutation, and this resulted in increased polymer yields but lower molecular weights for the P(3HB) polymers produced [61].

Based on the homology between *R. eutropha* PHA synthase and *Burkholderia glumae* lipase, the tertiary structure of which has been resolved using X-ray analysis, a threading model of the PhaC enzyme was proposed by Rehm *et al.* [13]. This was then used to map mutations generated by a single gene shuffling of four mutants that had reduced *in vivo* activity compared to the wild-type PHA synthase [13]. Although functional mapping of the mutant enzymes shown diagrammatically on this threading model may predict which amino acid residues are responsible for enzymatic properties such as activity or dimer formation, the increased activity achieved by Taguchi *et al.* [55] could not be predicted by a structural model.

The properties of PHAs are determined by the type of monomer. For example, it is known that P(3HB-*co*-3HA) random copolymers with a monomer ratio of 95 mol% 3HB to 5 mol% 3HA have properties similar to those of low-density polyethylene [62, 63]. Because these properties are desirable for the bulk use of PHAs as commodity plastics, the development of methods for the efficient production of PHA copolymers has become an important research topic. The *A. caviae* (*punctata*) PHA synthase is unique among type I PHA synthases as it is able to synthesize not only the P(3HB) homopolymer but also random copolyesters of 3HB and 3-hydroxyhexanoate (3HHx) [64]. Kichise *et al.* subsequently performed the first successful *in vitro* molecular evolution experiments on PhaC from *A. caviae* by applying an *in vitro* evolutionary technique to a limited region of the *phaC* and coexpressing those mutants with the monomer-supplying enzyme genes, *phaAB* from *R. eutropha* and *phaJ* from *A. caviae* to supply monomers from glucose or dodecanoate, respectively [65]. Two evolvants exhibited an increased activity towards 3HB-CoA of 56% and 21%, respectively, compared to the wild-type enzyme in *in vitro* assays. These mutations led to an enhanced accumulation (up to 6.5-fold higher than the wild-type PhaC) of P(3HB-*co*-3HHx) and increases in the 3HHx mol fraction (16–18 mol% compared to 10 mol% for the wild-type PHA synthase) in recombinant *E. coli* LS5218 strains grown on dodecanoate. Two single mutations, Asn149Ser and Asp171Gly for T3-11), were not highly conserved among PHA synthases. The success of these studies prompted the use of *in vitro* evolutionary technique to the entire *A. caviae* phaC gene in order to identify other beneficial mutations.

In a separate study, *A. caviae* PHA synthase was engineered *in vivo* using the mutator strain *E. coli* XL1-Red. This has a 5000-fold higher mutation rate than wild-type *E. coli*, and mutants were again screened for enhanced P(3HB) accumulation in recombinant *E. coli*. [66]. A higher *in vitro* specific activity (up to five fold) and a higher PHA content (up to 126% of wild-type) were observed for the overproducing mutants. In addition, mutants synthesized PHAs with increased average molecular weight but, in contrast to the previous study, the 3HHx fraction differed only slightly from the wild-type composition.

35.4.2 Engineering of the Type II *Pseudomonas* Species PHA Synthases

Unlike their type I counterparts, type II PHA synthases typically have substrate specificity towards mcl-3HA-CoA substrates but relatively poor substrate specificity towards scl-3HA-CoA substrates such as 3HB-CoA. An exception to this is the type II PHA synthase of *Pseudomonas* sp. 61-3, which has significant substrate specificity towards the 3HB-CoA (Table 35.2). In a landmark study conducted by Takase *et al.*, the *in vitro* evolutionary technique was applied to the PhaC1 PHA synthase from *Pseudomonas* sp. 61-3 to increase activity towards 3HB-CoA monomers [67]. Substitutions at two amino acid residues, Ser325 and Gln481, were found to dramatically affect the production of P(3HB) homopolymer in recombinant *E. coli* with glucose as the carbon source. The codons for these amino acids were subjected to site-specific saturation mutagenesis, and several individual substitutions [Ser325Cys, Ser325Thr, Gln481Lys, Gln481Met and Gln481Arg] were found that could dramatically increase the level of P(3HB) production. These mutations were combined as double mutants to further increase the level of P(3HB) production (340–400-fold higher than the wild-type enzyme) [67]. In a subsequent study, it was shown that these engineered PhaC1 synthases were able to accumulate higher amounts of scl-mcl PHAs compared to the wild-type PhaC1 synthase [68]. In addition, there was a shift in the monomer composition towards 3HB in the P(3HB-*co*-3HA) copolymer produced by the mutant synthases compared to the wild-type PHA synthase [68]. These changes in the *in vivo*-produced P(3HB-*co*-3HA) copolymer molar compositions correlated well with the *in vitro* biochemical data of the substrate specificity and activity of the enzymes, and represents one of the most well-rounded studies reported to date [68].

The results of these studies for the type II PHA synthase would be very useful for evaluating a similar evolution strategy for other types of PHA synthases, based on the amino acid sequence alignment of the PHA synthases. For example, position 481 in PhaC1 PHA synthase from *Pseudomonas* sp. 61-3 was found to be one of the residues which determined the substrate specificity of the enzyme, as described above. Interestingly, the amino acid residues corresponding to the position of this enzyme are conserved within each type of PHA synthases: Ala for type I, Gln for type II, Gly for type III, and Ser for type IV enzymes. Thus, the effects of mutating the highly conserved alanine (Ala510) of the *R. eutropha* PHA synthase (corresponding to position 481 in *Pseudomonas* sp. 61-3 PhaC1) were analyzed via site-specific saturation mutagenesis. Mutations, Ala510Met/Gln/Cys, at Ala510 were found to affect the substrate specificity of the *R. eutropha* PHA synthase, allowing a slightly higher 3HA incorporation compared to the wild-type PHA synthase in *R. eutropha* PHB^{-}4 (PHA-negative mutant), while other mutations either reduced or completely eliminated the amount of 3HA that could be incorporated into the polymer compared to the wild-type enzyme when grown on dodecanoate [69].

Glu130 was also identified during the *in vitro* evolution screening as a positive mutation. The Glu130Asp mutant was able to accumulate 10-fold higher P(3HB) from glucose compared to the wild-type PhaC1 enzyme [70]. The results of this study also showed that polymers produced by PHA synthases with mutations at Glu130Asp had higher molecular weights than those from other enzymes. Finally, a mutation at the Ser477 residue of PhaC1 further changed the substrate specificity towards scl-3HA-CoA monomers; however, when combined with a Ser325Cys or Ser325Thr mutation, this exhibited a synergistic enhancement of PHA production in addition to altered substrate specificity [71]. 'Mutation scrambling' among four beneficial positions (130, 325, 477, 481) for activity increase, change in substrate specificity and regulation of polymer molecular weight, would further create new 'super-enzymes'. Most recently, a possible mechanistic model for PHA polymerization has been proposed on the basis of the accumulated evolutionary studies [71]. The useful evolvants obtained in these systematic enzyme evolution have also been supplied to other bacteria [72–74] and plants [75, 76].

Additional engineering studies have been performed using type II PHA synthases. In a study reported by Solaiman, two isogenic PHA synthase genes (*phaC1* and *phaC2*) from *Pseudomonas resinovorans* were used by exchanging the α/β-hydrolase-fold coding regions to develop the hybrid genes *pha7* and *pha8*. The gene products of *pha7* and *pha8* were polymers with repeating unit compositions similar to the wild-type enzymes when grown on decanoate as the carbon source (23–27 mol% C8 and 73–78 mol% C10) [77]. However, two deletion mutants identified during construction of the *pha7* and *pha8* hybrid genes produced polymers with different repeating unit concentrations (40–45 mol% C8 and 55–60 mol% C10) compared to the wild-type PHA synthases and *pha7* and *pha8* genes [77]. The substrate specificity of PHA synthase 1 from *Pseudomonas putida* GPo1 was altered by localized semi-random mutagenesis, whereby the enzyme was evolved by using PCR-based gene fragmentation with degenerate primers followed by reassembly of the chimeric genes [78]. Based on multiple sequence alignments of PHA synthases, six conserved regions located in the predicted α/β-hydrolase fold were used to design degenerate primers corresponding to 23 amino acids distributed across the six regions. The fragments were mixed, and primerless reassembly PCR then used first to generate the recombined *phaC* genes and then to create 20 000 clones in *E. coli*. These plasmids were screened by transformation into *P. putida* GPp104, and then narrowed down to 13 candidates based on the opacity of 10 000 transformants. Among these 13 transformants, six candidates were chosen for transformation into *R. eutropha* PHB$^-$4, based on the mol% composition of the polymers they produced in *P. putida* GPp104. One of these candidates had the highest accumulation of PHA content and an increased 3HB mol% composition compared to the copolymer produced by the wild-type PhaC1. This mutant has several mutations from the semi-random mutagenesis that likely contributed to the changed substrate specificity and activity [78].

Another study was aimed at improving PHA synthases by a combinatorial strategy using the *P. oleovorans phaC1* gene as a scaffold from which to generate chimeras [79]. PCR products corresponding to the putative catalytic regions of PHA synthase enzymes were generated from soil DNA extracts and cloned into a synthetic *P. oleovorans phaC1* gene with a linker replacing the catalytic region of the enzyme. Transformants were screened with Nile blue A dye, and among 1478 clones obtained five were isolated that could produce more PHA than the native PhaC1 enzyme. Sequence analysis revealed that the active synthases all contained sequences corresponding to different species of *Pseudomonas*, and also contained 17 to 20 amino acid differences from the *P. oleovorans* PhaC1. On the other hand, the inactive chimeras contained sequences from type I synthases, indicating that this region could not be interchanged between the two types (I and II) of synthases [79].

35.5 Structure–Function Relationship of PHA Depolymerases

The biodegradation and biosynthesis of PHA have an interesting relationship. PHA degradation occurs in two forms, extracellular and intracellular, catalyzed by the actions of extracellular PHA depolymerase and intracellular PHA depolymerase, respectively. Of the two enzymes, the extracellular form has received much more attention in terms of mechanistic studies, notably of how the enzyme adsorbs to the PHA surface. In recent years the mechanisms of the PHA depolymerases have been the subject of many protein engineering studies, an overview of which is presented in the following sections and Table 35.6.

35.5.1 Domain Structure of Extracellular PHA Depolymerases

In the natural environment, the enzymatic degradation of PHA occurs by the action of specific hydrolyzing enzymes, extracellular PHA depolymerases (EC 3.1.1.75), which are secreted from various microbes independently of their PHA-accumulating ability. The extracellular PHA depolymerases are able to degrade partially crystalline PHA, but are inactive on amorphous PHA [8]. It has been proposed that degradation of the crystalline region of PHA proceeds in three steps: (i) adsorption of the enzyme to PHA; (ii) a nonhydrolytic disruption of the structure of PHA; and (iii) the hydrolysis.

The domain structure of bacterial extracellular depolymerases has been studied on the basis of amino acid sequence similarity and protein engineering [8]. As shown in Figure 35.5, bacterial extracellular P(3HB) depolymerases are typically comprised of three functional domains: catalytic (320–400 aa); linker (50–100 aa); and substrate-binding (40–60 aa). These three functional domains are essential for the enzymatic degradation of water-insoluble PHA through an adsorption–disruption–hydrolysis process, and the function and features of each domain are described below.

The catalytic domain contains a lipase-like catalytic triad (serine, aspartic acid, and histidine residues), as well as a signature sequence of a pentapeptide G-X-S-

Table 35.6 Sequence- and structure-based protein engineering studies of PHA depolymerases (PhaZ).

Enzyme	Position/domain	Purpose	Hydrolysis activity[a]	Reference
Extracellular PHA depolymerase				
P. lemoignei PhaZ4	55 aa deletion at C terminus	SBD identification	A	[91]
P. lemoignei PhaZ5	S138A/T	Active-site residue identification	I	[92]
	S195A	Active-site residue identification	A	[92]
P. lemoignei PhaZ7	H47A	Active-site residue identification	I	[93]
	A134G	Active-site residue identification	A	[93]
	A134G/H135L	Double mutation: active site residue identification	A	[93]
	S136A	Active-site residue identification	I	[93]
	S136T	Active-site residue identification	A	[93]
	D242A/N	Active-site residue identification	I	[93]
	A256A/N	Active-site residue identification	A	[93]
	H306A	Active-site residue identification	I	[93]
R. pickettii PhaZ	S139A	Active-site residue identification	I	[92, 94]
	S196A	Active-site residue identification	A	[92]
	D214G	Active-site residue identification	I	[94]
	H273D	Active-site residue identification	I	[94]
	Domain deletion, inversion chimera mutants	Identification of structure and function	A	[94]
	SBD	GST-SBD fusion: binding study	–	[95]
	SBD	His-tagged SBD	–	[90]
P. fluorescens PhaZ	S172A	Active-site residue identification	I	[96]

Table 35.6 *Continued*

Enzyme	Position/domain	Purpose	Hydrolysis activity[a]	Reference
P. funiculosum PhaZ	S39A	Active-site residue identification	I	[83]
P. stutzeri PhaZ	SBD	GST-SBD fusion: Binding study	–	[97]
	SBD deletion	Use for chemo-enzymatic PHA synthesis	A	[98]
C. testosterone PhaZ	SBD	GST-SBD fusion: binding study	–	[99]
C. acidovorans PhaZ	SBD	GST-SBD fusion: binding study	–	[100]
Marinobacter sp. NK-1	Domain deletion	GST-domain fusion: Identification of structure and function	–	[101]
Intracellular PHA depolymerase				
R. eutropha PhaZa1	C87A	Active-site residue identification	A	[85]
	S118A	Active-site residue identification	A	[85]
	H120Q	Active-site residue identification	A	[85]
	C183A	Active-site residue identification	I	[85]
	C183S	Active-site residue identification	A	[85]
	D355A	Active-site residue identification	I	[85]
	D356A	Active-site residue identification	A	[85]
	C370A	Active-site residue identification	A	[85]
	H388Q	Active-site residue identification	I	[85]
R. spheroids PhaZ1	C178A	Active-site residue identification	I	[102]
B. thuringiensis PhaZ	S102A	Active-site residue identification	I	[103]

a A: active mutant, I: inactive mutant which shows significantly reduced activity.

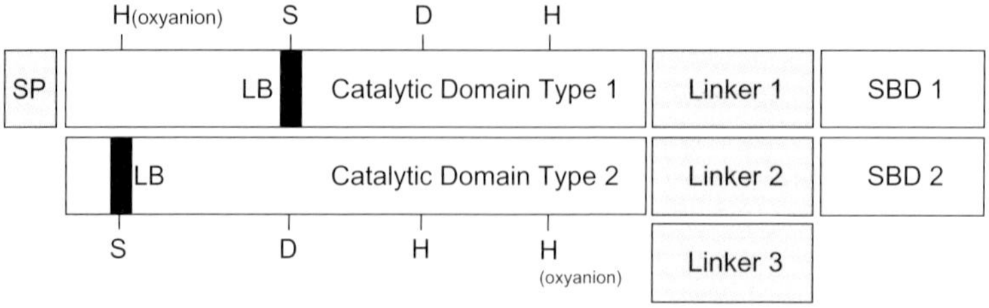

Figure 35.5 Domain model of extracellular PHA depolymerases [8]. SP = signal peptide; LB = lipase box; Catalytic domain type 1–2, one of two possible types of catalytic domains; Linker 1–3, one of three possible types of linking domains (Fn3, Thr, or Cad); SBD 1–2, one or two types of substrate-binding domain.

X-G, known as the lipase box [8]. Depending on the position of lipase box in PHA depolymerase, the catalytic domains have been classified into two types, I and II. In the type I domain, the lipase box is located in the center of the primary structure, whereas in the type II domain it is located at the N terminus of the structure.

The substrate-binding domain (SBD) is responsible for adsorption of the enzyme to the surface of the solid PHA; this helps the catalytic domain to interact with PHA chains, thus catalyzing their efficient hydrolysis. Based on sequence similarity, two different types of SBD, types I and II, can be differentiated, yet several amino acids – including positively charged amino acids such as histidine (His49), arginine (Arg44) and cysteine (Cys2) – are conserved in each type [8]. Recently, it has been proposed that the SBD has an additional and more active function of disrupting the structure of the PHA. This suggestion was based on the findings of Murase *et al.* [80], who reported that a catalytically inactive PHA depolymerase could change the morphology of a PHA single crystal, without changing the PHA molecular weight.

The role of the linker domain is not clearly understood, but it may function as a spacer that introduces a flexible region between the catalytic and SBDs to increase the hydrolytic efficiency of the catalytic domain. Based on sequence similarity, the linker domains are also divisible into three types: a cadherin-like type; a fibronection type III type; and a threonine-rich type. A recent study by Kataeva *et al.* [81] reported that a homologous protein to the linker domain of PHA depolymerase, the fibronectin type III homology domain of cellobiohydrolase ChbA from *Clostridium thermocellum*, also exhibited a disruptive function against crystalline substrates; this suggested that the linker domain may have an additional function of disrupting the structure of the PHA, similar to the SBD.

Some fungi are known to secrete PHA depolymerase [8]. For example, an extracellular PHA depolymerase from *Penicillium funiculosum* has been purified and well-characterized [82]. This has a molecular mass of about 33 000 Da, and is the smallest among the known extracellular depolymerases, the molecular masses of which typically range from 40 000 to 57 000 Da. The enzyme efficiently degrades

P(3HB) and a trimer of 3HB, although the degradation of P(3HB) was less efficient by two orders of magnitude than that of a multidomain enzyme from *R. pickettii*. Most recently, Hisano *et al.* [83] reported the crystal structure of the P(3HB) depolymerase from *P. funiculosum* at a resolution of 1.71 Å. The results showed a single-domain structure with a circularly permuted variant of the α/β-hydrolase fold. The same authors also determined the structure of the catalytically inactive mutant (Ser39Ala mutant)–3HB trimer methyl ester substrate complex at a resolution of 1.66 Å, as shown in Figure 35.6, and revealed the binding mode for a trimer substrate. In this depolymerase, solvent-exposed hydrophobic residues are aligned flat on the surface of the enzyme, serving as a polymer-adsorption site.

35.5.2
Intracellular PHA Depolymerase

Intracellular PHA depolymerase, which degrades PHA in an amorphous state, is a very important enzyme for the intracellular mobilization of PHA [9]. To date, only a few intracellular PHA depolymerase (PhaZa1) genes have been identified and characterized. *R. eutropha* H16 is the most well-characterized bacterium with

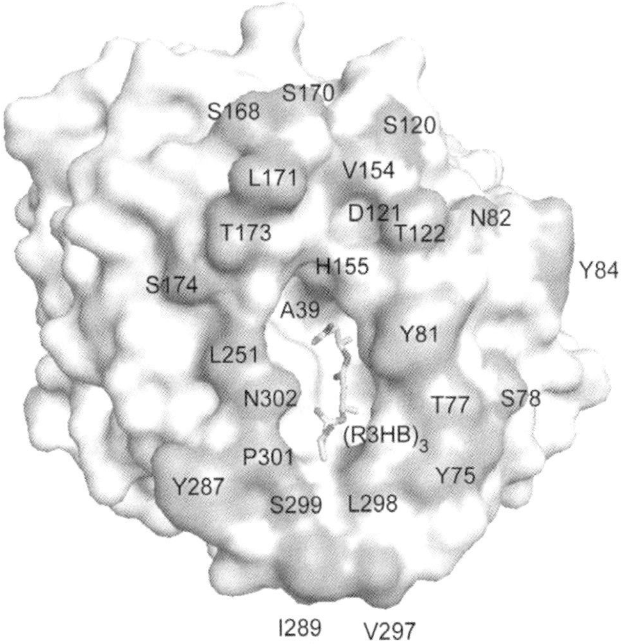

Figure 35.6 Crystal structure of PHA depolymerase from *P. funisulosum* (pdb: 2D80, 2D81) [83]. Positions of solvent-exposed hydrophobic residues, as well as polar and catalytic triad residues, are indicated. Residues Leu251, Leu298, Ser299, Pro301 and Asn302 also contribute to the substrate-binding site. A model of the (*R*)-3HB trimer bound in the crevice is shown as a yellow stick model.

respect to intracellular PHA degradation at the molecular level. The *phaZa1* gene of *R. eutropha* H16, which was the first intracellular PHA depolymerase gene to be cloned, encodes a protein with no classical lipase box (G-X-S-X-G) [84]. *R. eutropha* PhaZa1 was found to exist only as a P(3HB) granule-bound form in cells, its main hydrolytic products being 3HB oligomers. PhaZa1 shows a relatively high similarity to several proteins in databases, but no similarity with intracellular PHO depolymerase sequences of *P. oleovorans* or *P. aeruginosa*, nor with the extracellular P(3HB) depolymerases of *R. pickettii* T1 and *P. lemoignei*. Kobayashi *et al.* [85] constructed several mutants of PhaZa1 in order to identify the catalytic triad, and showed Cys183, Asp355 and His388 to be essential for PhaZa1 activity. Recently, a different type of intracellular P(3HB) depolymerase gene (*phaZd*) was cloned from *R. eutropha* His16 [86]. PhaZd is capable of hydrolyzing amorphous PHA into various 3HB oligomers, but the 3HB monomer is rarely detected as a hydrolytic product.

35.5.3 Amino Acid Residues Related to Binding Affinity

P(3HB) depolymerase from *R. pickettii* (PhaZ) adsorbs to P(3HB) via its SBD to enhance P(3HB) degradation. In order to evaluate the amino acid residues participating in P(3HB) adsorption, Hiraishi *et al.* [87] carried out a random mutagenesis study of the SBD from PhaZ. Genetic analysis of the isolated mutants with lowered binding activities showed that Ser, Tyr, Val, Ala and Leu residues in the SBD were replaced by other residues, and those residues were essential for full activity of both P(3HB) adsorption and degradation. These results suggested that PhaZ adsorbs onto the surface of P(3HB) not only via hydrogen bonds between hydroxyl groups of Ser in the enzyme and carbonyl groups in the P(3HB) polymer, but also via hydrophobic interaction between hydrophobic residues in the enzyme and methyl groups in the P(3HB) polymer.

Similarly, Jendrossek *et al.* [88] studied the function of poly(3-hydroxyoctanoate) (PHO) depolymerase from *P. fluorescens* GK13 by PCR random mutagenesis. Their results showed that the phenotypes of the recombinants depended on the PHO-degrading ability of their mutant enzymes, in which Leu and Phe residues were replaced and probably were involved in the interaction between the enzyme and PHO.

Fujita *et al.* [90] evaluated the adhesive forces between P(3HB) and P(3HB) depolymerase from *R. pickettii* by using atomic force microscopy (AFM)-based force–distance measurements. The SBD, fused with a histidine tag at the N terminus, was prepared and immobilized on the AFM tip surface via a self-assembled monolayer with a nitrilotriacetic acid group. By using these functionalized AFM tips, the single rupture force for P(3HB) was estimated at approximately 100 pN, which was much larger than that for the cleavage of single hydrogen bonding, as also revealed by AFM. Thus, the single rupture force estimated by these authors was due to the interaction between multiple amino acid residues rather than to one amino residue at the SBD and P(3HB) surface.

35.6
Application of PHA-Protein Binding Affinity

The surfaces of native PHA granules produced in bacterial cells are known to incorporate many proteins that show binding affinity to the granules' hydrophobic cores. The most abundant of these surface proteins, which are known as 'phasins' (PhaP, PhaF, PhaI), show a multibinding affinity not only to the PHA granules but also to specific DNA sequences in the genome [7]. As the binding of phasins is preferential towards the PHA granules rather than to the DNA, this property has been used to create a novel peptide-tag for immobilizing proteins *in vivo* on PHA granules. The same preferred binding has also been used to develop a convenient fusion protein recovery system, which employs the association between the PHA granules and the endogenous matrix (Figure 35.7). These novel protein immobilization and purification strategies have not been previously considered.

Moldes *et al.* [104] were first to demonstrate the feasibility of using phasin as a peptide-tag to anchor the fusion protein onto PHA granules in recombinant bacteria. Here, *P. putida* was used as a host for recombinant protein and PHA production. *P. putida* originally contains two phasins, PhaI (15.4 kDa) and PhaF (26.3 kDa); PhaF behaves not only as a structural protein but also as a transcriptional regulator of the biosynthetic *pha* cluster. The C-terminal region of PhaF is similar to that of a histone-like protein, whereas the N-terminal region has 57% similarity with the complete PhaI phasin. Thus, in these studies the N-terminal protein of PhaF was used as the peptide-tag (named BioF) for a fusion protein immobilization and purification system. Using β-galactosidase and amidase as examples, the PHA granules carrying the peptide-tag fusion protein could be isolated with a simple centrifugation step and used directly for certain applications. In addition, when requested, a practically pure preparation of the soluble peptide-tag fusion protein could be obtained by mild detergent treatment of the PHA granule.

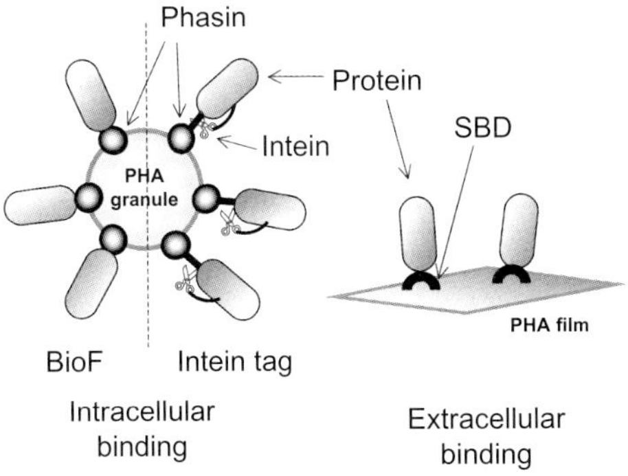

Figure 35.7 New technique using PHA-protein binding affinity. BioF is a tag using phasin (PhaF) and PHA binding affinity in the PHA-producing cells [104], and intein is a self-cleaving affinity tag based on protein splicing elements [105]. Under the extracellular condition, SBD from PHA depolymerase is a useful tag [110].

This protein recovery system has been improved by using an intein tag, which is a self-cleaving affinity tag based on protein splicing elements. In conventional affinity-based purification, the product protein must be cleaved from the affinity tag using proteases, when necessary. However, by using the intein as a linker between PHA-binding domain (phasin) and product protein, the latter can be released from the PHA-bound fusion protein simply by adjusting the temperature or pH to induce intein self-cleavage. This efficacy of this purification system was demonstrated by using PhaP as a protein tag in engineered *E. coli* [105] and *R. eutropha* [106]. It is expected that, by applying this procedure to the commercial situation, the costs associated with the purification of recombinant proteins would be greatly reduced.

Bäckström *et al.* [107] performed a study with the aim of displaying certain functional proteins on PHA granules by applying PHA-binding affinity. Here, by using PhaP as a PHA-binding tag, tailor-made PHA granules could be prepared that displayed diagnostically relevant antigens on their surfaces and then used as beads in fluorescence-activated cell sorting (FACS) analyses. The same group also reported a strategy for displaying functional proteins on PHA granules by covalently fusing them with PHA synthase (PhaC); notably, the extent of anchorage to the granules was shown not to be reliant on the PHA-protein binding affinity [108, 109].

The SBD of extracellular PHA depolymerases also represents a powerful option as a PHA-binding tag. Unfortunately, however, the binding ability of the SBD is limited to the crystalline state of PHA, which implies that, under intracellular conditions, the SBD tag would be unavailable for binding to the amorphous state of PHA. Nevertheless, the SBD would be more useful than phasins from the viewpoints of strong binding affinity, binding specificity and historical background of this protein. Park *et al.* [110] have developed a protein immobilization method by using a SBD-tagged fusion protein on PHA microbeads, and also succeeded in fabricating a microscale pattern of functional proteins on a PHA-coated glass substrate [111]. The latter technique could be applied in the generation of protein- and peptide-chips for use in biosensing systems.

35.7
Perspectives

Protein engineering is a discipline that incorporates the generation of novel protein molecules by means of genetic recombination. It appears that the first such studies of PHA-related enzymes were reported in 1987 by Thompson *et al.* [25], who identified the catalytic residue of the PhaA enzyme. Until 10 years ago, most of these studies had focused on understanding the structure–function relationships of enzymes, and many catalytically inactive and impaired mutants were generated on the basis of the rational design of enzyme. However, two additional directions have recently emerged with regards to PHA-related enzymes and proteins, namely: (i) the modification of enzyme properties in order to synthesize tailor-made PHA using directed evolution approaches; and (ii) the use of PHA–protein binding interactions for protein purification or patterning of functional proteins on PHA substrates as fusion proteins. Both approaches are directed mainly towards the application of PHA-related enzymes rather than to an understanding of the enzymes involved. The fusion-protein technology is especially attractive, and will undoubtedly provide new applications for PHA-related proteins in the future. In this respect, X-ray crystallographic structure determinations of PHA-related enzymes, including synthases, should lead to the creation of improved, high-performance evolvants.

Although PHA is today recognized as having a multifunctional role in environmental and medical fields, its commercialization as a bulk product remains problematic. Currently available, conventional technologies such as fermentation and genetic engineering may be incapable of overcoming the current problems of PHA production. Since it is essential that, from an industrial perspective, the material properties of PHA are improved and its production costs reduced, protein engineering – and especially directed evolution – will undoubtedly provide a means of creating super-enzymes to produce high-performance PHAs.

References

1 Lee, S.Y. (1996) Bacterial polyhydroxyalkanoates. *Biotechnology and Bioengineering*, **49**, 1–14.

2 Sudesh, K., Abe, H. and Doi, Y. (2000) Synthesis, structure and properties of polyhydroxyalkanoates: biological polyesters. *Progress in Polymer Science*, **25**, 1503–55.

3 Steinbüchel, A. and Valentin, H.E. (1995) Diversity of bacterial polyhydroxyalkanoic acids. *FEMS Microbiology Letters*, **128**, 219–28.

4 Steinbüchel, A., Hustede, H.E., Liebergell, M., Pieper, U., Timm, A. and Valentin, H. (1992) Molecular basis for biosynthesis and accumulation of polyhydroxyalkanoic acids in bacteria. *FEMS Microbiology Reviews*, **103**, 217–30.

5 Aldor, I.S. and Keasling, J.D. (2003) Process design for microbial plastic factories: metabolic engineering of polyhydroxyalkanoates. *Current Opinion in Biotechnology*, **14**, 475–83.

6 Anderson, A.J. and Dawes, E.A. (1990) Occurrence, metabolism, metabolic role, and industrial uses of bacterial polyhydroxyalkanoates. *Microbiological Reviews*, **54**, 450–72.

7 Madison, L.L. and Huisman, G.W. (1999) Metabolic engineering of poly(3-hydroxyalkanoates): from DNA to plastic. *Microbiology and Molecular Biology Reviews*, **63**, 21–53.

8 Jendrossek, D. and Handrick, R. (2002) Microbial degradation of polyhydroxyalkanoates. *Annual Review of Microbiology*, **56**, 403–32.

9 Sugiyama, A., Kobayashi, T., Shiraki, M. and Saito, T. (2004) Roles of poly(3-hydroxybutyrate) depolymerase and 3HB-oligomer hydrolase in bacterial PHB metabolism. *Current Microbiology*, **48**, 424–7.

10 Rehm, B.H. (2003) Polyester synthases: natural catalysts for plastics. *The Biochemical Journal*, **376**, 15–33.

11 Stubbe, J., Tian, J., Sinskey, A., He, A.J., Lawrence, A.G. and Liu, P. (2005) Nontemplate-dependent polymerization processes: polyhydroxyalkanoate synthases as a paradigm. *Annual Review of Biochemistry*, **74**, 433–80.

12 Zheng, Z., Li, M., Xue, X.J., Tian, H.L., Li, Z. and Chen, G.Q. (2006) Mutation on

N-terminus of polyhydroxybutyrate synthase of *Ralstonia eutropha* enhanced PHB accumulation. *Applied Microbiology and Biotechnology*, **72**, 896–905.

13 Rehm, B.H., Antonio, R.V., Spiekermann, P., Amara, A.A. and Steinbüchel, A. (2002) Molecular characterization of the poly(3-hydroxybutyrate) (PHB) synthase from *Ralstonia eutropha*: in vitro evolution, site-specific mutagenesis and development of a PHB synthase protein model. *Biochimica et Biophysica Acta*, **1594**, 178–90.

14 Tsuge, T., Imazu, S., Takase, K., Taguchi, S. and Doi, Y. (2004) An extra large insertion in the polyhydroxyalkanoate synthase from *Delftia acidovorans* DS-17: its deletion effects and relation to cellular proteolysis. *FEMS Microbiology Letters*, **231**, 77–83.

15 Gerngross, T.U., Snell, K.D., Peoples, O.P., Sinskey, A.J., Csuhai, E., Masamune, S. and Stubbe, J. (1994) Overexpression and purification of the soluble polyhydroxyalkanoate synthase from *Alcaligenes eutrophus*: evidence for a required posttranslational modification for catalytic activity. *Biochemistry*, **33**, 9311–20.

16 Hoppensack, A., Rehm, B.H. and Steinbüchel, A. (1999) Analysis of 4-phosphopantetheinylation of polyhydroxybutyrate synthase from *Ralstonia eutropha*: generation of β-alanine auxotrophic Tn5 mutants and cloning of the *panD* gene region. *Journal of Bacteriology*, **181**, 1429–35.

17 Jia, Y., Yuan, W., Wodzinska, J., Park, C., Sinskey, A.J. and Stubbe, J. (2001) Mechanistic studies on class I polyhydroxybutyrate (PHB) synthase from *Ralstonia eutropha*: class I and III synthases share a similar catalytic mechanism. *Biochemistry*, **40**, 1011–19.

18 Amara, A.A. and Rehm, B.H. (2003) Replacement of the catalytic nucleophile cysteine-296 by serine in class II polyhydroxyalkanoate synthase from *Pseudomonas aeruginosa*-mediated synthesis of a new polyester: identification of catalytic residues. *The Biochemical Journal*, **374**, 413–21.

19 Jia, Y., Kappock, T.J., Frick, T., Sinskey, A.J. and Stubbe, J. (2000) Lipases provide a new mechanistic model for polyhydroxybutyrate (PHB) synthases: characterization of the functional residues in *Chromatium vinosum* PHB synthase. *Biochemistry*, **39**, 3927–36.

20 Valappil, S.P., Boccaccini, A.R., Bucke, C. and Roy, I. (2007) Polyhydroxyalkanoates in Gram-positive bacteria: insights from the genera *Bacillus* and *Streptomyces*. *Antonie van Leeuwenhoek*, **91**, 1–17.

21 Müh, U., Sinskey, A.J., Kirby, D.P., Lane, W.S. and Stubbe, J. (1999) PHA synthase from *Chromatium vinosum*: cysteine 149 is involved in covalent catalysis. *Biochemistry*, **38**, 826–37.

22 Tian, J., Sinskey, A.J. and Stubbe, J. (2005) Class III polyhydroxybutyrate synthase: involvement in chain termination and reinitiation. *Biochemistry*, **44**, 8369–77.

23 Haywood, G.W., Anderson, A.J., Chu, L. and Dawes, E.A. (1988) Characterization of two 3-ketothiolases possessing differing substrate specificities in the polyhydroxyalkanoate synthesizing organism *Alcaligenes eutrophus*. *FEMS Microbiology Letters*, **52**, 91–6.

24 Slater, S., Houmiel, K.L., Tran, M., Mitsky, T.A., Taylor, N.B., Padgette, S.R. and Gruys, K.J. (1998) Multiple β-ketothiolases mediate poly(β-hydroxyalkanoate) copolymer synthesis in *Ralstonia eutropha*. *Journal of Bacteriology*, **180**, 1979–87.

25 Thompson, S., Mayerl, F., Peoples, O.P., Masamune, S., Sinskey, A.J. and Walsh, C.T. (1989) Mechanistic studies on β-ketoacyl thiolase from *Zoogloea ramigera*: identification of the active-site nucleophile as Cys89, its mutation to Ser89, and kinetic and thermodynamic characterization of wild-type and mutant enzymes. *Biochemistry*, **28**, 5735–42.

26 Palmer, M.A., Differding, E., Gamboni, R., Williams, S.F., Peoples, O.P., Walsh, C.T., Sinskey, A.J. and Masamune, S. (1991) Biosynthetic thiolase from *Zoogloea ramigera*. Evidence for a mechanism involving Cys-378 as the active site base. *Journal of Biological Chemistry*, **266**, 8369–75.

27 Williams, S.F., Palmer, M.A., Peoples, O.P., Walsh, C.T., Sinskey, A.J. and Masamune, S. (1992) Biosynthetic thiolase from *Zoogloea ramigera*. Mutagenesis of the putative active-site base Cys-378 to Ser-378 changes the partitioning of the acetyl *S*-enzyme intermediate. *Journal of Biological Chemistry*, **267**, 16041–3.

28 Modis, Y. and Wierenga, R.K. (1999) A biosynthetic thiolase in complex with a reaction intermediate: the crystal structure provides new insights into the catalytic mechanism. *Structure*, **7**, 1279–90.

29 Modis, Y. and Wierenga, R.K. (2000) Crystallographic analysis of the reaction pathway of *Zoogloea ramigera* biosynthetic thiolase. *Journal of Molecular Biology*, **297**, 1171–82.

30 Haywood, G.W., Anderson, A.J., Chu, L. and Dawes, E.A. (1988) The role of NADH- and NADPH-linked acetoacetyl-CoA reductases in the poly-3-hydroxybutyrate synthesizing organism *Alcaligenes eutrophus*. *FEMS Microbiology Letters*, **52**, 259–64.

31 Ploux, O., Masamune, S. and Walsh, C.T. (1988) The NADPH-linked acetoacetyl-CoA reductase from *Zoogloea ramigera*. Characterization and mechanistic studies of the cloned enzyme over-produced in *Escherichia coli*. *European Journal of Biochemistry*, **174**, 177–82.

32 Dennis, D., McCoy, M., Stangl, A., Valentin, H.E. and Wu, Z. (1998) Formation of poly(3-hydroxybutyrate-*co*-3-hydroxyhexanoate) by PHA synthase from *Ralstonia eutropha*. *Journal of Biotechnology*, **64**, 177–86.

33 Taguchi, K., Aoyagi, Y., Matsusaki, H., Fukui, T. and Doi, Y. (1999) Co-expression of 3-ketoacyl-ACP reductase and polyhydroxyalkanoate synthase genes induces PHA production in *Escherichia coli* HB101 strain. *FEMS Microbiology Letters*, **176**, 183–90.

34 Ren, Q., Sierro, N., Witholt, B. and Kessler, B. (2000) FabG, an NADPH-dependent 3-ketoacyl reductase of *Pseudomonas aeruginosa*, provides precursors for medium-chain-length poly-3-hydroxyalkanoate biosynthesis in *Escherichia coli*. *Journal of Bacteriology*, **182**, 2978–81.

35 Campos-Garcia, J., Caro, A.D., Najera, R., Miller-Maier, R.M., Al-Tahhan, R.A. and Soberon-Chavez, G. (1998) The *Pseudomonas aeruginosa rhlG* gene encodes an NADPH-dependent β-ketoacyl reductase which is specifically involved in rhamnolipid synthesis. *Journal of Bacteriology*, **180**, 4442–51.

36 Fukui, T. and Doi, Y. (1997) Cloning and analysis of the poly(3-hydroxybutyrate-*co*-3-hydroxyhexanoate) biosynthesis genes of *Aeromonas caviae*. *Journal of Bacteriology*, **179**, 4821–30.

37 Fukui, T., Shiomi, N. and Doi, Y. (1998) Expression and characterization of (*R*)-specific enoyl coenzyme A hydratase involved in polyhydroxyalkanoate biosynthesis by *Aeromonas caviae*. *Journal of Bacteriology*, **180**, 667–73.

38 Reiser, S.E., Mitsky, T.A. and Gruys, K.J. (2000) Characterization and cloning of an (*R*)-specific trans-2,3-enoylacyl-CoA hydratase from *Rhodospirillum rubrum* and use of this enzyme for PHA production in *Escherichia coli*. *Applied Microbiology and Biotechnology*, **53**, 209–18.

39 Tsuge, T., Fukui, T., Matsusaki, H., Taguchi, S., Kobayashi, G., Ishizaki, A. and Doi, Y. (2000) Molecular cloning of two (*R*)-specific enoyl-CoA hydratase genes from *Pseudomonas aeruginosa* and their use for polyhydroxyalkanoate synthesis. *FEMS Microbiology Letters*, **184**, 193–8.

40 Tsuge, T., Taguchi, K., Taguchi, S. and Doi, Y. (2003) Molecular characterization and properties of (*R*)-specific enoyl-CoA hydratases from *Pseudomonas aeruginosa*: metabolic tools for synthesis of polyhydroxyalkanoates via fatty acid β-oxidation. *International Journal of Biological Macromolecules*, **31**, 195–205.

41 Qin, Y.M., Haapalainen, A.M., Kilpelainen, S.H., Marttila, M.S., Koski, M.K., Glumoff, T., Novikov, D.K. and Hiltunen, J.K. (2000) Human peroxisomal multifunctional enzyme type 2. Site-directed mutagenesis studies show the importance of two protic residues for 2-enoyl-CoA hydratase 2 activity. *Journal of Biological Chemistry*, **275**, 4965–72.

42 Hisano, T., Tsuge, T., Fukui, T., Iwata, T., Miki, K. and Doi, Y. (2003) Crystal structure of the (*R*)-specific enoyl-CoA hydratase from *Aeromonas caviae* involved in polyhydroxyalkanoate biosynthesis. *Journal of Biological Chemistry*, **278**, 617–24.

43 Tsuge, T., Hisano, T., Taguchi, S. and Doi, Y. (2003) Alteration of chain length substrate specificity of *Aeromonas caviae* *R*-enantiomer-specific enoyl-coenzyme A hydratase through site-directed mutagenesis. *Applied and Environmental Microbiology*, **69**, 4830–6.

44 Hu, F., Cao, Y., Xiao, F., Zhang, J. and Li, H. (2007) Site-directed mutagenesis of *Aeromonas hydrophila* enoyl coenzyme A hydratase enhancing 3-hydroxyhexanoate fractions of poly(3-hydroxybutyrate-*co*-3-hydroxyhexanoate). *Current Microbiology*, **55**, 20–4.

45 Rehm, B.H., Kruger, N. and Steinbüchel, A. (1998) A new metabolic link between fatty acid de novo synthesis and polyhydroxyalkanoic acid synthesis. The *phaG* gene from *Pseudomonas putida* KT2440 encodes a 3-hydroxyacyl-acyl carrier protein-coenzyme a transferase. *Journal of Biological Chemistry*, **273**, 24044–51.

46 Matsumoto, K., Matsusaki, H., Taguchi, S., Seki, M. and Doi, Y. (2001) Cloning and characterization of the *Pseudomonas* sp. 61-3 *phaG* gene involved in polyhydroxyalkanoate biosynthesis. *Biomacromolecules*, **2**, 142–7.

47 Hoffmann, N., Amara, A.A., Beermann, B.B., Qi, Q., Hinz, H.J. and Rehm, B. H. (2002) Biochemical characterization of the *Pseudomonas putida* 3-hydroxyacyl ACP:CoA transacylase, which diverts intermediates of fatty acid de novo biosynthesis. *Journal of Biological Chemistry*, **277**, 42926–36.

48 Taguchi, K., Aoyagi, Y., Matsusaki, H., Fukui, T. and Doi, Y. (1999) Over-expression of 3-ketoacyl-ACP synthase III or malonyl-CoA-ACP transacylase gene induces monomer supply for polyhydroxybutyrate production in *Escherichia coli* HB101. *Biotechnology Letters*, **21**, 579–84.

49 Nomura, C.T., Taguchi, K., Taguchi, S. and Doi, Y. (2004) Coexpression of genetically engineered 3-ketoacyl-ACP synthase III (*fabH*) and polyhydroxyalkanoate synthase (*phaC*) genes leads to short-chain-length-medium-chain-length polyhydroxyalkanoate copolymer production from glucose in *Escherichia coli* JM109. *Applied and Environmental Microbiology*, **70**, 999–1007.

50 Nomura, C.T., Tanaka, T., Gan, Z., Kuwabara, K., Abe, H., Takase, K., Taguchi, K. and Doi, Y. (2004) Effective enhancement of short-chain-length-medium-chain-length polyhydroxyalkanoate copolymer production by coexpression of genetically engineered 3-ketoacyl-acyl-carrier-protein synthase III (*fabH*) and polyhydroxyalkanoate synthesis genes. *Biomacromolecules*, **5**, 1457–64.

51 Kourtz, L., Dillon, K., Daughtry, S., Madison, L.L., Peoples, O. and Snell, K. D. (2005) A novel thiolase-reductase gene fusion promotes the production of polyhydroxybutyrate in *Arabidopsis*. *Plant Biotechnology Journal*, **3**, 435–47.

52 Farinas, E.T., Bulter, T. and Arnold, F.H. (2001) Directed enzyme evolution. *Current Opinion in Biotechnology*, **12**, 545–51.

53 Tagauchi, S. and Doi, Y. (2004) Evolution of polyhydroxyalkanoate (PHA) production system by 'enzyme evolution': successful case studies of directed evolution. *Macromolecular Bioscience*, **4**, 146–56.

54 Nomura, C.T. and Taguchi, S. (2007) PHA synthase engineering toward superbiocatalysts for custom-made biopolymers. *Applied Microbiology and Biotechnology*, **73**, 969–79.

55 Taguchi, S., Maehara, A., Takase, K., Nakahara, M., Nakumura, H. and Doi, Y. (2001) Analysis of mutational effects of a polyhydroxybutyrate (PHB) polymerase on bacterial PHB accumulation using an in vivo assay system. *FEMS Microbiology Letters*, **198**, 65–71.

56 Taguchi, S., Ozaki, S., A. and Momose, H. (1998) Engineering of a cold-adapted protease by sequential random mutagenesis and a screening system. *Applied and Environmental Microbiology*, **126**, 689–93.

57 Taguchi S, S., Nakamura, H., Hiraishi, T., Yamato, I. and Doi, Y. (2002) In vitro evolution of a polyhydroxybutyrate synthase by intragenic suppression-type mutagenesis. *Journal of Biochemistry (Tokyo)*, **131**, 801–6.

58 Normi, Y.M., Hiraishi, T., Taguchi, S., Abe, H., Sudesh, K., Najimudin, N. and Doi, Y. (2005) Characterization and properties of G4X mutants of *Ralstonia eutropha* PHA synthase for poly(3-hydroxybutyrate) biosynthesis in *Escherichia coli*. *Macromolecular Bioscience*, **5**, 197–206.

59 Yano, T. and Kagamiyama, H. (2001) Directed evolution of ampicillin-resistant activity from a functionally unrelated DNA fragment: a laboratory model of molecular evolution. *Proceedings of the National Academy of Sciences of the United States of America*, **98**, 903–7.

60 Taguchi, S., Komada, S. and Momose, H. (2000) The complete amino acid substitutions at position 131 that is positively involved in cold adaptation of subtilisin BPN'. *Applied and Environmental Microbiology*, **66**, 1410–15.

61 Normi, Y.M., Hiraishi, T., Taguchi, S., Sudesh, K., Najimudin, N. and Doi, Y. (2005) Site-directed saturation mutagenesis at residue F420 and recombination with another beneficial mutation of *Ralstonia eutropha* polyhydroxyalkanoate synthase. *Biotechnology Letters*, **27**, 705–12.

62 Matsusaki, H., Abe, H. and Doi, Y. (2000) Biosynthesis and properties of poly(3-hydroxybutyrate-*co*-3-hydroxyalkanoates) by recombinant strains of *Pseudomonas* sp. 61-3. *Biomacromolecules*, **1**, 17–22.

63 Abe H, H. and Doi, Y. (2002) Side-chain effect of second monomer units on crystalline morphology, thermal properties, and enzymatic degradability for random copolyesters of (*R*)-3-hydroxybutyric acid with (*R*)-3-hydroxyalkanoic acids. *Biomacromolecules*, **3**, 133–8.

64 Doi, Y., Kitamura, S. and Abe, H. (1995) Microbial synthesis and characterization of poly(3-hydroxybutyrate-*co*-3-hydroxyhexanoate). *Macromolecules*, **28**, 4822–8.

65 Kichise, T., Taguchi, S. and Doi, Y. (2002) Enhanced accumulation and changed monomer composition in polyhydroxyalkanoate (PHA) copolyester by in vitro evolution of *Aeromonas caviae* PHA synthase. *Applied and Environmental Microbiology*, **68**, 2411–19.

66 Amara, A.A., Steinbüchel, A. and Rehm, B.H. (2002) *In vivo* evolution of the *Aeromonas punctata* polyhydroxyalkanoate (PHA) synthase: isolation and characterization of modified PHA synthases with enhanced activity. *Applied Microbiology and Biotechnology*, **59**, 477–82.

67 Takase K, K., Taguchi, S. and Doi, Y. (2003) Enhanced synthesis of poly(3-hydroxybutyrate) in recombinant *Escherichia coli* by means of error-prone PCR mutagenesis, saturation mutagenesis and *in vitro* recombination of the type II polyhydroxyalkanoate synthase gene. *Journal of Biochemistry (Tokyo)*, **133**, 139–45.

68 Takase, K., Matsumoto, K., Taguchi, S. and Doi Y, Y. (2004) Alteration of substrate chain-length specificity of type II synthase for polyhydroxyalkanoate biosynthesis by *in vitro* evolution: *in vivo* and in vitro enzyme assays. *Biomacromolecules*, **5**, 480–5.

69 Tsuge, T., Saito, Y., Narike, M., Muneta, K., Normi, Y.M., Kikkawa, Y., Hiraishi, T. and Doi, Y. (2004) Mutation effects of a conserved alanine (Ala510) in type I polyhydroxyalkanoate synthase from *Ralstonia eutropha* on polyester biosynthesis. *Macromolecular Bioscience*, **4**, 963–70.

70 Matsumoto, K., Takase, K., Aoki, E., Doi, Y. and Taguchi, S. (2005) Synergistic effects of Glu130Asp substitution in the type II polyhydroxyalkanoate (PHA) synthase: enhancement of PHA production and alteration of polymer molecular weight. *Biomacromolecules*, **6**, 99–104.

71 Matsumoto, K., Aoki, E., Takase, K., Doi, Y. and Taguchi, S. (2006) *In vivo* and *in vitro* characterization of Ser477X mutations in polyhydroxyalkanoate (PHA) synthase 1 from *Pseudomonas* sp. 61-3:

effects of beneficial mutations on enzymatic activity, substrate specificity, and molecular weight of PHA. *Biomacromolecules*, **7**, 2436–42.

72 Tsuge, T., Saito, Y., Kikkawa, Y., Hiraishi, T. and Doi, Y. (2004) Biosynthesis and compositional regulation of poly[(3-hydroxybutyrate)-*co*-(3-hydroxyhexanoate)] in recombinant *Ralstonia eutropha* expressing mutated polyhydroxyalkanoate synthase genes. *Macromolecular Bioscience*, **4**, 238–42.

73 Tsuge, T., Yano, K., Imazu, S., Numata, K., Kikkawa, Y., Abe, H., Taguchi, S. and Doi, Y. (2005) Biosynthesis of polyhydroxyalkanoate (PHA) copolymer from fructose using wild-type and laboratory-evolved PHA synthases. *Macromolecular Bioscience*, **4**, 963–70.

74 Tsuge, T., Watanabe, S., Sato, S., Hiraishi, T., Abe, H., Doi, Y. and Taguchi, S. (2007) Variation in copolymer composition and molecular weight of polyhydroxyalkanoate generated by saturation mutagenesis of *Aeromonas caviae* PHA synthase. *Macromolecular Bioscience*, **7**, 846–54.

75 Matsumoto, K., Nagao, R., Murata, T., Arai, Y., Kichise, T., Nakashita, H., Taguchi, S., Shimada, H. and Doi, Y. (2005) Enhancement of poly(3-hydroxybutyrate-*co*-3-hydroxyvalerate) production in the transgenic *Arabidopsis thaliana* by the in vitro evolved highly active mutants of polyhydroxyalkanoate (PHA) synthase from *Aeromonas caviae*. *Biomacromolecules*, **6**, 2126–30.

76 Matsumoto, K., Arai, Y., Nagao, R., Murata, T., Kakase, K., Nakashita, H., Taguchi, S., Shimada, H. and Doi, Y. (2006) Synthesis of short-chain-length/medium-chain-length polyhydroxyalkanoate (PHA) copolymers in peroxisome of the transgenic *Arabidopsis thaliana* harboring the PHA synthase gene from *Pseudomonas* sp. 61-3. *Journal of Polymers and the Environment*, **14**, 369–74.

77 Solaiman, D.K. (2003) Biosynthesis of medium-chain-length poly(hydroxyalkanoates) with altered composition by mutant hybrid PHA synthases. *Journal of Industrial Microbiology and Biotechnology*, **30**, 322–6.

78 Sheu, D.S. and Lee, C.Y. (2004) Altering the substrate specificity of polyhydroxyalkanoate synthase 1 derived from *Pseudomonas putida* GPo1 by localized semirandom mutagenesis. *Journal of Bacteriology*, **186**, 4177–84.

79 Niamsiri, N., Delamarre, S.C., Kim, Y.R. and Batt, C.A. (2004) Engineering of chimeric class II polyhydroxyalkanoate synthases. *Applied and Environmental Microbiology*, **70**, 6789–99.

80 Murase, T., Suzuki, Y., Doi, Y. and Iwata, T. (2002) Nonhydrolytic fragmentation of a poly[(*R*)-3-hydroxybutyrate] single crystal revealed by use of a mutant of polyhydroxybutyrate depolymerase. *Biomacromolecules*, **3**, 312–17.

81 Kataeva, I.A., Seidel, R.D.3rd , Shah, A., West, L.T., Li, X.L. and Ljungdahl, L.G. (2002) The fibronectin type 3-like repeat from the *Clostridium thermocellum* cellobiohydrolase CbhA promotes hydrolysis of cellulose by modifying its surface. *Applied and Environmental Microbiology*, **68**, 4292–300.

82 Brucato, C.L. and Wong, S.S. (1991) Extracellular poly(3-hydroxybutyrate) depolymerase from *Penicillium funiculosum*: general characteristics and active site studies. *Archives of Biochemistry and Biophysics*, **290**, 497–502.

83 Hisano, T., Kasuya, K., Tezuka, Y., Ishii, N., Kobayashi, T., Shiraki, M., Oroudjev, E., Hansma, H., Iwata, T., Doi, Y., Saito, T. and Miki, K. (2006) The crystal structure of polyhydroxybutyrate depolymerase from *Penicillium funiculosum* provides insights into the recognition and degradation of biopolyesters. *Journal of Molecular Biology*, **356**, 993–1004.

84 Saegusa, H., Shiraki, M., Kanai, C. and Saito, T. (2001) Cloning of an intracellular poly[D(-)-3-Hydroxybutyrate] depolymerase gene from *Ralstonia eutropha* H16 and characterization of the gene product. *Journal of Bacteriology*, **183**, 94–100.

85 Kobayashi, T. and Saito, T. (2003) Catalytic triad of intracellular poly(3-hydroxybutyrate) depolymerase (PhaZ1) in *Ralstonia eutropha* H16. *Journal of*

Bioscience and Bioengineering, **96**, 487–92.

86 Abe, T., Kobayashi, T. and Saito, T. (2005) Properties of a novel intracellular poly(3-hydroxybutyrate) depolymerase with high specific activity (PhaZd) in *Wautersia eutropha* H16. *Journal of Bacteriology*, **187**, 6982–90.

87 Hiraishi, T., Hirahara, Y., Doi, Y., Maeda, M. and Taguchi, S. (2006) Effects of mutations in the substrate-binding domain of poly[(*R*)-3-hydroxybutyrate] (PHB) depolymerase from *Ralstonia pickettii* T1 on PHB degradation. *Applied and Environmental Microbiology*, **72**, 7331–8.

88 Jendrossek, D., Schirmer, A. and Handrick, R. (1997) Recent advances in characterization of bacterial PHA depolymerases, in *1996 International Symposium on Bacterial Polyhydroxyalkanoates* (eds G. Eggink, A. Steinbüchel, Y. Poirier and B. Witholt), NRC Research Press, Ottawa, Canada, pp. 89–101.

90 Fujita, M., Kobori, Y., Aoki, Y., Matsumoto, N., Abe, H., Doi, Y. and Hiraishi, T. (2005) Interaction between poly[(*R*)-3-hydroxybutyrate] depolymerase and biodegradable polyesters evaluated by atomic force microscopy. *Langmuir*, **21**, 11829–35.

91 Behrends, A., Klingbeil, B. and Jendrossek, D. (1996) Poly(3-hydroxybutyrate) depolymerases bind to their substrate by a C-terminal located substrate binding site. *FEMS Microbiology Letters*, **143**, 191–4.

92 Shinohe, T., Nojiri, M., Saito, T., Stanislawski, T. and Jendrossek, D. (1996) Determination of the active sites serine of the poly (3-hydroxybutyrate) depolymerases of *Pseudomonas lemoignei* (PhaZ5) and of *Alcaligenes faecalis*. *FEMS Microbiology Letters*, **141**, 103–9.

93 Braaz, R., Handrick, R. and Jendrossek, D. (2003) Identification and characterisation of the catalytic triad of the alkaliphilic thermotolerant PHA depolymerase PhaZ7 of *Paucimonas lemoignei*. *FEMS Microbiology Letters*, **224**, 107–12.

94 Nojiri, M. and Saito, T. (1997) Structure and function of poly(3-hydroxybutyrate) depolymerase from *Alcaligenes faecalis* T1. *Journal of Bacteriology*, **179**, 6965–70.

95 Shinomiya, M., Iwata, T. and Doi, Y. (1998) The adsorption of substrate-binding domain of PHB depolymerases to the surface of poly(3-hydroxybutyric acid). *International Journal of Biological Macromolecules*, **22**, 129–35.

96 Schirmer, A., Matz, C. and Jendrossek, D. (1995) Substrate specificities of poly(hydroxyalkanoate)-degrading bacteria and active site studies on the extracellular poly(3-hydroxyoctanoic acid) depolymerase of *Pseudomonas fluorescens* GK13. *Canadian Journal of Microbiology*, **41**, 170–9.

97 Ohura, T., Kasuya, K. and Doi, Y. (1999) Cloning and characterization of the polyhydroxybutyrate depolymerase gene of *Pseudomonas stutzeri* and analysis of the function of substrate-binding domains. *Applied and Environmental Microbiology*, **65**, 189–97.

98 Suzuki, Y., Taguchi, S., Saito, T., Toshima, K., Matsumura, S. and Doi, Y. (2001) Involvement of catalytic amino acid residues in enzyme-catalyzed polymerization for the synthesis of polyesters. *Biomacromolecules*, **2**, 541–4.

99 Shinomiya, M., Iwata, T., Kasuya, K. and Doi, Y. (1997) Cloning of the gene for poly(3-hydroxybutyric acid) depolymerase of *Comamonas testosteroni* and functional analysis of its substrate-binding domain. *FEMS Microbiology Letters*, **154**, 89–94.

100 Kasuya, K., Ohura, T., Masuda, K. and Doi, Y. (1999) Substrate and binding specificities of bacterial polyhydroxybutyrate depolymerases. *International Journal of Biological Macromolecules*, **24**, 329–36.

101 Kasuya, K., Takano, T., Tezuka, Y., Hsieh, W.C., Mitomo, H. and Doi, Y. (2003) Cloning, expression and characterization of a poly(3-hydroxybutyrate) depolymerase from *Marinobacter* sp. NK-1. *International Journal of Biological Macromolecules*, **33**, 221–6.

102 Kobayashi, T., Nishikori, K. and Saito, T. (2004) Properties of an intracellular poly(3-hydroxybutyrate) depolymerase (PhaZ1) from *Rhodobacter spheroides*. *Current Microbiology*, **49**, 199–202.

103 Tseng, C.L., Chen, H.J. and Shaw, G.C. (2006) Identification and characterization of the *Bacillus thuringiensis phaZ* gene, encoding new intracellular poly-3-hydroxybutyrate depolymerase. *Journal of Bacteriology*, **188**, 7592–9.

104 Moldes, C., Garcia, P., Garcia, J.L. and Prieto, M.A. (2004) *In vivo* immobilization of fusion proteins on bioplastics by the novel tag BioF. *Applied and Environmental Microbiology*, **70**, 3205–12.

105 Banki, M.R., Gerngross, T.U. and Wood, D.W. (2005) Novel and economical purification of recombinant proteins: intein-mediated protein purification using in vivo polyhydroxybutyrate (PHB) matrix association. *Protein Science*, **14**, 1387–95.

106 Barnard, G.C., McCool, J.D., Wood, D.W. and Gerngross, T.U. (2005) Integrated recombinant protein expression and purification platform based on *Ralstonia eutropha*. *Applied and Environmental Microbiology*, **71**, 5735–42.

107 Bäckström, B.T., Brockelbank, J.A. and Rehm, B.H. (2007) Recombinant *Escherichia coli* produces tailor-made biopolyester granules for applications in fluorescence activated cell sorting: functional display of the mouse interleukin-2 and myelin oligodendrocyte glycoprotein. *BMC Biotechnology*, **7**, 3.

108 Peters, V. and Rehm, B.H. (2006) In vivo enzyme immobilization by use of engineered polyhydroxyalkanoate synthase. *Applied and Environmental Microbiology*, **72**, 1777–83.

109 Brockelbank, J.A., Peters, V. and Rehm, B.H. (2006) Recombinant *Escherichia coli* strain produces a ZZ domain displaying biopolyester granules suitable for immunoglobulin G purification. *Applied and Environmental Microbiology*, **72**, 7394–7.

110 Park, J.P., Lee, K.B., Lee, S.J., Park, T.J., Kim, M.G., Chung, B.H., Lee, Z.W., Choi, I.S. and Lee, S.Y. (2005) Micropatterning proteins on polyhydroxyalkanoate substrates by using the substrate binding domain as a fusion partner. *Biotechnology and Bioengineering*, **92**, 160–5.

111 Lee, S.J., Park, J.P., Park, T.J., Lee, S.Y., Lee, S. and Park, J.K. (2005) Selective immobilization of fusion proteins on poly(hydroxyalkanoate) microbeads. *Analytical Chemistry*, **77**, 5755–9.

36
Bioengineering of Sequence-Repetitive Polypeptides: Synthetic Routes to Protein-Based Materials of Novel Structure and Function

Sonha C. Payne, Melissa Patterson and Vincent P. Conticello

36.1
Introduction

Protein-based materials, which correspond to polymers of tandemly repeated oligopeptide sequence motifs, have been the focus of significant research interest over the past two decades [1]. The intellectual driving force for this process has come from two distinct directions: first, from interest in the fundamental polymer science of architecturally uniform macromolecules; and second from interest in the structural biology of native, protein-based materials. From the viewpoint of fundamental polymer science, protein-based materials represent an approach to understand the effect of polymer architectural parameters (composition, sequence and molar mass) on macromolecular properties. Ribosomal protein synthesis ensures a uniformity of polymer microstructure that is impossible to achieve using the conventional synthetic methods employed for organic polymerization reactions. Thus, non-natural polypeptide sequences can been synthesized with near-absolute control of architectural parameters, and these biologically synthesized poly(α-amino acids) can be considered as model uniform polymers. These synthetic protein-based materials may provide insight into the fundamental aspects of polymer physical chemistry both in solution and in the solid-state, potentially enabling the creation of material constructs that display novel behavior [1]. Furthermore, the observed control of polypeptide primary structure also implies the ability to define a higher-order structure through the progression of protein structural hierarchy. Secondary and super-secondary elements, and the interactions between them, can be specified through the sequence identity although, as with more conventional targets of protein design, the currently limited ability of theoretical approaches to reliably define the relationship between amino acid sequence and higher-order molecular and supramolecular structure is a significant constraint upon the design of novel polypeptide architectures. Nevertheless, genetic engineering methods have been employed to create artificial polypeptides of defined sequence that self-assemble into structurally defined supramolecular aggregates, including lamellar crystallites [2], structurally defined fibrils [3, 4],

Protein Engineering Handbook. Edited by Stefan Lutz and Uwe T. Bornscheuer

ISBN: 978-3-527-31850-6

surface-stabilizing coatings [5], smectic liquid crystalline mesophases [6, 7], thermoresponsive nanoparticles [8] and nanostructured hydrogels [9–11], on the basis of structural features programmed into the polypeptide sequences at the molecular level. These *de novo*-designed biomaterials provide an indication of the potential for biosynthesis to provide novel materials through the near-absolute control of macromolecular architecture.

The second factor that has motivated the investigation of protein-based materials lies in the desire to understand the chemical, biological and mechanical properties that underlie the native biological function of fibrous proteins [12]. Natural evolutionary processes have afforded an array of structurally diverse protein-based materials that are produced within organisms as a natural consequence of their life cycle. These native protein-based materials usually display low complexity sequences that consist of tandem repeats of a fundamental oligopeptide motif that displays limited plasticity in amino acid sequence, and thus they bear a nominal similarity to the repeat sequences of conventional organic polymers. The unique structural and functional properties of these native materials presumably arise as a consequence of their sequence specificity, which strongly influences the mode of self-assembly of the polymer chain into the supramolecular architectures that underlie their materials properties. Most notably, the materials properties of these native proteins often surpass the performance of synthetic materials within the relatively narrow compass of environmental conditions that define these biological systems [13]. Structural variants of these native proteins have been envisioned for technological applications as high-performance materials and, indeed, have provided the intellectual driving force for the development of conventional polymer science during the last century. Dragline silk fibers from the spider *Nephila clavipes* display a unique combination of high tensile and compressive strength that presumably originates in the segmented structure of the fibroin proteins that comprise the dragline fiber [14] (see also Chapter 37). The primary structures of these proteins consist of strictly alternating sequences of highly conserved alanine-rich and glycine-rich oligopeptide repeats [15]. These self-assemble into mesoscopic domains of distinctly different structural and mechanical properties that are covalently linked within the dragline fiber [16]. Similarly, the near-ideal resilience and extreme durability of the native bioelastomer elastin have been attributed to the primary structure of the protein [17], in which elastomeric sequences alternate with crosslinkable sequences to provide an elastomeric matrix that can endure a billion extension/relaxation cycles during the lifetime of an individual [18].

One of the main objectives of protein-based materials research has been directed toward understanding the sequence–structure–function relationships that define the biological role of these native protein-based materials, and which would provide the rationale for the design of synthetic analogues that emulate or expand upon the technologically useful properties of their natural counterparts. Often, the functionally critical structural properties of these native protein materials can be reproduced within synthetic polypeptides that comprise iterations of the canonical repeat sequences [19–21]. Thus, sequence-repetitive polypeptides may represent the best chemical models that are currently available to understand the critical structural features associated with the unique properties of native protein materials. However, in order to realize the full potential of protein-based materials, as revealed through the functional responses of native biological systems, it is necessary to attain a level of rational control over the self-assembly processes that govern the formation of the native structures. The foremost challenge in the bioengineering of novel protein-based materials resides in the correlation of sequence with higher-order structure, which is a general problem in *de novo* protein design. Although folded structures may occur frequently among random peptide sequence libraries [22], the extreme diversity of amino acid sequence space and limited knowledge of global (particularly long-range) sequence–structure correlations restrict the ability to select a specific structure from a random library of polypeptide sequences on the basis of first principles. Rather than *de novo* design, most synthetic approaches to the bioengineering of protein-based materials still rely on a permutation of conventional structural prototypes derived from native proteins.

The biosynthesis of artificial protein-based polymers derived from sequence-repetitive polypeptides has developed in conjunction with the fundamental advances in recombinant DNA cloning and protein expression techniques over the past 25 years. In 1980, Doel and coworkers reported the first example of the cloning and expression of an artificial gene based on a 150 sequence repeats of a simple (Asp-Phe) diad [23]. Although this technology was not developed for the synthesis of protein-based materials, these techniques were soon applied to the synthesis of sequence-repetitive polypeptides based on the canonical repeats observed for native fibrous proteins such as elastin, collagen, keratin and silk [24–26]. This approach met with mixed results in that, although significant knowledge was obtained with respect to cloning and expression of repetitive polypeptides, considerable challenges remained to be addressed, including the development of better methods to stabilize highly repetitive DNA sequences, to optimize recombinant protein yield, to promote appropriate post-translational modification, and to process the protein into a form that approximates that of the native state of the protein from which the sequence was originally derived. Of course, these difficulties are not specific to protein-based materials, but are generally observed for recombinant protein expression experiments in heterologous host systems and are usually addressed on a specific basis for the protein of interest. Nevertheless, recombinant protein expression has provided access to a significant number of synthetic protein-based materials, which have been employed as the subject of extensive structural characterization in solution and in the solid state. The information gleaned from these investigations has provided insight into the structural factors that underlie the macromolecular properties of native protein-based materials as well as into fundamental polymer physics. Moreover, the technological potential of protein-based polymers is beginning to be realized, particularly in the area of biomedical applications, in which the ability to tailor the biological, chemical and mechanical interface through control of the polypeptide sequence may represent a significant advantage over conventional polymeric biomaterials [27–30].

The aim of this chapter is to provide a description of current methodologies for the construction of sequence-repetitive polypeptides using the example of a synthetic elastin-mimetic protein material that is being developed for tissue-engineering applications. In addition to summarizing conventional methods for the assembly of sequence-repetitive genes, we also describe a strategy for the convergent synthesis of large genes encoding elastin-mimetic block copolymers in which the size and sequence of the individual domains can be easily controlled as well as the block–block interfaces and the terminal sequences.

36.2
Block Copolymers as Targets for Materials Design

Synthetic copolymers consisting of well-defined blocks of compositionally dissimilar monomers spontaneously self-assemble in the solid state into ordered domains of similar blocks [31]. These hybrid materials have been extensively studied and often have unique, technologically significant properties in comparison to blends of the respective homopolymers. For example, copolymers comprising distinct blocks of different mechanical and chemical properties have been employed as polymer surfactants, pressure-sensitive adhesives, blend compatibilizers, thermoplastic elastomers, mineralization templates and lithographic resists. In contrast, block *co*-polypeptides have not garnered as much attention. However, the recent development of biosynthetic and chemosynthetic methods for the preparation of well-defined block copolymers of peptide sequences promises the potential for rapid advancements [25, 32]. These materials could be potentially interesting based on the diverse structures and functions observed for naturally occurring protein materials, in which the repetitive sequence pattern induces a regular secondary structure within the individual domains of the block *co*-polypeptide that has an important effect upon the supramolecular organization of the material [33]. Segregation of the blocks into compositionally, structurally and spatially distinct domains occurs in analogy with synthetic block copolymers, affording ordered structures on the nanometer to micrometer size range. The sequence control and structural uniformity of these natural block copolymers is presumably responsible for their unique materials properties [34]. The genetic engineering of synthetic polypeptides enables the preparation of block copolymers composed of complex sequences in which the individual blocks may have different mechanical, chemical or biological properties [8–11]. The utility of these protein materials depends on the ability to functionally emulate or enhance the materials properties of conventional polymer systems, while retaining the benefits of greater control over the sequence and microstructure that protein engineering affords for the construction of materials. This precise control of macromolecular architecture provides an opportunity for tailoring technologically significant materials properties for directed applications, for example, in biomedicine.

Many native protein-based materials have macromolecular architectures that comprise repeating blocks of distinct amino acid sequence, as in the case, for example, of arachnid dragline silks. The alternating sequences of alanine-rich and glycine-rich domains define specific structural and functional properties in the final form of the material that are essential for its biological role. Domains within the sequence can display differences in structural order (i.e. crystallinity), in mechanical properties, in hydrophilic/hydrophobic balance, and in chemical reactivity. Often, these functional attributes act in concert to define the native function of the material. In addition, these block architectures have been hypothesized as being important to facilitate the processing of the materials under the relatively mild environmental conditions that are available to the organisms. Dragline silk proteins are spun under conditions far milder than those typically employed for the industrial spinning of conventional thermoplastic materials. The block architectures of the former materials promotes formation of liquid crystalline phases *in vivo* that greatly enhance the ease of fabrication of the fibers. This specialized mechanism for fiber formation represents an optimized match between polymer sequence and processing apparatus and conditions that would be desirable to emulate for synthetic polymeric materials [34].

36.2.1
Amphiphilic Block Copolymers

One notable characteristic of technological significance for many conventional block copolymers is the property of amphiphilicity, that is, a difference in hydrophilic versus lipophilic (hydrophobic) character between the respective blocks. Amphiphilic block copolymers represent a special class of materials that are composed of compositionally defined blocks that have significantly different interaction affinities for aqueous solutions [35]. These hybrid materials have attracted scientific interest due to their complex phase behavior in selective solvents, which parallels and complements that of small-molecule surfactant amphiphiles such as phospholipids. Amphiphilic diblock (**AB**) and triblock (**ABA** or **BAB**) copolymers undergo selective segregation of the hydrophobic domain in aqueous solvents to form micellar structures in which the corona of the micelle is derived from the hydrophilic block (**A**) and the core of the micelle from the hydrophobic block (**B**). The identity and sequence of the individual block units within the polymer dictates the nature of the supramolecular assembly. In contrast to small-molecule surfactants, the phase behavior of amphiphilic block copolymers can be modified conveniently through manipulation of the macromolecular architecture – that is, the length, composition and sequence of the individual blocks [36]. In addition, the hydrophilic/lipophilic balance can be adjusted systematically by variation of the relative lengths of the hydrophilic and hydrophobic blocks. These materials display several key features that may confer advantages over conventional surfactant amphiphiles in controlled delivery and release applications, namely, very low critical micelle concentrations, slow unimer–micelle exchange rates, high aggregate stabilities, and a controllable range of aggregate sizes and morphologies. The most commonly employed phases in medical applications consist of spherical micelles (polymer nanoparticles) that may occur in isolation or as an associated network, depending on the polymer architecture. As a consequence of their chimeric structure, amphiphilic block copolymers are ideally suited for applications involving the energetic and structural control of materials and biological interfaces, for example, as emulsifiers, delivery agents, dispersants, gelation agents, compatibilizers and foamants. Although most amphiphilic block copolymers are derived from conventional organic monomers, polypeptides that comprise hydrophilic/hydrophobic block structures may display many of the same technologically important physical attributes, particularly the ability to reversibly self-assemble from aqueous solution to form a range of well-defined supramolecular aggregates. It is envisioned that assemblies derived from these materials may be employed as biomaterials for *in vivo* applications, in which the greater control of polypeptide microstructure may afford advantages over similar synthetic polymer systems currently under investigation.

36.2.2
Elastin-Mimetic Block Copolymers

We have previously employed oligopeptide motifs derived from the elastin repeat sequence (Val-Pro-Xaa-Yaa-Gly) as substrates for the construction of biosynthetic block copolymers [8, 9, 37]. Elastin is a native protein-based material that is the primary structural component underlying the elastomeric mechanical response of compliant tissues in vertebrates and, therefore, has potential significance for human health as a medical biomaterial for the preparation of tissue-engineered analogues of native elastin-containing human systems [17]. Moreover, elastin-mimetic polypeptides display a well-defined correlation between repeat sequence and macromolecular properties (*vide infra*), which enables the creation of a wide variety of synthetic elastin analogues with tailorable biophysical properties [38]. The elastomeric domains of elastin comprise structurally similar oligopeptide motifs that are tandemly repeated in the native protein sequence. The local secondary structure and macromolecular thermodynamic and viscoelastic properties of the elastomeric domains can be emulated by synthetic polypeptides that are composed of a concatenated sequence of native oligopeptide motifs; the most common of which is the pentapeptide (Val-Pro-Gly-Val-Gly) [37]. Polypeptides based on these pentameric repeat sequences undergo reversible, temperature-dependent, hydrophobic assembly from aqueous solution in analogy to the phase behavior of native tropoelastin, the soluble precursor of crosslinked elastin. This process results in a spontaneous phase separation of the polypeptide above a critical solution temperature, T_t, which is near ambient temperature *in vitro*. This inverse temperature transition coincides with a conformational rearrangement of the local secondary structure within the pentapeptide motifs.

Structural investigations of synthetic elastin-based polypeptides have revealed that the materials' properties depend on the identity of the amino acids that occupy the third (Xaa) and fourth (Yaa) residues of the pentapeptide repeat sequence [37]. Alterations in the identity of the fourth residue (Yaa) modulate the position of the lower critical solution temperature of the polypeptide in aqueous solution in a manner commensurate with the effect of polarity of the amino acid side chain on the polymer–solvent interaction for the polypeptide series [(Val-Pro-Gly-Yaa-Gly)$_n$] [38]. In addition, the substitution of an Ala residue for the consensus Gly residue in the third (Xaa) position of the pentapeptide repeat results in a change in the mechanical response of the material from elastomeric to plastic [37]. The identity of the elastin-mimetic blocks can be varied between the elastomeric and plastic sequences to provide versatility in the mechanical properties of the synthetic construct. The transition temperature of the individual blocks can be easily adjusted by substitution of amino acid residues with the desired polarity profile for the canonical valine residue in position four of the pentapeptide repeat. Thus, the phase separation of elastin-mimetic domains within block copolymers can be employed as a general mechanism for the reversible self-assembly of protein-based materials on the mesoscopic scale [8, 9]. In addition, the combination of biocompatibility [39], physiological stability [37, 40], high recombinant protein yields [41] and tunable responsive properties of elastin-mimetic polypeptides [38] make synthetic protein-based materials derived from elastin-mimetic peptide sequences attractive candidates for the creation of *in vivo* biomaterials.

Previously, we have reported the construction of amphiphilic triblock (**BAB**) [9] copolymers derived from elastin-mimetic peptide sequences. These materials undergo reversible, temperature-dependent phase segregation of the hydrophobic block in aqueous solution to afford potentially biocompatible hydrogels under environmentally benign conditions. The sequence of the hydrophobic block can be chosen such that the lower critical solution temperature of the block occurs below 37 °C Collapse of the hydrophobic block results in the formation of microphase-separated protein domains under physiologically relevant conditions. The hydrophobic block (**B**) was derived from the plastic repeat sequence [(Val/Ile)-Pro-Ala-Val-Gly], while that of the hydrophilic block (**A**) is based on the elastomeric repeat sequence [(Val-Pro-Gly-Yaa-Gly)] in which polar glutamic acid residues are introduced at the Yaa position periodically throughout the repetitive sequence (Figure 36.1). Thus, the mechanical properties coincide with a change in polarity across the sequence such that the **BAB** triblock polypeptide behaves not only as an amphiphilic block copolymer, but also as the protein equivalent of a conventional polymeric thermoplastic elastomer [31]. Phase separation of the hydrophobic, plastic end-blocks (**B**) within the **BAB** polypeptide occurs above their critical solution temperature, while the hydrophilic, elastic mid-blocks (**A**) remain hydrated under these conditions and act as virtual crosslinks between the phase-separated plastic segments. The reversible self-assembly of the protein-based material from aqueous solution results in the formation of a hydrogel consisting of mechanically distinct domains on the mesoscopic scale. This hypothesis has been verified through a combination of calorimetric methods, spectroscopic analyses and mechanical testing of the elastin-mimetic polypeptides at temperatures below and above the critical solution temperature [9, 42, 43]. Thus, synthetic sequence-repetitive polypeptides based on elastin-mimetic sequences can be prepared which display potentially significant biological properties that emulate those of the native material. However, the physical properties of the polypeptide can be manipulated

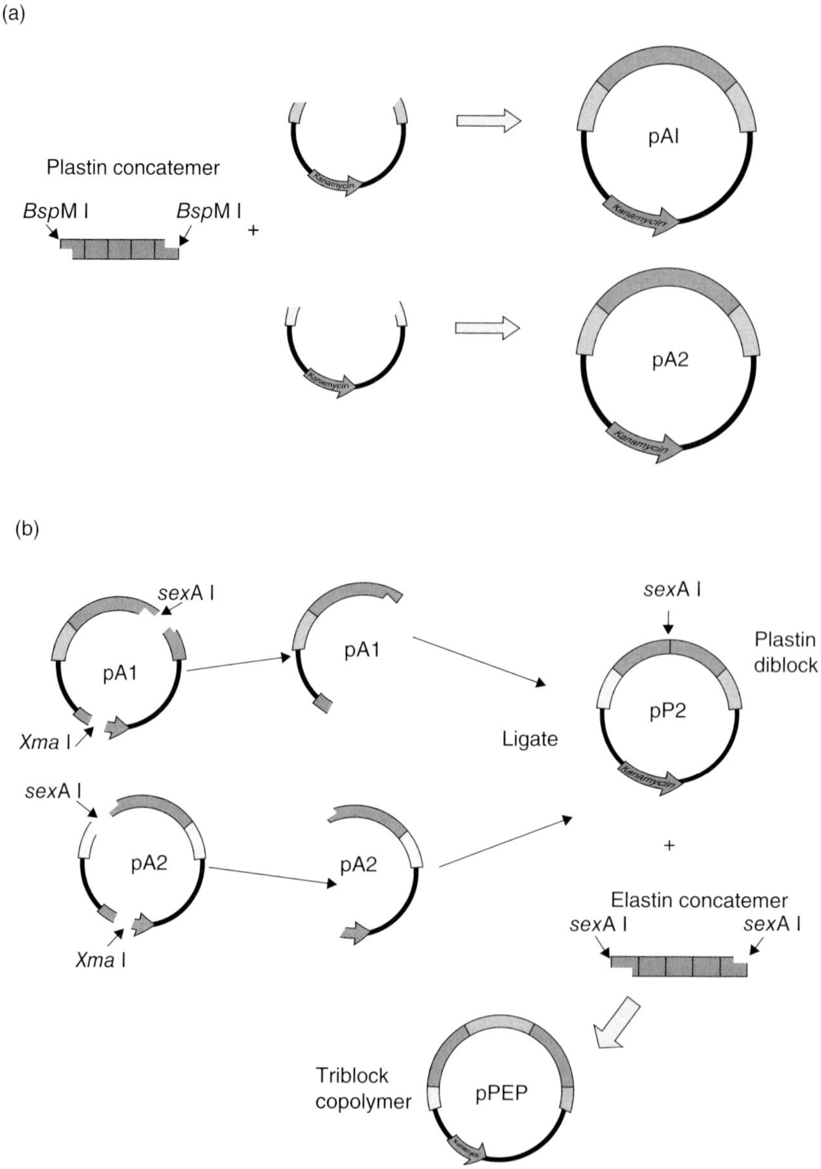

Figure 36.1 Previously reported strategy for the assembly of a block polypeptide based on elastin-mimetic sequence repeats. (a) DNA concatemers encoding the plastic cassettes were inserted into two modified expression vectors in which the positions of *Kpn*I and *Pst*I restriction sites flanking a central *Sex*AI site were reversed within the synthetic polylinker. Insertion of the concatemers afforded two plasmids, pA1 and pA2, which were subsequently joined together to form an acceptor plasmid for the elastin concatemer library; (b) Strategy for assembly of the triblock polypeptide through insertion of the elastin concatemer library into the compatible *Sex*AI site of the acceptor plasmid. Note that the size and sequence of the elastin concatemers can be easily varied, but the identity of the plastic cassettes is fixed in the acceptor plasmid, pP2, and cannot be easily varied without reconstruction of the entire expression cassette. In addition, modification of the terminal sequences and the block interfaces cannot easily be accomplished due to the incompatibility of the cohesive ends between concatemer libraries. This incompatibility also precludes the construction of more complex sequences using this approach and is essentially limited in scope to the creation of simple diblock and triblock constructs.

through the choice of sequence to facilitate processing of the materials using technologies similar to those employed for conventional, organic block copolymers.

36.3
Strategies for the Construction of Synthetic Genes Encoding Sequence-Repetitive Polypeptides

The synthesis of the protein-based materials based upon complex sequence repeats is best accomplished using the techniques of recombinant DNA (rDNA) technology and bacterial protein expression. The advantage of these methods lies in the ability to directly produce, with high fidelity, synthetic polypeptides of exact amino acid sequence and high molecular weight, as opposed to chemically synthesized oligopeptides, which are essentially limited to low degrees of polymerization (<60 residues) [1]. With regards to the discussion herein, the term 'protein-based material' implies a sequence-repetitive polypeptide, or a multidomain protein consisting of one or more sequence-repetitive polypeptides, that is encoded within a synthetic DNA expression cassette. As materials properties generally scale to some degree with chain length, the biosynthesis of these protein polymers usually requires the construction and expression of large, synthetic genes containing multiple direct repeats of a 'monomeric' DNA sequence of approximately 50 to 150 base pairs in length [44]. As automated DNA synthesis technology is currently limited to the production of oligodeoxynucleotides of lengths corresponding to about a hundred bases, sequences encoding medium to high-molecular-weight polypeptides cannot be obtained by direct synthesis of the entire gene. In addition, such repetitive DNA sequences may be unstable with respect to homologous recombination, and this may result in the structural instability of plasmid clones *in vivo*. Therefore, synthetic procedures for the cloning and expression of the repetitive genes may require special experimental considerations beyond conventional DNA manipulations (*vide infra*). However, a number of synthetic approaches have been described that represent significant technical improvements over the synthetic route described by Doel *et al.* [23], and their immediate successors [24, 45]. A recent review has comprehensively summarized the scope of experimental methods that have been developed for the synthesis of repetitive genes encoding sequence-repetitive polypeptides [46]. Two main approaches have been described that are complementary in experimental methodology: (i) DNA cassette concatemerization [24, 45]; and (ii) recursive directional ligation [47] (Figure 36.2). Both strategies involve the chemical synthesis of the corresponding DNA sequence encoding the desired peptide repeat motif, enzymatically induced concatemerization, ligation of the concatemer into a plasmid vector, propagation in a bacterial host and, finally, expression of the repetitive polypeptides. However, the two strategies differ significantly in the method that is employed for generation of the concatemers and subsequent manipulation of the cloning vectors.

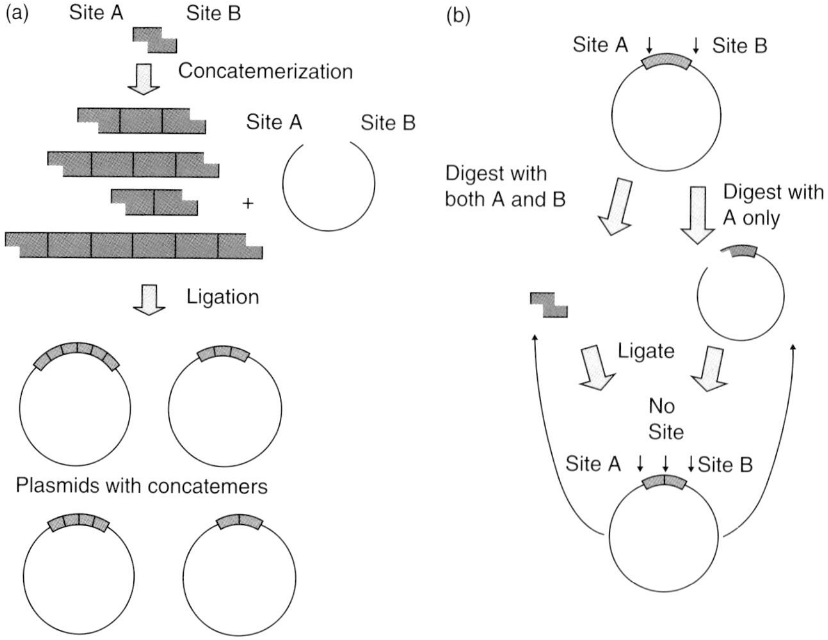

Figure 36.2 Schematic representations of commonly applied methods for the synthesis of concatemeric genes encoding sequence-repetitive polypeptides. (a) DNA cassette concatemerization; (b) Recursive directional ligation.

36.3.1
DNA Cassette Concatemerization

Historically, DNA cassette concatemerization was the initial strategy to be developed for the synthesis of artificial genes encoding sequence-repetitive polypeptides, and variants of this technology are probably still the most commonly employed procedure for the assembly of genes encoding large, repetitive polypeptides (Figure 36.2a). This experimental protocol involves the construction of double-stranded oligonucleotide segments (DNA 'monomers') containing nonpalindromic, cohesive ends. Generation of the cohesive-ended DNA monomers is generally accomplished through the use of restriction endonucleases capable of recognizing and cleaving nonpalindromic sequences. The sizes of the oligopeptide repeats are usually chosen such that they could be conveniently encoded within single DNA cassettes of approximately 50 to 150 base pairs in length prior to concatemerization. Self-ligation of the DNA monomers proceeds in a head-to-tail fashion to generate a library of concatemers which differ in length by increments of the monomer. Preparative agarose gel electrophoresis is used to fractionate the concatemers according to the degree of concatemerization. Concatemers within the desired size range are extracted from the gel and used directly in subsequent cloning steps. This fractionation process provides an effective strategy for enriching the library in concatemers of larger size that may be under-represented within the general population resulting from the ligation reaction. A critical consideration for the successful application of this procedure is the efficacy of cloning and screening a population of concatemers to identify a construct of appropriate size. Unless concatemeric DNA cassettes corresponding to individual bands are excised from the gel, it is difficult to isolate and clone concatemers of determinate size using this approach. Usually, the sizes of individual concatemers are identified through screening a population of clones in parallel using either colony screening polymerase chain reaction (PCR) or restriction digestion of isolated plasmid-based constructs. Often, this process may require the screening of a large number of clones to identify a cassette of the desired size. DNA cassettes corresponding to very high degrees of concatemerization have been isolated using this procedure, although it is typically challenging to isolate a clone corresponding to a specific size. Although laborious, these protocols have been widely employed for the synthesis of artificial genes encoding sequence-repetitive polypeptides based on natural sequences, as well as artificial proteins having no natural parallel [1]. However, difficulties have been reported in obtaining long concatemers as cloned inserts using this approach [48]. Modified concatemerization strategies have been described in which DNA adaptors have been appended to the termini to facilitate cloning into conventional plasmid-based vectors [4, 49], although these approaches do not necessarily address the problems associated with low yields of long concatemers.

36.3.2
Recursive Directional Ligation

In contrast to the DNA cassette concatemerization approach, recursive directional ligation permits the isolation of concatemers of determinate size through a controlled oligomerization process that is facilitated by the DNA manipulation experiments (Figure 36.2b). Although several variations of the basic protocol have been described [46], the general procedure involves iterative directional insertions in a plasmid-based vector in which smaller concatemers are joined together recursively to form larger ones [47]. The size of the DNA product that results from the cloning procedure corresponds to the sum of the initial DNA reactants. Thus, two DNA monomers can be joined together in a plasmid to form a dimeric construct. Two equivalents of the resulting dimeric construct can be joined to form a tetrameric construct, and so on. Repetitive application of this process, in which the products from a prior step are employed as the reactants in a successive step, can afford large concatemeric cassettes of determinate size. This procedure relies on the judicious choice of restriction sites at the termini of the DNA cassettes to facilitate the directional cloning process. As for DNA cassette concatemerization, restriction endonucleases that recognize and cleave nonpalindromic sites are very useful for the generation of cohesive-ended DNA fragments that are competent for selective ligation. Recursive directional ligation has the advantage that synthetic genes of determinate size and sequence can be obtained, although the process can be labor-intensive for the assembly of large genes. For example, the assembly of a concatemer that encodes 32 (2^5) repeats of the basic sequence motif requires at least five iterations of the directional cloning process in which the size of the concatemeric construct is doubled at the end of each step. Nevertheless, a significant number of sequence-repetitive polypeptides have been produced from synthetic genes assembled in this fashion. Recursive directional ligation is no doubt the technique of choice for the creation of synthetic genes of defined size and sequence. In theory, neither gene assembly strategy places any restriction on the size of the cloned DNA concatemers, although practical considerations (e.g. the efficiency of transformation of plasmid-based constructs and genetic instability of repetitive DNA sequences) may limit the effective size of cloned DNA inserts. It has been found that the ease of isolation of long DNA concatemers depends heavily on the identity of the DNA sequence. For many of the elastin-derived constructs, very large synthetic genes (≥8000 base pairs) can be obtained that encode sequence-repetitive polypeptides (see below), whereas for other polypeptide sequences the isolation of long DNA concatemers becomes very difficult due to genetic instability leading to a recombinative loss of the majority of the coding sequence.

36.3.3
Genetic Assembly of Synthetic Genes Encoding Block Architectures

Previous preparations of elastin-based block copolymers have employed DNA cassette concatemerization to generate synthetic genes encoding sequence-repetitive polypeptides [9, 37]. The hydrophilic and hydrophobic blocks were encoded within separate DNA cassettes, concatemerized independently, and joined together via sequential ligation into a multifunctional polylinker sequence to create genetic fusions encoding the block copolymers. The construction of a synthetic DNA cassette encoding the elastin-mimetic triblock polymer illustrates the general approach that was employed for the synthesis of these materials (see Figure 36.1). The two DNA monomers **E** and **P** encode the elastic and plastic repeat sequences, respectively (Figure 36.3). The restriction endonuclease *Sex*AI is employed for the generation of nonpalindromic DNA monomers encoding the elastin block. The repeat sequence, (Val-Pro-Gly-Xaa-Gly), of the elastic block contains a strictly conserved 'Pro-Gly' unit, which forms the corners of a type II β-turn structure that is essential for the development of elastomeric behavior within the pentapeptide repeats. The restriction endonuclease *Sex*AI recognizes and cleaves a nucleotide heptanucleotide sequence [5′-ACC(T/A)GGT-3′] that contains the coding sequence of the Pro-Gly element common to these elastin repeat sequences. Moreover, the cleavage pattern generates nonpalindromic 5′-cohesive ends with five-base, single-stranded extensions, which ensures exclusive head-to-tail concatemerization of the DNA monomers that encode the elastomeric repeat sequences. A similar procedure is employed for the concatemerization of DNA monomers based on the plastic repeat sequence [Val/Ile-Pro-Ala-Val-Gly], which employs the restriction endonuclease *Bsp*MI to generate DNA monomers competent for self-ligation in an exclusive head-to-tail manner. This endonuclease recognizes the hexanucleotide sequence

```
(a)
       Val Pro Gly Val Gly Ile Pro Gly Val Gly Ile Pro Gly Val Gly Ile Pro Gly
          ↓
AG CTT GTA CCT GGT GTT GGC ATC CCG GGT GTA GGT ATC CCA GGC GTT GGT ATT CCG GGT
     A CAT GGA CCA CAA CCG TAG GGC CCA CAT CCA TAG GGT CCG CAA CCA TAA GGC CCA
HinD III SexA I ↑

Val Gly Ile Pro Gly Val Gly Val Pro Gly
                                ↓
GTA GGC ATC CCA GGC GTT GGC GTA CCT GGT G
CAT CCG TAG GGT CCG CAA CCG CAT GGA CCA CCT AG
                              SexA I ↑  BamH I

(b)
       Val Pro Ala Val Gly Ile Pro Ala Val Gly Ile Pro Ala Val Gly Ile Pro Ala
                      ↓
AG CTT GTA CCT GCT GTT GGT ATC CCG GCT GTT GGT ATC CCA GCT GTT GGC ATT CCG GCT
     A CAT GGA CGA CAA CCA TAG GGC CGA CAA CCA TAG GGT CGA CAA CCG TAA GGC CGA
HinD III BspM I                ↑

Val Gly Ile Pro Ala Val Gly Val Pro Ala Val Gly Ile
                                            ↓
GTA GGT ATC CCG GCT GTT GGT GTA CCT GCT GTT GGT ATC G
CAT CCA TAG GGC CGA CAA CCA CAT GGA CGA CAA CCA TAG CCT AG
                              BspM I             ↑ BamH I
```

Figure 36.3 Oligonucleotide cassettes encoding the elastin **E** (a) and plastin **P** (b) repeat sequences that were employed in a previously reported synthesis of an elastin-mimetic triblock polypeptide. The recognition sites for the restriction endonucleases that are employed for generation of the respective DNA monomers are highlighted. Arrows indicate the cleavage positions on the sense and anti-sense strands for the respective DNA monomers. Note that the cohesive ends generated from restriction cleavage are compatible and nonpalindromic for each monomer, but are not compatible between monomers. The cleavage sites for restriction endonuclease *Bsp*MI occur at well-defined positions downstream of the recognition site, as expected for type IIs restriction endonucleases.

5′-ACCTGC-3′, which occurs within the DNA monomers encoding the plastic repeat. Cleavage occurs at a downstream position (+4/+8) on the DNA strand, which generates 5′ cohesive ends with four-base, single-stranded extensions. As *Bsp*MI is a type IIs restriction endonuclease, the sequence of the extensions depends solely on the position of the hexanucleotide cleavage site. The sequences of the cohesive ends can be made nonpalindromic through the choice of the downstream DNA sequence, which permits exclusive formation of extensions that are compatible with head-to-tail concatemerization of the DNA monomer. Self-ligation of each DNA cassette (as depicted in Figure 36.2a) afforded a population of concatemers encoding repeats of the elastic and plastic sequences, respectively.

In order to construct a gene encoding the triblock polymer, an expression plasmid must be modified such that the polylinker can accommodate concatemers of the appropriate size and sequence in the desired order. A convergent strategy was employed that would enable the insertion of a variety of central elastin blocks. Two related plasmids were constructed from the expression vector pET-24a such that the order of the restriction sites employed for the insertion of the elastin and plastin repeats (*Sex*AI and *Bsp*MI, respectively) were reversed in the corresponding polylinker sequences. Plasmid pA1 encodes the *Sex*AI site downstream of the *Bsp*MI site, and plasmid pA2 encodes the sites in the opposite orientation. Plastin concatemers were inserted into the *Bsp*MI site of the modified polylinkers in plasmids pA1 and pA2. Repetitive genes encoding 16 repeats of the plastic sequence were isolated and identified via restriction cleavage with the endonucleases *Kpn*I and *Pst*I, which flank the *Bsp*MI insertion sites. The pair of recombinant plasmids, pPA1 and pPA2, encoded the N-terminal and C-terminal plastin domains of triblock polymer **1**, respectively. Restriction cleavage of each plasmid with *Sex*AI and *Xma*I afforded two fragments, which were separated with preparative agarose gel electrophoresis. Enzymatic ligation of the plastin-containing cleavage fragments of pPA1 and pPA2 afforded the recombinant plasmid pP2, which encoded a single *Sex*AI restriction site between two identical plastin blocks. The *Xma*I site occurs within the antibiotic resistance marker, which ensures the propagation of productive ligation products under antibiotic selection. The pool of concatemers was inserted into the compatible *Sex*AI site of pP2 to afford a series of expression constructs, pPEP, that encoded triblock copolymers containing central blocks of variable length. Individual clones were cleaved with *Kpn*I/*Pst*I to liberate gene fragments corresponding to the plastic and elastic domains and the parent plasmid, which were analyzed by agarose gel electrophoresis to determine the size of the elastin domains. This protocol has afforded synthetic genes larger than 5000 base pairs that encode elastin-mimetic block polypeptides with molar masses as great as 200 kDa, which could be expressed in high yield in a bacterial host system derived from *E. coli*.

36.4 A Hybrid Approach to the Controlled Assembly of Complex Architectures of Sequence-Repetitive Polypeptides

Despite the successful production of the target polypeptides, the original approach had several disadvantages. First, a laborious series of cloning steps was necessary to prepare the plasmid containing the expression construct encoding the triblock polypeptide. Second, while the size of the central elastin block could be varied easily using this strategy, the size and sequence of the end block domains could not be altered unless the entire expression cassette was reconstructed. Finally, the peptide sequences at the block termini or interfaces could not be altered using this approach, thus limiting the ability to add further modifications to the polypeptide sequence. In order to facilitate the modification of synthetic DNA constructs encoding elastin-mimetic block polypeptides, a more versatile cloning strategy was devised that combined the favorable attributes of DNA cassette concatemerization and recursive directional ligation. Individual concatemers corresponding to polypeptide blocks of defined composition were generated through the former procedure, while the concatemers encoding separate blocks were joined together sequentially through the use of the latter process. This approach permits the creation of synthetic genes that encode a structurally diverse range of sequence-repetitive polypeptides in which the size and sequences of individual blocks could be varied independently and the termini and interfaces could be modified easily.

The successful application of this procedure was greatly facilitated through the use of a seamless cloning strategy [50]. This process utilizes the recognition/cleavage characteristics of the type IIs restriction endonucleases, in which cleavage of a DNA duplex occurs at a specific position that is *downstream* of its recognition site. This cleavage pattern usually generates synthetic duplexes with 5′-cohesive ends in which the identity of bases within the overhangs is independent of the recognition site. Thus, the position of the recognition site determines the identity of the cohesive ends such that the sequence can be defined to facilitate directional cloning without the need for an array of endonucleases with unique internal recognition/cleavage patterns that match the internal sequence of the DNA duplex. Moreover, the restriction sites for type IIs endonucleases are cleaved from the DNA cassette during the procedure and, hence, are not incorporated into the coding sequence of the DNA monomer as extraneous sequences. Synthetic DNA duplexes flanked by inverted type IIs recognition sites can be enzymatically cleaved to generate ligation-competent DNA monomers with nonpalindromic, complementary cohesive ends. Moreover, if different type IIs restriction endonucleases are employed, then selective cleavage can occur at either end of the DNA cassette such that the termini can be functionally distinguished.

Initially, we described the use of seamless cloning [50] to prepare DNA monomers that could be enzymatically joined in an exclusively head-to-tail manner to generate concatemer libraries and to facilitate the insertion of concatemeric genes directly into the cloning site of an expression plasmid [51]. However, this technique can be applied not only for the generation and cloning of concatemers, but also for the directional assembly of concatemers into blocks of defined size and sequence through the selective use of two distinct type IIs restriction endonucleases. Currently, a large number (>100) of type IIs restriction endonucleases are commercially available that differ in the size and sequence of the recognition site and the position, orientation and restriction pattern of the cleavage site. In all cases, we prefer to use endonucleases that cleave downstream of the recognition site at a defined position to generate 5′ cohesive-ended fragments with single-stranded overhangs of three or four bases. Previous studies in our laboratory employed either *Eam1104*I or *Sap*I, which generate three-base, single-stranded ends. Currently, the enzymes of choice for this procedure are those that have at least six nucleotides in the recognition sequence and cleave at a defined position to generate four-base overhangs. A number of endonucleases that meet these criteria are commercially available (Table 36.1) such that at least two enzymes can be identified that do not cleave at internal sites within conventional cloning plasmids and are suitable for use in this procedure. Generally, restriction endonucleases are chosen that cleave proximally to the recognition site to avoid the necessity of synthesizing longer regions of superfluous DNA. The enzymes that are employed

Table 36.1 Recognition and cleavage characteristics of commercially available type IIs restriction endonucleases.

Enzyme	Recognition sequence	Cleavage pattern
*Aar*I	5′-CACCTGC-3′	(+4/+8)
*Acc*36I	5′-ACCTGC-3′	(+4/+8)
*Ace*III	5′-CAGCTC-3′	(+7/+11)
*Bau*I	5′-CACGAG	(−5/−1)
*Bbs*I	5′-GAAGAC-3′	(+2/+6)
*Bfu*AI	5′-ACCTGC-3′	(+4/+8)
*Bsa*I	5′-GGTCTC-3′	(+1/+5)
*Bse*Y1	5′-CCCAGC-3′	(−5/−1)
*Bsm*BI	5′-CGTCTC	(+1/+5)
*Bsp*MI	5′-ACCTGC-3′	(+4/+8)
*Bte*ZI	5′-GCGATG-3′	(+10/+14)
*Eco*31I	5′-GGTCTC-3′	(+1/+5)

in the procedure that we describe are *Bbs*I, *Bsm*BI and *Bsa*I, which were chosen on the basis of commercial availability, differences in recognition site, and compatibility with the cloning and expression plasmids employed for the construction of the DNA cassettes.

Two pairs of synthetic oligonucleotides were annealed to create two DNA monomers, S1 and S3, that encoded the elastic and plastic peptide repeat sequences, respectively (Figure 36.4). These DNA cassettes share several features that are critical for the gene assembly process. First, *Hin*DIII and *Bam*HI restriction sites were placed at the 5′ and 3′ termini, respectively, to facilitate insertion of the DNA monomer into the polylinker of the cloning plasmid pZErO-2. Second, recognition sites for the type IIs restriction endonucleases *Bbs*I and *Bsm*BI were placed at the termini of the DNA cassette at internal positions that define the beginning and end, respectively, of the coding sequence of peptide repeat motif. Sequential cleavage with these two enzymes releases the DNA monomer from the cloning plasmid. Cleavage with either enzyme generates compatible, nonpalindromic 5′-cohesive ends derived from the four-base sequences, 5′-TCCA-3′ in the forward (sense) direction and 5′-TGGA-3′ in the reverse (anti-sense) direction. Thus, the

(a)

```
                Val Pro Gly Ala Gly Val Pro Gly Ala Gly Val Pro Gly Glu Gly Val Pro Gly
                 ↓
AG CTT GAA GAC GTT CCA GGT GCA GGC GTA CCG GGT GCT GGC GTT CCG GGT GAA GGT GTT CCA GGC
     A CTT CTG CAA GGT CCA CGT CCG CAT GGC CCA CGA CCG CAA GGC CCA CTT CCA CAA GGT CCG
HinD III Bbs I          ↑

Ala Gly Val Pro Gly Ala Gly Val Pro
                          ↓
GCA GGT GTA CCG GGT GCG GGT GTT CCA AGA GAC GG
CGT CCA CAT GGC CCA CGC CCA CAA GGT TCT CTG CCC TAG
                                  ↑ BsmB I  BamH I
```

(b)

```
                Ile Pro Ala Val Gly Ile Pro Ala Val Gly Ile Pro Ala Val Gly Ile Pro Ala
                 ↓
AG CTT GAA GAC ATT CCA GCT GTT GGT ATC CCG GCT GTT GGT ATC CCA GCT GTT GGC ATT CCG GCT
     A CTT CTG TAA GGT CGA CAA CCA TAG GGC CGA CAA CCA TAG GGT CGA CAA CCG TAA GGC CGA
HinD III Bbs I          ↑

Val Gly Ile Pro Ala Val Gly Ile Pro
                          ↓
GTA GGT ATC CCG GCT GTT GGT ATT CCA AGA GAC GG
CAT CCA TAG GGC CGA CAA CCA TAA GGT TCT CTG CCC TAG
                                  ↑ BsmB I  BamH I
```

(c)

```
      ↓                                     ↓
C ATG GTT CCA AGA GAC CAG GTA CCG GTC TCG TCC AGG TGT AGG CTA ATA
      CAA GGT TCT CTG GTC CAT GGC CAG AGC AGG TCC ACA TCC GAT TAT TCG A
Nco I       ↑ Bsa I      Kpn I  Bsa I          ↑                HinD III
```

Figure 36.4 (a,b) Oligonucleotide cassettes encoding the elastic **S1** (a) and plastic **S3** (b) repeat sequences. The recognition sites for the relevant restriction endonucleases that are employed for generation of the DNA monomer are highlighted. Arrows indicate the cleavage positions on the sense and anti-sense strands for the respective endonucleases. The cleavage sites for the type IIs restriction endonuclease *Bbs*I and *Bsm*BI occur at well-defined positions downstream of the recognition site. Note that the cohesive ends generated from restriction cleavage are compatible and nonpalindromic for both types of monomer and the resulting concatemers; (c) DNA sequence of the expression plasmid adaptor. The translational start sequence coincides with the position of the *Nco*I cleavage site. Note that the recognition sequences for the enzyme *Bsa*I occur internally with respect to the cleavage sites. Restriction digestion with *Bsa*I affords cohesive ends that are compatible with those of the concatemeric cassettes when oriented in the sense direction.

DNA monomers are competent for self-ligation (as in Figure 36.2a) to generate a library of concatemers that differ by increments of 75 base pairs that corresponds to the coding sequence of the DNA monomer (Figure 36.5a). Furthermore, the cloning plasmids retain the restriction sites for *Bbs*I and *Bsm*BI in the modified polylinker after excision of the monomer and the resulting cohesive ends are compatible with directional insertion of the DNA concatemers. Thus, after electrophoretic size fractionation, the pool of concatemers of desired size range can be recloned into the modified cloning plasmid. The cloned concatemers are screened for the size of the insert and clones containing concatemers of appropriate size are propagated for further use in the construction of the DNA block cassettes (Figure 36.5b). Using this procedure, concatemers can be routinely isolated

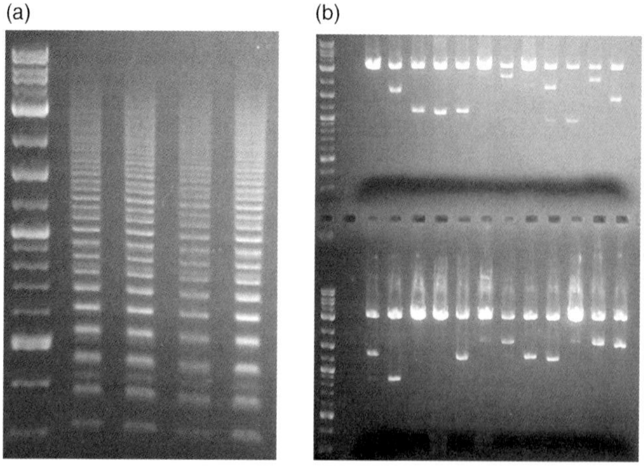

Figure 36.5 (a) Agarose gel electrophoretogram of the pool of DNA concatemers resulting from self-ligation of DNA monomers S1 and S3. The difference in size between successive concatemers corresponds to the 75 base pairs of the DNA monomer; (b) Agarose gel electrophoretogram of DNA fragments resulting from restriction cleavage of the plasmid clones containing the concatemeric DNA inserts. The larger size bands in each lane correspond to the plasmid pZErO-2. The smaller bands correspond to the sizes of the DNA concatemers, which range from approximately 1500 to 2200 base pairs. The far left lane in each case represents the DNA size standards.

that are at least 2000 base pair in size and, once recloned, have the same relationship to the *Bbs*I and *Bsm*BI restrictions sites as were established in the original monomeric DNA cassette. In addition, it is critically important to note that the cohesive ends of the two different sets of concatemers are identical and are, therefore, compatible for ligation to each other in the appropriate context.

The second step of the assembly process involves selective ligation of pairs of concatemers to create synthetic DNA fusions that encode blocks of sequence-repetitive polypeptides (Figure 36.6). The two concatemers may be identical as in conventional recursive directional ligation, or correspond to different sequence repeats, as required for the construction of block polypeptides. For the synthesis of artificial genes encoding BAB-type triblock polypeptides, two successive ligation steps are performed to generate initially an AB-diblock and a BAB-triblock. This gene assembly involves not only the selective use of either *Bbs*I or *Bsm*BI, but also the participation of a third enzyme *Nco*I, which cleaves at an internal position in the plasmid pZErO-2 that is within the kanamycin resistance gene. *Nco*I is a conventional type II restriction endonuclease that recognizes and cleaves at staggered (1/5) positions within the palindromic hexanucleotide site to generate 5′-cohesive ends with four-base, single-stranded extensions. Initially, the concatemeric genes are contained within the separate plasmid clones from which they were initially isolated. One clone contains the proximal concatemer and the other contains the distal one in terms of position relative to the anticipated N terminus

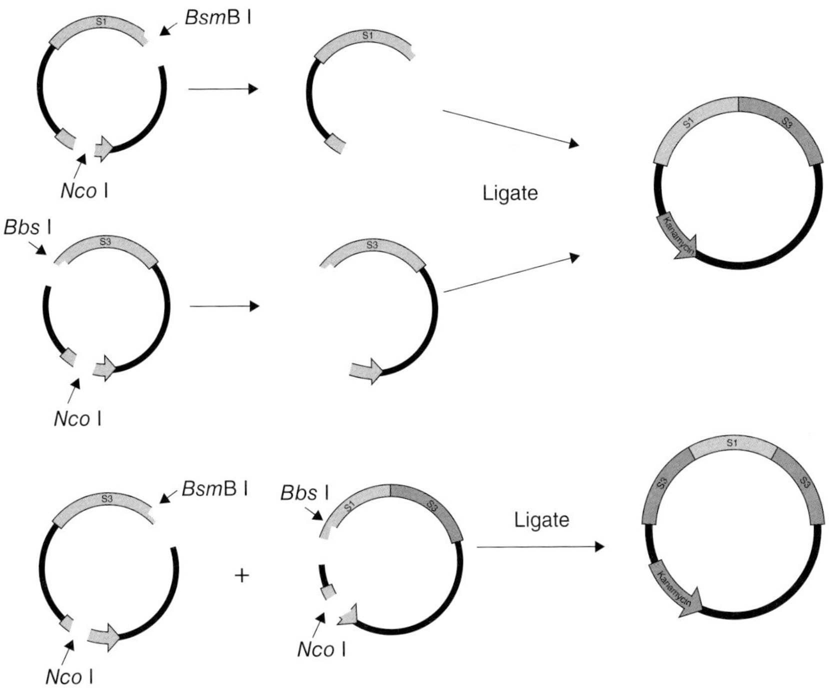

Figure 36.6 Directional strategy for assembly of synthetic genes encoding block polypeptides of elastin-mimetic sequence repeats. Plasmid clones corresponding to elastic (S1) and plastic (S3) concatemers are assembled iteratively to create initially an AB-diblock construct that serves as the input for the subsequent synthesis of a BAB-triblock cassette. A seamless cloning approach is employed in which two independent type IIs restriction endonucleases, *Bbs*I and *Bsm*BI, flank the 5′ and 3′ positions of both concatemeric cassettes in an inverted orientation. Selective cleavage generates cohesive ends that are nonpalindromic but compatible, which allows the two concatemers to be joined in a specific orientation. Although this strategy is iterative rather than recursive, the procedure has no intrinsic limitations that would prohibit the latter approach.

within the coding sequence of the product. Each plasmid clone is subjected to restriction digestion with *Nco*I and one of the type IIs restriction endonucleases, *Bsm*BI for the proximal (S1) clone and *Bbs*I for the distal (S3) clone. This process cleaves each plasmid into two pieces that can be separated using agarose gel electrophoresis to isolate the fragments that contain the respective concatemers. Subsequent ligation of these two fragments not only reconstitutes the kanamycin resistance marker, but also generates a direct fusion of the proximal concatemer to the distal one, affording a construct that encodes the AB-diblock polypeptide. This process typically proceeds with very high efficiency due to reconstitution of the selectable marker, and also affords a diblock cassette that is flanked by unique *Bbs*I and *Bsm*BI sites that can be employed in further rounds of direction ligation. Indeed, the initial diblock construct is employed as the distal block in a second round of directional ligation in which the proximal block corresponds to the S3-type concatemer. Artificial genes encoding multiple blocks of sequence-repetitive polypeptides of defined block order and block size can be constructed straightforwardly in this manner. In addition, the presence of two distinct type IIs restriction endonuclease sites provides a mechanism to easily introduce modifications between blocks or at either end of the sequence if compatible termini are present in the modifying sequences. In all cases, the original restriction endonuclease recognition sites are retained at the respective termini of the ligation products, while seamless junction points are created at the new interfaces between the blocks. The assembly of the DNA construct, including all modifications, is completed within the cloning plasmid and the final expression cassette is liberated from the plasmid by sequential restriction digestion with *Bbs*I and *Bsm*BI. Using this procedure, we have created block-like sequences of artificial genes as large as 8000 base pairs that encode elastin-mimetic polypeptides of molar mass greater than 250 kDa.

The final step of the procedure involves directional cloning of the concatemeric DNA cassette into an appropriately modified expression vector. However, the nonpalindromic cohesive ends of the concatemeric cassettes are not compatible with cohesive ends that would be generated by restriction digestion at the endonuclease cleavage sites within the polylinker of conventional expression plasmids. Therefore, the pET-24d expression plasmid was modified through the introduction of a synthetic adaptor sequence (Figure 36.6), which was inserted between the *Nco*I/*Hin*DIII sites within the polylinker. Several critical features have been introduced into the adaptor sequence; the most important of which is the introduction of two *Bsa*I recognition sites. This type IIs restriction endonuclease cleaves at a defined position downstream of its recognition site to generate 5′-cohesive ends that can be made to be compatible with those of the DNA concatemers. Note that the recognition sites are internally oriented within the synthetic polylinker with respect to the cleavage sites. Thus, restriction cleavage at both sites should liberate a small stuffer fragment that contains the recognition sites for *Bsa*I, effectively removing these sites from the plasmid and leaving only the cohesive ends. The enzyme *Bsa*I was chosen for use in this procedure, as it does not cleave at internal sites within the expression plasmid, while both *Bbs*I and *Bsm*BI cleave at one or more internal positions. In addition, a recognition/cleavage site for *Kpn*I was placed between the two *Bsa*I sites as a mechanism to prevent unproductive side reactions such as re-ligation of the polylinker into the plasmid or ligation of a concatemeric DNA into a plasmid in which only one of the *Bsa*I sites was cleaved. Thus, restriction digestion with the second enzyme should reduce background levels of undesired cloning products, some of which can only be identified by double-stranded DNA sequence analysis. Finally, the sequence of the adaptor is configured such that concatemer cassettes that have been inserted into the appropriate sites would be presented in the appropriate reading frame to produce the desired polypeptide upon induction of gene expression. The target genes are placed under the control of a T7*lac* promoter in this modified pET-24a expression vector, which has been used effectively to direct overexpression of highly repetitive, heterologous polypeptides in *E. coli* strains such as BL21(DE3) or BLR(DE3) [52]. High yields (multigram quantities) of the elastin-mimetic polypeptides have been obtained using this expression system in terrific broth medium, despite the structural incongruity of these proteins to the endogenous proteins of the host bacterium. Elastin-mimetic polypeptides, including block copolymers, are easily purified from the endogenous host proteins via repetitive cycling through the phase transition, which results in selective precipitation of the polypeptide at high salt concentration (500 m*M* NaCl) [53]. The final isolated yields of the elastin-mimetic polypeptides are usually in the range from 200 to 800 mg of purified protein l^{-1} of the shake-flask culture. SDS–PAGE and reverse-phase HPLC indicated a high level of purity of the recombinant polypeptides (>95%). Amino acid compositional analysis, N-terminal Edman degradation and matrix-assisted laser desorption/ionization-time of flight (MALDI-TOF) mass spectrometry confirmed the integrity of the protein constructs.

36.5 Future Outlook

We envision that this hybrid cloning approach should be amenable to the construction of artificial genes encoding a wide range of sequence-repetitive polypeptides, including those of highly complex macromolecular architecture. Many experimental studies of native protein-based materials have been predicated on the assumption that the physical behavior of these materials in solution and in the solid state could be approximated using protein polymers derived from tandem repeats of the canonical sequence motif. However, many native protein-based materials have unique N- and C-terminal sequences that strongly deviate in composition from the repeat sequence. These sequences have been proposed to have multiple roles *in vivo*, including to facilitate the processing of the materials or to participate in interactions with other proteins to form supramolecular assemblies. One well-known example of the former process is the involvement of the C-terminal propeptides in directing the folding and assembly of fibrous collagens [54], while an example of the latter process is the interaction of the C terminus of tropoelastin with fibrillin to form elastic fibers [55]. The approach that we describe here can be employed easily to introduce native terminal sequences to examine the effect of their presence on the physical behavior of the sequence-repetitive domain of these polypeptides. This approach has been recently employed in the biosynthesis of a synthetic derivative of *Araneus diadematus* dragline fibroin [56], in which the native, non-repetitive C-terminal sequence [57, 58] was appended to the repetitive domain. Distinct differences in assembly characteristics were noted between two silk variants that differed only in the presence of the C-terminal domain [59]. These results suggest that uniquely reactive peptide sequences may also be added to the termini of synthetic protein materials to facilitate their use within directed applications, for example, terminal derivatization of polypeptides to enhance surface attachment through introduction of specific chemisorptive interactions [5, 60]. In addition to the presence of unique terminal sequences, the internal repetitive sequences of native protein-based materials often contain larger repeat motifs that are observed at length scales beyond the simple oligopeptide repeat motifs. For example, the amino acid sequences of silk proteins across different genera and species often display amphiphilic repeats of longer hydrophobic and shorter hydrophilic domains are superimposed at greater length scales upon the crystalline repeats [61]. This macro-repeat pattern has been proposed as an essential element to the processing of silk proteins, which typically occurs at high protein concentrations (20–40% wt/vol.) within the silk glands of these organisms. In contrast, protein-based materials based on repetition of the simple sequence repeats of native silks are often difficult to process due to limited solubility *in vitro*, perhaps as a consequence of the absence of these native sequence features. The synthesis of more complex protein-based materials has been hindered to a great extent by the difficulty in assembling the corresponding genes using previously described cloning strategies. We believe that the technology described in this chapter may provide relatively easy access to synthetic genes of sufficient complexity to encode sequence variations similar to those observed for native protein-based materials. These novel types of material provide an opportunity to further examine the effect of sequence complexity on the physical behavior of protein-based materials, as well as the capacity to design synthetic multifunctional polypeptides for development as high-performance materials for specialized applications.

Acknowledgments

The authors acknowledge the financial support of the Herman Frasch Foundation (418–97HF), NIH (5R01HL071136–04) and the NSF (EEC-9731643).

References

1 van Hest, J.C. and Tirrell, D.A. (2001) *Chemical Communications*, 1897–904.

2 Krejchi, M.T., Atkins, E.D., Waddon, A.J., Fournier, M.J., Mason, T.L. and Tirrell, D.A. (1994) *Science*, **265**, 1427–32.

3 West, M.W., Wang, W., Patterson, J., Mancias, J.D., Beasley, J.R. and Hecht, M.H. (1999) *Proceedings of the National Academy of Sciences of the United States of America*, **96**, 11211–16.

4 (a) Topilina, N.I., Higashiya, S., Rana, N., Ermolenkov, V.V., Kossow, C., Carlsen, A, Ngo, S. C., Wells, C.C., Eisenbraun, E.T., Dunn, K.A., Lednev, I.K., Geer, R.E., Kaloyeros, A.E. and Welch, J.T. (2006) *Biomacromolecules*, **7**, 1104–11.
(b) Higashiya, S., Topilina, N.I., Ngo, S.C., Zagorevskii, D. and Welch, J.T. (2007) *Biomacromolecules*, **8**, 1487–97.

5 (a) Henderson, D.B., Davis, R.M., Ducker, W.A. and Van Cott, K.E. (2005) *Biomacromolecules*, **6**, 1912–20.
(b) Tulpar, A., Henderson, D.B., Mao, M., Caba, B., Davis, R.M., Van Cott, K.E. and Ducker, W.A. (2005) *Langmuir*, **21**, 1497–506.

6 Yu, S.M., Conticello, V.P., Zhang, G., Kayser, C., Fournier, M.J., Mason, T.L. and Tirrell, D.A. (1997) *Nature*, **389**, 167–70.

7 Yu, S.M., Soto, C.M. and Tirrell, D.A. (2000) *Journal of the American Chemical Society*, **122**, 6522–9.

8 Lee, T.A.T., Cooper, A., Apkarian, R.P. and Conticello, V.P. (2000) *Advanced Materials*, **12**, 1105–10.

9 Wright, E.R., McMillan, R.A., Cooper, A., Apkarian, R.P. and Conticello, V.P. (2002) *Advanced Functional Materials*, **2**, 149–54.

10 Petka, W.A., Harden, J.L., McGrath, K.P., Wirtz, D. and Tirrell, D.A. (1998) *Science*, **281**, 389–92.

11 Qu, Y., Payne, S.C., Apkarian, R.P. and Conticello, V.P. (2000) *Journal of the American Chemical Society*, **122**, 5014–15.

12 Fraser, R.D.B. and MacRae, T.P. (1973) *Conformation in Fibrous Proteins*, Academic Press, New York, NY.

13 Vollrath, F. and Porter, D. (2006) *Soft Matter*, **2**, 377–85.

14 Oroudjev, E., Soares, J., Arcidiacono, S., Thompson, J.B., Fossey, S.A. and Hansma, H.G. (2002) *Proceedings of the National Academy of Sciences of the United States of America*, **99** (Suppl 2), 6460–5.

15 Hinman, M.B. and Lewis, R.V. (1992) *The Journal of Biological Chemistry*, **267**, 19320–4.
(b) Xu, M. and Lewis, R.V. (1990) *Proceedings of the National Academy of Sciences of the United States of America*, **87**, 7120–4.

16 Simmons, A.H., Michal, C.A. and Jelinski, L.W. (1996) *Science*, **271**, 84–7.

17 Rosenbloom, J., Abrams, W.R. and Mecham, R. (1993) *The FASEB Journal*, **7**, 1208–18.

18 Gosline, J., Lillie, M., Carrington, E., Guerette, P., Ortlepp, C. and Savage, K. (2002) *Philosophical Transactions of the Royal Society of London. Series B, Biological Sciences*, **357**, 121–32.

19 Lewis, R.V., Hinman, M., Kothakota, S. and Fournier, M.J. (1996) *Protein Expression and Purification*, **7**, 400–6.

20 Prince, J.T., McGrath, K.P., DiGirolamo, C.M. and Kaplan, D.L. (1995) *Biochemistry*, **34**, 10879–85.

21 Fahnestock, S.R., Yao, Z. and Bedzyk, L.A. (2000) *Journal of Biotechnology*, **74**, 105–19.

22 Davidson, A.R. and Sauer, R.T. (1994) *Proceedings of the National Academy of Sciences of the United States of America*, **91**, 2146–50.

23 Doel, M.T., Eaton, M., Cook, E.A., Lewis, H., Patel, T. and Carey, N.H. (1980) *Nucleic Acids Research*, **8**, 4575–92.

24 Cappello, J., Crissman, J.W., Dorman, M., Mikolajczak, M., Textor, G., Marquet, M. and Ferrari, F. (1990) *Biotechnology Progress*, **6**, 198–202.

25 Goldberg, I., Salerno, A.J., Patterson, T. and Williams, J.I. (1989) *Gene*, **80**, 305–14.

26 McPherson, D.T., Morrow, C., Minehan, D.S., Wu, J., Hunter, E. and Urry, D.W. (1992) *Biotechnology Progress*, **8**, 347–52.

27 Langer, R. and Tirrell, D.A. (2004) *Nature*, **428**, 487–92.

28 Chilkoti, A., Christensen, T. and MacKay, J.A. (2006) *Current Opinion in Chemical Biology*, **10**, 652–7.

29 Hofmann, S., Knecht, S., Langer, R., Kaplan, D.L., Vunjak-Novakovic, G., Merkle, H.P. and Meinel, L. (2006) *Tissue Engineering*, **12**, 2729–38.

30 Herrero-Vanrell, R., Rincon, A.C., Alonso, M., Reboto, V., Molina-Martinez, I.T. and Rodriguez-Cabello, J.C. (2005) *Journal of Controlled Release*, **102**, 113–22.

31 Noshay, A., McGrath, J.E. and Copolymers, B. (1977) *Overview and Critical Survey*, Academic Press, New York.

32 Deming, T.J. (1997) *Nature*, **390**, 386–9.

33 (a) Bellomo, E.G., Wyrsta, M.D., Pakstis, L., Pochan, D.J. and Deming, T.J. (2004) *Nature Materials*, **3**, 244–8.
(b) Nowak, A.P., Breedveld, V., Pakstis, L., Ozbas, B., Pine, D.J., Pochan, D. and Deming, T.J. (2002) *Nature*, **417**, 424–8.

34 (a) Knight, D.P. and Vollrath, F. (2002) *Philosophical Transactions of the Royal Society of London. Series B, Biological Sciences*, **357**, 155–63.
(b) Vollrath, F. and Knight, D.P. (2001) *Nature*, **410**, 541–8.

35 (a) Mortensen, K. (1998) *Current Opinion in Colloid and Interface Science*, **3**, 12–19.
(b) Alexandridis, P. (1996) *Current Opinion in Colloid and Interface Science*, **1**, 490–501.

36 Hajduk, D.A., Kossuth, M.B., Hillmyer, M.A. and Bates, F.S. (1998) *The Journal of Physical Chemistry B*, **102**, 4269–76.

37 Wright, E.R. and Conticello, V.P. (2002) *Advanced Drug Delivery Reviews*, **54**, 1057–73.

38 (a) Urry, D.W., Luan, C.H., Parker, T.M., Gowda, D.C., Prasad, K.U., Reid, M.C. and Safavy, A. (1991) *Journal of the American Chemical Society*, **113**, 4346–7. (b) Urry, D.W., Gowda, D.C., Parker, T., Luan, C.H., Reid, M.C., Harris, C.M., Pattanaik, A. and Harris, R.D. (1992) *Biopolymers*, **32**, 1243–50.

39 Urry, D.W., Parker, T.M., Reid, M.C. and Gowda, D.C. (1991) *The Journal of Bioactive and Compatible Polymers*, **6**, 263–82.

40 Mecham, R.P., Broekelman, T.J., Fliszar, C.J., Shapiro, S.D., Welgus, H.G. and Senior, R.M. (1997) *The Journal of Biological Chemistry*, **272**, 18071–6.

41 (a) Daniell, H., Guda, C., McPherson, D.T., Zhang, X., Xu, J. and Urry, D.W. (1997) *Methods in Molecular Biology*, **63**, 359–71.
(b) Chow, D.C., Dreher, M.R., Trabbic-Carlson, K. and Chilkoti, A. (2006) *Biotechnology Progress*, **22**, 638–46.

42 Wu, X., Sallach, R., Haller, C.A., Caves, J.A., Nagapudi, K., Conticello, V.P., Levenston, M.E. and Chaikof, E.L. (2005) *Biomacromolecules*, **6**, 3037–44.

43 Nagapudi, K., Brinkman, W.T., Leisen, J., Thomas, B.S., Wright, E.R., Haller, C., Wu, X., Apkarian, R.P., Conticello, V.P. and Chaikof, E.L. (2005) *Macromolecules*, **38**, 345–54.

44 (a) Cappello, J. and Ferrari, F. (1994) *Plastics from Microbes: Microbial Synthesis of Polymers and Polymer Precursors* (ed. D.P. Mobley), Hanser/Gardner Publications, Munich, Germany, pp. 35–92.
(b) Ferrari, F.A. and Cappello, J. (1997) *Protein–Based Materials* (eds K. McGrath and D. Kaplan), Birkhauser, Boston, MA, pp. 37–60.

45 McGrath, K.P., Fournier, M.J., Mason, T.L. and Tirrell, D.A. (1992) *Journal of the American Chemical Society*, **114**, 727–33.

46 Lixin, M. (2006) *Biomacromolecules*, **7**, 2099–107.

47 Meyer, D.E. and Chilkoti, A. (2002) *Biomacromolecules*, **3**, 357–67.

48 Won, J. and Barron, A.E. (2002) *Macromolecules*, **35**, 8281–7.

49 Junger, A., Kaufman, D., Scheibel, T. and Weberskirch, R. (2005) *Macromolecular Bioscience*, **5**, 494–501.

50 Padgett, K.A. and Sorge, J.A. (1996) *Gene*, **168**, 31–5.

51 (a) McMillan, R.A., Lee, T.A.T. and Conticello, V.P. (1999) *Macromolecules*, **32**, 3643–8.
(b) Goeden-Wood, N.L., Conticello, V.P., Muller, S.J. and Keasling, J.D. (2002) *Biomacromolecules*, **3**, 874–9.

52 Studier, F.W., Rosenberg, A.H., Dunn, J.J. and Dubendorff, J.W. (1989) *Methods in Enzymology*, **185**, 60–89.

53 McPherson, D.T., Xu, J. and Urry, D.W. (1996) *Protein Expression and Purification*, **7**, 51–7.

54 McLaughlin, S.H. and Bulleid, N.J. (1998) *Matrix Biology*, **16**, 369–77.

55 Kozel, B.A., Wachi, H., Davis, E.C. and Mecham, R.P. (2003) *The Journal of Biological Chemistry*, **278**, 18491–8.

56 Huemmerich, D., Helsen, C.W., Quedzuweit, S., Oschmann, J., Rudolph, R. and Scheibel, T. (2004) *Biochemistry*, **43**, 13604–12.

57 Beckwitt, R. and Arcidiacono, S. (1994) *The Journal of Biological Chemistry*, **269**, 6661–3.

58 Sponner, A., Vater, W., Rommerskirch, W., Vollrath, F., Unger, E., Grosse, F. and Weisshart, K. (2005) *Biochemical and Biophysical Research Communications*, **338**, 897–902.

59 Exler, J.H., Hümmerich, D. and Scheibel, T. (2007) *Angewandte Chemie – International Edition in English*, **46**, 3559–62.

60 Dalsin, J.L., Hu, B.H., Lee, B.P. and Messersmith, P.B. (2003) *Journal of the American Chemical Society*, **125**, 4253–8.

61 Bini, E., Knight, D.P. and Kaplan, D.L. (2004) *The Journal of Biological Chemistry*, **335**, 27–40.

37
Silk Proteins – Biomaterials and Bioengineering

Xiaoqin Wang, Peggy Cebe and David. L. Kaplan

37.1
Silk Protein Polymers – An Overview

Silk proteins represent one group of an important set of proteins termed fibrous proteins. Fibrous proteins provide the 'materials' of nature, whether in the form of fibers (orb webs of spiders), tissue structures (collagens in tissues), hair (keratin) or as ceramics (seashell organic–inorganic composites), among many other examples (see also Chapter 36). This general grouping of proteins is characterized by the presence of highly repetitive primary sequences [1] which lead to long stretches of regular secondary structures, unlike the highly mixed secondary structures seen in globular proteins. This feature leads to the formation of complex long-range ordered materials, such as fibers, basement membranes and related structures, with key information 'encoded' into the protein structure at the primary sequence and secondary structural levels. This chemistry and structure leads to fibrous proteins forming mechanically robust but flexible structures, while also providing recognition sites in regular patterns for cell binding, for mineralization, and for interactions with extracellular matrix (ECM) components such as glycosaminoglycans (GAGs) and matrix-associated proteins to generate different levels of tissue hydration or mechanical function. The highly repetitive primary sequences allow key features of the sequences of these large proteins to be abbreviated as short synthetic genetic variants.

Silk proteins represent a unique family of the class of these fibrous proteins due to their novel structure and resulting functions [2, 3]. From a materials science perspective, silks spun by spiders and silkworms represent the strongest and toughest natural fibers known [2, 4], and on a weight basis, are stronger than steel. These properties derive in part from the highly oriented, numerous and small beta-sheet crystals in combination with less-crystalline regions that play an important role in modulating water, hydrogen bonding and elasticity. The novel material features of silks have recently been extended due to insights into self-assembly, the role of water in assembly, and the formation of liquid crystalline features [5–8]. These insights have led to new processing modes to generate novel materials from

Protein Engineering Handbook. Edited by Stefan Lutz and Uwe T. Bornscheuer

ISBN: 978-3-527-31850-6

silk proteins, including hydrogels, ultrathin films, thick films, conformal coatings, three-dimensional (3D) porous matrices, nanoscale-diameter fibers and large-diameter fibers [9–14] (Figure 37.1). Importantly, the content of beta sheet can be controlled based on the mode of processing and postprocessing, resulting in an ability to modulate the mechanical properties as well as the enzymatic degradability of these materials [13, 15, 16].

The molecular structure of many silks consists of large regions (blocks) of hydrophobic amino acids, segregated by relatively short, more hydrophilic, regions (spacers). The hydrophobic domains organize into protein crystals (beta sheets). These beta sheets form thermally nonreversible physical crosslinks that stabilize the silk structures, and generally are dominated by repeats of alanine, glycine–alanine or glycine–alanine–serine. The less-crystalline regions of some silks consist of: (i) β-spirals similar to a β-turn composed of glycine–proline–glycine–XX repeats (where X is mostly glutamine); (ii) helical structures composed of glycine–glycine–X, which give rise to the elasticity; and (iii) tyrosine-containing sequences [2, 17, 18]. At the N and C termini, most silks also contain nonrepetitive regions that appear to be instrumental in controlling solubility and may also play a role in the controlled assembly of silk proteins as related to the mechanical properties

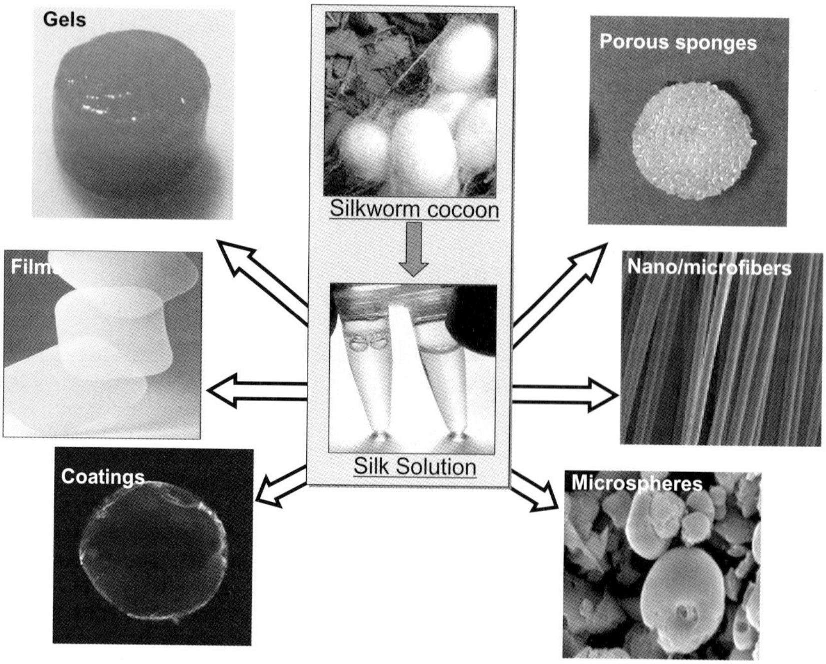

Figure 37.1 Silk materials in diverse formats formed from solubilized silk from silkworm cocoons: hydrogels, two-dimensional films, nanoscale layer-by-layer coatings, microspheres, nano/microfibers and porous sponges.

[19, 20]. Most silk proteins have a high molecular weight (in excess of 300 kDa), and the amino acid chemistry of most silks involved in orb webs or cocoon structures is dominated by hydrophobic side chains. A variety of silk genes have been isolated from different silkworms and spider species to provide specific insight into the above features relevant to the structure–function relationships in this family of proteins. For example, genetic sequence data have been obtained for silkworm species such as *Bombyx mori* (the commercialized source of silk for textiles via sericulture), *Antheraea pernyi, Antheraea mylitta, Antheraea assama* and *Antheraea yamamai* [21–24], and spider species including *Nephila clavipes, Araneus diadematus,* the Araneomorphae and Mygalomorphae (tarantulas), and recently even the black widow spider (*Latrodectus hesperus*) [25–27].

Native silk fibers are the strongest known natural fibers that also exhibit impressive toughness and rival even high-performance synthetic fibers in terms of mechanical performance [3, 4, 28, 29]. For example, recent data with reprocessed silkworm silk thin films demonstrated a modulus of 100 nm-thick films of between 6 and 8 GPa, with a toughness of 328 kJ m^{-3} [29]. Recent data with reprocessed silk fibroin 3D porous matrices also showed that the resistance to compression of these systems exceeded that of commonly used biomaterial polymeric scaffolds (e.g. polyesters, collagen) [30]. Thermal stability is also a hallmark of silks, as they can be steam-sterilized without any loss of structure or mechanical integrity. These properties have been exploited in various forms in recent studies in tissue engineering. Importantly, silks achieve their impressive mechanical properties with physical crosslinks (beta sheets). Unlike collagens and many other polymeric biomaterials, silk beta-sheet crosslinks are created without a need for chemical or photoinitiated crosslinking, thus facilitating easier processing and materials fabrication. These physical crosslinks form through hydrogen bonding and hydrophobic interactions via inter- and intra-chain interactions.

The thermal properties of silk fibroin and silk fibers have been investigated for a long time [31, 32]. The glass transition (T_G) is 178 °C for dry *B. mori* fibroin [32–34], and the thermal degradation temperature is 230 °C [32, 34]. Most studies use differential scanning calorimetry (DSC) [31, 32, 35–38] for qualitative studies of the temperature location and breadth of the glass transition relaxation process. The modern variant, temperature-modulated DSC, provides quantitative measurement of the heat capacity step at the glass transition, which can be used to determine the absolute determination of the degree of crystallinity in silk films [39, 40]. The role of water on the glass transition can also be quantitatively investigated [40, 41].

The thermal degradation of silks is studied using thermogravimetric analysis [40, 42] or thermomechanical analyzers [34] to study the impact of beta-sheet crystals on the glass transition, crystallization and thermal stability of fibroin and other silks. The impact of proteolytic degradation on crystallinity was studied using DSC [43]. Recently, the blending of silk with polymers and biopolymers has been shown to improve thermal degradation behavior, including silk blends with chitosan [38], gelatin [42, 44, 45], nylon 66 [46] and nylon 6 [47], fucoidan [48], poly(vinyl alcohol) or epoxy [49, 50], or grafting with methacrylamide [51]. The effects of hydration on thermal properties have been investigated for blends [52] and for silk fibroin [40].

The stability of silks to a wide range of environmental insults is particularly impressive, and is derived in a large part from the beta-sheet crystalline structures. Silks are stable to: (i) most organic solvents; (ii) relatively high temperatures for proteins (>200 °C); and (iii) most acids/bases. Importantly, spun silks containing beta-sheet crystals are stable in water. In order to solubilize silk, concentrated salts (e.g. 50% LiBr, calcium nitrate), concentrated acids (which degrade the material), or solvents [e.g. *N*-methylenemorpholine-*N*-oxide (NMMO)] can be used. Once solubilized, the protein can be processed into aqueous solutions without the salts, or lyophilized for subsequent solubilization in solvents such as hexafluoroisopropanol (HFIP) or ionic liquids. The presence of diverse amino acid side-chain chemistries on silk protein chains facilitates coupling chemistry to functionalize silks, such as with cytokines, morphogens or cell-binding domains [53, 54]. For example, cell growth factors can be coupled to these proteins with carbodimide chemistry (1-ethyl-3-(3-dimethylaminopropyl), to enhance functional utility, without any negative impact on biological function. This has been demonstrated with arginine–glycine–aspartic acid peptides (RGD), parathyroid hormone (PTH) and bone morphogenetic protein-2 (BMP-2) (e.g. [54, 55]). Enzymes and fluorescent labels have been coupled to silk using similar strategies [56, 57]. Furthermore, the predominantly hydrophobic nature of the protein can be exploited to functionalize the interfaces of bulk silk materials with less-hydrophobic additives or modifications, allowing facile approaches to modifying the material surfaces such as with polyaspartic acid for mineralization, or peptide domains for nucleation and polymerization of silica precursors for glassification [58, 59]. Despite the dominant hydrophobic feature of these proteins, all aqueous modes of processing have been developed to permit the direct incorporation of labile chemicals, biochemicals or biologicals into the silk aqueous processing stream, and this has resulted in functionalized silk-based materials without any loss of the biological activity of the additives [60, 61].

The solubility of silk proteins has also been addressed via genetic modifications of engineered variants with environmentally regulated molecular triggers. These designs were based on either chemical or biochemical reactions, and exploited spider silk sequence features as a starting point [62–65]. In one system, methionine residues were added to the consensus sequence to serve as redox triggers; here, oxidation of the methionines improved the solubility in water, while reduction reduced the solubility. In another design, an enzymatic site for cyclic AMP-dependent protein kinase phosphorylation was encoded in the silk sequence to exploit reversible phosphorylation/dephosphorylation biochemical reactions related to solubilization (phosphorylation to promote solubility, dephosphorylation to reduce solubility).

Water plays a critical and central role in the assembly of silk polymer chains into functional materials. Mechanisms underlying silk protein polymer assembly have been elucidated and exploited towards a new family of novel biomaterials [5–7] (Figure 37.2). Silks exhibit many features common to block copolymer

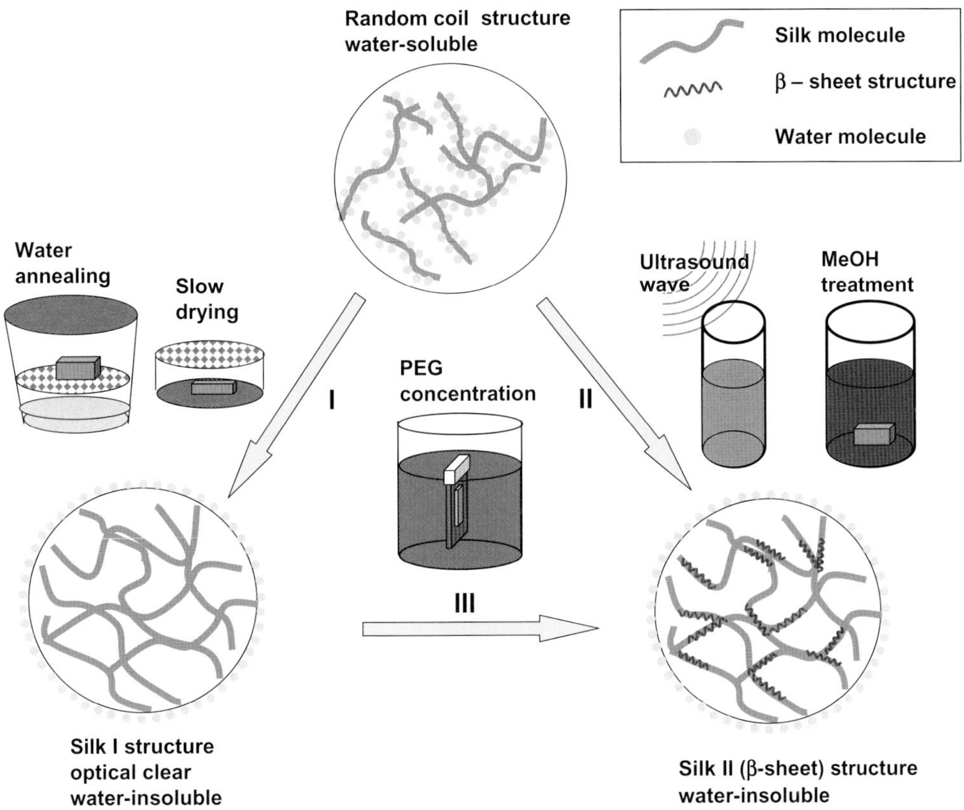

Figure 37.2 Silk structural changes upon material processing. Silk exists mainly in random coils when solubilized in water and quickly dried in air. Water annealing and slow drying (days) exclude water and induce the formation of hydrated silk structures; silk I has an increase in α-helices and β-turns (route I). At this stage, the protein materials are water-insoluble and optically clear after treatment. The ultrasonication of silk solution and methanol treatment of quickly dried silk materials also changes the hydrophobic hydration of silk and induces the formation of stable silk II structures, due to extensive stacking of β-sheets of hydrophobic domains (route II). The material after treatment is water-insoluble and chemically stable. Osmotic pressure by dialysis against polyethylene glycol (PEG) solution can be used to regulate the loss of water and therefore control silk I and silk II structural transitions, depending on the concentration of PEG used and the dialysis time. Furthermore, the silk I structure may transit to the more stable silk II structure upon material shearing, heating, and exposure to polar solvent such as methanol (route III).

systems, including long subsequences of amino acids defining the chemistry of large hydrophobic domains or blocks, with hydrophilic domains interspersed. The tendency to form long stretches of homogeneous secondary structure leads to flexible amphiphilic features in solution, with surfactancy, micelle formation and micelle-based lyotropic liquid crystalline phase behavior. A model for silk processing in silkworms and spiders was developed and validated with experimental results. Silk spinning begins with protein chain folding at lower concentrations of protein; proceeds through the formation of soft micelles and then micellar aggregates (globules) through water loss and increasing protein concentration; this leads to fiber formation due to physical shear during fiber spinning. Many aspects of this process can be mimicked *in vitro* and, in doing so, traditional fiber outcomes are now morphed into electrospun nanoscale diameter fibers, porous 3D matrices, flexible films and nanoscale coatings. Remarkably, these processes, from protein solubilization to assembly and materials formation, can all be conducted in water. The structural stability of the materials can be locked in through the formation of beta sheets, the crystalline domains present in the silk proteins. The remarkable aspect of this process is that the control of the protein-processing environment (*in vitro* or *in vivo*) and subsequent formation into solid-state materials (gels, fibers, films, sponges) is regulated by the content of water–how much is present and how quickly it is removed. Thus, techniques such as rapid dehydration (methanol treatment), shear thinning (natural spinning process), increased chain mobility (ultrasonication, heating) to drive changes in hydrophobic hydratation, and osmotic stress [dialysis against a hygroscopic polymer such as polyethylene glycol (PEG)] can be used to regulate the process. The rate of water removal from the hydrophobic domains of the silk dictates the extent of beta-sheet formation (numbers, sizes, distribution), and in turn the mechanical properties of the silk material formed and the degradation profiles. An unexpected observation from the above-described processes is the ability to use water in the processing to regulate material clarity. Thus, with slow water annealing, optically transparent films can be formed from silk, via regulation of the crystallization process [5, 13].

Silks are being explored as biomaterial matrices for a wide range of cell and tissue studies. Silk proteins can be reprocessed into a wide range of material morphologies and structures (high or low beta-sheet content) depending on the mode of solution preparation and handling [5, 11–14, 30, 66]. Silk materials have been used in many formats for tissue formation, including ligament-cell-based reconstruction due to the alignment and mechanical strength of the twisted silk structure [3, 67], and spider silks for Schwann cells related to peripheral nerve injuries and reconstruction [68].

The high surface area of small-diameter fibers, which is considered useful for improved cell interactions, has prompted interest in electrospinning silk proteins into fibers for biomaterial applications. Nanoscale-diameter fibers have been generated from silkworm silk solutions via electrospinning [11] and studied for nanomechanical properties using atomic force microscopy (AFM) nanoindentation techniques [69]. The biological compatibility of these materials in terms of adhesion and growth of human bone marrow-derived stromal cells was also reported [66]. In addition, native silk fibers have been prepared and chemically modified on the surfaces with cell adhesion peptides (RGD) to assess human bone marrow stromal cell responses and tissue formation [67, 70]. Fibers have also been formed by electrospinning genetically engineered spider silks in aqueous and solvent (HFIP) systems [71–73].

Films have been generated from reprocessed silk fibroin in a variety of modes to control structure and morphology [12, 13]. In addition, these films have been chemically modified with cell adhesion peptides (RGD) or morphogens/cytokines (BMP-2, PTH) to study cell adherence and differentiation toward tissue-specific outcomes such as bone [54, 55, 70]. Films have also been formed from genetically engineered spider silks. Cloning strategies and the ability to modify film surfaces for the attachment of functional molecules suggest utility for wound dressings, anti-adhesion barriers and enzyme immobilization scaffolds [56, 74].

Hydrogels were formed from aqueous silk solutions through osmotic stress, leading to control of the gel features based on the concentration of silk protein, the concentration of the osmotic stress-inducing polymer (PEG), pH, temperature, divalent or monovalent cations, or sonication [9, 30, 75]. Porous 3D scaffolds were formed from regenerated silk fibroin using freeze-drying, salt leaching and gas-foaming techniques with porosities up to 99% and pore sizes controllable from tens of microns to 1000 microns, depending on need [14, 30]. These scaffolds have been used in a variety of tissue-engineering studies, including those with cartilage [76] and bone [77, 78]. Genetically engineered spider silks have been formed into hydrogels, with and without crosslinking, by using visible light after applying ammonium peroxodisulfate and tris(2,2′-bipyridyl)dichlororuthenium (II) [79].

Sponge-like, porous 3D structures are important in tissue engineering to support cell and tissue ingrowth, to allow the transport of nutrients and metabolic wastes, and to promote tissue development [14, 30]. Sponges were formed from regenerated silk fibroin using freeze-drying, salt leaching and gas-foaming techniques [14, 16] and have been used in a variety of tissue-engineering studies including those with cartilage [77, 80] and bone [76, 78, 81].

Small-sized vesicles have been formed from both silkworm and spider silks, both of which offer options in the delivery of components such as pharmaceuticals, cell growth factors and other materials. Genetically engineered spider silks were formed with β-sheet-rich thin polymer shells with mechanical stability [82]. Silkworm silk has been processed into microvesicles using a variety of strategies, and successfully used to delivery small- and large-molecule drugs, as well as growth factors [61] (Figure 37.3).

The controlled deposition of ultrathin layers (nanometer-length scale) can be achieved by exploiting the self-assembly of silk proteins and solution control of protein assembly [10]. An aqueous, stepwise deposition process was established, and the structural control of the silk protein locked in the features of the coatings due to the beta sheets formed. This process allows the control of conformal thin films down to nanometer thickness, and can be used for the highly controlled build-up of deposited layers, such as for the entrapment and delivery of drugs or other compounds [83] (Figure 37.3).

Silks have been extensively studied for biocompatibility as well as enzymatic degradability. Silks have been used as sutures for decades, are FDA approved, biocompatible, and are less immunogenic and inflammatory than collagens or polyesters such as poly(lactic-*co*-glycolic acid) (PLGA) [3, 76, 84]. The data from silk biomaterials studied *in vitro* and *in vivo* suggest that purified degradable silk

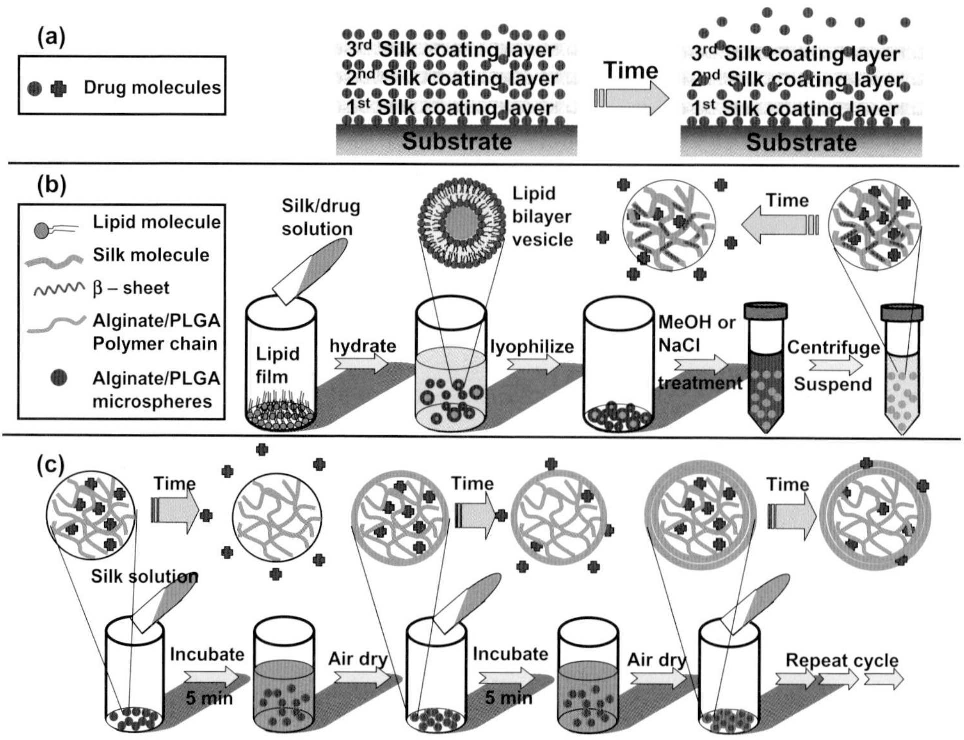

Figure 37.3 Silk materials for drug delivery. Extensive β-sheet structural networks in silk materials provide a microenvironment capable of maintaining drug activity and retarding release. The release can be controlled by: (a) layer-by-layer nanocoatings on surfaces; the drugs can be deposited between silk nanolayers or encapsulated in the layers, and release rate is controlled by the number of layers; (b) silk microspheres prepared with liposome-templating, depending on amount of β-sheet structure formed due to methanol or sodium chloride treatments, the drug release rates can be controlled; (c) alginate and polylactic/glycolic acid (PLGA) microspheres coated with silk using a layer-by-layer deposition method. Drugs are pre-encapsulated in microspheres before coating. Similar to (a), the drug release can be controlled by the number of coated layers.

fibroin exhibits biocompatibility commensurate or better than collagen or polyester-based biomaterials (PLGA copolymers) [70, 76]. There is also no evidence for bioburdens in silk-based biomaterials [3]. Silks are slowly degrading protein biomaterials, with the degradation rates being dependent on the form, structure and implant site utilized [15, 16, 67]. The early-observed adverse reactions to silks were due to the presence of residual sericin, a family of glue-like proteins that coat the core silk fibers, and not the fibers (fibroin) themselves [3, 85]. Polyester fibers coated with silks and implanted *in vivo* did not generate any thrombogenic response [86]. A detailed study of the inflammatory potential of silks showed that C3 activation, fibrinogen adsorption and mononuclear cell activation, among other measures in comparison to polystyrene and poly(2-hydroxyethyl methacrylate), indicated that silks were no different than these model surfaces in terms of humoral responses related to inflammation, and the degree of activation and adhesion of the immunocompetent cells was less of a problem [87]. The injection of solubilized silk fibroin and a series of other proteins was reported to induce amyloidosis in animal studies [88]. However, the mode of delivery was different from that in biomaterial matrix forms, and it has been widely shown that many proteins under these conditions will induce amyloid-like features. There is no documented evidence for these concerns from a clinical perspective, despite silk materials having been used for hundreds of years [3].

37.2 Silk Protein Polymers – Methods of Preparation

Two methods will be reviewed here to provide a sense of the different procedures involved in the preparation or processing of silk proteins for the types of study outlined in Section 37.1. First, the techniques used to create genetically engineered silks will be described, providing a route to spider silks. Second, the techniques used to reverse engineer silkworm silk from *B. mori* are detailed, for generating proteins for biomaterial scaffolds.

37.2.1 Preparation of Spider Silks

Two approaches are used to clone silks (Figure 37.4): (i) olignonucleotide synthesis and multimerization based on consensus repeats from silk sequence data; and (ii) cDNA library formation and the selection of clones encoding spider silk from mRNA isolated from silk-producing glands. As silks are highly repetitive in terms of their primary amino acid sequence, the former approach has been the most widely utilized. Further, synthetic oligonucleotides can be optimized for different expression systems in an attempt to improve final protein yield. cDNA clones have remained problematic in terms of stability during cloning, further limiting their utility in the generation of silks useful for the types of study summarized above. We first successfully generated synthetic silk genes to express spider silk consensus repeats [89] and then isolated, cloned and expressed a native partial cDNA clone from spiders, *Nephila clavipes* dragline silk [90]. Subsequent to these earlier studies, a variety of recombinant silk protein variants have been generated, mainly via synthetic oligonucleotide approaches [63, 64, 91, 92]. Structural variants of silks have been generated through the use of cDNA libraries and the use of synthetic gene technology, including silks with novel features (e.g. inorganic nucleating domains, cell-binding domains, molecular triggers to control assembly) [63, 64, 91, 93, 94]. As silks generally do not have significant levels of post-translational modifications, *Escherichia coli* can be used as a host system, as can higher organisms. A variety of host systems have been utilized to express silkworm or spider silks, yeast (e.g. *Pichia pastoris*), insect cells (e.g. sf9), bovine mammary epithelial

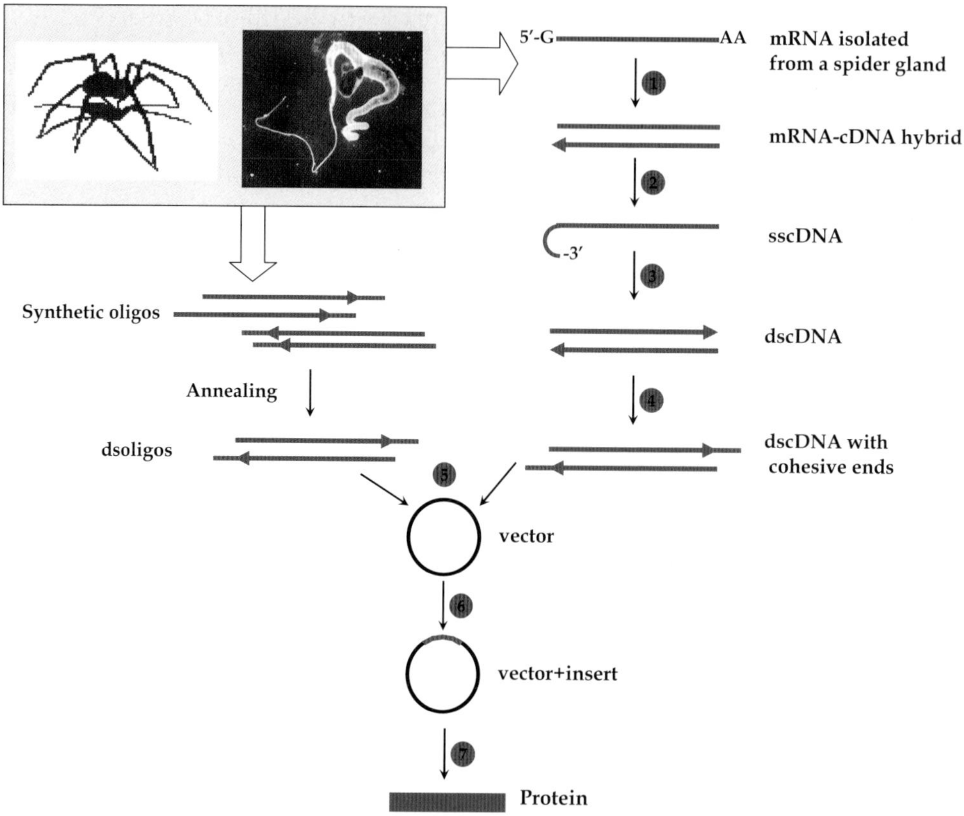

Figure 37.4 Cloning silk genes – from silkworms and spiders. Two basic paths are shown. On the left-hand side the synthetic genes are constructed and cloned based on consensus sequences; on the right-hand side the cDNA libraries are used to isolate and clone targeted silk genes. (①) mRNA from a gland is incubated with reverse transcriptase and oligo(dT); (②) The resulting mRNA–cDNA hybrid is treated with alkali to hydrolyze the RNA, leaving the single-stranded (ss) cDNA; (③) DNA polymerase synthesizes the complementary DNA strand; S1 nuclease is then used to cleave the loop; (④) the double-stranded (ds) cDNA is incubated with terminal transferase to generate cohesive ends; (⑤) Both cDNA/synthetic oligos and a cloning vector are digested with restriction enzymes and (⑥) ligated into a vector; (⑦) the cloning vector is transformed into the expression host, where the protein is expressed.

alveolar cells, baby hamster kidney cells, transgenic plants such as *Arabidopsis,* soybean, potato and tobacco, transgenic mammals such as mouse and goat, and finally transgenic silkworms [72, 95–104]. The sizes of the expressed silk proteins remain low (usually 50 kDa or less) when compared to the native proteins, due to challenges such as gene stability.

Synthetic oligonucleotide gene construction techniques are most commonly used to construct silk clones for subsequent expression. The repeat unit selected and used in the design of spider silk clones can originate, for example from the consensus sequence from the main dragline silk of the spider *N. clavipes* (Accession #P19 837) [105]. The consensus sequence design was used to form multimers (combining the single consensus sequence many times to build larger genes to encode longer proteins), leading to the formation of a 15mer recombinant protein where one repeat of this consensus sequence contains 33 amino acids [SGRGGLGGQGAGAAAAAGGAGQGGYGGLGSQGT] [61, 92, 94]. The monomer building block was designed and constructed for cloning using synthetic oligonucleotides and then amplified using the polymerase chain reaction (PCR). The hydrophobic beta-sheet-forming domain is underlined. Multimers (in this case a 15mer) encoding the repeat were cloned through the transfer of cloned inserts between two shuttle vectors based on pUC19 and pCR-Script, which were ampicillin and chloramphenicol resistant, respectively [64, 89]. The expression vector pET-30a was used to house the clones. The construct pET30–15mer was obtained by subcloning the *Nco*I–*Not*I fragment of the 15mer in pCR-Script directly into the pET-30a vector. This system houses a His-tag to facilitate purification, and the multimer silk systems, and system can be further modified through the addition of other sequences, such as cell-binding domains, mineralization domains and other fusions, by insertion into the linker site in the vector.

Constructs such as pET-30a(+)–15mer were transformed into the *E. coli* host strains RY-3041, a mutant strain defective in the expression of SlyD protein, for protein expression [91]. This allows for easier purification by reducing background bands that emerge after nickel column purification. The proteins are purified on nickel-affinity columns, utilizing the His-tag fusion. Purified samples are extensively dialyzed against Millipore H_2O to remove any final contaminating proteins. For dialysis, Snake Skin membranes (Pierce, Rockford, IL) with a molecular weight cut-off (MWCO) of 7000 or lower are generally used. The dialyzed samples are lyophilized and characterized for amino acid composition and protein concentration, as well as N-terminal sequence, to assure that the correct protein is generated for subsequent use in structure–function studies.

37.2.2 Preparation of Scaffolds

Some of the basic approaches and uses of porous silk protein scaffolds were briefly reviewed earlier in the chapter. In order to generate these types of structure, silkworm cocoons or raw silkworm fiber (from textile sources) must be: (i) purified, to remove the contaminating sericins and other debris; (ii) destructured (loss of beta-sheet physical crosslinks) to provide solubilization in water; and (iii) further processed into the targeted material format (for the methods described here, porous 3D sponges) (Figure 37.5). Initially, cocoons of *B. mori* silkworm are boiled for 20 min in an aqueous solution of 0.02 *M* Na_2CO_3, and then rinsed thoroughly with distilled water to extract the glue-like sericin proteins. The extracted silk fibroin is then dissolved in 9.3 *M* LiBr solution at 60 °C for 4 h, yielding a 20% (w/v) fibroin solution. This solution is dialyzed against distilled water using Slide-a-Lyzer dialysis cassettes (MWCO 3500, Pierce) for 2 days. The final concentration

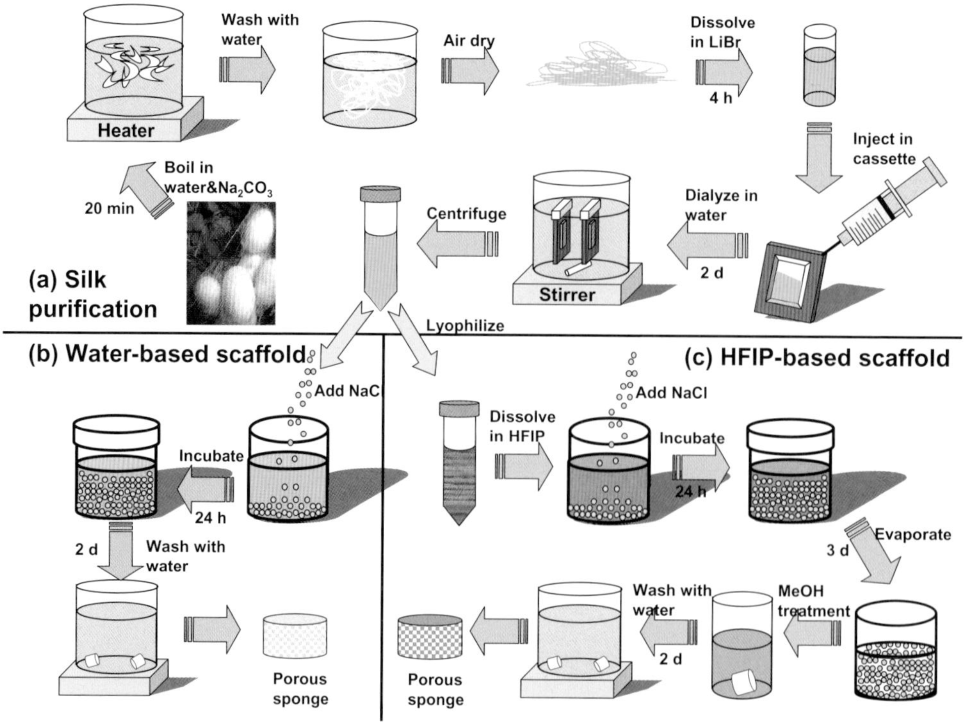

Figure 37.5 Silk porous scaffold processing. (a) All-aqueous silk purification. Lithium salt used in the process is removed during exhaustive dialysis; (b) Porous sponge-like silk scaffold prepared with water and sodium chloride granules. The sodium chloride is removed by extraction at the end of the process; (c) Porous sponge-like silk scaffold prepared with hexafluoroisopropanol (HFIP), methanol and sodium chloride granules. The organic solvents and salt are removed by evaporation and extraction.

of silk fibroin aqueous solution is approximately 8% (w/v). For the preparation of scaffolds using an all-aqueous process, the solutions are used directly in combination with NaCl crystals to form stable 3D matrices with control of pore size (crystal size used) and degradability (fibroin concentration), both of which impact directly on the mechanical properties and enzymatic degradability [16]. For the solvent fabrication process, the dialysis solution is lyophilized and the silk powder is then dissolved directly in the HFIP to generate different solution concentrations which are used in the formation of 3D porous spongy scaffolds with control of pore size (salt crystal size) and mechanical properties (based on fibroin concentration used) [14].

HFIP-derived silk fibroin scaffolds are prepared by adding 4 g of granular NaCl (sieved to a desired particle size to match the target pore size) into 2 ml of 8 wt% silk fibroin in HFIP. The containers are covered overnight to reduce the evaporation of HFIP to form more homogeneous structures. Subsequently, the solvent is evaporated at room temperature for 3 days. The silk/porogen matrix is then treated in methanol for variable periods of time to control the extent of beta-sheet formation. The matrices are then immersed in water for 2 days to remove the NaCl, and air-dried. For the aqueous-based silk fibroin scaffolds, 4 g of granular NaCl (particle size based on the target pore size but increased by 10% as the outer layer dissolves initially to saturate the aqueous solution) is added to 2 ml of 8 wt% silk fibroin solution in disk-shaped Teflon containers (1.8 cm diameter × 2 cm height). The containers are covered, left at room temperature for 24 h, and then immersed in water to extract the NaCl for 2 days. In the above processes, there is no residual lithium salt or organic solvent (HFIP) [16, 30].

Both, the aqueous- or organic solvent-derived scaffolds have been used to study stem cell growth and differentiation toward bone and cartilage tissue [77, 78, 80, 81]. For example, tissue engineering of cartilage has been carried out using human bone marrow-derived mesenchymal stem cells (hMSCs) and porous silk scaffolds [78, 80]. hMSCs were isolated, expanded in culture, and characterized with respect to the expression of surface markers and ability for chondrogenic and osteogenic differentiation [76–78]. With the HFIP-derived silk scaffolds, the expanded cells were seeded on the scaffolds and RGD-coupled silk scaffolds (modified to enhance cell attachment, $3.5 \pm 0.5\,\mathrm{p}M$ RGD cm^{-2}) with scaffold pores ~200 μm in size. The resulting constructs were cultured for 4 weeks in either control medium (DMEM with 10% fetal bovine serum) or chondrogenic medium. The hMSCs became attached to silk scaffolds via cellular extensions, formed cell networks after 2 weeks, and continuous sheets after 4 weeks of culture. With time in culture, the GAG content increased. The deposition of GAGs and type II collagen were markedly higher for hMSCs cultured on silk than on control collagen scaffolds. hMSC differentiation into chrondrogenic tissue *in vitro* was also studied using the aqueous-derived silk fibroin scaffolds [80]. Chondrogenesis was assessed by real time RT–PCR for cartilage-specific ECM markers, and with histological and immunohistochemical evaluations of cartilage-specific ECM components. Dexamethasone and transforming growth factor-beta3 (TGF-β3) were essential for the survival, proliferation and chondrogenesis of hMSCs in the scaffolds in serum-free media. After 3 weeks of cultivation, the spatial cell arrangement and collagen type II distribution in the constructs was cartilage-like, and a good homogeneity of the new ECM was observed. Overall, the high porosity, slow degradation and structural integrity of the silk scaffolds provides important benefits to cartilage tissue formation *in vitro* when compared with other commonly used degradable polymeric matrices, such as collagen and PLGA.

37.3
Silk Protein Polymers – Future Perspectives and Challenges

The above review and brief examples have laid the groundwork for future opportunities with the study and use of silk proteins for a wide range of potential interests.

These unique insights into the self-assembly and control of material structure and morphology offer important and useful options to regulate material features, such as mechanical properties and biological responses (rates of degradation, remodeling, environmental compatibility). The ability to control primary sequence via genetic engineering, allows the option of novel variants in silk protein domain distributions to further explore structure–function relationships, such as those analogous to synthetic block copolymer studies related to phase diagrams to regulate morphology. Based on the information provided above, we are in a position to exploit this understanding towards well-defined and predictable material structures with the control of primary sequence, surface chemistry, structure and morphology at length scales ranging from nanometer to macroscopic, to gain insight into sequence–structure relationships. The rich hydrogen bonding, peptide backbone rotational restrictions and complex chemistries in these proteins are, for the most part, absent from the synthetic polymer block design systems. Thus, new and important insights should be forthcoming on the role of these small, but numerous, forces on protein polymer assembly and function.

The ability to generate mechanically robust materials from silks also provides unique and important benefits for biomaterials and scaffolds for tissue engineering. Although, today these processes are still at an early stage of development, they suggest future opportunities for soft-tissue reconstruction, hard-tissue regeneration and drug delivery options. New families of biomaterials which slowly (but fully) degrade and so persist for years are envisioned, and may prove valuable for systems such as sustained release and surface coatings. The incorporation of enzymes, sensing components, inorganic phases, conducting polymers and related features, coupled with silk to create functionalized assembled protein systems, may offer useful options for the future. These systems may prove valuable in a range of applications, including components of nondestructive materials assessments, in self-decontamination material systems, regenerative biomaterials and many related technological opportunities involving controlled release. In Nature, silks – unlike collagens – are not generally mineralized, such that there is no significant control over silk–inorganic interfaces. However, techniques to control the mineralization of silks have been developed, offering new pathways to stiffer, silk-composite materials. The ability to genetically engineer silk chimeras to encode mineralizing domains may also offer insight into new molecular-level composites.

Silk processing and materials features fit well with the focal point of 'green chemistry' and environmental compatibility. As the assembly of silk into new materials can be conducted entirely in water, options to incorporate sensitive biological components become feasible. The metallization of these well-defined surfaces on these length scales would also offer templating opportunities for stabilized catalysts, with the silk protein being removed after templating function. These new options for using all-aqueous methods to process silks into functional materials should also be applicable to a broader range of polymer science needs. As silks represent the most hydrophobic proteins known to mankind, concepts in synthetic polymer block designs which embrace design rules from silks may lead to new methods for the aqueous processing of these polymers.

Silks exist in Nature as a diverse set of chemistries related to a diverse set of functional features; examples include different types of fiber with different mechanical properties, cocoon composites and sticky silks for adhesion. This ability to embrace such diversity in novel materials designs awaits the use of combinatorial approaches, and with sufficient sequence data from different spider and silkworm silks now available the opportunity has been presented to begin this process. The genetic tools described in this chapter may be applied to this goal, with the main challenge being more in the screening and selection process to identify the targeted functional features required in the new silk variants. The results of recent studies have also shown that nontraditional silks (nonbeta-sheet systems) are present in Nature, and offer new options in silk-based materials processing, assembly and functional features. One such example is honeybee silk, which is composed of coiled-coil proteins [106]. Moreover, when these systems are combined with traditional silk processes it is likely that an expanded set of material properties will become apparent.

As noted above, chimeric silk proteins have now been created that combine silk with mineralization domains to generate organic–inorganic composites. The challenge here will be to develop this approach in order to explore size variants and to incorporate full-length functional proteins, such as for cell signaling and related needs.

Despite the optimism of these areas, there remain many challenges. Silks are particularly amenable to genetic approaches (as outlined above), due to an absence of significant post-translational modifications, in direct contrast to collagens. However, whilst cloning and expression techniques for silks have been improved and self-assembly and processing into many material formats is now better understood, many problems persist. Further improvements are required both in the yield of protein, and in the expression of full-length silks in heterologous hosts to fully embrace the complex sequence structure required for the accurate assembly and function of silks. Clearly, the use of silks in commodity materials, including high-performance composites, biomaterials and durable materials, will require the creation of robust, low-cost expression systems.

Even with the correct silk protein, the ability to fully recapitulate the remarkable mechanical properties of silk fibers spun by spiders and silkworms remains an unmet goal. Further insight into the processing of these proteins in aqueous environments, to mimic the native processes of silk-spinning organisms, and to organize these proteins into functional materials will be required. Ultimately, this goal will only be met through a better control of beta-sheet characteristics, including size, distribution and orientation, as these impact directly on the materials' properties. Such features are also directly amenable to aqueous processing, notably the rate and extent of water removal during processing.

Acknowledgments

Our deep thanks go to our many colleagues over the years, who have contributed to the study of silk proteins, as well as to the funding agencies that have provided the essential fiscal support for the studies, including the NSF, NIH and AFOSR. We thank Olena Rabotyagova for the preparation of Figure 37.4.

References

1 McGrath, K. and Kaplan, D.L. (1998) *Protein-Based Materials*, Birkahuser, Boston.

2 Kaplan, D.L., Adams, W., Farmer, B. and Viney, C. (1994) Silk polymers, in *Materials Science and Biotechnology*, Vol. **544** (ed. American Chemical Society Symposium Series), American Chemical Society, Washington, D.C, pp. 1–385.

3 Altman, G.H., Diaz, F., Jakuba, C., Calabro, T., Horan, R.L., Chen, J., Lu, H., Richmond, J. and Kaplan, D.L. (2003) Silk-based biomaterials. *Biomaterials*, **24**, 401–16.

4 Gosline, J.M., DeMont, M.E. and Denny, M.W. (1986) The structure and properties of spider silk. *Endeavour*, **10**, 37–43.

5 Jin, H.J. and Kaplan, D.L. (2003) Mechanism of silk processing in insects and spiders. *Nature*, **424**, 1057–61.

6 Bini, E., Knight, D.P. and Kaplan, D.L. (2004) Mapping domain structures in silks from insects and spiders related to protein assembly. *Journal of Molecular Biology*, **335**, 27–40.

7 Wong, C., Bini, E., Hensman, J., Knight, D.P., Lewis, R.V. and Kaplan, D.L. (2006) Role of pH and charge on silk protein assembly in insects and spiders. *Applied Physics A – Materials Science & Processing*, **82**, 223–33.

8 Knight, D.P. and Vollrath, F. (2001) Liquid crystalline spinning of spider silk. *Nature*, **410**, 541–8.

9 Kim, U.J., Park, J., Li, C., Jin, H.J., Valluzzi, R. and Kaplan, D.L. (2004) Structure and properties of silk hydrogels. *Biomacromolecules*, **5**, 786–92.

10 Wang, X., Kim, H.J., Xu, P., Matsumoto, A. and Kaplan, D.L. (2005) Biomaterial coatings by stepwise deposition of silk fibroin. *Langmuir*, **21**, 11335–41.

11 Jin, H.J., Fridrikh, S.V., Rutledge, G.C. and Kaplan, D.L. (2002) Electrospinning *Bombyx mori* silk with poly (ethylene oxide). *Biomacromolecules*, **3**, 1233–9.

12 Jin, H.J., Park, J., Kim, U.J., Valluzzi, R., Cebe, P. and Kaplan, D.L. (2004) Biomaterial films of *Bombyx mori* silk with poly(ethylene oxide). *Biomacromolecules*, **5**, 711–17.

13 Jin, H.J., Park, J., Karageorgiou, V., Kim, U.J., Valluzzi, R., Cebe, P. and Kaplan, D.L. (2005) Water-stable silk films with reduced β-sheet content. *Advanced Functional Materials*, **15**, 1241–7.

14 Nazarov, R., Jin, H.J. and Kaplan, D.L. (2004) Porous 3D scaffolds from regenerated silk fibroin. *Biomacromolecules*, **5**, 718–26.

15 Horan, R.L., Antle, K., Collette, A.L., Wang, Y., Huang, J., Moreau, J.E., Volloch, V., Kaplan, D.L. and Altman, G.H. (2005) In vitro degradation of silk fibroin. *Biomaterials*, **26**, 3385–93.

16 Kim, H.J., Kim, U.J., Vunjak-Novakovic, G., Min, B.H. and Kaplan, D.L. (2005) Influence of macroporous protein scaffolds on bone tissue engineering from bone marrow stem cells. *Biomaterials*, **26**, 4442–52.

17 Kaplan, D.L., Mello, C.M., Arcidiacono, S., Fossey, S., Sencal, K. and Muller, W. (1997) Silk, in *Protein-Based Materials* (eds K. McGrath and D.L. Kaplan), Birkhauser, pp. 103–31.

18 Zhou, C.Z., Confalonieri, F., Jacquet, M., Perasso, R., Li, Z.G. and Janin, J. (2001) Silk fibroin: Structural implications of a

remarkable amino acid sequence. *Proteins Structure Function and Genetics*, **44**, 119–22.

19 Scheibel, T. (2004) Spider silks: recombinant synthesis, assembly, spinning, and engineering of synthetic proteins. *Microbial Cell Factories*, **3**, Art, no. 14.

20 Van Beek, J.D., Hesst, S., Vollrath, F. and Meier, B.H. (2002) The molecular structure of spider dragline silk: folding and orientation of the protein backbone. *Proceedings of the National Academy of Sciences of the United States of America*, **99**, 10266–71.

21 Chatterjee, S.N. and Tanushree, T. (2004) Molecular profiling of silkworm biodiversity in India: an overview. *Russian Journal of Genetics*, **40**, 1339–47.

22 Hwang, J.S., Lee, J.S., Goo, T.W., Yun, E.Y., Lee, K.S., Kim, Y.S., Jin, B.R., Lee, S.M., Kim, K.Y., Kang, S.W. and Suh, D.S. (2001) Cloning of the fibroin gene from the oak silkworm, *Antheraea yamamai* and its complete sequence. *Biotechnology Letters*, **23**, 1321–6.

23 Li, W., Terenius, O., Hirai, M., Nilsson, A.S. and Faye, I. (2005) Cloning, expression and phylogenetic analysis of Hemolin, from the Chinese oak silkmoth, *Antheraea pernyi*. *Developmental and Comparative Immunology*, **29**, 853–64.

24 Goldsmith, M.R., Shimada, T. and Abe, H. (2005) The genetics and genomics of the silkworm, *Bombyx mori*. *Annual Review of Entomology*, **50**, 71–100.

25 Grab, J.E., DiMauro, T., Lewis, R.V. and Hayashi, C.Y. (2007) Expansion and intragenic homogenization of spider silk genes since the triassic: evidence from mygalomorphae (tarantulas and their kin) spidroins. *Molecular Biology and Evolution*, **24**, 2454–64.

26 Ayoub, N.A., Garb, J.E., Tinghitella, R.M., Collin, M.A. and Hayashi, C.Y. (2007) Blueprint for a high-performance biomaterial: full-length spider dragline silk genes. *PLoS ONE*, **2**, e514.

27 Ayoub, N.A. and Hayashi, C.Y. (2008) Multiple recombining loci encode MaSp1, the primary constituent of dragline silk, in widow spiders (Latrodectus: Theridiidae). *Molecular Biology and Evolution*, **25**, 277–86.

28 Cunniff, P.M., Fossey, S.A., Auerbach, M.A., Song, J.W., Kaplan, D.L., Adams, W.W., Eby, R.K., Mahoney, D. and Vezie, D.L. (1994) Mechanical and thermal properties of dragline silk from the spider *Nephila clavipes*. *Polymers for Advanced Technology*, **5**, 401–10.

29 Jiang, C., Wang, X., Gunawidjaja, R., Lin, Y.H., Gupta, M.K., Kaplan, D.L., Naik, R.R. and Tsukruk, V.V. (2007) Mechanical properties of robust ultrathin silk fibroin films. *Advanced Functional Materials*, **17**, 2229–37.

30 Kim, U.J., Park, J., Kim, H.J., Wada, M. and Kaplan, D.L. (2005) Three-dimensional aqueous-derived biomaterial scaffolds from silk fibroin. *Biomaterials*, **26**, 2775–85.

31 Magoshi, J. and Nakamura, S. (1975) Studies on physical properties and structure of silk. Glass transition and crystallization of silk fibroin. *Journal of Applied Polymer Science*, **19**, 1013–15.

32 Magoshi, J., Magoshi, Y., Nakamura, S., Kasai, N. and Kakudo, M. (1977) Physical-properties and structure of silk. 5. Thermal-behavior of silk fibroin in random-coil conformation. *Journal of Polymer Science Part B: Polymer Physics*, **15**, 1675–84.

33 Agarwal, N., Hoagland, D.A. and Farris, R.J. (1997) Effect of moisture absorption on the thermal properties of *Bombyx mori* silk fibroin films. *Journal of Applied Polymer Science*, **63**, 401–10.

34 Motta, A., Fambri, L. and Migliaresi, C. (2002) Regenerated silk fibroin films: Thermal and dynamic mechanical analysis. *Macromolecular Chemistry and Physics*, **203**, 1658–65.

35 Magoshi, J., Magoshib, Y., Beckerc, M.A., Katod, M., Hane, Z., Tanaka, T., Inoue, S. and Nakamura, S. (2000) Crystallization of silk fibroin from solution. *Thermochimica Acta*, **352–3**, 165–9.

36 Zhang, H., Magoshi, J., Becker, M., Chen, J.Y. and Matsunaga, R. (2002) Thermal properties of *Bombyx mori* silk fibers. *Journal of Applied Polymer Science*, **86**, 1817–20.

37 Tanaka, T., Magoshi, J., Magoshi, Y., Inoue, S., Kobayashi, M., Tsuda, H., Becker, M.A. and Nakamura, S. (2002) Thermal properties of *Bombyx mori* and several wild silkworm silks – Phase transition of liquid silk. *Journal of Thermal Analysis & Calorimetry*, **70**, 825–32.

38 Sashina, E.S., Janowska, G., Zaborski, M. and Vnuchkin, A.V. (2007) Compatibility of fibroin/chitosan and fibroin/cellulose blends studied by thermal analysis. *Journal of Thermal Analysis & Calorimetry*, **89**, 887–91.

39 Hu, X., Kaplan, D. and Cebe, P. (2006) Determining beta sheet crystallinity in fibrous proteins by thermal analysis and infrared spectroscopy. *Macromolecules*, **39**, 6161–70.

40 Hu, X., Kaplan, D. and Cebe, P. (2007) Effect of water on thermal properties of silk fibroin. *Thermichimica Acta*, **461**, 137–44.

41 Hu, X., Kaplan, D. and Cebe, P. (2008). Dynamic protein-water relationships during beta sheet formation. *Macromolecules*, **41**, 3939–48.

42 Gil, E.S., Frankowski, D.J., Bowman, M.K., Gozen, A.O., Hudson, S.M. and Spontak, R.J. (2006) Mixed protein mends composed of gelatin and *Bombyx mori* silk fibroin: Effects of solvent-induced crystallization and composition. *Biomacromolecules*, **7**, 728–35.

43 Taddei, P. and Monti, P. (2005) Vibrational infrared conformational studies of model peptides representing the semicrystalline domains of *Bombyx mori* silk fibroin. *Biopolymers*, **78**, 249–58.

44 Gil, E.S., Spontak, R.J. and Hudson, S.M. (2005) Effect of beta-sheet crystals on the thermal and rheological behavior of protein-based hydrogels derived from gelatin and silk fibroin. *Macromolecular Bioscience*, **5**, 702–9.

45 Gil, E.S., Frankowski, D.J., Hudson, S.M. and Spontak, R.J. (2007) Silk fibroin membranes from solvent-crystallized silk fibroin/gelatin blends: effects of blend and solvent composition. *Materials Science and Engineering C*, **27**, 426–31.

46 Liu, Y., Shao, Z.Z., Zhou, P. and Chen, X. (2004) Thermal and crystalline behavior of silk fibroin/nylon 66 blend films. *Polymer*, **45**, 7705–10.

47 Chen, H., Hu, X. and Cebe, P. (2008) Thermal properties and phase transitions in nylon-6 and silk fibroin blends. *Journal of Thermal Analysis*, in press.

48 Cheng, Z.L., Wang, S. and Zhu, H.S. (2004) Structural and thermal characteristics of silk fibroin/fucoidan blend films. *Acta Polymerica Sinica*, **3**, 446–9.

49 Tsukada, M., Nagura, M., Ishikawa, H. and Shiozaki, H. (1991) Structural characteristics of silk fibers treated with epoxides. *Journal of Applied Polymer Science*, **43**, 643–9.

50 Tsukada, M., Freddi, G. and Crighton, J.S. (1994) Structure and compatibility of poly (vinyl alcohol) -silk fibroin (pva/sf) blend films. *Journal of Polymer Science Part B: Polymer Physics*, **32**, 243–8.

51 Kameda, T. and Tsukada, M. (2006) Structure and thermal analyses of MAA-grafted silk fiber using DSC and C-13 solid-state NMR. *Macromolecular Materials and Engineering*, **291**, 877–82.

52 Lee, K.Y. and Ha, W.S. (1999) DSC studies on bound water in silk fibroin/S-carboxymethyl kerateine blend films. *Polymer*, **40**, 4131–4.

53 Karageorgiou, V., Meinel, L., Hofmann, S., Malhotra, A., Volloch, V., Schwob, J., Kaplan, D.L. (2004) Bone morphogenetic protein-2 decorated fibroin films induce osteogenic differentiation of human bone marrow stromal cells. *Journal of Biomedical Materials Research*, **71A**, 528–37.

54 Sofia, S., McCarthy, M.B., Gronowicz, G. and Kaplan, D.L. (2001) Functionalized silk-based biomaterials for bone formation. *Journal of Biomedical Materials Research*, **54**, 139–48.

55 Karageorgiou, V. and Kaplan, D.L. (2005) Porosity of 3D biomaterial scaffolds and osteogenesis. *Biomaterials*, **26**, 5474–91.

56 Huemmerich, D., Slotta, U. and Scheibel, T. (2006) Processing and modification of films made from recombinant spider silk proteins. *Applied Physics Letters Part A*, **82**, 219–22.

57 Vepari, C. and Kaplan, D.L. (2007) Silk as a biomaterial. *Progress in Polymer Science*, **32**, 991–1007.
58 Li, C., Jin, H.J., Botsaris, B.D. and Kaplan, D.L. (2005) Silk apatite composites from electrospun fibers. *Journal of Materials Research*, **20**, 3374–84.
59 Wong, P.F.C., Bini, E., Henseman, J., Knight, D.P., Lewis, R.V. and Kaplan, D.L. (2006) Solution behavior of synthetic silk peptides and modified recombinant silk proteins. *Applied Physics A Materials Science & Processing*, **82**, 193–203.
60 Li, C., Vepari, C., Jin, H.J., Kim, H.J. and Kaplan, D.L. (2006) Electrospun silk-BMP-2 scaffolds for bone tissue engineering. *Biomaterials*, **27**, 3115–24.
61 Wang, X., Wenk, E., Matsumoto, A., Meinel, L., Li, C. and Kaplan, D.L. (2007) Silk microspheres for encapsulation and controlled release. *Journal of Controlled Release*, **117**, 360–70.
62 Valluzzi, R., Szela, S., Avtges, P., Kirschner, D. and Kaplan, D. (1999) Methionine redox controlled crystallization of biosynthetic silk spidroin. *Journal of Physical Chemistry B*, **103**, 11382–92.
63 Szela, S., Avtges, P., Valluzzi, R., Winkler, S., Wilson, D., Kirschner, D. and Kaplan, D.L. (2000) Reduction-oxidation control of beta-sheet assembly in genetically engineered silk. *Biomacromolecules*, **1**, 534–42.
64 Winkler, S., Wilson, D. and Kaplan, D.L. (2000) Controlling beta-sheet assembly in genetically engineered silk by enzymatic phosphorylation/dephosphorylation. *Biochemistry*, **39**, 12739–46.
65 Foo, C.W.P., Bini, E., Huang, J., Lee, S.Y. and Kaplan, D.L. (2006) Solution behavior of synthetic silk peptides and modified recombinant silk proteins. *Applied Physics Letters Part A*, **82**, 193–203.
66 Jin, H.J., Chen, J., Karageorgiou, V., Altman, G.H. and Kaplan, D.L. (2004) Human bone marrow stromal cell responses on electrospun silk fibroin mats. *Biomaterials*, **25**, 1039–47.
67 Altman, G., Horan, R., Lu, H., Moreau, J., Martin, I., Richmond, J. and Kaplan, D.L. (2002) Silk matrix for tissue engineered anterior cruciate ligaments. *Biomaterials*, **23**, 4131–41.
68 Allmeling, C., Jokuszies, A., Reimers, K., Kall, S. and Vogt, P.M. (2006) Use of spider silk fibres as an innovative material in a biocompatible artificial nerve conduit. *Journal of Cellular and Molecular Medicine*, **10**, 770–7.
69 Wang, M., Jin, H.J., Kaplan, D.L. and Rutledge, G.C. (2004) Mechanical properties of electrospun silk fibers. *Macromolecules*, **37**, 6856–64.
70 Chen, J., Altman, G.H., Karageorgiou, V., Horan, R., Collette, A., Volloch, V., Colabro, T. and Kaplan, D.L. (2003) Human bone marrow stromal cell and ligament fibroblast responses on RGD-modified silk fibers. *Journal of Biomedical Materials Research Part A*, **67**, 559–70.
71 Arcidiacono, S., Mello, C.M., Butler, M., Welsh, E., Soares, J.W., Allen, A., Ziegler, D., Laue, T. and Chase, S. (2002) Aqueous processing and fiber spinning of recombinant spider silks. *Macromolecules*, **35**, 1262–6.
72 Lazaris, A., Arcidiacono, S., Huang, Y., Zhou, J.F., Duguay, F., Chretien, N., Welsh, E.A., Soares, J.W. and Karatzas, C.N. (2002) Spider silk fibers spun from soluble recombinant silk produced in mammalian cells. *Science*, **295**, 472–6.
73 Stephens, J.S., Fahnestock, S.R., Farmer, R.S., Kiick, K.L., Chase, D.B. and Rabolt, J.F. (2005) Effects of electrospinning and solution casting protocols on the secondary structure of a genetically engineered dragline spider silk analog investigated via Fourier transform Raman spectroscopy. *Biomacromolecules*, **6**, 1405–13.
74 Junghans, F., Morawietz, M., Conrad, U., Scheibel, T., Heilmann, A. and Spohn, U. (2006) Preparation and mechanical properties of layers made of recombinant spider silk proteins and silk from silk worm. *Applied Physics Letters Part A*, **82**, 253–60.
75 Wang, X., Kluge, J.A., Leisk, G.G. and Kaplan, D.L. (2008) Sonication-induced gelation of silk fibroin for cell encapsulation. *Biomaterials*, **29**, 1054–64.

76 Meinel, L., Hoffmann, S., Karageorgiou, V., Zichner, L., Langer, R. and Kaplan, D.L. (2004) Engineering cartilage-like tissue using human mesenchymal stem cells and silk protein scaffolds. *Biotechnology and Bioengineering*, **88**, 379–91.

77 Meinel, L., Karageorgiou, V., Fajardo, R., Snyder, B., Shinde-Patil, V., Zichner, L., Kaplan, D.L., Langer, R. and Vunjak-Novakovic, G. (2004) Bone tissue engineering using human mesenchymal stem cells; effects of scaffold material and medium flow. *Annals of Biomedical Engineering*, **32**, 112–22.

78 Meinel, L., Kargeorgiou, V., Hofmann, S., Fajardo, R., Snyder, B., Li, C., Zichner, L., Langer, R., Vunjak-Novakovic, G. and Kaplan, D.L. (2004) Engineering bone-like tissue in vitro using human bone marrow stem cells and silk scaffolds. *Journal of Biomedical Materials Research*, **71A**, 25–34.

79 Rammensee, S., Huemmerich, D., Hermanson, K.D., Scheibel, T. and Bausch, A.R. (2006) Rheological characterization of hydrogels formed by recombinantly produced spider silk. *Applied Physics Letters Part A*, **82**, 261–4.

80 Wang, Y., Kim, U.J., Blasioli, D.J., Kim, H.J. and Kaplan, D.L. (2005) In vitro cartilage tissue engineering with 3D porous aqueous-derived silk scaffolds and mesenchymal stem cells. *Biomaterials*, **26**, 7082–94.

81 Meinel, L.R., Fajardo Hofmann, S., Langer, R., Chen, J., Snyder, B., Vunjak-Novakovic, G. and Kaplan, D.L. (2005) Silk-implants for the healing of critical size bone defects. *Bone*, **37**, 688–98.

82 Hermanson, K.D., Huemmerich, D., Scheibel, T. and Bausch, A.R. (2007) Engineered microcapsules fabricated from reconstituted spider silk. *Advanced Materials*, **19**, 1810–15.

83 Wang, X., Wenk, E., Hu, X., Castro, G.R., Meinel, L., Wang, X., Li, C., Merkle, H. and Kaplan, D.L. (2007) Silk coatings on PLGA and alginate microspheres for protein delivery. *Biomaterials*, **28**, 4161–9.

84 Panilaitis, B., Altman, G.H., Chen, J., Jin, H.J., Karageorgiou, V. and Kaplan, D.L. (2003) Macrophage responses to silk. *Biomaterials*, **24**, 3079–85.

85 Soong, H.K. and Kenyon, K.R. (1984) Adverse reactions to virgin silk sutures in cataract surgery. *Ophthalmology*, **91**, 479–83.

86 Sakabe, H., Ito, H., Miyamoto, T., Noishiki, Y. and Ha, W.S. (1989) In vivo blood compatibility of regenerated silk fibroin. *Sen-I Gakkaishi*, **45**, 487–90.

87 Santin, M., Motta, A., Freddi, G. and Cannas, M. (1999) In vitro evaluation of the inflammatory potential of the silk fibroin. *Journal of Biomedical Materials Research Part A*, **46**, 382–9.

88 Lundmark, K., Westermark, G.T., Olsén, A. and Westermark, P. (2005) Protein fibrils in nature can enhance amyloid protein A amyloidosis in mice: cross-seeding as a disease mechanism. *Proceedings of the National Academy of Sciences of the United States of America*, **102**, 6098–102.

89 Prince, J.P., McGrath, K.P., DiGirolamo, C.M. and Kaplan, D.L. (1995) Construction, cloning and expression of synthetic spider dragline silk DNA. *Biochemistry*, **34**, 10879–85.

90 Arcidiacono, S., Mello, C., Kaplan, D., Cheley, S. and Bayley, H. (1998) Purification and characterization of recombinant spider silk expressed in *Escherichia coli. Applied Microbiology and Biotechnology*, **49**, 31–8.

91 Huang, J., Valluzzi, R., Bini, E., Vernaglia, B. and Kaplan, D.L. (2003) Cloning, expression and assembly of sericin-like protein. *Journal of Biological Chemistry*, **278**, 46117–23.

92 Bini, E., Foo, C.W.P., Huang, J., Karageorgiou, V., Kitchel, B. and Kaplan, D.L. (2006) RGD-functionalized bioengineered spider dragline silk biomaterial. *Biomacromolecules*, **7**, 3139–45.

93 Wong, C.P.F., Kitchel, B., Huang, J., Patwardhan, S.V., Belton, D., Perry, C. and Kaplan, D.L. (2006) Novel nanocomposites from spider silk-silica fusion (chimeric) proteins. *Proceedings of the National Academy of Sciences of the United States of America*, **103**, 9428–33.

94 Huang, J., Wong, C., George, A. and Kaplan, D.L. (2007) The effect of genetically engineered spider silk-dentin matrix protein 1 chimeric protein on hydroxyapatite nucleation. *Biomaterials*, **28**, 2358–67.

95 Fahnestock, S.R. and Irwin, S.L. (1997) Synthetic spider dragline silk proteins and their production in *Escherichia coli*. *Applied Microbiology and Biotechnology*, **47**, 23–32.

96 Fahnestock, S.R. and Bedzyk, L.A. (1997) Production of synthetic spider dragline silk protein in *Pichia pastoris*. *Applied Microbiology and Biotechnology*, **47**, 33–9.

97 Scheller, J.K.H, Gührs, F., Grosse, U. and Conrad, U. (2001) Production of spider silk proteins in tobacco and potato. *Nature Biotechnology*, **19**, 573–7.

98 Karatzas, C.N. *et al.* (2003) High-toughness spider silk fibers spun from soluble recombinant silk produced mammalian cells, in *Biopolymers. Polyamides and Complex Proteinaceous Materials II* (eds S.R. Fahnestock and A. Steinbüchel), Wiley-VCH Verlag GmbH, Weinheim, pp. 500–10.

99 Huemmerich, D., Scheibel, T., Vollrath, F., Cohen, S., Gat, U. and Ittah, S. (2004) Novel assembly properties of recombinant spider dragline silk proteins. *Current Biology*, **14**, 2070–4.

100 Menassa, R., Zhu, H., Karatzas, C.N., Lazaris, A., Richman, A. and Brandle, J. (2004) Spider dragline silk proteins in transgenic tobacco leaves: accumulation and field production. *Plant Biotechnology Journal*, **2**, 431–8.

101 Barr, L.A., Fahnestock, S.R. and Yang, J. (2004) Production and purification of recombinant DP1B silk-like protein in plants. *Molecular Breeding*, **13**, 345–56.

102 Yang, J., Barr, L.A., Fahnestock, S.R. and Liu, Z.B. (2005) High yield recombinant silk-like protein production in transgenic plants through protein targeting. *Transgenic Research*, **14**, 313–24.

103 Miao, Y., Miao, Y., Zhang, Y., Nakagaki, K., Zhao, T., Zhao, A., Meng, Y., Nakagaki, M., Park, E.Y. and Maenaka, K. (2006) Expression of spider flagelliform silk protein in *Bombyx mori* cell line by a novel Bac-to-Bac/BmNPV baculovirus expression system. *Applied Microbiology and Biotechnology*, **71**, 192–9.

104 Kwang, S.L., Bo, Y.K., Yeon, H.J., Soo, D.W., Hung, D.S. and Byung, R.J. (2007) Molecular cloning and expression of the C-terminus of spider flagelliform silk protein from *Araneus ventricosus*. *Journal of Biosciences*, **32**, 705–12.

105 Xu, M. and Lewis, R.V. (1990) Structure of a protein superfiber: spider dragline silk. *Proceedings of the National Academy of Sciences of the United States of America*, **87**, 7120–4.

106 Sutherland, T.D., Weisman, S., Trueman, H.E., Sriskantha, A., Trueman, J.W.H. and Haritos, V.S. (2007) Conservation of essential design features in coiled coil silks. *Molecular Biology and Evolution*, **24**, 2424–32.

Index

Protein Engineering Handbook. Edited by Stefan Lutz and Uwe T. Bornscheuer

ISBN: 978-3-527-31850-6

d